INTERNATIONAL SERIES OF MONOGRAPHS IN AERONAUTICS AND ASTRONAUTICS

DIVISION II: AERODYNAMICS
EDITORS: R. T. JONES AND W. P. JONES

VOLUME 7

WING THEORY IN SUPERSONIC FLOW

WING THEORY IN SUPERSONIC FLOW

by

ELIE CARAFOLI

Professor at the Polytechnical Institute
Director of the Institute of Fluid Mechanics,
Bucharest

with the cooperation of

DAN MATEESCU *and* ADRIANA NASTASE

from the Institute of Fluid Mechanics,
Bucharest

THE QUEEN'S AWARD TO INDUSTRY 1966

PERGAMON PRESS

OXFORD · LONDON · EDINBURGH · NEW YORK
TORONTO · SYDNEY · PARIS · BRAUNSCHWEIG

Pergamon Press Ltd., Headington Hill Hall, Oxford
4 & 5 Fitzroy Square, London W.1
Pergamon Press (Scotland) Ltd., 2 & 3 Teviot Place, Edinburgh 1
Pergamon Press Inc., 44-01 21st Street, Long Island City, New York 11101
Pergamon of Canada Ltd., 207 Queen's Quay West, Toronto 1
Pergamon Press (Aust.) Pty. Ltd., 19a Boundary Street, Rushcutters Bay, N.S.W. 2011, Australia
Pergamon Press S.A.R.L., 24 rue des Écoles, Paris 5e
Vieweg & Sohn GmbH, Burgplatz 1, Braunschweig

First edition 1969

Library of Congress Catalog Card No. 67-27481

PRINTED IN HUNGARY

08 012330 9

CONTENTS

IV. STUDY OF ORDINARY DELTA WINGS

V. YAWED DELTA WINGS. DELTA WINGS WITH GENERAL VARIATION OF INCIDENCE

VI. TRAPEZOIDAL AND RECTANGULAR WINGS

VII. STUDY OF POLYGONAL WINGS

VIII. CRUCIFORM WINGS

IX. SIMPLE OR CRUCIFORM WING FITTED WITH A CONICAL BODY

X. MISCELLANEOUS

XI. DETERMINATION OF PRESSURES AND AERODYNAMIC CHARACTERISTICS OF WINGS BEYOND THE ASSUMPTION OF SMALL DISTURBANCES, IN SUPERSONIC AND MODERATE HYPERSONIC FLOWS. EXPERIMENTAL VERIFICATIONS

XII. SOME THEORETICAL CONSIDERATIONS ON WINGS WITH CURVED LEADING EDGES AND WING–FUSIFORM BODY SYSTEMS

PREFACE

FOR a long time now we have been concerned with the study of triangular and polygonal wings in supersonic flow, basing this study on the theory of high order conical motions. Several of my collaborators and I have examined the cases of isolated simple wings, cruciform wings, wings with vertical plane tail, simple or cruciform wings fitted with a body, and so on.

For the purposes of our study we have tried, on the one hand, to establish a unified theory that will permit a simple solution to all these problems and, on the other hand, to set up a corresponding method for the actual calculation of isolated simple wings or of more complex systems, thereby obtaining explicit and simple formulae for each real application that should be of use in practical aeronautics. This method is used throughout this work, since it is highly suitable for application to the cases under discussion. Consequently we have considered it unnecessary to deal with any other methods of approach, particularly since they might have prejudiced the unity and method of this exposition, based on two simple geometrical parameters of the wings which are described below.

Chapter I is a recapitulation of the fundamental equations in the mechanics of inviscid fluids; by linearizing these formulae we have obtained a series of corresponding equations which form the basis of the theory of high order conical motions. In Chapter II a unified theory is established for angularly shaped wings in supersonic flow, assuming that the height or thickness of the wing surface is given by a superposition of homogeneous polynomials of various orders as functions of coordinates of the plane in which the wing lies. The whole theory is based on just two geometrical elements: the subsonic leading edge of a thin wing, and the ridge that separates two régimes of variation of the wing surface or of the pressures on the wing. The main formulae which give the pressures and aerodynamic characteristics of the wing when the equation of the surface is known, or determine the wing surface when the pressures are given on it, have been established in this way for various forms of wings—triangular, trapezoidal or polygonal. Chapter III gives the method for calculating these characteristics (lift, drag and moments) for various shapes of the trailing edge line.

Wings of ordinary shapes, delta wings, arrow-head or swept-back wings, yawed wings, trapezoidal wings, polygonal wings, and so on, are studied in Chapters IV, V, VI and VII, by means of the application of the formulae obtained. These chapters give practical data of use to specialists, and describe simple methods for carrying out new investigations for solving certain special problems that appear in practical supersonic aeronautics, particularly regarding the shape of the wing surface and the way it varies.

In order to give concrete solutions to certain problems of particular practical importance, as for instance the problem of cruciform wings, with applications to control surfaces of rockets or to aircraft tails, the whole of Chapter VIII is devoted to such problems, which are treated in this way for the first time. We have dwelt at some length on these problems, assuming that they are of particular interest in practice. For this purpose the problem of wing-body interference, which forms the subject of Chapter IX, has been given a concrete solution and a correspondingly extensive development. For the same reason, and particularly considering the importance of the ailerons and periodic oscillations of wings during flight, we have briefly considered these problems in Chapter X. In that chapter we also introduced the study of flaps situated at the leading edges of wings, in order to eliminate the theoretically infinite velocity that would otherwise occur there.

Following the same idea, we have treated at greater length the problem of flow separation at leading edges, mainly in order to represent the phenomenon of the detachment of vortices from the upper surface in the shape of a spiral, and of checking experimentally the theoretical forecasts.

In the last chapter there is an extension of the theory of conical motions to some very similar problems which are termed *quasi-conical* motions, with the aid of which we can study wings with curved leading edges instead of straight ones. We can thus deal with aerodynamic problems of wings of general shape which occur on modern aeroplanes.

In fact, an essential problem which appears in connection with supersonic phenomena and the corresponding theories for calculating the pressures on the wing is whether the theories are confirmed in practice, thereby ensuring the thorough validity of the derived formulae for the calculation of aerodynamic characteristics of wings or combined lifting systems.

For this reason, it was thought necessary to devote the eleventh chapter to experimental verifications; developing the theory still further, we have tried to extend the applicability of the formulae beyond the supersonic régime up to the moderate hypersonic régime, this applicability being also checked experimentally.

As regards the relative amount of space devoted to the various topics, we must confess that it has not been easy to make a fair division. We decided to deal more fully with common problems of current application in supersonic aeronautics, and to give a more concise and general exposition of those problems which, although important, are more specialized and also necessitate difficult and laborious calculations, so that the complete treatment of certain concrete examples would have acquired an unjustifiable length.

A unified theory and a simple method of calculation based on the assumption of small disturbances was applied to all the topics, but because of the possibilities of extending the results obtained to the moderate hypersonic régime (as pointed out above) the present book offers, to those concerned with engineering applications, useful equations for calculating the lift, drag and moments of wings in supersonic flow, even in cases where the incidences and Mach numbers involved are quite large.

Although the treatment of the problems is made at an advanced level and corresponds to certain specialist lectures in the graduate courses of engineering sciences, the approach to the subject is rational and complete, so that it can be followed easily and used successfully by teachers and students as well as by specialists in modern aerodynamics.

On the other hand, because of its initial recapitulative and introductory section, this book can stand by itself and may be used *ab initio* as an advanced textbook in the subject, even by those who are not already acquainted with it.

We wish to express our appreciation to all those collaborators who helped us in achieving the aim of the present book; among them we mention Beatrice Horovitz, Cornel Berbente and Stefan Staicu who have assisted us in the elaboration of certain chapters. We also wish to extend our warmest thanks to Elena Lazu for having translated this book into English and having typed the manuscript.

Finally, we take this opportunity of giving our special thanks to the editorial board of Pergamon Press for all their kindness in the publishing of this book.

ELIE CARAFOLI

ACKNOWLEDGEMENTS

IN THE present book we have reproduced some experimental results, diagrams and photographs from several authors' works, with a view to comparing them with the theory under discussion. Consequently, we should like to express our grateful appreciation to the following publishers and individuals:

The Controller of Her Britannic Majesty's Stationery Office; William F. Hausmann, Deputy Assistant Administrator for International Affairs, National Aeronautics and Space Administration; Irwin Hersey, Director of Publications, American Institute of Aeronautics and Astronautics; Antonio Ferri; J. E. Garrick; M. Landhal; G. Drougge and B. Beane; C. H. McLellan, L. C. Squire and P. G. Wilby.

References are made to the respective authors in the bibliography.

Should there be any omissions from the present acknowledgement, we wish to apologize in advance.

CHAPTER I

RECAPITULATIVE SURVEY OF THE EQUATIONS OF MOTION OF PERFECT FLUIDS IN SUPERSONIC FLOW

IN THIS first chapter we shall try to present briefly some of the fundamental results concerning the flow of perfect compressible fluids, confining ourselves to those results which will be made use of in the following chapters for discussing various problems which constitute the object of the present work.

Proceeding from the usual equations, which are generally well known by those concerned with fluid mechanics, we shall try to obtain in the simplest way the most characteristic and useful formulae for applications in the field we are going to deal with.

In view of the fact that this work is concerned with a branch of applied aerodynamics, we shall pass over the details representing the purely mathematical aspects of the formulae, and give generally only their simplest expressions which can be more easily understood and applied.

The main tasks that we are aiming at in this chapter concentrate on some fundamental problems concerning in particular the velocity potential equation, the applications of the theory of small disturbances to supersonic flow, the theory of high order conical flows and, of course, the usual relations and formulae given for the motion of perfect fluids in supersonic flow, which are absolutely necessary for discussing and understanding these fundamental problems.

I. 1. General Equations

We shall first briefly establish some fundamental equations of motion.

I. 1.1. Equations of Motion for a Perfect Fluid

Assuming a fluid considered as an inviscid continuum medium, and denoting by **V** the speed of a fluid particle at any point, where the density si ϱ, while p is the pressure, we have to write, if the fluid's own weight is neglected (this fluid being the air or another gas) and an irrotational motion is considered, the following equations:

the irrotationality condition (the curl of the velocity is zero),

$$\nabla \wedge \mathbf{V} = 0, \tag{1.1}$$

the continuity equation,

$$\frac{\partial \varrho}{\partial t} + \nabla(\varrho \mathbf{V}) = 0, \tag{1.2}$$

and *the momentum equation* (Euler's dynamic equation),

$$\frac{d\mathbf{V}}{dt} = \frac{\partial \mathbf{V}}{\partial t} + \nabla\left(\tfrac{1}{2} V^2\right) + (\nabla \wedge \mathbf{V}) \wedge \mathbf{V} = \frac{\partial \mathbf{V}}{\partial t} + \nabla\left(\tfrac{1}{2} V^2\right) = -\frac{1}{\varrho} \nabla p. \tag{1.3}$$

where ∇ is the operator $\left(\frac{\partial}{\partial x}\mathbf{i} + \frac{\partial}{\partial y}\mathbf{j} + \frac{\partial}{\partial z}\mathbf{k}\right)$, $\frac{d}{dt}$ is the total derivative with respect to time moving with the fluid.

Considering further an isentropic transformation of the moving gas, the general physical equation — which is given in the form

$$\frac{p}{\varrho} = RT, \tag{1.4}$$

where T is the temperature and R is the universal gas constant—is replaced in this case by the isentropic equation

$$\frac{p}{\varrho^{\gamma}} = \text{const.}, \tag{1.5}$$

where γ is the ratio of specific heats, under constant pressure and constant volume, respectively ($\gamma = 1{\cdot}4$ for air).

I. 1.2. Pressure Equation

From the irrotationality condition (1.1), we deduce that the velocity is the gradient of a velocity potential φ, that is we have

$$\mathbf{V} = \nabla \varphi. \tag{1.6}$$

On the other hand, observing that taking into account (1.5), we may write

$$\int \frac{dp}{\varrho} = \frac{\gamma}{\gamma - 1} \frac{p}{\varrho}, \qquad \frac{\nabla p}{\varrho} = \nabla\left(\frac{\gamma}{\gamma-1} \frac{p}{\varrho}\right); \tag{1.7}$$

equation (1.3) therefore becomes in this case

$$\frac{\partial \mathbf{V}}{\partial t} + \nabla\left(\tfrac{1}{2} V^2\right) + \frac{\nabla p}{\varrho} = \nabla\left(\frac{\partial \varphi}{\partial t} + \tfrac{1}{2} V^2 + \frac{\gamma}{\gamma-1} \frac{p}{\varrho}\right) = 0, \tag{1.8}$$

whence we obtain the pressure equation in a differential form and a finite form, called the Bernoulli–Lagrange equation,

$$d\left(\frac{\partial\varphi}{\partial t}\right)+V\,dV+\frac{dp}{\varrho}=0, \tag{1.9a}$$

$$\frac{\partial\varphi}{\partial t}+\tfrac{1}{2}V^2+\frac{\gamma}{\gamma-1}\frac{p}{\varrho}=\text{const.} \tag{1.9b}$$

In the case of a steady motion, the potential does not vary with time and the above equation becomes

$$\tfrac{1}{2}V^2+\frac{\gamma}{\gamma-1}\frac{p}{\varrho}=\frac{\gamma}{\gamma-1}\frac{p_0}{\varrho_0}=\text{const.}, \tag{1.10}$$

where we have denoted by p_0 and ϱ_0 the pressure and density, respectively, at a stagnation point ($V=0$).

I. 1.3. Momentum Theorems

Let us consider a volume τ bounded by a surface σ and let $\varrho\,d\tau$ be an element of mass. If we neglect the external (mass) forces, the equilibrium between the inertia forces and the overall resultant of the pressure, which will be denoted by **P**, will be given by the relation

$$\frac{d\mathbf{I}}{dt}=\int_\tau\frac{d\mathbf{V}}{dt}\varrho\,d\tau=\int_\sigma\mathbf{n}p\,d\sigma=\mathbf{P}, \tag{1.11}$$

where **n** is the normal unit vector on the surface σ, directed towards the interior of the fluid, while **I** represents the momentum.

Taking into account the continuity equation (1.2) and denoting by **i**, **j**, **k** the unit vectors in the direction of the rectangular coordinate axes, we may write successively

$$\varrho\frac{d\mathbf{V}}{dt}=\frac{\partial(\varrho\mathbf{V})}{\partial t}+\frac{\partial(\varrho u\mathbf{V})}{\partial x}+\frac{\partial(\varrho v\mathbf{V})}{\partial y}+\frac{\partial(\varrho w\mathbf{V})}{\partial z}$$

$$=\frac{\partial(\varrho\mathbf{V})}{\partial t}+\mathbf{i}\nabla(\varrho u\mathbf{V})+\mathbf{j}\nabla(\varrho v\mathbf{V})+\mathbf{k}\nabla(\varrho w\mathbf{V}) \tag{1.12a}$$

$$\mathbf{V}=\mathbf{i}u+\mathbf{j}v+\mathbf{k}w. \tag{1.12b}$$

Inserting into (1.11), transforming the volume integrals into surface integrals (in compliance with Gauss' theorem) and denoting by Q the rate of flow across the surface σ (or the flux of the vector **V** through σ), it follows readily that

$$\mathbf{P} = \int_\tau \varrho \frac{d\mathbf{V}}{dt}\, d\tau = \int_\tau \frac{\partial(\varrho\mathbf{V})}{\partial t}\, d\tau - \mathbf{i}\int_\sigma \varrho u(\mathbf{n}\cdot\mathbf{V})\, d\sigma$$

$$-\mathbf{j}\int_\sigma \varrho v(\mathbf{n}\cdot\mathbf{V})\, d\sigma - \mathbf{k}\int_\sigma \varrho w(\mathbf{n}\cdot\mathbf{V})\, d\sigma \tag{1.13}$$

$$= \int_\tau \frac{\partial(\varrho\mathbf{V})}{\partial t}\, d\tau - \int_\sigma \varrho\mathbf{V}\, dQ,$$

which represents the momentum theorem and which will be of use later on for calculating the resultant force.

Similarly, we may calculate the moment of the pressure about any point O, which will be denoted by $\mathbf{M}_p$.

Denoting by $\mathbf{r}$ the position vector of the elementary mass $\varrho\, d\tau$, we write

$$\mathbf{r}\wedge\frac{d\mathbf{V}}{dt} = \frac{d(\mathbf{r}\wedge\mathbf{V})}{dt} \tag{1.14}$$

and consequently, the moment of pressure will be given by the integral

$$\frac{d\mathbf{K}}{dt} = \mathbf{M}_p = \int_\tau \varrho \frac{d(\mathbf{r}\wedge\mathbf{V})}{dt}\, d\tau, \tag{1.15}$$

where $\mathbf{K}$ represents the moment of momentum.

Setting further

$$\mathbf{m} = \mathbf{r}\wedge\mathbf{V} \tag{1.16}$$

and noting that, analogously to (1.12), we have

$$\varrho\frac{d\mathbf{m}}{dt} = \frac{\partial(\varrho\mathbf{m})}{\partial t} + \frac{\partial(\varrho u\mathbf{m})}{\partial x} + \frac{\partial(\varrho v\mathbf{m})}{\partial y} + \frac{\partial(\varrho w\mathbf{m})}{\partial z} \tag{1.17}$$

and hence the place of $\mathbf{V}$ in the preceding calculation is taken by $\mathbf{m}$, we shall write finally, replacing $\mathbf{V}$ in (1.13) by $\mathbf{m}$,

$$\mathbf{M}_p = \int_\tau \frac{\partial(\varrho\mathbf{r}\wedge\mathbf{V})}{\partial t}\, d\tau - \int_\sigma \varrho(\mathbf{r}\wedge\mathbf{V})\, dQ, \tag{1.18}$$

which represents the moment of momentum theorem.

I. 2. Propagation of Small Disturbances

An important parameter entering in our equations is the speed of sound, which will be denoted by a and which is equal to the velocity of propagation of small disturbances.

Let us consider a one-dimensional motion along any axis, say Ox; designating the velocity in this direction by u and applying equations (1.2) and (1.3), we obtain successively

$$\frac{1}{\varrho}\frac{\partial\varrho}{\partial t}+\frac{\partial u}{\partial x}+\frac{u}{\varrho}\frac{\partial\varrho}{\partial x}=\frac{\partial(\ln\varrho)}{\partial t}+\frac{\partial u}{\partial x}+u\frac{\partial(\ln\varrho)}{\partial x}=0 \quad (1.19a)$$

$$\frac{\partial u}{\partial t}+u\frac{\partial u}{\partial x}=-\frac{1}{\varrho}\frac{\partial p}{\partial x}=-\frac{\partial p}{\partial\varrho}\frac{1}{\varrho}\frac{\partial\varrho}{\partial x}=-\frac{dp}{d\varrho}\frac{\partial(\ln\varrho)}{\partial x}, \quad (1.19b)$$

in which ∂ has been replaced by d, since the pressure varies only as a function of ϱ (1.5).

If we consider slow motions, the velocity u and the density and pressure variations as functions of t and x are small, so that the products of these magnitudes can be neglected.

In this case, we may write

$$\frac{\partial u}{\partial x}=-\frac{\partial(\ln\varrho)}{\partial t}, \quad \frac{\partial u}{\partial t}=-\frac{dp}{d\varrho}\frac{\partial(\ln\varrho)}{\partial x}. \quad (1.20)$$

Setting further

$$\frac{dp}{d\varrho}=a^2 \quad (1.21)$$

and considering that in the interval of pressure variation with respect to density, a^2 may be considered constant, we may differentiate the first equation (1.12) with respect to t, and the second one with respect to x, thus obtaining the equation of propagation for the small disturbances,

$$\frac{\partial^2 u}{\partial t^2}=a^2\frac{\partial^2 u}{\partial x^2}, \quad (1.22)$$

whose solution may be written in d'Alembert's general form

$$u=u_1(x-at)+u_2(x+at). \quad (1.23)$$

Assuming that we are dealing only with a displacement in the direction $x>0$, then only the first term in the right-hand side is valid, showing that the disturbance at any point, represented for instance by the velocity u of a particle at that point, propagates with the speed a, since we always find the same velocity u for $x=at$.

Equation (1.21) is the acoustic equation, while a is the speed of the sound propagation; for this reason a is called *the speed of sound.*

Further, taking account of (1.4) and (1.5), we write

$$a^2=\frac{dp}{d\varrho}=\gamma\frac{p}{\varrho}=\gamma RT, \quad (1.24)$$

whence it may be seen that the speed of sound of the same gas depends on the gas temperature.

For the sake of simplicity, we have considered above an example of one-dimensional motion (1.18); however, we can similarly discuss without any difficulty the case of spherical disturbances, where a solution similar to the above (1.22) also shows us that a is the velocity of propagation for these disturbances [C−2].

In the following, we shall give a graphical representation of the motion, in order to indicate the role played by the speed of sound in defining the different character of subsonic and supersonic flows.

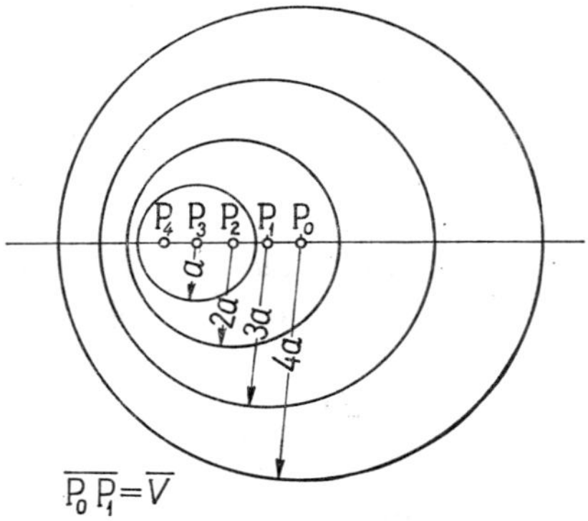

FIG. 1.1

For this purpose, we shall assume that at any point P a disturbance takes place, caused by a body moving with velocity **V** with respect to the undisturbed fluid. This disturbance is transmitted in all directions with the speed of sound a.

If $V < a$, then the disturbance propagates faster than the body, while the latter displaces in a medium previously *warned*, that is in a medium which is disturbed everywhere (Fig. 1.1). The space around the body is more and more affected by this disturbance, so that, after a relatively short time, we may say that, at least in the immediate proximity of the body, the flow is very close to the asymptotic one corresponding to *the impulsive motion*, in which it is supposed that the characteristics of the motion propagate instantaneously throughout the entire space. This characteristic of affecting the entire fluid space defines *subsonic flow*.

If $V > a$, the disturbing body moves faster than the propagation velocity (i.e. the speed of sound), so that a part of the surrounding space is not warned of this disturbance, thus remaining undisturbed. It is only in the interior of a cone *(the Mach cone)* that the space is disturbed (Fig. 1.2). The semi-angle at the vertex, which will be designated by μ and is called the Mach angle, is given by the relation

$$\frac{1}{\sin \mu} = \frac{V}{a} = M, \tag{1.25}$$

where the ratio M is called *the Mach number*.

In calculations, the tangent of the angle μ and a coefficient B, the inverse of the tangent, occur very frequently, being defined by the relations

$$B = \sqrt{M_\infty^2 - 1} = \frac{1}{\tan \mu}, \tag{1.26}$$

where M_∞ represents the Mach number of the free stream.

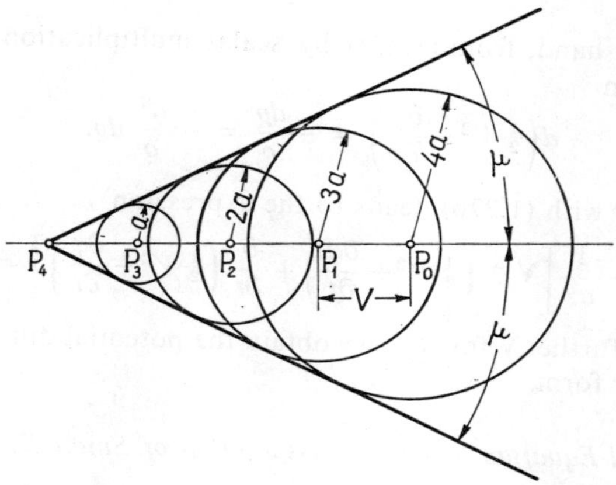

FIG. 1.2

This property of only partially affecting the fluid space characterizes *supersonic flow*.

Throughout the disturbed space of a subsonic motion, there may exist restricted regions in which the fluid particles have higher speeds than the speed of sound; likewise, in the disturbed space of a supersonic motion may also exist such restricted regions, in which the fluid particles have lower speeds than the speed of sound.

In both cases, in the regions close to the separation boundaries, a mixed motion occurs of transonic type.

In the following, we shall deal with the problems concerning the basic formulae applied to the supersonic flow.

I. 3. Velocity Potential Equation

Under the assumption of an irrotational motion, there exists a velocity potential which has been denoted above by φ and whose differential equation will be established below.

I. 3.1. General Equation of the Velocity Potential

For this purpose, we shall proceed from equations (1.2) and (1.3) or (1.8), put in the form

$$\nabla\left(\tfrac{1}{2}V^2+\frac{\partial\varphi}{\partial t}\right) = -\frac{\nabla p}{\varrho} = -a^2\frac{\nabla\varrho}{\varrho}, \tag{1.27a}$$

$$\nabla\mathbf{V}+\mathbf{V}\frac{\nabla\varrho}{\varrho}+\frac{1}{\varrho}\frac{\partial\varrho}{\partial t} = 0. \tag{1.27b}$$

On the other hand, from (1.27a) by scalar multiplication of each side by $d\mathbf{r}$, we obtain

$$d\left(\tfrac{1}{2}V^2+\frac{\partial\varphi}{\partial t}\right) = -\frac{dp}{\varrho} = -\frac{a^2}{\varrho}\,d\varrho, \tag{1.28}$$

which, together with (1.27b), leads to the expression

$$\nabla\mathbf{V}-\frac{1}{a^2}\left[\mathbf{V}\nabla\left(\tfrac{1}{2}V^2+\frac{\partial\varphi}{\partial t}\right)+\frac{\partial}{\partial t}\left(\tfrac{1}{2}V^2+\frac{\partial\varphi}{\partial t}\right)\right] = 0. \tag{1.29}$$

Substituting further $\mathbf{V}$ for $\nabla\varphi$, we obtain the potential differential equation in a vector form.

I. 3.2. Potential Equation under the Assumption of Small Disturbances

The potential equation (1.29) can be easily expressed in rectangular, cartesian or other coordinates, but the expressions are cumbersome; on the other hand, in the following chapters, this equation will be employed only under the assumption of small disturbances; that is the reason why we shall try to give a simplified form. To this end, we shall assume that the free stream U_∞ is disturbed (for instance, in the presence of a slender or thin body) due to an additional small velocity whose projections on a system of orthogonal axes $Oxyz$ are u, v, w, respectively.

Denoting by $\Phi(x, y, z, t)$ the disturbance velocity potential, the total velocity components will be

$$U = \frac{\partial\varphi}{\partial x} = U_\infty+u = U_\infty+\frac{\partial\Phi}{\partial x}, \quad v = \frac{\partial\Phi}{\partial y}, \quad w = \frac{\partial\Phi}{\partial z}, \tag{1.30a}$$

$$\varphi = U_\infty x+\Phi, \tag{1.30b}$$

respectively.

Neglecting the second-order terms, we can write successively:

$$V^2 = (U_\infty+u)^2+v^2+w^2 \approx U_\infty^2+2U_\infty u = U_\infty^2+2U_\infty\frac{\partial\Phi}{\partial x}, \tag{1.31a}$$

$$\nabla(\tfrac{1}{2}V^2) \approx \nabla(U_\infty u) = \nabla\left(U_\infty\frac{\partial\Phi}{\partial x}\right), \tag{1.31b}$$

$$\frac{\partial}{\partial t}(\tfrac{1}{2}V^2) \approx U_\infty \frac{\partial u}{\partial t} = U_\infty \frac{\partial^2 \Phi}{\partial x \partial t}, \tag{1.31c}$$

$$\mathbf{V}\nabla(\tfrac{1}{2}V^2) \approx \mathbf{V}\nabla(U_\infty u) \approx U_\infty^2 \frac{\partial u}{\partial x} = U_\infty^2 \frac{\partial^2 \Phi}{\partial x^2}, \tag{1.31d}$$

$$\mathbf{V}\nabla\frac{\partial \varphi}{\partial t} \approx (U_\infty + u)\frac{\partial^2 \Phi}{\partial t \partial x} + v\frac{\partial^2 \Phi}{\partial t \partial y} + w\frac{\partial^2 \Phi}{\partial t \partial z} \approx U_\infty \frac{\partial^2 \Phi}{\partial x \partial t}. \tag{1.31e}$$

On the other hand, taking account of (1.24) and denoting by a_∞ and a_0 the speeds of sound in the free stream (where $\partial\varphi/\partial t$ may be taken equal to zero) and at the stagnation points ($V = 0$), the Bernoulli equation (1.9b) becomes

$$\frac{\partial \Phi}{\partial t} + \tfrac{1}{2}V^2 + \frac{a^2}{\gamma - 1} = \tfrac{1}{2}U_\infty^2 + \frac{a_\infty^2}{\gamma - 1} = \frac{a_0^2}{\gamma - 1} = \text{const.}, \tag{1.32}$$

whence

$$a^2 = a_\infty^2 - \frac{\gamma - 1}{2}\left(V^2 - U_\infty^2 + 2\frac{\partial \Phi}{\partial t}\right) \approx a_\infty^2. \tag{1.33}$$

Substituting the above in (1.29) and denoting further the free stream Mach number by M_∞,

$$M_\infty = \frac{U_\infty}{a_\infty}, \tag{1.34}$$

we finally obtain the differential equation of the velocity potential under the assumption of small disturbances:

$$(1 - M_\infty^2)\frac{\partial^2 \Phi}{\partial x^2} + \frac{\partial^2 \Phi}{\partial y^2} + \frac{\partial^2 \Phi}{\partial z^2} = \frac{1}{a_\infty^2}\left(\frac{\partial^2 \Phi}{\partial t^2} + 2U_\infty \frac{\partial^2 \Phi}{\partial x \partial t}\right). \tag{1.35}$$

In the case of steady flows, this expression is simplified and becomes

$$-B^2\frac{\partial^2 \Phi}{\partial x^2} + \frac{\partial^2 \Phi}{\partial y^2} + \frac{\partial^2 \Phi}{\partial z^2} = 0, \tag{1.36}$$

where B has been defined by (1.26).

For incompressible flows ($M_\infty \approx 0$), expression (1.35) as well as (1.36) are reduced to the simple *Laplacian*

$$\nabla\mathbf{V} = \nabla^2\varphi = \frac{\partial^2 \Phi}{\partial x^2} + \frac{\partial^2 \Phi}{\partial y^2} + \frac{\partial^2 \Phi}{\partial z^2} = 0, \tag{1.37}$$

which represents just the continuity equation in this case.

I. 3.3. Expression of the Potential Equation in Cylindrical Coordinates

It is more convenient to make use of cylindrical coordinates for the study of axially-symmetrical motions; therefore, we shall give below the

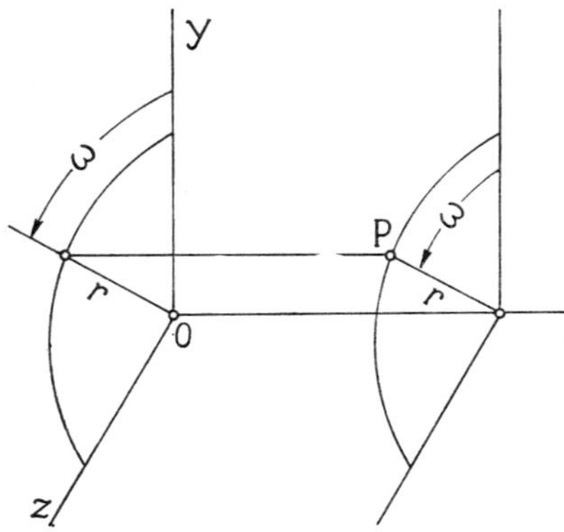

FIG. 1.3

expression of the potential equation in these coordinates (Fig. 1.3), which are related to the cartesian ones by the relations

$$x = x, \quad y = r\cos\omega, \quad z = r\sin\omega. \tag{1.38}$$

Since x is unchanged, we have to write as a function of r and ω, the following expression

$$\frac{\partial^2\Phi}{\partial y^2}+\frac{\partial^2\Phi}{\partial z^2} = \frac{\partial v}{\partial y}+\frac{\partial w}{\partial z}, \tag{1.39}$$

which represents the divergence in the plane Oyz of a vector having the components v and w. In cylindrical coordinates, the velocities are V_r and V_ω respectively (Fig. 1.4); considering an element dr and $r\,d\omega$, we shall successively write, according to the true definition of divergence,

$$\frac{1}{r\,dr\,d\omega}\left[\frac{\partial}{\partial r}(v_r r\,d\omega)\,dr+\frac{\partial}{\partial\omega}(V_\omega\,dr)\,d\omega\right]$$
$$= \frac{1}{r}\left[\frac{\partial(rV_r)}{\partial r}+\frac{\partial V_\omega}{\partial\omega}\right] = \frac{1}{r}\left[\frac{\partial}{\partial r}\left(r\frac{\partial\Phi}{\partial r}\right)+\frac{\partial}{\partial\omega}\left(\frac{\partial\Phi}{r\,\partial\omega}\right)\right]. \tag{1.40}$$

Substituting in (1.35), we obtain

$$-B^2\frac{\partial^2\Phi}{\partial x^2}+\frac{\partial^2\Phi}{\partial r^2}+\frac{\partial^2\Phi}{r^2\,\partial\omega^2}+\frac{1}{r}\frac{\partial\Phi}{\partial r} = \frac{1}{a^2}\left(\frac{\partial^2\Phi}{\partial t^2}+2U_\infty\frac{\partial^2\Phi}{\partial x\,\partial t}\right), \tag{1.41}$$

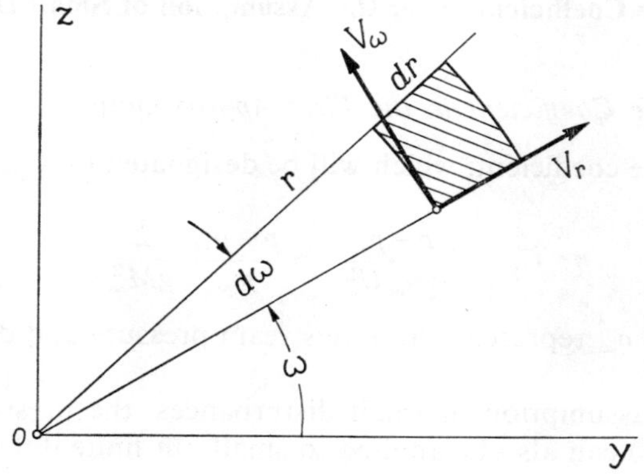

FIG. 1.4

which becomes for steady flows,

$$-B^2\frac{\partial^2\Phi}{\partial x^2}+\frac{\partial^2\Phi}{\partial r^2}+\frac{\partial^2\Phi}{r^2\,\partial\omega^2}+\frac{1}{r}\frac{\partial\Phi}{\partial r}=0, \tag{1.42}$$

and for axially-symmetrical steady motions

$$-B^2\frac{\partial^2\Phi}{\partial x^2}+\frac{\partial^2\Phi}{\partial r^2}+\frac{1}{r}\frac{\partial\Phi}{\partial r}=0. \tag{1.43}$$

Let us notice that for the incompressible motions, where

$$M_\infty\approx 0,\quad B^2=M_\infty^2-1\approx -1, \tag{1.44}$$

expressions (1.42) and (1.43) become simple *Laplacians,*

$$\nabla^2\Phi=\frac{\partial^2\Phi}{\partial x^2}+\frac{1}{r}\frac{\partial}{\partial r}\left(r\frac{\partial\Phi}{\partial r}\right)+\frac{1}{r^2}\frac{\partial^2\Phi}{\partial\omega^2}=0 \tag{1.45}$$

for general motions and

$$\nabla^2\Phi=\frac{\partial^2\Phi}{\partial x^2}+\frac{1}{r}\frac{\partial}{\partial r}\left(r\frac{\partial\Phi}{\partial r}\right)=0 \tag{1.46}$$

for axially-symmetrical motions. As is seen, the function Φ is harmonic in this case, while the above equations represent the continuity equation.

If the velocity components in the direction Ox vanish, then the motion is a plane one, represented by the potential equation in polar coordinates,

$$\nabla^2\Phi=\frac{1}{r}\frac{\partial}{\partial r}\left(r\frac{\partial\Phi}{\partial r}\right)+\frac{1}{r^2}\frac{\partial^2\Phi}{\partial\omega^2}=0. \tag{1.47}$$

I. 4. Pressure Coefficient under the Assumption of Small Disturbances

I. 4.1. Pressure Coefficient in the First Approximation

The pressure coefficient, which will be designated by C_p, is given by the relation

$$C_p = \frac{p-p_\infty}{\frac{1}{2}\varrho_\infty U_\infty^2} = \frac{p-p_\infty}{p_\infty}\frac{2}{\gamma M_\infty^2}, \tag{1.48}$$

where p_∞ and ϱ_∞ represent the free stream pressure and density respectively.

Under the assumption of small disturbances, the pressure differential equation (1.28) can also be applied to small but finite differences; we can thus write

$$\frac{\delta p}{\varrho_\infty} = -\delta\left(\frac{1}{2}V^2+\frac{\partial\Phi}{\partial t}\right). \tag{1.49}$$

Setting further

$$\delta p = p-p_\infty, \tag{1.50a}$$

$$\delta\left(\frac{1}{2}V^2+\frac{\partial\Phi}{\partial t}\right) = \frac{1}{2}(V^2-U_\infty^2)+\frac{\partial\Phi}{\partial t} \approx U_\infty u+\frac{\partial\Phi}{\partial t}, \tag{1.50b}$$

since $\partial\Phi/\partial t$ is zero at infinite, and replacing in (1.48), we finally obtain the pressure coefficient under the assumption of small disturbances,

$$C_p = -\frac{2}{U_\infty^2}\left(U_\infty u+\frac{\partial\Phi}{\partial t}\right). \tag{1.51}$$

For steady flows, this formula becomes simpler,

$$C_p = -2\frac{u}{U_\infty}. \tag{1.52}$$

I. 4.2. Pressure Coefficient in the Second Approximation

For wings with greater incidence and higher Mach numbers, but mainly for bodies of revolution, even for very slender ones, the above formula (1.52) is no longer sufficient.

Indeed, it may be demonstrated that the second degree terms of the disturbance velocity become very important and cannot be neglected.

Therefore, we shall try to complete the above approximate formula, so that it might also be applied to the case of slender bodies of revolution and eventually to the study in a second approximation of the wing characteristics.

From (1.32) we have

$$a^2 = a_\infty^2 - \frac{\gamma-1}{2}\left(V^2 - U_\infty^2 + 2\frac{\partial \Phi}{\partial t}\right), \tag{1.53}$$

which, dividing by a_∞^2 and taking account of (1.24), becomes

$$\frac{T}{T_\infty} = 1 - \frac{\gamma-1}{2a_\infty^2}\left[2uU_\infty + q^2 + 2\frac{\partial \Phi}{\partial t}\right], \tag{1.54}$$

where we set

$$q^2 = u^2 + v^2 + w^2. \tag{1.55}$$

On the other hand, taking account of (1.4) and (1.5), we deduce readily

$$\frac{T}{T_\infty} = \frac{p}{\varrho}\frac{\varrho_\infty}{p_\infty} = \left(\frac{p}{p_\infty}\right)^{(\gamma-1)/\gamma}, \tag{1.56}$$

whence

$$\frac{p}{p_\infty} = \left(\frac{T}{T_\infty}\right)^{\gamma/(\gamma-1)} = \left[1 - \frac{\gamma-1}{2}\,\frac{2uU_\infty + q^2 + 2\dfrac{\partial \Phi}{\partial t}}{a_\infty^2}\right]^{\gamma/(\gamma-1)}. \tag{1.57}$$

In this case, the pressure coefficient can be calculated to a better approximation by taking three terms in the expansion of the square bracket and neglecting the secondary terms,

$$\begin{aligned} C_p &= \frac{p-p_\infty}{p_\infty}\frac{2}{\gamma M_\infty^2} \\ &= \frac{2}{\gamma M_\infty^2}\left\{\left[1 - \frac{\gamma-1}{2}\left(2\frac{u}{U_\infty} + \frac{q^2}{U_\infty^2} + \frac{2}{U_\infty^2}\frac{\partial \Phi}{\partial t}\right)M_\infty^2\right]^{\gamma/(\gamma-1)} - 1\right\} \\ &\approx -\left(2\frac{u}{U_\infty} + \frac{v^2+w^2}{U_\infty^2} + \frac{2}{U_\infty^2}\frac{\partial \Phi}{\partial t}\right) + \frac{u^2}{U_\infty^2}(M_\infty^2 - 1) \\ &\quad + \frac{1}{a_\infty^2}\left(2\frac{u}{U_\infty} + \frac{1}{U_\infty^2}\frac{\partial \Phi}{\partial t}\right)\frac{\partial \Phi}{\partial t}. \end{aligned} \tag{1.58}$$

If the motion is steady, we obtain a simpler expression,

$$C_p = -\left(2\frac{u}{U_\infty} + \frac{v^2+w^2}{U_\infty^2}\right) + \frac{u^2}{U_\infty^2}(M_\infty^2 - 1), \tag{1.59}$$

where, for the axially-symmetrical motions, we can replace v^2+w^2 by the square of the radial velocity,

$$V_r^2 = v^2 + w^2. \tag{1.60}$$

For bodies of revolution, where V_r^2 is indeed of the same order of magnitude as u, this formula can be applied to the calculation of the axial drag due to pressures.

I. 5. Theory of Wings of Infinite Aspect Ratio in Supersonic Flow under the Assumption of Small Disturbances

We are dealing with a steady plane motion around a wing of infinite aspect ratio, whose normal section constitutes the wing profile.

I. 5.1. Pressure Distribution on the Profile (Ackeret's Theory)

We shall assume that the thickness of the profile and its incidence relative to the free stream velocity U_∞ are small, so that the method of small disturbances is applicable.

Let us refer the profile to a system of axes Oxy, having the axis Ox parallel to the free stream velocity (Fig. 1.5). Let $\Phi(x, y)$ denote the dis-

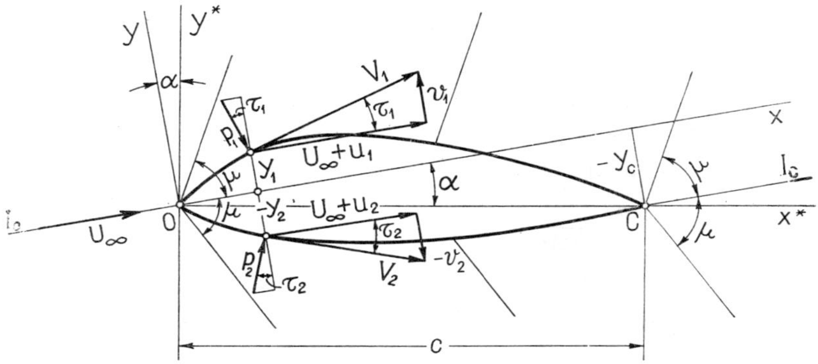

FIG. 1.5

turbance velocity potential and put

$$\varphi = U_\infty x + \Phi, \quad u = \frac{\partial \Phi}{\partial x}, \quad v = \frac{\partial \Phi}{\partial y}; \tag{1.61}$$

in this case, the potential equation (1.36) becomes

$$\frac{\partial^2 \Phi}{\partial y^2} = B^2 \frac{\partial^2 \Phi}{\partial x^2}, \tag{1.62}$$

having the well-known solution

$$\Phi = \Phi_1(x - By) + \Phi_2(x + By). \tag{1.63}$$

Referring to Fig. 1.5, it is readily seen that the fluid space is divided into two parts by the line I_0OCI_c, the potential Φ_1 being valid in the upper part and the potential Φ_2 in the lower part.

On the lines

$$x-By = C_1, \quad x+By = C_2 \tag{1.64}$$

(where C_1 and C_2 are constants), which are called *simple Mach lines* because they make the angle μ, $-\mu$, respectively, with the axis Ox,

$$\left(\frac{dy}{dx}\right)_1 = \frac{1}{B} = \tan\mu, \quad \left(\frac{dy}{dx}\right)_2 = -\frac{1}{B} = -\tan\mu, \tag{1.65}$$

the values of the potential are constants.

Writing further that the equations of the contour are given by the equations

$$y_1 = y_1(x), \quad y_2 = y_2(x), \tag{1.66}$$

the subscripts 1 and 2 referring to the upper and lower surfaces respectively, we obtain on *the profile contour*,

$$\frac{dy_1}{dx} = \frac{v_1}{U_\infty+u_1} \approx \frac{v_1}{U_\infty} = \frac{1}{U_\infty}\frac{\partial\Phi_1}{\partial y}, \tag{1.67a}$$

$$\frac{dy_2}{dx} = \frac{v_2}{U_\infty+u_2} \approx \frac{v_2}{U_\infty} = \frac{1}{U_\infty}\frac{\partial\Phi_2}{\partial y}. \tag{1.67b}$$

From the potential expression (1.63), we deduce, on the upper and lower surfaces,

$$u_1 = \frac{\partial\Phi_1}{\partial x} = -\frac{1}{B}\frac{\partial\Phi_1}{\partial y} = -\frac{v_1}{B} = -\frac{U_\infty}{B}\frac{dy_1}{dx}, \tag{1.68a}$$

$$u_2 = \frac{\partial\Phi_2}{\partial x} = \frac{1}{B}\frac{\partial\Phi_2}{\partial y} = \frac{v_2}{B} = \frac{U_\infty}{B}\frac{dy_2}{dx}, \tag{1.68b}$$

respectively, whence it results for the pressure coefficient on the profile contour, on the upper and lower surfaces,

$$C_{p_1} = -2\frac{u_1}{U_\infty} = \frac{2}{B}\frac{dy_1}{dx}, \quad C_{p_2} = -2\frac{u_2}{U_\infty} = -\frac{2}{B}\frac{dy_2}{dx}. \tag{1.69}$$

I. 5.2. *Aerodynamic Characteristics*

Lift coefficient. Denoting by τ the angle made by the tangent to the profile with the axis Ox,

$$\frac{dy}{dx} = \tan\tau \approx \tau, \tag{1.70}$$

and by ds a surface element of the contour, we shall write for the lift coefficient

$$C_l = \frac{1}{c}\int_{0c} C_{p_2} \cos \tau_2 \, ds_2 - \frac{1}{c}\int_{0c} C_{p_1} \cos \tau_1 \, ds_1$$

$$= -\frac{2}{Bc}\left(\int_0^c \frac{dy_1}{dx}\,dx + \int_0^c \frac{dy_2}{dx}\,dx\right) = -\frac{4}{B}\frac{y_c}{c} = \frac{4\alpha}{B}, \qquad (1.71)$$

where we have replaced y_c/c by the incidence made by the profile chord with the direction of the free stream

$$\alpha \approx -\frac{y_c}{c}. \qquad (1.72)$$

Drag coefficient. For the drag coefficient due to pressures, we shall successively write

$$C_d = \frac{1}{c}\int_{0c} C_{p_1} \sin \tau_1 \, ds_1 - \frac{1}{c}\int_{0c} C_{p_2} \sin \tau_2 \, ds_2$$

$$= \frac{2}{Bc}\left(\int_0^c \left(\frac{dy_1}{dx}\right)^2 dx + \int_0^c \left(\frac{dy_2}{dx}\right)^2 dx\right). \qquad (1.73)$$

If we refer the profile to a system of axes Ox^*y^*, where the profile chord lies along the axis Ox^* (Fig. 1.5), we shall write approximately

$$x^* \approx x, \quad y_1 \approx y_1^* - \alpha x, \quad -y_2 \approx -y_2^* + \alpha x; \qquad (1.74)$$

differentiating and substituting in (1.73), we finally obtain

$$C_d = \frac{2}{Bc}\left(2\alpha^2 c + \int_0^c \left(\frac{dy_1^*}{dx}\right)^2 dx + \int_0^c \left(\frac{dy_2^*}{dx}\right)^2 dx\right). \qquad (1.75)$$

Moment coefficient. If we examine again Fig. 1.5, we can write for the moment coefficient, with respect to the leading edge O,

$$\mu = \frac{1}{c^2}\int_{0c} C_{p_2}(x \cos \tau_2 + y_2 \sin \tau_2)\, ds_2 - \frac{1}{c^2}\int_{0c} C_{p_1}(x \cos \tau_1 + y_1 \sin \tau_1)\, ds_1$$

$$\approx -\frac{2}{Bc^2}\int_{0c} x(dy_1 + dy_2). \qquad (1.76)$$

Taking account of (1.74) and (1.71), we obtain

$$\mu = \tfrac{1}{2}\frac{4\alpha}{B} - \frac{2}{Bc^2}\int_{0c} x(dy_1^* + dy_2^*) = \tfrac{1}{2} C_l + \frac{2}{B}\frac{S_1 - S_2}{c^2} \qquad (1.77)$$

where S_1 represents the area of the upper section of the profile and S_2 the area of the lower section (considered with respect to the profile chord).

I. 5.3. Yawed Wing of Infinite Aspect Ratio

Let it be assumed that a wing of infinite aspect ratio makes an angle χ with the normal to the free stream velocity; then the wing is said to be *yawed*.

Let U_l and U_n denote the velocity components in the direction of the span and normal to it respectively (Fig. 1.6),

$$U_l = U_\infty \sin\chi, \quad U_n = U_\infty \cos\chi. \tag{1.78}$$

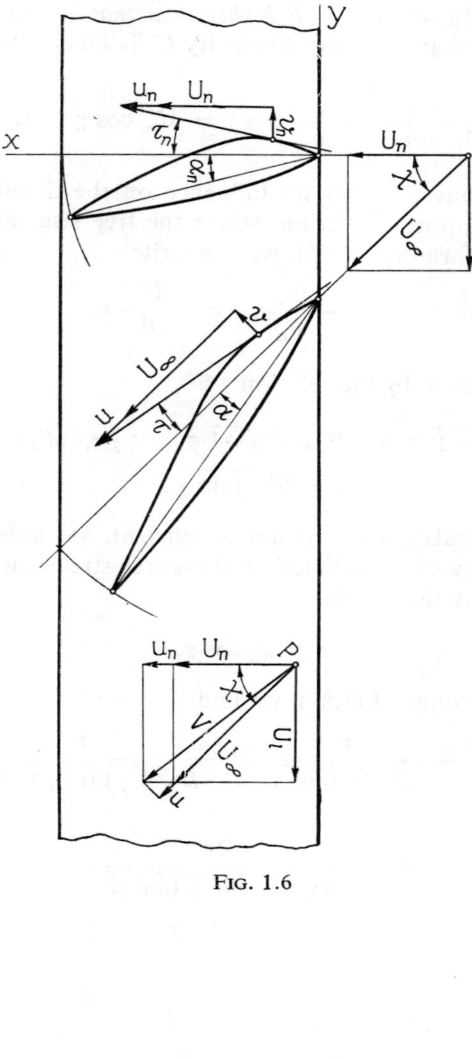

FIG. 1.6

As we can see, the longitudinal velocity U_l plays no part in the motion of an inviscid perfect fluid, by giving the wing a spanwise translation motion which produces no new effect. It readily results that the normal component U_n alone has a direct action upon the wing, thus playing the role of the velocity U_∞ in a non-yawed wing, that is a wing which moves in a direction normal to the direction of its span.

Considering two sections of the wing, a normal one and the other in the velocity direction, we obtain two profiles having the chords c_n, c, respectively, the angles of incidence α_n, α, respectively, and the tangents to the profile τ_n, τ at corresponding points, connected by the relations

$$c_n = c \cos \chi, \quad \alpha = \alpha_n \cos \chi, \quad \tau = \tau_n \cos \chi. \tag{1.79}$$

The case when the velocity U_n is higher than that of sound ($U_n > a_\infty$). Let it be assumed that the normal velocity U_n is higher than the speed of sound and set

$$M_n = \frac{U_n}{a_\infty} = \frac{U_\infty}{a_\infty} \cos \chi = M_\infty \cos \chi > 1. \tag{1.80}$$

As indicated above, U_l has no influence on the disturbance velocity distribution in the normal section, where the free stream velocity is U_n, so that, applying formulae (1.68), we can write

$$u_n = -\frac{U_n}{B_n} \tau_n = -\frac{U_\infty}{B_n} \tau, \tag{1.81}$$

where B_n will be given by the relation

$$\begin{aligned} B_n = \sqrt{M_n^2 - 1} &= \sqrt{M_\infty^2 \cos^2 \chi - 1} = \cos \chi \sqrt{M_\infty^2 - 1 - \tan^2 \chi} \\ &= \cos \chi \sqrt{B^2 - \tan^2 \chi}. \end{aligned} \tag{1.82}$$

In order to calculate the pressure coefficient, we must calculate the disturbance velocity u in the direction of the free stream, which, as seen in Fig. 1.6, is given by the relation

$$u = u_n \cos \chi, \tag{1.83}$$

whence, taking account of (1.81), we find

$$\frac{u}{U_\infty} = -\frac{\tau}{\sqrt{B^2 - \tan^2 \chi}} = -\frac{\tau}{B\sqrt{1 - \tan^2 \chi \tan^2 \mu}} \tag{1.84}$$

and consequently

$$C_p = \frac{2\tau}{B\sqrt{1 - \tan^2 \chi \tan^2 \mu}}. \tag{1.85}$$

Starting from this formula for the pressure coefficient, we can calculate the lift, drag and moment coefficients which have the same expression as for the normal wing ($\chi = 0$), but with B in the denominator of these formulae replaced by B', where

$$B' = B\sqrt{1-\tan^2\chi\tan^2\mu}\,. \tag{1.86}$$

We note, however, that in this case the corresponding integrals must be taken along the profile of the wing section parallel to the free stream velocity, as is usually done for swept-back and delta wings, which will be later dealt with.

The case when the velocity U_n is lower than the speed of sound ($U_n < a_\infty$). In this case, the motion over the wing is effectively subsonic, with all the consequences related to such a motion, as, for instance, the velocity behaviour at the leading edge of a thin wing which tends towards infinity according to a law deduced from the equations of motion of perfect fluids.

For certain subsequent applications, we shall study particularly the behaviour of the disturbance velocities in a compressible fluid with respect to those in an incompressible fluid.

If u_n and v_n are the disturbance velocities in a plane Oxy normal to the span, velocities that are derived from a potential Φ_n, then applying equation (1.36) of the potential (restricted of course to the two-dimensional problem) we may write

$$(1-M_n^2)\frac{\partial^2\Phi_n}{\partial x^2}+\frac{\partial^2\Phi_n}{\partial y^2} = (1-M_n^2)\frac{\partial u_n}{\partial x}+\frac{\partial v_n}{\partial y} = 0. \tag{1.87}$$

Within the assumption of small disturbances, the boundary conditions are given by the relation

$$\left(\frac{dy}{dx}\right)_{\text{pr}} = \left(\frac{v_n}{U_\infty+u_n}\right)_{\text{pr}} \approx \left(\frac{v_n}{U_\infty}\right)_{\text{pr}}. \tag{1.88}$$

Let us now consider a system of axes $O\xi y$ representing a plane in which the motion is incompressible and let us make a transformation defined by the relations

$$x = \sqrt{1-M_n^2}\,\xi, \quad y = y, \tag{1.89a}$$

$$u_n = \sqrt{1-M_n^2}\,u_{ni}, \quad v_n = v_{ni}, \quad \Phi_n = \Phi_{ni}, \tag{1.89b}$$

where u_{ni}, v_{ni} and Φ_{ni} represent the velocities and the potential of an incompressible motion in the plane $O\xi y$, respectively, which are derived from the transformations (1.89).

Indeed, replacing in (1.87) Φ_n by Φ_{ni} and x by ξ, we obtain

$$\frac{\partial^2 \Phi_{ni}}{\partial \xi^2} + \frac{\partial^2 \Phi_{ni}}{\partial y^2} = \frac{\partial u_{ni}}{\partial \xi} + \frac{\partial v_{ni}}{\partial y} = 0, \tag{1.90}$$

which is just the potential equation in an incompressible motion, for which the boundary conditions on the profile will be, similarly to (1.88),

$$\left(\frac{dy}{d\xi}\right)_{\text{pr}} \approx \left(\frac{v_{ni}}{U_\infty}\right)_{\text{pr}} = \left(\frac{v_n}{U_\infty}\right)_{\text{pr}} \approx \left(\frac{dy}{dx}\right)_{\text{pr}}; \tag{1.91}$$

hence the profiles are similar in both planes.

Since the velocity u_{ni} is known (from the incompressible motion), the velocity u_n follows readily as a function of u_{ni}; on the other hand, taking account of (1.83), which corresponds for the incompressible flow to $u_i = u_{ni} \cos \chi$, we can write successively

$$\frac{u}{\cos \chi} = u_n = \frac{\partial \Phi_n}{\partial x} = \frac{\partial \Phi_{ni}}{\partial \xi} \frac{d\xi}{dx} = \frac{u_{ni}}{\sqrt{1-M_n^2}} = \frac{u_i}{\sqrt{1-M_n^2}} \frac{1}{\cos \chi}. \tag{1.92}$$

Likewise, we obtain for the pressure coefficient

$$C_p = -2 \frac{u}{U_\infty} = -2 \frac{u_i}{U_\infty} \frac{1}{\sqrt{1-M_n^2}} = \frac{C_{pi}}{\sqrt{1-M_n^2}}, \tag{1.93}$$

where C_{pi} is the pressure coefficient in the incompressible flow.

Although a treatment of the subsonic motion is not included in this work, which deals exclusively with supersonic motion under the assumption of small disturbances, we shall have to apply the above formulae later to the calculation of the *suction* force, for thin wings with subsonic leading edges, where this suction phenomenon occurs.

I. 6. Theory of Axially-symmetric Bodies under the Assumption of Small Disturbances (Application to the Circular Cone)

The problem of the flow around bodies of revolution does not constitute as such the subject of our investigation. Nevertheless, in one of our next chapters, we discuss the problem of the wing–conical body interference, a problem which implies the knowledge of the flow around this axially-symmetrical body, considered alone; this will be examined below.

I. 6.1. *Resolution of the General Motion*

Let us consider a system of axes of cylindrical coordinates x, r, ω; under the assumption of small disturbances, the general motion around a *slender body* of circular cross-section, having an *incidence* α with respect to the free stream, may be resolved into two parts: an axially-symmetrical one due to the free stream having the velocity U_∞, parallel to the axis of the body of revolution, and the second due to the velocity v_∞, much smaller than U_∞, normal to this one and parallel to the axis Oy (Fig. 1.7).

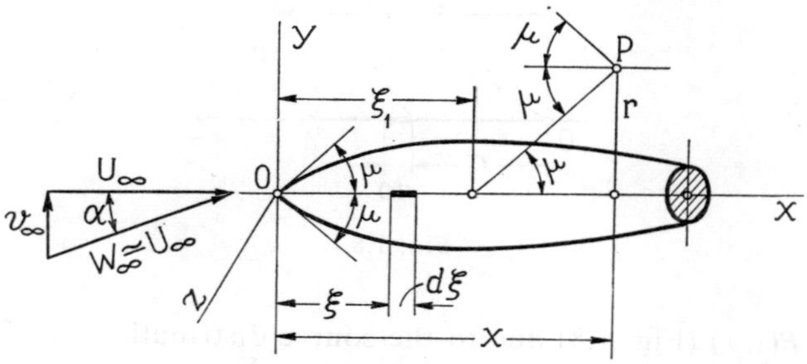

FIG. 1.7

The angle of incidence, which is given by the relation

$$\frac{v_\infty}{U_\infty} = \tan\alpha \approx \alpha, \tag{1.94}$$

is assumed to be very small, as well as the thickness of the slender body, in order that the assumption of small disturbances may be applicable.

Denoting by Φ_1 the perturbation velocity potential in the axially-symmetrical motion, and by Φ_2 the velocity potential in the lateral motion, we shall write for the total potential

$$\varphi = U_\infty x + \Phi_1 + \Phi_2. \tag{1.95}$$

We shall discuss separately below the axially-symmetric motion and the lateral motion.

I. 6.2. *Axially-symmetric Flow*

The potential Φ_1 of the supersonic axially-symmetric flow around bodies of revolution will satisfy equation (1.43):

$$-B^2\frac{\partial^2\Phi_1}{\partial x^2} + \frac{\partial^2\Phi_1}{\partial r^2} + \frac{1}{r}\frac{\partial\Phi_1}{\partial r} = 0. \tag{1.96}$$

In order to find the solution of this equation, let us first consider the incompressible subsonic motion, for which the slender bodies may be generated through the method of sources distributed upon the slender body axis, considered as axis of a system of coordinates s, r.

In this system, let us consider an elementary source $q(\sigma)\,d\sigma$, $q(\sigma)$ being the intensity of the source per unit length, at point σ. The velocity potential at

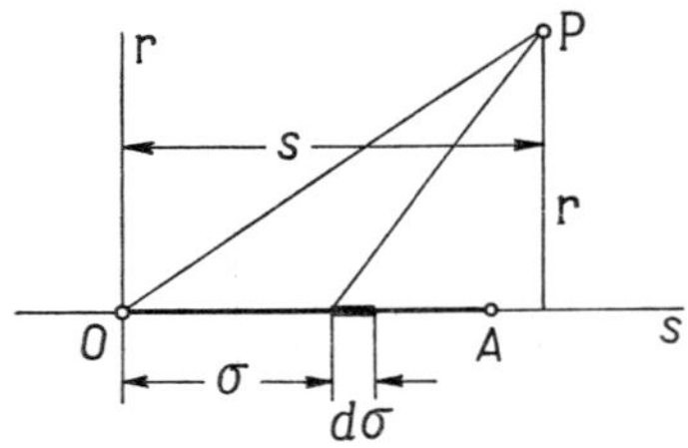

FIG. 1.8

the point $P(s, r)$ (Fig. 1.8) due to the source distribution on the segment OA, will be

$$\Phi_1 = -\frac{1}{4\pi}\int_{OA}\frac{q(\sigma)\,d\sigma}{\sqrt{(s-\sigma)^2+r^2}}, \tag{1.97}$$

which satisfies the equation of the incompressible motion (1.46), in which x has been replaced by s,

$$\frac{\partial^2\Phi_1}{\partial s^2}+\frac{\partial^2\Phi_1}{\partial r^2}+\frac{1}{r}\frac{\partial\Phi_1}{\partial r} = 0. \tag{1.98}$$

If we perform the change of variables

$$x = iBs, \quad \xi = iB\sigma, \tag{1.99}$$

we can write in these new coordinates, with the introduction of a coefficient of proportionality,

$$\Phi_1 = \int_0^{\xi_1}\frac{f(\xi)\,d\xi}{\sqrt{(x-\xi)^2-B^2r^2}}. \tag{1.100}$$

On the other hand, introducing the new variables x, r into (1.98), we obtain for Φ_1 the equation (1.96) for the potential of a supersonic axially-symmetric motion under the assumption of small disturbances.

In this way, by analogy with the incompressible motion, we have found the formula with the aid of which we can construct the potential for the motion around a slender body in supersonic flow.

Choosing the function $f(\xi)$, in the manner indicated by Kármán and Moore [K-2], various forms of slender bodies can be obtained. We first notice that the integral cannot extend beyond a value ξ_1, beyond which the above expression (1.100) would become imaginary (Fig. 1.7); hence, the expression under the square root sign should be always positive, that is

$$(x-\xi)^2 - B^2r^2 \geqslant 0, \tag{1.101}$$

whence it follows, for the limit of integration,

$$\xi_1 = x - Br. \tag{1.102}$$

Further, we have to perform the integration (1.100) for various forms of the function $f(\xi)$.

In order to simplify the evaluation of this integral, we shall introduce a new variable z defined by the relation

$$\xi = x - Br \cosh z, \quad d\xi = -Br \sinh z \, dz, \tag{1.103}$$

whence it follows that

$$z = \cosh^{-1}\frac{x}{Br}, \quad \text{for} \quad \xi = 0,$$

$$z = 0, \quad \text{for} \quad \xi = \xi_1.$$

Substituting into (1.100), we obtain on

$$\Phi_1 = -\int_{\cosh^{-1}\frac{x}{Br}}^{0} f(x - Br\cosh z)\, dz, \tag{1.104}$$

and

$$u_1 = \frac{\partial \Phi_1}{\partial x} = -\int_{\cosh^{-1}\frac{x}{Br}}^{0} f'(x - Br\cosh z)\, dz + \frac{f(0)}{\sqrt{x^2 - B^2r^2}}, \tag{1.105a}$$

$$v_1 = \frac{\partial \Phi_1}{\partial r} = B\int_{\cosh^{-1}\frac{x}{Br}}^{0} f'(x - Br\cosh z)\cosh z\, dz - \frac{x}{r}\frac{f(0)}{\sqrt{x^2 - B^2r^2}}, \tag{1.105b}$$

where $f'(x - Br\cosh z) = f'(\xi)$ represents the derivative with respect to $\xi = x - Br\cosh z$; on the other hand, we notice that $f(0) = 0$, since the nose of the slender bodies considered in supersonic flow is supposed to be sharp. Applying the above formulae, it is easy to consider various applications.

Application to the circular cone. If we choose for $f(x-Br\cosh z)$ the simplest expression, that is $f = -C_1(x-Br\cosh z)$, we obtain then the axially-symmetrical flow around a circular cone (Fig. 1.9); in fact we have successively

$$\Phi_1 = C_1 \int_{\cosh^{-1}\frac{x}{Br}}^{0} (x-Br\cosh z)\,dz = C_1\left(\sqrt{x^2-B^2r^2} - x\cosh^{-1}\frac{x}{Br}\right), \tag{1.106a}$$

$$u_1 = C_1 \int_{\cosh^{-1}\frac{x}{Br}}^{0} dz = -C_1 \cosh^{-1}\frac{x}{Br}, \tag{1.106b}$$

$$v_1 = -C_1 B \int_{\cosh^{-1}\frac{x}{Br}}^{0} \cosh z\,dz = \frac{C_1}{r}\sqrt{x^2-B^2r^2}. \tag{1.106c}$$

As is seen, the velocities u_1 and v_1 are the same on any radius vector

$$\frac{r}{x} = \text{const.} \tag{1.107}$$

This motion is called a *conical motion*, which will be generalized and extended to triangular wings later. From the above expressions it is easily seen that u_1 and v_1 are homogeneous functions of zero order, in x and r.

Let us notice now that the velocities are real only if

$$\frac{r}{x} < \frac{1}{B} = \tan\mu, \tag{1.108}$$

which means that the motion is included in the Mach cone (Fig. 1.9).

We can find the value of the velocity on the cone by

$$\frac{r}{x} = \tan\delta, \tag{1.109}$$

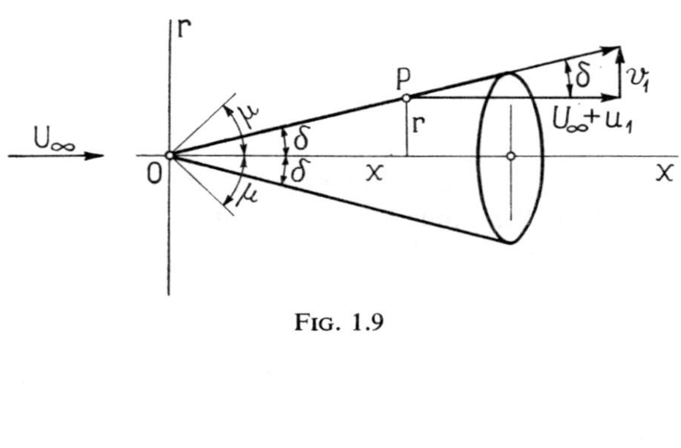

FIG. 1.9

the angle δ being the semi-angle at the cone vertex. Further, writing down the ratio of the total velocities on the cone, we obtain

$$\tan\delta = \frac{v_1}{U_\infty + u_1} = \frac{C_1\sqrt{\cot^2\delta - B^2}}{U_\infty - C_1\cosh^{-1}\left(\dfrac{\cot\delta}{B}\right)}, \tag{1.110}$$

whence we deduce the value of the constant C_1,

$$C_1 = \frac{U_\infty \tan\delta}{\sqrt{\cot^2\delta - B^2} + \tan\delta\cosh^{-1}\dfrac{\cot\delta}{B}} = \frac{U_\infty}{\cot\delta\sqrt{\cot^2\delta - B^2} + \cosh^{-1}\dfrac{\cot\delta}{B}}. \tag{1.111}$$

Taking account of (1.60), where V_r corresponds to the above v_1, and applying the formula (1.59) obtained previously, we obtain the pressure coefficient,

$$\begin{aligned} C_p &= -2\left(\frac{u_1}{U_\infty} + \frac{1}{2}\frac{v_1^2}{U_\infty^2}\right) + B^2\frac{u_1^2}{U_\infty^2} \\ &= 2\frac{C_1}{U_\infty}\cosh^{-1}\frac{\cot\delta}{B} - \frac{C_1^2}{U_\infty^2}(\cot^2\delta - B^2) + B^2\left(\frac{C_1}{U_\infty}\cosh^{-1}\frac{\cot\delta}{B}\right), \end{aligned} \tag{1.112}$$

where the constant C_1 is replaced by its value given above (1.111).

I. 6.3. *Lateral Motion around Bodies of Revolution*

The general lateral motion around a slender body, due to the velocity v_∞ directed towards the axis oy, is defined by a potential Φ_2 of the form

$$\Phi_2 = \left(rv_\infty + \frac{\partial\Phi_1}{\partial r}\right)\cos\omega, \tag{1.113}$$

where Φ_1 represents an arbitrary axially-symmetrical potential that satisfies equation (1.96). Indeed, introducing this expression into the general equation of small disturbances in the form (1.42), it may readily be seen that Φ_2 satisfies this equation; that is we obtain

$$-B^2\frac{\partial^2\Phi_2}{\partial x^2} + \frac{\partial^2\Phi_2}{\partial r^2} + \frac{1}{r}\frac{\partial\Phi_2}{\partial r} + \frac{1}{r^2}\frac{\partial^2\Phi_2}{\partial\omega^2} = 0 \tag{1.114}$$

and hence we can construct Φ_2 in the following way.

Taking account of (1.105 b), where $f(0) = 0$, we can write for the general form of the potential

$$\Phi_2 = \cos\omega\left[rv_\infty + \int_{\cosh^{-1}\frac{x}{Br}}^{0} g(x - Br\cosh z)\cosh z\,dz\right], \quad (1.115)$$

$g(x - Br\cosh z)$ being an arbitrary function of $(x - Br\cosh z)$.

This is the way in which the lateral motions of slender bodies have been studied by C. Ferrari [F–8] and H. S. Tsien [T–8].

Application to the circular cone. Taking the simplest expression for $g(x - Br\cosh z)$, that is $g = 2C_2(x - Br\cosh z)$, we obtain the potential for motion around the circular cone,

$$\Phi_2 = rv_\infty\cos\omega - 2C_2\cos\omega\left[Br\cosh^{-1}\frac{x}{Br} - x\sqrt{\frac{x^2}{B^2r^2} - 1}\right] \quad (1.116)$$

and, by differentiation,

$$u_2 = \frac{\partial\Phi_2}{\partial x} = 2C_2\cos\omega\sqrt{\frac{x^2}{B^2r^2} - 1}, \quad (1.117a)$$

$$v_2 = \frac{\partial\Phi_2}{\partial r} = v_\infty\cos\omega - C_2\cos\omega\left[B\cosh^{-1}\frac{x}{Br} + \frac{x}{r}\sqrt{\frac{x^2}{B^2r^2} - 1}\right], \quad (1.117b)$$

where the constant C_2 will be determined as above, by imposing the condition that the ratio of the velocities on the cone shall be equal to the tangent of the semi-angle at the vertex,

$$\tan\delta = \frac{v_2}{u_2} = \frac{v_\infty - C_2\left[B\cosh^{-1}\frac{\cot\delta}{B} + \cot\delta\sqrt{\frac{\cot^2\delta}{B^2} - 1}\right]}{2C_2\sqrt{\frac{\cot^2\delta}{B^2} - 1}}; \quad (1.118)$$

whence

$$C_2 = \frac{v_\infty}{B\cosh^{-1}\frac{\cot\delta}{B} + \cot\delta\sqrt{\frac{\cot^2\delta}{B^2} - 1 + 2\sqrt{\frac{1}{B^2} - \tan^2\delta}}}. \quad (1.119)$$

The third velocity component, which will be denoted by V_ω, can easily be deduced from (1.116), by differentiating with respect to ω,

$$V_\omega = \frac{\partial\Phi_2}{r\partial\omega} = -v_\infty\sin\omega + C_2\sin\omega\left(B\cosh^{-1}\frac{x}{Br} - \frac{x}{r}\sqrt{\frac{x^2}{B^2r^2} - 1}\right). \quad (1.120)$$

Taking into account that v_∞ is small, and $u_2 \approx v_\infty\tan\delta\cos\omega$ on the cone is hence very small, one can easily see that this lateral motion introduces only secondary terms as far as the pressure drag in the direction of the cone axis is concerned: on the one hand because u_2 assumes equal values of opposite sign for ω and $\frac{1}{2}\pi-\omega$, and on the other hand because u_2^2, v_2^2 and V_ω^2 are secondary terms.

For the same reason, as far as u_2 is concerned, we obtain a normal (or lateral) component.

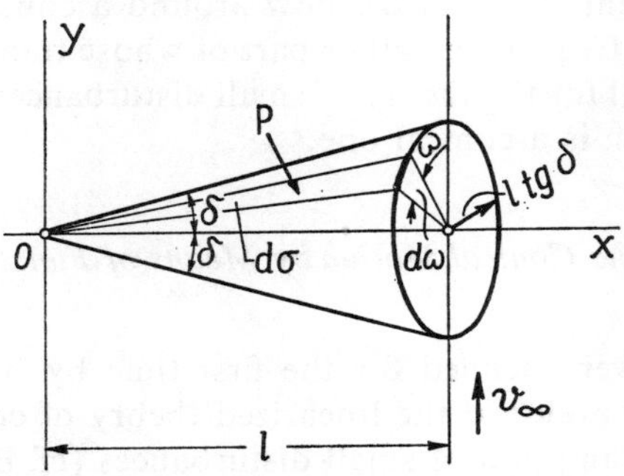

FIG. 1.10

Indeed, considering Fig. 1.10, we shall denote a triangular element $d\sigma$ on a length l of the cone

$$d\sigma = \tfrac{1}{2}\frac{l}{\cos\delta}\, l \tan\delta\, d\omega, \tag{1.121}$$

where the same pressure p is exerted; that is where we have the same pressure coefficient C_p,

$$C_p = -\,2\frac{u_2}{U_\infty} = -4\frac{C_2}{U_\infty}\cos\omega\sqrt{\frac{\cot^2\delta}{B^2}-1}\,. \tag{1.122}$$

Projecting then the pressure once on a normal to the cone axis (multiplying by $\cos\delta$) and once on a parallel to Oy (multiplying by $\cos\omega$), and taking on the other hand the circular section of the cone as a reference surface, we finally obtain

$$C_l = \frac{-2\frac{1}{2}l^2\tan\delta}{\pi\, l^2\tan^2\delta}\int_0^\pi C_p\cos\omega\, d\omega = 2\frac{C_2}{U_\infty}\cot\delta\sqrt{\frac{\cot^2\delta}{B^2}-1}\,. \tag{1.123}$$

I. 7. Conical Motions in Supersonic Flow

If all characteristics of the motion, that is velocity, pressure, density, etc., are the same on a radius vector issuing from a point O, but may differ from one radius vector to another, then this motion is called *conical motion*.

For instance, the motion around a cone of revolution, having the axis directed streamwise, is a rigorous conical motion if the vertex angle and the incidence of the cone axis with respect to the stream do not exceed a limit, beyond which the shock wave at the cone vertex detaches, while the motion ceases to be a conical one.

Another more characteristic example, which will be considered particularly in the present study, is the flow around a conical body, lying in the direction of the free stream, all or part of whose transverse dimensions are sufficiently small for the theory of small disturbances to be applicable; the resulting motion is a conical one.

I. 7.1. Solution of the Conical Motion by Means of Functions of a Complex Variable

These motions were defined for the first time by Busemann in 1935, who subsequently developed the linearized theory of conical motions on the basis of the assumption of small disturbances (cf. Busemann (1943)).

A series of other authors, most of them cited in the general references, have discussed several aspects of this kind of motion using various methods and obtaining very interesting results, which have been of use in the calculation of supersonic wings.

From the definition of the conical motion, it follows that the velocity is a function of r_1/x_1; hence it is homogeneous of zero order with respect to these variables.

Let q be the projection of the velocity upon any direction l and let Φ designate the velocity potential; then

$$q = \frac{\partial \Phi}{\partial l}\,. \tag{1.124}$$

This velocity satisfies an equation of the form (1.42), in which, however, Φ will be replaced by q.

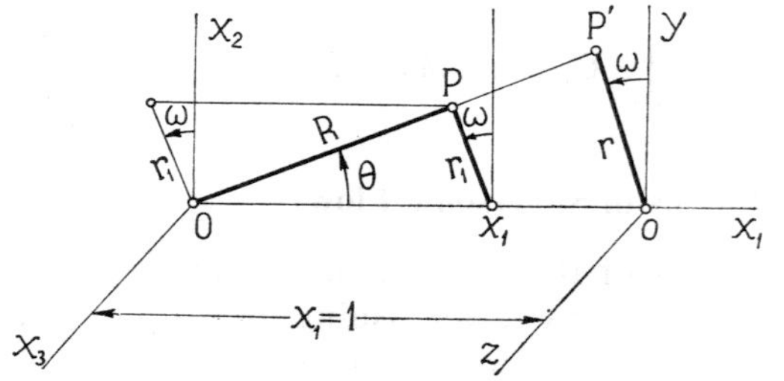

FIG. 1.11

Indeed, in a system of cylindrical coordinates x_1, r_1, ω, we have

$$-B^2\frac{\partial^2 q}{\partial x_1^2}+\frac{1}{r_1}\frac{\partial}{\partial r_1}\left(r_1\frac{\partial q}{\partial r_1}\right)+\frac{1}{r_1^2}\frac{\partial^2 q}{\partial\omega^2}$$
$$=\frac{\partial}{\partial l}\left[-B^2\frac{\partial^2 \Phi}{\partial x_1^2}+\frac{1}{r_1}\frac{\partial}{\partial r_1}\left(r_1\frac{\partial \Phi}{\partial r_1}\right)+\frac{1}{r_1^2}\frac{\partial^2 \Phi}{\partial\omega^2}\right]=0. \quad (1.125)$$

For reasons which will be seen later, the usual coordinates x, r have been replaced by x_1, r_1.

Let us now consider another system of coordinates which is derived from the first one and in which only r_1 is changed, that is

$$x_1 = x_1, \quad \mathbf{r} = ar_1, \quad \omega = \omega. \quad (1.126)$$

Let the velocity q be assumed to be a harmonic function with respect to this system, that is having the *Laplacian* equal to zero, as established above for the harmonic functions (1.46),

$$\triangle q=\nabla^2 q=\frac{\partial^2 q}{\partial x_1^2}+\frac{1}{\mathbf{r}}\frac{\partial}{\partial \mathbf{r}}\left(\mathbf{r}\frac{\partial q}{\partial \mathbf{r}}\right)+\frac{1}{\mathbf{r}^2}\frac{\partial^2 q}{\partial\omega^2}=0. \quad (1.127)$$

As is seen, in order that this equation may be derived from the first, we must take

$$a = iB. \quad (1.128)$$

In order to facilitate the calculation further, we shall use spherical coordinates R, θ, ω, by putting

$$x_1 = R\cos\theta, \quad \mathbf{r} = R\sin\theta, \quad \omega=\omega, \quad (1.129)$$

which will be introduced into (1.127). For this purpose, we put

$$x_1^2+\mathbf{r}^2 = R^2, \quad \frac{\mathbf{r}}{x_1}=\tan\theta, \quad (1.130)$$

whence

$$\frac{\partial\theta}{\partial x_1}=-\frac{\sin\theta}{R}, \quad \frac{\partial\theta}{\partial\mathbf{r}}=\frac{\cos\theta}{R}, \quad \frac{\partial R}{\partial x_1}=\frac{x_1}{R}. \quad (1.131)$$

Further, taking into account that q does not vary with R, the motion being conical, we shall find successively:

$$\frac{\partial^2 q}{\partial x_1^2}=\frac{\sin\theta}{R}\frac{\partial}{\partial\theta}\left(\frac{\sin\theta}{R}\frac{\partial q}{\partial\theta}\right)+\frac{\sin\theta\cos\theta}{R^2}\frac{\partial q}{\partial\theta}, \quad (1.132a)$$

$$\frac{1}{\mathbf{r}}\frac{\partial}{\partial\mathbf{r}}\left(\mathbf{r}\frac{\partial q}{\partial\mathbf{r}}\right)=\frac{\cos^2\theta}{R^2\sin\theta}\frac{\partial}{\partial\theta}\left(\sin\theta\frac{\partial q}{\partial\theta}\right)-\frac{\sin\theta\cos\theta}{R^2}\frac{\partial q}{\partial\theta}, \quad (1.132b)$$

which we introduce into (1.127) and obtain the condition that the velocity be a harmonic function:

$$\frac{\partial}{\partial \theta}\left(\sin\theta\frac{\partial q}{\partial \theta}\right)+\frac{\partial}{\partial \omega}\left(\frac{1}{\sin\theta}\frac{\partial q}{\partial \omega}\right) = 0. \tag{1.133}$$

Performing the following transformation according to Gurevich [G—30],

$$\theta = i\Theta, \quad \tanh\Theta = \frac{2\varrho}{1+\varrho^2}, \tag{1.134}$$

the above equation becomes

$$\frac{1}{\varrho}\frac{\partial}{\partial \varrho}\left(\varrho\frac{\partial q}{\partial \varrho}\right)+\frac{1}{\varrho^2}\frac{\partial^2 q}{\partial \omega^2} = 0, \tag{1.135}$$

which is similar to (1.47), if we replace Φ by q and r by ϱ, which shows that q is in fact a harmonic function in the plane of polar coordinates ϱ and ω.

Further we shall write

$$\frac{\mathbf{r}}{x_1} = iB\frac{r_1}{x_1} = \tan\theta = i\tanh\Theta = i\frac{2\varrho}{1+\varrho^2}. \tag{1.136}$$

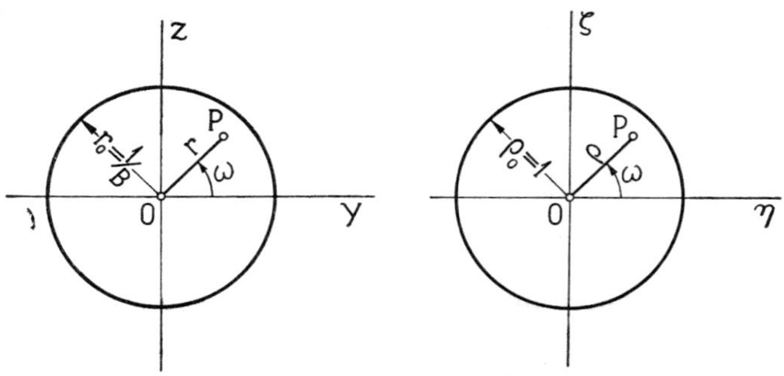

FIG. 1.12

Considering the system of coordinates x_1, r_1, ω or x_1, x_2, x_3 (Fig. 1.11), this relation given by Busemann [B—24] is valid in any plane normal to Ox_1 and hence to a certain plane

$$x_1 = 1\,; \tag{1.137}$$

in this case, relation (1.130) may be written under the simple form

$$Br = \frac{2\varrho}{1+\varrho^2}, \quad \omega = \omega, \quad \left(r = \frac{r_1}{x_1}\right), \tag{1.138}$$

which transforms the radius vectors of the physical plane Oyz into those of the auxiliary plane $O\eta\varsigma$ (Fig. 1.12).

The coordinates in these planes are defined by the following relations:

$$y = \frac{x_2}{x_1} = r\cos\omega, \quad z = \frac{x_3}{x_1} = r\sin\omega, \tag{1.139a}$$

$$\eta = \varrho\cos\omega, \quad \zeta = \varrho\sin\omega. \tag{1.139b}$$

It is clearly seen that the coordinates y, z, r, respectively, are similar to the real ones x_2, x_3, r_1, respectively, in the plane $x_1 = 1$.

This physical plane plays a particular part since the knowledge of the characteristics of the motion (velocities, pressures, etc.) at the points of this plane determines the whole motion; indeed, on each radius vector issuing from the vertex of the Mach cone and passing through a point P, we have the same characteristics.

The Mach cone intersects the physical plane that we are considering ($x_1 = 1$) in a circle of radius

$$r_0 = \tan\mu = \frac{1}{B}; \tag{1.140}$$

in the auxiliary plane this circle corresponds to another circle of radius

$$\varrho_0 = 1. \tag{1.141}$$

In this auxiliary plane, the disturbance velocity components, that is u, v, w, as well as the arbitrary component q, are harmonic functions and they can be related with the corresponding associated harmonic functions u', v', w', thus obtaining analytical functions of the complex variable

$$\xi = \eta + i\zeta = \varrho e^{i\omega}; \tag{1.142}$$

in this sense, we shall define and designate these functions in the following manner

$$\mathcal{U} = u + iu', \quad \mathcal{V} = v + iv', \quad \mathcal{W} = w + iw'. \tag{1.143}$$

I. 7.2. Compatibility Relations in the Conical Motion

As shown above, in Busemann's auxiliary plane, which will be referred to in the following as the ξ-plane and which was obtained through the geometrical transformation (1.138), the components u, v, w of the disturbance velocity represent the real parts of the analytical functions $\mathcal{U}, \mathcal{V}, \mathcal{W}$. These functions will have the same properties in any other plane, x for instance (not to be taken as an arbitrary coordinate along the axis Ox_1),

$$x = y + i\mathfrak{z} = x(\xi), \tag{1.144}$$

the conformal transformation of the ξ-plane (Fig. 1.13). In this sense, we shall consider a certain special transformation

$$Bx = \frac{2\xi}{1+\xi^2}, \quad \left(\xi = \frac{1-\sqrt{1-B^2x^2}}{Bx}\right), \tag{1.145}$$

which has the property of reproducing in its actual magnitude the trace of the wing inside the Mach cone in the physical plane Oyz, on the abscissa of this new auxiliary x-plane. Indeed, for the abscissae y ($x = y$, $\mathfrak{z} = 0$), we obtain

$$By = \frac{2\eta}{1+\eta^2}, \tag{1.146}$$

which is just the geometrical transformation (1.138) applied to the radius vectors r, of the plane ($x_1 = 1$), hence of the abscissae y, η in the physical and auxiliary planes, which is represented by relation (1.146). For this purpose, we have retained the same notation y for the abscissa, but we have denoted the ordinate by $\mathfrak{z}$, in order not to be confounded with z.

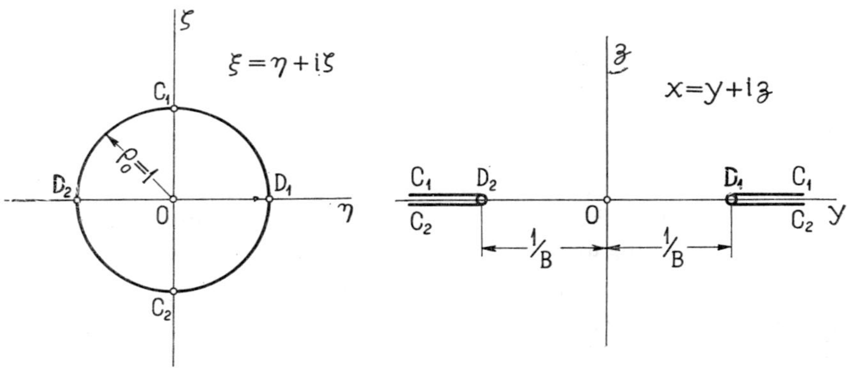

FIG. 1.13

This property of the conformal transformation (1.145) presents some advantages as regards the computation and the understanding of the various formulae and therefore this auxiliary plane will be currently adopted, in which the upper semicircle in the ξ-plane is represented on the upper semi-plane in the x-plane, the contour of the circle of radius $\varrho_0 = 1$ being situated on the abscissa Oy at C_1D_1 and D_2C_1. The same thing is also true for the lower semicircle which is transformed into the lower semi-plane, the circle contour being situated on an abscissa at D_1C_2 and D_2C_2. The points C_1 and C_2 in the plane are situated at infinity.

The analytical functions $\mathcal{U}(\xi)$, $\mathcal{V}(\xi)$ and $\mathcal{W}(\xi)$ defined by (1.143), which become $\mathcal{U}(x)$, $\mathcal{V}(x)$ and $\mathcal{W}(x)$ in the auxiliary plane x, are related to each other by perfectly defined relations resulting from the irrotationality conditions (1.1); that is, from the fact that the vorticity is zero at any point of the space considered ($\nabla \wedge \mathbf{V} = 0$),

$$\frac{\partial u}{\partial x_2} = \frac{\partial v}{\partial x_1}, \quad \frac{\partial v}{\partial x_3} = \frac{\partial w}{\partial x_2}, \quad \frac{\partial w}{\partial x_1} = \frac{\partial u}{\partial x_3}. \tag{1.147}$$

Let us assume that $\mathcal{U} = u+iu'$ is known; then $\mathcal{V} = v+iv'$ and $\mathcal{W} = w+iw'$ can be readily deduced from the above relations. Indeed, we have

$$\frac{\partial v}{\partial x_1} = \frac{\partial v}{\partial y}\frac{\partial y}{\partial x_1} + \frac{\partial v}{\partial z}\frac{\partial z}{\partial x_1} = -\frac{1}{x_1}\left(\frac{\partial v}{\partial y}y + \frac{\partial v}{\partial z}z\right), \tag{1.148a}$$

$$\frac{\partial u}{\partial x_2} = \frac{\partial u}{\partial y}\frac{\partial y}{\partial x_2} = \frac{1}{x_1}\frac{\partial u}{\partial y}, \tag{1.148b}$$

whence

$$-\frac{\partial u}{\partial y} = y\frac{\partial v}{\partial y} + z\frac{\partial v}{\partial z}. \tag{1.149}$$

Now, taking into consideration the coordinates r and ω in the plane Oyz, we have

$$r^2 = y^2+z^2, \quad \frac{\partial r}{\partial y} = \frac{y}{r} = \cos\omega, \quad \frac{\partial r}{\partial z} = \frac{z}{r} = \sin\omega, \tag{1.150a}$$

$$\tan\omega = \frac{z}{y}, \quad \frac{\partial \omega}{\partial y} = -\frac{z}{r^2} = -\frac{\sin\omega}{r}, \quad \frac{\partial \omega}{\partial z} = \frac{\cos\omega}{r}. \tag{1.150b}$$

On the other hand, from Busemann's geometrical transformation (1.138), we likewise have

$$\frac{dr}{d\varrho} = \frac{2}{B}\frac{1-\varrho^2}{(1+\varrho^2)^2}. \tag{1.151}$$

Introducing in (1.149) we have

$$\frac{\partial u}{\partial y} = \frac{\partial u}{\partial \varrho}\frac{d\varrho}{dr}\frac{\partial r}{\partial y} + \frac{\partial u}{\partial \omega}\frac{\partial \omega}{\partial y} = \frac{B}{2}\left[\frac{\partial u}{\partial \varrho}\frac{(1+\varrho^2)^2}{1-\varrho^2}\cos\omega - \frac{\partial u}{\partial \omega}\frac{1+\varrho^2}{\varrho}\sin\omega\right], \tag{1.152a}$$

$$y\frac{\partial v}{\partial y} + z\frac{\partial v}{\partial z} = r\frac{\partial v}{\partial r} = r\frac{dv}{d\varrho}\frac{d\varrho}{dr} = \frac{\varrho(1+\varrho^2)}{1-\varrho^2}\frac{\partial v}{\partial \varrho}, \tag{1.152b}$$

respectively.

Finally, the following equation is obtained

$$\frac{\partial v}{\partial \varrho} = -\frac{B}{2}\left[\frac{1+\varrho^2}{\varrho}\cos\omega\frac{\partial u}{\partial \varrho} - \frac{1-\varrho^2}{\varrho}\sin\omega\frac{\partial u}{\varrho\,\partial \omega}\right]. \tag{1.153}$$

Taking into account the fact that the associated harmonic functions u and u' satisfy the conditions of monogeneity,

$$\frac{\partial u}{\partial \varrho} = \frac{\partial u'}{\varrho\,\partial \omega}, \quad \frac{\partial u'}{\varrho\,\partial \omega} = -\frac{\partial u'}{\partial \varrho}, \tag{1.154}$$

the above equation (1.153) can be written

$$\begin{aligned} \frac{\partial v}{\partial \varrho} &= -\frac{B}{2}\left(\frac{1+\varrho^2}{\varrho}\cos\omega\,\frac{\partial u}{\partial \varrho} + \frac{1-\varrho^2}{\varrho}\sin\omega\,\frac{\partial u'}{\partial \varrho}\right) \\ &= -\operatorname{Re}\frac{B}{2}\left(\frac{1+\varrho^2}{\varrho}\cos\omega\,\frac{\partial \mathcal{U}}{\partial \varrho} - i\,\frac{1-\varrho^2}{\varrho}\sin\omega\,\frac{\partial \mathcal{U}}{\partial \varrho}\right) \\ &= -\operatorname{Re}\frac{B}{2}\left(\xi+\frac{1}{\xi}\right)\frac{\partial \mathcal{U}}{\partial \varrho}. \end{aligned} \tag{1.155}$$

On the other hand, we note that we can write

$$\xi = \varrho e^{i\omega}, \quad \frac{\partial}{\partial \varrho} = \frac{\partial \xi}{\partial \varrho}\,\frac{d}{d\xi} = e^{i\omega}\,\frac{d}{d\xi}, \tag{1.156a}$$

$$\operatorname{Re} e^{i\omega}\frac{d\mathcal{V}}{d\xi} = \operatorname{Re}\frac{\partial \mathcal{V}}{\partial \varrho} = \frac{\partial v}{\partial \varrho}, \tag{1.156b}$$

$$\frac{\partial \mathcal{U}}{\partial \varrho} = e^{i\omega}\,\frac{d\mathcal{U}}{d\xi}. \tag{1.156c}$$

Introducing into (1.155) and taking also account of (1.145), we obtain the following *compatibility relation*:†

$$d\mathcal{U} = -\frac{2}{B}\,\frac{\xi}{1+\xi^2}\,d\mathcal{V} = -x\,d\mathcal{V} \tag{1.157}$$

in the two auxiliary planes ξ and x.

Using the same method of approach, we likewise obtain for $\mathcal{W} = w+iw'$ a similar *compatibility relation*,

$$d\mathcal{U} = \frac{2i}{B}\,\frac{\xi}{1-\xi^2}\,d\mathcal{W} = \frac{ix}{\sqrt{1-B^2x^2}}\,d\mathcal{W}. \tag{1.158}$$

These compatibility relations will be made use of widely in the study of conical motions, particularly expressed in the plane x, under their unified form,

$$d\mathcal{U} = -x\,d\mathcal{V} = \frac{ix}{\sqrt{1-B^2x^2}}\,d\mathcal{W}. \tag{1.159}$$

† A very interesting direct demonstration of the compatibility relations was given by L. M. Milne-Thomson in his treatise *Theoretical Aerodynamics* (2nd ed., Macmillan and Co. Ltd., St. Martin's Street, London, 1952, p. 365).

I. 8. High-order Conical Flows

It should be recalled that, for systematizing the notations used in the following chapters, we have adopted a system of cartesian coordinates x_1, x_2, x_3 (Fig. 1.14). The same coordinates as above (1.139) are retained for the physical plane, that is $y = x_2/x_1$ and $z = x_3/x_1$ which correspond to x_2 and x_3 in the plane $x_1 = 1$.

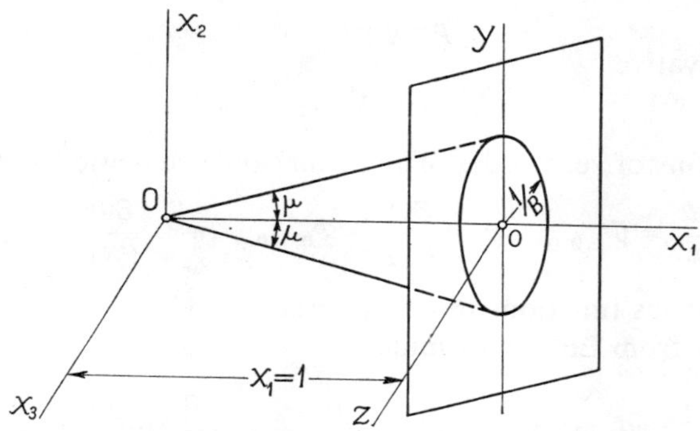

FIG. 1.14

Therefore, for a clear understanding of the notions and for avoiding any confusion, the same physical plane Oyz normal to the axis Ox_1, at point $x_1 = 1$, will always be considered.

I. 8.1. Considerations on the Potential and Velocities as High-order Homogeneous Functions

A conical flow of n order with respect to the vertex O, is the flow for which the disturbance velocity potential $\Phi(x_1, x_2, x_3)$ is an nth order homogeneous function with respect to x_1, x_2, x_3. It follows that the nth order derivatives of the potential are constant on any radius vector issuing from O, hence they are zero order homogeneous functions constituting the actual *conical field*.

The study of the high order conical motions, which is closely related to the analysis of the homogeneous functions of the same order, has been comprehensively treated in this manner, the basic principles being also established, by Paul Germain [G—10, 11] and many other authors, from whom we cite Beshkine [B—10]. These authors have obtained extremely valuable results in this field and are indicated in our references.

If $n = 1$, we obtain the true conical flow defined in the previous paragraph.

Let $\Phi_{p,q,r,}$ denote the derivative of order p with respect to x_1, of order q with respect to x_2 and of order r with respect to x_3:

$$\Phi_{p,q,r,} = \frac{\partial^{p+q+r}\Phi}{\partial x_1^p \, \partial x_2^q \, \partial x_3^r} . \tag{1.160}$$

If

$$p+q+r = n, \tag{1.161}$$

then the derivative

$$\Phi_{p,q,r} = \Phi_{n-q-r,q,r} \tag{1.162}$$

is homogeneous of zero order. The disturbance velocities written as

$$u = \frac{\partial\Phi}{\partial x_1} = \Phi_{1,0,0}, \quad v = \frac{\partial\Phi}{\partial x_2} = \Phi_{0,1,0}, \quad w = \frac{\partial\Phi}{\partial x_3} = \Phi_{0,0,1} \tag{1.163}$$

are homogeneous functions of $(n-1)$ order.

Proceeding from Euler's formula,

$$n\Phi = \left(x_1 \frac{\partial}{\partial x_1} + x_2 \frac{\partial}{\partial x_2} + x_3 \frac{\partial}{\partial x_3}\right)\Phi \tag{1.164}$$

and successively applying the same formula to the derivatives of order one, we shall finally express the potential Φ as a function of its derivatives of nth order:

$$n!\Phi = \left(x_1 \frac{\partial}{\partial x_1} + x_2 \frac{\partial}{\partial x_2} + x_3 \frac{\partial}{\partial x_3}\right)^{(n)} \Phi. \tag{1.165}$$

Introducing the notations (1.133), the following equation is obtained:

$$\frac{n!}{x_1{}^n}\Phi = \left(\frac{\partial}{\partial x_1} + y \frac{\partial}{\partial x_2} + z \frac{\partial}{\partial x_3}\right)^{(n)} \Phi, \tag{1.166}$$

whence it can be seen that Φ/x_1^n is a zero order homogeneous function.

Similarly, for the disturbance velocity components u, v, w, we may write

$$\frac{(n-1)!}{x_1^{n-1}} u = \left(\frac{\partial}{\partial x_1} + y \frac{\partial}{\partial x_2} + z \frac{\partial}{\partial x_3}\right)^{(n-1)} u, \tag{1.167a}$$

$$\frac{(n-1)!}{x_1^{n-1}} v = \left(\frac{\partial}{\partial x_1} + y \frac{\partial}{\partial x_2} + z \frac{\partial}{\partial x_3}\right)^{(n-1)} v, \tag{1.167b}$$

$$\frac{(n-1)!}{x_1^{n-1}} w = \left(\frac{\partial}{\partial x_1} + y \frac{\partial}{\partial x_2} + z \frac{\partial}{\partial x_3}\right)^{(n-1)} w, \tag{1.167c}$$

respectively, which are also zero order homogeneous functions and where u, v, w will be replaced by their expressions in (1.163). Developing further and denoting by C_m^s the coefficients of the binomial expansion, the above expressions will be put in the form

$$\frac{(n-1)!}{x_1^{n-1}} u = \sum_{q=0}^{n-1} \sum_{r=0}^{q} C_{n-1}^q C_q^r y^{q-r} z^r \Phi_{n-q,\, q-r,\, r}, \tag{1.168a}$$

$$\frac{(n-1)!}{x_1^{n-1}} v = \sum_{q=0}^{n-1} \sum_{r=0}^{q} C_{n-1}^q C_q^r y^{q-r} z^r \Phi_{n-q-1,\, q-r+1,\, r}, \tag{1.168b}$$

$$\frac{(n-1)!}{x_1^{n-1}} w = \sum_{q=0}^{n-1} \sum_{r=0}^{q} C_{n-1}^q C_q^r y^{q-r} z^r \Phi_{n-q-1,\, q-r,\, r+1}, \tag{1.168c}$$

If we consider the plane $x_3 = 0$, and set $z = 0$ in relations (1.167), we obtain, after performing the expansion,

$$\frac{(n-1)!}{x_1^{n-1}} u = \sum_{q=0}^{n-1} C_{n-1}^q y^q \Phi_{n-q,\, q,\, 0}, \tag{1.169a}$$

$$\frac{(n-1)!}{x_1^{n-1}} v = \sum_{q=0}^{n-1} C_{n-1}^q y^q \Phi_{n-q-1,\, q+1,\, 0}, \tag{1.169b}$$

$$\frac{(n-1)!}{x_1^{n-1}} w = \sum_{q=0}^{n-1} C_{n-1}^q y^q \Phi_{n-q-1,\, q,\, 1}, \tag{1.169c}$$

which can also be deduced from (1.168), setting $r = 0$.

Similarly, in view of some special applications, if we consider the plane $x_2 = 0$ and set $y = 0$ in (1.167), we obtain

$$\frac{(n-1)!}{x_1^{n-1}} u = \sum_{r=0}^{n-1} C_{n-1}^r z^r \Phi_{n-r,\, 0,\, r}, \tag{1.170a}$$

$$\frac{(n-1)!}{x_1^{n-1}} v = \sum_{r=0}^{n-1} C_{n-1}^r z^r \Phi_{n-r-1,\, 1,\, r}, \tag{1.170b}$$

$$\frac{(n-1)!}{x_1^{n-1}} w = \sum_{r=0}^{n-1} C_{n-1}^r z^r \Phi_{n-r-1,\, 0,\, r+1}, \tag{1.170c}$$

respectively.

All the above expressions are zero order homogeneous functions; they enjoy the same properties as the velocities in the true conical motion.

I. 8.2. Representation of High-order Conical Motions by Complex Variable Functions

Let us observe that the homogeneous functions of zero order $\Phi_{n-q-r,q,r}$ play the same part as the velocities u, v, w in the true conical motion, which are also zero order homogeneous functions. Exactly like these ones, the functions $\Phi_{n-q-r,q,r}$ are harmonic in the auxiliary plane $O\eta\zeta$ (Fig. 1.13) or in any other plane derived from this one by the conformal transformation (for instance x in Fig. 1.13) and hence a corresponding conjugate harmonic function $\Phi'_{n-q-r,q,r}$ can be associated with them; thus we may obtain an analytic function $F_{n-q-r,q,r}$,

$$F_{n-q-r,q,r} = \Phi_{n-q-r,q,r} + i\Phi'_{n-q-r,q,r}, \tag{1.171}$$

in any of the auxiliary planes defined above. Considering for example the plane $x = y + i\mathfrak{z}$ (1.145), we may write instead of expressions (1.161) and (1.170),

$$\frac{(n-1)!}{x_1^{n-1}} u = \operatorname{Re} \sum_{q=0}^{n-1} C_{n-1}^{q} x^q F_{n-q,q,0}, \tag{1.172a}$$

$$\frac{(n-1)!}{x_1^{n-1}} v = \operatorname{Re} \sum_{q=0}^{n-1} C_{n-1}^{q} x^q F_{n-q-1,q+1,0}, \tag{1.172b}$$

$$\frac{(n-1)!}{x_1^{n-1}} w = \operatorname{Re} \sum_{q=0}^{n-1} C_{n-1}^{q} x^q F_{n-q-1,q,1}, \tag{1.172c}$$

respectively, for the case $z = 0$ ($x_3 = 0$), and

$$\frac{(n-1)!}{x_1^{n-1}} u = \operatorname{Re} \sum_{r=0}^{n-1} C_{n-1}^{r} \left(\frac{-ix}{\sqrt{1-B^2x^2}}\right)^r F_{n-r,0,r} \tag{1.173a}$$

$$\frac{(n-1)!}{x_1^{n-1}} v = \operatorname{Re} \sum_{r=0}^{n-1} C_{n-1}^{r} \left(\frac{-ix}{\sqrt{1-B^2x^2}}\right)^r F_{n-r-1,1,r}, \tag{1.173b}$$

$$\frac{(n-1)!}{x_1^{n-1}} w = \operatorname{Re} \sum_{r=0}^{n-1} C_{n-1}^{r} \left(\frac{-ix}{\sqrt{1-B^2x^2}}\right)^r F_{n-r-1,0,r+1}, \tag{1.173c}$$

for the case $y = 0$ ($x_2 = 0$), where z in formula (1.170), in compliance with Busemann's geometrical transformation (1.138), has been replaced by a function of $\mathfrak{z}$, as found from the following successive operations:

$$z = \frac{1}{B}\frac{2\zeta}{1+\zeta^2} = \frac{1}{B}\left(\frac{-2i\xi}{1-\xi^2}\right)_{\eta=0} = \left(\frac{-ix}{\sqrt{1-B^2x^2}}\right)_{y=0} = \frac{\mathfrak{z}}{\sqrt{1+B^2\mathfrak{z}^2}}. \tag{1.174}$$

I. 8.3. Compatibility Relations

Let us consider one of the $(n-1)$ order derivatives of the function Φ, for example $\Phi_{n-1,0,0}$, and observe that this first order function may represent the potential of a conical flow proper, as it has been defined above (§ I.7), for which the respective velocities

$$u_n = \frac{\partial \Phi_{n-1,0,0}}{\partial x_1} = \Phi_{n,0,0} \quad v_n = \frac{\partial \Phi_{n-1,0,0}}{\partial x_2} = \Phi_{n-1,1,0},$$

$$w_n = \frac{\partial \Phi_{n-1,0,0}}{\partial x_3} = \Phi_{n-1,0,1} \tag{1.175}$$

are zero order homogeneous functions and, as shown above, play the part of the disturbance velocities u, v, w of a conical motion proper, representing the real part of the analytical functions $\mathcal{U}$, $\mathcal{V}$, $\mathcal{W}$. Thus, we have to consider the similar components u_n, v_n, w_n or $\Phi_{n,0,0}$, $\Phi_{n-1,1,0}$, $\Phi_{n-1,0,1}$, which represent the real part of the analytical functions $F_{n,0,0}$, $F_{n-1,1,0}$, $F_{n-1,0,1}$, the latter taking the place of $\mathcal{U}$, $\mathcal{V}$, $\mathcal{W}$. In this case, the compatibility relations established above (1.159) will also be applied, that is,

$$dF_{n,0,0} = -x\, dF_{n-1,1,0} = \frac{ix}{\sqrt{1-B^2x^2}}\, dF_{n-1,0,1}. \tag{1.176}$$

Let us consider the relation between the first side and the second side and apply it successively to the functions $F_{n-1,1,0}$, $F_{n-2,2,0}$, $\ldots$, $F_{n-q+1,q-1,0}$, $\ldots$, instead of the function $F_{n,0,0}$; multiplying then the first sides and then the second sides, we finally obtain

$$dF_{n,0,0} = (-x)^q\, dF_{n-q,q,0} = (-x)^{q-1}\, dF_{n-q+1,q-1,0} = \ldots \tag{1.177}$$

In an analogous manner, taking into consideration the relation between the first and the third side and proceeding as above for the functions $F_{n-1,0,1}$, $F_{n-2,0,2}$, $\ldots$, $F_{n-r+1,0,r-1}$, $\ldots$, we obtain a similar relation

$$dF_{n,0,0} = \left(\frac{ix}{\sqrt{1-B^2x^2}}\right)^r dF_{n-r,0,r} = \left(\frac{ix}{\sqrt{1-B^2x^2}}\right)^{r-1} dF_{n-r+1,0,r-1} = \ldots \tag{1.178}$$

Similarly, we can start from $F_{n-q,q,0}$ or $F_{n-r,0,r}$ and apply the same relations; we obtain

$$dF_{n-q,q,0} = \left(\frac{ix}{\sqrt{1-B^2x^2}}\right)^r dF_{n-q-r,q,r}, \tag{1.179a}$$

$$dF_{n-r,0,r} = (-x)^q\, dF_{n-q-r,q,r}, \tag{1.179b}$$

respectively, which, replacing in (1.177) or (1.178), leads to the general relation

$$dF_{n,0,0} = (-x)^q \left(\frac{ix}{\sqrt{1-B^2x^2}}\right)^r dF_{n-q-r,\,q,\,r}, \tag{1.180}$$

valid for any q, r, which satisfy the condition

$$0 \leqslant q+r \leqslant n. \tag{1.181}$$

The general expressions (1.180), obtained by Paul Germain [G—10,11], termed *the compatibility relations*, will be later on applied to certain concrete problems.

CHAPTER II

UNIFIED THEORY OF ANGULAR WINGS BASED ON HIGH-ORDER CONICAL FLOW. DETERMINATION OF THE AXIAL DISTURBANCE VELOCITY (u) AND OF THE DOWNWASH (w) IN THE DIRECT, INDIRECT AND MIXED PROBLEMS

IN THE following we shall try to give a unified method for the study of triangular, rectangular and trapezoidal wings, or, generally, of polygonal wings of usual forms, as they are made use of in supersonic aeronautics.

This method is based upon two characteristic features of such wings, namely: *the subsonic leading edge* of a thin wing and *the separation ridge* (situated on the wing), which will be defined below from the geometric and aerodynamic point of view.

II. 1. General Considerations

II. 1.1. Definitions

Let us consider an orthogonal system $Ox_1x_2x_3$, having the axis Ox_1 parallel to the free stream velocity U_∞, and a *flattened angular wing*, with its vertex at the origin, approximately contained in the plane Ox_1x_2 (Fig. 2.1). The edges OA_1 and OA_2 are called *leading edges*, while the enclosing line $CDFG$ is called the *trailing edge*.

The Mach cone is defined by the angle μ and the circle representing the intersection of this cone with the *physical plane*, defined in the previous chapter as being the plane normal to Ox_1, at the distance $x_1 = 1$. The radius of this circle is

$$r_0 = \tan \mu = \frac{1}{B}. \tag{2.1}$$

The coordinates y, z in the physical plane represent, as previously shown (1.139), the ratios

$$y = \frac{x_2}{x_1}, \quad z = \frac{x_3}{x_1}. \tag{2.2}$$

With respect to the Mach cone, the leading edge may be situated in the interior of this cone, in which case it is termed a *subsonic leading edge* (OA_2 for example), or may be external, in which case it is termed a *supersonic leading edge* (OA_1 for example), and enjoys the properties established in the preceding chapter for the yawed wing of infinite aspect ratio (§I. 5.3).

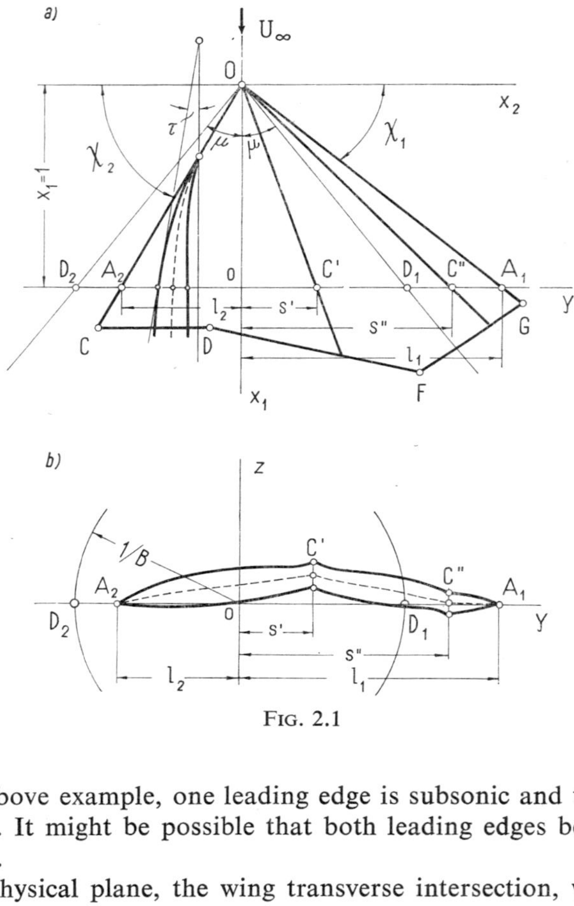

FIG. 2.1

In the above example, one leading edge is subsonic and the other one supersonic. It might be possible that both leading edges be subsonic or supersonic.

In the physical plane, the wing transverse intersection, which will be called *the wing trace* in the physical plane, lies on the abscissa axis and coincides with it when the assumption of small disturbances is used. This trace can be situated partially in the interior of the Mach cone and partially in the exterior, or totally in the interior, or totally in the exterior.

The position of the leading edges will be defined in the wing plane by the angles χ_1 and χ_2, and their traces in the physical plane ($x_1 = 1$) by l_1 and

l_2 (Fig. 2.1), which represent in fact the real dimensions of the wing trace in this plane, and which are given by the following relations:

$$l_1 = \cot \chi_1, \quad l_2 = \cot \chi_2. \tag{2.3}$$

Similarly, the separation line issuing from the vertex O and situated on the wing at those points where the wing surface presents certain discontinuities, will be termed a *ridge*. It is hence a ridge separating two parts of the wing, in which the surfaces vary in a different manner. As in the case of the leading edges, a ridge may be *subsonic* or *supersonic*, according as it is situated in the interior or exterior of the Mach cone. The ridge OC' in Fig. 2.1 is a *subsonic ridge*, while OC'' is a *supersonic ridge*.

The leading edge and the ridges will play a fundamental role in the theory of *angular* wings in supersonic flow.

If the Mach cones issuing from any point of the trailing edge do not cross the wing, then the latter is called an *angular wing*, owing to the fact that its properties do not alter and do not depend on the enclosing line, which can be displaced very far downstream, in which case it will be termed an *indefinite wing*.

Generally, the wing will also have a thickness, so that in a cross section normal to the axis Ox_1, or in a longitudinal section normal to the axis Ox_2, we obtain special *flattened* contours.

The contour obtained in an arbitrary section $x_2 =$ const. represents the *wing profile* in that section.

The upper surface of the wing is called the *upper face* or *extrados*, while the lower surface is called the *lower face* or *intrados*. The same terms are also retained for wing profiles of various sections.

The distance between the two faces at any point, approximately measured in the direction of the axis Ox_3, represents *the thickness of the wing* at that point.

The surface generated by the middles of the segments made up of the distances between the two faces, represents a thin wing and is termed *the wing frame;* in the longitudinal section which defines the wing profile, we shall similarly have *the profile frame*.

Let us now denote by Z the equations of the wing surface; they are functions of x_1 and x_2,

$$Z = Z(x_1, x_2). \tag{2.4}$$

The tangent to the profile in a section $x_2 =$ const., at any of its points, makes the angle τ with the direction of the free stream velocity, which we have currently denoted by U_∞; under the assumption of a sufficiently thin wing and contained approximately in the plane Ox_1x_2, a condition which corresponds in fact to the small disturbances assumption, we have

$$\frac{\partial Z}{\partial x_1} = -\tan \tau \approx -\tau \tag{2.5}$$

and this angle will be termed the *deflexion* of the surface or of the profile at the point considered, with respect to the stream direction, if this is in the direction of the axis Ox_1.

Let us now consider the points P_1, P_2, P_s, situated on the same vertical, both on the profile and on its frame (Fig. 2.2); denoting by τ_1, τ_2 and α the respective deflexions, we may write

$$\frac{\partial Z_1}{\partial x_1} = -\tau_1, \quad \frac{\partial Z_2}{\partial x_1} = -\tau_2, \quad \frac{\partial Z_s}{\partial x_1} = \frac{\partial}{\partial x_1}\left(\frac{Z_1+Z_2}{2}\right) = -\frac{\tau_1+\tau_2}{2} = -\alpha, \tag{2.6}$$

in which Z_1 and Z_2 define the points P_1 and P_2, while Z_s defines the point P_s of the frame.

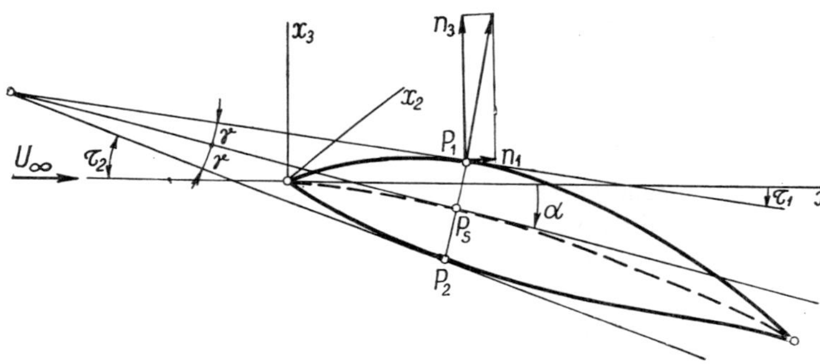

FIG. 2.2

On the other hand, denoting by γ the angle made by the tangent to the frame with each of the tangents to the two points on the profile, within the assumed approximations, we set

$$\tau_1 = \alpha - \gamma, \quad \tau_2 = \alpha + \gamma. \tag{2.7}$$

It follows from these relations that under the assumption of small disturbances and hence of the linearized equations, on the basis of the principle of the superposition of effects, the supersonic flow around the wing, defined by (2.4), may result from the superposition of motions around two wings: a thin wing constituted of its frame, having the deflexion α with respect to the stream (which may be called *incidence*), and a thick wing (within the limit admissible in our assumption regarding the cross dimensions of the wing) having the deflexion $\pm\gamma$ with respect to the stream (which may be called *slope*), hence symmetrical to the stream.

Beginning from now, each of these wings may be treated separately.

Since the lift can be given only by the thin wing, this one will be called *lifting wing*.

The second wing will be called *wing of symmetrical thickness*, and the complete wing *thick lifting wing*.

II. 1.2. Conditions of Flow on the Wing. Boundary Conditions

Denoting by n_1, n_2, n_3 the direction cosines of the normals to the surface at any point, where the velocities are $U_\infty+u$, v, w, respectively, in the direction of the three axes of coordinates, from the condition that the velocity component along the normal be zero, it follows that

$$n_1(U_\infty+u)+n_2v+n_3w = 0. \tag{2.8}$$

Let us now observe that as the wing is thin and contained in the plane Ox_1x_2, we have (Fig. 2.2),

$$-n_1 \approx \frac{\partial Z}{\partial x_1} \approx -\tau, \quad n_2 \approx \frac{\partial Z}{\partial x_2}, \quad n_3 \approx 1. \tag{2.9}$$

On the other hand, u, v and w being small disturbances, and n_1 and n_2 being also very small, we can neglect the terms of second order ($n_1u\approx 0$, $n_2v \approx 0$) and finally write the general condition on the wing,

$$w \approx -\tau U_\infty, \tag{2.10}$$

a relation which will be separately applied to the thin wing and the wing of symmetrical thickness, both on the lower and upper faces,

$$w \approx -\alpha U_\infty, \quad w \approx \pm\gamma U_\infty. \tag{2.11}$$

As a consequence of the fact that we have assumed the wing to lie in the plane Ox_1x_2 and to have a small thickness, the above conditions should be considered valid in the first approximation, just on this plane, although they refer to the points on the wing surface, hence to a certain distance from the plane Ox_1x_2, being of course assumed as very small, according to our assumption, made at the beginning of this study.

By separately treating these two cases, the *thin lifting wing* and the *wing of symmetrical thickness*, the boundary conditions (2.11) regarding the velocities w are completed with the conditions that must satisfy the other components u and v.

Denoting the upper face by subscript 1 and the lower face by subscript 2, and taking account of the antisymmetry of the motion we have, hence, for the *thin lifting wing*,

$$w_1 = w_2 = -\alpha U_\infty, \quad u_1 = -u_2, \quad v_1 = -v_2, \tag{2.12}$$

while outside the wing, in the plane Ox_1x_2,

$$u = v = 0. \tag{2.13}$$

For the wing of symmetrical thickness, taking account of the symmetry with respect to the plane Ox_1x_2 and retaining the same notations as above, we have on the wing

$$w_1 = -w_2 = \gamma U_\infty, \quad u_1 = u_2, \quad v_1 = v_2, \tag{2.14}$$

while outside the wing, in the plane Ox_1x_2,

$$w = 0. \tag{2.15}$$

II. 1.3. Definition of Problems referring to the Geometry and Aerodynamics of Angular Wings

The aerodynamic study of wings consists of three main problems, which will be defined below:

Direct problem, in which the pressure distribution on a given (known) wing is looked for. Now a wing of triangular or polygonal planform is given by the equation of its surface, which can have a very general expression. Within our treatment of the high order conical flows, the wing will be defined by its equation Z, which is assumed to be a superposition of homogeneous polynomials in x_1, x_2, of the form

$$Z = \sum_{n=1}^{N} \sum_{q=0}^{n} C_{n-q,\,q} x_1^{n-q} x_2^q = \sum_{n=1}^{N} x_1^n \sum_{q=0}^{n} C_{n-q,\,q} y^q, \tag{2.16}$$

N being the degree of the homogeneous polynomial of the highest order, considered in the expression of Z, where the coefficients $C_{n-q,\,q}$ are constants and y has the same significance as above (2.2). The surface deflexion with respect to the general stream will be given by relation (2.5); hence, taking account of (2.10) and (2.9), we may write

$$\begin{aligned} w \approx -\tau U_\infty \approx U_\infty \frac{\partial Z}{\partial x_1} &= \sum_{n=1}^{N} \sum_{q=0}^{n-1} w_{n-q-1,\,q} x_1^{n-q-1} x_2^q \\ &= \sum_{n=1}^{N} x_1^{n-1} \sum_{q=0}^{n-1} w_{n-q-1,\,q} y^q, \end{aligned} \tag{2.17}$$

where the coefficients $w_{n-q-1,\,q}$ are constants.

Within the linearized theory, in order to solve the general problem, it is sufficient to solve the case of a single polynomial of $(n-1)$ order, in the form of which the downwash w may occur,

$$w = \sum_{q=0}^{n-1} w_{n-q-1,\,q} x_1^{n-q-1} x_2^q = x_1^{n-1} \sum_{q=0}^{n-1} w_{n-q-1,\,q} y^q. \tag{2.18}$$

However, it should be noted that, as shown below, portions of the wing, separated for instance by the lines OC_1 and OC_2 (Fig. 2.3), may be represented by different polynomials. As shown below, the separation lines OC_1 and OC_2 will be called *ridges* and will play a particular part in the treatment of the problem.

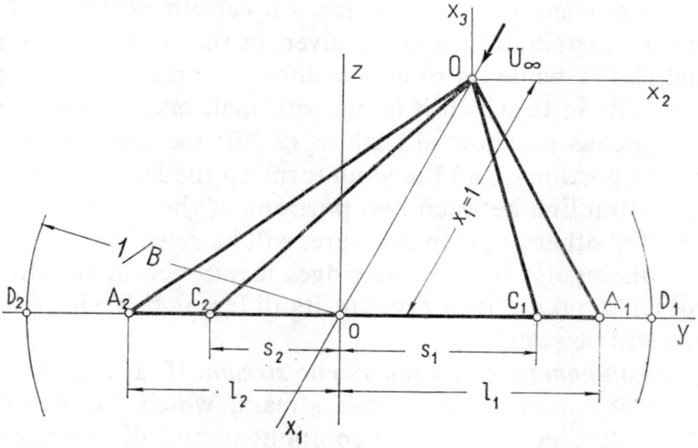

FIG. 2.3

In order to find the pressure at any point or, in other words, *the pressure coefficient*, it may be easily seen from formula (1.52), previously established, that is,

$$C_p = -2\frac{u}{U_\infty}, \tag{2.19}$$

that it is necessary to find the *axial disturbance velocity u*.

The *indirect problem* consists in finding the equations of a wing which has a given pressure distribution, that is, as follows from (2.19) a given distribution of the velocity u, which will also be given in the form of a sum of homogeneous polynomials of various orders; similarly to the direct problem, we shall consider a single homogeneous polynomial of $(n-1)$ order,

$$u = \sum_{q=0}^{n-1} u_{n-q-1,\,q} x_1^{n-q-1} x_2^q = x_1^{n-1} \sum_{q=0}^{n-1} u_{n-q-1,\,q} y^q. \tag{2.20}$$

In order to find the equations of the wing surface, we shall first look for the velocity w, from which we shall then successively deduce Z; thus

$$\frac{\partial Z}{\partial x_1} = -\tau = \frac{w}{U_\infty}, \tag{2.21a}$$

$$Z = \frac{1}{U_\infty}\int w\, dx_1 + f(x_2), \tag{2.21b}$$

where $f(x_2)$ is a function in x_2, which must satisfy certain enclosing conditions of the wing surfaces.

The same remark as above will be also made, namely that the distribution (2.20) can be different on portions separated by lines, which will be also called *ridges*, corresponding to the indirect problem.

The *mixed problem* is one in which, on certain portions on the wing, the downwash distribution may be given in the form of a homogeneous polynomial (2.18), while on other portions the pressure distribution, that is of the velocity u, to which it is proportional, may be given in the form of a homogeneous polynomial such as (2.20); the pressure is looked for on the former portions, and the wing form on the latter ones.

The separation line between two portions of the wing, one of given incidence and the other of given pressure, will be referred to as *mixed ridge*, in order to distinguish it from the ridges mentioned in the direct and indirect problems, taking into account its different behaviour with respect to them, as will be seen later on.

Case of a non-homogeneous supersonic stream. If, at a given instant, the thin wing is in a non-homogeneous stream, which, besides the general velocity U_∞, also has the normal components w_∞ distributed according to an arbitrary law, the stream exercises on this wing a pressure, as if at each point it had a deflexion τ with respect to the general direction U_∞ of the stream,

$$\tau = \frac{w_\infty}{U_\infty}. \tag{2.22}$$

This effect is equivalent to that of a homogeneous stream (in the absence of the vertical component w), if the same deflexion as above is given to the wing.

Hence, besides the motion corresponding to the initial wing, as it actually is, another motion corresponding to a *deformed wing* with the deflexion τ or the downwash $w = -\tau U_\infty$ (given by (2.22)) will be superposed and decomposed according to a sum of homogeneous polynomials of various orders, the mathematical treatment being as for the direct problem.

II. 2. Behaviour of the Axial Disturbance Velocity u in the Neighbourhood of the Subsonic Leading Edges or the Subsonic Ridges on the Thin Wing (Direct Problem)

II. 2.1. Motion round a Thin Angular Plate with Subsonic Edges

Let us consider a thin plane wing, having an angle of incidence α with respect to the free stream U_∞ (Fig. 2.4). The downwash of the whole wing is the same and equal to

$$w = -\alpha U_\infty = \text{const.}; \tag{2.23}$$

we obtain true conical motion (of order one).

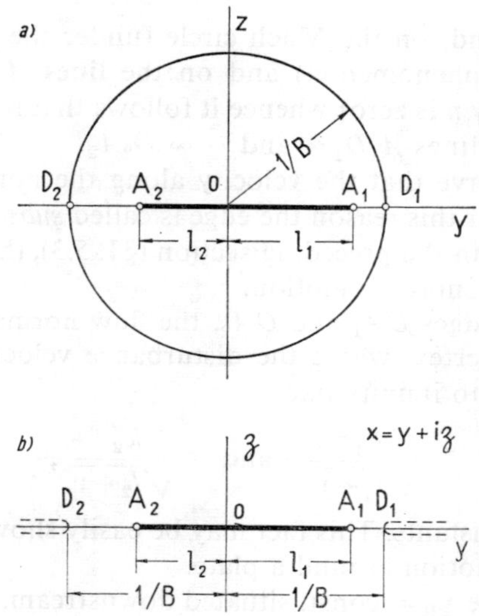

FIG. 2.4

In the physical plane, the wing trace A_2A_1 lies on the abscissa axis, A_1 being at $y = l_1$ and A_2 at $y = -l_2$. In the auxiliary plane $x = y+i\mathfrak{z}$, defined previously by the transformation of Busemann's plane (1.145), this trace is exactly reproduced on the abscissa axis. The problem consists in finding the disturbance velocities u, v, w, which represent the real parts of the functions $\mathcal{U}$, $\mathcal{V}$, $\mathcal{W}$, defined previously (1.143),

$$\mathcal{U} = u+iu'; \quad \mathcal{V} = v+iv'; \quad \mathcal{W} = w+iw'. \tag{2.24}$$

Of particular interest is u, which appears in the pressure formula (1.52), from which the other components are derived by applying the compatibility relations (1.159). In this sense, we shall write only the relation between $\mathcal{U}$ and $\mathcal{W}$

$$d\mathcal{U} = \frac{ix}{\sqrt{1-B^2x^2}}\, d\mathcal{W}, \tag{2.25}$$

which becomes on the wing ($\mathfrak{z} = 0$)

$$du+i\,du' = \frac{iy}{\sqrt{1-B^2y^2}}\,(dw+i\,dw'). \tag{2.26}$$

Further, taking the imaginary part on both sides and remembering that w is constant on the whole wing ($dw = 0$),

we have
$$du' = \frac{y}{\sqrt{1-B^2y^2}}\, dw = 0. \tag{2.27}$$

On the other hand, on the Mach circle (under the assumption due to the nature of the phenomenon) and on the lines A_1D_1 and D_2A_2, the disturbance velocity u is zero; whence it follows that in the auxiliary plane x, u is zero on the lines $A_1D_1\infty$ and $-\infty D_2A_2$.

Let us now observe that the velocity along the normal to the leading edge is subsonic (for this reason the edge is called *subsonic*) and that along this line, as shown in the preceding section (§ I. 5.3), the motion enjoys all the properties of a subsonic motion.

On the leading edges OA_1 and OA_2, the flow normal to these lines encounters a sharp vertex, where the disturbance velocity components *(in particular u)*, tend to infinity like

$$\frac{k_1}{\sqrt{l_1-y}} \quad \text{and} \quad \frac{k_2}{\sqrt{l_2+y}},$$

k_1 and k_2 being constants. This fact may be easily shown from consideration of the plane motion around a plate.

Indeed, in a plane $x_1 = \text{const.}$ situated downstream, the vertical motion around the plate, neglecting for the sake of simplicity the effect of boundaries due to the Mach cone (these boundaries can be considered, but the

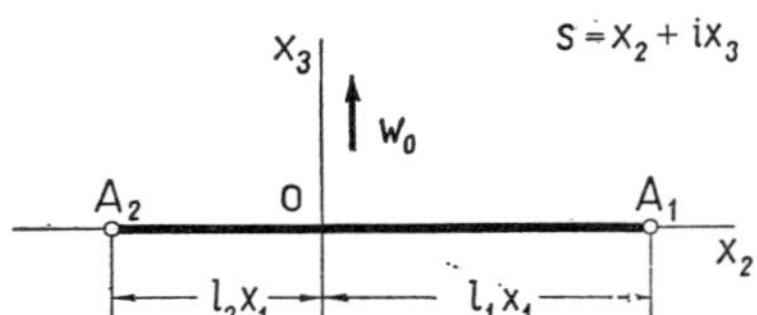

FIG. 2.5

problem would be laborious), will be given approximately by the following potential

$$f(s) = -iw_0(s-\sqrt{(s+x_1l_2)(s-x_1l_1)}\,), \tag{2.28}$$

a function of the complex variable (Fig. 2.5)

$$s = x_2+ix_3. \tag{2.29}$$

The approximation consists in the fact that in an arbitrary section of the triangular plate (for a specified x_1), the flow due to the vertical stream w_0 is assumed to be plane.

The longitudinal (axial) velocity u on the plate ($s = x_2$) will be given by

$$u = \left(\frac{\partial f}{\partial x_1}\right)_{x_3=0} = -\frac{w_0[2l_1l_2x_1+(l_1-l_2)x_2]}{2\sqrt{(x_1l_2+x_2)(x_1l_1-x_2)}} = -\frac{w_0[2l_1l_2+(l_1-l_2)y]}{2\sqrt{(l_2+y)(l_1-y)}}, \tag{2.30}$$

whence one may see the behaviour of this velocity for $y \to l_1$ and $y \to -l_2$.

Obviously, $\mathcal{U}$ will also tend to infinity at the same points as

$$\mathcal{U}_{x \to l_1} = \frac{k_1}{\sqrt{l_1-x}}, \quad \mathcal{U}_{x \to -l_2} = \frac{k_2}{\sqrt{l_2+x}}. \tag{2.31}$$

Further, let us use a *hydrodynamic analogy*, by considering $\mathcal{U}$ as a potential in which the imaginary part u', which is constant on the trace A_2A_1, according to (2.30), plays the part of a stream function; the real part u, which is equal to zero on the remaining part of the abscissa, plays the part of a potential.

It may be easily seen that $\mathcal{U}$ has the form

$$\mathcal{U} = \frac{q_0+q_1x}{\sqrt{(l_1-x)(l_2+x)}} = u+iu', \tag{2.32}$$

which satisfies all the above conditions (Fig. 2.4): the imaginary part u' is equal to zero on A_2A_1 (the wing trace), the real part u is equal to zero on $-\infty D_2A_2$ and $A_1D_1\infty$, and at points $A_1 (y = l_1)$ and $A_2 (y = -l_2)$ has the behaviour given by (2.31).

From the compatibility relation (2.25), it is seen that, since w is finite at $x = 0$, in order to avoid a discontinuity we must have

$$[d\mathcal{U}]_{x=0} = 0, \tag{2.33}$$

whence, putting the derivative of $\mathcal{U}$ equal to zero, we get

$$q_1 = q_0 \frac{l_1-l_2}{2l_1l_2}. \tag{2.34}$$

The manner of determining the constant q_0 will be indicated subsequently.

II. 2.2. Case of the Subsonic Leading Edges of a Thin Wing in High-order Conical Motion

Let us now consider the same angular wing, having, however, the incidence α or the downwash $w = -\alpha U_\infty$ in the form of a homogeneous polynomial of $(n-1)$ order, as it has been defined above (2.18).

At an arbitrary point on one of the leading edges, by analogy with what has been indicated above (2.31), we may assume that the velocity tends towards infinity as

$$\frac{k_1(x_1, x_2)}{\sqrt{l_1 - y}}, \quad \frac{k_2(x_1, x_2)}{\sqrt{l_2 + y}},$$

respectively, where $k_1(x_1, x_2)$ and $k_2(x_1, x_2)$ depend on the points considered on the leading edges.

For this reason it follows that the $(n-1)$ order derivative of the velocity u, with respect to x_1 and x_2, the derivative which has been denoted by $\Phi_{n-q, q, 0}$, is a zero order homogeneous function which will contain all singularities of the form

$$\frac{1}{(l_1 - y)^{n-\frac{1}{2}}}, \quad \frac{1}{(l_2 + y)^{n-\frac{1}{2}}}.$$

It follows that $F_{n-q, q, 0}$, which represents an analytical function of x (§ I. 8.2), will also contain all singularities in the complex form

$$\frac{1}{(l_1 - x)^{\frac{1}{2}}}, \quad \frac{1}{(l_1 - x)^{\frac{3}{2}}}, \ldots, \quad \frac{1}{(l_1 - x)^{n-\frac{1}{2}}}, \tag{2.35}$$

for the leading edge A_1 and

$$\frac{1}{(l_2 + x)^{\frac{1}{2}}}, \quad \frac{1}{(l_2 + x)^{\frac{3}{2}}}, \ldots, \quad \frac{1}{(l_2 + x)^{n-\frac{1}{2}}}, \tag{2.36}$$

for the leading edge A_2.

These results will be used subsequently in order to find the solution of the various problems we are going to treat in this book.

Formulae (2.35) and (2.36), showing the behaviour of the derivatives of the velocity u (written in complex form) near the subsonic leading edges of a thin wing, constitute *the first fundamental condition* for the calculation of the disturbance velocity u on a thin wing.

We may observe now that expressions (2.35) and (2.36) represent multipoles of an odd order (mono-pole, tri-pole, etc.).

II. 2.3. Behaviour of the Disturbance Velocity near the Separation Ridges

Let us again consider a plane wing with subsonic edges (Fig. 2.6) and let OC be a separation line, which has been termed above a *ridge* (§ II. 1); let it be assumed that the incidence is different on each side, w' from C

to A_1 and w'' from A_2 to C. The jump made by the downwash w along the ridge will be

$$w'-w'' = \Delta w = -(\alpha'-\alpha'')U_\infty. \tag{2.37}$$

In the immediate neighbourhood of the ridge OC, hence round the point C ($y = s$) in the auxiliary plane, but only on the upper side of the wing surface, this jump introduces a logarithmic term into the function $\mathcal{U}$, or, in

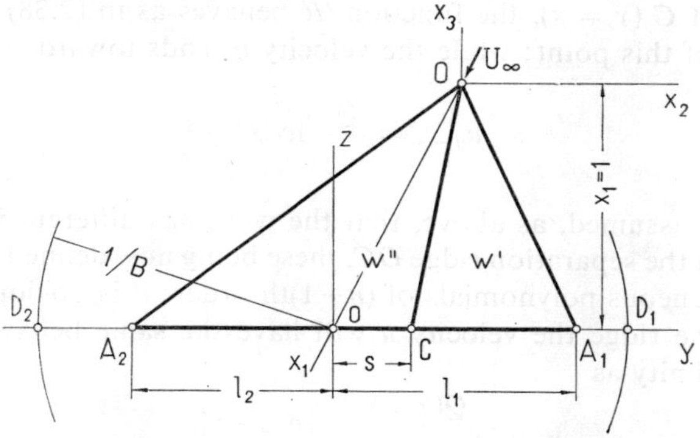

FIG. 2.6

the *hydrodynamic sense*, a *source* of intensity $2Q$ for instance, which is easily determined as follows. In this sense, we write

$$\mathcal{U}_{x\to s} = \frac{Q}{\pi}\ln(s-x), \tag{2.38}$$

and consider a small semicircle of radius r round the point C ($y = s$), by putting

$$s-x = re^{i(\theta+\pi)}. \tag{2.39}$$

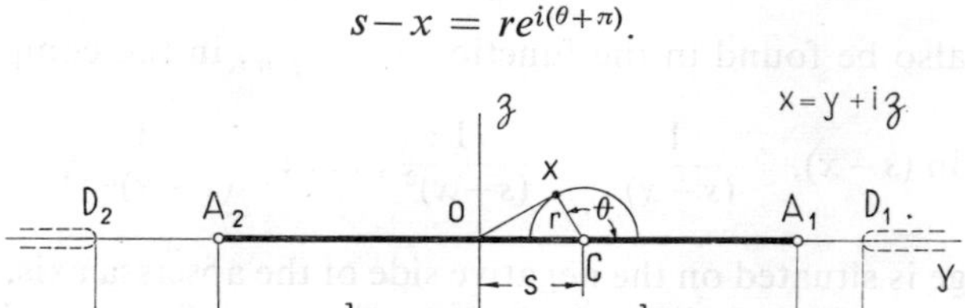

FIG. 2.7

Applying the compatibility relation (2.25) and integrating round the semicircle of radius r (Fig. 2.7), we obtain successively

$$(w''-w')\frac{is}{\sqrt{1-B^2s^2}} = \frac{Q}{\pi}\ln re^{i(\pi+\theta)}\Big|_0^\pi = iQ, \tag{2.40}$$

whence the source intensity, a function of the downwash jump and the position of the point C ($y = s$), follows:

$$Q = -\frac{s\Delta w}{\sqrt{1-B^2s^2}}\,. \tag{2.41}$$

Hence it is seen that in order to have a downwash jump $\Delta w = w' - w''$ at the point C ($y = s$), the function $\mathcal{U}$ behaves as in (2.38) in the neighbourhood of this point; while the velocity u tends towards infinity as

$$u_{y \to s} = \frac{Q}{\pi} \ln |s-y|. \tag{2.42}$$

Let it be assumed, as above, that the wing has different incidences on each side of the separation ridge OC, these being now defined by two different homogeneous polynomials of $(n-1)$th order; it is obvious that at any point on the ridge the velocity u will have the same behaviour, tending towards infinity as

$$\frac{Q(x_1, x_2)}{\pi} \ln |s-y|,$$

where $Q(x_1, x_2)$ depends now on the position of the point on the ridge OC. Hence, the $(n-1)$th order derivatives with respect to x_1 and x_2 of the disturbance velocity u, which has been denoted by $\Phi_{n-q,\,q,\,0}$, will obviously contain all singularities of the form

$$\ln |s-y|, \quad \frac{1}{(s-y)}, \quad \frac{1}{(s-y)^2}, \ldots, \quad \frac{1}{(s-y)^{n-1}},$$

which will also be found in the function $F_{n-q,\,q,\,0}$, in the complex form

$$\ln (s-x), \quad \frac{1}{(s-x)}, \quad \frac{1}{(s-x)^2}, \ldots, \quad \frac{1}{(s-x)^{n-1}}\,. \tag{2.43}$$

If the ridge is situated on the negative side of the abscissa axis, at $y = -s$, for instance, we shall similarly find in the function $F_{n-q,\,q,\,0}$ singularities of the complex form

$$\ln (s+x), \quad \frac{1}{(s+x)}, \quad \frac{1}{(s+x)^2}, \ldots, \quad \frac{1}{(s+x)^{n-1}}\,. \tag{2.44}$$

In this sense, it should be noted that in expression $(s \pm x)$ which is found in formulae (2.43) and (2.44), the minus sign represents ridges situated on the positive side of the abscissa axis, while the plus sign represents ridges situated on the negative side of the abscissa axis.

These forms of the behaviour of the velocity u in the neighbourhood of the separation ridges constitute *the second fundamental condition* for the calculation of u on any thick lifting wing, as will be shown below.

We observe that expressions (2.43) and (2.44) represent sources and multipoles of even order (bi-pole, quadri-pole, etc.).

II. 3. Contribution of the Subsonic Leading Edges of Thin Wings and of the Separation Ridges to the Construction of the Velocity u (Direct Problem)

II. 3.1. Preliminary Considerations

Let us consider Euler's formulae (1.169a, 1.169c) and (1.172a, 1.172c), which for $z = 0$ are put in the form

$$(n-1)!\frac{u}{x_1^{n-1}} = \sum_{q=0}^{n-1} C_{n-1}^q y^q \Phi_{n-q,\,q,\,0} = \mathrm{Re}\sum_{q=0}^{n-1} C_{n-1}^q x^q F_{n-q,\,q,\,0} \tag{2.45a}$$

$$(n-1)!\frac{w}{x_1^{n-1}} = \sum_{q=0}^{n-1} C_{n-1}^q y^q \Phi_{n-q-1,\,q,\,1} = \mathrm{Re}\sum_{q=0}^{n-1} C_{n-1}^q x^q F_{n-q-1,\,q,\,1}, \tag{2.45b}$$

and observe that the latter formula is equal on the wing to the downwash w (2.18), which we write in the form

$$\frac{w}{x_1^{n-1}} = \sum_{q=0}^{n-1} w_{n-q-1,\,q}\, y^q. \tag{2.46}$$

It follows that

$$C_{n-1}^q \Phi_{n-q-1,\,q,\,1} = (n-1)!\, w_{n-q-1,\,q} = \text{const.} \tag{2.47a}$$

$$d\Phi_{n-q-1,\,q,\,1} = 0. \tag{2.47b}$$

Further, let us consider the corresponding analytical functions $F_{n-q-1,\,q,\,1}$ and $F_{n-q,\,q,\,0}$, which we introduce into the compatibility relations (1.179); at the points situated on the *wing trace* ($\mathfrak{z} = 0$), these become, for $r = 1$,

$$d(\Phi_{n-q,\,q,\,0} + i\Phi'_{n-q,\,q,\,0}) = \frac{iy}{\sqrt{1-B^2y^2}}\, d(\Phi_{n-q-1,\,q,\,1} + i\Phi'_{n-q-1,\,q,\,1}) \tag{2.48}$$

whence, taking into account (2.47b), we have

$$d\Phi_{n-q-1,\,q,\,1} = d\Phi'_{n-q,\,q,\,0} = 0. \tag{2.49}$$

In the sense of the *hyrodynamic analogy* we may assume that $F_{n-q,\,q,\,0}$ would represent the potential of a fictitious incompressible motion, in which $\Phi'_{n-q,\,q,\,0}$ is a stream function, in which case we deduce from (2.49) that *the wing trace is a stream line*. This result will be considered a *fundamental characteristic*. With this characteristic and with the condition established in the preceding sections (§§ II. 2.2 and II. 2.3), we shall proceed to the calculation of the disturbance velocity u, through the summation of the contributions of subsonic leading edges and separation ridges.

In this sense, let us consider the trace of the thin subsonic wing A_2A_1, having its edges at A_1 ($y = l_1$) and A_2 ($y = -l_2$) and having one ridge at C_1 ($y = s_1$) and one ridge at C_2 ($y = -s_2$), as is shown in Fig. 2.8.

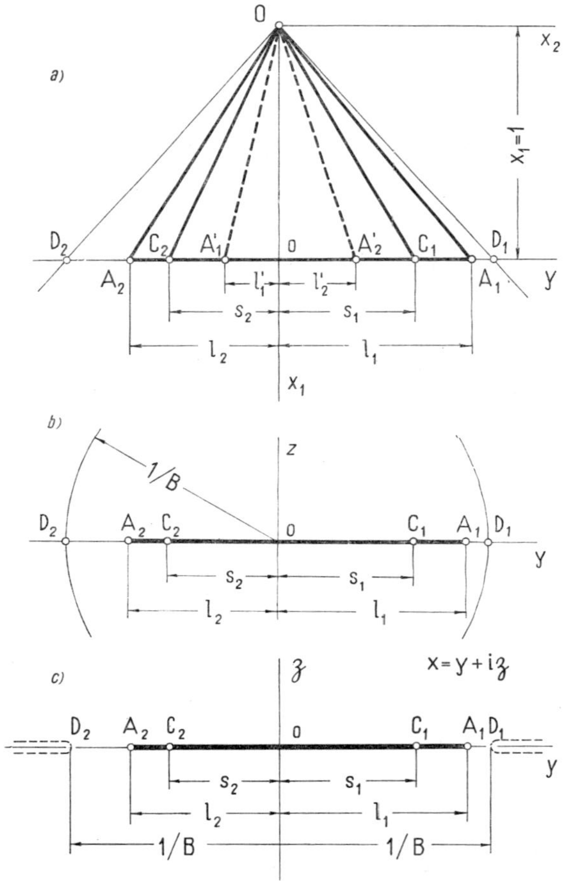

FIG. 2.8

By the conformal transformation

$$\chi^2 = \frac{l_1 - x}{l_2 + x}, \quad (\chi = \lambda + i\nu), \tag{2.50}$$

we observe that the wing trace becomes the whole abscissa axis with the origin at A_1, representing both the upper and lower surfaces of the wing (Fig. 2.9). This being now a stream line, let us represent the singularities

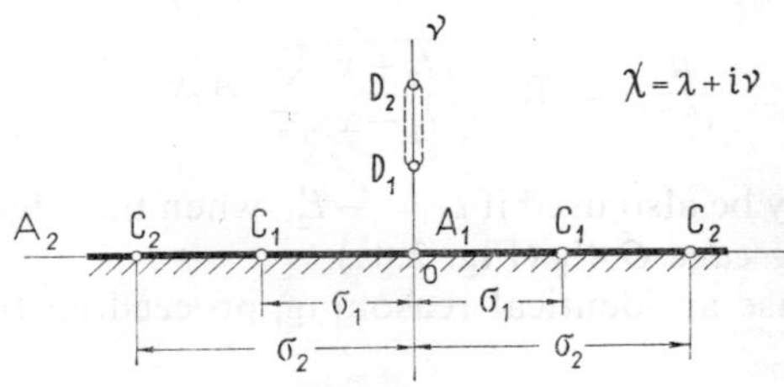

FIG. 2.9

at the leading edges and ridges satisfying this condition. The ordinate axis of the new plane represents the lines $A_1D_1\infty$ and $-\infty\, D_2A_2$ in the plane x, and hence is an equipotential line ($u = 0$).

II. 3.2. Contribution of the Subsonic Leading Edges of a Thin Wing

We situate at A_1 all singularities (2.35) which occur in the χ plane in the form of multipoles with odd exponents, having their axis on the true abscissa, in order to satisfy the condition that this should be a stream line, and the ordinate axis an equipotential. Accordingly, the function $F_{n-q,\,q,\,0}$ will include all terms of the form

$$\frac{1}{\chi}, \quad \frac{1}{\chi^3}, \quad \frac{1}{\chi^5}, \ldots, \quad \frac{1}{\chi^{2n-1}}$$

or, in the x plane, it will include terms of the form

$$\sqrt{\frac{l_2+x}{l_1-x}}, \quad \sqrt{\frac{l_2+x}{l_1-x}}\,\frac{(l_2+x)}{(l_1-x)}, \ldots, \quad \sqrt{\frac{l_2+x}{l_1-x}}\left(\frac{l_2+x}{l_1-x}\right)^{n-1},$$

so that we may write for the leading edge A_1

$$F_{n-q,\,q,\,0} = \sqrt{\frac{l_2+x}{l_1-x}} \sum_{p=0}^{n-1} \frac{a_p x^p}{(l_1-x)^{n-1}}. \tag{2.51}$$

Applying Euler's formula (2.45a) and denoting by u the contribution of the leading edge A_1, and observing on the other hand, as shown above (§ II. 2.2), that this velocity tends towards infinity at A_1 like

$$u_{y \to l_1} = \frac{k}{\sqrt{l_1 - y}}, \tag{2.52}$$

hence the denominator can have only this radical; we write finally for the contribution of the subsonic leading edge A_1 of the thin wing

$$\frac{u}{x_1^{n-1}} = \operatorname{Re} \sqrt{\frac{l_2+x}{l_1-x}} \sum_{q=0}^{n-1} A_q x^q. \tag{2.53}$$

This formula may be also used if $l_2 = -l_2'$, when the edge OA_2 becomes the *subsonic trailing edge* OA'_2 (Fig. 2.8a).

Similarly, if we use an identical reasoning, proceeding from a transformation

$$\chi^2 = \frac{l_2+x}{l_1-x}, \tag{2.54}$$

we have for the leading edge A_2,

$$F_{n-q,\, q,\, 0} = \sqrt{\frac{l_1-x}{l_2+x}} \sum_{p=0}^{n-1} \frac{a_p x^p}{(l_2+x)^{n-1}} \tag{2.55}$$

whence, applying Euler's formula, we obtain the contribution of the subsonic leading edge A_2 of the thin wing,

$$\frac{u}{x_1^{n-1}} = \operatorname{Re} \sqrt{\frac{l_1-x}{l_2+x}} \sum_{q=0}^{n-1} A_q x^q. \tag{2.56}$$

This expression is also valid for the case in which the edge OA_1 becomes the *subsonic trailing edge* OA'_1, replacing l_1 by $-l_1'$.

For the contribution of the two subsonic leading edges of a thin wing, we add the two expressions (2.53) and (2.56) and, after performing some simple calculations, we may write

$$\frac{u}{x_1^{n-1}} = \operatorname{Re} \frac{1}{\sqrt{(l_1-x)(l_2+x)}} \sum_{q=0}^{n} A_q x^q. \tag{2.57}$$

Let us now return to expressions (2.53) and (2.56) which represent the contribution of each leading edge taken separately.

If the edge A_2 is supersonic, then it is seen from Fig. 2.8 that the transformation (2.50) is valid only if $l_2 = 1/B$. It follows that in (2.53) l_2 must be replaced by $1/B$, obtaining thus for the contribution of the leading edge A_1 the following expression:

$$\frac{u}{x_1^{n-1}} = \text{Re} \sqrt{\frac{1+Bx}{B(l_1-x)}} \sum_{q=0}^{n-1} A_q x^q. \tag{2.58}$$

In a similar way, for the contribution of the edge A_2 when the edge A_1 is supersonic, l_1 is replaced by $1/B$ in (2.56) and we obtain

$$\frac{u}{x_1^{n-1}} = \text{Re} \sqrt{\frac{1-Bx}{B(l_2+x)}} \sum_{q=0}^{n-1} A_q x^q. \tag{2.59}$$

Note. Since the expressions for u indicate only the form of representation, the same symbol A_q has been used for the various constants, although they are not identical in all these expressions.

II. 3.3. Contribution of Ridges

We have observed that in the χ plane (Fig. 2.9), the abscissa axis is a *stream line* and the ordinate axis an *equipotential* for the *fictitious complex potential* $F_{n-q,\,q,\,0}$. The singularities (2.43), representing sources and multipoles, will satisfy these conditions if they are situated on each side of the origin on the abscissa axis at points $\chi = \pm\sigma_1$, corresponding to the ridge C_1 $(y = s_1)$ in the physical and auxiliary planes (Fig. 2.8).

It may be easily seen that the contribution of the ridge C_1 to the fictitious potential will be

$$F_{n-q,\,q,\,0} = c_0 \ln \frac{\chi-\sigma_1}{\chi+\sigma_1} + \sum_{k=1}^{n-1} c_k \left[\frac{1}{(\chi-\sigma_1)^k} - \frac{(-1)^k}{(\chi+\sigma_1)^k} \right], \tag{2.60}$$

where $c_0, \ldots, c_k$ are constant coefficients.

Let us consider again the x plane by applying relation (2.50), setting for σ_1, which results from the same relation,

$$\sigma_1 = \sqrt{\frac{l_1-s_1}{l_2+s_1}}, \tag{2.61}$$

and transform the logarithmic expression given in (2.60) into a hyperbolic expression

$$\ln \frac{\chi-\sigma_1}{\chi+\sigma_1} = -2 \tanh^{-1} \frac{\sigma_1}{\chi} = -2 \cosh^{-1} \frac{\chi}{\sqrt{\chi^2-\sigma_1^2}}$$

$$= -2 \cosh^{-1} \sqrt{\frac{(l_2+s_1)(l_1-x)}{(l_1+l_2)(s_1-x)}}, \tag{2.62}$$

finally

$$F_{n-q,\, q,\, 0} = c \cosh^{-1} \sqrt{\frac{(l_2+s_1)(l_1-x)}{(l_1+l_2)(s_1-x)}} + \frac{\sqrt{(l_1-x)(l_2+x)}}{(s_1-x)^{n-1}} \sum_{p=0}^{n-2} c_p x^p. \tag{2.63}$$

Further, applying Euler's formula (2.45a) and considering that the velocity u behaves like $\ln(s_1-x)$ in the immediate neighbourhood of the ridge C_1, we may write as the expression for u:

$$\frac{u}{x_1^{n-1}} = \mathrm{Re}\Bigg[\sum_{q=0}^{n-1} C_q x^q \cosh^{-1} \sqrt{\frac{(l_2+s_1)(l_1-x)}{(l_1+l_2)(s_1-x)}} + \sum_{q=0}^{n-2} D_q x^q \sqrt{(l_1-x)(l_2+x)}\Bigg], \tag{2.64}$$

where C_q, D_q are constant coefficients.

If the leading edge A_1 is supersonic, then, for the reasons indicated above (i.e. the transformation (2.50) is valid only if $l_1 = 1/B$), we shall replace l_1 by $1/B$; if the leading edge A_2 is supersonic, then we shall replace l_2 by $1/B$; finally, if both edges are supersonic, then we shall replace l_1 and l_2 by $1/B$. Hence, instead of the above formula, we may write for each case:

$$\frac{u}{x_1^{n-1}} = \mathrm{Re}\Bigg(\sum_{q=0}^{n-1} C_q x^q \cosh^{-1} \sqrt{\frac{(l_2+s_1)(1-Bx)}{(1+Bl_2)(s_1-x)}} + \sum_{q=0}^{n-2} D_q x^q \sqrt{\frac{(1-Bx)(l_2+x)}{B}}\Bigg), \tag{2.65}$$

for the case when the leading edge OA_1 is supersonic;

$$\frac{u}{x_1^{n-1}} = \mathrm{Re}\Bigg(\sum_{q=0}^{n-1} C_q x^q \cosh^{-1} \sqrt{\frac{(1+Bs_1)(l_1-x)}{(1+Bl_1)(s_1-x)}} + \sum_{q=0}^{n-2} D_q x^q \sqrt{\frac{(l_1-x)(1+Bx)}{B}}\Bigg), \tag{2.66}$$

for the case when the leading edge OA_2 is supersonic;

$$\frac{u}{x_1^{n-1}} = \mathrm{Re}\Bigg(\sum_{q=0}^{n-1} C_q x^q \cosh^{-1} \sqrt{\frac{(1+Bs_1)(1-Bx)}{2B(s_1-x)}} + \sum_{q=0}^{n-2} D_q x^q \sqrt{\frac{1-B^2x^2}{B^2}}\Bigg), \tag{2.67}$$

for the case when both edges are supersonic.

When the edges are supersonic, it might be possible that the ridges situated on the same side are also supersonic ($s > 1/B$); in this case, since the above expressions remain valid, as will be seen later (§ II. 4.5), we may perform the following replacements, deduced from the relations between the trigonometric and hyperbolic functions,

$$\cosh^{-1}\beta = i\cos^{-1}\beta. \tag{2.68}$$

When the ridge is situated on the left side of the origin, on the negative side of the abscissa axis, then proceeding from the transformation (2.54), and using a similar approach, we find the same formulae as above, in which however s_1 will be replaced by $-s_2$, l_1 by $-l_2$, l_2 by $-l_1$; thus we have

$$F_{n-q,\,q,\,0} = g\cosh^{-1}\sqrt{\frac{(l_1+s_2)(l_2+x)}{(l_1+l_2)(s_2+x)}} + \sqrt{\frac{(l_1-x)(l_2+x)}{(s_2+x)^{n-1}}}\sum_{p=0}^{n-2} g_p x^p, \tag{2.69a}$$

$$\frac{u}{x_1^{n-1}} = \mathrm{Re}\left(\sum_{q=0}^{n-1} G_q x^q \cosh^{-1}\sqrt{\frac{(l_1+s_2)(l_2+x)}{(l_1+l_2)(s_2+x)}} + \sum_{q=0}^{n-2} D_q x^q \sqrt{(l_1-x)(l_2+x)}\right). \tag{2.69b}$$

Here, the same remark may be made in connection with the position of the leading edges. If A_1 is supersonic, we replace l_1 by $1/B$; if A_2 is supersonic, we replace l_2 by $1/B$; and if A_1 and A_2 are both supersonic, then we replace l_1 and l_2 by $1/B$. We obtain three formulae similar to (2.65, 2.66, 2.67),

$$\frac{u}{x_1^{n-1}} = \mathrm{Re}\left(\sum_{q=0}^{n-1} G_q x^q \cosh^{-1}\sqrt{\frac{(1+Bs_2)(l_2+x)}{(1+Bl_2)(s_2+x)}} + \sum_{q=0}^{n-2} D_q x^q \sqrt{\frac{(1-Bx)(l_2+x)}{B}}\right), \tag{2.70a}$$

$$\frac{u}{x_1^{n-1}} = \mathrm{Re}\left(\sum_{q=0}^{n-1} G_q x^q \cosh^{-1}\sqrt{\frac{(l_1+s_2)(1+Bx)}{(1+Bl_1)(s_2+x)}} + \sum_{q=0}^{n-2} D_q x^q \sqrt{\frac{(l_1-x)(1+Bx)}{B}}\right), \tag{2.70b}$$

$$\frac{u}{x_1^{n-1}} = \mathrm{Re}\left(\sum_{q=0}^{n-1} G_q x^q \cosh^{-1}\sqrt{\frac{(1+Bs_2)(1+Bx)}{2B(s_2+x)}} + \sum_{q=0}^{n-2} D_q x^q \sqrt{\frac{1-B^2x^2}{B^2}}\right). \tag{2.70c}$$

Note. For the sake of simplicity, we have again generally used the same symbols for denoting the constants, even though they are not identical in the various expressions of u.

II. 3.4. Case in which the Downwash Distribution is an Arbitrary Homogeneous Function

For the sake of simplicity, we consider a single term of the distribution of the downwash w, given in general form by (2.18), namely the $(n-1)$th order polynomial,

$$w = \sum_{q=0}^{n-1} w_{n-q-1,\,q}\, x_1^{n-q-1}\, x_2^q, \tag{2.71}$$

where $w_{n-q-1,\,q}$ is an arbitrary function of $y = x_2/x_1$, continuous and differentiable; hence the whole expression above is no longer an $(n-1)$th order homogeneous polynomial, but an arbitrary homogeneous function of the same order. In this case, with certain additions, results previously obtained may be applied.

Let us first consider the effect of separation ridges and subsonic leading edges of thin wings; employing the same reasoning as in sections (§ II. 3.3), we observe that the disturbance velocity u behaves in the immediate neighbourhood of these ridges like

$$\frac{k_1(x_1, x_2)}{\sqrt{l_1 - y}}, \quad \frac{k_2(x_1, x_2)}{\sqrt{l_2 + y}}, \quad \frac{Q(x_1, x_2)}{\pi} \ln(s - y).$$

As in the preceding cases referring to the high-order homogeneous polynomials, the functions $F_{n-q,\,q,\,0}$ will contain the same singularities. By applying Euler's formula, the formulae for u/x_1^{n-1} retain the same expression.

The contributions of ridges and edges do not, however, give the complete solution of the problem. From one ridge to another, or from one edge to a ridge, etc., an infinity of elementary ridges, hence an infinity of sources, each of them producing a contribution du, given by an expression similar to (2.64), will be considered. In this expression, the coefficients C_q are linear functions of the differences between the downwashes on the right-hand and left-hand sides of the ridge, that is

$$\Delta w_{n-q-1,\,q} = w'_{n-q-1,\,q} - w''_{n-q-1,\,q}, \tag{2.72}$$

as will be seen later in determining the constants.

In the case of continuous variation of the coefficient $w_{n-q-1,\,q}$ between two ridges situated say on the right-hand side of the origin at $y = s_r$ and $y = s'_r$, then the above finite differences are transformed into differentials of the form

$$dw_{n-q-1,\,q} = \frac{dw_{n-q-1,\,q}}{ds}\,ds = f_q(s)\,ds, \tag{2.73}$$

with respect to which the coefficients C_q are linear functions, where the usual variable y has been replaced by s. In this case, we may write for the contribution of the elementary ridge lying at $y = s$,

$$\frac{1}{x_1^{n-1}}\frac{du}{ds} = \mathrm{Re}\left(\sum_{q=0}^{n-1} C_q(s)x^q \cosh^{-1}\sqrt{\frac{(l_2+s)(l_1-x)}{(l_1+l_2)(s-x)}} + \sum_{q=0}^{n-2} D_q x^q \sqrt{(l_1-x)(l_2+x)}\right), \tag{2.74}$$

and the contribution of the whole segment, from $y = s_r$ to $y = s_r'$, between the two ridges considered, as well as their own contribution, will be given by the expression

$$u = \mathrm{Re}\int_{s_r}^{s_r'} \frac{du}{ds}\,ds + u_r + u_r', \tag{2.75}$$

u_r and u_r' being the contributions of the two ridges.

Similarly, if the coefficient $w_{n-q-1,\,q}$ varies between two left ridges situated at $y = -s_l$ and $y = -s_l'$, we shall have for the contribution of an elementary ridge and for the whole segment from $y = -s_l$ to $y = -s_l'$,

$$\frac{1}{x_1^{n-1}}\frac{du}{ds} = \mathrm{Re}\left(\sum_{q=0}^{n-1} G_q(s)x^q \cosh^{-1}\sqrt{\frac{(l_1+s)(l_2+x)}{(l_1+l_2)(s+x)}} + \sum_{q=0}^{n-2} D_q x^q \sqrt{(l_1-x)(l_2+x)}\right), \tag{2.76a}$$

$$u = \mathrm{Re}\int_{s_l}^{s_l'} \frac{du}{ds}\,ds + u_l + u_l', \tag{2.76b}$$

u_l and u_l' being the contributions of the two ridges.

Let us consider a conical motion proper ($n = 1$), for which however the downwash varies with respect to the abscissa y, considered now as a usual variable, dependent on a zero order homogeneous function,

$$w = w(y). \tag{2.77}$$

Instead of Q in formula (2.41), we have now

$$dQ = -\frac{s}{\sqrt{1-B^2s^2}}\left(\frac{dw}{dy}\right)_{y=s} ds, \tag{2.78}$$

whence, taking account of the sign change by passing from ln to $\cosh^{-1}$ (2.62),

$$\frac{du}{ds} = \text{Re}\,\frac{2s}{\pi\sqrt{1-B^2s^2}}\cosh^{-1}\sqrt{\frac{(l_2+s)(l_1-x)}{(l_1+l_2)(s-x)}}\cdot\left(\frac{dw}{dy}\right)_{y=s}, \tag{2.79a}$$

$$\frac{du}{ds} = -\,\text{Re}\,\frac{2s}{\pi\sqrt{1-B^2s^2}}\cosh^{-1}\sqrt{\frac{(l_1+s)(l_2+x)}{(l_1+l_2)(s+x)}}\cdot\left(\frac{dw}{dy}\right)_{y=-s}, \tag{2.79b}$$

for the right-hand side of the origin and the left-hand side, respectively; hence, for the whole wing, we integrate from 0 to l_1 using (2.75), and from $-l_2$ to 0 using (2.76b), thus performing a simple quadrature, more or less laborious. The contribution of the edges lying at A_1 and A_2 will be then added, thus obtaining the complete solution.

II. 4. Examples of Application to Ordinary Wings in the Direct Problem

In the preceding section we gave the fundamental formulae with the aid of which the disturbance velocity u may be constructed, formulae which represent two effects: *the effect of the leading edges* and *the effect of the separation ridges*.

With the aid of these formulae it is possible to solve all the problems of the thin triangular wing and of the triangular wing of symmetrical thickness, which, by superposition, give us the actual wings, defined at the beginning as *thick lifting wings*. We admit the wing to be perfectly determined as concerns its *planform* by its trailing edge A_2A_1 and its length ($x_1=1$).

We give the results for the following problems:

II. 4.1. Thin Wing with Subsonic Edges and Two Separation Ridges

For the wing without ridges, the disturbance velocity u is given by formula (2.57). If the wing has in addition a separation ridge in the right side of the origin and another one on the left side, expressions (2.64) and (2.69b) will be added to this formula, representing the effect of the ridges C_1 and C_2 (Fig. 2.8). In addition, if we note that the second terms of the right-hand side of these last expressions reduce to (2.57) through combination, then, for the thin wing with subsonic edges and two ridges on the right and left side of the origin, we can write the formula for the disturbance velocity in the form

$$\frac{u}{x_1^{n-1}} = \mathrm{Re}\left[\frac{\sum_{q=0}^{n} A_q x^q}{\sqrt{(l_1-x)(l_2+x)}} + \sum_{q=0}^{n-1} C_q x^q \cosh^{-1}\sqrt{\frac{(l_2+s_1)(l_1-x)}{(l_1+l_2)(s_1-x)}}\right.$$

$$\left. + \sum_{q=0}^{n-1} G_q x^q \cosh^{-1}\sqrt{\frac{(l_1+s_2)(l_2+x)}{(l_1+l_2)(s_2+x)}}\right]. \tag{2.80}$$

Case of thin wing with subsonic trailing edge. If one of the subsonic edges, A_2 for example, is on the right side of the axis Ox_1 (Fig. 2.8a), then this line becomes a *subsonic trailing edge*. Analogous to Joukowski's hypothesis about the conditions of motion at the trailing edge in incompressible flow, the hypothesis that the velocity is zero at the trailing edge will be also made here.

Let us consider a simple wing without a separation ridge. In this case, the previous formula (2.53), which contains only the influence of the subsonic leading edge A_1, is also valid for the case given in Fig. 2.8a; l_2 is, however, to be replaced by $-l_2'$. Similarly, formula (2.56) is also valid for the case when the edge A_1 becomes a *trailing edge* (OA_1'), but then we must put $l_1 = -l_1'$ (Fig. 2.8a).

Thus we obtain

$$\frac{u}{x_1^{n-1}} = \mathrm{Re}\sum_{q=0}^{n-1} A_q x^q \sqrt{\frac{x-l_2'}{l_1-x}}, \tag{2.81a}$$

$$\frac{u}{x_1^{n-1}} = \mathrm{Re}\sum_{q=0}^{n-1} A_q x^q \sqrt{\frac{-l_1'-x}{l_2+x}}. \tag{2.81b}$$

Note. In the above formulae no account has been taken of the influence of the vortices which separate from the centre up to $y = -l_2'$, $y = -l_1'$ respectively, owing to the variation of circulation round the wing and to the fact that the trailing edge is contained in the Mach cone.

II. 4.2. Thin Wing with One Subsonic Edge and One Supersonic Edge

Let us assume that the edge A_2 is subsonic, while A_1 is supersonic (Fig. 2.10); in this case, formula (2.59) should be applied. In addition, let us consider a ridge lying at C_1, for which formula (2.65) should be applied, in which the second term in the right-hand side can be combined with (2.59). Thus we have

$$\frac{u}{x_1^{n-1}} = \mathrm{Re}\left[\sum_{q=0}^{n-1} A_q x^q \sqrt{\frac{1-Bx}{B(l_2+x)}} + \sum_{q=0}^{n-1} C_q x^q \cosh^{-1}\sqrt{\frac{(s_1+l_2)(1-Bx)}{(1+Bl_2)(s_1-x)}}\right]. \tag{2.82}$$

We have assumed that beyond the ridge C_1, the wing is continued up to A_1 by a plate without incidence (Fig. 2.10).

If the ridge C_1 is situated outside the Mach cone then it will constitute the true leading edge A_1 (we superpose OC_1 on OA_1 on the real wing). The contribution of the ridge C_1 represents now the contribution of the leading edge A_1, and we put $s_1 = l_1$; finally, taking into account (2.68) and replacing C_q by K_q to indicate that we are dealing with a supersonic leading

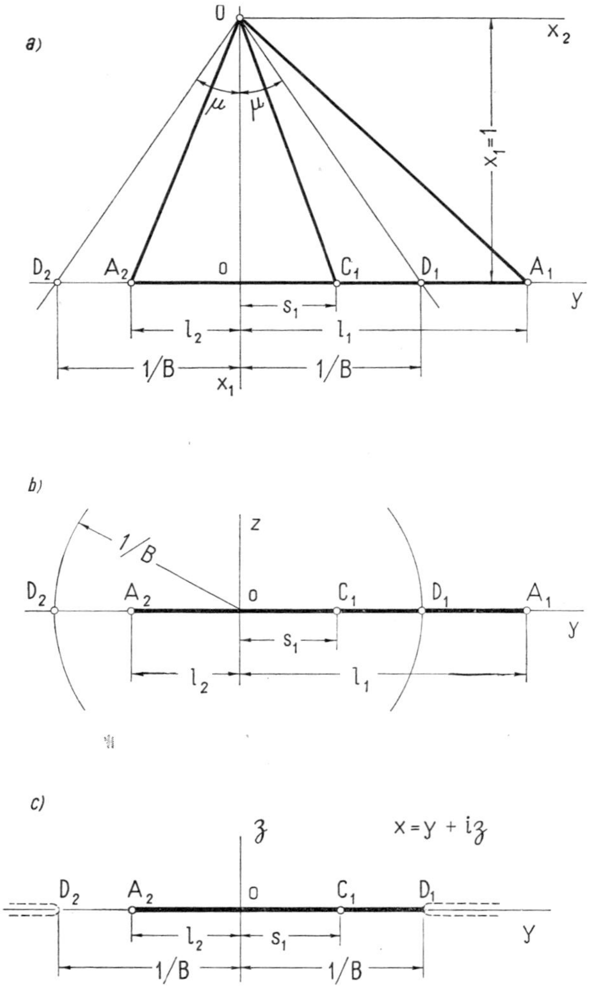

FIG. 2.10

edge, we obtain

$$\frac{u}{x_1^{n-1}} = \operatorname{Re}\left[\sum_{q=0}^{n-1} A_q x^q \sqrt{\frac{1-Bx}{B(l_2+x)}} + \sum_{q=0}^{n-1} K_q x^q \cos^{-1}\sqrt{\frac{(l_1+l_2)(1-Bx)}{(1+Bl_2)(l_1-x)}}\right]. \tag{2.83}$$

If the edge A_1 is subsonic and A_2 is supersonic, then we combine (2.58) and (2.66), replacing s_2 from its latter formula by l_2; then, taking account of (2.68) and replacing G_q by H_q as above, we obtain the expression of the contribution of the supersonic leading edge OA_2,

$$\frac{u}{x_1^{n-1}} = \operatorname{Re}\left[\sum_{q=0}^{n-1} A_q x^q \sqrt{\frac{1+Bx}{B(l_1-x)}} + \sum_{q=0}^{n-1} H_q x^q \cos^{-1}\sqrt{\frac{(l_1+l_2)(1+Bx)}{(1+Bl_1)(l_2+x)}}\right]. \tag{2.84}$$

Case of trailing edges. If we assume that the leading edge A_2 becomes a trailing edge, then this has no longer any effect and the solution is given by (2.64) with $l_1 = 1/B$, as in (2.65), and with l_2 replaced by $-l_2'$; we obtain thus,

$$\frac{u}{x_1^{n-1}} = \operatorname{Re}\left[\sum_{q=0}^{n-1} C_q x^q \cosh^{-1}\sqrt{\frac{(s_1-l_2')(1-Bx)}{(1-Bl_2')(s_1-x)}} + \sum_{q=0}^{n-2} D_q x^q \sqrt{\frac{(1-Bx)(x-l_2')}{B}}\right]. \tag{2.85}$$

From this expression, we deduce the case of the wing having a subsonic leading edge A_2' and a supersonic leading edge A_1, if the ridge superposes on OA_1, putting for this purpose $s_1 = l_1$ and replacing C_q by K_q,

$$\frac{u}{x_1^{n-1}} = \operatorname{Re}\left[\sum_{q=0}^{n-1} K_q x^q \cos^{-1}\sqrt{\frac{(l_1-l_2')(1-Bx)}{(1-Bl_2')(l_1-x)}} + \sum_{q=0}^{n-2} D_q x^q \sqrt{\frac{(1-Bx)(x-l_2')}{B}}\right]. \tag{2.86}$$

These formulae remain also valid for the case when OA_2' superposes on Ox_1 ($l_2' = 0$).

In a similar way, proceeding from formula (2.69b), if OA_1 becomes a trailing edge, we put $l_1 = -l_1'$; since OA_2 is supersonic, we put $l_2 = 1/B$, obtaining finally

$$\frac{u}{x_1^{n-1}} = \operatorname{Re}\left[\sum_{q=0}^{n-1} G_q x^q \cosh^{-1}\sqrt{\frac{(s_2-l_1')(1+Bx)}{(1-Bl_1')(s_2+x)}} + \sum_{q=0}^{n-2} D_q x^q \sqrt{\frac{(1+Bx)(-x-l_1')}{B}}\right]. \tag{2.87}$$

If the ridge OC_2 becomes the true supersonic leading edge, then, replacing G_q by H_q, we have

$$\frac{u}{x_1^{n-1}} = \mathrm{Re}\left[\sum_{q=0}^{n-1} H_q x^q \cos^{-1}\sqrt{\frac{(l_2-l_1')(1+Bx)}{(1-Bl_1')(l_2+x)}} + \sum_{q=0}^{n-2} D_q x^q \sqrt{\frac{(1+Bx)(-x-l_1')}{B}}\right]. \tag{2.88}$$

These formulae are also valid for the case when OA_1' superposes on the axis Ox_1 ($l_1' = 0$).

II. 4.3. Wing of Symmetrical Thickness and Thin Wing with Supersonic Edges

Since in the case of the wing of symmetrical thickness, the abscissa axis is an axis of symmetry, this wing has no leading edge, in the sense of thin wing theory. We could consider two ridges OC_1 and OC_2 (Fig. 2.11), and assume that on the right side of OC_1 or on the left side of OC_2, the wing is continued by a thin plane parallel to the free stream, the trace being thus situated on the abscissa axis in the x plane.

In this way, the two ridges OC_1 and OC_2 are thus leading edges of the wing of symmetrical thickness.

The summation of the contributions of the two ridges given by (2.67) and (2.70c), once for the ridge C_1 ($y = s_1$) and once for the ridge C_2 ($y = -s_2$)—ridges that represent now the leading edges of the wing of symmetrical thickness—gives

$$\frac{u}{x_1^{n-1}} = \mathrm{Re}\left[\sum_{q=0}^{n-1} C_q x^q \cosh^{-1}\sqrt{\frac{(1+Bs_1)(1-Bx)}{2B(s_1-x)}} + \sum_{q=0}^{n-1} G_q x^q \cosh^{-1}\sqrt{\frac{(1+Bs_2)(1+Bx)}{2B(s_2+x)}} + \sum_{q=0}^{n-2} D_q x^q \sqrt{\frac{1-B^2x^2}{B^2}}\right]. \tag{2.89}$$

This expression would also correspond to a *thin wing*, which had on the right side of the ridge OC_1 and on the left side of the ridge OC_2 plane surfaces at zero incidence prolonged beyond the Mach cone up to the leading edges.

If s_1 and s_2 are greater than $1/B$, that is they cross the Mach cone and become the true leading edges ($s_1 = l_1$, $s_2 = l_2$), the above formula also corresponds to the case of a thin wing having both its edges supersonic, since in such a case the upper surface of the thin wing behaves like the upper surface of a wing of symmetrical thickness, while on the lower surface the axial disturbance velocities are equal and of opposite sign.

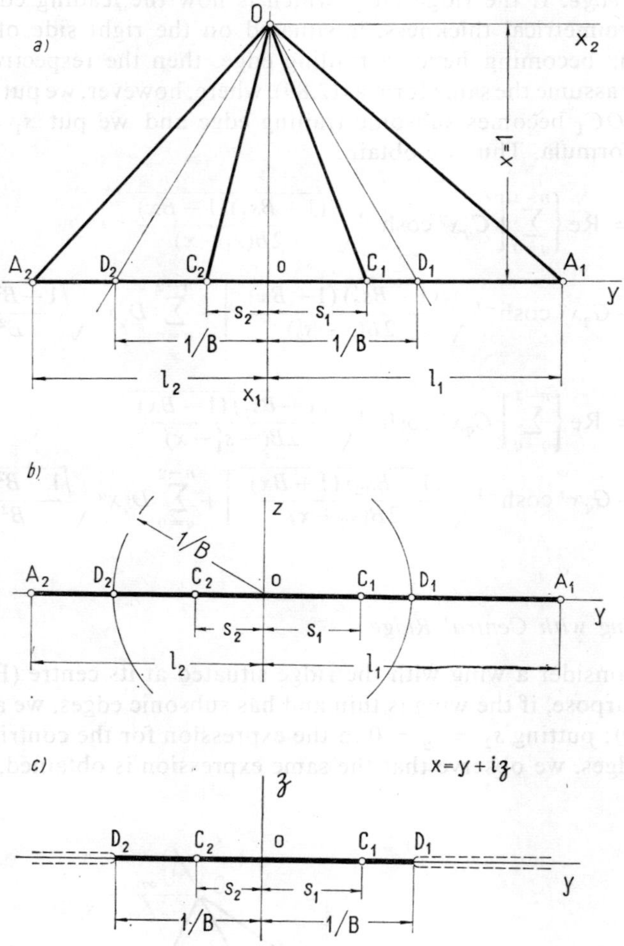

FIG. 2.11

Considering then (2.68) and replacing C_q by K_q, we may write

$$\frac{u}{x_1^{n-1}} = \text{Re}\left[\sum_{q=0}^{n-1} K_q x^q \cos^{-1}\sqrt{\frac{(1+Bl_1)(1-Bx)}{2B(l_1-x)}} + \sum_{q=0}^{n-1} H_q x^q \cos^{-1}\sqrt{\frac{(1+Bl_2)(1+Bx)}{2B(l_2+x)}} + \sum_{q=0}^{n-2} D_q x^q \sqrt{\frac{1-B^2x^2}{B^2}}\right]. \tag{2.90}$$

Trailing edge. If the ridge OC_2, which is now the leading edge of the wing of symmetrical thickness, is situated on the right side of the axis ($s_2 = -s_2'$), becoming hence a trailing edge, then the respective contribution will assume the same form as (2.89), where, however, we put $s_2 = -s_2'$; similarly, OC_1 becomes subsonic trailing edge and we put $s_1 = -s_1'$ in the same formula. Thus we obtain,

$$\frac{u}{x_1^{n-1}} = \mathrm{Re}\left\{\sum_{q=0}^{n-1}\left[C_q x^q \cosh^{-1}\sqrt{\frac{(1+Bs_1)(1-Bx)}{2B(s_1-x)}} + G_q x^q \cosh^{-1}\sqrt{\frac{(1-Bs_2')(1+Bx)}{2B(x-s_2')}}\right] + \sum_{q=0}^{n-2} D_q x^q \sqrt{\frac{1-B^2x^2}{B^2}}\right\}, \tag{2.91a}$$

$$\frac{u}{x_1^{n-1}} = \mathrm{Re}\left\{\sum_{q=q}^{n-1}\left[C_q x^q \cosh^{-1}\sqrt{\frac{(1-Bs_1')(1-Bx)}{2B(-s_1'-x)}} + G_q x^q \cosh^{-1}\sqrt{\frac{(1+Bs_2)(1+Bx)}{2B(s_2+x)}}\right] + \sum_{q=0}^{n-2} D_q x^q \sqrt{\frac{1-B^2x^2}{B^2}}\right\}. \tag{2.91b}$$

II. 4.4. Wing with Central Ridge

Let us consider a wing with the ridge situated at its centre (Fig. 2.12). For this purpose, if the wing is thin and has subsonic edges, we apply formula (2.80); putting $s_1 = s_2 = 0$ in the expression for the contribution of the two ridges, we observe that the same expression is obtained, since we

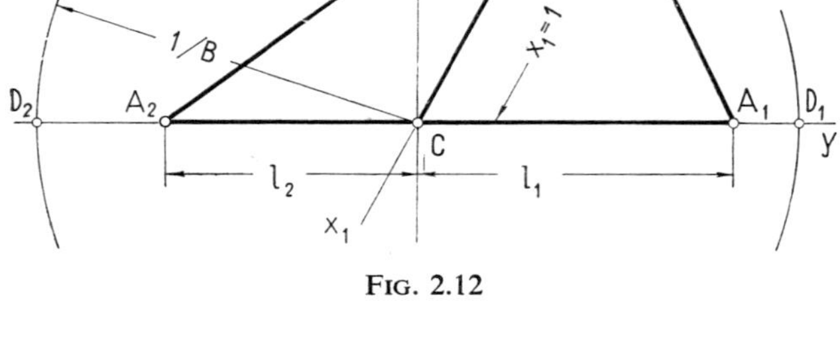

FIG. 2.12

have

$$\operatorname{Re}\cosh^{-1}\sqrt{\frac{l_2(l_1-x)}{-(l_1+l_2)x}} = \operatorname{Re}\cosh^{-1}\sqrt{\frac{l_1(l_2+x)}{(l_1+l_2)x}}\,. \tag{2.92}$$

On the other hand, replacing C_q or G_q by Q_q, in order to indicate that the two ridges coalesce at the origin, we may write

$$\frac{u}{x_1^{n-1}} = \operatorname{Re}\left\{\sum_{q=0}^{n}\frac{A_q x^q}{\sqrt{(l_1-x)(l_2+x)}} + \sum_{q=0}^{n-1} Q_q x^q \cosh^{-1}\sqrt{\frac{l_1(l_2+x)}{(l_1+l_2)x}}\right\}. \tag{2.93}$$

If the wing is of *symmetrical thickness*, we set $l_1 = l_2 = 1/B$ in (2.92) and obtain

$$\operatorname{Re}\cosh^{-1}\sqrt{\frac{1+Bx}{2Bx}} = \tfrac{1}{2}\operatorname{Re}\cosh^{-1}\sqrt{\frac{1}{B^2x^2}} \tag{2.94}$$

which will be introduced into expressions (2.89) and (2.90) in order to obtain the contribution of the central ridge. We have thus

$$\begin{aligned}\frac{u}{x_1^{n-1}} = \operatorname{Re}\Bigg\{\sum_{q=0}^{n-1}\Bigg[& C_q x^q \cosh^{-1}\sqrt{\frac{(1+Bs_1)(1-Bx)}{2B(s_1-x)}} \\ &+ G_q x^q \cosh^{-1}\sqrt{\frac{(1+Bs_2)(1+Bx)}{2B(s_2+x)}} + Q_q x^q \cosh^{-1}\sqrt{\frac{1}{B^2x^2}}\Bigg] \\ &+ \sum_{q=0}^{n-2} D_q x^q \sqrt{\frac{1-B^2x^2}{B^2}}\Bigg\},\end{aligned} \tag{2.95a}$$

$$\begin{aligned}\frac{u}{x_1^{n-1}} = \operatorname{Re}\Bigg\{\sum_{q=0}^{n-1}\Bigg[& K_q x^q \cos^{-1}\sqrt{\frac{(1+Bl_1)(1-Bx)}{2B(l_1-x)}} \\ &+ H_q x^q \cos^{-1}\sqrt{\frac{(1+Bl_2)(1+Bx)}{2B(l_2+x)}} + Q_q x^q \cosh^{-1}\sqrt{\frac{1}{B^2x^2}}\Bigg] \\ &+ \sum_{q=0}^{n-2} D_q x^q \sqrt{\frac{1-B^2x^2}{B^2}}\Bigg\}.\end{aligned} \tag{2.95b}$$

These expressions are valid both for the wing of symmetrical thickness and for the thin wing with two ridges, prolonged beyond the Mach cone by plane surfaces at zero incidence, or for the thin wing with supersonic edges respectively.

II. 4.5. Validity of Formulae outside the Mach Cone

If we are given a wing of symmetrical thickness or a thin wing with one or both edges supersonic, then, as long as the part of the wing outside the Mach cone has neither incidence nor slope, but only a plane parallel to the stream, it is obvious that the velocity w on these external portions is equal to zero, as is also shown by all formulae established above.

However, if a ridge situated on the side of the supersonic leading edge crosses the Mach cone ($s_1 > 1/B$, $s_2 > 1/B$), all formulae retain their validity both inside and outside the Mach cone. Indeeed, for conical motions of order one, applying formula (2.65) and taking account of (2.41), (2.37) and (2.62), we may write for the contribution of the right side ridge, for instance,

$$\mathcal{U}_0 = -\frac{2s_1 w}{\pi\sqrt{1-B^2 s_1^2}} \cosh^{-1} \sqrt{\frac{(l_2+s_1)(1-Bx)}{(1+Bl_2)(s_1-x)}}, \tag{2.96}$$

where w takes the place of w'' ($w' = 0$) in (2.37).

Taking $s_1 > 1/B$ and considering (2.68), we put

$$\mathcal{U}_0 = -\frac{2s_1 w}{\pi\sqrt{B^2 s_1^2-1}} \cos^{-1} \sqrt{\frac{(l_2+s_1)(1-Bx)}{(1+Bl_2)(s_1-x)}}. \tag{2.96 bis}$$

It is readily seen that for the points situated outside the Mach cone ($s_1 > x > 1/B$), the expression under the square root, whose absolute value will be indicated by β, becomes negative and hence we have

$$\cos^{-1}\sqrt{-\beta} = \cos^{-1} i\sqrt{\beta} = \tfrac{1}{2}\pi \pm i \ln(\sqrt{\beta}+\sqrt{1+\beta}), \tag{2.97}$$

whence

$$u = \operatorname{Re}\mathcal{U}_0 = \frac{-w}{B\sqrt{1-\dfrac{1}{B^2 s_1^2}}} = \frac{\tau U_\infty}{B\sqrt{1-\tan^2\chi\tan^2\mu}}, \tag{2.98}$$

the same expression as that found for the yawed wing with angle of yaw χ (1.84).

Moreover, it may be directly remarked that the supersonic ridge is the line of a wedge, tangent to the Mach cone inside which u is constant, hence also on the Mach circle, on the portion contained in the interior of the wedge tangent to the cone (Fig. 2.13).

Further, using the same reasoning, the functions $F_{n-q,q,0}$ play the same part as $\mathcal{U}_0$ in the true conical motion ($n = 1$), as may also be seen from the expression (2.63), where the first term is similar to the above one

(2.96), and the second term becomes imaginary beyond the Mach cone and hence plays no part.

Applying Euler's formula (2.45a), which does not alter in the least the validity of the function $F_{n-q,q,0}$ on the whole interval inside and outside the Mach cone, since these functions are multiplied by y^q alone we obtain

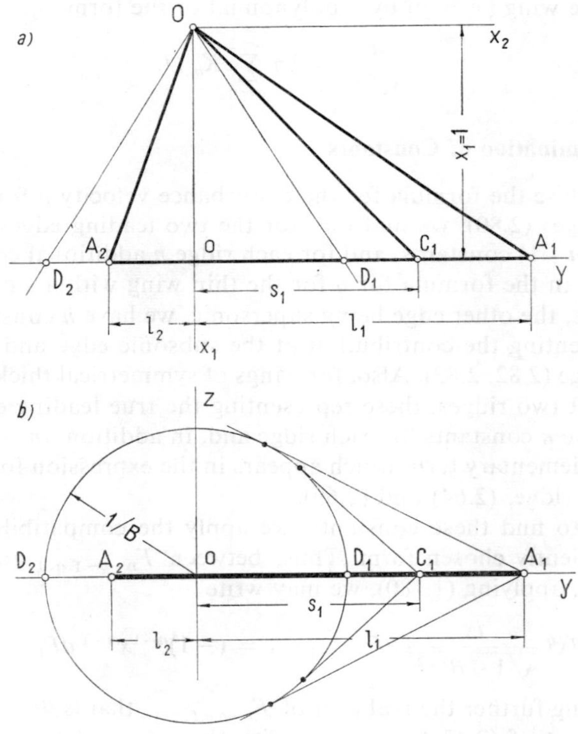

FIG. 2.13

the general expression for u (2.64), which therefore also retains its validity on the wing inside and outside the Mach cone.

A solution of this problem has also been given by another method of approach,† by looking for a solution of the hyperbolic equation

$$-B^2\frac{\partial^2 u}{\partial x_1^2}+\frac{\partial^2 u}{\partial x_2^2}+\frac{\partial^2 u}{\partial x_3^2}=0, \tag{2.99}$$

† Sobolev and Smirnov, quotation from the book *Course of Higher Mathematics*, by V.I. Smirnov, vol. III, 2nd part, Moscow, 1953. Rumanian translation.

where the function u is of zero order, thus corresponding to conical motions of order one (the potential Φ is a homogeneous function of order one).

Note on the expression for the axial velocity u outside the Mach cone. When the leading edge or even some ridges are outside the Mach cone, $\mathrm{Re}\cos^{-1}\sqrt{\beta}$ corresponding to these reduces to $\frac{1}{2}\pi$ in the region outside the Mach cone but inside the wedge, as indicated by (2.97). In this case, u is given on the wing ($x = y$) by a polynomial of the form

$$\frac{u}{x_1^{n-1}} = \tfrac{1}{2}\pi \sum_{q=0}^{n-1} K_q\, y^q. \tag{2.100}$$

II. 5. Determination of Constants

If we analyse the formula for the disturbance velocity u for wings with subsonic edges (2.80), we find that for the two leading edges we have to determine $(n+1)$ constants, and for each ridge n additional constants.

Likewise, in the formula for u for the thin wing with a single subsonic leading edge, the other edge being supersonic, we have n constants for the term representing the contribution of the subsonic edge and n constants for each ridge (2.82, 2.83). Also, for wings of symmetrical thickness, which have at least two ridges, these representing the true leading edges in this case, we have n constants for each ridge and, in addition, $(n-1)$ constants of the complementary term which appears in the expression for the contribution of a ridge, (2.64) and (2.69).

In order to find these constants, we apply the compatibility relations in a conveniently chosen form. Thus, between $F_{n-q-1,q,1}$ and $F_{1,n-1,0}$, for instance, applying (1.180), we may write

$$(-1)^q x^q \frac{ix}{\sqrt{1-B^2x^2}}\, dF_{n-q-1,\,q,\,1} = (-1)^{n-1} x^{n-1}\, dF_{1,\,n-1,\,0}. \tag{2.101}$$

Considering further the real part of $F_{n-q-1,q,1}$, that is $\Phi_{n-q-1,q,1}$, and, taking account of (2.47a), we may write the same relation as above in the form

$$\begin{aligned}\Phi_{n-q-1,\,q,\,1} &= \frac{(n-1)!}{C_{n-1}^q}\, w_{n-q-1,\,q}\\ &= \mathrm{Im}\,(-1)^{n-q-1}\int x^{n-q-2}\sqrt{1-B^2x^2}\, dF_{1,\,n-1,\,0}\end{aligned} \tag{2.102}$$

to which an arbitrary constant is added.

Let us now put the homogeneous function of zero order u/x_1^{n-1} in the form

$$\frac{u}{x_1^{n-1}} = \mathrm{Re}\,\mathcal{U}_{n-1}, \tag{2.103}$$

where $\mathcal{U}_{n-1}$ is an analytic function of the complex variable x, and notice that we have†

$$F_{1,\,n-1,\,0} = \frac{d^{n-1}\mathcal{U}_{n-1}}{dx^{n-1}}\,; \tag{2.104}$$

in this case expression (2.93) assumes the following form,

$$\frac{(n-1)!}{C^q_{n-1}}\,w_{n-q-1,\,q} = \mathrm{Im}\,(-1)^{n-q-1}\int x^{n-q-2}\,\sqrt{1-B^2x^2}\,\frac{d^n\mathcal{U}_{n-1}}{dx^n}\,dx, \tag{2.105}$$

to which an arbitrary constant is also added.

On each side of each ridge, we have $w'_{n-q-1,\,q}$ on the right and $w''_{n-q-1,q}$, on the left, these values being different (whence the term 'separation ridge').

If we perform the above integral along a semicircle σ of very small radius, having as centre the ridge trace ($y = s$), we may write the relation

$$\mathrm{Im}\,(-1)^{n-q-1}\int_{\sigma} x^{n-q-2}\,\sqrt{1-B^2x^2}\,\frac{d^n\mathcal{U}_{n-1}}{dx^n}\,dx$$

$$= \frac{(n-1)!}{C^q_{n-1}}\,(w''_{n-q-1,\,q}-w'_{n-q-1,\,q}). \tag{2.106}$$

On the other hand, considering two points $x = y'$ and $x = y''$ on the right and left sides of the ridge and a third point x_0, where the velocity w is zero, we also have the following two relations,

$$\frac{(n-1)!}{C^q_{n-1}}\,w'_{n-q-1,\,q} = \mathrm{Im}\,(-1)^{n-q-1}\int_{x_0}^{y'} x^{n-q-2}\,\sqrt{1-B^2x^2}\,\frac{d^n\mathcal{U}_{n-1}}{dx^n}\,dx\,, \tag{2.107a}$$

$$\frac{(n-1)!}{C^q_{n-1}}\,w''_{n-q-1,\,q} = \mathrm{Im}\,(-1)^{n-q-1}\int_{x_0}^{y''} x^{n-q-2}\,\sqrt{1-B^2x^2}\,\frac{d^n\mathcal{U}_{n-1}}{dx^n}\,dx. \tag{2.107b}$$

By varying q from 0 to $(n-1)$, we obtain a number of equations which is generally smaller than the number of unknowns. The necessary additional relations will be deduced from the following additional conditions:

(1) $\Phi_{n-q-1,\,q,\,1}$ is continuous when crossing the axis Ox_1, being hence continuous for $x = 0$;

(2) $\Phi_{n-q-1,\,q,\,1}$ is also continuous when crossing the Mach cone, being hence continuous for $x = \pm 1/B$.

† Indeed, we have successively

$$\Phi_{100} = u, \quad \Phi_{1,\,n-1,\,0} = \frac{\partial^{n-1}u}{\partial x_2^{n-1}} = \frac{1}{x_1^{n-1}}\frac{\partial^{n-1}u}{\partial y^{n-1}} = \mathrm{Re}\,\frac{d^{n-1}\mathcal{U}_{n-1}}{dx^{n-1}}\,.$$

Applying the relation (2.102),

$$d\Phi_{n-q-1,\,q,\,1} = \text{Re}\, dF_{n-q-1,\,q,\,1} = \text{Im}\,(-1)^{n-q-1}\, x^{n-q-2}\sqrt{1-B^2x^2}\, dF_{1,\,n-1}$$

$$= \text{Im}\,(-1)^{n-q-1}\, x^{n-q-2}\sqrt{1-B^2x^2}\,\frac{d^n\mathcal{U}_{n-1}}{dx^n}\,dx, \qquad (2.108)$$

we observe that, for $q = n-1$, $\Phi_{n-q-1,\,q,\,1}$ would be discontinuous at the origin; in order to avoid this, we must necessarily have

$$\left(\frac{d^n\mathcal{U}_{n-1}}{dx^n}\right)_{x=0} = 0, \qquad (2.109)$$

which satisfies the first condition set above.

For wings of symmetrical thickness, or for thin wings with one or both edges supersonic, the function $\mathcal{U}_{n-1}$ contains in one form or another the terms $\sqrt{1-Bx}$ and $\sqrt{1+Bx}$, as may be seen in formulae (2.89), (2.91), etc. The derivatives of $\mathcal{U}_{n-1}$ will necessarily contain these terms in the denominator, to the power $(2p-1)/2$, p being the order of the derivative.

In order to avoid this, it is necessary that $\mathcal{U}_{n-1}$ should satisfy the condition that

$$\frac{d^n\mathcal{U}_{n-1}}{dx^n}$$

should contain at most $\sqrt{1-Bx}$, $\sqrt{1+Bx}$, or $\sqrt{1-B^2x^2}$ in the denominator.

It is possible to ensure this, by writing that there exists no jump when the Mach cone is crossed and hence that we must have the condition

$$\Delta\left(\frac{d^p\mathcal{U}_{n-1}}{dx^p}\right)_{x=\pm\frac{1}{B}} = 0, \quad (p = 1, 2, \ldots, n-1). \qquad (2.110)$$

We shall first assume that the ridges are situated inside the Mach cone and are then put outside the cone, in the case when they will constitute supersonic leading edges or ridges, after the determination of the constants.

On the other hand, we note that the derivation order p is at most $p = n-1$, since the last derivative could contain $\sqrt{1-B^2x^2}$ in the denominator, which cancels after simplification, as seen from formula (2.107).

Even if it had not been eliminated, the term $\sqrt{1-B^2x^2}$ in the denominator is integrable.

The relations to be obtained are linear and homogeneous equations with respect to the coefficients of u, this being linear and homogeneous with respect to its own coefficients.

We notice that relations (2.110) are automatically satisfied for thin wings, on the Mach cone, towards the side of the subsonic edge.

Note. For simplifying the operations referring to the determination of constants, from examination of the integral (2.106), which represent a semi-residue round the ridge $y = s$, we notice that those terms of $\mathcal{U}_{n-1}$ which are analytic in the neighbourhood of the point $y = s$, may be disregarded. Expanding this function in the neighbourhood of this point, it is easily seen that $\mathcal{U}_{n-1}$, which represents the contribution of the right hand side ridge (2.64), and the similar function which represents the contribution of the left-hand side ridge (2.69b) respectively, reduce to the following simple expressions:

$$\mathcal{U}_{\substack{n-1 \\ x \to s_1}} = -\tfrac{1}{2} \sum_{q=0}^{n-1} C_q x^q \ln (s_1 - x), \tag{2.111a}$$

$$\mathcal{U}_{\substack{n-1 \\ x \to -s_2}} = -\tfrac{1}{2} \sum_{q=0}^{n-1} G_q x^q \ln (s_2 + x), \tag{2.111b}$$

which are obtained by expressing $\cosh^{-1}$ in logarithmic form and making x tend towards $s_1, -s_2$, respectively, retaining hence only the logarithmic term.

When the ridge is situated at the centre (2.93), the condition (2.109) is not applicable; it should be eventually replaced by the condition that u be continuous for $x = 0$. In order that this may be possible, we must have $Q_0 = 0$, as can also be clearly seen in the above formulae (2.111), where we put $s_1 = s_2 = 0$ and whence it may be seen in fact that the discontinuity of u at the origin cannot be avoided unless we put $C_0 = G_0 = 0$. It follows that, for $n = 1$, formulae (2.64) and (2.69) are no longer valid, if they are applied at the origin.

If, however, a discontinuity exists, this may be imagined as a broken surface of the wing in the direction Ox_1, having different incidences on each side of the origin. But, in this case, we have assumed a zero velocity, similarly with Jukovski's hypothesis, since the line Ox_1 plays now the part of a trailing edge (§ II. 4.2).

II. 6. Examples of the Determination of Constants

II. 6.1. Constants for Subsonic Edges

We shall treat below a complete example of determining the constants which occur in the formulae for the disturbance velocity u, in order to establish the technique of this operation.

We did in fact indicate in the preceding section the general method for determining the constants, without showing, however, the whole range of operations which are involved in a concrete case.

In the following we shall give the technique for the calculations for a particular example, retaining, however, a certain generality. In this sense, we shall consider a thin wing with subsonic edges and two separation ridges in a conical flow of second order, for which, by applying formula (2.80), we find the general expression

$$\frac{u}{x_1} = \operatorname{Re} \mathcal{U}_1 = \operatorname{Re}\left\{\frac{A_0+A_1x+A_2x^2}{\sqrt{(l_1-x)(l_2+x)}}\right.$$
$$+(C_0+C_1x)\cosh^{-1}\sqrt{\frac{(l_2+s_1)(l_1-x)}{(l_1+l_2)(s_1-x)}}$$
$$\left.+(G_0+G_1x)\cosh^{-1}\sqrt{\frac{(l_1+s_2)(l_2+x)}{(l_1+l_2)(s_2+x)}}\right\}, \tag{2.112}$$

where, for the sake of generality, the symmetry was not considered.

The incidence or the axial velocity w, which is assumed to be given, has the form

$$w = w_{10}x_1+w_{01}x_2 = x_1(w_{10}+w_{01}y), \tag{2.113}$$

the coefficients w_{10} and w_{01} being different on the three sides of the wing, A_2C_2, C_2OC_1, C_1A_1 (Fig. 2.14), namely:

$$w''_{10},\ w''_{01};\quad w^0_{10},\ w^0_{01};\quad w'_{10},\ w'_{01}.$$

The three constants A_0, A_1, A_2 corresponding to the subsonic edges, will be given by the following three distinct relations, obtained by applying formulae (2.107),

$$w'_{10} = \operatorname{Im}\int_{y'}^{1/B}\sqrt{1-B^2x^2}\,\frac{d^2\mathcal{U}_1}{dx^2}\,dx, \tag{2.114a}$$

$$w'_{01} = -\operatorname{Im}\int_{y'}^{1/B}\frac{\sqrt{1-B^2x^2}}{x}\,\frac{d^2\mathcal{U}_1}{dx^2}\,dx, \tag{2.114b}$$

$$\left(\frac{d^2\mathcal{U}_1}{dx^2}\right)_{x=0} = 0, \tag{2.114c}$$

in which y' is situated on the segment C_1A_1, while the limit $1/B$ (or any other limit $y_0 > 1/B$) corresponds to $w = 0$.

It may be easily seen that we have $\operatorname{Im}\mathcal{U}_{n-1} = 0$ (or const.), on the wing, hence the above integrals (2.114a, b) are zero at all the wing points. Denoting just by $F(x)$ such an indefinite integral and following the path of integration (Fig. 2.14), where the semicircle γ of small radius ε round the

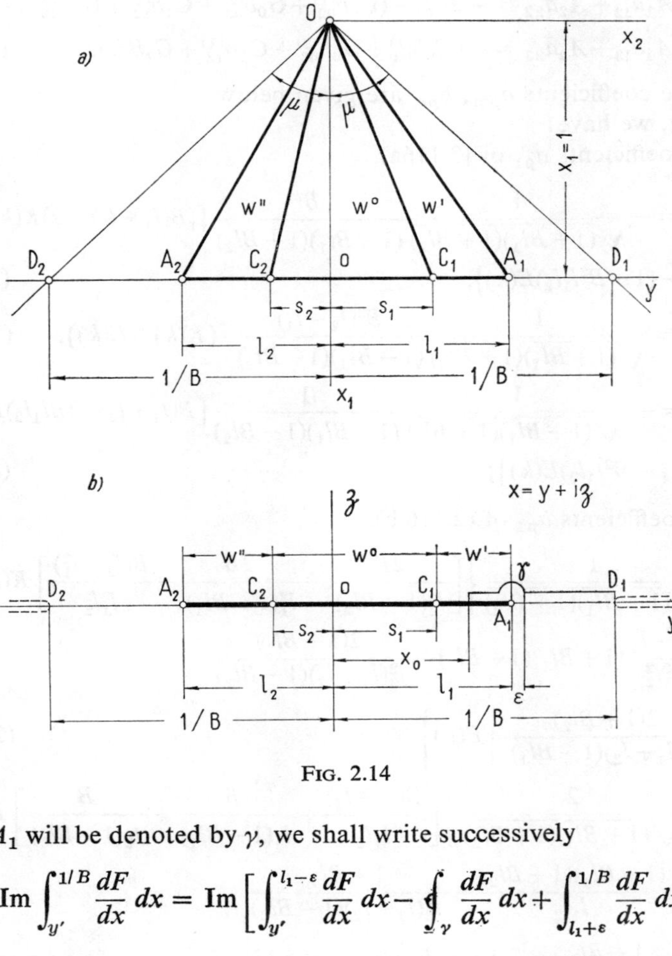

FIG. 2.14

point A_1 will be denoted by γ, we shall write successively

$$\operatorname{Im}\int_{y'}^{1/B}\frac{dF}{dx}\,dx = \operatorname{Im}\left[\int_{y'}^{l_1-\varepsilon}\frac{dF}{dx}\,dx - \oint_{\gamma}\frac{dF}{dx}\,dx + \int_{l_1+\varepsilon}^{1/B}\frac{dF}{dx}\,dx\right]$$

$$= \operatorname{Im}\left[F(l_1-\varepsilon)-F(y')+F(l_1+\varepsilon)-F(l_1-\varepsilon)+F(1/B)-F(l_1+\varepsilon)\right]$$

$$= \operatorname{Im}\left[F(1/B)-F(y')\right] = \operatorname{Im} F(1/B)$$

because $F(y')$ is real.

It should, however, be noticed that we can apply the first two relations (2.114) to the side of the edge A_2; the respective integrals are not distinct from those considered above.

Further, from the three relations (2.114), we obtain a system of three equations for the three constants A_0, A_1, A_2:

$$A_0a_{01}+A_1a_{11}+A_2a_{21} = w'_{10}-(C_0b_{01}^{(1)}+G_0b_{01}^{(2)}+C_1b_{11}^{(1)}+G_1b_{11}^{(2)}), \quad (2.116a)$$

$$A_0a_{02}+A_1a_{12}+A_2a_{22} = -2w'_{01}-(C_0b_{02}^{(1)}+G_0b_{02}^{(2)}+C_1b_{12}^{(1)}+G_1b_{12}^{(2)}), \quad (2.116b)$$

$$A_0a_{03}+A_1a_{13}+A_2a_{23} = -(C_0b_{03}^{(1)}+G_0b_{03}^{(2)}+C_1b_{13}^{(1)}+G_1b_{13}^{(2)}), \quad (2.116c)$$

where the coefficients a_{pm}, b_{pm} are given below.

In fact, we have:

for the coefficients a_{p1} of (2.116a),

$$a_{01} = -\frac{1}{\sqrt{(1+Bl_1)(1+Bl_2)}}\frac{B^2}{(1-Bl_1)(1+Bl_2)}\{[B(l_1+l_2)-2]K(k) -(1-B^2l_1l_2)E(k)\}, \quad (2.117a)$$

$$a_{11} = \frac{1}{\sqrt{(1+Bl_1)(1+Bl_2)}}\frac{B^2(l_2-l_1)}{(1-Bl_1)(1-Bl_2)}\cdot(K(k)-E(k)), \quad (2.117b)$$

$$a_{21} = -\frac{1}{\sqrt{(1+Bl_1)(1+Bl_2)}}\frac{1}{(1-Bl_1)(1-Bl_2)}[B(l_1+l_2-2Bl_1l_2)K(k) -(1-B^2l_1l_2)E(k)]; \quad (2.117c)$$

for the coefficients a_{p2} of (2.116 b),

$$a_{02} = \frac{1}{\sqrt{(1+Bl_1)(1+Bl_2)}}\left\{\left[\frac{2B}{l_2^2(1-Bl_2)}-\frac{2B}{l_1^2(1-Bl_1)}-\frac{B(l_2^2-l_1^2)}{l_1^2l_2^2}\right]K(k) +\left[\frac{l_2-l_1}{l_1^2l_2^2}(1+Bl_1)(1+Bl_2)-\frac{2(1+Bl_1)}{l_2^2(l_1+l_2)(1-Bl_2)} +\frac{2(1+Bl_2)}{l_1^2(l_1+l_2)(1-Bl_1)}\right]E(k)\right\} \quad (2.118a)$$

$$a_{12} = -\frac{2}{\sqrt{(1+Bl_1)(1+Bl_2)}}\left\{\left[\frac{B(l_1+l_2)}{l_1l_2}+\frac{B}{l_1(1-Bl_1)}+\frac{B}{l_2(1-Bl_2)}\right]K(k) -\left[\frac{(1+Bl_1)(1+Bl_2)}{l_1l_2}+\frac{1+Bl_1}{l_2(l_1+l_2)(1-Bl_2)} +\frac{1+Bl_2}{l_1(l_1+l_2)(1-Bl_1)}\right]E(k)\right\}, \quad (2.118b)$$

$$a_{22} = \frac{2}{\sqrt{(1+Bl_1)(1+Bl_2)}}\frac{B^2(l_2-l_1)}{(1-Bl_1)(1-Bl_2)}[K(k)-E(k)]; \quad (2.118c)$$

for the coefficients a_{p3} of (2.116c),

$$a_{03} = 3l_1^2+3l_2^2-2l_1l_2, \quad (2.119a)$$

$$a_{13} = 4l_1l_2(l_2-l_1), \quad (2.119b)$$

$$a_{23} = 8l_1^2l_2^2. \quad (2.119c)$$

Likewise, for the coefficients $b_{p1}^{(i)}$ of (2.116a), we have

$$b_{01}^{(i)} = \frac{r_i}{\sqrt{(1+Bl_1)(1+Bl_2)}}\left[\frac{l_2 K(k)}{l_2+g_i}+\frac{g_i(l_1+l_2)}{r_i^2}\Pi(\varrho, k)\right], \quad (2.120a)$$

$$b_{11}^{(i)} = -\frac{r_i}{\sqrt{(1+Bl_1)(1+Bl_2)}}\left[\frac{K(k)}{l_2+g_i}-\frac{l_1+l_2}{r_i^2}\Pi(\varrho, k)\right]; \quad (2.120b)$$

for the coefficients $b_{p1}^{(i)}$ of (2.116b),

$$b_{02}^{(i)} = \frac{r_i}{\sqrt{(1+Bl_1)(1+Bl_2)}}\left\{\left[\frac{2}{g_i^2(l_2+g_i)}-\frac{B(l_1+l_2)}{g_i l_1 l_2}\right]K(k) + \frac{(1+Bl_1)(1+Bl_2)}{g_i l_1 l_2}E(k)-\frac{2(l_1+l_2)}{g_i^2 r_i^2}\Pi(\varrho, k)\right\}, \quad (2.121a)$$

$$b_{12}^{(i)} = \frac{2r_i}{\sqrt{(1+Bl_1)(1+Bl_2)}}\left[\frac{2+B^2 g_i l_2}{g_i(l_2+g_i)}K(k)-\frac{(2-B^2 g_i^2)(l_1+l_2)}{g_i r_i^2}\Pi(\varrho, k)\right]; \quad (2.121b)$$

for the coefficients $b_{p3}^{(i)}$ of (2.116c),

$$b_{03}^{(i)} = \frac{l_1 l_2 r_i}{g_i}\left(l_2-l_1+\frac{2l_1 l_2}{g_i}\right), \quad (2.122a)$$

$$b_{13}^{(i)} = \frac{4l_1^2 l_2^2 r_i}{g_i}. \quad (2.122b)$$

In these formulae, $K(k)$, $E(k)$ and $\Pi(\varrho, k)$ are complete elliptic integrals of the first, second and third kind, where we put

$$k = \sqrt{\frac{(1-Bl_1)(1-Bl_2)}{(1+Bl_1)(1+Bl_2)}}, \quad (2.123a)$$

$$\varrho = \frac{(1-Bl_1)(l_2+g_i)}{(1+Bl_2)(l_1-g_i)}. \quad (2.123b)$$

For the sake of simplicity, the following notation

$$r_i = \sqrt{(l_1-g_i)(l_2+g_i)}, \quad g_i = (-1)^{i+1}s_i, \quad (i = 1, 2), \quad (2.124)$$

has been used above.

II. 6.2. Constants of Ridges

For the four constants due to ridges, we take two semi-residues for each constant round the ridges s_1 and s_2. For this purpose, expressing $\cosh^{-1}$ in logarithmic form, we take the respective reduced expressions (2.111),

$$\underset{x\to s_1}{\mathcal{U}_1'} = -\tfrac{1}{2}(C_0+C_1 x)\ln(s_1-x), \quad (2.125a)$$

$$\underset{z\to -s_2}{\mathcal{U}_1''} = -\tfrac{1}{2}(G_0+G_1 x)\ln(s_2+x). \quad (2.125b)$$

In this sense, the four necessary relations will be

$$w'_{10}-w^0_{10} = \operatorname{Im}\oint_{\sigma_1} \sqrt{1-B^2x^2}\,\frac{d^2\mathcal{U}'_1}{dx^2}\,dx = \frac{\pi(C_1+B^2s_1C_0)}{2\sqrt{1-B^2s_1^2}}, \tag{2.126a}$$

$$w'_{01}-w^0_{01} = -\operatorname{Im}\oint_{\sigma_1} \frac{\sqrt{1-B^2x^2}}{x}\,\frac{d^2\mathcal{U}'_1}{dx^2}\,dx = -\frac{\pi[C_0+s_1(2-B^2s_1^2)C_1]}{2s_1^2\sqrt{1-B^2s_1^2}}, \tag{2.126b}$$

$$w^0_{10}-w''_{10} = \operatorname{Im}\oint_{\sigma_2} \sqrt{1-B^2x^2}\,\frac{d^2\mathcal{U}''_1}{dx^2}\,dx = \frac{\pi(G_1-B^2s_2G_0)}{\sqrt{21-B^2s_2^2}}, \tag{2.126c}$$

$$w^0_{01}-w''_{01} = -\operatorname{Im}\oint_{\sigma_2} \frac{\sqrt{1-B^2x^2}}{x}\,\frac{d^2\mathcal{U}''_1}{dx^2}\,dx = -\frac{\pi[G_0-s_2(2-B^2s_1^2)G_1]}{2s_2^2\sqrt{1-B^2s_2^2}}, \tag{2.126d}$$

where σ_1 and σ_2 are semicircles of very small radii round the points s_1 and s_2.

From the above equations we obtain directly:

$$C_0 = \frac{2s_1}{\pi(1-B^2s_1^2)^{3/2}}\left[(2-B^2s_1^2)(w'_{10}-w^0_{10})+s_1(w'_{01}-w^0_{01})\right], \tag{2.127a}$$

$$C_1 = -\frac{2}{\pi(1-B^2s_1^2)^{3/2}}\left[(w'_{10}-w^0_{10})+B^2s_1^3(w'_{01}-w^0_{01})\right], \tag{2.127b}$$

$$G_0 = -\frac{2s_2}{\pi(1-B^2s_2^2)^{3/2}}\left[(2-B^2s_2^2)(w^0_{10}-w''_{10})-s_2(w^0_{01}-w''_{01})\right], \tag{2.127c}$$

$$G_1 = -\frac{2}{\pi(1-B^2s_2^2)^{3/2}}\left[(w^0_{10}-w''_{10})-B^2s_2^3(w^0_{01}-w''_{01})\right]\cdot \tag{2.127d}$$

Note. The results obtained above are also valid for the case of supersonic ridges.

II.6.3. Case of the Wing of Symmetrical Planform (Delta Wing)

For the case of the wing of symmetrical planform (delta wing), we put, as above, $l_1 = l_2 = l$ and $s_1 = s_2 = s$, considering two cases: *symmetrical incidence* and *antisymmetrical incidence*.

Symmetrical incidence. In this case we put

$$w''_{10} = w'_{10}, \quad w''_{01} = -w'_{01}, \quad w^0_{10}, \quad w^0_{01} = 0. \tag{2.128}$$

From considerations of geometrical symmetry on the one hand and from the examination of relations (2.127) on the other hand, it follows that

$$A_1 = 0, \quad C_0 = G_0, \quad C_1 = -G_1. \tag{2.129}$$

Antisymmetrical incidence. In this case we may write

$$w''_{10} = -w'_{10}, \quad w''_{01} = w'_{01}, \quad w^0_{10} = 0, \quad w^0_{01} \tag{2.130}$$

and from similar considerations it follows that

$$A_0 = A_2 = 0, \quad C_0 = -G_0, \quad C_1 = G_1. \tag{2.131}$$

Of course, both for the symmetrical and antisymmetrical wing, the constants may be determined directly in a simpler way, but we have aimed in the above at treating the general case, which is more laborious and involves complex operations.

II. 7. Contribution of the Ridges and the Leading Edge to the Calculation of the Downwash w, when the Pressures are given on the Wing (Indirect Problem)

II. 7.1. Analogy between u in the Direct Problem and w in the Indirect Problem

As already defined previously (§ II. 1.3), this problem consists in finding the equations of the wing surface in the case when the pressure distribution is given. Since the pressure coefficient is proportional to the longitudinal

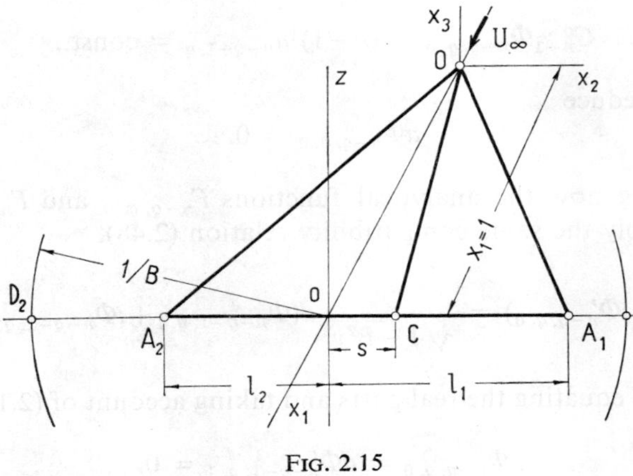

FIG. 2.15

disturbance velocity u (2.19), the pressure distribution will be defined by the expression for u, which will be given by a homogeneous polynomial of $(n-1)$th order of the form indicated by (2.20).

From this condition there results a motion in which the downwash w, appearing in the calculation of the wing equations (2.21b), is to be determined. This problem is the inverse of the preceding one, in which w was given and u was determined (direct problem); that is the reason why, for simplifying our exposition, we have termed it the *indirect problem*.

Let us consider exactly the same wing as for the direct problem, with a separation ridge OC (Fig. 2.15), on which, however, the distribution of longitudinal velocities u is given instead of that of the downwash w, as in the preceding cases of the direct problem.

From Euler's formulae (2.45), which we write once again

$$\frac{(n-1)!u}{x_1^{n-1}} = \sum_{q=0}^{n-1} C_{n-1}^q y^q \Phi_{n-q,\,q,\,0} = \operatorname{Re} \sum_{q=0}^{n-1} C_{n-1}^q x^q F_{n-q,\,q,\,0}\,, \tag{2.132a}$$

$$\frac{(n-1)!w}{x_1^{n-1}} = \sum_{q=0}^{n-1} C_{n-1}^q y^q \Phi_{n-q-1,\,q,\,1} = \operatorname{Re} \sum_{q=0}^{n-1} C_{n-1}^q x^q F_{n-q-1,\,q,\,1}\,, \tag{2.132b}$$

we notice that the first relation corresponds to the velocity u on the wing (2.20), which will be given in the form

$$\frac{u}{x_1^{n-1}} = \sum_{q=0}^{n-1} u_{n-q-1,\,q}\, y^q\,. \tag{2.133}$$

It follows at once, by equating this expression with the second term in (2.132a), that

$$C_{n-1}^q \Phi_{n-q,\,q,\,0} = (n-1)!u_{n-q-1,\,q} = \text{const.}, \tag{2.134}$$

whence we deduce

$$d\Phi_{n-q,\,q,\,0} = 0. \tag{2.135}$$

Considering now the analytical functions $F_{n-q,\,q,\,0}$ and $F_{n-q-1,\,q,\,1}$, to which we apply the same compatibility relation (2.48),

$$d(\Phi_{n-q,\,q,\,0} + i\Phi'_{n-q,\,q,\,0}) = \frac{iy}{\sqrt{1-B^2y^2}}\, d(\Phi_{n-q-1,\,q,\,1} + i\Phi_{n-q-1,\,q,\,1}), \tag{2.136}$$

we obtain, by equating the real parts and taking account of (2.135),

$$d\Phi_{n-q,\,q,\,0} = d\Phi'_{n-q-1,\,q,\,1} = 0. \tag{2.137}$$

If, following the *hydrodynamic analogy*, we consider a fictitious potential $F_{n-q-1,q,1}$, then the wing trace is a stream line for this fictitious potential.

From Euler's relations (2.132b), it follows that we may construct w if we know the functions $F_{n-q-1,q,1}$, which may be determined from the above condition (2.137) and from the contribution of the leading edges and ridges.

Hence, as regards the boundary conditions, this *indirect* problem, in which we are looking for the functions $F_{n-q-1,q,1}$, is equivalent to the *direct* problem, in which we are looking for the functions $F_{n-q,q,0}$.

In order to make clear the character of this equivalence and to enumerate the different cases that occur, we shall juxtapose the indirect problems with the equivalent direct problems.

The following table can be given, which may be easily explained through the corresponding figures.

A. *Indirect problems*	B. *Equivalent direct problems*
1. *Thin* wing with subsonic or supersonic leading edges (Fig. 2.16a).	1. Wing of *symmetrical thickness* with subsonic or supersonic leading edges (Fig. 2.16b).

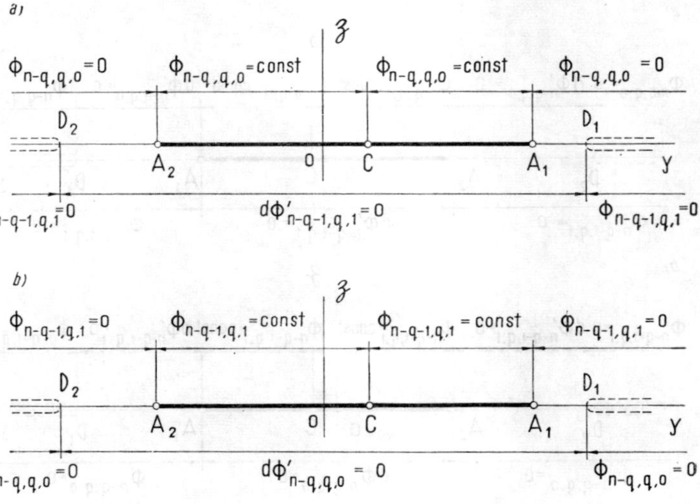

FIG. 2.16

2. Wing of *symmetrical thickness* with one subsonic leading edge and one supersonic leading edge (Fig. 2.17a).

2. *Thin* wing with one subsonic leading edge and one supersonic leading edge (Fig. 2.17b).

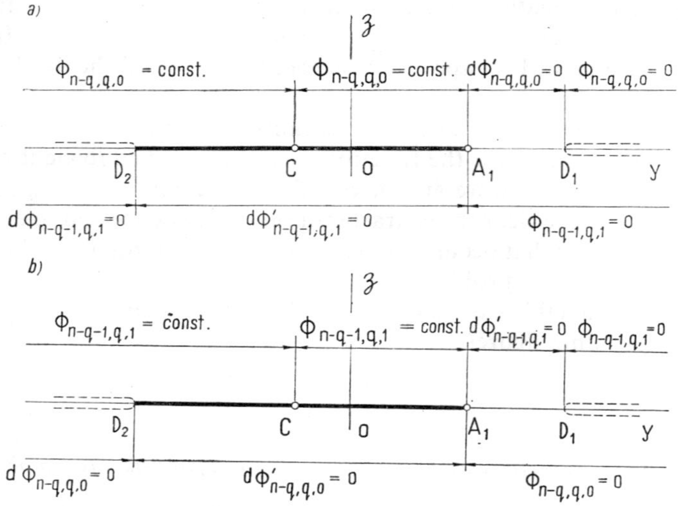

FIG. 2.17

3. Wing of *symmetrical thickness* with subsonic leading edges (Fig. 2.18a).

3. *Thin* wing with subsonic leading edges (Fig. 2.18b).

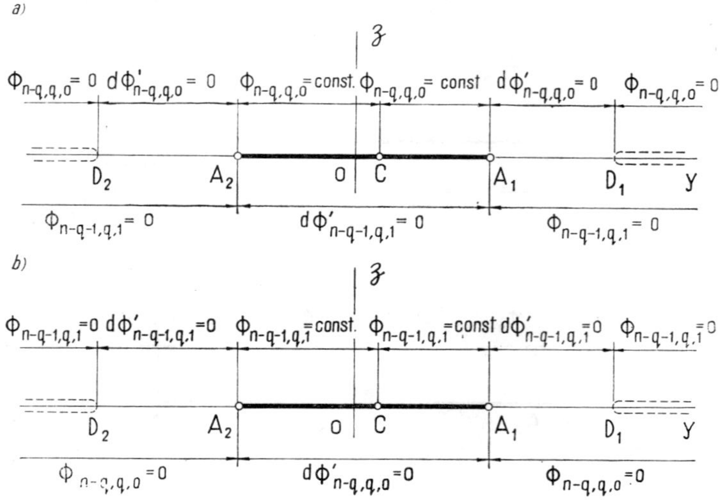

FIG. 2.18

The identity of the boundary conditions which the functions $F_{n-q,q,0}$ in the indirect problem and $F_{n-q-1,q,1}$ in the direct problem respectively satisfy is obvious and results from the corresponding figures.

In view of the correspondences shown above, the evaluation of the function

$$\frac{w}{x_1^{n-1}} = w_{n-1} = \operatorname{Re} \mathcal{W}_{n-1} \tag{2.138}$$

reduces to the evaluation of the equivalent function which refers to u. From this point, the solution should be looked for in the expressions for u in the preceding sections. However, in the case of the indirect problem, there occurs a singularity at the origin, which will be defined in the following section.

II. 7.2. Logarithmic Singularity at the Origin

Let us proceed from the compatibility relation in the form (2.101), for instance, and note that for $q = n-1$, hence for $dF_{0,n-1,1}$, x appears in the denominator of the right-hand side:

$$dF_{0,n-1,1} = -\frac{i\sqrt{1-B^2x^2}}{x} d(\Phi_{1,n-1,0}+i\Phi'_{1,n-1,0}). \tag{2.139}$$

According to the data of the problem, $\Phi_{1,n-1,0}$ is constant (2.137), while $\Phi'_{1,n-1,0}$ is a function whose differential does not vanish at the origin. Hence, it follows that, at the origin, a single function $F_{n-q-1,q,1}$, that is $F_{0,n-1,1}$, contains a logarithmic singularity, as if we had a ridge at $x = s = 0$. Setting hence $s_1 = s_2 = 0$ in an expression similar to (2.63) and (2.69a), we obtain for the logarithmic term

$$\operatorname{Re} F_{0,n-1,1} = \operatorname{Re} Q_0 \cosh^{-1} \sqrt{\frac{l_2(x-l_1)}{(l_1+l_2)x}} = \operatorname{Re} Q_0 \cosh^{-1} \sqrt{\frac{l_1(l_2+x)}{(l_1+l_2)x}}, \tag{2.140}$$

where the subscript (0) refers to the singularity at the origin. Noticing further that the other functions $F_{n-q-1,q,1}$ contain no singularity at the origin, by applying Euler's formula (2.132b), we obtain

$$\frac{(n-1)!\,w}{x_1^{n-1}} = \operatorname{Re} \sum_{q=0}^{n-1} C_{n-1}^q x^q F_{n-q-1,q,1}$$
$$= \operatorname{Re} Q x_1^{n-1} \cosh^{-1} \sqrt{\frac{l_1(x+l_2)}{(l_1+l_2)x}}, \tag{2.141}$$

where Q without the subscript (0) has been introduced, since this is different from Q_0. This term will be added to the expression for w, which is to be found in the same way as for u for the corresponding equivalent wing.

II. 7.3. Expressions for the Downwash w

Considering a *wing of symmetrical thickness* having subsonic leading edges (this wing is playing the part of the thin wing in the direct problem), we obtain the following solutions:

(a) The contribution of a ridge OC_1 situated at $y = s_1$, by analogy with (2.64), will be

$$\frac{w}{x_1^{n-1}} = \operatorname{Re}\left[\sum_{q=0}^{n-1} C_q x^q \cosh^{-1}\sqrt{\frac{(l_2+s_1)(l_1-x)}{(l_1+l_2)(s_1-x)}} + Qx^{n-1}\cosh^{-1}\sqrt{\frac{l_1(x+l_2)}{(l_1+l_2)x}} + \sum_{q=0}^{n-2} D_q x^q \sqrt{(l_1-x)(l_2+x)}\right], \tag{2.142}$$

while in the case when the wing is *thin* (the formula being similar to u corresponding to the equivalent thick wing in the direct problem, according to Fig. 2.16), we have to put $l_1 = l_2 = 1/B$ in the above formulae, and thus obtain

$$\frac{w}{x_1^{n-1}} = \operatorname{Re}\left[\sum_{q=0}^{n-1} C_q x^q \cosh^{-1}\sqrt{\frac{(1+Bs_1)(1-Bx)}{2B(s_1-x)}} + Qx^{n-1}\cosh^{-1}\sqrt{\frac{1+Bx}{2Bx}} + \sum_{q=0}^{n-2} D_q x^q \sqrt{\frac{1-B^2x^2}{B^2}}\right]. \tag{2.143}$$

Likewise, for the contribution of a ridge situated on the thick wing at $y = -s_2$, by analogy with (2.69b), we have

$$\frac{w}{x_1^{n-1}} = \operatorname{Re}\left[\sum_{q=0}^{n-1} G_q x^q \cosh^{-1}\sqrt{\frac{(l_1+s_2)(l_2+x)}{(l_1+l_2)(s_2+x)}} + Qx^{n-1}\cosh^{-1}\sqrt{\frac{l_1(l_2+x)}{(l_1+l_2)x}} + \sum_{q=0}^{n-2} D_q x^q \sqrt{(l_1-x)(l_2+x)}\right]; \tag{2.144}$$

and for the thin wing, by putting $l_1 = l_2 = 1/B$

$$\frac{w}{x_1^{n-1}} = \operatorname{Re}\left[\sum_{q=0}^{n-1} G_q x^q \cosh^{-1}\sqrt{\frac{(1+Bs_2)(1+Bx)}{2B(s_2+x)}} + Qx^{n-1}\cosh^{-1}\sqrt{\frac{1+Bx}{2Bx}} + \sum_{q=0}^{n-2} D_q x^q \sqrt{\frac{1-B^2x^2}{B^2}}\right]. \tag{2.145}$$

(b) The contribution of the leading edge OA_1, by analogy with (2.53), will be

$$\frac{w}{x_1^{n-1}} = \mathrm{Re}\left[\sum_{q=0}^{n-1} A_q x^q \sqrt{\frac{l_2+x}{l_1-x}} + Qx^{n-1}\cosh^{-1}\sqrt{\frac{l_1(l_2+x)}{(l_1+l_2)x}}\right]. \tag{2.146}$$

When the edge OA_2 becomes supersonic, l_2 will be replaced by $1/B$.

(c) The contribution of the leading edge OA_2, by analogy with (2.56) will be

$$\frac{w}{x_1^{n-1}} = \mathrm{Re}\left[\sum_{q=0}^{n-1} A_q x^q \sqrt{\frac{l_1-x}{l_2+x}} + Qx^{n-1}\cosh^{-1}\sqrt{\frac{l_1(l_2+x)}{(l_1+l_2)x}}\right]. \tag{2.147}$$

When the edge OA_1 becomes supersonic, l_1 will be replaced by $1/B$.

(d) The contribution of the two leading edges OA_1 and OA_2, by analogy with (2.57), is

$$\frac{w}{x_1^{n-1}} = \mathrm{Re}\left[\sum_{q=0}^{n} \frac{A_q x^q}{\sqrt{(l_1-x)(l_2+x)}} + Qx^{n-1}\cosh^{-1}\sqrt{\frac{l_1(l_2+x)}{(l_1+l_2)x}}\right]. \tag{2.148}$$

(e) The contribution of a ridge situated at the origin cannot be easily deduced from the first term of the right-hand side of expression (2.142) by putting $s_1 = 0$, owing to some very laborious indeterminacies, which cannot be simplified with indetermined coefficients.

That is the reason why we shall use a direct approach, proceeding from the true conical motion ($n = 1$), for which the compatibility relation (2.25) will be put in the form

$$d\mathcal{W} = -\frac{i\sqrt{1-B^2x^2}}{x}\, d\mathcal{U}. \tag{2.149}$$

It should be noted now that if at the origin the actual value of $\mathcal{U}$ presents a jump, say K (only in this case, however, will the ridge be situated at the origin), then $\mathcal{U}$ has a logarithmic singularity at the origin, that is a vortex in the hydrodynamic sense, $-i(K/\pi)\ln x$; on the other hand, $\mathcal{U}$ may also contain another term iax, compatible with the data of the problem, which produces another singularity at the origin for $\mathcal{W}$, so that we may write

$$\mathcal{U}_{x\to 0} = -i\left(\frac{K}{\pi}\ln x - ax\right) + b, \tag{2.150}$$

where b is a constant. Introducing in (2.149) and integrating in the neighbourhood of the origin ($x \to 0$), we obtain

$$\mathcal{W}_{x\to 0} = \frac{K}{\pi}\frac{1}{x} + a\ln x + \ldots, \tag{2.151}$$

whence it is seen that, besides the logarithmic singularity, which has the same effect as that at point $y = s_1$, the central ridge also produces a doublet.

Proceeding further in a manner analogous to that used for the direct problem (§ II. 2), we can state that the functions $F_{n-q-1,q,1}$ contain at the origin all singularities from $1/x$ to $1/x^n$, hence $1/x^n$ additional to the case of a ridge situated at any point $y = s$.

Taking account of the boundary conditions, in the same way as in (§ II. 2), an additional term in $1/x$ is obtained, namely

$$D_{-1}\frac{\sqrt{(l_1-x)(l_2+x)}}{x}$$

so that, instead of expression (2.144), we have

$$\frac{w}{x_1^{n-1}} = \mathrm{Re}\left(\sum_{q=0}^{n-1} Q_q x^q \cosh^{-1}\sqrt{\frac{l_1(l_2+x)}{(l_1+l_2)x}} + \sum_{q=-1}^{n-2} D_q x^q \sqrt{(l_1-x)(l_2+x)}\right) \tag{2.152}$$

II. 8. Some Considerations on the Indirect Problem

II. 8.1. Examples of Application

In order to illustrate the simplicity of the above method for constructing the function w, we shall give some examples of common applications.

Wing of symmetrical thickness with subsonic leading edges and two separation ridges (Fig. 2.19). We combine (2.142), (2.144) and (2.148)

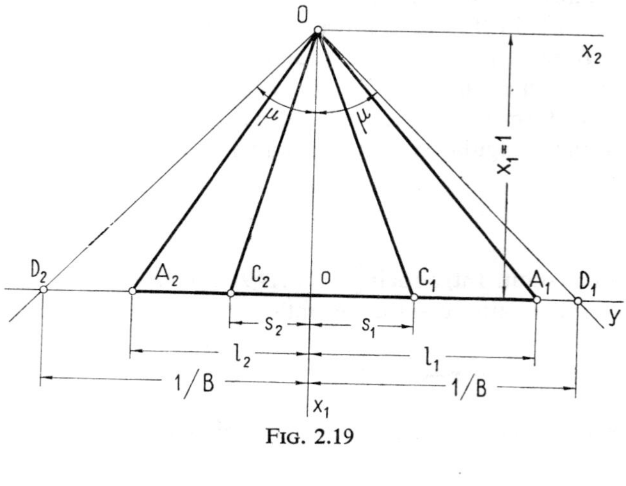

FIG. 2.19

and obtain, by combining the similar expressions,

$$\frac{w}{x_1^{n-1}} = \mathrm{Re}\left[\sum_{q=0}^{n} \frac{A_q x^q}{\sqrt{(l_1-x)(l_2+x)}} + \sum_{q=0}^{n-1} C_q x^q \cosh^{-1}\sqrt{\frac{(l_2+s_1)(l_1-x)}{(l_1+l_2)(s_1-x)}}\right.$$
$$\left. + \sum_{q=0}^{n-1} G_q x^q \cosh^{-1}\sqrt{\frac{(l_1+s_2)(l_2+x)}{(l_1+l_2)(s_2+x)}} + Q x^{n-1} \cosh^{-1}\sqrt{\frac{l_1(x+l_2)}{(l_1+l_2)x}}\right]. \tag{2.153}$$

Wing of symmetrical thickness with the edge OA_1 subsonic and OA_2 supersonic and a ridge OC_2 (Fig. 2.20). We shall combine (2.146), in which l_2 is replaced by $1/B$; then (2.144), in which l_2 is replaced by $1/B$; then again (2.144), in which l_2 is replaced by $1/B$ and s_2 by l_2.

$$\frac{w}{x_1^{n-1}} = \mathrm{Re}\left[\sum_{q=0}^{n-1}\left(A_q x^q \sqrt{\frac{1+Bx}{B(l_1-x)}} + G_q x^q \cosh^{-1}\sqrt{\frac{(l_1+s_2)(1+Bx)}{(1+Bl_1)(s_2+x)}}\right.\right.$$
$$\left.\left. + H_q x^q \cos^{-1}\sqrt{\frac{(l_1+l_2)(1+Bx)}{(1+Bl_1)(l_2+x)}}\right) + Q x^{n-1}\cosh^{-1}\sqrt{\frac{l_1(1+Bx)}{(1+Bl_1)x}}\right], \tag{2.154}$$

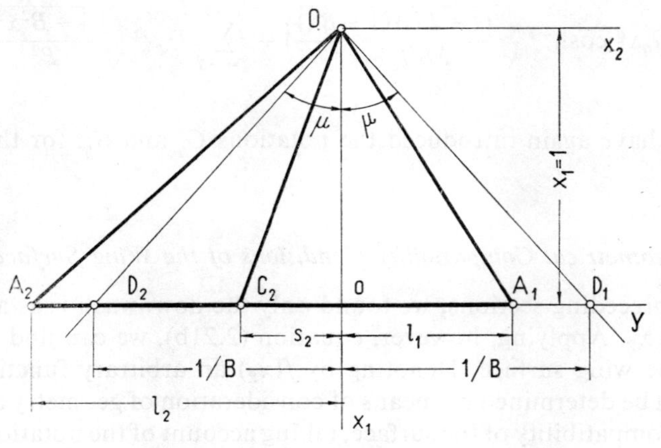

FIG. 2.20

in which we have introduced the constant H_q to indicate that the respective term refers to the supersonic leading edge OA_2.

Thin wing with both subsonic leading edges and a ridge OC_1. We shall apply (2.143) once for the ridge situated at $y = s_1$ and once again for $s_1 = l_1$

and (2.145), setting $s_2 = l_2$; we obtain thus

$$\frac{w}{x_1^{n-1}} = \mathrm{Re}\left[\sum_{q=0}^{n-1}\left(C_q' x^q \cosh^{-1}\sqrt{\frac{(1+Bs_1)(1-Bx)}{2B(s_1-x)}}\right.\right.$$
$$+ C_q x^q \cosh^{-1}\sqrt{\frac{(1+Bl_1)(1-Bx)}{2B(l_1-x)}} + G_q x^q \cosh^{-1}\sqrt{\frac{(1+Bl_2)(1+Bx)}{2B(l_2+x)}}\Bigg)$$
$$\left. + Qx^{n-1}\cosh^{-1}\sqrt{\frac{1+Bx}{2Bx}} + \sum_{q=0}^{n-2} D_q x^q \sqrt{\frac{1-B^2x^2}{B^2}}\right], \qquad (2.155)$$

where we have retained the notations C_q and G_q for the leading edges, since these play the same part as any other ridge. In exchange, we have denoted by C_q' the coefficient referring to the ridge OC_1.

Thin wing with central ridge. This is an application of the preceding case, in which, however, the ridge is situated at the centre.

From this fact, taking account of results previously obtained (2.152), the first term in the right-hand side of expression (2.155) and the last one modify as in (2.152), so that we may write for this case

$$\frac{w}{x_1^{n-1}} = \mathrm{Re}\left[\sum_{q=0}^{n-1}\left(Q_q x^q \cosh^{-1}\sqrt{\frac{1+Bx}{2Bx}} + C_q x^q \cosh^{-1}\sqrt{\frac{(1+Bl_1)(1-Bx)}{2B(l_1-x)}}\right.\right.$$
$$\left.\left. + G_q x^q \cosh^{-1}\sqrt{\frac{(1+Bl_2)(1+Bx)}{2B(l_2+x)}}\right) + \sum_{q=-1}^{n-2} D_q x^q \sqrt{\frac{1-B^2x^2}{B^2}}\right], \qquad (2.156)$$

where we have again introduced the notations C_q and G_q, for the leading edges.

II. 8.2. Geometrical Compatibility Conditions of the Wing Surface

In the preceding sections, we found only the downwash w as a function of x_1 and x_2. Applying, however, equation (2.21b), we can find the equation of the wing surface. Denoting by $f(x_2)$ an arbitrary function of x_2, which will be determined by means of consideration of geometry and of the physical compatibility of the surface, taking account of the notation (2.138), we may in fact write

$$Z(x_1, x_2) - f(x_2) = \frac{1}{U_\infty}\int w\,dx_1 = \frac{1}{U_\infty}\int x_1^{n-1} w_{n-1}\,dx_1$$
$$= \frac{x_2^n}{U_\infty}\int \frac{w_{n-1}}{y^{n+1}}\,dy. \qquad (2.157)$$

The function w_{n-1} is real on the wing, and if an imaginary term due to an imaginary constant or to a logarithmic term appears, then this must be eliminated.

In the neighbourhood of the ridge or of the subsonic leading edges (for wings of symmetrical thickness) the singularities $\ln (y-s)$ and $(y-s)^{-1/2}$ appear, being integrable, and hence the wing equation is finite at these points. At the origin, however, ($y = 0$), where w_{n-1} behaves like $y^{n-1} \ln y$, as follows from the general term

$$Qx^{n-1} \cosh^{-1} \sqrt{\frac{l_1(l_2+x)}{(l_1+l_2)x}},$$

included in the above formulae, we may write for the behaviour of Z,

$$\underset{y\to 0}{Z} = -x_2^n \int \frac{y^{n-1} \ln y}{y^{n+1}}\, dy = -x_2^n \int \frac{\ln y}{y^2}\, dy = x_2^n \frac{1+\ln y}{y} =$$

$$= x_1^n y^{n-1}(1+\ln y), \tag{2.158}$$

whence it follows that Z is finite at the origin if $n > 1$. For the conical motion ($n = 1$), Z becomes infinite at the origin and hence, in such a case, the solution is not compatible from the geometrical and physical standpoints.

In the case when there exists a ridge at the centre, then, in the proximity of the origin we have

$$\underset{x\to 0}{\mathcal{W}_{n-1}} = -\tfrac{1}{2} \sum_{q=0}^{n-1} Q_q x^q \ln x + \frac{D_{-1}}{x}, \tag{2.159}$$

whence we obtain the form of the surface around the origin, by putting $x = y$ and applying the integral (2.157),

$$\underset{y\to 0}{U_\infty Z} = -x_2^n \int \frac{w_{n-1}}{y^{n+1}}\, dy = x_2^n \Bigg[\frac{D_{-1}}{(n+1)y^{n+1}} -$$

$$- \tfrac{1}{2} \sum_{q=0}^{n-1} \frac{Q_q}{n-q} \left(\frac{\ln y}{y^{n-q}} - \int \frac{dy}{y^{n-q+1}} \right) \Bigg]. \tag{2.160}$$

Performing this integration and replacing x_2 by $x_1 y$, it follows finally that

$$\underset{y\to 0}{U_\infty Z} = x_1^n \left[\frac{D_{-1}}{(n+1)y} - \tfrac{1}{2} \sum_{q=0}^{n-1} \frac{Q_q y^q}{n-q} \left(\ln y + \frac{1}{n-q} \right) \right]. \tag{2.161}$$

Therefore, it is seen that Z is finite at the origin if $D_1 = Q_0 = 0$, that is if the coefficients $u_{n-1,0}$ and $u_{n-2,1}$ do not have discontinuities at the origin, as shown later (2.170). It also follows that $\Phi_{n,0,0}$ and $\Phi_{n-1,1,0}$, to which they are proportional, (2.134), must be continuous at the origin;

which means, applying Euler's formulae (1.172a) and (1.172b), that u and v are continuous at the origin, since in both formulae the term in $q = 0$ vanishes.

It may be seen that for $n = 2$ too, the equation Z becomes infinite at the origin if a ridge exists at the centre.

For $n \geqslant 3$, the ridge can be situated at the centre, without Z becoming infinite at the origin, since the condition that $u_{n-1,0}$ and $u_{n-2,1}$ should not have discontinuites at the origin, may be satisfied and yet the ridge may exist, since the other coefficients $u_{n-3,2}$, etc., may be discontinuous at the origin.

II. 8.3. Determination of Constants in the Indirect Problem

The procedure is similar to that used for the direct problem; therefore, only certain ideas necessary for the technique of establishing the constants in this indirect problem will be given.

A brief examination of the formulae in the preceding section indicates the number of constants: $3n+2$ for the wing of symmetrical thickness with two ridges and subsonic leading edges; $3n+1$ for the wing of symmetrical thickness with a single ridge and one subsonic leading edge and the other supersonic; $4n$ for the wing of symmetrical thickness with one ridge and both leading edges supersonic (or for the thin wing).

From the compatibility relations, which we write again in the form

$$(-x)^q \, dF_{n-q,q,0} = (-x)^{n-1} \frac{ix}{\sqrt{1-B^2x^2}} \, dF_{0,n-1,1}, \tag{2.162}$$

we obtain the general formula for the determination of the constants,

$$\Phi_{n-q,q,0} = \operatorname{Im} (-1)^{n-q} \int \frac{x^{n-q}}{\sqrt{1-B^2x^2}} \frac{d^n \mathcal{W}_{n-1}}{dx^n} \, dx + \text{const.} \tag{2.163}$$

where the function $\mathcal{W}_{n-1}$ is defined by (2.138); in a manner similar to that for $\mathcal{U}_{n-1}$, it may be shown that

$$F_{0,n-1,1} = \frac{d^n \mathcal{W}_{n-1}}{dx^n}. \tag{2.164}$$

Denoting by superscripts (′) and (″) the values on the right- and left-hand sides of the ridge respectively, and taking into consideration (2.134), we may write

$$\Phi'_{n-q,q,0} = \frac{(n-1)!u'_{n-q-1,q}}{C^q_{n-1}} = \operatorname{Im} (-1)^{n-q} \int_{x_0}^{y'} \frac{x^{n-q}}{\sqrt{1-B^2x^2}} \frac{d^n \mathcal{W}_{n-1}}{dx^n} \, dx, \tag{2.165a}$$

$$\Phi''_{n-q,\,q,\,0} = \frac{(n-1)!\,u''_{n-q-1,\,q}}{C^q_{n-1}} = \operatorname{Im}\,(-1)^{n-q}\int_{x_0}^{y''} \frac{x^{n-q}}{\sqrt{1-B^2x^2}}\,\frac{d^n\mathcal{W}_{n-1}}{dx^n}\,dx, \tag{2.165b}$$

x_0 being a point where the axial disturbance velocity is zero ($u = 0$), while y' and y'' are points of the wing on the right- and left-hand sides of the ridge OC_1 ($y' > s_1$, $y'' < s_1$).

For the case of a ridge (or for the case of leading edges of the thin wing or supersonic leading edges of the wing of symmetrical thickness, when these edges play the role of a ridge), the integration is performed on a semi-circle σ, of very small radius round the ridge, such that from the above expressions we obtain a simpler relation

$$\frac{(n-1)!}{C^q_{n-1}}\,(u''_{n-q-1,\,q} - u'_{n-q-1,\,1,\,q}) = \operatorname{Im}\,(-1)^{n-q}\int_{\sigma_1} \frac{x^{n-q}}{\sqrt{1-B^2x^2}}\,\frac{d^n\mathcal{W}_{n-1}}{dx^n}\,dx. \tag{2.166}$$

We have therefore to calculate the semi-residue of the expression under the integral sign. In this case however, it is sufficient to expand $\mathcal{W}_{n-1}$ around the points $y = s_1$ and $y = -s_2$ respectively, and to obtain thus the *reduced* expression of $\mathcal{W}_{n-1}$,

$$\underset{x\to s_1}{\mathcal{W}_{n-1}} = -\tfrac{1}{2}\sum_{q=0}^{n-1} C_q x^q \ln\,(s_1 - x), \tag{2.167a}$$

$$\underset{x\to -s_2}{\mathcal{W}_{n-1}} = -\tfrac{1}{2}\sum_{q=0}^{n-1} G_q x^q \ln\,(s_2 + x), \tag{2.167b}$$

which will be introduced above, since the other terms are analytic around s_1 and s_2, and have no contribution.

Determination of constants through the application of formulae (2.165) will be made inside the Mach circle and the results will be then transposed outside the Mach cone since, as shown for the direct problem, it may be also proved here, in the same manner, that the results are also valid beyond the Mach cone.

With this remark, the other continuity conditions at the crossing of the Mach cone will be written, similarly to (2.110),

$$\Delta\left(\frac{d^p\mathcal{W}_{n-1}}{dx^p}\right)_{x=\pm 1/B} = 0. \tag{2.168}$$

These are all the equations for determining the constants: $4n$ equations in all.

Remarks concerning the determination of constants. For the indirect problem, the number of equations obtained does not always correspond to the number of constants to be determined. Thus, for instance:

(a) In the case of the thin wing with subsonic edges and a single ridge (2.155), the number of constants is equal to the number of equations ($4n$ equations in the case of the preceding examples).

(b) In the case of a wing of symmetrical thickness with one subsonic edge and the other supersonic and a single ridge (2.154), one of the equations (2.168) referring to $x = \pm 1/B$, are automatically satisfied, according as the subsonic edge is situated on the right- or left-hand side of the origin. Hence, we have only $3n$ auxiliary relations for $3n+1$ coefficients. In this case, we have a simple indeterminacy which must be eliminated, by imposing on the problem one more condition related to physical and geometrical considerations.

(c) In the case of the wing of symmetrical thickness with both edges subsonic and two ridges (2.153), relations (2.168) are automatically satisfied, so that the $3n$ equations are not sufficient for determining $3n+2$ constants. In this case, we have a double indeterminacy, which will be eliminated by imposing two additional physical or geometrical conditions on the wing.

(d) In the case of the thin wing with a central ridge (2.156), the doublet should also be introduced in the simplified reduced expression of $\mathcal{W}_{n-1}$; that is, we have to write

$$\mathcal{W}_{n-1} = -\tfrac{1}{2}\sum_{q=0}^{n-2} Q_q x^q \ln x + \frac{D_{-1}}{x} \tag{2.169}$$

since the term $Q_{n-1} x^{n-1} \ln x$ has no residue in the integrals included in (2.166).

In exchange, in these integrals, we shall have a residue for $q = 0$, due to the doublet and a residue for $q = 1$, respectively, due to the doublet and the term $Q_0 \ln x$. From this fact, applying equation (2.169), it follows that

$$D_{-1} = \frac{u''_{n-1,0} - u'_{n-1,0}}{n\pi}, \quad Q_0 = -\frac{2(u''_{n-2,1} - u'_{n-2,1})}{(n-1)\pi}. \tag{2.170}$$

(e) Finally, it should be remarked that, as for the direct problem, the auxiliary relations between the coefficients of the axial velocity u and those of the downwash w, which are looked for, are linear and homogeneous.

II. 9. Wing whose Incidence Variation and Pressure Distribution are given Alternatively on Separate Portions (Mixed Problem)

II. 9.1. Data of the Problem

Let us consider a triangular wing OA_1A_2 with subsonic edges, having a ridge OT which divides it into two distinct regions (Fig. 2.21); on one region, TOA_1 for instance, the incidence variation is given, on the other one TOA_2, the pressure distribution is given. Both distributions will be considered as given in the form of homogeneous polynomials of $(n-1)$th order. The pressure distribution and the shape of the whole wing surface are to be determined.

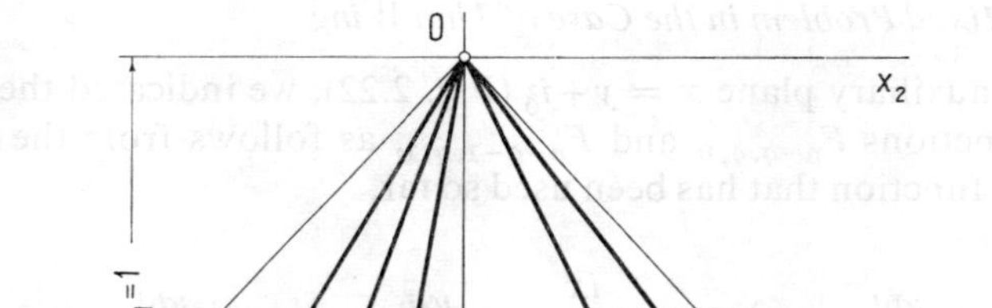

FIG. 2.21

The boundary conditions in the region TOA_1 are similar to those for the direct problem, and in the region TOA_2 to those for the indirect problem. In the mixed problem, the boundary conditions for both regions coexist.

From the standpoint of our method, which is based on the contribution of ridges and leading edges, the new element in the mixed problem is the separation ridge OT of the two regions, which has been called a mixed-ridge.

In the mixed problem, a distinction between the wing leading edges should also be made, since the boundary conditions in their vicinity are different and their contributions to the expressions for the disturbance velocities (the axial disturbance velocity or the downwash) will be different. We shall call the edge OA_1 (Fig. 2.21) a *direct leading edge*, in the vicinity of which the boundary conditions are similar to those for the direct problem, and the edge OA_2, for similar reasons, will be called an *indirect leading edge*.

Likewise, the ridges in the "direct region" OTA_1 will be called *direct ridges*, and those situated in the "indirect region" OTA_2 will be called *indirect ridges*.

In the following, we shall establish separately the contributions of these aerodynamic elements to the expressions of the disturbance velocities u and w, for the thin wing and the wing of symmetrical thickness.

II. 9.2. Mixed Problem in the Case of Thin Wing

In the auxiliary plane $x = y + i\mathfrak{z}$ (Fig. 2.22), we indicated the behaviour of the functions $F_{n-q,\,q,\,0}$ and $F_{n-q-1,\,q,\,1}$, as follows from the same compatibility function that has been used so far,

$$d(\Phi_{n-q,\,q,\,0}+i\Phi'_{n-q,\,q,\,0}) = \frac{ix}{\sqrt{1-B^2x^2}}\, d(\Phi_{n-q-1,\,q,\,1}+i\Phi'_{n-q-1,\,q,\,1}). \quad (2.171)$$

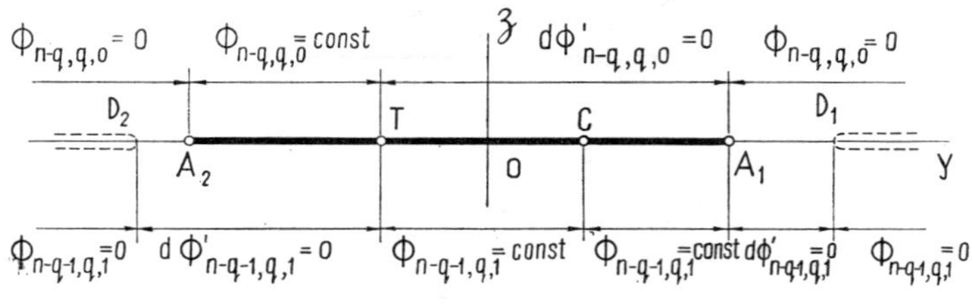

FIG. 2.22

It is readily seen that the portion TA_1 of the wing trace is a stream line for the functions $F_{n-q,\,q,\,0}$ (the same as for the direct problem); on the remaining part of the wing trace, that is on TA_2, the whole abscissa is an equipotential line for the same functions.

Hence, in the case of the mixed problem for a thin wing, $F_{n-q,\,q,\,0}$ and hence the axial velocity u may be determined through the hydrodynamic analogy. We are confronted with a direct problem in which the portion TA_1 plays the role of a wing, that is the direct leading edge A_1 ($y = l_1$) and the mixed ridge T ($y = -t$) behave like two *subsonic leading edges* in the direct problem, for which, taking account of the previous formulae

(§ II.4), we can write separately

$$\frac{u}{x^{n-1}} = \mathrm{Re}\sqrt{\frac{t+x}{l_1-x}}\sum_{q=0}^{n-1} A_q x^q, \tag{2.172}$$

for the *direct subsonic leading edge* OA_1,

$$\frac{u}{x^{n-1}} = \mathrm{Re}\sqrt{\frac{l_1-x}{t+x}}\sum_{q=0}^{n-1} A_q x^q, \tag{2.173}$$

for the *mixed ridge OT* and

$$\frac{u}{x_1^{n-1}} = \mathrm{Re}\sum_{q=0}^{n} \frac{A_q x^q}{\sqrt{(l_1-x)(t+x)}}, \tag{2.174}$$

for the *direct edge* OA_1 and the *mixed ridge OT*.

If the direct leading edge OA_1 becomes supersonic, l_1 will be replaced by $1/B$ in the formula (2.173); if the mixed ridge becomes supersonic, t will be replaced by $1/B$ in formula (2.172).

Contribution of a direct ridge. Further, if we have a ridge OC_1 on the same portion TA_1, situated at the point $y = s_1$ on the right side of the origin ($s_1 > 0$), which we have called a *direct ridge*, we shall simply apply the general formula of the ridge on the thin wing OTA_1 (2.64) in which l_2 will be replaced by t, and thus obtain the *contribution of a ridge* in this complex case,

$$\frac{u}{x_1^{n-1}} = \mathrm{Re}\left[\sum_{q=0}^{n-1} C_q x^q \cosh^{-1}\sqrt{\frac{(t+s_1)(l_1-x)}{(l_1+t)(s_1-x)}} + \sum_{q=0}^{n-2} D_q x^q \sqrt{(l_1-x)(t+x)}\right]. \tag{2.175}$$

If the direct ridge is situated on the left side of the origin ($y = -s_2 < 0$), then l_2 will be replaced by t in expression (2.69) and we obtain

$$\frac{u}{x_1^{n-1}} = \mathrm{Re}\left[\sum_{q=0}^{n-1} G_q x^q \cosh^{-1}\sqrt{\frac{(l_1+s_2)(t+x)}{(l_1+t)(s_2+x)}} + \sum_{q=0}^{n-2} D_q x^q \sqrt{(l_1-x)(t+x)}\right]. \tag{2.176}$$

The same remarks as above will also be made here, namely: if the direct leading edge OA_1 becomes supersonic, l_1 will be replaced by $1/B$ in (2.175) or (2.176) respectively; if the mixed ridge becomes supersonic, t will be replaced by $1/B$; and if both of them become supersonic, we put $l_1 = t = 1/B$. Likewise, if the ridge OC_1, after the replacement of l_1 by $1/B$, becomes in its turn supersonic, then we also put $C_q \cosh^{-1}\beta = K_q \cos^{-1}\beta$.

In this latter case, if we also put $s_1 = l_1$, the ridge becomes a supersonic leading edge, while the above formula represents the contribution of the direct supersonic leading edge.

Likewise, the contribution of the mixed supersonic ridge is also obtained from formula (2.176) in which, after the replacement of $t = 1/B$, we put $s_1 = t$. Thus we obtain:

$$\frac{u}{x_1^{n-1}} = \text{Re}\left[\sum_{q=0}^{n-1} T_q x^q \cos^{-1}\sqrt{\frac{(t+l_1)(1+Bx)}{(1+Bl_1)(t+x)}}\right.$$
$$\left.+\sum_{q=0}^{n-2} D_q x^q \sqrt{\frac{(1+Bx)(l_1-x)}{B}}\right]; \qquad (2.177)$$

if $l_1 > 1/B$, put $1/B$ instead of l_1.

In the following, we shall have to study in particular the contributions of the indirect leading edge and indirect ridge to the expression of u, for which we have no equivalent relations in the direct problem.

Contribution of the indirect leading edge A_2. For this purpose we proceed in the usual way, first considering the true conical motion ($n = 1$), in which case the pressure on the portion A_2T is given by the value of the velocity $u = u_2 =$ const., which passes from $u = 0$ on the left side of A_2, to $u = u_2$ on the right side of A_2. This jump is obtained through a logarithmic singularity, a vortex in the hydrodynamic sense,

$$\underset{x\to l_2}{\mathcal{U}} = i\,\frac{k}{\pi}\ln(l_2+x), \qquad (2.178)$$

where $\mathcal{U}$ has always the same significance given at the beginning. For the nth order conical motion, at the point A_2 we have all the singularities

$$i\ln(x+l_2), \quad \frac{i}{x+l_2}, \ldots, \quad \frac{i}{(x+l_2)^n}$$

Performing a transformation similar to (2.54),

$$\chi^2 = \frac{t+x}{l_1-x}, \qquad (2.179)$$

we shall transpose all these singularities in the new auxiliary plane (Fig. 2.23), singularities that will be situated at points A_2 on the ordinate axis, having the ordinates

$$\nu_2 = \pm\sqrt{\frac{t-l_2}{l_1+l_2}}, \qquad (2.180)$$

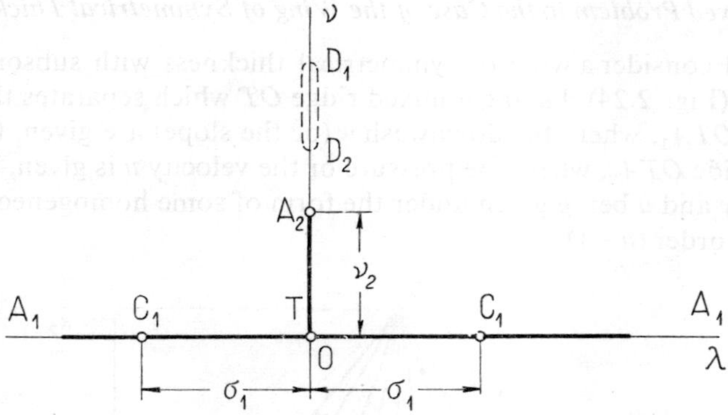

FIG. 2.23

so that the abscissa is a "stream line". Hence, similarly to (2.60), we have

$$F_{n-q,\,q,\,0} = ig \ln \frac{\chi - i\nu_2}{\chi + i\nu_2} + \sum_{p=1}^{n-1} g_p \left[\frac{i^p}{(\chi - i\nu_2)^p} - \frac{(-i)^p}{(\chi + i\nu_2)^p} \right]; \tag{2.181}$$

in the auxiliary plane x, performing the necessary transformations, we obtain

$$F_{n-q,\,q,\,0} = c \cos^{-1} \sqrt{\frac{(l_1+l_2)(t+x)}{(l_1+t)(l_2+x)}} + \frac{\sqrt{(l_1-x)(t+x)}}{(x+l_2)^{n-1}} \sum_{p=0}^{n-2} c_p x^p . \tag{2.181 bis}$$

Applying further Euler's formula, we finally obtain the *contribution of the indirect leading edge A_2*,

$$\frac{u}{x_1^{n-1}} = \operatorname{Re} \left[\sum_{q=0}^{n-1} H_q x^q \cos^{-1} \sqrt{\frac{(l_1+l_2)(t+x)}{(l_1+t)(l_2+x)}} + \sum_{q=0}^{n-2} D_q x^q \sqrt{(l_1-x)(t+x)} \right]. \tag{2.182}$$

If the direct leading edge becomes supersonic, we put $l_1 = 1/B$; if the mixed ridge also becomes supersonic, we also put $t = 1/B$.

Note. Any ridge which has been called an *indirect* ridge, situated on the portion A_2T, behaves exactly like the indirect leading edge which has been studied above. Thus, if this *indirect ridge* is situated at the point $y = -s_2'$, then l_2 from the above formula will be replaced by s_2', thus obtaining its contribution.

II. 9.3. Mixed Problem in the Case of the Wing of Symmetrical Thickness

We shall consider a wing of symmetrical thickness with subsonic leading edges (Fig. 2.24), having a mixed ridge OT which separates the right-hand side OTA_1, where the downwash w (or the slope) are given, from the left-hand side OTA_2, where the pressure or the velocity u is given, both the functions w and u being given under the form of some homogeneous polynomials of order $(n-1)$.

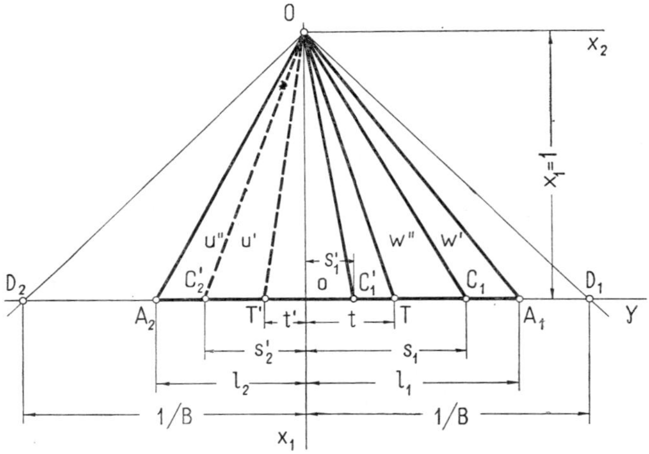

FIG. 2.24

In addition, we may consider a direct ridge OC_1 on the right-hand par (direct region) and eventually an indirect ridge (indirect region) on the left hand side.

In this case too, analysing the boundary conditions in the plane $x = y+i\mathfrak{z}$ on the basis of the compatibility relations written in the form (2.171), we can state that for the motion defined by the fictitious potential $F_{n-q-1,q,1}$, the wing portion A_2T is a stream line, while the remaining part of the axis Oy is an equipotential line. Therefore we may determine through the hydrodynamic analogy the functions $F_{n-q-1,q,1}$, and hence w. We are thus dealing with an indirect problem, in which the portion TA_2 plays the role of a wing. Consequently, the contributions of the fundamental elements (edges and ridges) will contain a term similar to (2.141) at the origin (in which $l_1 = t$ was set beforehand). This term is characteristic of the indirect problem.

With these remarks, the indirect leading edge OA_2 $(y = -l_2)$ and the mixed ridge OT $(y = t)$ behave like two subsonic edges in the indirect problem. Hence, replacing l_1 by t, we have:

Contribution of the indirect subsonic leading edge OA_2 from (2.147),

$$\frac{w}{x_1^{n-1}} = \mathrm{Re}\left[\sqrt{\frac{t-x}{l_2+x}}\sum_{q=0}^{n-1} A_q x^q + Q x^{n-1}\cosh^{-1}\sqrt{\frac{t(x+l_2)}{(t+l_2)x}}\right]; \qquad (2.183)$$

if $t > 1/B$, one replaces t by $1/B$.

Contribution of the mixed ridge OT from (2.146),

$$\frac{w}{x_1^{n-1}} = \mathrm{Re}\left[\sqrt{\frac{l_2+x}{t-x}}\sum_{q=0}^{n-1} A_q x^q + Q x^{n-1}\cosh^{-1}\sqrt{\frac{t(x+l_2)}{(t+l_2)x}}\right]; \qquad (2.184)$$

if $l_2 > 1/B$, we also put $1/B$ instead of l_2.

Contribution of the mixed ridge OT and the indirect leading edge, taken together, from (2.148),

$$\frac{w}{x_1^{n-1}} = \mathrm{Re}\left[\frac{\sum_{q=0}^{n} A_q x^q}{\sqrt{(l_2+x)(t-x)}} + Q x^{n-1}\cosh^{-1}\sqrt{\frac{t(x+l_2)}{(t+l_2)x}}\right]. \qquad (2.185)$$

Contribution of the indirect ridge OC_1'. If the ridge is situated on the right side of the origin ($y = s_1' > 0$), from (2.142) (where s_1 is also replaced by s_1'),

$$\frac{w}{x_1^{n-1}} = \mathrm{Re}\left[\sum_{q=0}^{n-1} C_q' x^q \cosh^{-1}\sqrt{\frac{(l_2+s_1')(t-x)}{(t+l_2)(s_1'-x)}}\right.$$
$$\left. + \sum_{q=0}^{n-2} D_q x^q \sqrt{(t-x)(l_2+x)} + Q x^{n-1}\cosh^{-1}\sqrt{\frac{t(x+l_2)}{(t+l_2)x}}\right]. \qquad (2.186)$$

If the indirect ridge OC_2' is situated on the left side of the origin ($y = -s_2' < 0$), from (2.144) we deduce

$$\frac{w}{x_1^{n-1}} = \mathrm{Re}\left[\sum_{q=0}^{n-1} G_q' x^q \cosh^{-1}\sqrt{\frac{(t+s_2')(l_2+x)}{(t+l_2)(s_2'+x)}}\right.$$
$$\left. + \sum_{q=0}^{n-2} D_q x^q \sqrt{(t-x)(l_2+x)} + Q x^{n-1}\cosh^{-1}\sqrt{\frac{t(l_2+x)}{(t+l_2)x}}\right]. \qquad (2.187)$$

We mention that the contribution of an indirect supersonic leading edge OA_2 to the expression of w is obtained by putting $l_2 = 1/B$ and then $s_2 = l_2$ and $G_q' x^q \cosh^{-1}\beta = H_q x^q \cos^{-1}\beta$ in (2.187).

Further, it is also necessary to study the contributions of the direct leading edge and the direct ridge to the expression of w, for which we have no equivalent relations in the direct problem.

In order that, in the case of conical motion, the jump of the downwash w from a value $w = w'$ to another value $w = w''$ be ensured, it is necessary that w tend towards infinity like

$$\mathcal{W}_{x\to l_1} = i\frac{k}{\pi}\ln(l_1-x). \tag{2.188}$$

For high-order conical motion, we consider that w tends towards infinity, as in the case of the conical motion.

Performing the transformation in the auxiliary plane

$$\chi^2 = \frac{t-x}{l_2+x}, \tag{2.189}$$

we remark that in this plane we have to solve for $F_{n-q-1,\,q,\,1}$ the same boundary problem as for $F_{n-q,\,q,\,0}$ in the case of the thin wing. Proceeding exactly as in this latter case, taking account of Euler's formulae (2.132b) and including also the term referring to the origin—obtained by putting $l_1 = t$ in (2.141)—the contribution of the direct leading edge OA_1 ($y = l_1$) will be of the form

$$\frac{w}{x_1^{n-1}} = \operatorname{Re}\left[\sum_{q=0}^{n-1} K_q x^q \cos^{-1}\sqrt{\frac{(l_1+l_2)(t-x)}{(t+l_2)(l_1-x)}} + \sum_{q=0}^{n-2} D_q x^q \sqrt{(l_2+x)(t-x)} + Q x^{n-1}\cosh^{-1}\sqrt{\frac{t(x+l_2)}{(t+l_2)x}}\right]. \tag{2.190}$$

The contribution of a direct ridge OC_1 is obtained by replacing l_1 by s_1 in the preceding formula.

Further, it is to be noted that in the case when the indirect edge OA_2 or the mixed ridge OT become supersonic, we replace l_2 by $1/B$ or t by $1/B$, respectively, in the above formula (2.190).

II. 9.4. Additional Considerations on the Mixed Problem

We shall give below some additional considerations necessary for making clear some aspects of the mixed problem.

Important note on the contribution of the mixed ridge. In the direct problems of triangular wings, it has been shown that in the case when the leading edge becomes trailing edge, this makes no contribution to the expression for u. This is justified by the fact that the phenomenon is subsonic in the vicinity of the leading edge; hence, analagous to Jukovsky's hypothesis, the disturbance velocity must be zero along the corresponding edge. Since in the case of a thin wing, the mixed ridge has a behaviour similar to that of a direct leading edge, it follows that, if the mixed subsonic ridge is situated on the same side of the origin as a direct leading edge and hence in the "aerodynamic shadow" of this edge, its contribution to the axial disturbance velocity will also be zero.

In a similar way, if in the case of a wing of symmetrical thickness, the mixed subsonic ridge is situated in the aerodynamic shadow of an indirect leading edge, then the ridge will no longer make any contribution to the disturbance downwash w.

Case of the wing with two mixed ridges. In the above considerations, the case of the wing with a mixed ridge has been treated. By means of the hydrodynamic analogy, the expression for u has been given for the thin wing, and the expression for w for the wing of symmetrical thickness.

If the number of ridges is greater, the problem gets complicated, since, generally, it becomes multiconnected in the hydrodynamic sense. There are also here certain cases in which the problem remains simply connected from this point of view. These cases will be specified below. We shall consider only the practical example of the symmetrical wing with two mixed ridges situated at the points $y = \pm t$. For this type of wing, two possible cases are to be considered:

(a) The velocity w is given on the central portion of the wing and the velocity u on the two lateral parts.

In this situation, for the thin wing, the boundary conditions correspond to a simply connected domain from the hydrodynamic point of view, as regards the velocity u; we may hence write

$$\frac{u}{x_1^{n-1}} = \mathrm{Re}\left\{\frac{\sum\limits_{q=0}^{n} A_q x^q}{\sqrt{t^2-x^2}} + \sum_{q=0}^{n-1}\left[K_q x^q \cos^{-1}\sqrt{\frac{(t+l)(t-x)}{2t(l-x)}}\right.\right.$$

$$\left.\left. + H_q x^q \cos^{-1}\sqrt{\frac{(t+l)(t+x)}{2t(l+x)}}\right]\right\}. \tag{2.191}$$

In exchange, for the wing of symmetrical thickness, the boundary conditions correspond to a simply connected domain also for u, only if the wing has both leading edges supersonic. In this case, the expression of u (2.191) indicated above is also valid.

(b) The velocity u is given on the central portion of the wing, and the velocity w on the two lateral parts.

In this situation, for the wing of symmetrical thickness, the boundary conditions also correspond to a simply connected domain from the hydrodynamic point of view, this time as regards the velocity w; we may therefore write

$$\frac{w}{x_1^{n-1}} = \text{Re}\left\{\frac{\sum_{q=0}^{n} A_q x^q}{\sqrt{t^2-x^2}} + \sum_{q=0}^{n-1}\left[K_q x^q \cos^{-1}\sqrt{\frac{(l+t)(t-x)}{2t(l-x)}}\right.\right.$$
$$\left.\left. + H_q x^q \cos^{-1}\sqrt{\frac{(l+t)(t+x)}{2t(l+x)}}\right] + Q x^{n-1}\cosh^{-1}\sqrt{\frac{x+t}{2x}}\right\}. \tag{2.192}$$

In exchange, for the thin wing, the boundary conditions are those of a simply connected domain, also for w, only if the wing has both leading edges supersonic. In this case, the expression of w (2.192) is also valid.

Likewise, it should be remarked that no restriction should be made concerning the number of supersonic mixed ridges, since they bring no modifications in the order of connection of the problems treated.

Note. In order to determine w in the mixed problems in which u on the wing was determined through the hydrodynamic analogy, we start from the compatibility relations put in the particular form

$$\frac{w}{x_1^{n-1}} = \frac{1}{(n-1)!}\,\text{Im}\sum_{q=0}^{n-1} C_{n-1}^q(-1)^{n-q-1}x^q\int x^{n-q-2}\sqrt{1-B^2x^2}\,\frac{d^n\mathcal{U}_{n-1}}{dx^n}\,dx, \tag{2.193}$$

to which an arbitrary constant is added.

Conversely, in order to determine u when w on the wing is known, we use the same compatibility relations, put, however, in the form

$$\frac{u}{x_1^{n-1}} = \frac{1}{(n-1)!}\,\text{Im}\sum_{q=0}^{n-1} C_{n-1}^q(-1)^{n-q}x^q\int\frac{x^{n-q}}{\sqrt{1-B^2x^2}}\,\frac{d^n\mathcal{W}_{n-1}}{dx^n}\,dx, \tag{2.194}$$

with the approximation of a constant.

II. 9.5. Determination of Constants

We shall consider the case of a triangular wing having a direct ridge OC_1, an indirect ridge OC_2', and a mixed ridge OT. The case of the thin wing and that of the wing of symmetrical thickness will be treated separately. For determining the constants, the same method as in the preceding cases will be employed.

For the thin wing we determine u through the hydrodynamic analogy. In this case, the conditions for the direct problem will be used. These conditions must be completed with those which ensure the correct jumps of the velocity u near the indirect ridge and edge. Thus, denoting by σ_1' and λ_1 the very small semicircles round the indirect ridge and edge, we may write

$$\mathrm{Re}(-1)^{n-q-1}\oint_{\sigma_1'} x^{n-q-1}\frac{d^n\mathcal{U}_{n-1}}{dx^n}\,dx = \frac{(n-1)!\,(u''_{n-q-1,q}-u'_{n-q-1,q})}{C^q_{n-1}} \tag{2.195a}$$

$$\mathrm{Re}(-1)^{n-q-1}\oint_{\lambda_1'} x^{n-q-1}\frac{d^n\mathcal{U}_{n-1}}{dx^n}\,dx = \frac{(n-1)!\,u''_{n-q-1,q}}{C^q_{n-1}}. \tag{2.195b}$$

For the wing of symmetrical thickness, where w is determined through the hydrodanymic analogy, we employ the relations for the indirect problem. In addition, in order to ensure the jumps of the velocity w near the direct leading edge and ridge, it is necessary that the following conditions

$$\mathrm{Re}(-1)^{n-q-1}\oint_{\lambda_1} x^{n-q-1}\frac{d^n\mathcal{W}_{n-1}}{dx^n}\,dx = \frac{(n-1)!\,w'_{n-q-1,q}}{C^q_{n-1}}, \tag{2.196a}$$

$$[(q = 0, 1, \ldots (n-1)]$$

$$\mathrm{Re}(-1)^{n-q-1}\oint_{\sigma_1} x^{n-q-1}\frac{d^n\mathcal{W}_{n-1}}{dx^n}\,dx = \frac{(n-1)!\,(w''_{n-q-1,q}-w'_{n-q-1,q})}{C^q_{n-1}} \tag{2.196b}$$

be fulfiled, where λ_1 and σ_1 are semicircles of very small radii around the the respective edges and ridges.

CHAPTER III

THE CALCULATION OF THE OVERALL FORCES AND MOMENTS ON WINGS WITH TRAILING EDGES OF ARBITRARY PLANFORM SHAPE

IN THE preceding chapter, we have established the axial disturbance velocities u for different positions and cases of an angular wing in supersonic flow. Since the pressure coefficient C_p is proportional to the axial velocity u, the expression for the latter will be used for integrating the pressures on the wing in order to calculate the aerodynamic characteristics of the wing, i.e. lift, wave drag, rolling, pitching and yawing moments.

All these depend, of course, on the planform of the wing and particularly on its enclosing line, which constitutes in fact the trailing edge proper of the wing.

On the other hand, in the case of the subsonic leading edges of thin wings, suction forces localized on these edges can appear, producing a forward force which diminishes the drag; lateral forces and yawing moments also appear.

In the following, we shall examine different practical cases, in order to establish convenient formulae for calculating the aerodynamic characteristics.

III. 1. Preliminary Considerations

III. 1.1. Geometric and Aerodynamic Definition of an Actual Wing

The equation of the surface of any wing may be represented by an n order homogeneous polynomial in x_1, x_2,

$$Z_n = \sum_{q=0}^{n} C_{n-q,\,q} x_1^{n-q} x_2^{q}; \tag{3.1}$$

this simple wing of order n may be considered *alone* or as a *component wing* of an actual wing.

An actual wing is generally made up of a superposition of polynomials of the form (3.1),

$$Z = \sum_{n=1}^{N} \sum_{q=0}^{n} C_{n-q,\,q}\, x_1^{n-q}\, x_2^q, \tag{3.2}$$

which is equivalent to a superposition of component wings of orders $1, 2, \ldots, N$ and which will be referred to in the following as a *complex wing*.

By differentiation with respect to x_1, we obtain the downwash w, which defines this wing from the point of view of the boundary conditions; for the component and complex wing, we have

$$w^{(n-1)} = U_\infty \frac{\partial Z_n}{\partial x_1} = \sum_{q=0}^{n-1} w_{n-q-1,\,q}\, x_1^{n-q-1}\, x_2^q = x_1^{n-1} \sum_{q=0}^{n-1} w_{n-q-1,\,q}\, y^2. \tag{3.3a}$$

$$w_{n-1} = \frac{w^{(n-1)}}{x_1^{n-1}} = \sum_{q=0}^{n-1} w_{n-q-1,\,q}\, y^q, \tag{3.3b}$$

$$w = U_\infty \frac{\partial Z}{\partial x_1} = \sum_{n=1}^{N} x_1^{n-1} \sum_{q=0}^{n-1} w_{n-q-1,\,q}\, y^q = \sum_{n=1}^{N} x_1^{n-1}\, w_{n-1}, \tag{3.3c}$$

respectively.

Each of the component wings generates a conical motion of order n, whence the velocity $u^{(n-1)}$ results; the motion round the complex wing consists of the superposition of these motions, from which u is obtained; employing similar notations, we have for the component and complex wing

$$u^{(n-1)} = x_1^{n-1}\, u_{n-1}, \quad u = \sum_{n=1}^{N} x_1^{n-1}\, u_{n-1}, \tag{3.4}$$

respectively.

Replacing y by $x = y+i\mathfrak{z}$ in the expressions for w_{n-1} and u_{n-1}, we introduce the notations $\mathcal{W}_{n-1}$ and $\mathcal{U}_{n-1}$—both having been already employed previously (2.103), (2.138)—defined by the relations

$$\frac{w^{(n-1)}}{x_1^{n-1}} = w_{n-1} = \mathrm{Re}\,\mathcal{W}_{n-1}, \tag{3.5a}$$

$$\frac{u^{(n-1)}}{x_1^{n-1}} = u_{n-1}\mathrm{Re}\,\mathcal{U}_{n-1}, \tag{3.5b}$$

which will be of use later in our future calculations.

Note. In order that the constants of the function $\mathcal{U}_{n-1}$ in the component motions of different orders should not be confused, each of these constants will be designated by two subscripts, the first one indicating the order of the motion (n for instance) and the second one being the subscript of the constant, equal to the exponent of x, by which it is multiplied. Thus, for the nth order conical motion, the respective constants in the expression of $\mathcal{U}_{n-1}$ will be given in the form

$$\begin{array}{lllll} A_{n,0}, & A_{n,1}, & A_{n,2}, \dots, & A_{n,q}, \dots \\ C_{n,0}, & C_{n,1}, & C_{n,2}, \dots, & C_{n,q}, \dots \\ K_{n,0}, & K_{n,1}, & K_{n,2}, \dots, & K_{n,q}, \dots \\ D_{n,0}, & D_{n,1}, & D_{n,2}, \dots, & D_{n,q}, \dots \\ \cdot\ \cdot\ \cdot & \cdot\ \cdot\ \cdot & \cdot\ \cdot\ \cdot & \cdot\ \cdot\ \cdot & \cdot \end{array}$$

Examples of a complex wing. An actual or complex wing, as it has been termed above, is perfectly defined from the geometrical point of view. If the wing is thin, generally no closure condition on the leading or trailing edge is necessary. For wings of symmetrical thickness, however, the closure conditions of the two surfaces—the upper and the lower one, on all edges of the wing—are necessary; otherwise, the wing could be no longer dealt with within the theory of small disturbances or would not correspond to reality. That is the reason why the examples given below refer to the wing of symmetrical thickness, and they may also represent the thin wing; the conditions which have to be introduced for the former wing are not, however, absolutely necessary for the latter wing.

Let us consider a wing of symmetrical thickness and of triangular form also symmetrical, called a *delta wing;* the two simplest forms of its representation will have the following expressions as the equations of their respective surfaces,

$$Z = t(1-x_1)(l^2x_1^2-x_2^2), \tag{3.6a}$$

$$Z' = t'(1-x_1)(lx_1 \mp x_2), \tag{3.6b}$$

where the length of the wing or the middle chord has been taken equal to unity, its span equal to $2l$, and its maximum thickness proportional to t and t', respectively.

For the first wing the respective surface contains no separation ridge; for the second one, however, where the minus sign corresponds to $x_2 > 0$, and the plus sign to $x_2 < 0$, the surface contains a central separation ridge.

The downwashes w on the wing, which define these wings as regards the high-order conical motions that they generate, will be

$$w = \frac{\partial Z}{\partial x_1} U_\infty = tU_\infty(2l^2x_1 - 3l^2x_1^2 + x_2^2), \tag{3.7a}$$

$$w' = \frac{\partial Z'}{\partial x_1} U_\infty = t'U_\infty(l - 2lx_1 \pm x_2). \tag{3.7b}$$

It is readily seen now that the first wing is made up of two components of orders $n = 2$ and $n = 3$, while the second wing is also made up of two components, having, however, the orders $n = 1$ and $n = 2$. Employing the usual notation, we may write for these two wings

$$w_1 = w_{10} + w_{01} y = 2tl^2 U_\infty, \quad (w_{01} = 0, \quad n = 2) \tag{3.8a}$$

$$w_2 = w_{20} + w_{02} y^2 = tU_\infty(-3l^2 + y^2), \quad (n = 3) \tag{3.8b}$$

$$w_0' = t' l U_\infty, \quad (n = 1) \tag{3.8c}$$

$$w_1' = w_{10}' \pm w_{01}' y = t' U_\infty(-2l \pm y), \quad (n = 2) \tag{3.8d}$$

respectively.

From the above examples it is seen that the simplest complex delta wing of symmetrical thickness will have $N \geqslant 3$, if the wing has natural symmetry as regards the downwashes (3.7 a) and $N \geqslant 2$, respectively, if the wing has natural and forced symmetry (3.7 b).

We cannot have a wing of symmetrical thickness with N less than 3 in the first case or less than 2 in the second. For thin wings there is no restriction as regards the construction of the wing. In the following chapters we shall take account of this fact.

For other planforms, as, for instance, the right trapezium, we may construct an enclosed wing of symmetrical thickness only with minimum $N = 4$, if we wish to have natural continuity of downwashes on the wing. In certain simpler cases, we may have a complex wing with $N < 4$, but with a forced continuity of the velocities. This problem will be dealt with in the appropriate chapter on trapezoidal wings.

The planform. A wing may have a polygonal form; starting from the simplest form, we shall first consider a triangular wing, whence we shall then deduce other forms: trapezoidal, double-trapezoidal wings, etc., having any enclosing line, which constitutes the *trailing edge* (Fig. 3.1)

We shall generally denote by S the wing total surface area and by h a reference length (the length or the root chord along the longitudinal axis Ox_1, or the base upon the axis Ox_2, etc.).

If we denote by $C_p^{(n)}$ and C_p the pressure coefficients of the component wing of n order and of the complex wing, applying (1.52) and taking also account of (3.4), we obtain

$$C_p^{(n)} = -2\frac{u^{(n-1)}}{U_\infty} = -2x_1^{n-1}\frac{u_{n-1}}{U_\infty},$$
$$C_p = -2\frac{u}{U_\infty} = -2\sum_{n=1}^{N} x_1^{n-1}\frac{u_{n-1}}{U_\infty}. \tag{3.9}$$

Let us denote by $L^{(n)}$ the lift of the component wing of n order and by L the lift of a complex (actual) wing; the following dimensionless lift coefficients will be defined:

$$C_l^{(n)} = \frac{2L^{(n)}}{\varrho_\infty U_\infty^2 S}, \quad C_l = \frac{2L}{\varrho_\infty U_\infty^2 S} = \sum_{n=1}^{N} C_l^{(n)}, \tag{3.10}$$

Similarly, denoting by $D^{(n)}$ and D the wave drag of a simple wing of order n and of the complex wing, we shall also define the dimensionless coefficients of the wave drag

$$C_{dn}^{(n)} = \frac{2D^{(n)}}{\varrho_\infty U_\infty^2 S}, \quad C_d = \frac{2D}{\varrho_\infty U_\infty^2 S}, \tag{3.11}$$

where we have employed the double indices for reasons that will be seen later.

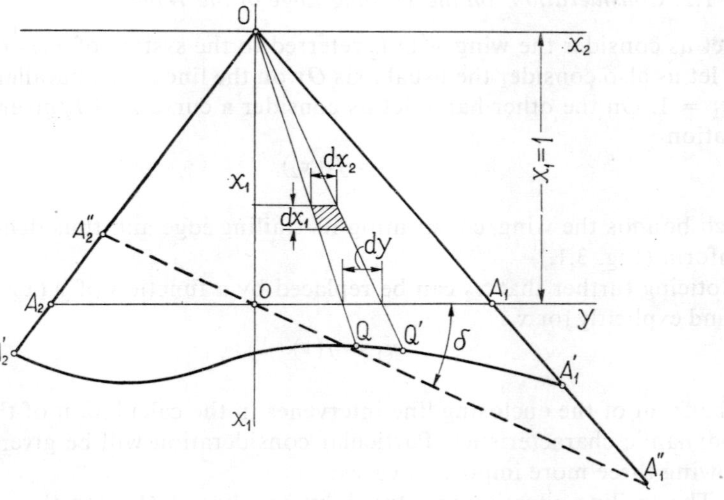

FIG. 3.1

Further, passing to the aerodynamic moments with respect to the axes Ox_1, Ox_2, Ox_3 and using the notations M_1 (for the rolling moment), M_2 (for the pitching moment) and M_3 (for the yawing moment) for the complex wing and correspondingly $M_1^{(n)}$, $M_2^{(n)}$, $M_3^{(n)}$, for the moments referring to the simple wing of order n (or the nth order component of a complex wing), we shall define the following dimensionless coefficients:

$$\mu_1^{(n)} = \frac{2M_1^{(n)}}{\varrho_\infty U_\infty^2 Sh}, \quad \mu_1 = \frac{2M_1}{\varrho_\infty U_\infty^2 Sh} = \sum_{n=1}^{N} \mu_1^{(n)} \tag{3.12}$$

for the rolling moments,

$$\mu_2^{(n)} = \frac{2M_2^{(n)}}{\varrho_\infty U_\infty^2 Sh}, \quad \mu_2 = \frac{2M_2}{\varrho_\infty U_\infty^2 Sh} = \sum_{n=1}^{N} \mu_2^{(n)} \tag{3.13}$$

for the pitching moments, and

$$\mu_{3n}^{(n)} = \frac{2M_3^{(n)}}{\varrho_\infty U_\infty^2 Sh}, \quad \mu_3 = \frac{2M_3}{\varrho_\infty U_\infty^2 Sh} \tag{3.14}$$

for the yawing moments.

In the following we shall establish the expressions for all the above coefficients.

III. 1.2. Considerations on the Trailing Edge of the Wing

Let us consider the wing $A_1'OA_2'$ referred to the system of axes $Ox_1x_2x_3$ and let us also consider the usual axis Oy on the line A_2A_1, parallel to Ox_2 at $x_1 = 1$. On the other hand, let us consider a curve $A_1'QA_2'$ given by the equation

$$x_1 = f_1(x_2), \tag{3.15}$$

which bounds the wing, constituting its trailing edge and thus defining its planform (Fig. 3.1.)

Noticing further that x_2 can be replaced by a function of y ($x_2 = x_1y$), we find explicitly for x_1,

$$x_1 = f(y). \tag{3.16}$$

The form of the enclosing line intervenes in the calculation of the wing aerodynamic characteristics. Particular consideration will be given to the following three more important cases:

1. The trailing edge is constituted by the line A_2OA_1 on the Oy axis. In such a case

$$x_1 = f(y) = 1. \tag{3.17}$$

2. The trailing edge is a straight line $A_2''OA_1''$, which makes the angle $-\delta$ with the Oy axis; from the relation

$$\frac{x_2}{x_1-1} = \frac{x_1 y}{x_1-1} = \cot\delta \tag{3.18}$$

we deduce

$$x_1 = f(y) = \frac{1}{1-y\tan\delta}. \tag{3.19}$$

3. The enclosing curve is arbitrary, but may be expressed by a polynomial of the form

$$x_1 = f(y) = P(y) = a_0+a_1y+a_2y^2+a_3y^3+a_4y^4+\dots, \tag{3.20}$$

which reduces to an even polynomial if the wing is symmetrical with respect to the Ox_1 axis

$$x_1 = f(y) = P(y) = a_0+a_2y^2+a_4y^4+\dots. \tag{3.21}$$

Of course, other shapes of enclosing lines can be also obtained, but expression (3.20) is particularly advantageous for calculations. Thus, for instance, a hyperbola with the vertex at the point $x_1 = 1$, $x_2 = 0$ may be written in the form

$$\left(x_1-\tfrac{1}{2}\right)^2-k^2x_2^2 = \tfrac{1}{4}, \quad x_1 = f(y) = \frac{1}{1-k^2y^2}. \tag{3.22}$$

Other combinations between these forms can also be made, by considering eventually a more complicated rational function, such as to obtain the most convenient enclosing curve from the aerodynamic and geometrical standpoints.

In the following, we shall refer particularly to the leading edges defined by (3.17), (3.19) and (3.21) which represent cases ordinarily met with in practice.

III. 2. Calculation of Forces

III. 2.1. Lift Coefficient

Let us consider an element $d\sigma$ of the component wing of order n, say

$$d\sigma = dx_1\,dx_2 = x_1\,dx_1\,dy; \tag{3.23}$$

on both faces of this element, a force is exerted which is approximately normal to the plane Ox_1x_2 and equal to the *lift* of the element, this lift being proportional to

$$-2C_p^{(n)}d\sigma = 4\frac{u_{n-1}}{U_\infty}\,x_1^n\,dx_1\,dy. \tag{3.24}$$

Integrating over the elementary triangle OQQ' (Fig. 3.1) from $x_1 = 0$ to $x_1 = f(y)$, we obtain an elementary "action"

$$4\frac{u_{n-1}}{U_\infty}\,dy\int_0^{f(y)} x_1^n\,dx_1 = 4\frac{[f(y)]^{n+1}}{n+1}\,\frac{u_{n-1}}{U_\infty}\,dy, \tag{3.25}$$

which we integrate along the line A_2OA_1, obtaining thus the lift coefficient corresponding to the n order component wing,

$$C_l^{(n)} = \frac{4}{(n+1)U_\infty S}\int_{-l_2}^{l_1}[f(y)]^{n+1}u_{n-1}\,dy; \tag{3.26}$$

for the complex wing, the total coefficient is

$$C_l = \sum_{n=1}^{N} C_l^{(n)}. \tag{3.27}$$

If we consider the three enclosing lines of the wing, expression (3.26) may be particularized by replacing $f(y)$ by its expression given in (3.17) when the enclosing line is A_2A_1, or by its expression given in (3.19), when the enclosing line is the straight line $A_2''OA_1''$, or finally by the expression given in (3.21).

Application of the residue method. Further, let us consider the wing trace A_2A_1 in the physical plane (Fig. 3.2 b). The above integral (3.26) may be performed on a closed circuit around the thin wing, once on the upper surface, from A_2 to A_1, and once on the lower surface, from A_1 to A_2, since the pressure coefficient C_p (or u) is equal and of opposite sign on the two faces. In this case, considering a closed contour consisting of the upper and lower surface lines, we may write

$$C_l^{(n)} = -\frac{2}{(n+1)SU_\infty}\oint_{A_1A_2A_1}[f(y)]^{n+1}u_{n-1}\,dy. \tag{3.28}$$

For the sake of generality, let us assume now that the leading edge OA_1 is subsonic and OA_2 is supersonic (Fig. 3.2a), and that the portion OCA_2 is a flat plate without incidence. The line OC is hence a separation ridge which will be made to tend towards OA_2, constituting thus the true leading edge. In the auxiliary plane $x = y+i\mathfrak{z}$, we have the configuration shown in Fig. 3.2c, where on the portion from A_1 to $+\infty$, $u = 0$, while on the portion from D_2 to $-\infty$, $u = 0$ too.

As the above integral may be performed on the wing, y may be replaced by x in all expressions under the integral sign: we have thus $f(x)$ instead of $f(y)$, $\mathcal{U}_{n-1}$ instead of u_{n-1} and, of course, dx instead of dy.

Further, we observe that $\mathcal{U}_{n-1}$ is analytic throughout the entire space outside the wing trace, except the point situated at infinity; then we assume that $f(x)$ has the same value on the upper and on lower surfaces and contains eventually an arbitrary number p of poles. In this case, the above integral may be carried out on any contour K, which does not include the singularities of the function $f(x)$, bounded by the circles γ_i of very small radii (Fig. 3.2c); thus we have

$$C_l^{(n)} = -\frac{2}{(n+1)U_\infty S}\left\{\oint_K [f(x)]^{n+1}\mathcal{U}_{n-1}\,dx - \sum_{i=1}^{p}\oint_{\gamma_i}[f(x)]^{n+1}\mathcal{U}_{n-1}\,dx\right\}, \quad (3.29)$$

where the contour K will lie at infinity.

Accordingly, everything reduces to the calculation of the residue at infinity and around the poles of the function $f(x)$.

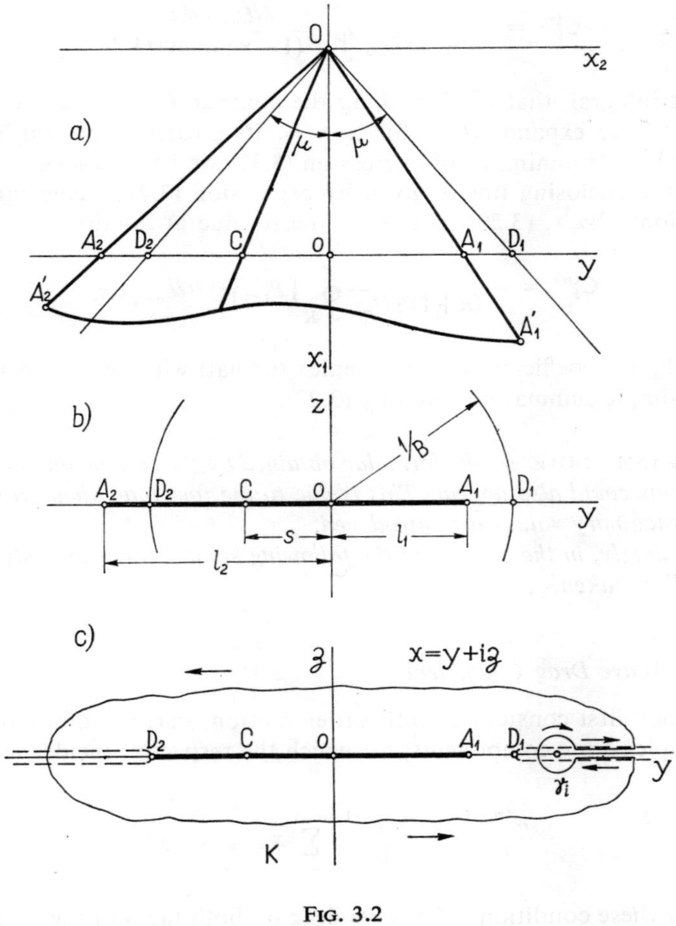

FIG. 3.2

In this sense, the cases indicated above will have the following distinguishing formulae:

1. When the trailing edge is the straight line A_2A_1, taking account of (3.17), the integral (3.29) reduces to a simple residue at infinity

$$C_l^{(n)} = -\frac{2}{(n+1)SU_\infty}\oint_K \mathcal{U}_{n-1}\,dx. \tag{3.30}$$

2. When the trailing edge is the straight line $A_2''OA_1''$ (Fig. 3.1), replacing y by x in (3.19), expression (3.29) reduces only to the residue at the point $y = \cot\delta$, on the circle γ of very small radius,

$$C_l^{(n)} = \frac{2}{(n+1)SU_\infty}\oint_\gamma \frac{\mathcal{U}_{n-1}\,dx}{(1-x\tan\delta)^{n+1}}, \tag{3.31}$$

the first integral, that is, that along the contour K, being equal to zero. Indeed, if we expand $\mathcal{U}_{n-1}$ for $x \to \infty$, its greatest exponent is $(n-1)$, while in the denominator of expression (3.31) we have $(n+1)$.

3. If the enclosing line is given by expression (3.21), replacing in this expression y by x, (3.29) reduces to the residue at infinity,

$$C_l^{(n)} = -\frac{2}{(n+1)SU_\infty}\oint_K [P_{(x)}]^{n+1}\mathcal{U}_{n-1}\,dx. \tag{3.32}$$

For the lift coefficient of the complex (actual) wing, we are confronted with a simple summation given by (3.27).

IMPORTANT NOTE. *In the formulae obtained by the residue method, imaginary terms could also appear. This is due to the fact that when crossing the ridges imaginary values are introduced.*

Accordingly, in the above and the following similar formulae only the real part will be taken.

III. 2.2. Wave Drag Coefficient

We shall first consider an mth order motion, corresponding to one of the complex wing components, for which the respective incidence is given by

$$\frac{w^{(m-1)}}{U_\infty} = \frac{x_1^{m-1}}{U_\infty}\sum_{\varkappa=0}^{m-1} x_{m-\varkappa-1,\,\varkappa}\, y^{\varkappa}. \tag{3.33}$$

Under these conditions, the wave drag on both faces of the element $d\sigma$, corresponding to the nth order component wing, will be proportional to

$$2\frac{w^{(m-1)}}{U_\infty} C_p^{(n)}\, d\sigma = -\frac{4}{U_\infty^2} x_1^{m+n-1}\, dx_1 \left(\sum_{\varkappa=0}^{m-1} w_{m-\varkappa-1,\varkappa}\, y^\varkappa\right) u_{n-1}\, dy, \tag{3.34}$$

whence, integrating over the whole elementary triangle OQQ' (Fig. 3.1) between the limits $x_1 = 0$ and $x_1 = f(y)$, we obtain

$$-\frac{4[f(y)]^{m+n}}{(m+n)U_\infty^2}\left(\sum_{\varkappa=0}^{m-1} w_{m-\varkappa-1,\varkappa}\, y^\varkappa\right) u_{n-1}\, dy. \tag{3.35}$$

Further, denoting by $C_{dm}^{(n)}$ the drag coefficient on the component wing which generates the nth order motion, due to the downwash corresponding to the mth order wing given by (3.33), we have

$$C_{dm}^{(n)} = -\frac{4}{(m+n)SU_\infty^2}\int_{-l_2}^{l_1}\left(\sum_{\varkappa=0}^{m-1} w_{m-\varkappa-1,\varkappa}\, y^\varkappa\right)[f(y)]^{m+n}\, u_{n-1}\, dy. \tag{3.36}$$

For the nth order wing considered alone, putting $m = n$ in the above expression, we obtain

$$C_{dn}^{(n)} = -\frac{2}{nSU_\infty^2}\int_{-l_2}^{l_1}\left(\sum_{\varkappa=0}^{n-1} w_{n-\varkappa-1,\varkappa}\, y^\varkappa\right)[f(y)]^{2n}\, u_{n-1}\, ay. \tag{3.37}$$

In view of practical applications, for the three cases considered above in connection with the enclosing line defined by $f(y)$, expression (3.36) becomes respectively

$$C_{dm}^{(n)} = -\frac{4}{(m+n)SU_\infty^2}\int_{-l_2}^{l_1}\left(\sum_{\varkappa=0}^{m-1} w_{m-\varkappa-1,\varkappa}\, y^\varkappa\right) u_{n-1}\, dy, \tag{3.38a}$$

$$C_{dm}^{(n)} = -\frac{4}{(m+n)SU_\infty^2}\int_{-l_2}^{l_1}\left(\sum_{\varkappa=0}^{m-1} w_{m-\varkappa-1,\varkappa}\, y^\varkappa\right)\frac{u_{n-1}\, dy}{(1-y\tan\delta)^{m+n}}, \tag{3.38b}$$

$$C_{dm}^{(n)} = -\frac{4}{(m+n)SU_\infty^2}\int_{-l_2}^{l_1}\left(\sum_{\varkappa=0}^{m-1} w_{m-\varkappa-1,\varkappa}\, y^\varkappa\right)[P(y)]^{m+n}\, u_{n-1}\, dy. \tag{3.38c}$$

Passing now to the total drag of the complex wing, and observing that m and n vary from 1 to N, we may write

$$C_d = \sum_{m=1}^{N}\sum_{n=1}^{N} C_{dm}^{(n)}. \tag{3.39}$$

These formulae are applicable to all cases, but the integrations are very laborious. For this reason, for very many cases of practical applications, one may apply the residue method given below.

The residue method. Let us now assume that the downwashes vary continuously over the whole wing according to the expression (3.33); we may replace y by x, obtaining thus

$$\frac{w^{(m-1)}}{x_1^{m-1}} = w_{m-1} = \operatorname{Re} \sum_{\varkappa=0}^{m-1} w_{m-\varkappa-1,\varkappa}\, x^{\varkappa} \tag{3.40}$$

This function corresponds identically to (3.33) at the points of the wing ($x = y$), has the same value on the upper and lower surfaces and is uniform throughout the space around the wing trace A_2A_1. In this case, employing the same reasoning as for the lift and proceeding from (3.36), we may write

$$C_{dm}^{(n)} = \frac{2}{(m+n)SU_\infty^2}\left\{\oint_K \left(\sum_{\varkappa=0}^{m-1} w_{m-\varkappa-1,\varkappa}\, x^{\varkappa}\right)[f(x)]^{m+n}\mathcal{U}_{n-1}\, dx - \sum_{i=1}^{p}\oint_{\gamma_i}\left(\sum_{\varkappa=0}^{m-1} w_{m-\varkappa-1,\varkappa}\, x^{\varkappa}\right)[f(x)]^{m+n}\mathcal{U}_{n-1}\, dx\right\}. \tag{3.41}$$

Further, we shall consider the various forms of the above integral for the three cases mentioned, and write:

$$C_{dm}^{(n)} = \frac{2}{(m+n)SU_\infty^2}\oint_K \left(\sum_{\varkappa=0}^{m-1} w_{m-\varkappa-1,\varkappa}\, x^{\varkappa}\right)\mathcal{U}_{n-1}\, dx, \tag{3.42a}$$

$$C_{dm}^{(n)} = -\frac{2}{(m+n)SU_\infty^2}\oint_\gamma \left(\sum_{\varkappa=0}^{m-1} w_{m-\varkappa-1,\varkappa}\, x^{\varkappa}\right)\frac{\mathcal{U}_{n-1}\, dx}{(1-x\tan\delta)^{m+n}}, \tag{3.42b}$$

$$C_{dm}^{(n)} = \frac{2}{(m+n)SU_\infty^2}\oint_K \left(\sum_{\varkappa=0}^{m-1} w_{m-\varkappa-1,\varkappa}\, x^{\varkappa}\right)[P(x)]^{m+n}\mathcal{U}_{n-1}\, dx. \tag{3.42c}$$

For the nth order wing considered alone, we put $m = n$ in the above integrals.

We observe that the first and third integrals represent the residue of the point at infinity, while the second integral represents the residue at the point $y = \cot\delta$ (Fig. 3.2c).

The total drag coefficient will be given in the same form (3.39), which is also the calculation procedure adopted for performing in a convenient order the summations involved by the drag calculation.

Note. The above formulae are valid for thin wings in all cases, as well as for wings of symmetrical thickness with supersonic leading edges. If one or both leading edges of the wing of symmetrical thickness are subsonic, then the residue method is no longer applicable and the integrals (3.38) performed along the wing trace must be employed directly. The same remark holds for the wings fitted with one or several separation ridges. The difficulty arises from the fact that in these cases the downwash can no longer be put in the form of a simple analytic function, continuous over the whole wing.

III. 3. Calculation of Moments

III. 3.1. Rolling Moment Coefficient

If the elementary lift corresponding to the element $d\sigma = x_1\, dx_1\, dy$ is multiplied by $x_2 = x_1 y$, then we obtain an expression proportional to the elementary rolling moment

$$-2C_p^{(n)}\, x_2\, d\sigma = 4\frac{u_{n-1}}{U_\infty} x_1^{n+1}\, dx_1 y\, dy. \tag{3.43}$$

Integrating then over the whole elementary triangle OQQ' (Fig. 3.1), from 0 to $x_1 = f(y)$, we obtain

$$\frac{4}{U_\infty}\frac{x_1^{n+2}}{n+2}\bigg|_0^{f(y)} u_{n-1} y\, dy = \frac{4}{U_\infty}\frac{[f(y)]^{n+2}}{n+2} u_{n-1} y\, dy, \tag{3.44}$$

the nth order wing alone being of course considered here.

Performing the integral over the whole wing, we obtain the corresponding rolling moment coefficient

$$\mu_1^{(n)} = \frac{4}{(n+2)ShU_\infty}\int_{-l_2}^{l_1} [f(y)]^{n+2}\, u_{n-1} y\, dy, \tag{3.45}$$

where $f(y)$ assumes different values for the three enclosing lines considered above, indicated by (3.17), (3.19) and (3.21).

Employing the same reasoning as in the preceding cases and making use of the same general notations, the rolling moment coefficient is also given by the expression

$$\mu_1^{(n)} = -\frac{2}{(n+2)ShU_\infty}\left\{\oint_K [f(x)]^{n+2}\, \mathcal{U}_{n-1}\, x\, dx \right.$$
$$\left. - \sum_{i=1}^{p} \oint_{\gamma_i} [f(x)]^{n+2}\, \mathcal{U}_{n-1}\, x\, dx\right\}. \tag{3.46}$$

For the three classes of enclosing lines we have

$$\mu_1^{(n)} = -\frac{2}{(n+2)ShU_\infty}\oint_K \mathcal{U}_{n-1}\, x\, dx, \tag{3.47a}$$

$$\mu_1^{(n)} = \frac{2}{(n+2)ShU_\infty}\oint_\gamma \frac{\mathcal{U}_{n-1}\, x\, dx}{(1-x\tan\delta)^{n+2}}, \tag{3.47b}$$

$$\mu_1^{(n)} = -\frac{2}{(n+2)ShU_\infty}\oint_K [P(x)]^{n+2}\, \mathcal{U}_{n-1} x\, dx. \tag{3.47c}$$

When the line of the trailing edge is a straight line which makes an angle $-\delta$ with the Oy axis, corresponding hence to the second formula (3.47), taking into consideration the obvious relation

$$\frac{1}{n+1}\frac{\partial}{\partial z}\left[\frac{1}{(1-xz)^{n+1}}\right] = \frac{x}{(1-xz)^{n+2}}, \tag{3.48}$$

and writing tan δ instead of z and comparing (3.31) with (3.47b), we may write

$$h\mu_1^{(n)} = \frac{n+1}{n+2}\frac{\partial C_l^{(n)}}{\partial(\tan\delta)}. \tag{3.49}$$

If we put $n = 1$, we obtain a formula similar to that given by Milne-Thomson† for the true conical motion.

This expression greatly facilitates the problem, since it requires only the expression for $C_l^{(n)}$ for finding the rolling moment coefficient.

For the complex wing we have obviously

$$\mu_1 = \sum_{n=1}^{N} \mu_1^{(n)}. \tag{3.50}$$

III. 3.2. Pitching Moment Coefficient

In order to obtain the moment corresponding to the element $d\sigma$, we multiply the elementary lift (3.24) by x_1 and obtain

$$2C_p^{(n)}x_1\,d\sigma = -4\frac{u_{n-1}}{U_\infty}x_1^{n+1}\,dx_1\,dy, \tag{3.51}$$

from which, integrating over the whole elementary triangle OQQ' and referring to the nth order component wing, we have

$$\frac{4}{U_\infty}u_{n-1}\,dy\int_0^{f(y)} x_1^{n+1}\,dx_1 = \frac{4}{U_\infty}\frac{[f(y)]^{n+2}}{(n+2)}u_{n-1}\,dy. \tag{3.52}$$

Further, integrating over the whole wing, we obtain for the nth order component wing the pitching moment coefficient,

$$\mu_2^{(n)} = -\frac{4}{(n+2)ShU_\infty}\int_{-l_2}^{l_1}[f(y)]^{n+2}u_{n-1}\,dy, \tag{3.53}$$

where $f(y)$ can be replaced by the corresponding expressions for the three enclosing lines.

† L. M. Milne-Thomson, *Theoretical Aerodynamics*, p. 377, 2nd ed., Macmillan & Co. Ltd., London, 1952.

Employing the same reasoning as in the preceding cases, we may apply the residue method, obtaining thus

$$\mu_2^{(n)} = \frac{2}{(n+2)ShU_\infty}\left\{\oint_K [f(x)]^{n+2}\,\mathcal{U}_{n-1}\,dx - \sum_{i=1}^{p}\oint_{\gamma_i}[f(x)]^{n+2}\,\mathcal{U}_{n-1}\,dx\right\}, \tag{3.54}$$

from which we have the three cases of enclosing edge:

$$\mu_2^{(n)} = \frac{2}{(n+2)ShU_\infty}\oint_K \mathcal{U}_{n-1}\,dx = -\frac{1}{h}\frac{n+1}{n+2}C_l^{(n)}, \tag{3.55a}$$

$$\mu_2^{(n)} = -\frac{2}{(n+2)ShU_\infty}\oint_\gamma \frac{\mathcal{U}_{n-1}\,dx}{(1-x\tan\delta)^{n+2}} = -\frac{1}{h}\frac{n+1}{n+2}\frac{\partial}{\partial(\tan\delta)}[C_l^{(n)}\tan\delta], \tag{3.55b}$$

$$\mu_2^{(n)} = \frac{2}{(n+2)ShU_\infty}\oint_K [P(x)]^{n+2}\,\mathcal{U}_{n-1}\,dx. \tag{3.55c}$$

The total coefficient μ_2 for the complex wing is easily obtained as

$$\mu_2 = \sum_{n=1}^{N}\mu_2^{(n)}. \tag{3.56}$$

Note. The right-hand side of (3.55a) is obvious. As regards formula (3.55 b), the final result to be obtained from its right-hand side may be easily deduced through an approach similar to that employed for $\mu_1^{(n)}$. Indeed, taking into account (3.49), we may write

$$\begin{aligned} h\mu_2^{(n)} &= -\frac{2}{(n+2)SU_\infty}\oint_\gamma \frac{\mathcal{U}_{n-1}[(1-x\tan\delta)+x\tan\delta]}{(1-x\tan\delta)^{n+2}}\,dx \\ &= -\frac{n+1}{n+2}C_l^{(n)} - h\mu_1^{(n)}\tan\delta = -\frac{n+1}{n+2}\left[C_l^{(n)} + \frac{\partial C_l^{(n)}}{\partial(\tan\delta)}\cdot\tan\delta\right] \\ &= -\frac{n+1}{n+2}\frac{\partial}{\partial(\tan\delta)}[C_l^{(n)}\tan\delta]. \end{aligned} \tag{3.57}$$

III. 3.3. Yawing Moment Coefficient

It should be observed first that the elementary drag, corresponding to the element $d\sigma$ of the nth order wing, due to the motion of the mth order component wing, must be multiplied by $x_2 = x_1 y$ in order to obtain the elementary yawing moment, proportional to

$$-2\frac{w^{(m-1)}}{U_\infty}C_p^{(n)}\,x_2\,d\sigma=\frac{4}{U_\infty^2}\,x_1^{m+n}\,dx_1\left(\sum_{\varkappa=0}^{m-1}w_{m-\varkappa-1,\,\varkappa}\,y^{\varkappa}\right)u_{n-1}\,y\,dy, \tag{3.58}$$

which, integrated over the whole triangle OQQ', becomes

$$\frac{4x_1^{m+n+1}}{(m+n+1)\,U_\infty^2}\Bigg|_0^{f(y)}\left(\sum_{\varkappa=0}^{m-1}w_{m-\varkappa-1,\,\varkappa}\,y^{\varkappa}\right)u_{n-1}\,y\,dy$$
$$=\frac{4}{(m+n+1)U_\infty^2}\left(\sum_{\varkappa=0}^{m-1}w_{m-\varkappa-1,\,\varkappa}\,y^{\varkappa}\right)[f(y)]^{m+n+1}\,u_{n-1}\,y\,dy. \tag{3.59}$$

By integration, we obtain the corresponding coefficient on the entire nth order component wing due to the downwash of the mth order one

$$\mu_{3m}^{(n)}=\frac{4}{(m+n+1)U_\infty^2Sh}\int_{-l_2}^{l_1}\left(\sum_{\varkappa=0}^{m-1}w_{m-\varkappa-1,\,\varkappa}\,y^{\varkappa}\right)[f(y)]^{m+n+1}\,u_{n-1}\,y\,dy, \tag{3.60}$$

where $f(y)$ will be replaced by its expressions corresponding to the three main enclosing lines.

For the yawing coefficient of the nth order wing alone, we put $m = n$ in the above formula.

For the complex wing, proceeding from (3.60) and summing from 1 to N both for n and for m, we obtain

$$\mu_3=\sum_{m=1}^{N}\sum_{n=1}^{N}\mu_{3m}^{(n)}. \tag{3.61}$$

These formulae are valid for all cases and for any enclosing line.

If, however, the downwash is of the form (3.40) and fulfils the conditions previously indicated, by following the same reasoning as in the preceding cases, we may write directly, without passing through intermediate stages, the residue formula:

$$\mu_{3m}^{(n)}=-\frac{2}{(m+n+1)ShU_\infty^2}\left\{\oint_K\left(\sum_{\varkappa=0}^{m-1}w_{m-\varkappa-1,\,\varkappa}x^{\varkappa}\right)[f(x)]^{m+n+1}\,\mathcal{U}_{n-1}x\,dx\right.$$
$$\left.-\sum_{i=1}^{p}\oint_{\gamma_i}\left(\sum_{\varkappa=0}^{m-1}w_{m-\varkappa-1,\,\varkappa}\,x^{\varkappa}\right)[f(x)]^{m+n+1}\,\mathcal{U}_{n-1}\,x\,dx\right\}, \tag{3.62}$$

where we put $m = n$ for the n order wing alone.

Further, considering the three cases of enclosing line, we may write

$$\mu_{3m}^{(n)} = -\frac{2}{(m+n+1)ShU_\infty^2}\oint_K \left(\sum_{\varkappa=0}^{m-1} w_{m-\varkappa-1,\varkappa}\, x^\varkappa\right) \mathcal{U}_{n-1}x\, dx, \tag{3.63a}$$

$$\mu_{3m}^{(n)} = \frac{2}{(m+n+1)ShU_\infty^2}\oint_\gamma \left(\sum_{\varkappa=0}^{m-1} w_{m-\varkappa-1,\varkappa}\, x^\varkappa\right) \frac{\mathcal{U}_{n-1}x\, dx}{(1-x\tan\delta)^{m+n+1}}, \tag{3.63b}$$

$$\mu_{3m}^{(n)} = -\frac{2}{(m+n+1)ShU_\infty^2}\oint_K \left(\sum_{\varkappa=0}^{m-1} w_{m-\varkappa-1,\varkappa}\, x^\varkappa\right)[P(x)]^{m+n+1}\mathcal{U}_{n-1}x\, dx. \tag{3.63c}$$

For the complex wing, the general formula (3.61) may be applied.

It may be now observed that $(n+1)$ may be replaced by $(m+n)$ in formula (3.48), from which, taking also account of (3.42), we obtain a new expression for (3.63b.),

$$\mu_{3m}^{(n)} = \frac{2}{(m+n+1)ShU_\infty^2}\oint_\gamma \left(\sum_{\varkappa=0}^{m-1} w_{m-\varkappa-1,\varkappa}\, x^\varkappa\right) \frac{\mathcal{U}_{n-1}x\, dx}{(1-x\tan\delta)^{m+n+1}}$$
$$= -\frac{1}{h}\,\frac{m+n}{m+n+1}\,\frac{\partial}{\partial(\tan\delta)}[C_{dm}^{(n)}]. \tag{3.64}$$

Note. As in the wave drag calculation, the same remark will be made here in conjunction with the above formulae which are valid for all cases pertaining to thin wings, as well as for the case of the wing of symmetrical thickness with supersonic leading edges. If one or both leading edges of a wing of symmetrical thickness are subsonic, then the residue method is no longer applicable and the general formula (3.60) should be employed.

III. 4. Some Special Applications of the Formulae for Calculating the Forces and Moments

III. 4.1. Definition of a Wing Bounded by a Line. Situation Outside the Mach Cone

Let us consider a wing $OA_1'DA_2'$ (Fig. 3.3) having the edge OA_1 supersonic and examine only the portion $OCDA_2'$, bounded by the straight line CD, parallel to the Ox_1 axis; we must calculate the forces and moments on this portion of the wing.

Let us denote by G the dimensionless coefficient of forces or moments for the whole wing and by G_1 and G_2 the coefficients corresponding to the

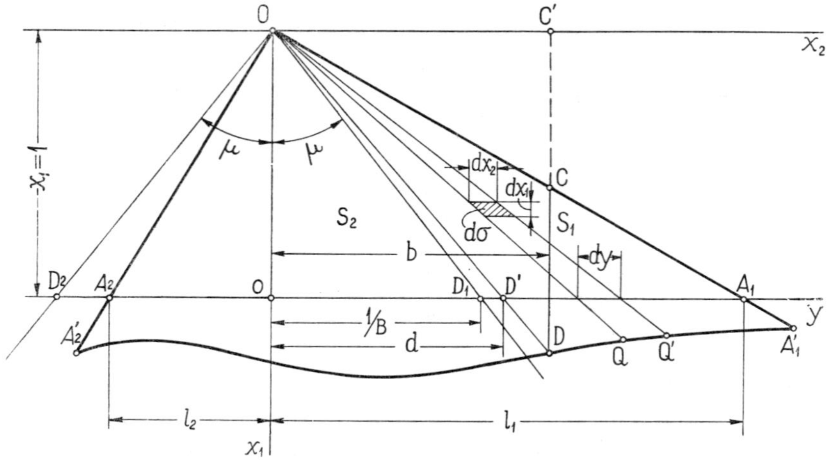

FIG. 3.3

two wing portions, $CA_1'D$ and $OCDA_2'$; if S, S_1, S_2 represent the surfaces of the whole wing and of the two portions, we have

$$S_2G_2 = SG - S_1G_1. \tag{3.65}$$

In this expression G is known from the formulae established in the preceding paragraphs for the whole wing; for calculating G_2, G_1 must be also known. This coefficient may be easily calculated if the portion $CA_1'D$ is outside the Mach cone, that is if we have

$$d > \frac{1}{B}.$$

In fact, on the surface S_1, situated beyond the Mach cone, the expression for u_{n-1} corresponding to the nth order component wing is given at the points on the wing ($x = y$) by a polynomial of the form

$$u_{n-1} = \frac{\pi}{2} \sum_{q=0}^{n-1} K_q y^q, \tag{3.66}$$

which, as mentioned previously (2.100), represents the contribution of the supersonic leading edge OA_1.

Let us now return to the element of surface $d\sigma$, on which forces and moments act, proportional to the following expressions

$$\Phi(y)\,dy \cdot x_1^p\,dx_1, \quad \Psi(y)y\,dy \cdot x_1^r\,dx_1, \tag{3.67}$$

as established above for the aerodynamic forces and moments of wings with different trailing edge lines.

If our considerations are confined to the surface S_1 only, then the limits of integration on the elementary triangle OQQ' are the following

$$x_1 = \frac{b}{y}, \quad x_1 = f(y), \tag{3.68}$$

where $f(y)$ has the same significance as in (3.16).

In this case, expressions (3.67) become

$$\frac{\Phi(y)\, dy}{p+1}\left\{[f(y)]^{p+1} - \left(\frac{b}{y}\right)^{p+1}\right\}, \tag{3.69a}$$

$$\frac{\Psi(y) y\, dy}{r+1}\left\{[f(y)]^{r+1} - \left(\frac{b}{y}\right)^{r+1}\right\}. \tag{3.69b}$$

As may be seen, these expressions differ from those established for calculating the aerodynamic coefficients on the whole wing only by the second term in the curly brackets. They may be integrated further along the Oy axis between the limits $y = d$ and $y = l_1$ (Fig. 3.3).

If the enclosing curve is known, we may write in the general case

$$x_{1D} = \frac{b}{d} = f(d), \tag{3.70}$$

from which d will be obtained in the general case. For the three particular cases of the curve $f(y)$, we have

$$d = b, \quad d = \frac{b}{1+b \tan \delta}, \quad dP(d) = b, \tag{3.71}$$

for the lines A_2A_1, $A_2''OA_1''$ and $A_2'QA_1'$ (Fig. 3.1), respectively.

III.4.2. Calculation of the Dimensionless Coefficients on Portions outside the Mach Cone

As indicated above, the formulae for establishing the coefficients C_1, corresponding to the surface S_1 of the wing, differ from those of the general coefficients over the whole wing only by the second term in the brackets of expressions (3.69) and by the lower limit of integration d.

Therefore, considering the formulae established previously for the coefficients of forces and moments on the whole wing, we may easily write the formulae which refer to the same coefficients corresponding to the surface S_1. In order to differentiate these coefficients from those of the whole wing, the additional subscript (1) will be employed.

Thus, by analogy with (3.26) and (3.27), we may write for the lift coefficients

$$S_1 C_{1l}^{(n)} = \frac{4}{(n+1)U_\infty} \int_d^{l_1} \left\{ [f(y)]^{n+1} - \left(\frac{b}{y}\right)^{n+1} \right\} u_{n-1}\, dy, \quad (3.72\text{a})$$

$$S_1 C_{1l} = S_1 \sum_{n=1}^{N} C_{1l}^{(n)}. \quad (3.72\text{b})$$

Likewise, for the drag coefficient of the nth order component wing, which is due to the downwash of the motion of the mth order component wing, as well as for the total drag coefficient of the complex wing, by analogy with (3.36) and (3.39), we have

$$S_1 C_{1dm}^{(n)} = -\frac{4}{(m+n)U_\infty^2} \int_d^{l_1} \left(\sum_{\varkappa=0}^{m-1} w_{m-\varkappa-1,\varkappa}\, y^\varkappa \right) \left\{ [f(y)]^{m+n} - \left(\frac{b}{y}\right)^{m+n} \right\} u_{n-1}\, dy, \quad (3.73\text{a})$$

$$S_1 C_{1d} = S_1 \sum_{m=1}^{N} \sum_{n=1}^{N} C_{1dm}^{(n)}, \quad (3.73\text{b})$$

respectively. For the nth order wing alone, we put $m = n$ in the formula (3.73a).

In a similar manner, for the rolling moment coefficient, taking into consideration (3.45) and (3.50), we may write

$$S_1 \mu_{11}^{(n)} = \frac{4}{(n+2)hU_\infty} \int_d^{l_1} \left\{ [f(y)]^{n+2} - \left(\frac{b}{y}\right)^{n+2} \right\} u_{n-1}\, y\, dy, \quad (3.74\text{a})$$

$$S_1 \mu_{11} = S_1 \sum_{n=1}^{N} \mu_{11}^{(n)} \quad (3.74\text{b})$$

For the pitching moment coefficient, we have also terms similar to (3.53) and (3.56),

$$S_1 \mu_{12}^{(n)} = -\frac{4}{(n+2)hU_\infty} \int_d^{l_1} \left\{ [f(y)]^{n+2} - \left(\frac{b}{y}\right)^{n+2} \right\} u_{n-1}\, dy, \quad (3.75\text{a})$$

$$S_1 \mu_{12} = S_1 \sum_{n=1}^{N} \mu_{12}^{(n)}. \quad (3.75\text{b})$$

Finally, for the yawing moment coefficient, considering (3.60) and (3.61) we may write

$$S_1\mu_{13m}^{(n)} = \frac{4}{(m+n+1)hU_\infty^2}\int_d^{l_1}\left(\sum_{\varkappa=0}^{m-1} w_{m-\varkappa-1,\varkappa}\, y^\varkappa\right)\left\{[f(y)]^{m+n+1} - \left(\frac{b}{y}\right)^{m+n+1}\right\} u_{n-1}\, y\, dy, \tag{3.76a}$$

$$S_1\mu_{13} = S_1\sum_{m=1}^{N}\sum_{n=1}^{N}\mu_{13m}^{(n)}, \tag{3.76b}$$

respectively.

For the nth order wing alone, we put $m = n$ in (3.76a).

All the above formulae are simple quadratures, which may be easily integrated since the expressions under the integral sign are elementary.

As regards the surface S_1, its expression may be easily deduced from the three leading edge cases considered above. Thus we have

$$S_1 = \frac{(l_1-b)^2}{2l_1}, \tag{3.77a}$$

$$S_1 = \frac{[l_1-b(1-l_1\tan\delta)]^2}{2l_1(1-l_1\tan\delta)}, \tag{3.77b}$$

$$S_1 = \tfrac{1}{2}\int_d^{l_1}\left\{[P(y)]^2-\frac{b^2}{y^2}\right\}dy. \tag{3.77c}$$

Note. The application of the formulae established above is convenient, particularly if we can easily calculate the aerodynamic characteristics of the whole wing $A_2'OA_1'$ by the residue method (a fact that greatly facilitates the problem), and then subtract the characteristics corresponding to the surface S_1, which is in fact very simple (Fig. 3.3).

If, however, the leading edge OA_1 becomes parallel to Ox_2 ($l_1 \to \infty$), and the line DC' borders, in this case, one half of the wing with straight leading edge ($OC'DA_2'$), then, besides the above general method, it might be more convenient to calculate directly the aerodynamic characteristics of this half of the wing.

In this case, we refer to previous formulae, where the limits of integration should be altered as follows (Fig. 3.3):

We shall first integrate from $y = -l_2$ to $y = 1/B$; then from $y = 1/B$ to $y = d$, the velocities u and w being represented in this interval by homogeneous polynomials in x_1 and x_2. For the rest, we shall integrate along the line DC (or along the line DC', if the edge is straight), where the velocities u and w are also represented by homogeneous polynomials in x_1 and x_2.

It is obvious that these two latter integrals are very simple.

III. 4.3. Observations on the Application of the Residue Method

The formulae that give the residue at infinity include terms under the integral sign of the form

$$P(x)\cdot \mathcal{U}_{n-1}\, dx$$

where $P(x)$ is a polynomial in x, and $\mathcal{U}_{n-1}$ also includes the contribution of the separation ridges under the form of $\cosh^{-1} X$, X being an analytical function of x.

Consequently, the formulae indicated above will contain integrals of the form

$$\begin{aligned}\oint_K x^p \cosh^{-1}X\, dx &= \left(\frac{x^{p+1}}{p+1}\cosh^{-1}X\right)_K \\ &\quad - \frac{1}{p+1}\oint_K x^{p+1}\frac{d}{dx}(\cosh^{-1}X)\, dx \\ &= -\frac{1}{p+1}\oint_K x^{p+1}\frac{d}{dx}(\cosh^{-1}X)\, dx, \end{aligned} \tag{3.78}$$

since the first term in the right-hand side is equal to zero. In fact, for the ridges situated on the right- and left-hand sides of the origin, we may write

$$\begin{aligned}\cosh^{-1}X_1 &= \cosh^{-1}\sqrt{\frac{(l_2+s_1)(l_1-x)}{(l_1+s_2)(s_1-x)}}, \\ \cosh^{-1}X_2 &= \cosh^{-1}\sqrt{\frac{(l_1+s_2)(l_2+x)}{(l_1+l_2)(s_2+x)}}, \end{aligned} \tag{3.79}$$

respectively. It may be seen that these functions are uniform at infinity; the above relation (3.78), which is advantageous in that the calculation of the residue can be performed more easily, is therefore justified.

In this connection, in order to simplify the integrals such as those indicated in (3.78), some characteristic developments are given below:

$$\begin{aligned}&\frac{d}{dx}\left(\cosh^{-1}\sqrt{\frac{(l_2+s_1)\,(l_1-x)}{(l_1+l_2)\,(s_1-x)}}\right) \\ &= \frac{-i\sqrt{(l_1-s_1)(l_2+s_1)}}{2x^2}\left(1-\frac{l_1}{x}\right)^{-1/2}\left(1+\frac{l_2}{x}\right)^{-1/2}\left(1-\frac{s_1}{x}\right)^{-1} \\ &= -\frac{i\sqrt{(l_1-s_1)\,(l_2+s_1)}}{2x^2}\left[\sum_{\nu=0}^{\infty}\left(\frac{s_1}{x}\right)^{\nu}+\sum_{\nu=1}^{\infty}\sum_{\mu=1}^{\nu}\frac{(2\mu)!}{2^{2\mu}(\mu!)^2}\left[l_1^{\mu}\right.\right. \\ &\left.\left.+(-l_2)^{\mu}\right]\frac{s_1^{\nu-\mu}}{x^{\nu}}+\sum_{\nu=2}^{\infty}\sum_{\mu=2}^{\nu}\sum_{\lambda=1}^{\mu-1}\frac{(2\lambda)!(2\mu-2\lambda)!(-1)^{\mu-\lambda}}{2^{2\mu}(\lambda!)^2[(\mu-\lambda)!]^2}\frac{l_1^{\lambda}\,l_2^{\mu-\lambda}\,s_1^{\nu-\mu}}{x^{\nu}}\right], \end{aligned} \tag{3.80a}$$

$$\frac{d}{dx}\left(\cosh^{-1}\sqrt{\frac{(l_1+s_2)(l_2+x)}{(l_1+l_2)(s_2+x)}}\right)$$

$$= -\frac{i\sqrt{(l_1+s_2)(l_2-s_2)}}{2x^2}\left(1-\frac{l_1}{x}\right)^{-1/2}\left(1+\frac{l_2}{x}\right)^{-1/2}\left(1+\frac{s_2}{x}\right)^{-1}$$

$$= -\frac{i\sqrt{(l_1+s_2)(l_2-s_2)}}{2x^2}\left[\sum_{\nu=0}^{\infty}\left(\frac{-s_2}{x}\right)^{\nu}+\sum_{\nu=1}^{\infty}\sum_{\mu=1}^{\nu}\frac{(2\mu)!}{2^{2\mu}(\mu!)^2}[l_1^{\mu}\right.$$

$$+(-l_2)^{\mu}]\frac{(-s_2)^{\nu-\mu}}{x^{\nu}}$$

$$\left.+\sum_{\nu=2}^{\infty}\sum_{\mu=2}^{\nu}\sum_{\lambda=1}^{\mu-1}\frac{(-1)^{\mu-\lambda}(2\lambda)!(2\mu-2\lambda)!}{2^{2\mu}(\lambda!)^2[(\mu-\lambda)!]^2}\frac{l_1^{\lambda}l_2^{\mu-\lambda}(-s_2)^{\nu-\mu}}{x^{\nu}}\right]. \qquad (3.80\text{b})$$

If one or both leading edges are supersonic, we replace l_1 or l_2 by $1/B$ and $l_1 = l_2 = 1/B$, respectively. Then, after performing these operations, if s_1 or s_2 become the actual supersonic leading edges, we put $s_1 = l_1$ or $s_2 = l_2$, according to the sign convention for the leading edge abscissae.

If the wing has a symmetrical planform, we put $l_1 = l_2 = l$ and obtain

$$\frac{d}{dx}\left(\cosh^{-1}\sqrt{\frac{(l+s_1)(l-x)}{2l(s_1-x)}}\right) = -\frac{i\sqrt{l^2-s_1^2}}{2x^2}\left(1-\frac{l^2}{x^2}\right)^{-1/2}\left(1-\frac{s_1}{x}\right)^{-1}$$

$$= -\frac{i\sqrt{l^2-s_1^2}}{2x^2}\left[\sum_{\mu=0}^{\infty}\left(\frac{s_1}{x}\right)^{\mu}+\sum_{\mu=1}^{\infty}\sum_{\lambda=1}^{\mu}\frac{(2\lambda)!\,l^{2\lambda}}{2^{2\lambda}(\lambda!)^2}\left(\frac{s_1^{2(\mu-\lambda)}}{x^{2\mu}}+\frac{s_1^{2(\mu-\lambda)+1}}{x^{2\mu+1}}\right)\right], \qquad (3.81\text{a})$$

$$\frac{d}{dx}\left(\cosh^{-1}\sqrt{\frac{(l+s_2)(l+x)}{2l(s_2+x)}}\right) = -\frac{i\sqrt{l^2-s_2^2}}{2x^2}\left(1-\frac{l^2}{x^2}\right)^{-1/2}\left(1+\frac{s_2}{x}\right)^{-1}$$

$$= -\frac{i\sqrt{l^2-s_2^2}}{2x^2}\left[\sum_{\mu=0}^{\infty}\left(-\frac{s_2}{x}\right)^{\mu}+\sum_{\mu=1}^{\infty}\sum_{\lambda=1}^{\mu}\frac{(2\lambda)!\,l^{2\lambda}}{2^{2\lambda}(\lambda!)^2}\left(\frac{s_2^{2(\mu-\lambda)}}{x^{2\mu}}-\frac{s_2^{2(\mu-\lambda)+1}}{x^{2\mu+1}}\right)\right]. \qquad (3.81\text{b})$$

Likewise, if we are confronted with a central ridge, we write

$$\frac{d}{dx}\left(\cosh^{-1}\sqrt{\frac{l^2}{x^2}}\right) = -\frac{il}{x^2}\left(1-\frac{l^2}{x^2}\right)^{-1/2} = -\frac{il}{x^2}\sum_{\lambda=0}^{\infty}\frac{(2\lambda)!}{2^{2\lambda}(\lambda!)^2}\left(\frac{l}{x}\right)^{2\lambda}. \qquad (3.82)$$

If the edges are supersonic, we replace l by $1/B$ in the above formulae. In this case s_1 or s_2 may be simple ridges or leading edges.

In this connection, we note that in the construction of the analytic function $\mathcal{U}_{n-1}$, terms of the form

$$\sqrt{\frac{l_2+x}{l_1-x}}, \quad \sqrt{\frac{l_1-x}{l_2+x}}, \quad \frac{1}{\sqrt{(l_1-x)(l_2+x)}}, \quad \sqrt{(l_1-x)(l_2+x)}$$

multiplied by a polynomial $P(x)$ also appear. In order to simplify the evaluation of integrals appearing in the calculation of aerodynamic characteristics by the residue method, the expansions of the above expressions in powers of $1/x$ will be given below, when x tends infinity:

$$\sqrt{\frac{l_2+x}{l_1-x}} = i\left(1+\frac{l_2}{x}\right)^{1/2}\left(1-\frac{l_1}{x}\right)^{-1/2}$$

$$= i\left[1+\sum_{\mu=1}^{\infty}\frac{(2\mu-2)!}{2^{2\mu-1}\,\mu[(\mu-1)!]^2}\,\frac{(-1)^{\mu-1}l_2^{\mu}+(\mu-1)l_1^{\mu}}{x^{\mu}}\right.$$

$$\left.+\sum_{\mu=2}^{\infty}\sum_{\nu=1}^{\mu-1}\frac{(-1)^{\nu-1}(2\nu-2)!\,(2\mu-2\nu)!\,l_1^{\mu-\nu}\,l_2^{\nu}}{2^{2\mu-1}\,\nu[(\nu-1)!]^2\,[(\mu-\nu)!]^2\,x^{\mu}}\right], \tag{3.83a}$$

$$\sqrt{\frac{l_1-x}{l_2+x}} = -i\left(1-\frac{l_1}{x}\right)^{1/2}\left(1+\frac{l_2}{x}\right)^{-1/2}$$

$$= -i\left[1-\sum_{\mu=1}^{\infty}\frac{(2\mu-2)!}{2^{2\mu-1}\,\mu[(\mu-1)!]^2}\,\frac{l_1^{\mu}+(-1)^{\mu-1}(\mu-1)l_2^{\mu}}{x^{\mu}}\right.$$

$$\left.+\sum_{\mu=2}^{\infty}\sum_{\nu=1}^{\mu-1}\frac{(-1)^{\nu-\mu-1}(2\nu-2)!\,(2\mu-2\nu)!\,l_1^{\nu}\,l_2^{\mu-\nu}}{2^{2\mu-1}\,\nu[(\nu-1)!]^2\,[(\mu-\nu)!]^2\,x^{\mu}}\right], \tag{3.83b}$$

$$\frac{1}{\sqrt{(l_1-x)(l_2+x)}} = \frac{i}{x}\left(1-\frac{l_1}{x}\right)^{-1/2}\left(1+\frac{l_2}{x}\right)^{-1/2}$$

$$= \frac{i}{x}\left[1+\sum_{\mu=1}^{\infty}\frac{(2\mu)!}{2^{2\mu}\,(\mu!)^2}\,\frac{l_1^{\mu}+(-l_2)^{\mu}}{x^{\mu}}\right.$$

$$\left.+\sum_{\mu=2}^{\infty}\sum_{\nu=1}^{\mu-1}\frac{(-1)^{\mu-\nu}(2\nu)!\,(2\mu-2\nu)!\,l_1^{\nu}\,l_2^{\mu-\nu}}{2^{2\mu}\,[(\mu-\nu)!]^2(\nu!)^2\,x^{\mu}}\right], \tag{3.83c}$$

$$\sqrt{(l_1-x)(l_2+x)} = -ix\left(1-\frac{l_1}{x}\right)^{1/2}\left(1+\frac{l_2}{x}\right)^{1/2}$$

$$= -ix\left[1-\sum_{\mu=1}^{\infty}\frac{(2\mu-2)!}{2^{2\mu-1}\,\mu[(\mu-1)!]^2}\,\frac{l_1^{\mu}+(-l_2)^{\mu}}{x^{\mu}}\right.$$

$$\left.-\sum_{\mu=2}^{\infty}\sum_{\nu=1}^{\mu-1}\frac{(-1)^{\mu-\nu-1}(2\nu-2)!\,(2\mu-2\nu-2)!}{2^{2\mu-2}\,\nu(\mu-\nu)!\,[(\nu-1)!]^2[(\mu-\nu-1)!]^2}\,\frac{l_1^{\mu-\nu}\,l_2^{\nu}}{x^{\mu}}\right]. \tag{3.83d}$$

The first two expansions are also valid when one of the leading edges becomes supersonic. The abscissa of the respective edge, l_2 or l_1, will be replaced by $1/B$. In the last expression, if one or both leading edges become supersonic, the respective edge abscissa will be replaced by $1/B$.

III. 5. Calculation of the Suction Force induced by the Subsonic Leading Edges of a Thin Wing

III. 5.1. Theoretical Considerations on the Suction Force

Besides the aerodynamic forces and moments induced by pressures, as they were defined above, *suction forces* also appear at the subsonic leading edges of a thin wing; such forces will be determined by following a previous approach $C-2$].

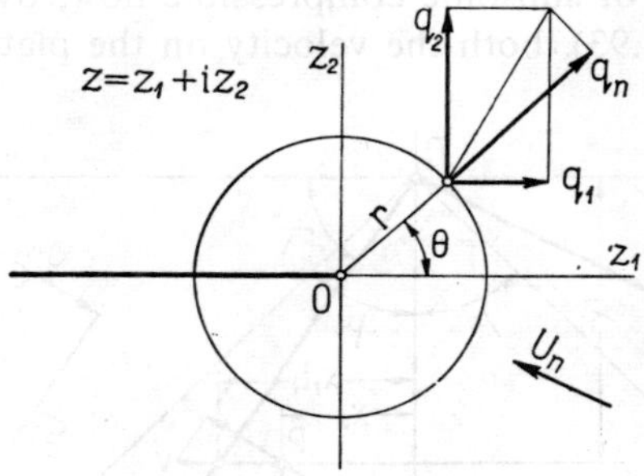

FIG. 3.4

For this purpose, let Oz_1z_2 be a system of axes having the origin on the leading edge, in an arbitrary section (Fig. 3.4). Denoting by z and q the complex variable and the complex velocity, respectively,

$$z = z_1+iz_2 = re^{i\theta}, \tag{3.84a}$$

$$q = q_1-iq_2, \tag{3.84b}$$

this velocity tends to infinity in the vicinity of the leading edge ($z \to 0$) exactly as $z^{-1/2}$, that is

$$\underset{z\to 0}{q} = \frac{iA}{\sqrt{z}}\frac{iA}{\sqrt{r}e^{i(\theta/2)}} = \frac{A}{\sqrt{r}}\left(\sin\frac{\theta}{2}+i\cos\frac{\theta}{2}\right), \tag{3.85}$$

and on the plate ($\theta = \pm\pi$), the absolute value of the velocity is

$$V = \frac{A}{\sqrt{r}}. \tag{3.86}$$

Applying the momentum theorem to a control volume contained in the circular cylinder of a small radius r and length equal to unity, we may write

$$F = -\varrho\int_{-\pi}^{\pi} q_1q_nr\,d\theta = -\varrho\int_{-\pi}^{\pi}(q_1\cos\theta+q_2\sin\theta)q_1r\,d\theta = \pi\varrho A^2, \tag{3.87}$$

where F denotes the *suction force* on the unit length, exerted in the wing plane in the *direction normal* to the leading edge, which, for this purpose, is assumed to be slightly rounded off, and q_n is the radial component.

The velocity component along this normal (Fig. 3.5), which will be denoted by U_{n1} is subsonic and hence the corresponding Mach number is smaller than unity,

$$M_n = \frac{U_{n1}}{a_\infty} = \frac{U_\infty}{a_\infty}\cos\chi_1 = M_\infty \cos\chi_1 < 1. \tag{3.88}$$

As a general result of subsonic compressible flow, by application of formulae (1.92) and (1.93), both the velocity on the plate and the suction F

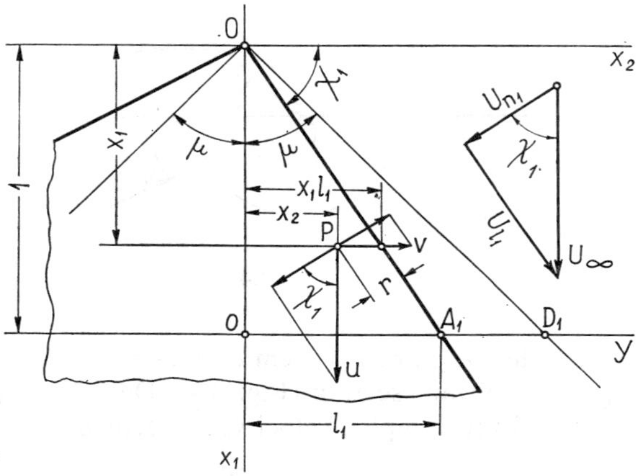

FIG. 3.5

increase in the ratio $(1-M_n)^{-1/2}$, becoming thus

$$V_c = \frac{A}{\sqrt{1-M_n^2}\sqrt{r}} = \frac{A_c}{\sqrt{r}}, \tag{3.89a}$$

$$F_c = \frac{F}{\sqrt{1-M_n^2}} = \pi\varrho\frac{A^2}{\sqrt{1-M_n^2}} = \pi\varrho A_c^2\sqrt{1-M_n^2}, \tag{3.89b}$$

where V_c is to be determined; noting that $\cot\chi_1 = l_1$, the term $\sqrt{1-M_n^2}$ becomes

$$\sqrt{1-M_n^2} = \sqrt{1-M_\infty^2\cos^2\chi_1} = \sin\chi_1\sqrt{1-(M_\infty^2-1)\cot^2\chi_1}$$
$$= \sin\chi_1\sqrt{1-B^2l_1^2}. \tag{3.90}$$

Let us consider a point P in the vicinity of the leading edge A_1 (Fig. 3.5), in any one section x_1; it may be easily seen that the total velocity, in the direction normal to the line OA_1, is

$$V_c = u \cos \chi_1 - v \sin \chi_1 = \frac{ul_1 - v}{\sqrt{1+l_1^2}}. \tag{3.91}$$

Let us now consider the two zero-order homogeneous functions $F_{n-q,q,0}$ and $F_{n-q-1,q+1,0}$; they are related by the compatibility relation (1.180) and consequently we have

$$dF_{n-q-1,\,q+1,\,0} = -\frac{1}{x}\, dF_{n-q,\,q,\,0}, \tag{3.92a}$$

$$F_{n-q-1,\,q+1,\,0} = -\int \frac{1}{x}\, dF_{n-q,\,q,\,0} \approx -\frac{1}{l_1} F_{n-q,\,q,\,0}, \tag{3.92b}$$

since on the Oy axis, $x \approx l_1$ for the points in the vicinity of the leading edge A_1, as was the case of the point P. From Euler's relations (1.168), it may be easily seen that the above formula (3.92) leads to the relation

$$v = -\frac{u}{l_1} \tag{3.93}$$

which, inserted into (3.91),

$$V_c = u \frac{\sqrt{1+l_1^2}}{l_1} = \frac{u}{\cos \chi_1}. \tag{3.94}$$

III. 5.2. Calculation of the Suction Force and Moment

Thrust force. On an element $dx_1/\sin \varkappa_1$ of the leading edge, the suction force (3.89 b), normal to this, becomes

$$\begin{aligned} dF_c &= \pi \varrho A_c^2 \sqrt{1-M_n^2}\, \frac{dx_1}{\sin \chi_1} = \pi \varrho \sqrt{1-B^2 l_1^2}\, (V_c^2 r)_{r\to 0}\, dx_1 \\ &= \pi \varrho \frac{\sqrt{1-B^2 l_1^2}}{\cos^2 \chi_1} (u^2 r)_{\substack{y\to l_1 \\ r\to 0}}\, dx_1. \end{aligned} \tag{3.95}$$

On examining Fig. 3.5, we may write

$$r = (x_1 l_1 - x_2) \sin \chi_1 = x_1 (l_1 - y) \sin \chi_1, \tag{3.96}$$

and accordingly (3.95) becomes

$$\begin{aligned} dF_c &= \pi \varrho \frac{\sqrt{1-B^2 l_1^2}}{\cos^2 \chi_1} [u^2 (l_1 - y)]_{y\to l_1} \sin \chi_1 x_1\, dx_1 \\ &= \pi \varrho \frac{\sqrt{1-B^2 l_1^2}}{l_1 \cos \chi_1} [u^2 (l_1 - y)]_{y\to l_1} x_1\, dx_1. \end{aligned} \tag{3.97}$$

As indicated above, this elementary force is normal to the leading edge OA_1; its projection in the direction Ox_1 is a *thrust force*,

$$dR_1 = -\cos\chi_1\, dF_c = -\pi\varrho\,\frac{\sqrt{1-B^2l_1^2}}{l_1}\,[u^2(l_1-y)]_{y\to l_1} x_1\,dx_1, \tag{3.98}$$

Let us now observe that for a complex wing we may write

$$u = \sum_{n=1}^{N} x_1^{n-1}u_{n-1} = \sum_{m=1}^{N} x_1^{m-1}u_{m-1}, \tag{3.99a}$$

$$u^2 = \sum_{m=1}^{N}\sum_{n=1}^{N} x_1^{m+n-2}u_{n-1}u_{m-1}. \tag{3.99b}$$

Inserting into (3.98) and integrating from $x_1 = 0$ to

$$x_1 = f(l_1) = a_1, \tag{3.100}$$

where a_1 is the projection of the edge OA_1 on the Ox_1 axis, we obtain the suction force for the complex wing

$$R_1 = -\pi\varrho\,\frac{\sqrt{1-B^2l_1^2}}{l_1}\left[\sum_{m=1}^{N}\sum_{n=1}^{N}\frac{u_{m-1}\,u_{n-1}}{m+n}(l_1-y)\right]_{y\to l_1} a_1^{m+n}, \tag{3.101}$$

which, as it may be seen, is a *thrust force*.

Considering $\varrho \approx \varrho_\infty$, the respective axial suction coefficient is given by the relation

$$\sigma_1 = \frac{R_1}{\frac{1}{2}\varrho_\infty U_\infty^2 S} = -\frac{2\pi\sqrt{1-B^2l_1^2}}{U_\infty^2 S l_1}\left[\sum_{m=1}^{N}\sum_{n=1}^{N}\frac{u_{m-1}\,u_{n-1}}{m+n}(l_1-y)\right]_{y\to l_1} a_1^{m+n}$$

$$= \sum_{m=1}^{N}\sum_{n=1}^{N}\sigma_{1m}^{(n)}, \tag{3.102}$$

where, by analogy with the wave drag, an intermediate coefficient $\sigma_{1m}^{(n)}$ was introduced, due to the interference of two component wings of orders n and m,

$$\sigma_{1m}^{(n)} = -\frac{2\pi\sqrt{1-B^2l_1^2}}{U_\infty^2 S l_1}\left[\frac{u_{m-1}\,u_{n-1}}{m+n}(l_1-y)\right]_{y\to l_1} a_1^{m+n}. \tag{3.103}$$

This coefficient simplifies the calculations for the total drag coefficient of the complex wing.

For an nth wing considered alone, we put $m = n$ in formula (3.103) and obtain

$$\sigma_{1n}^{(n)} = -\frac{2\pi\sqrt{1-B^2l_1^2}}{U_\infty^2 S l_1}\left[\frac{u_{n-1}^2}{2n}(l_1-y)\right]_{y\to l_1} a_1^{2n}. \tag{3.104}$$

Similarly, for the suction force due to the leading edge OA'_2, if this edge is also subsonic, we denote by a_2 its projection on the Ox_1 axis and write

$$\sigma_{2m}^{(n)} = -\frac{2\pi\sqrt{1-B^2l_2^2}}{U_\infty^2 S l_2}\left[\frac{u_{m-1}\,u_{n-1}}{m+n}(l_2+y)\right]_{y\to -l_2} a_2^{m+n}, \qquad (3.105)$$

which is deduced from (3.103), replacing l_1 by $-l_2$.

For the nth order wing considered alone, we put $m = n$ in the above expression and obtain

$$\sigma_{2n}^{(n)} = -\frac{2\pi\sqrt{1-B^2l_2^2}}{U_\infty^2 S l_2}\left[\frac{u_{n-1}^2}{2n}(l_2+y)\right]_{y\to -l_2} a_2^{2n}. \qquad (3.106)$$

Similarly with (3.102), the total suction coefficient for the complex wing will be

$$\sigma_2 = \sum_{m=1}^{N}\sum_{n=1}^{N}\sigma_{2m}^{(n)}. \qquad (3.107)$$

Owing to an eventual asymmetry in the general case, a fact which happens for instance in the case of a yawed delta wing, we shall have a lateral force and a yawing moment.

Lateral suction force. Denoting by T the lateral suction force and by τ the lateral suction coefficient, we may write

$$\tau = \frac{2T}{\varrho_\infty U_\infty^2 S}. \qquad (3.108)$$

Further, it may be easily seen that the lateral force is the projection of the suction force in the direction Ox_2, while the axial force (thrust force) is the projection of the same force in the direction Ox_1. The following relations between the coefficients of these forces follow readily:

$$\tau_{1m}^{(n)} = -\frac{1}{l_1}\sigma_{1m}^{(n)}, \quad \tau_{2m}^{(n)} = \frac{1}{l_2}\sigma_{2m}^{(n)}, \quad \tau_m^{(n)} = -\frac{1}{l_1}\sigma_{1m}^{(n)} + \frac{1}{l_2}\sigma_m^{(n)}, \qquad (3.109)$$

and for the nth order wing considered alone we put $m = n$; we have further,

$$\tau_1 = -\frac{1}{l_1}\sigma_1, \quad \tau_2 = \frac{1}{l_2}\sigma_2 \quad \tau = -\frac{1}{l_1}\sigma_1 + \frac{1}{l_2}\sigma_2, \qquad (3.110)$$

for the complex wing.

Yawing moment. Denoting the yawing moment of the suction forces by G and the respective coefficient by γ, we may write

$$\gamma = \frac{2G}{\varrho_\infty U_\infty^2 Sh}. \qquad (3.111)$$

In order to find the elementary yawing moment, the elementary force dF_c given by (3.97) is multiplied by its distance to the wing vertex $x_1/\sin \varkappa_1$;

employing the same notations for the two leading edges, the corresponding moment will be given by the relation

$$dG_1 = \frac{x_1}{\sin \chi_1} dF_c = \frac{x_1 \cos \chi_1}{\sin \chi_1 \cos \chi_1} dF_c. \tag{3.112}$$

From this relation it may be seen that x_1 has an exponent greater by one than the thrust force due to suction, and that in addition the term

$$\sin \chi_1 \cos \chi_1 = \frac{l_1}{1+l_1^2}, \tag{3.113}$$

exists in the denominator of the above expression. In this case, the coefficients of the yawing moments can be easily made functions of those of the thrust forces; taking account of (3.103) and (3.102) and retaining the same indices, we may write

$$\gamma_{1m}^{(n)} = \frac{2\pi\sqrt{1-B^2l_1^2}\,(1+l_1^2)}{U_\infty^2 S l_1^2 h}\left[\frac{u_{m-1}u_{n-1}}{m+n+1}(l_1-y)\right]_{y\to l_1} a_1^{m+n+1}$$

$$= -\frac{m+n}{m+n+1}\,\frac{a_1(1+l_1^2)}{hl_1}\,\sigma_{1m}^{(n)}, \tag{3.114a}$$

$$\gamma_{1n}^{(n)} = -\frac{2n}{2n+1}\,\frac{a_1(1+l_1^2)}{hl_1}\,\sigma_{1n}^{(n)}, \tag{3.114b}$$

$$\gamma_1 = \sum_{m=1}^{N}\sum_{n=1}^{N}\gamma_{1m}^{(n)}. \tag{3.114c}$$

Similarly, we may write for the coefficients of the yawing moments which refer to the leading edge OA_2, the following formulae, deduced from the preceding ones by replacing l_1 by $-l_2$:

$$\gamma_{2m}^{(n)} = -\frac{2\pi\sqrt{1-B^2l_2^2}\,(1+l_2^2)}{U_\infty^2 S l_2^2 h}\left[\frac{u_{m-1}u_{n-1}}{m+n+1}(l_2+y)\right]_{y\to -l_1} a_2^{m+n+1}$$

$$= \frac{m+n}{m+n+1}\,\frac{a_2(1+l_2^2)}{hl_2}\,\sigma_{2m}^{(n)}, \tag{3.115a}$$

$$\gamma_{2n}^{(n)} = \frac{2n}{2n+1}\,\frac{a_2(1+l_2^2)}{hl_2}\,\sigma_{2n}^{(n)}, \tag{3.115b}$$

$$\gamma_2 = \sum_{m=1}^{N}\sum_{n=1}^{N}\gamma_{2m}^{(n)}. \tag{3.115c}$$

The moment coefficients, for the nth order wing considered alone and for the complex wing, will be

$$\gamma_n^{(n)} = \gamma_{1n}^{(n)}+\gamma_{2n}^{(n)}, \quad \gamma = \gamma_1+\gamma_2. \tag{3.116}$$

CHAPTER IV

STUDY OF ORDINARY DELTA WINGS

THIS chapter examines wings of symmetrical triangular form called *delta wings*, ordinarily used in supersonic aviation.

It is for this reason that we shall treat more comprehensively delta wings with a given surface, for which the axial velocity u will be determined in various practical cases, in order to calculate the pressures and aerodynamic characteristics *(direct problem)*. On the other hand, since cases occur when a certain pressure distribution on the wing is given, and it is necessary to calculate the corresponding surface, several examples of the *indirect* and mixed *problem* will also be treated.

The expression for the axial velocity u in the direct problem or that for the downwash w in the indirect problem will be deduced from the general relations obtained in Chapter II by putting $l_1 = l_2 = l$, very simple symmetrical forms being thus obtained by means of these substitutions. As shown in the same chapter, two indices will be used for the constants in the expressions for u and w, the first indicating the order of the motion and the second the exponent of x ($A_{n,q}x^q$).

It should be remarked that a good deal of the material included in this chapter has been approached differently by various authors. We have tried to employ the same unified method, applicable to all problems.

IV. 1. Introductory Considerations on the Delta Wing

Let us consider a *delta wing*, that is a triangular wing of symmetrical planform with respect to the Ox_1 axis, which is also the wing axis (Fig. 4.1).

The downwash w on the wing is given by the following expressions which characterize the *n-th order component wing*:

$$\frac{w^{(n-1)}}{x_1^{n-1}} = w_{n-1} = \sum_{k=0}^{n-1} (\pm 1)^k w_{n-k-1,\,k}\, y^k$$

$$= \sum_{\varkappa=0}^{E\left(\frac{n-1}{2}\right)} w_{n-2\varkappa-1,\,2\varkappa}\, y^{2\varkappa} \pm \sum_{\varkappa=0}^{E\left(\frac{n-2}{2}\right)} w_{n-2\varkappa-2,\,2\varkappa+1}\, y^{2\varkappa+1}, \tag{4.1}$$

for the case of symmetrical distribution of incidence (relative to the thin wing) or slope (relative to the wing of symmetrical thickness) and

$$\frac{w^{(n-1)}}{x_1^{n-1}} = w_{n-1} = \sum_{k=0}^{n-1} (\pm 1)^{k+1} w_{n-k-1,\,k}\, y^k$$

$$= \pm \sum_{\varkappa=0}^{E\left(\frac{n-1}{2}\right)} w_{n-2\varkappa-1,\,2\varkappa}\, y^{2\varkappa} + \sum_{\varkappa=0}^{E\left(\frac{n-2}{2}\right)} w_{n-2\varkappa-2,\,2\varkappa+1}\, y^{2\varkappa+1}, \qquad (4.2)$$

for the case of antisymmetrical distribution of incidence or slope, where $E(t)$ represents the greatest integer which does not exceed t ($t = \frac{n-1}{2}, \frac{n-2}{2}$).

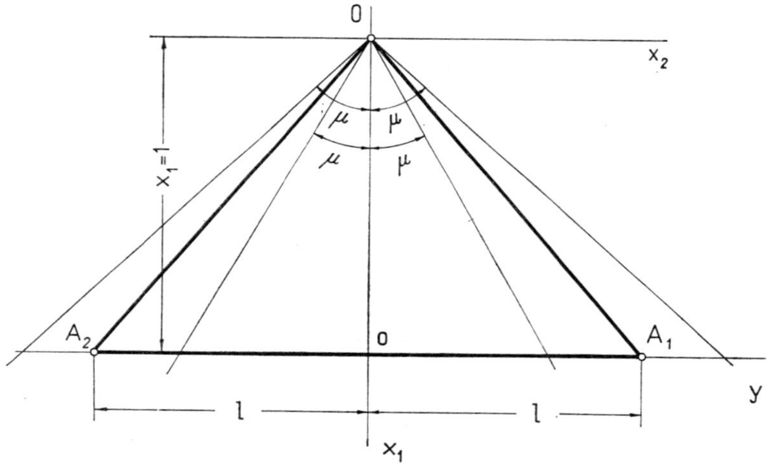

FIG. 4.1

In the last side of (4.1), the even terms which lead to a *natural symmetry* of incidence or slope have been separated from the odd terms which lead to a *forced symmetry*.

Also, in the last side of (4.2), the even terms which lead to a *forced antisymmetry* of incidence have been separated from the odd terms which lead to a *natural antisymmetry*.

On the other hand, it should be mentioned that the plus sign (+) refers to $y > 0$, and the minus sign (−) to $y < 0$.

The total velocity w of the *complex wing* will be represented by a superposition of homogeneous polynomials of the same form as above, from $n = 1$ up to $n = N$, as defined in the preceding chapter (3.3c), N being the highest order:

$$w = \sum_{n=1}^{N} w^{(n-1)} = \sum_{n=1}^{N} x_1^{n-1} w_{n-1}. \tag{4.3}$$

As shown in Chapter II, to the downwash $w^{(n-1)}$ defined by (4.1) or (4.2), there corresponds an nth order conical motion, having the axial velocity $u^{(n-1)}$, which will be put in the form

$$u^{(n-1)} = x_1^{n-1} u_{n-1} \tag{4.4}$$

and which represents the real part of the analytical function $\mathcal{U}_{n-1}$,

$$\mathcal{U}_{n-1} = u_{n-1} + iu'_{n-1}, \tag{4.5}$$

where u'_{n-1} is the associated harmonic function of u_{n-1}.

The total velocity u of the complex wing will be

$$u = \sum_{n=1}^{N} u^{(n-1)} = \sum_{n=1}^{N} x_1^{n-1} u_{n-1} = \mathrm{Re} \sum_{n=1}^{N} x_1^{n-1} \mathcal{U}_{n-1}. \tag{4.6}$$

In conjunction with the aerodynamic characteristics, the notations employed in Chapter III will be recalled.

Thus, denoting by $C_l^{(n)}$ the lift coefficient of the nth order component wing, the total coefficient C_l of the complex wing will be

$$C_l = \sum_{n=1}^{N} C_l^{(n)}. \tag{4.7}$$

It should be noticed that for the wing with antisymmetrical incidence and for the wing of symmetrical thickness whose incidence is equal to zero, the lift coefficient is zero.

Likewise, denoting by $C_{dm}^{(n)}$ the drag coefficient of the nth order component wing, due to the downwash of the mth order component wing, the total drag coefficient C_d will be yielded by the relation

$$C_d = \sum_{m=1}^{N} \sum_{n=1}^{N} C_{dm}^{(n)}. \tag{4.8}$$

In an analogous manner, if $\mu_1^{(n)}$ is the rolling moment coefficient of the nth order component wing, differing from zero only in the case of the thin wing with antisymmetry of incidence, the total moment is

$$\mu_1 = \sum_{n=1}^{N} \mu_1^{(n)}; \tag{4.9}$$

if we denote by $\mu_2^{(n)}$ the pitching moment coefficient of the nth order component wing, differing from zero only in the case of the thin wing with symmetry of incidence, which may be also written as a function of $C_l^{(n)}$ (3.55a),

$$\mu_2^{(n)} = -\frac{n+1}{n+2} C_l^{(n)}, \tag{4.10}$$

then, for the total pitching coefficient, we have

$$\mu_2 = \sum_{n=1}^{N} \mu_2^{(n)} = -\sum_{n=1}^{N} \frac{n+1}{n+2} C_l^{(n)}. \tag{4.11}$$

It should be further noticed that the yawing moment coefficient is equal to zero owing to the symmetrical distribution of the wave drag; hence

$$\mu_{3m}^{(n)} = \mu_3 = 0. \tag{4.12}$$

Finally, taking account of the intermediate suction coefficient $\sigma_m^{(n)}$, which is due to the interference of two component wings of orders n and m, the total suction coefficient will be given by the relation

$$\sigma = \sum_{m=1}^{N} \sum_{n=1}^{N} \sigma_m^{(n)}. \tag{4.13}$$

IV. 2. Thin Delta Wing with Subsonic Leading Edges having Symmetrical Incidences

IV. 2.1. General Formulae for the Axial Velocity

If the downwash w is put in the form (4.1), where the sign (+) refers to the positive part of the abscissa and the sign (−) to the negative part, then the nth order conical motion is also symmetrical; the expression for u also contains the contribution of a ridge at the centre and is obtained from (2.93) by putting $l_1 = l_2 = l$. Using the new double notations, we have for the analytic function $\mathcal{U}_{n-1}$,

$$\mathcal{U}_{n-1} = \sum_{q=0}^{E\left(\frac{n}{2}\right)} \frac{A_{n,\,2q}\, x^{2q}}{\sqrt{l^2 - x^2}} + \sum_{q=1}^{E\left(\frac{n-1}{2}\right)} Q_{n,\,2q}\, x^{2q} \cosh^{-1} \sqrt{\frac{l^2}{x^2}}, \tag{4.14}$$

where the even index $2q$ ensures the symmetry of the motion, while $E\left(\frac{n}{2}\right)$, $E\left(\frac{n-1}{2}\right)$ represent the greatest integers which do not exceed $\frac{n}{2}$ and $\frac{n-1}{2}$, respectively, and where the contribution of the central ridge, deduced from (2.93) by setting $l_1 = l_2 = l$, is written above in a simpler form, taking into account the following relation:

$$\operatorname{Re} \cosh^{-1} \sqrt{\frac{l+x}{2x}} = \operatorname{Re} \cosh^{-1} \sqrt{\frac{l-x}{-2x}} = \tfrac{1}{2} \operatorname{Re} \cosh^{-1} \sqrt{\frac{l^2}{x^2}}. \tag{4.15}$$

On the other hand, $Q_{n,0} = 0$ was considered in the above expression, in order to avoid certain physical incompatibilities, u having to be finite at the centre ($y = 0$), as it is shown in § II. 5.

The central ridge is the effect of the change of sign, which implies two different functions for w, on the right and left-hand sides of the origin.

For the complex wing we obtain w and u by applying (4.3) and (4.6).

For the first four conical motions $n = 1, 2, 3, 4$, the values of the constants $A_{n,2q}$ are functions of the coefficients of w, as follows:

$$A_{10} = -\frac{w_{00}l^2}{E(k)}, \tag{4.16}$$

for $n = 1$;

$$A_{20} = -\frac{2l^2}{\pi}\frac{\pi(1-B^2l^2)w_{10}+[E(k)-B^2l^2K(k)]\,lw_{01}}{(1-2B^2l^2)E(k)+B^2l^2K(k)}. \tag{4.17a}$$

$$A_{22} = \frac{1}{\pi}\frac{\pi(1-B^2l^2)w_{10}+2[E(k)-K(k)]\,B^2l^3w_{01}}{(1-2B^2l^2)E(k)+B^2l^2K(k)}, \tag{4.17b}$$

for $n = 2$;

$$Q_{32} = -\frac{w_{11}}{\pi}, \quad Q_{42} = -\frac{2w_{21}}{\pi}, \tag{4.18a}$$

$$A_{n,2q} = -\frac{l^{2-2q}}{\pi\Omega_n}\sum_{k=0}^{n-1}\Omega_{n,2q}^{(k)}\,l^k\,w_{n-k-1,k}, \tag{4.18b}$$

for $n = 3$ and $n = 4$.

In the above formulae, $K(k)$ and $E(k)$ are complete elliptical integrals of the first and second kind, having as modulus

$$k = \sqrt{1-B^2l^2}, \tag{4.19}$$

which differs from the index k utilized in the expressions for w (4.1 and 4.2).

The coefficients $\Omega_{n,2q}^{(k)}$ and Ω_n appearing in the above formulae are given in Appendix I.1.

On the basis of the above values of the constants, we give for the motion around the component wing of the second order, the variation of $u^{(1)}/w_{10}$, which, as seen from the relation

$$C_p = -2\frac{u}{U_\infty} \tag{4.20}$$

has the same form as the pressure coefficient.

Thus, the diagrams of Figs. 4.2 and 4.3 represent the variation of $u^{(1)}/w_{10}$ in the cross-section $x_1 = 1$ and in the longitudinal section $x_2 = 0{\cdot}2$, respectively.

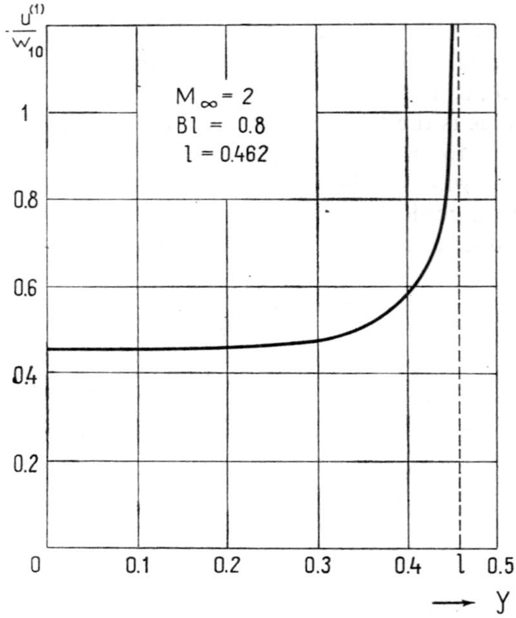
$-\frac{u^{(1)}}{w_{10}}$
$M_\infty = 2$
$Bl = 0.8$
$l = 0.462$
1
0.8
0.6
0.4
0.2
0
0.1
0.2
0.3
0.4
l
0.5
→ y

FIG. 4.2

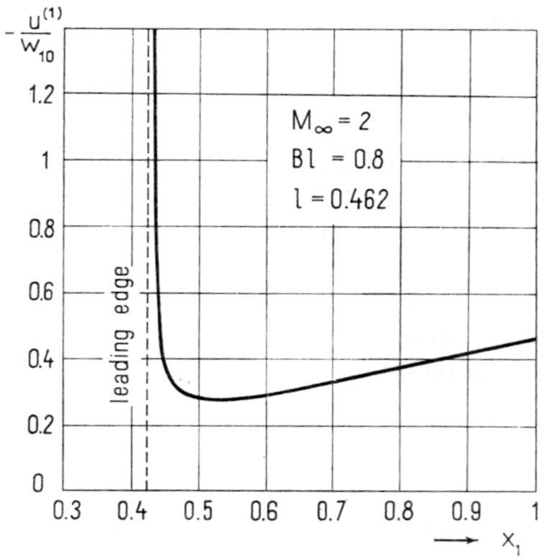
$-\frac{u^{(1)}}{w_{10}}$
1.2
1
0.8
0.6
0.4
0.2
0
$M_\infty = 2$
$Bl = 0.8$
$l = 0.462$
leading edge
0.3
0.4
0.5
0.6
0.7
0.8
0.9
1
→ x_1

FIG. 4.3

In these diagrams we have taken $M_\infty = 2$ and $Bl = 0{\cdot}8$, $l = 0{\cdot}462$, $n = 2$, $w_{01} = 0$.

It should be remarked now that the true conical motion ($n = 1$) represents the flow around a triangular flat plate having the incidence

$$\alpha_{00} = -\frac{w_{00}}{U_\infty}. \tag{4.21}$$

For the complex wing, this *component* could be taken as a *reference wing*, which will yield the position with respect to the free stream U_∞ by means of the incidence α_{00}.

IV. 2.2. Lift Coefficient

For the case of a single component wing of order n, taking into account (3.30), where we put $S = l$, we may write

$$C_l^{(n)} = \frac{4\pi}{(n+1)lU_\infty}\left[\sum_{q=0}^{E\left(\frac{n}{2}\right)} \frac{(2q)!\, l^{2q}}{2^{2q}(q!)^2} A_{n,\,2q} + \sum_{q=1}^{E\left(\frac{n-1}{2}\right)} \frac{(2q)!\, l^{2q+1}}{(2q+1)2^{2q}(q!)^2} Q_{n,\,2q}\right]. \tag{4.22}$$

For the total coefficient we apply (4.7).

For the first four wings considered alone ($n = 1, 2, 3, 4$), we have

$$C_l^{(1)} = \frac{2\pi}{lU_\infty} A_{10}, \tag{4.23a}$$

$$C_l^{(2)} = \frac{4\pi}{3lU_\infty}\left(A_{20} + \frac{l^2}{2} A_{22}\right), \tag{4.23b}$$

$$C_l^{(3)} = \frac{\pi}{lU_\infty}\left(A_{30} + \frac{l^2}{2} A_{32} + \frac{l^3}{6} Q_{32}\right), \tag{4.23c}$$

$$C_l^{(4)} = \frac{4\pi}{5lU_\infty}\left(A_{40} + \frac{l^2}{2} A_{42} + \frac{3l^4}{8} A_{44} + \frac{l^3}{6} Q_{42}\right), \tag{4.23d}$$

respectively.

If these four wings are the constituent parts of a complex wing, the lift coefficient C_l of this wing will be the sum of the four coefficients given above.

Note. THE PITCHING MOMENT COEFFICIENT is related to lift by the same formulae as in (4.10) and (4.11), utilizing the lift coefficients calculated above.

IV. 2.3. Wave Drag Coefficient

As regards the even terms of the expression (4.1) for w, we shall apply the residue method for calculating the wave drag. The odd terms of w cannot be put in a convenient simple analytical form; for this reason, formula (3.42 a) will be applied for even terms and formula (3.38) for odd terms (as well as for even ones too).

For symmetry reasons, we integrate only on one-half of the wing, from 0 to l.

First we put

$$\int_0^l y^t \cosh^{-1}\sqrt{\frac{l^2}{y^2}}\,dy = \frac{l}{t+1}\int_0^l \frac{y^t\,dy}{\sqrt{l^2-y^2}} = \frac{l}{t+1}I_t, \qquad (4.24)$$

where

$$I_{2r} = \int_0^l \frac{y^{2r}\,dy}{\sqrt{l^2-y^2}} = \frac{\pi}{2}\frac{(2r)!\,l^{2r}}{2^{2r}(r!)^2}, \quad (t = 2r), \qquad (4.25a)$$

$$I_{2r+1} = \int_0^l \frac{y^{2r+1}\,dy}{\sqrt{l^2-y^2}} = \frac{2^{2r}(r!)^2\,l^{2r+1}}{(2r+1)!}, \quad (t = 2r+1). \qquad (4.25b)$$

The wave drag coefficient $C_{dm}^{(n)}$, corresponding to the nth order component wing due to the vertical velocity w of the mth order component motion, taking into consideration the above formulae, will be

$$C_{dm}^{(n)} = -\frac{8}{(m+n)lU_\infty^2}\sum_{k=0}^{m-1} w_{m-k-1,\,k}\left[\sum_{q=0}^{E\left(\frac{n}{2}\right)} I_{k+2q}A_{n,\,2q} + \sum_{q=1}^{E\left(\frac{n-1}{2}\right)} \frac{l}{k+2q+1} I_{k+2q}Q_{n,\,2q}\right], \qquad (4.26)$$

where I_{k+2q} is given by (4.25 a) if $k+2q = 2r$, that is if k is an even number ($k = 2\varkappa$), or by (4.25 b), if $k+2q = 2r+1$, that is if k is an odd number ($k = 2\varkappa+1$).

For the whole complex wing, we have the same summation (4.8) as usual.

As examples, we have:

$$C_{d2}^{(3)} = -\frac{8}{5lU_\infty^2}\left[w_{10}\frac{\pi}{2}\left(A_{30}+\frac{l^2}{2}A_{32}+\frac{l^3}{6}Q_{32}\right) + w_{01}l\left(A_{30}+\frac{2l^2}{3}A_{32}+\frac{l^3}{6}Q_{32}\right)\right], \qquad (4.27a)$$

$$C_{d3}^{(3)} = -\frac{4}{3lU_\infty^2}\left[w_{20}\frac{\pi}{2}\left(A_{30}+\frac{l^2}{2}A_{32}+\frac{l^3}{6}Q_{32}\right)\right.$$

$$+w_{11}l\left(A_{30}+\frac{2l^2}{3}A_{32}+\frac{l^3}{6}Q_{32}\right)$$

$$\left.+w_{02}\frac{\pi l^2}{4}\left(A_{30}+\frac{3l^2}{4}A_{32}+\frac{3l^3}{20}Q_{32}\right)\right]. \tag{4.27b}$$

For the remaining coefficients $C_{dm}^{(n)}$ of the complex wing consisting of four component wings, we apply of course (4.26); in this connection, the Integrals I_{k+2q} for $n = 1, 2, 3, 4$ and $m = 1, 2, 3, 4$ are given in Appendix *I*.2.

IV. 2.4. Suction Coefficient

For reasons of symmetry, the two leading edges have the same suction force and therefore, in the expressions for the suction coefficient, both leading edges will be considered.

The coefficient $\sigma_m^{(n)}$, which is due to the interference of two component wings of orders *n* and *m*, will be given by (3.103); we multiply this formula by 2 and put $l_1 = l$, $S = l$, $a_1 = 1$, and obtain

$$\sigma_m^{(n)} = -\frac{2\pi\sqrt{1-B^2l^2}}{(m+n)l^3U_\infty^2}\left(\sum_{q=0}^{E\left(\frac{n}{2}\right)} A_{n,2q}\,l^{2q}\right)\left(\sum_{k=0}^{E\left(\frac{m}{2}\right)} A_{m,2k}\,l^{2k}\right). \tag{4.28}$$

The suction coefficient of the *n*th order wing considered alone is obtained from the above formula by putting $m = n$,

$$\sigma_n^{(n)} = -\frac{\pi\sqrt{1-B^2l^2}}{nl^3U_\infty^2}\left(\sum_{q=0}^{E\left(\frac{n}{2}\right)} A_{n,2q}\,l^{2q}\right)^2. \tag{4.29}$$

As an example, we put $m = n = 4$ in (4.29) and obtain

$$\sigma_4^{(4)} = -\frac{\pi\sqrt{1-B^2l^2}}{4l^3U_\infty^2}(A_{40}+A_{42}l^2+A_{44}l^4)^2. \tag{4.30}$$

The total coefficient for the whole wing follows by applying formula (4.13) for $N = 4$.

IV. 3. Thin Delta Wing with Subsonic Edges having Antisymmetrical Incidences

IV. 3.1. Definition of the Wing and Determination of the Axial Velocity

The downwash corresponding to the nth order wing is put in the form (4.2), which represents an antisymmetrical distribution; the plus sign refers to the positive part of the abscissa and the minus sign to the negative part.

For the axial disturbance velocity we likewise obtain an antisymmetrical expression for the function $\mathcal{U}_{n-1}$, in which the real part represents the axial velocity u_{n-1}:

$$\mathcal{U}_{n-1} = \sum_{q=0}^{E\left(\frac{n-1}{2}\right)} \frac{A_{n,\,2q+1}\, x^{2q+1}}{\sqrt{l^2-x^2}} + \sum_{q=0}^{E\left(\frac{n-2}{2}\right)} Q_{n,\,2q+1}\, x^{2q+1} \cosh^{-1} \sqrt{\frac{l^2}{x^2}}. \tag{4.31}$$

The expression for the central ridge is determined by the jump of the constants $(\pm 1)^k w_{n-k-1,\,k}$.

For the four component wings, proceeding from $n = 1$ to $n = 4$, the values of the constants are

$$A_{11} = -\frac{2lw_{00}}{\pi} \tag{4.32}$$

for $n = 1$;

$$A_{21} = -\frac{l}{\pi} \frac{2(1-B^2l^2)[2E(k)-B^2l^2K(k)]w_{10}+\pi(1-B^2l^2)lw_{01}}{(2-B^2l^2)E(k)-B^2l^2K(k)}, \tag{4.33a}$$

$$Q_{21} = -\frac{2w_{10}}{\pi}, \tag{4.33b}$$

for $n = 2$;

$$A_{n,\,2q+1} = -\frac{l^{1-2q}}{\pi\Omega_n^*} \sum_{k=0}^{n-1} \Omega_{n,\,2q+1}^{(k)}\, l^k\, w_{n-k-1,\,k}, \tag{4.34a}$$

$$Q_{31} = -\frac{4w_{20}}{\pi}, \quad Q_{41} = -\frac{6w_{30}}{\pi}, \quad Q_{43} = -\frac{3B^2w_{30}+2w_{12}}{3\pi}, \tag{4.34b}$$

for $n = 3$ and $n = 4$, where $\Omega_{n,\,2q+1}^{(k)}$ and Ω_n^* are given in Appendix I.1. Referring to $K(k)$ and $E(k)$, their significance is the same as in § IV. 2.1 and the modulus k is given by (4.19).

The variation of $u^{(1)}/w_{01}$, which has the same form as the variation of C_p (4.20), is plotted at the cross-section $x_1 = 1$ (Fig. 4.4) and at the longitudinal section $x_2 = 0{\cdot}2$ (Fig. 4.5), respectively.

As in the case of the wing with symmetry of incidence, we have considered $M_\infty = 2$ and $Bl = 0{\cdot}8$, $l = 0{\cdot}462$, $n = 2$ and $w_{10} = 0$.

IV. 3.2. Wave Drag Coefficient

As in the preceding case (§ IV. 2.3), formula (3.38) is applied. Using the same notations (4.24) and (4.25), we may write:

$$C_{dm}^{(n)} = -\frac{8}{(m+n)lU_\infty^2}\sum_{k=0}^{m-1} w_{m-k-1,\,k}\left[\sum_{q=0}^{E\left(\frac{n-1}{2}\right)} I_{k+2q+1}A_{n,\,2q+1} + \sum_{q=0}^{E\left(\frac{n-2}{2}\right)} \frac{l}{k+2q+2} I_{k+2q+1}Q_{n,\,2q+1}\right], \tag{4.35}$$

in which I_{k+2q+1} is given by (4.25a) if $k+2q+1 = 2r$, that is, if k is an odd number ($k = 2\varkappa+1$), or by (4.25b) if $k+2q+1 = 2r+1$, that is, if k is an even number ($k = 2\varkappa$).

For the whole complex wing we shall apply the same summation (4.8) as usual.

As an example we put $m = n = 3$ and obtain:

$$C_{d3}^{(3)} = -\frac{4}{3U_\infty^2}\left[w_{20}\left(A_{31}+\frac{2l^2}{3}A_{33}+\frac{l}{2}Q_{31}\right)+w_{11}\frac{\pi l}{4}\left(A_{31}+\frac{3l^2}{4}A_{33}+\frac{l}{3}Q_{31}\right) + w_{02}\frac{2l^2}{3}\left(A_{31}+\frac{4l^2}{5}A_{33}+\frac{l}{4}Q_{31}\right)\right]. \tag{4.36}$$

In order to calculate the drag coefficients $C_{dm}^{(n)}$ for the four component wings ($n = 1,2,3,4$), the integrals I_{k+2q+1} are given in Appendix I.2.

IV. 3.3. Suction Coefficient

We proceed as in the case of the wing with symmetry of incidence (§ IV. 1.4), by writing first the expression for the coefficient $\sigma_m^{(n)}$ due to the axial velocities of the two constituent conical motions of orders m and n, according to the relation (3.103) established previously, where we put $l_1 = l$, $S = l$, $a_1 = 1$ and multiply by 2 for both leading edges:

$$\sigma_m^{(n)} = -\frac{2\pi\sqrt{1-B^2l^2}}{(m+n)\,l^3U_\infty^2}\left(\sum_{q=0}^{E\left(\frac{n-1}{2}\right)} A_{n,\,2q+1}\,l^{2q+1}\right)\left(\sum_{k=0}^{E\left(\frac{m-1}{2}\right)} A_{m,\,2k+1}\,l^{2k+1}\right). \tag{4.37}$$

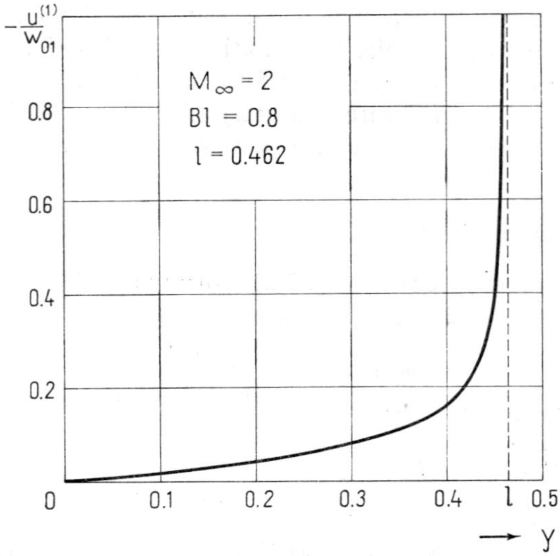
$-\frac{u^{(1)}}{w_{01}}$
M∞ = 2
Bl = 0.8
l = 0.462
0.8
0.6
0.4
0.2
0
0.1
0.2
0.3
0.4
l
0.5
→ Y

FIG. 4.4

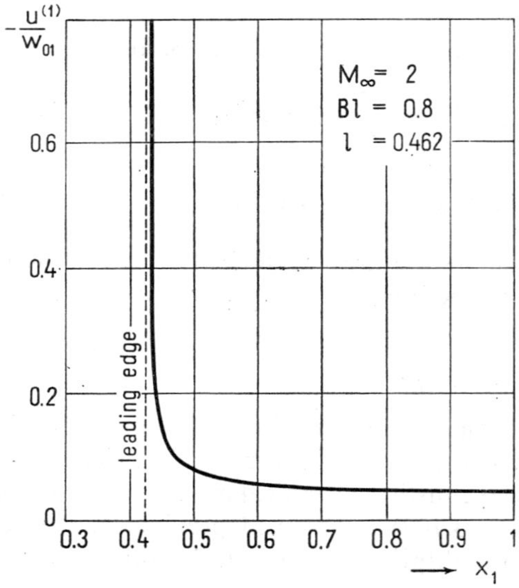
$-\frac{u^{(1)}}{w_{01}}$
M∞ = 2
Bl = 0.8
l = 0.462
0.6
0.4
0.2
0
leading edge
0.3
0.4
0.5
0.6
0.7
0.8
0.9
1
→ x1

FIG. 4.5

For the nth order wing considered alone, the suction coefficient is defined by the expression

$$\sigma_n^{(n)} = -\frac{\pi\sqrt{1-B^2l^2}}{nl^3U_\infty^2}\left(\sum_{q=0}^{E\left(\frac{n-1}{2}\right)} A_{n,2q+1}\, l^{2q+1}\right)^2, \tag{4.38}$$

which may be readily obtained from (4.37) by putting $m = n$.

The total suction coefficient of the complex wing follows from formula (4.13).

Further, for a component wing of order $n = 4$, we have to apply (4.38) which yields

$$\sigma_4^{(4)} = -\frac{\pi\sqrt{1-B^2l^2}}{4lU_\infty^2}(A_{41}+A_{43}l^2)^2. \tag{4.39}$$

IV. 3.4. Rolling Moment Coefficient

The rolling moment coefficient is given by the following expression, from formula (3.47a), where $S = l$ and $h = 1$,

$$\mu_1^{(n)} = \frac{4\pi}{(n+2)\,lU_\infty}\left\{\sum_{q=0}^{E\left(\frac{n-1}{2}\right)} \frac{(2q+2)!\,l^{2(q+1)}}{2^{2(q+1)}[(q+1)!]^2} A_{n,\,2q+1}\right.$$

$$\left. + \sum_{q=0}^{E\left(\frac{n-2}{2}\right)} \frac{(2q+2)!\,l^{2q+3}}{(2q+3)2^{2(q+1)}[(q+1)!]^2} Q_{n,\,2q+1}\right\}. \tag{4.40}$$

We find thus for each component wing,

$$\mu_1^{(1)} = \frac{2\pi l}{3U_\infty} A_{11}, \tag{4.41a}$$

$$\mu_1^{(2)} = \frac{\pi l}{2U_\infty}\left(A_{21}+\frac{l}{3}Q_{21}\right), \tag{4.41b}$$

$$\mu_1^{(3)} = \frac{2\pi l}{5U_\infty}\left(A_{31}+\frac{3l^2}{4}A_{33}+\frac{l}{3}Q_{31}\right), \tag{4.41c}$$

$$\mu_1^{(4)} = \frac{\pi l}{3U_\infty}\left(A_{41}+\frac{3l^2}{4}A_{43}+\frac{l}{3}Q_{41}+\frac{3l^3}{20}Q_{43}\right). \tag{4.41d}$$

The total moment μ_1 is obtained by the summation of these coefficients.

IV. 4. Thin Delta Wing with Supersonic Edges and Symmetry of Incidence

IV. 4.1. Expression for the Axial Velocity

The downwash w has the same expression as for the wing with subsonic leading edges (4.1). The axial velocity u is, however, deduced from (2.95b), by putting $l_1 = l_2 = l$ and taking into account the symmetry conditions. Thus, using throughout double indices for constants, we obtain an expression which includes the effect of a central ridge:

$$\mathcal{U}_{n-1} = \sum_{q=0}^{n-1} K_{n,q}\, x^q \left[\cos^{-1}\sqrt{\frac{(1+Bl)(1-Bx)}{2B(l-x)}} + (-1)^q \cos^{-1}\sqrt{\frac{(1+Bl)(1+Bx)}{2B(l+x)}}\right] + \sum_{q=1}^{E\left(\frac{n-1}{2}\right)} Q_{n,2q} x^{2q} \cosh^{-1}\sqrt{\frac{1}{B^2x^2}}$$
$$+ \sum_{q=0}^{E\left(\frac{n-2}{2}\right)} D_{n,2q}\, x^{2q}\sqrt{\frac{1-B^2x^2}{B^2}}\,. \tag{4.42}$$

where $Q_{n,0} = 0$ was considered, as shown in § IV. 2.1.

For the first four component wings, the constants of these expressions are given as functions of the coefficients of the downwashes (4.1); we have thus

$$K_{10} = -\frac{2w_{00}}{\pi B}\frac{Bl}{\sqrt{B^2l^2-1}} \tag{4.43}$$

for $n = 1$;

$$K_{20} = -\frac{2l}{\pi(B^2l^2-1)^{3/2}}\left[(B^2l^2-2)\,w_{10} - lw_{01}\right], \tag{4.44a}$$

$$K_{21} = -\frac{2}{\pi(B^2l^2-1)^{3/2}}\left[w_{10} + B^2l^3w_{01}\right], \tag{4.44b}$$

$$D_{20} = -\frac{2Bl}{\pi(B^2l^2-1)}\left[w_{10} + lw_{01}\right] \tag{4.44c}$$

for $n = 2$;

$$K_{30} = -\frac{l}{\pi(B^2l^2-1)^{5/2}}\left[(2B^4l^4 - 5B^2l^2 + 6)w_{20} + 3lw_{11} + (B^2l^2+2)l^2w_{02}\right], \tag{4.45a}$$

$$K_{31} = \frac{2}{\pi(B^2l^2-1)^{5/2}}\left[(B^2l^2+2)w_{20} - (B^2l^2-4)B^2l^3w_{11} + 3B^2l^4w_{02}\right], \tag{4.45b}$$

$$K_{32} = -\frac{1}{\pi l(B^2l^2-1)^{5/2}}[3B^2l^2w_{20}+(4B^2l^2-1)lw_{11}+(2B^2l^2+1)B^2l^4w_{02}], \quad (4.45c)$$

$$Q_{32} = -\frac{w_{11}}{\pi}, \quad (4.45d)$$

$$D_{30} = -\frac{Bl}{\pi(B^2l^2-1)^2}[(2B^2l^2-5)w_{20}+(B^2l^2-4)lw_{11}-3l^2w_{02}] \quad (4.45e)$$

for $n = 3$, and

$$K_{40} = -\frac{l}{3\pi(B^2l^2-1)^{7/2}}[3(2B^6l^6-7B^4l^4+8B^2l^2-8)w_{30}-3(B^2l^2+4)lw_{21}$$
$$+(B^4l^4-8B^2l^2-8)l^2w_{12}-3(3B^2l^2+2)l^3w_{03}], \quad (4.46a)$$

$$K_{41} = -\frac{1}{\pi(B^2l^2-1)^{7/2}}[3(3B^2l^2+2)w_{30}+(2B^4l^4-7B^2l^2+20)B^2l^3w_{21}$$
$$+15B^2l^4w_{12}+3(B^2l^2+4)B^2l^5w_{03}], \quad (4.46b)$$

$$K_{42} = \frac{1}{\pi l(B^2l^2-1)^{7/2}}[3(B^2l^2+4)B^2l^2w_{30}+(4B^4l^4+13B^2l^2-2)lw_{21}$$
$$-(2B^4l^4-13B^2l^2-4)B^2l^4w_{12}+3(4B^2l^2+1)B^2l^5w_{03}], \quad (4.46c)$$

$$K_{43} = -\frac{1}{3\pi l^2(B^2l^2-1)^{7/2}}[3(4B^2l^2+1)B^2l^2w_{30}+15B^4l^5w_{21}$$
$$+(20B^4l^4-7B^2l^2+2)l^2w_{12}+3(2B^2l^2+3)B^4l^7w_{03}], \quad (4.46d)$$

$$Q_{42} = -\frac{2w_{21}}{\pi}, \quad (4.46e)$$

$$D_{40} = -\frac{Bl}{3\pi(B^2l^2-1)^3}[(6B^4l^4-17B^2l^2+26)w_{30}+(2B^4l^4-6B^2l^2+19)lw_{21}$$
$$+(B^2l^2+14)l^2w_{12}+(4B^2l^2+11)l^3w_{03}], \quad (4.46f)$$

$$D_{42} = -\frac{Bl}{3\pi l^2(B^2l^2-1)^3}[(2B^2l^2+13)B^2l^2w_{30}-(2B^4l^4-9B^2l^2-8)B^2l^3w_{21}$$
$$+(6B^4l^4+11B^2l^2-2)l^2w_{12}-(11B^2l^2+4)B^2l^5w_{03}] \quad (4.46g)$$

for $n = 4$.

With the aid of these values we have plotted curves of $u^{(1)}/w_{10}$ (which is proportional to C_p) in the cross-section $x_1 = 1$, Fig. 4.6, and in the longitudinal section $x_2 = 0{\cdot}2$, Fig. 4.7.

For calculating these curves we have taken $M_\infty = 2$ and $Bl = 1{\cdot}3$, $l = 0{\cdot}75$, $n = 2$, $w_{01} = 0$.

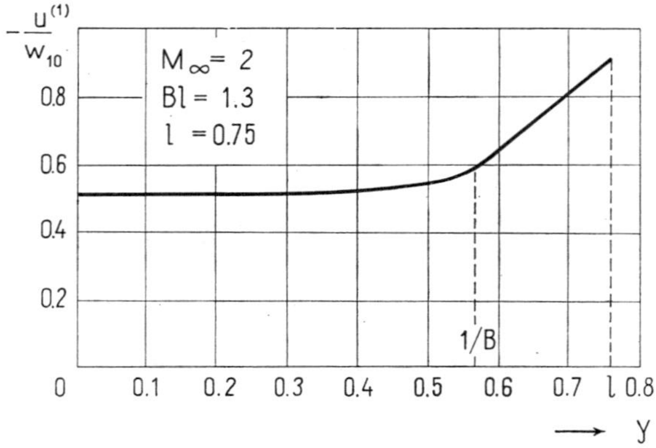

FIG. 4.6

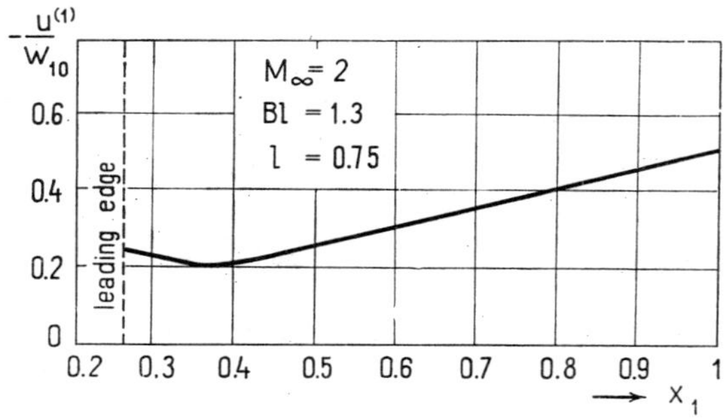

FIG. 4.7

IV. 4.2. Lift Coefficient

Applying formula (3.30) and considering (4.42), we obtain the lift coefficient of the single wing or of the nth order component wing:

$$C_l^{(n)} = \frac{4\pi}{(n+1)lU_\infty}\left[\sum_{q=0}^{n-1} K_{n,q}\frac{\sqrt{B^2l^2-1}}{(q+1)B^{q+1}}\sum_{r=0}^{E\left(\frac{q}{2}\right)}\frac{(2r)!}{2^{2r}(r!)^2}(Bl)^{q-2r}\right.$$
$$+\sum_{q=1}^{E\left(\frac{n-1}{2}\right)}\frac{(2q)!}{(2q+1)2^{2q}(q!)^2B^{2q+1}}Q_{n,2q}$$
$$\left.+\sum_{q=0}^{E\left(\frac{n-2}{2}\right)}\frac{(2q)!}{(2q+2)2^{2q}(q!)^2B^{2(q+1)}}D_{n,2q}\right]. \tag{4.47}$$

Applying this formula to each of the four wings considered above, we obtain

$$C_l^{(1)} = \frac{2\pi}{U_\infty}\frac{\sqrt{B^2l^2-1}}{Bl}K_{10} = -\frac{4w_{00}}{BU_\infty}, \tag{4.48a}$$

$$C_l^{(2)} = \frac{4\pi}{3BlU_\infty}\left[\sqrt{B^2l^2-1}\left(K_{20}+\frac{l}{2}K_{21}\right)+\frac{1}{2B}D_{20}\right], \tag{4.48b}$$

$$C_l^{(3)} = \frac{\pi}{BlU_\infty}\left[\sqrt{B^2l^2-1}\left(K_{30}+\frac{l}{2}K_{31}+\frac{2B^2l^2+1}{6B^2}K_{32}\right)\right.$$
$$\left.+\frac{1}{6B^2}Q_{32}+\frac{1}{2B}D_{30}\right], \tag{4.48c}$$

$$C_l^{(4)} = \frac{4\pi}{5BlU_\infty}\left[\sqrt{B^2l^2-1}\left(K_{40}+\frac{l}{2}K_{41}+\frac{2B^2l^2+1}{6B^2}K_{42}\right.\right.$$
$$\left.\left.+\frac{2B^2l^2+1}{8B^2}lK_{43}\right)+\frac{1}{6B^2}Q_{42}+\frac{1}{2B}D_{40}+\frac{1}{8B^3}D_{42}\right], \tag{4.48d}$$

respectively.

The total coefficient of the complex wing is given by the sum of the above coefficients.

Note. The *pitching moment* coefficient has the same expression as (4.10), hence it is deduced from the lift coefficient defined above.

IV. 4.3. Wave Drag Coefficient

First we shall write

$$H_{k+q,k} = \text{Re}\int_0^l y^{k+q}\left[\cos^{-1}\sqrt{\frac{(1+Bl)(1-By)}{2B(l-y)}}\right.$$
$$\left.+(-1)^q\cos^{-1}\sqrt{\frac{(1+Bl)(1+By)}{2B(l+y)}}\right]dy$$
$$= \frac{1}{k+q+1}\left\{\frac{1+(-1)^{k+1}}{2}l^{k+q+1}\cos^{-1}\frac{1}{Bl}\right.$$
$$\left.+\sqrt{B^2l^2-1}\sum_{t=0}^{k+q}\frac{1+(-1)^{k-t}}{2}l^{k+q-t}\mathcal{J}_t\right\}, \tag{4.49}$$

where we have successively

$$\mathcal{J}_{2r} = \text{Re}\int_0^l\frac{y^{2r}\,dy}{\sqrt{1-B^2y^2}} = \frac{\pi}{2}\frac{(2r)!}{2^{2r}(r!)^2B^{2r+1}}, \quad (t=2r), \tag{4.50a}$$

$$\mathcal{J}_{2r+1} = \text{Re}\int_0^l\frac{y^{2r+1}\,dy}{\sqrt{1-B^2y^2}} = \frac{2^{2r}(r!)^2}{(2r+1)!B^{2(r+1)}}, \quad (t=2r+1), \tag{4.50b}$$

that is, a different value for even numbers of t from that for odd numbers of t.

With the aid of these notations, we may write the drag coefficient $C_{dm}^{(n)}$, by applying formula (3.38):

$$C_{dm}^{(n)} = -\frac{8}{(m+n)lU_\infty^2} \sum_{k=0}^{m-1} w_{m-k-1,\,k} \left[\sum_{q=0}^{n-1} H_{k+q,\,k} K_{n,\,q} \right.$$
$$+ \sum_{q=1}^{E\left(\frac{n-1}{2}\right)} \frac{1}{k+2q+1} \mathcal{J}_{k+2q} Q_{n,\,2q}$$
$$\left. + \sum_{q=0}^{E\left(\frac{n-2}{2}\right)} \frac{1}{(k+2q+2)B} \mathcal{J}_{k+2q} D_{n,\,2q} \right], \tag{4.51}$$

where $H_{k+q,\,k}$ is defined in the above expression (4.49), while $\mathcal{J}_{k+2q}$ is given by formula (4.50a) if $k+2q = 2r$ (that is $k = 2\varkappa$), or by formula (4.50b) if $k+2q = 2r+1$ (that is $k = 2\varkappa+1$).

For the coefficient $C_{dn}^{(n)}$ of the nth order wing considered alone, we put $m = n$ in the above formula.

As an example, we obtain the following expression for the component wing of order $n = 2$ considered alone:

$$C_{d2}^{(2)} = -\frac{2}{BlU_\infty^2} \left\{ w_{10} \frac{\pi}{2} \left[\sqrt{B^2l^2-1} \left(K_{20} + \frac{l}{2} K_{21} \right) + \frac{1}{2B} D_{20} \right] \right.$$
$$\left. + w_{01} \frac{1}{B} \left[\left(B^2l^2 \cos^{-1} \frac{1}{Bl} + \sqrt{B^2l^2-1} \right) \left(\tfrac{1}{2} K_{20} + \frac{l}{3} K_{21} \right) + \frac{1}{3B} D_{20} \right] \right\}. \tag{4.52}$$

For the comlex wing we perform the summation (4.8), where $N = 4$; for calculating the coefficients $C_{dm}^{(n)}$, integrals $H_{k+q,\,k}$ and $\mathcal{J}_{k+2q}$ are given in Appendix I.3.

IV. 5. Wing of Symmetrical Thickness

IV. 5.1. Usual Expression for the Function $\mathcal{U}_{n-1}$

Let us consider a wing of symmetrical thickness, for which downwash w is given by formula (4.1), corresponding to a symmetrical slope distribution.

If the wing of symmetrical thickness has supersonic leading edges ($l > 1/B$), the expression for the axial velocity is identical to that obtained above for the thin wing with supersonic leading edges. Indeed, since both for the thin wing and the wing of symmetrical thickness the upper and lower surfaces are independent as regards the air flow, and the downwashes are identical, it follows that the axial velocities are also identical. However, for the wing of symmetrical thickness, these axial velocities are equal and of the same sign on the upper and lower faces, while for thin wings they are equal and of opposite sign. The wave drag coefficient is hence identical to the one found for the thin wing with supersonic edges (4.51).

What happens, however, when the leading edges of the wing of symmetrical thickness are subsonic ($l<1/B$)?

In this case, the leading edges work as two symmetrical ridges and the wing may be considered as being continued by two flat plates without incidence, which are in fact the prolongation of the plane of symmetry of the wing of symmetrical thickness.

The axial velocity u_{n-1}, or the corresponding function $\mathcal{U}_{n-1}$, retains exactly the same form as (4.42) by virtue of the validity inside and outside of the Mach cone of the formulae which refer to the contribution of ridges (§ II. 4.5). Denoting the respective coefficients by $C_{n,q}$ for the subsonic leading edges (which play the same part as the ridges themselves) and by $K_{n,q}$ for the supersonic edges, and observing that all the latter coefficients have $\sqrt{B^2l^2-1}$ at the denominator, which becomes imaginary when $Bl<1$, one may deduce from the known relation $\cosh^{-1}\beta = i\cos^{-1}\beta$ (2.68) the following formula for determining $C_{n,q}$,

$$C_{n,q} = K_{n,q}\,\frac{\sqrt{B^2l^2-1}}{\sqrt{1-B^2l^2}}\,. \tag{4.53}$$

Accordingly, for the wing of symmetrical thickness with subsonic leading edges (ridges), the same expression (4.42) can be retained for the function $\mathcal{U}_{n-1}$. However, in order to distinguish the two problems by their representation form, i.e. the wing of symmetrical thickness with subsonic leading edges and the thin wing with supersonic leading edges, $\mathcal{U}_{n-1}$ will be set under the following form, employing the ordinary notations $C_{n,q}$ and $\cosh^{-1}$;

$$\begin{aligned}\mathcal{U}_{n-1} &= \sum_{q=0}^{n-1} C_{n,q}\,x^q\left[\cosh^{-1}\sqrt{\frac{(1+Bl)(1-Bx)}{2B(l-x)}}\right.\\ &\quad\left.+(-1)^q\cosh^{-1}\sqrt{\frac{(1+Bl)(1+Bx)}{2B(l+x)}}\right]\\ &\quad+\sum_{q=1}^{E\left(\frac{n-1}{2}\right)} Q_{n,2q}\,x^{2q}\cosh^{-1}\sqrt{\frac{1}{B^2x^2}}+\sum_{q=0}^{E\left(\frac{n-2}{2}\right)} D_{n,2q}\,x^{2q}\sqrt{\frac{1-B^2x^2}{B^2}}\,,\end{aligned} \tag{4.54}$$

where $E\left(\frac{n-1}{2}\right)$ and $E\left(\frac{n-2}{2}\right)$ are the greatest integers contained in $\frac{n-1}{2}$ and $\frac{n-2}{2}$

For the first four component wings, the constants $C_{n,q}$ are deduced from $K_{n,q}$ (4.43, 4.44, 4.45, 4.46), as indicated in formula (4.53).

IV. 5.2. Wave Drag Coefficient

In an analogous way to the preceding case, we shall write

$$H'_{k+q,k} = \operatorname{Re}\int_0^l y^{k+q}\left[\cosh^{-1}\sqrt{\frac{(1+Bl)(1-By)}{2B(l-y)}}\right.$$
$$\left.+(-1)^q\cosh^{-1}\sqrt{\frac{(1+Bl)(1+By)}{2B(l+y)}}\right]dy$$
$$= \frac{l^{k+q+1}}{k+q+1}\left\{\frac{1+(-1)^{k+1}}{2}\cosh^{-1}\frac{1}{Bl}\right.$$
$$+(-1)^q[1-(-1)^{k+q+1}]\cosh^{-1}\frac{1+Bl}{2\sqrt{Bl}}$$
$$\left.+\sqrt{1-B^2l^2}\sum_{t=0}^{k+q}\frac{1+(-1)^{k-t}}{2l^{t+1}}\mathcal{J}'_t\right\}, \qquad (4.55)$$

where $\mathcal{J}'_t$ has the following forms:

$$\mathcal{J}'_{2r} = \operatorname{Re}\int_0^l \frac{y^{2r}\,dy}{\sqrt{1-B^2y^2}}$$
$$= \frac{(2r)!}{2^{2r}(r!)^2B^{2r+1}}\left[\sin^{-1}Bl-\sqrt{1-B^2l^2}\sum_{p=0}^{r-1}\frac{2^{2p}(p!)^2}{(2p+1)!}(Bl)^{2p+1}\right], \quad (t=2r) \qquad (4.56a)$$

$$\mathcal{J}'_{2r+1} = \operatorname{Re}\int_0^l \frac{y^{2r+1}\,dy}{\sqrt{1-B^2y^2}}$$
$$= \frac{2^{2r}(r!)^2}{(2r+1)!B^{2(r+1)}}\left[1-\sqrt{1-B^2l^2}\sum_{p=0}^{r}\frac{(2p)!}{2^{2p}(p!)^2}(Bl)^{2p}\right], \quad (t=2r+1), \qquad (4.56b)$$

the first integral for even numbers of t and the second one for odd numbers of t.

Thus, the intermediate drag coefficient $C_{dm}^{(n)}$, corresponding to the two conical motions of order m and n, is expressed by

$$C_{dm}^{(n)} = -\frac{8}{(m+n)lU_\infty^2}\sum_{k=0}^{m-1} w_{m-k-1,k}\left[\sum_{q=0}^{n-1} H'_{k+q,k}\, C_{n,q} + \sum_{q=1}^{E\left(\frac{n-1}{2}\right)} \frac{1}{k+2q+1}\left(l^{k+2q+1}\cosh^{-1}\frac{1}{Bl}+\mathcal{J}'_{k+2q}\right)Q_{n,2q} + \sum_{q=0}^{E\left(\frac{n-2}{2}\right)} \frac{1}{(k+2q+2)B}\left(l^{k+2q+1}\sqrt{1-B^2l^2}+\mathcal{J}'_{k+2q}\right)D_{n,2q}\right], \tag{4.57}$$

where $H'_{k+q,k}$ is defined in the above expression (4.55), while $\mathcal{J}'_{k+2q}$ is given by (4.56a) if $k+2q = 2r$, that is, $k = 2\varkappa$ (even number), or has the form (4.56b) if $k+2q = 2r+1$, that is, $k = 2\varkappa+1$ (odd number).

As an example putting $m = n = 2$ in (4.57), we find the drag coefficient of a second-order single wing:

$$C_{d2}^{(2)} = -\frac{2}{BlU_\infty^2}\left\{w_{10}\left[\left(C_{20}+\frac{l}{2}C_{21}\right)\sqrt{1-B^2l^2}\sin^{-1}Bl + 2C_{20}\,Bl\cosh^{-1}\frac{1+Bl}{2\sqrt{Bl}}+\frac{1}{2B}D_{20}\left(\sin^{-1}Bl+Bl\sqrt{1-B^2l^2}\right)\right] + w_{01}\frac{1}{B}\left[\left(\tfrac{1}{2}C_{20}+\frac{l}{3}C_{21}\right)\left(\sqrt{1-B^2l^2}-1+B^2l^2+B^2l^2\cosh^{-1}\frac{1}{Bl}\right) - \tfrac{2}{3}C_{21}B^2l^3\cosh^{-1}\frac{1+Bl}{2\sqrt{Bl}}+\frac{1}{3B}D_{20}\left(1-(1-B^2l^2)^{3/2}\right)\right]\right\}. \tag{4.58}$$

For the complex wing, consisting of four wings of order $n = 1, 2, 3, 4$ and $m = 1, 2, 3, 4$, the coefficients $C_{dm}^{(n)}$ will be calculated by means of the integrals $H'_{k+q,k}$ and $\mathcal{J}'_{k+2q}$, which are included in Appendix I.4.

IV. 6. Thin Delta Wing with Supersonic Edges and Antisymmetry of Incidence

IV. 6.1. Determination of the Axial Velocity Function

The downwashes w on the wing have the same expression (4.2) as for the corresponding wing with subsonic edges, but the axial velocity u_{n-1} has a structure differing from the axial velocity of this latter wing. Indeed, putting $l_1 = l_2 = l$ in (2.95b), and, on the other hand, considering the antisymmetry conditions, we finally obtain

$$\mathcal{U}_{n-1} = \sum_{q=0}^{n-1} K_{n,q} x^q \Big[\cos^{-1} \sqrt{\frac{(1+Bl)(1-Bx)}{2B(l-x)}} - (-1)^q \cos^{-1} \sqrt{\frac{(1+Bl)(1+Bx)}{2B(l+x)}} + \sum_{q=0}^{E\left(\frac{n-2}{2}\right)} Q_{n,2q+1} x^{2q+1} \times \cosh^{-1} \sqrt{\frac{1}{B^2x^2}} + \sum_{q=0}^{E\left(\frac{n-3}{2}\right)} D_{n,2q+1} x^{2q+1} \sqrt{\frac{1-B^2x^2}{B^2}}, \tag{4.59}$$

where $E(\frac{n-2}{2})$ and $E(\frac{n-3}{2})$ represent the greatest integers contained in $\frac{n-2}{2}$ and $\frac{n-3}{2}$, respectively.

For the first four component wings ($n = 1, 2, 3, 4$), the constants $K_{n,q}$ have the same values indicated in (4.43), (4.44a, b), (4.45a, b, c), (4.46a, b, c, d), while the remaining constants have the following values:

$$Q_{21} = -\frac{2w_{10}}{\pi} \tag{4.60}$$

for $n = 2$;

$$Q_{31} = -\frac{4w_{20}}{\pi}, \tag{4.61a}$$

$$D_{31} = \frac{B}{\pi(B^2l^2-1)}\left[(2B^2l^2-5)B^2l^2w_{20} - (2B^2l^2+1)lw_{11} - 3B^2l^4w_{02}\right], \tag{4.61b}$$

for $n = 3$;

$$Q_{41} = -\frac{6w_{30}}{\pi}, \tag{4.62a}$$

$$Q_{43} = -\frac{3B^2w_{30}+2w_{12}}{3\pi}, \tag{4.62b}$$

$$D_{41} = \frac{B}{\pi(B^2l^2-1)^3}\left[3(5B^4l^4-16B^2l^2+21)B^2l^2w_{30} - 2(B^4l^4-5B^2l^2-1)lw_{21} - (5B^4l^4-34B^2l^2-1)l^2w_{12} - 10B^2l^5w_{03}\right], \tag{4.62c}$$

for $n = 4$.

Employing relation (4.58) and putting $n = 2$, we have plotted the diagrams of $u^{(1)}/w_{01}$ in the cross-section $x_1 = 1$, Fig. 4.8, and in the longitudinal section $x_2 = 0{\cdot}2$, Fig. 4.9.

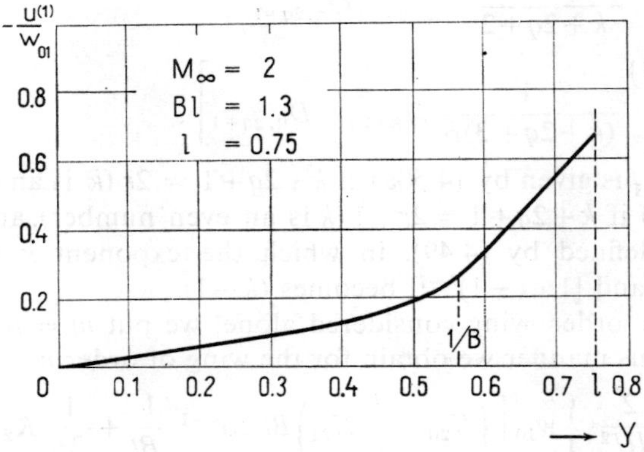

FIG. 4.8

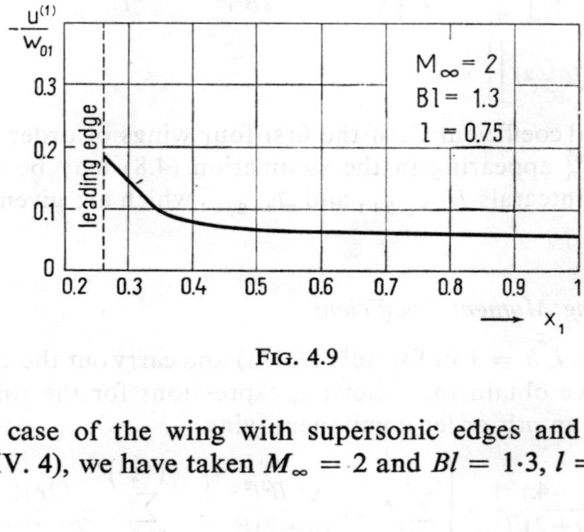

FIG. 4.9

As in the case of the wing with supersonic edges and symmetry of incidence (§ IV. 4), we have taken $M_\infty = 2$ and $Bl = 1{\cdot}3$, $l = 0{\cdot}75$, $n = 2$, $w_{10} = 0$.

IV. 6.2. Wave Drag Coefficient

Applying formula (3.38a) we may calculate the coefficient $C_{dm}^{(n)}$, integrating from 0 to l for reasons of symmetry and multiplying by two. For the integrals we employ the notations from formulae (4.49), and (4.50). For this coefficient we may write

$$C_{dm}^{(n)} = -\frac{8}{(m+n)lU_\infty^2} \sum_{k=0}^{m-1} w_{m-k-1,\,k} \left[\sum_{q=0}^{n-1} H_{k+q,\,k+1} K_{n,\,q} \right.$$

$$+ \sum_{q=0}^{E\left(\frac{n-2}{2}\right)} \frac{1}{k+2q+2} \mathcal{J}_{k+2q+1} Q_{n,\,2q+1}$$

$$\left. + \sum_{q=0}^{E\left(\frac{n-3}{2}\right)} \frac{1}{(k+2q+3)B} \mathcal{J}_{k+2q+1} D_{n,\,2q+1} \right], \tag{4.63}$$

where $\mathcal{J}_{k+2q+1}$ is given by (4.50a) if $k+2q+1 = 2r$ (k is an odd number) or by (4.50 b) if $k+2q+1 = 2r+1$ (k is an even number), and also, where $H_{k+q,\,k+1}$ is defined by (4.49), in which the exponent k of the terms $[1+(-1)^{k+1}]$ and $[1+(-1)^{k-t}]$ becomes $(k+1)$.

For the nth order wing considered alone, we put $m = n$ in the above formula; in this manner we obtain for the wing of order $n = 2$:

$$C_{d2}^{(2)} = -\frac{2}{BlU_\infty^2} \left\{ w_{10} \left[\left(K_{20} + \frac{l}{2} K_{21} \right) Bl \cos^{-1} \frac{1}{Bl} + \frac{1}{2B} K_{21} \sqrt{B^2 l^2 - 1} \right. \right.$$

$$\left. + \frac{1}{2B} Q_{21} \right] + w_{01} \frac{\pi l}{4} \left[\left(K_{20} + \frac{2B^2 l^2 + 1}{3B^2 l^2} l K_{21} \right) \sqrt{B^2 l^2 - 1} \right.$$

$$\left. \left. + \frac{l}{3B^2 l^2} Q_{21} \right] \right\}. \tag{4.64}$$

For the total coefficient C_d of the first four wings of order $n = 1, 2, 3, 4$, the terms $C_{dm}^{(n)}$ appearing in the summation (4.8) may be calculated by means of the integrals $H_{k+q,\,k+1}$ and $\mathcal{J}_{k+2q+1}$ which are given in Appendix I.3.

IV. 6.3. Rolling Moment Coefficient

We put $S = l$, $h = 1$ in formula (3.47a) and carry out the corresponding integration; we obtain the following expressions for the rolling moment coefficient of an nth order component wing:

$$\mu_1^{(n)} = \frac{4\pi}{(n+2)lU_\infty} \left[\sum_{q=0}^{n-1} K_{n,\,q} \frac{\sqrt{B^2 l^2 - 1}}{(q+2)B^{q+2}} \sum_{r=0}^{E\left(\frac{q+1}{2}\right)} \frac{(2r)!}{2^{2r}(r!)^2} (Bl)^{q-2r+1} \right.$$

$$+ \sum_{q=0}^{E\left(\frac{n-2}{2}\right)} \frac{(2q+2)!}{(2q+3)2^{2(q+1)}[(q+1)!]^2 B^{2q+3}} Q_{n,\,2q+1}$$

$$\left. + \sum_{q=0}^{E\left(\frac{n-3}{2}\right)} \frac{(2q+2)!}{(2q+4)2^{2(q+1)}[(q+1)!]^2 B^{2(q+2)}} D_{n,\,2q+1} \right]. \tag{4.65}$$

For each single wing we have:

$$\mu_1^{(1)} = \frac{2\pi}{3BU_\infty} K_{10}\sqrt{B^2l^2-1}\,, \tag{4.66a}$$

$$\mu_1^{(2)} = \frac{\pi}{2BU_\infty}\left[\left(K_{20}+\frac{2B^2l^2+1}{3B^2l^2}\,lK_{21}\right)\sqrt{B^2l^2-1}+\frac{l}{3B^2l^2}\,Q_{21}\right], \tag{4.66b}$$

$$\mu_1^{(3)} = \frac{2\pi}{5BU_\infty}\left[\left(K_{30}+\frac{2B^2l^2+1}{3B^2l^2}\,lK_{31}+\frac{2B^2l^2+1}{4B^2l^2}\,l^2K_{32}\right)\sqrt{B^2l^2-1}\right.$$
$$\left.+\frac{l}{3B^2l^2}\,Q_{31}+\frac{l^2}{4B^3l^3}\,D_{31}\right], \tag{4.66c}$$

$$\mu_1^{(4)} = \frac{\pi}{3BU_\infty}\left[\left(K_{40}+\frac{2B^2l^2+1}{3B^2l^2}\,lK_{41}+\frac{2B^2l^2+1}{4B^2l^2}\,l^2K_{42}\right.\right.$$
$$\left.+\frac{8B^4l^4+4B^2l^2+3}{20B^4l^4}\,l^3K_{43}\right)\sqrt{B^2l^2-1}+\frac{l}{3B^2l^2}\,Q_{41}$$
$$\left.+\frac{3l^3}{20B^4l^4}\,Q_{43}+\frac{l^2}{4B^3l^3}\,D_{41}\right]. \tag{4.66d}$$

IV. 7. Delta Wing with Sonic Leading Edges

IV. 7.1. Determination of the Axial Velocities

In the preceding considerations we have treated the wings with subsonic leading edges ($l<1/B$) and those with supersonic leading edges ($l>1/B$). If l is made to tend towards $1/B$ in the expressions of the axial velocities u, obtained in the examples treated earlier, we obtain the axial velocities u corresponding to wings with *sonic* leading edges ($l=1/B$, Fig. 4.10). If the transition from the expressions for the axial velocities corresponding to wings with supersonic leading edges, to those with sonic edges is more laborious since it implies an *a priori* knowledge of the constants, this transition is performed automatically if we proceed from the expression for u corresponding to wings with subsonic leading edges. Indeed, putting $l = 1/B$ in expression (4.14) and (4.31) for $\mathcal{U}_{n-1}$, corresponding to wings with symmetrical and antisymmetrical incidences, we obtain instead of these expressions:

$$\mathcal{U}_{n-1} = \sum_{q=0}^{E\left(\frac{n}{2}\right)} \frac{BA'_{n,2q}\,x^{2q}}{\sqrt{1-B^2x^2}} + \sum_{q=1}^{E\left(\frac{n-1}{2}\right)} Q'_{n,\,2q}\,x^{2q}\cosh^{-1}\sqrt{\frac{1}{B^2x^2}}\,, \tag{4.67a}$$

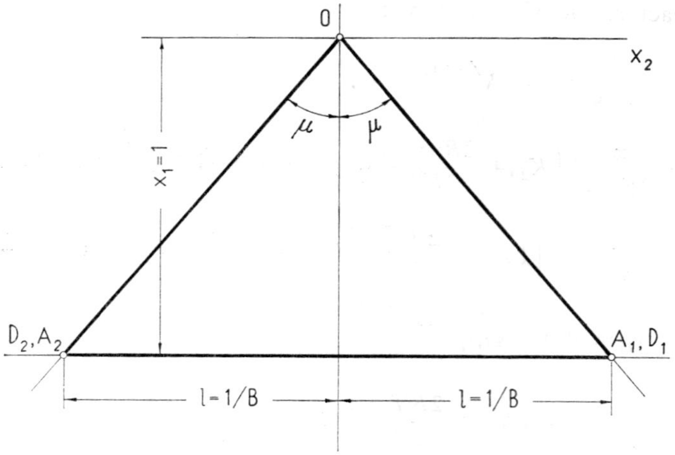

FIG. 4.10

in the case of symmetry, and

$$\mathcal{U}_{n-1} = \sum_{q=0}^{E\left(\frac{n-1}{2}\right)} \frac{BA'_{n,\,2q+1}x^{2q+1}}{\sqrt{1-B^2x^2}} + \sum_{q=0}^{E\left(\frac{n-2}{2}\right)} Q'_{n,\,2q+1}\,x^{2q+1}\cosh^{-1}\sqrt{\frac{1}{B^2x^2}}, \tag{4.67b}$$

in the case of antisymmetry.

In these expressions $E(t)$ is the greatest integer contained in $t\left(t=\frac{n}{2},\ \frac{n-1}{2},\ \frac{n-2}{2}\right)$.

All the above coefficients have been marked by (′) in order not to be confused with those in the original formulae mentioned above. The former coefficients are deduced from the latter but have different and very simple forms. Several examples are given below.

IV. 7.2. Determination of Constants

(a) *Symmetry of incidence.* For the four component wings ($n = 1, 2, 3, 4$), instead of the coefficients $A_{n,2q}$ given by (4.16), (4.17) and (4.18), we deduce the corresponding coefficients by simple passing to limit operations:

$$A'_{10} = -\frac{2w_{00}}{\pi B^2} \tag{4.68}$$

for $n = 1$;

$$A'_{20} = -\frac{2}{3\pi B^2}\left(4w_{10}+\frac{1}{B}w_{01}\right), \tag{4.69a}$$

$$A'_{22} = \frac{2}{3\pi}\left(2w_{10}-\frac{1}{B}w_{01}\right) \tag{4.69b}$$

for $n = 2$;

$$A'_{30} = -\frac{1}{15\pi B^2}\left(44w_{20}+\frac{9}{B}w_{11}+\frac{4}{B^2}w_{02}\right), \tag{4.70a}$$

$$A'_{32} = \frac{1}{15\pi}\left(28w_{20}-\frac{}{B}w_{11}-\frac{12}{B^2}w_{02}\right), \tag{4.70b}$$

$$Q'_{32} = -\frac{w_{11}}{\pi} \tag{4.70c}$$

for $n = 3$;

$$A'_{40} = -\frac{2}{105\pi B^2}\left(160w_{30}+\frac{27}{B}w_{21}+\frac{32}{3B^2}w_{12}+\frac{6}{B^3}w_{03}\right), \tag{4.71a}$$

$$A'_{42} = \frac{2}{105\pi}\left(104w_{30}-\frac{1}{B}w_{21}-\frac{248}{3B^2}w_{12}-\frac{36}{B^3}w_{03}\right), \tag{4.71b}$$

$$A'_{44} = \frac{4B^2}{105\pi}\left(4w_{30}-\frac{10}{B}w_{21}+\frac{68}{3B^2}w_{12}-\frac{3}{B^3}w_{03}\right), \tag{4.71c}$$

$$Q'_{42} = -\frac{2w_{21}}{\pi} \tag{4.71d}$$

for $n = 4$.

(b) *Antisymmetry of incidence*. In an analogous way, for the first four component wings ($n = 1, 2, 3, 4$,) the coefficients $A'_{n,2q+1}$ are obtained by simple passing to limit operations from the coefficients $A_{n,2q+1}$ given by expressions (4.32), (4.33) and (4.34); thus we have:

$$A'_{11} = -\frac{2w_{00}}{\pi B} \tag{4.72}$$

for $n = 1$;

$$A'_{21} = -\frac{4}{3\pi B}\left(w_{10}+\frac{1}{B}w_{01}\right), \tag{4.73a}$$

$$Q'_{21} = -\frac{2w_{10}}{\pi} \tag{4.73b}$$

for $n = 2$;

$$A'_{31} = \frac{4}{15\pi B}\left(2w_{20}-\frac{8}{B}w_{11}-\frac{3}{B^2}w_{02}\right), \tag{4.74a}$$

$$A'_{33} = -\frac{4B}{15\pi}\left(6w_{20}-\frac{4}{B}w_{11}+\frac{1}{B^2}w_{02}\right), \tag{4.74b}$$

$$Q'_{31} = -\frac{4w_{20}}{\pi} \tag{4.74c}$$

for $n = 3$;

$$A'_{41} = \frac{1}{35\pi B}\left(117w_{30}-\frac{272}{3B}w_{21}-\frac{30}{B^2}w_{12}-\frac{16}{B^3}w_{03}\right), \tag{4.75a}$$

$$A'_{43} = -\frac{B}{35\pi}\left(149w_{30}-\frac{176}{3B}w_{21}+\frac{2}{B^2}w_{12}+\frac{48}{B^3}w_{03}\right), \tag{4.75b}$$

$$Q'_{41} = -\frac{6w_{30}}{\pi}, \tag{4.75c}$$

$$Q'_{43} = -\frac{3B^2w_{30}-2w_{12}}{3\pi} \tag{4.75d}$$

for $n = 4$.

IV. 7.3. Aerodynamic Characteristics

The aerodynamic characteristics of wings with sonic leading edges will not be given here, as they are of the same form as those for wings with subsonic edges. Only the constants in the expressions for the axial velocities u differ and, in this respect, we have given the above examples for the sake of comparison. These constants will be introduced in the formulae for the lift, drag and moment coefficients.

As regards the suction forces, these are equal to zero, as may be readily seen from the fact that sonic edges constitute the limit of supersonic edges, where the axial velocity u is zero and consequently the suction force does not exist.

IV. 8. Thin Delta Wing with given Pressure on Surface (Indirect Problem)

IV. 8.1. General Expression for the Downwash

It should first be recalled that the pressure coefficient, given by the relation

$$C_p = -2\frac{u}{U_\infty},$$

is proportional to the axial velocity u. Accordingly, the pressure distribution may be indicated by the axial velocity u, which will be expressed by a superposition of homogeneous polynomials of the form (2.20).

For the component or single wing of nth order, the axial velocity, designated by $u^{(n-1)}$, will be represented by a homogeneous polynomial of order $(n-1)$ which, taking into consideration only the natural symmetry, will be expressed in the following form:

$$\frac{u^{(n-1)}}{x_1^{n-1}} = \sum_{\varkappa=0}^{E\left(\frac{n-1}{2}\right)} u_{n-2\varkappa-1,\, 2\varkappa}\, y^{2\varkappa}. \tag{4.76}$$

Further, applying formula (2.155), in which we put $l_1 = l_2 = l$ and where the ridge OC_1 does not exist ($C'_q = 0$), considering the symmetry and introducing double indices, we obtain the following expression for the corresponding downwash function $\mathcal{W}_{n-1}$:

$$\mathcal{W}_{n-1} = \sum_{q=0}^{n-1} C_{n,\,q}\, x^q \left[\cosh^{-1}\sqrt{\frac{(1+Bl)(1-Bx)}{2B(l-x)}} + (-1)^q \cosh^{-1}\sqrt{\frac{(1+Bl)(1+Bx)}{2B(l+x)}}\right] + \frac{1-(-1)^n}{2} Q_n\, x^{n-1} \cosh^{-1}\sqrt{\frac{1}{B^2x^2}} + \sum_{q=0}^{E\left(\frac{n-2}{2}\right)} D_{n,\,2q}\, x^{2q} \sqrt{\frac{1-B^2x^2}{B^2}}, \quad (\mathcal{W}_{n-1} = w_{n-1} + i w'_{n-1}), \tag{4.77}$$

where $E\left(\frac{n-2}{2}\right)$ is the greatest integer contained in $\frac{n-2}{2}$.

This formula is also valid if the leading edges are supersonic, when we consider the relation

$$C_{n,\,q} \cosh^{-1}\beta = K_{n,\,q} \cos^{-1}\beta, \quad \left(C_{n,\,q}\sqrt{1-B^2l^2} = K_{n,\,q}\sqrt{B^2l^2-1}\right). \tag{4.78}$$

On the other hand, it should be noticed that for $n = 1$, w tends towards infinity at the origin, as may be seen from the above expression (4.77); we have also seen that the wing surface Z tends towards infinity at the origin (2.161). Hence, the case $n = 1$ should be eliminated. Further we consider the component wings of order $n = 2$ and $n = 3$. Applying the relations established previously for determining the constants (§ II. 8.3), we have calculated the corresponding constants:

$$C_{20} = -\frac{2B^2l^2u_{10}}{\pi\sqrt{1-B^2l^2}}, \quad C_{21} = \frac{2u_{10}}{\pi l^2\sqrt{1-B^2l^2}}, \tag{4.79a}$$

$$D_{20} = -\frac{2u_{10}B}{\pi l} \tag{4.79b}$$

for $n = 2$;

$$C_{30} = -\frac{B^2l[(3-2B^2l^2)u_{20}+l^2u_{02}]}{\pi(1-B^2l^2)^{3/2}}, \tag{4.80a}$$

$$C_{31} = \frac{2[B^2u_{20}-(2-3B^2l^2)u_{02}]}{\pi(1-B^2l^2)^{3/2}}, \tag{4.80b}$$

$$C_{32} = \frac{(2-3B^2l^2)u_{20}+(6-9B^2l^2+2B^4l^4)l^2u_{02}}{\pi l^3(1-B^2l^2)^{3/2}}, \tag{4.80c}$$

$$D_{30} = \frac{-B[(1-2B^2l^2)u_{20}-l^2u_{02}]}{\pi l^2(1-B^2l^2)}, \tag{4.80d}$$

$$Q_3 = -\frac{u_{20}+3l^2u_{02}}{\pi l^3} \tag{4.80e}$$

for $n = 3$.

IV. 8.2. Determination of the Equations for the Wing Surface

The surface of the complex wing consists of a superposition of constituent wings, represented by homogeneous polynomials of order $n = 1, 2, \ldots, N$.

The surface equation Z_n of the nth order component wing, with the addition of an arbitrary function of x_2, is given by formula (2.157), which will be applied to expression (4.77); thus we obtain for the case of subsonic edges ($l < 1/B$):

$$\begin{aligned} Z_n = &-\frac{x_2^n}{U_\infty}\int\frac{w_{n-1}\,dy}{y^{n+1}} = -\frac{x_1^n}{U_\infty}y^n\Bigg\{\sum_{q=0}^{n-1}C_{n,q}E_{n,q}(y) \\ &-\sum_{q=0}^{E\left(\frac{n-2}{2}\right)}D_{n,2q}\frac{1}{(n-2q)B}\left[\frac{\sqrt{1-B^2y^2}}{y^{n-2q}}+B^2\Gamma_{n-2q-1}(y)\right] \\ &+Q_n\frac{1-(-1)^n}{2y}\left[\sqrt{1-B^2y^2}-\cosh^{-1}\sqrt{\frac{1}{B^2y^2}}\right]\Bigg\}. \end{aligned} \tag{4.81}$$

The new notations introduced in this formula will be explained in the following formulae:

$$E_{n,q}(y) = \int \frac{1}{y^{n+1-q}}\left[\cosh^{-1}\sqrt{\frac{(1+Bl)(1-By)}{2B(l-y)}}\right.$$

$$\left.+(-1)^q\cosh^{-1}\sqrt{\frac{(1+Bl)(1+By)}{2B(l+y)}}\right]dy$$

$$= -\frac{1}{n-q}\left\{\left[\frac{1}{y^{n-q}}-\frac{1}{l^{n-q}}\right]\cosh^{-1}\sqrt{\frac{(1+Bl)(1-By)}{2B(l-y)}}\right.$$

$$+\left[\frac{(-1)^q}{y^{n-q}}-\frac{(-1)^n}{l^{n-q}}\right]\cosh^{-1}\sqrt{\frac{(1+Bl)(1+By)}{2B(l+y)}}$$

$$\left.-\frac{\sqrt{1-B^2l^2}}{2l^{n-q}}\sum_{t=1}^{n-q}[1-(-1)^{n-t}]\,l^{t-1}\,\Gamma_t(y)\right\}, \tag{4.82a}$$

$$\Gamma_t(y) = \int \frac{dy}{y^t\sqrt{1-B^2y^2}}, \tag{4.82b}$$

$$\Gamma_{2r+1}(y) = \int \frac{dy}{y^{2r+1}\sqrt{1-B^2y^2}} = -\frac{(2r)!}{2^{2r}(r!)^2y^{2r}}\left[(By)^{2r}\cosh^{-1}\sqrt{\frac{1}{B^2y^2}}\right.$$

$$\left.+\sqrt{1-B^2y^2}\sum_{p=0}^{r-1}\frac{2^{2p}(p!)^2}{(2p+1)!}(By)^{2(r-p-1)}\right], \quad (t = 2r+1), \tag{4.82c}$$

$$\Gamma_{2(r+1)}(y) = \int \frac{dy}{y^{2(r+1)}\sqrt{1-B^2y^2}}$$

$$= -\frac{2^{2r}(r!)^2}{(2r+1)!\,y^{2r+1}}\sqrt{1-B^2y^2}\sum_{p=0}^{r}\frac{(2p)!}{2^{2p}(p!)^2}(By)^{2(r-p)},$$

$$(t = 2(r+1)). \tag{4.82d}$$

Inserting these expressions into (4.81), we obtain successively for each of the two wings considered ($n = 2, 3$):

$$Z_2 = \frac{x_1^2}{U_\infty}\left\{\tfrac{1}{2}C_{20}\left[\left(1-\frac{y^2}{l^2}\right)\left(\cosh^{-1}\sqrt{\frac{(1+Bl)(1-By)}{2B(l-y)}}\right.\right.\right.$$

$$\left.\left.+\cosh^{-1}\sqrt{\frac{(1+Bl)(1+By)}{2B(l+y)}}\right)+\frac{y^2}{l^2}\sqrt{1-B^2l^2}\cosh^{-1}\sqrt{\frac{1}{B^2y^2}}\right]$$

$$+C_{21}y\left[\left(1-\frac{y}{l}\right)\cosh^{-1}\sqrt{\frac{(1+Bl)(1-By)}{2B(l-y)}}\right.$$

$$\left.-\left(1+\frac{y}{l}\right)\cosh^{-1}\sqrt{\frac{(1+Bl)(1+By)}{2B(l+y)}}+\frac{y}{l}\sqrt{1-B^2l^2}\cosh^{-1}\sqrt{\frac{1}{B^2y^2}}\right]$$

$$\left.+\frac{D_{20}}{2B}\left[\sqrt{1-B^2y^2}-B^2y^2\cosh^{-1}\sqrt{\frac{1}{B^2y^2}}\right]\right\}, \tag{4.83a}$$

$$
\begin{aligned}
Z_3 = \frac{x_1^3}{U_\infty}\Bigg\{ & \tfrac{1}{3} C_{30}\Bigg[\left(1-\frac{y^3}{l^3}\right)\cosh^{-1}\sqrt{\frac{(1+Bl)(1-By)}{2B(l-y)}} \\
& +\left(1+\frac{y^3}{l^3}\right)\cosh^{-1}\sqrt{\frac{(1+Bl)(1+By)}{2B(l+y)}}+\frac{y^2}{l^2}\sqrt{(1-B^2l^2)(1-B^2y^2)}\Bigg] \\
& +\tfrac{1}{2} C_{31}y\Bigg[\left(1-\frac{y^2}{l^2}\right)\left(\cosh^{-1}\sqrt{\frac{(1+Bl)(1-By)}{2B(l-y)}}\right. \\
& \left.-\cosh^{-1}\sqrt{\frac{(1+Bl)(1+By)}{2B(l+y)}}\right)+\frac{y}{l}\sqrt{(1-B^2l^2)(1-B^2y^2)}\Bigg] \\
& +C_{32}y^2\Bigg[\left(1-\frac{y}{l}\right)\cosh^{-1}\sqrt{\frac{(1+Bl)(1-By)}{2B(l-y)}} \\
& +\left(1+\frac{y}{l}\right)\cosh^{-1}\sqrt{\frac{(1+Bl)(1+By)}{2B(l+y)}}\Bigg]+\frac{D_{30}}{3B}(1-B^2y^2)^{3/2} \\
& -Q_3y^2\left[\sqrt{1-B^2y^2}-\cosh^{-1}\sqrt{\frac{1}{B^2y^2}}\right]\Bigg\}. \qquad (4.83b)
\end{aligned}
$$

From the above relations it may be observed that the wing surface is finite in the vicinity of the edges OA_1 and OA_2 ($y = \pm l$).

However, in the vicinity of the origin ($y = 0$), the wing surface behaves like

$$\underset{y\to 0}{Z_n} = -x_1^n\, y^{n-1}\ln y, \qquad (4.84)$$

from which it results that the height is finite at the centre if $n > 1$ and infinite if $n = 1$.

In order to give an indication of the surface shape, such a surface of the component wing of order $n = 2$ was sketched in Fig. 4.11. In these diagrams we have taken

$$M_\infty = 2,\ Bl = 0{\cdot}8,\ l = 0{\cdot}462,\ u_{10} = 0{\cdot}05\ U_\infty.$$

Case of supersonic edges. It is easy to show that, in the case of supersonic edges, the expression for Z_n remains exactly the same as (4.81), putting, if need be, $K_{n,q}\, F_{n,q}\,(y)$ instead of $C_{n,q}\, E_{n,q},\,(y)$ where $F_{n,q}\,(y)$ is deduced from $E_{n,q}(y)$, by changing $\cosh^{-1}\beta$ and $\sqrt{1-B^2l^2}$ into $\cos^{-1}\beta$ and $\sqrt{B^2l^2-1}$ in its expression (4.82d). The imaginary terms, if they occur, are of course cancelled.

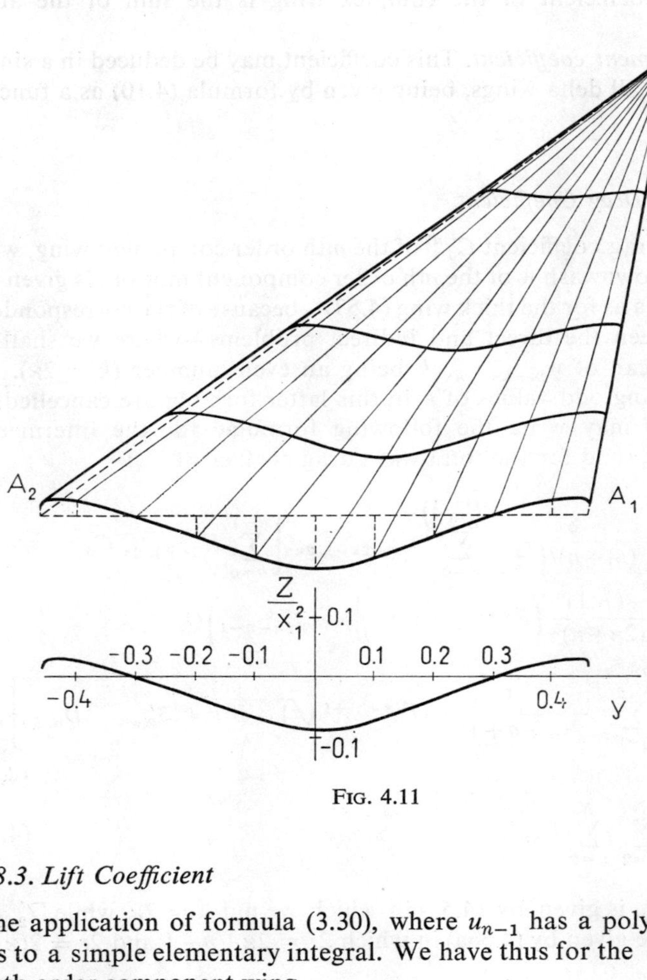

FIG. 4.11

IV. 8.3. Lift Coefficient

The application of formula (3.30), where u_{n-1} has a polynomial form, leads to a simple elementary integral. We have thus for the coefficient of the nth order component wing,

$$C_l^{(n)} = \frac{8}{U_\infty} \sum_{\varkappa=0}^{E\left(\frac{n-1}{2}\right)} \frac{u_{n-2\varkappa-1,\, 2\varkappa} l^{2\varkappa}}{(n+1)(2\varkappa+1)}, \tag{4.85}$$

and for the complex wing we apply (4.7).

For each single or component wing of order $n = 2, 3$, we have

$$C_l^{(2)} = \frac{8}{3} \frac{u_{10}}{U_\infty}, \quad C_l^{(3)} = \frac{2}{3} \frac{3u_{20} + l^2 u_{02}}{U_\infty}. \tag{4.86}$$

The total coefficient of the complex wing is the sum of the above coefficients.

Pitching moment coefficient. This coefficient may be deduced in a similar manner as for all delta wings, being given by formula (4.10) as a function of $C_l^{(n)}$.

IV. 8.4. Wave Drag Coefficient

The wave drag coefficient $C_{dn}^{(m)}$ of the mth order component wing, which is due to the downwash w of the nth order component motion, is given by a similar formula as for the thick wing (4.57)—because of the correspondence existing between the direct and indirect problems—where we shall put $u_{m-k-1,\,k}$ instead of $w_{m-k-1,\,k}$, k being an even number ($k = 2\varkappa$). The terms containing odd values of k in this latter formula are cancelled.

Finally, we may write the following formulae for the intermediate coefficient $C_{dn}^{(m)}$ and for the total wave drag coefficient C_d:

$$C_{dn}^{(m)} = -\frac{8}{(m+n)lU_\infty^2} \sum_{\varkappa=0}^{E\left(\frac{m-1}{2}\right)} u_{m-2\varkappa-1,\,2\varkappa} \left[\sum_{q=0}^{n-1} H'_{2\varkappa+q,\,2\varkappa}\, C_{n,\,q} \right.$$
$$+\frac{1-(-1)^n}{2(2\varkappa+n)} \left(l^{2\varkappa+n} \cosh^{-1} \frac{1}{Bl} + \mathcal{J}'_{2\varkappa+n-1} \right) Q_n$$
$$\left. + \sum_{q=0}^{E\left(\frac{n-2}{2}\right)} \frac{1}{2(\varkappa+q+1)B} \left(l^{2(\varkappa+q)+1} \sqrt{1-B^2l^2} + \mathcal{J}'_{2(\varkappa+q)} \right) D_{n,\,2q} \right], \tag{4.87a}$$

$$C_d = \sum_{m=2}^{N} \sum_{n=2}^{N} C_{dn}^{(m)}, \tag{4.87b}$$

where $H'_{2\varkappa+q,\,2\varkappa}$ is given by (4.55) in which we put $k = 2\varkappa$, while $\mathcal{J}'_{2\varkappa+n-1}$ and $\mathcal{J}'_{2(\varkappa+q)}$ are given by (4.56a) in which $2r = 2\varkappa+n-1$ and $2r = 2(\varkappa+q)$, respectively.

As an example, putting $m = n = 3$ in the above formula, we find

$$C_{d3}^{(3)} = -\frac{4}{3BlU_\infty^2} \left\{ u_{20} \left[C_{30} \left(\sqrt{1-B^2l^2} \sin^{-1} Bl + 2Bl \cosh^{-1} \frac{1+Bl}{2\sqrt{Bl}} \right) \right. \right.$$
$$+ C_{31} \frac{l}{2} \sqrt{1-B^2l^2} \sin^{-1} Bl + C_{32} \frac{1}{3B^2} \left(\frac{1+2B^2l^2}{2} \sqrt{1-B^2l^2} \sin^{-1} Bl \right.$$
$$\left. + 2B^3l^3 \cosh^{-1} \frac{1+Bl}{2\sqrt{Bl}} - \tfrac{1}{2} Bl(1-B^2l^2) \right) + Q_3 \frac{1}{6B^2} \left(\sin^{-1} Bl \right.$$

$$+2B^3l^3 \cosh^{-1}\frac{1}{Bl} - Bl\sqrt{1-B^2l^2}\Big) + D_{30}\frac{1}{2B^2}(\sin^{-1}Bl$$

$$+Bl\sqrt{1-B^2l^2})\Big] + \frac{1}{B^2}u_{02}\Big[\tfrac{1}{3}C_{30}\Big(\frac{1+2B^2l^2}{2}\sqrt{1-B^2l^2}\sin^{-1}Bl$$

$$+2B^3l^3\cosh^{-1}\frac{1+Bl}{2\sqrt{Bl}} - Bl\sqrt{1-B^2l^2}\Big)$$

$$+C_{31}\frac{l}{4}\Big(\frac{1+2B^2l^2}{2}\sqrt{1-B^2l^2}\sin^{-1}Bl - \frac{1-B^2l^2}{2}Bl\Big)$$

$$+C_{32}\frac{1}{5B^2}\Big(\frac{3+4B^2l^2+8B^4l^4}{8}\sqrt{1-B^2l^2}\sin^{-1}Bl$$

$$+2B^5l^5\cosh^{-1}\frac{1+Bl}{2\sqrt{Bl}} - \tfrac{3}{8}Bl(1+B^2l^2-2B^4l^4)\Big)$$

$$+Q_3\frac{1}{40B^2}\Big(3\sin^{-1}Bl + 8B^5l^5\cosh^{-1}\frac{1}{Bl}$$

$$-(3+2B^2l^2)Bl\sqrt{1-B^2l^2}\Big) + D_{30}\frac{1}{8B^2}(\sin^{-1}Bl$$

$$-(1-2B^2l^2)Bl\sqrt{1-B^2l^2})\Big]\Big\}. \tag{4.88}$$

For the total drag coefficient, we put $N = 3$ in (4.87). Apart from the coefficients indicated above, the remaining coefficients may be calculated by means of the integrals given in Appendix I. 4, in which $k = 2\varkappa$ too.

Case of supersonic edges. At the beginning we have shown that formula (4.77) is also valid for the case of supersonic edges and may be put in the usual form by replacing the terms $C_{n,q}\cosh^{-1}\beta$ by $K_{n,q}\cos^{-1}\beta$. In such a case, however, we may also apply the residue method, finding thus the following general expression, which must be similar to (4.87a) if in the latter we take $l > 1/B$. In fact we have to take the real part of this expression. We obtain

$$C_{dn}^{(m)} = -\frac{8}{(m+n)lU_\infty^2}\sum_{\varkappa=0}^{E\left(\frac{m-1}{2}\right)} u_{m-2\varkappa-1,\,2\varkappa}\Big[\sum_{q=0}^{n-1} H_{2\varkappa+q,\,2\varkappa}K_{n,q}$$

$$+\frac{1-(-1)^n}{2(2\varkappa+n)}\mathcal{J}_{2\varkappa+n-1}Q_n + \sum_{q=0}^{E\left(\frac{n-2}{2}\right)}\frac{1}{2(\varkappa+q+1)}\mathcal{J}_{2(\varkappa+q)}D_{n,\,2q}\Big] \tag{4.89}$$

where $H_{2\varkappa+q,\,2\varkappa}$ is given by (4.49) in which we put $k = 2\varkappa$, while $\mathcal{J}_{2\varkappa+n-1}$ and $\mathcal{J}_{2(\varkappa+q)}$ are given by (4.50a) in which $2r = 2\varkappa+n-1$ and $2r = 2(\varkappa+p)$, respectively.

Particularizing these formulae for $m = n = 3$, we have likewise

$$C_{d3}^{(3)} = -\frac{2\pi}{3BlU_\infty^2}\left\{u_{20}\left[\left(K_{30}+\frac{l}{2}K_{31}+\frac{2B^2l^2+1}{2B^2l^2}\frac{l^2}{3}K_{32}\right)\sqrt{B^2l^2-1}\right.\right.$$
$$\left.+\frac{1}{6B^2}Q_3+\frac{1}{2B}D_{30}\right]+u_{02}\frac{1}{2B^2}\left[\left(\frac{2B^2l^2+1}{3}K_{30}\right.\right.$$
$$\left.+\frac{2B^2l^2+1}{4}lK_{31}+\frac{8B^4l^4+4B^2l^2+3}{5B^2l^2}\frac{l^2}{4}K_{32}\right)\sqrt{B^2l^2-1}$$
$$\left.\left.+\frac{3}{20B^2}Q_3+\frac{1}{4B}D_{30}\right]\right\}. \tag{4.90}$$

The above expression is also deduced from (4.88) by putting $l > 1/B$ and taking its real part.

It is to be noticed in this case that, when $\text{Re}\,Bx > 1$, we have

$$\text{Re}\left[\sin^{-1}Bx\right]_{x=y} = \frac{\pi}{2}.$$

The total drag coefficient of the complex wing is given by (4.87b) with $N = 3$. The remaining coefficients occurring in this expression may be calculated by means of the values of the integrals $H_{2\varkappa+q,\,2\varkappa}$ and $\mathcal{J}_t$ given in Appendix I.3, in which we also put $k = 2\varkappa$.

IV. 9. Examples of Delta Wing of Symmetrical Thickness with a given Pressure Distribution on the Surface (Indirect Problem)

IV. 9.1. Determination of the Downwashes

For the mth order component wing we consider the same pressure distribution as in the preceding case, expressed by the axial velocity u_{n-1}(4.76).

Applying again formula (2.153), where the ridges OC_1 and OC_2 do not exist ($C_q = G_q = 0$) and where we put $l_1 = l_2 = l$, taking into account the symmetry, we obtain the following expression for the downwash function $\mathcal{W}_{n-1}$:

$$\mathcal{W}_{n-1} = \sum_{q=0}^{E\left(\frac{n}{2}\right)}\frac{A_{n,\,2q}\,x^{2q}}{\sqrt{l^2-x^2}}+\frac{1-(-1)^n}{2}Q_n\,x^{n-1}\cosh^{-1}\sqrt{\frac{l^2}{x^2}}\,,$$
$$(\mathcal{W}_{n-1} = w_{n-1}+iw'_{n-1}), \tag{4.91}$$

where $E\left(\frac{n}{2}\right)$ is the greatest integer contained in $\frac{n}{2}$.

This expression will be particularized only for a single wing of order $n = 2$, since we are treating only an example of calculation, the problem being generally very laborious and, on the other hand, not very important for practical applications. The case $n = 1$ has been avoided since, owing to the singularity at the centre, it presents a geometrical incompatibility.

Let us now observe that the logarithmic term at the origin is equal to zero ($Q_2 = 0$) for $n = 2$, a fact that greatly simplifies the problem.

Thus for the case considered we have

$$\mathcal{W}_1 = \frac{A_{20}+A_{22}x^2}{\sqrt{l^2-x^2}}. \tag{4.92}$$

Applying further the relations established earlier for determining the constants (§ II. 8.3), it is to be noted that in such a case one of the constants remains arbitrary. In order to determine this constant, an additional condition may be imposed, which will be assumed to be of the general form

$$-k^*A_{20}+A_{22}l^2 = 0 \tag{4.93}$$

where k^* is arbitrary.

If, for example, we consider that the tangent is zero at the leading edges of the wing, then we have to take $k^* = -1$ in the above relation.

In this latter case, the following values are obtained for the constants:

$$A_{20} = \frac{6(1-B^2l^2)^2u_{10}}{3B^2l^2(1+2B^2l^2)K(k)-3(4-B^2l^2)E(k)}, \tag{4.94a}$$

$$A_{22} = -\frac{6(1-B^2l^2)^2u_{10}}{l^2[3B^2l^2(1+2B^2l^2)K(k)-3(4-B^2l^2)E(k)]}. \tag{4.94b}$$

In these formulae $K(k)$ and $E(k)$ are complete elliptical integrals of the first and second kind, having as modulus

$$k = \sqrt{1-B^2l^2}. \tag{4.95}$$

IV. 9.2. Determination of the Equation of the Wing Surface

We consider the general expression for the downwash given by (4.92); applying formula (2.157) we find the equation of the surface of the wing of symmetrical thickness which, with the addition of an arbitrary function of x_2, may be written as

$$Z_n = -\frac{x_1^n}{U_\infty}y^n\Bigg\{\sum_{q=0}^{E\left(\frac{n}{2}\right)} A_{n,2q}\Gamma'_{n-2q+1}(y) + Q_n\frac{1-(-1)^n}{2y}\left(\frac{\sqrt{l^2-y^2}}{l}-\cosh^{-1}\sqrt{\frac{l^2}{y^2}}\right)\Bigg\}. \tag{4.96}$$

In this formula we have employed the following notations, similar to those of the preceding case:

If $n-2q+1 = 2r+1$, that is, if n is an even number, then

$$\Gamma'_{2r+1}(y) = -\frac{(2r)!}{2^{2r}(r!)^2 l^{2(r+1)}}\left[l\cosh^{-1}\sqrt{\frac{l^2}{y^2}} + \sqrt{l^2-y^2}\sum_{p=0}^{r-1}\frac{2^{2p}(p!)^2}{(2p+1)!}\left(\frac{l}{y}\right)^{2(p+1)}\right]. \tag{4.97}$$

If $n-2q+1 = 2(r+1)$, that is, if n is an odd number, then

$$\Gamma'_{2(r+1)}(y) = -\frac{2^{2r}(r!)^2}{(2r+1)!\,l^{2r+3}}\sqrt{l^2-y^2}\sum_{p=0}^{r}\frac{(2p)!}{2^{2p}(p!)^2}\left(\frac{l}{y}\right)^{2p+1} \tag{4.98}$$

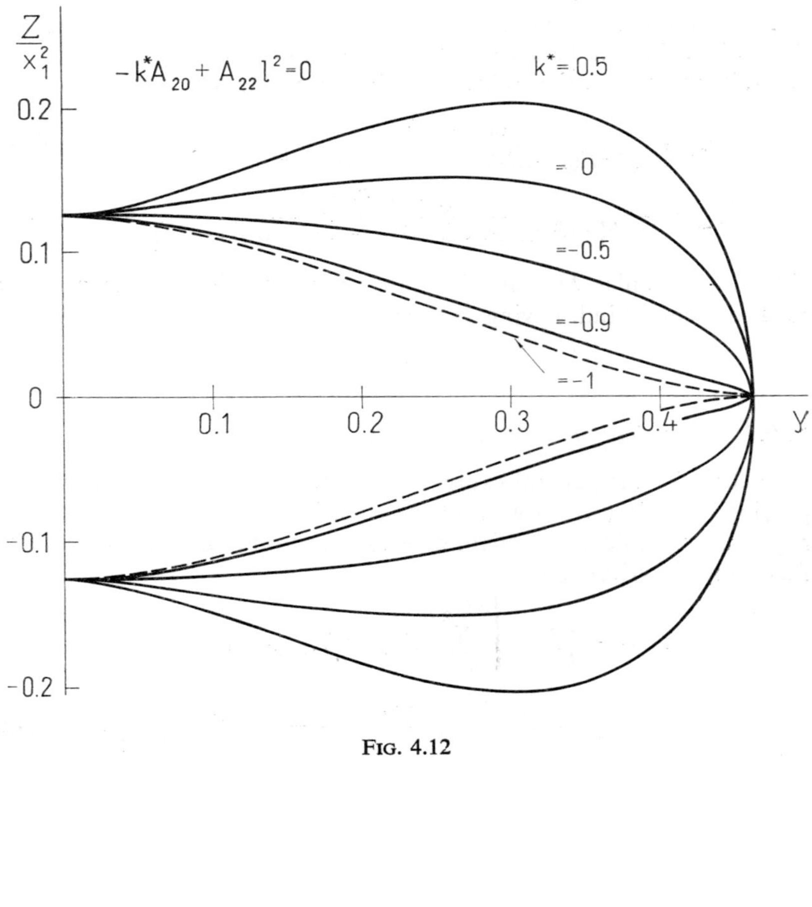

FIG. 4.12

Applying these formulae for $n = 2$, we obtain the following surface:

$$Z_2 = \frac{x_1^2}{lU_\infty}\left[A_{20}\frac{1}{2l^2}\left(l\sqrt{l^2-y^2}+y^2\cosh^{-1}\sqrt{\frac{l^2}{y^2}}\right)\right.$$
$$\left.+A_{22}y^2\cosh^{-1}\sqrt{\frac{l^2}{y^2}}\right]. \tag{4.99}$$

It is seen that the surface Z_2 is finite at the origin and equal to zero on the leading edges OA_1 and OA_2 ($y = \pm l$, $Z_2 = 0$). Figure 4.12 gives the forms of such surfaces of order $n = 2$ with different k^*, for $M_\infty = 2$, $Bl = 0{\cdot}8$, $l = 0{\cdot}462$, $u_{10} = 0{\cdot}05\, U_\infty$.

IV. 9.3. Wave Drag Coefficient

The wave drag coefficient $C_{dn}^{(m)}$ of the mth order component wing, due to the downwash w of the nth order component wing, as well as the total coefficient C_d of the complex wing, are given by the following expressions obtained by applying formula (3.42a):

$$C_{dn}^{(m)} = -\frac{8}{(m+n)lU_\infty^2}\sum_{\varkappa=0}^{E\left(\frac{m-1}{2}\right)} u_{m-2\varkappa-1,\,2\varkappa}\left[\sum_{q=0}^{E\left(\frac{n}{2}\right)} I_{2(\varkappa+q)}\, A_{n,\,2q}\right.$$
$$\left.+\frac{1-(-1)^n}{2(2\varkappa+n)}\, I_{2\varkappa+n-1}\, Q_n\right], \tag{4.100a}$$

$$C_d = \sum_{m=2}^{N}\sum_{n=2}^{N} C_{dn}^{(m)}, \tag{4.100b}$$

where $I_{2(\varkappa+q)}$ and $I_{2\varkappa+n-1}$ are given by (4.25a) in which $2r = 2(\varkappa+q)$ and $2r = 2\varkappa+n-1$, respectively.

Applying this formula for the single wing of order $n = m = 2$, we obtain

$$C_{d2}^{(2)} = -\frac{\pi u_{10}}{lU_\infty^2}\left[A_{20}+\tfrac{1}{2}\, l^2 A_{22}\right]. \tag{4.101}$$

IV. 10. Mixed Problem applied to a Thin Delta Wing in a True Conical Motion ($n = 1$)

It should be noted from the beginning that to treat a single or component wing of nth order implies very laborious calculations; therefore, our considerations will be confined to a simple case, i.e. that of true conical motion ($n = 1$).

IV. 10.1. Determination of the Velocities u and w

Let us consider a thin delta wing separated by two ridges OT_1 and OT_2 into three parts (Fig. 4.13): the central part OT_1T_2 having constant incidence and consequently constant downwash w ($w = w_0$), and the two lateral parts OA_1T_1 and OA_2T_2 having constant pressure and consequently constant u ($u = u_0$).

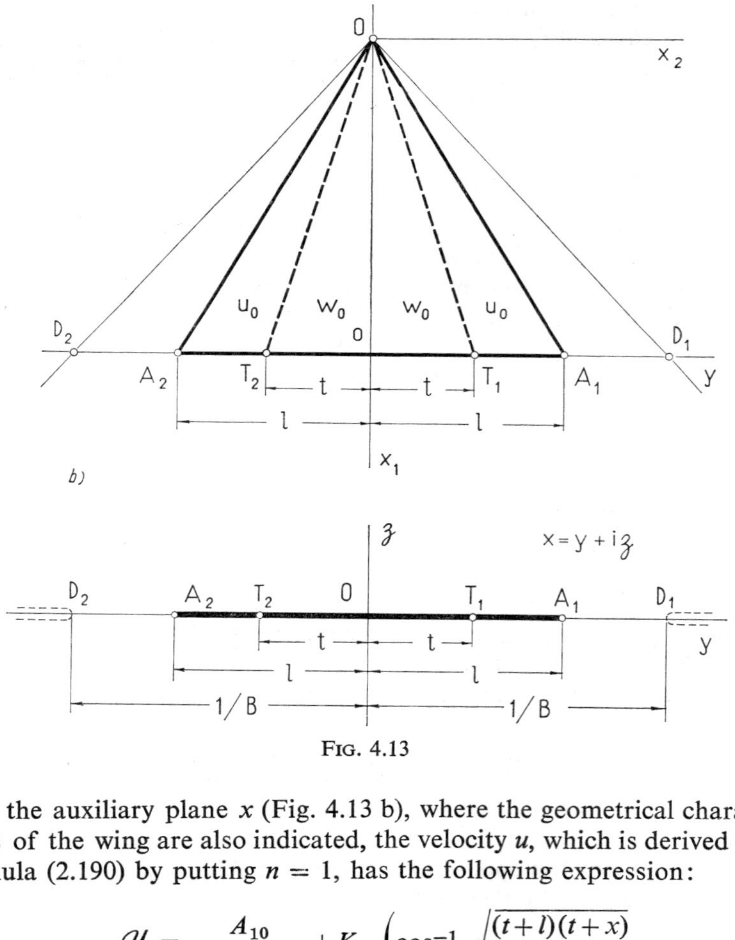

FIG. 4.13

In the auxiliary plane x (Fig. 4.13 b), where the geometrical characteristics of the wing are also indicated, the velocity u, which is derived from formula (2.190) by putting $n = 1$, has the following expression:

$$\mathcal{U} = \frac{A_{10}}{\sqrt{t^2 - x^2}} + K_{10}\left(\cos^{-1}\sqrt{\frac{(t+l)(t+x)}{2t(l+x)}}\right.$$

$$\left. + \cos^{-1}\sqrt{\frac{(t+l)(t-x)}{2t(l-x)}}\right) \tag{4.102}$$

which is valid on the whole wing and in which the constants, determined in the ordinary way, have the values given below:

$$K_{10} = \frac{2u_0}{\pi} \tag{4.103a}$$

$$A_{10} = -\frac{w_0 t^2}{E(k)} + \frac{2u_0 t^2 \sqrt{l^2-t^2}}{\pi E(k)} \left\{ -\frac{1-B^2t^2}{l^2-t^2} K(k) + \frac{2t(1-B^2l^2)}{l(1+Bt)} \left[\Pi\left(-\frac{(l-t)k_1}{l+t},\ k_1\right) - \Pi\left(-\frac{(l+t)k_1}{l-t},\ k_1\right) \right] \right\} \tag{4.103b}$$

where $K(k)$, $E(k)$ and $\Pi(\alpha, k_1)$ are complete elliptic integrals of the first, second and third kind, having respectively the modulus

$$k = \sqrt{1-B^2t^2}; \quad k_1 = \frac{1-Bt}{1+Bt}. \tag{4.104}$$

For determining the downwash w, we proceed from the compatibility relation (2.25),

$$d\mathcal{W} = -\frac{i\sqrt{1-B^2x^2}}{x}\, d\mathcal{U}. \tag{4.105}$$

To obtain a complex general solution is, however, more laborious as regards practical applications; therefore, we give the expression for w only on the portion T_1A_1 (A_2T_2, respectively) which is of the form

$$w = \frac{A_{10}}{2t^2(1+Bt)} \{4Bt(1+2Bt)F(\varphi,k_1) + (1+Bt)^2[2E(\varphi,k_1) + \cot\varphi\sqrt{1-k_1^2\sin^2\varphi}]\} - \frac{2K_{10}\sqrt{l^2-t^2}}{(1+Bt)(l^2-t^2)} \left\{ -(1-B^2t^2)F(\varphi,k_1) + \frac{t(1-B^2l^2)}{l} \left[\Pi\left(\varphi, -\frac{(l-t)k_1}{l+t}, k_1\right) - \Pi\left(\varphi, -\frac{(l+t)k_1}{l-t}, k_1\right) \right] \right\} \tag{4.106}$$

in which $F(\varphi, k_1)$, $E(\varphi, k_1)$ and $\Pi(\varphi, \alpha, k_1)$ are elliptic integrals of the first, second and third kind, of modulus k_1 given in the above formula (4.104) and the argument φ is related to y according to the relation

$$\varphi = \sin^{-1}\sqrt{\frac{(y-t)(1+Bt)}{(y+t)(1-Bt)}}, \quad y = -\frac{t(k_1\sin^2\varphi+1)}{k_1\sin^2\varphi-1}. \tag{4.107}$$

IV. 10.2. Surface of the Wing

On the portion T_1T_2, the wing consists of a flat plate; on the portion T_1A_1 (as well as on the portion T_2A_2), we apply the general formula (2.157) which in our case can be written in the following form

$$Z = -\frac{x_2}{U_\infty}\int \frac{w}{y^2}\,dy + f(x_2) = -\frac{x_1}{U_\infty}\left[w + y\int \frac{dw}{y}\right] + f(x_2). \qquad (4.108)$$

Applying this formula we obtain the following expression for Z,

$$\begin{aligned}
Z = &-\frac{x_1}{U_\infty}\Bigg(\frac{A_{10}}{2t^2(1+Bt)}\Bigg\{4t(1+Bt)y\tan^{-1}\frac{2\tan\varphi}{(1+Bt)\sqrt{1-k_1^2\sin^2\varphi}}\\
&+4Bt[1+2Bt+2ty(1+Bt)]F(\varphi,k_1)+(1+Bt)^2(1-2ty)(2F(\varphi,k_1)\\
&+\cot\varphi\sqrt{1-k_1^2\sin^2\varphi})\Bigg\}+\frac{K_{10}\sqrt{l^2-t^2}}{2l^2t(1+Bt)(l^2-t^2)}\\
&\times\Bigg\{2y(1+Bt)(l^2-t^2)\tan^{-1}\frac{2\tan\varphi}{(1+Bt)\sqrt{1-k_1^2\sin^2\varphi}}\\
&+4t[(1-B^2t^2)l^2-(1-B^2l^2)ty]\,F(\varphi,k_1)\\
&+4t^2(1-B^2l^2)(y-l)\Bigg[\Pi\left(\varphi,-\frac{(l+t)k_1}{l-t},k_1\right)\\
&+\Pi\left(\varphi,-\frac{(l-t)k_1}{l+t},k_1\right)\Bigg]\Bigg\}\Bigg)+f(x_2). \qquad (4.109)
\end{aligned}$$

CHAPTER V

YAWED DELTA WINGS. DELTA WINGS WITH GENERAL VARIATION OF INCIDENCE

THE wings dealt with in the preceding chapter have the shape of an isosceles triangle with the axis in the direction of the free stream and the base normal to it.

The present chapter also treats delta wings, differing from the preceding ones by the position of their axis with respect to the velocity direction (yawed delta wings). For yawed delta wings the flows become asymmetrical and the axial velocities u have forms differing from the former ones.

This chapter closes with an examination of delta wings whose incidence or slope varies according to a homogeneous function instead of a homogeneous polynomial, the field of application being thus broadened.

V. 1. Study of Yawed Delta Wings

V. 1.1. Preliminary Considerations

Let us consider any delta wing referred to a system of axes $O\xi_1\xi_2\xi_3$, having height h and base $2b$ (Fig. 5.1). Let it be assumed that its axis makes an angle δ with the direction of the stream parallel to the axis Ox_1 of the current system $Ox_1x_2x_3$ so far considered. We say that the wing is *yawed*, having the angle of yaw δ.

Any one point P of coordinates ξ_1, ξ_2 with respect to the initial system $O\xi_1\xi_2$, will have as new coordinates x_1, x_2, defined by the following formulae:

$$x_1 = \xi_1 \cos\delta + \xi_2 \sin\delta, \quad x_2 = -\xi_1 \sin\delta + \xi_2 \cos\delta. \tag{5.1}$$

Conversely, the initial coordinates ξ_1, ξ_2 with respect to the new coordinates, will be given by the expression

$$\xi_1 = x_1 \cos\delta - x_2 \sin\delta, \quad \xi_2 = x_1 \sin\delta + x_2 \cos\delta. \tag{5.2}$$

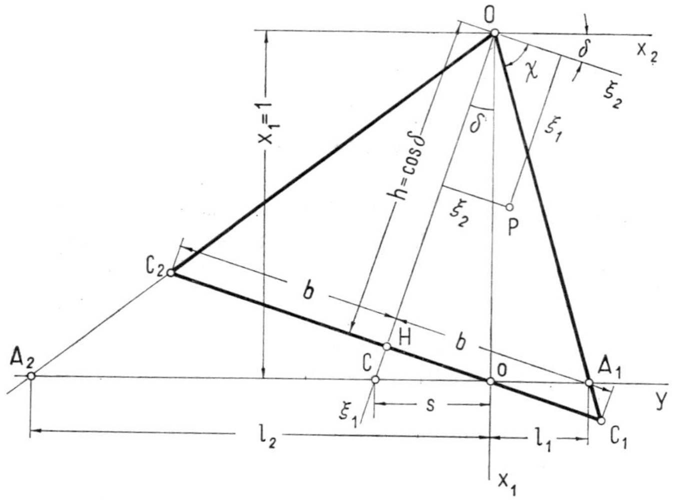

FIG. 5.1

Further, let us write the equation of the surface of the delta wing considered, which is put in the form

$$Z = Z(\xi_1, \xi_2); \tag{5.3}$$

if ξ_1, ξ_2 are replaced by their expressions given in (5.2), the equation of the same wing is found as a function of the new coordinates x_1, x_2.

In order to calculate the slope τ of the wing surface at any point, as it has been defined in a previous chapter (2.5), we may proceed from (5.3) and find successively

$$\tau = -\frac{\partial Z}{\partial x_1} = -\frac{\partial Z}{\partial \xi_1}\frac{\partial \xi_1}{\partial x_1} - \frac{\partial Z}{\partial \xi_2}\frac{\partial \xi_2}{\partial x_1} = \tau' \cos\delta + \tau'' \sin\delta \tag{5.4}$$

where

$$\tau' = -\frac{\partial Z}{\partial \xi_1}, \quad \tau'' = -\frac{\partial Z}{\partial \xi_2} \tag{5.5}$$

are the initial wing slopes (for $\delta = 0$) at the same point P, the former in the direction $O\xi_1$, and the latter in the direction $O\xi_2$.

In its general form (5.4), the slope τ consists, in the new position of the yawed wing, both of the longitudinal slope τ' and the transverse one τ''; the respective downwashes w' and w'', given by

$$w' = -\tau' U_\infty, \quad w'' = -\tau'' U_\infty, \tag{5.6}$$

will enter the formula for the total velocity w which refers to the new boundary conditions of the yawed wing,

$$w = -\tau U_\infty = w' \cos\delta + w'' \sin\delta = w(\xi_1, \xi_2). \tag{5.7}$$

Replacing ξ_1 and ξ_2 by their expressions given in (5.2), we obtain w as a function of x_1 and x_2, which is no longer symmetrical with respect to the Ox_1 axis. Indeed, for the component wing of order n and for the complex wing, we find the same general expressions which have been established previously (3.3),

$$w = \sum_{n=1}^{N} w^{(n-1)} = \sum_{n=1}^{N} x_1^{n-1} w_{n-1}, \tag{5.8}$$

where

$$w_{n-1} = \sum_{k=0}^{n-1} w_{n-k-1,\,k}\, y^k = \mathrm{Re} \sum_{k=0}^{n-1} w_{n-k-1,\,k}\, x^k = \mathrm{Re}\, \mathcal{W}_{n-1}. \tag{5.9}$$

In order to make use of the results obtained previously as regards the axial velocity u, we adopt the same physical plane as before, that is the plane normal to Ox_1, at a distance $x_1 = 1$. The intersections of the axis Oy with the edges OC_1 and OC_2 yield l_1 and l_2. Examining (Fig. 5.1) we may write

$$l_1 = \frac{b-\sin\delta}{\cos\delta + b\tan\delta}, \quad l_2 = \frac{b+\sin\delta}{\cos\delta - b\tan\delta}, \tag{5.10a}$$

$$h = \cos\delta, \quad S = bh = b\cos\delta \tag{5.10b}$$

where S stands for the wing surface.

Note on the wing position with respect to the Mach cone. Examining Fig. 5.2 we observe that the leading edges of the wing may have the following three positions with respect to the Mach cone:

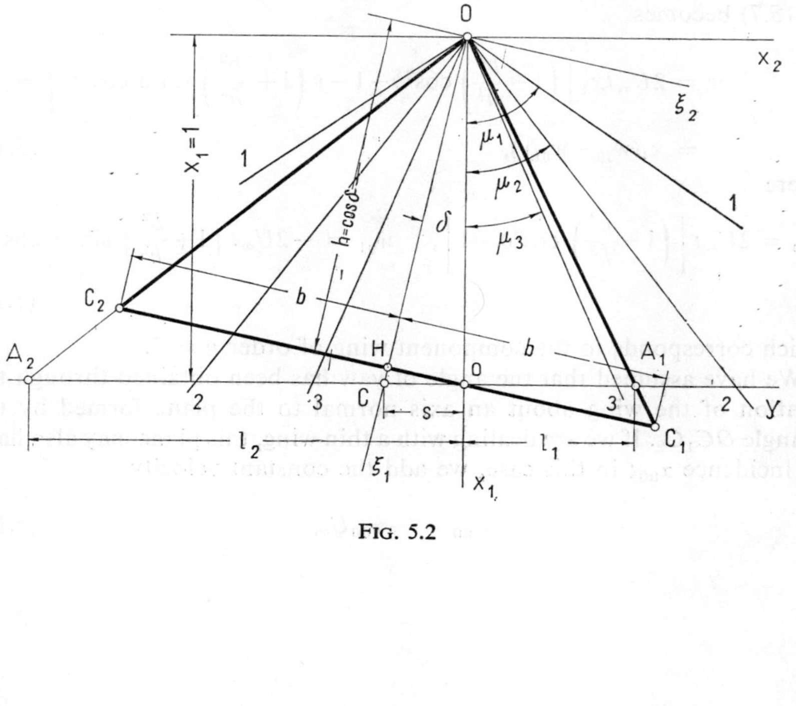

FIG. 5.2

1. If the angle of the cone is μ_1, both leading edges are subsonic;
2. If the angle of the cone is μ_2, the edge OA_1 is subsonic and OA_2 is supersonic;
3. If the angle of the cone is μ_3, both leading edges are supersonic.

Each of these positions will be analysed below.

V. 1.2. Examples

THIN WING. A thin wing may consist of one single wing of order n or of a system of more components of different orders, thus constituting a complex wing. For the sake of simplicity, let us consider a thin wing (Fig. 5.1), having a natural symmetry of incidence, for which the surface equation is given by

$$Z = t\left(\frac{b^2}{h^2}\xi_1^2-\xi_2^2\right). \tag{5.11}$$

t being proportional to the maximum camber of the thin wing; it follows readily that

$$w' = \frac{\partial Z}{\partial \xi_1}U_\infty = 2U_\infty t\frac{b^2}{h^2}\xi_1, \quad w'' = \frac{\partial Z}{\partial \xi_2}U = -2U_\infty t\xi_2. \tag{5.12}$$

After performing the necessary substitutions, the total downwash given by (5.7) becomes

$$w = 2U_\infty tx_1\left[\left(1+\frac{b^2}{h^2}\right)\cos^2\delta-1-y\left(1+\frac{b^2}{h^2}\right)\sin\delta\cos\delta\right] =$$

$$= x_1(w_{10}+w_{01}y), \tag{5.13}$$

where

$$w_{10} = 2U_\infty t\left[\left(1+\frac{b^2}{h^2}\right)\cos^2\delta-1\right], \quad w_{01} = -2U_\infty t\left(1+\frac{b^2}{h^2}\right)\sin\delta\cos\delta, \tag{5.14}$$

which corresponds to the component wing of order $n = 2$.

We have assumed that the angle of yaw has been obtained through the rotation of the wing about an axis normal to the plane formed by the triangle OC_1C_2. If we are dealing with a thin wing, this plane may also have an incidence α_{00}; in this case, we add the constant velocity

$$w_{00} = -\alpha_{00}U_\infty \tag{5.15}$$

to (5.14), a fact that will add in its turn another conical motion ($n = 1$) to the motion considered above. Therefore, the thin complex wing is constituted of two component wings ($n = 1, 2$).

THICK WING. Since a wing of symmetrical thickness should be closed at the leading edges and at the trailing edge, this wing cannot consist only of a single wing of order n, but of more components of different orders which satisfy this condition. That is the reason why the example given below represents such a complex wing of symmetrical thickness, considering of course the simplest cases under the assumption of natural and forced symmetry.

Therefore, let us consider the following equation of the surface of the wing

$$Z = t(h-\xi_1)\left(\frac{b}{h}\xi_1 \mp \xi_2\right) \tag{5.16}$$

from which we deduce

$$w' = \frac{\partial Z}{\partial \xi_1} U_\infty = U_\infty t\left(b - 2\frac{b}{h}\xi_1 \pm \xi_2\right). \tag{5.17a}$$

$$w'' = \frac{\partial Z}{\partial \xi_2} U_\infty = \mp U_\infty t(h-\xi_1), \tag{5.17b}$$

the upper signs being valid for $\xi_2 > 0$, and the lower ones for $\xi_2 < 0$. It is hence seen that there is a separation ridge at the centre, as a result of forced symmetry, which now becomes the ridge OH with respect to the system Ox_1x_2 (Fig. 5.1). The total downwash becomes

$$w = U_\infty t\left\{(b\cos\delta \mp h\sin\delta) + x_1\left[\pm\sin 2\delta - \frac{b}{h}(1+\cos 2\delta) + y\left(\frac{b}{h}\sin 2\delta \pm \cos 2\delta\right)\right]\right\} = w_{00} + x_1(w_{10} + w_{01}y) \tag{5.18}$$

whence

$$w_{00} = U_\infty t(b\cos\delta \mp h\sin\delta) \tag{5.19}$$

results for $n = 1$, and

$$w_{10} = U_\infty t\left[\pm\sin 2\delta - \frac{b}{h}(1+\cos 2\delta)\right], \tag{5.20a}$$

$$w_{01} = U_\infty t\left(\pm\cos 2\delta + \frac{b}{h}\sin 2\delta\right), \tag{5.20b}$$

for $n = 2$.

In the above expressions (5.19) and (5.20), the upper sign is valid for the side OHC_1 on the right of the ridge, and the lower sign is valid for the side OHC_2 on the left of the ridge.

Note. The above examples have been given in order to be eventually considered for the complete study of a yawed complex wing. The computations being very laborious, we restricted ourselves to giving the simplest examples by taking $n = 1, 2$, these examples being, however, complete for illustrating the method.

V. 1.3. Specific Forms of the General Formulae for Calculating the Aerodynamic Coefficients

The general formulae for calculating the lift and wave drag coefficients, as well as the moment coefficients, have been given in Chapter III. When the residue method is applicable (at the point $y = \cot \delta$), these coefficients may be put in the following forms:

Lift coefficient, for which formula (3.31) has been applied:

$$C_l^{(n)} = -\operatorname{Im} \frac{4\pi(-\cot \delta)^{n+1}}{(n+1)!SU_\infty} \left[\frac{d^n}{dx^n} (\mathcal{U}_{n-1}) \right]_{x = \cot \delta} . \tag{5.21}$$

Wave drag coefficient, obtained by applying formula (3.42b), where $\mathcal{W}_{m-1}$ is defined by (5.9), replacing n by m in the latter formula,

$$C_{dm}^{(n)} = \operatorname{Im} \frac{4\pi(-\cot \delta)^{m+n}}{(m+n)!SU_\infty^2} \left[\frac{d^{m+n-1}}{dx^{m+n-1}} (\mathcal{W}_{m-1}\mathcal{U}_{n-1}) \right]_{x = \cot \delta} . \tag{5.22}$$

Rolling moment coefficient, obtained by applying formula (3.47b), where the true height of the delta wing may be taken as reference length ($h = \cos \delta$),

$$\mu_1^{(n)} = -\operatorname{Im} \frac{4\pi(-\cot \delta)^{n+2}}{(n+2)!ShU_\infty} \left[\frac{d^{n+1}}{dx^{n+1}} (x\mathcal{U}_{n-1}) \right]_{x = \cot \delta} . \tag{5.23}$$

Pitching moment coefficient, obtained by applying formula (3.55b),

$$\mu_2^{(n)} = \operatorname{Im} \frac{4\pi(-\cot \delta)^{n+2}}{(n+2)!ShU_\infty} \left[\frac{d^{n+1}}{dx^{n+1}} \mathcal{U}_{n-1} \right]_{x = \cot \delta} . \tag{5.24}$$

Yawing moment coefficient, obtained from (3.63b),

$$\mu_{3m}^{(n)} = -\operatorname{Im} \frac{4\pi(-\cot \delta)^{m+n+1}}{(m+n+1)!ShU_\infty^2} \left[\frac{d^{m+n}}{dx^{m+n}} (x\mathcal{U}_{n-1}\mathcal{W}_{m-1}) \right]_{x = \cot \delta} . \tag{5.25}$$

Note on the rolling and pitching moments. With respect to the wing axes $O\xi_1$ and $O\xi_2$, the rolling and pitching moment coefficients are as follows

$$\mu_1' = \mu_1 \cos\delta - \mu_2 \sin\delta, \quad \mu_2' = \mu_1 \sin\delta + \mu_2 \cos\delta, \tag{5.26}$$

respectively.

Suction effect

If one or both leading edges are subsonic, then we have to take further into account the following:

Suction coefficients, where formulae (3.103) and (3.105) have been applied:

$$\sigma_{1m}^{(n)} = -\frac{2\pi\sqrt{1-B^2l_1^2}\,a_1^{m+n}}{(m+n)l_1(l_1+l_2)SU_\infty^2}\left[\left(\sum_{k=0}^{m} A_{m,k}l_1^k\right)\left(\sum_{q=0}^{n} A_{n,q}l_1^q\right)\right], \tag{5.27a}$$

$$\sigma_{2m}^{(n)} = -\frac{2\pi\sqrt{1-B^2l_2^2}\,a_2^{m+n}}{(m+n)l_2(l_1+l_2)SU_\infty^2}\left[\left(\sum_{k=0}^{m} A_{m,k}(-l_2)^k\right)\left(\sum_{q=0}^{n} A_{n,q}(-l_2)^q\right)\right], \tag{5.27b}$$

for wings with both subsonic leading edges, and

$$\sigma_{1m}^{(n)} = -\frac{2\pi\sqrt{1-B^2l_1^2}(1+Bl_1)}{(m+n)Bl_1SU_\infty^2}\left[\left(\sum_{k=0}^{m-1} A_{m,k}\,l_1^k\right)\left(\sum_{q=0}^{n-1} A_{n,q}\,l_1^q\right)\right]a_1^{m+n}, \tag{5.28}$$

for wings with one subsonic edge ($l_1 < 1/B$) and the other supersonic ($l_2 > 1/B$).

In the above formulae we have denoted by a_1 and a_2 the projections of the leading edges OC_1 and OC_2 on the Ox_1 axis,

$$a_1 = h\cos\delta + b\sin\delta, \quad a_2 = h\cos\delta - b\sin\delta. \tag{5.29}$$

Yawing moment coefficients due to suction, for which formulae (3.114a) and (3.115a) have been applied:

$$\gamma_{1m}^{(n)} = -\frac{m+n}{m+n+1}\,\frac{(1+l_1^2)a_1}{hl_1}\,\sigma_{1m}^{(n)}, \tag{5.30a}$$

$$\gamma_{2m}^{(n)} = \frac{m+n}{m+n+1}\,\frac{(1+l_2^2)a_2}{hl_2}\,\sigma_{2m}^{(n)}. \tag{5.30b}$$

V. 2. Thin Yawed Delta Wing with Subsonic Edges

V. 2.1. Expressions for the Axial Velocities

Generally, the initial delta wing has both natural and forced symmetry of incidence; it follows that the axial disturbance velocities u will contain the contribution of a ridge OC. For the sake of simplicity we shall deal only with the *natural symmetry of incidence.*

In the case of a yawed wing, the motions are asymmetrical and formula (2.57) may be applied, from which the general expression of the respective axial velocity function $\mathcal{U}_{n-1} = u_{n-1} + iu'_{n-1}$ is deduced:

$$\mathcal{U}_{n-1} = \sum_{q=0}^{n} \frac{A_{n,\,q}\, x^q}{\sqrt{(l_1 - x)(l_2 + x)}}\,. \tag{5.31}$$

In this case, considering $n = 1$ and $n = 2$, the coefficients w_{00}, w_{10}, w_{01} of the downwash are given by (5.15) and (5.14), deduced, of course, by proceeding from a perfectly designed wing (5.11).

The corresponding constants $A_{n,q}$ are given by the following relations:

$$A_{10} = -\frac{w_{00} l_1 l_2 \sqrt{(1+Bl_1)(1+Bl_2)}}{(1+Bl_1)(1+Bl_2)E(k) - B(l_1+l_2)K(k)}, \quad A_{11} = \frac{l_1 - l_2}{2l_1 l_2} A_{10}, \tag{5.32a}$$

$$k = \sqrt{\frac{(1-Bl_1)(1-Bl_2)}{(1+Bl_1)(1+Bl_2)}}, \tag{5.32b}$$

for $n = 1$, where $K(k)$ and $E(k)$ are complete elliptical integrals of the first and second kind having the modulus k given by relations (5.32b);

$$A_{20} = \frac{1}{\Delta_2} \begin{vmatrix} w_{10} & a_{11} & a_{21} \\ -2w_{01} & a_{12} & a_{22} \\ 0 & a_{13} & a_{23} \end{vmatrix}, \tag{5.33a}$$

$$A_{21} = \frac{1}{\Delta_2} \begin{vmatrix} a_{01} & w_{10} & a_{21} \\ a_{02} & -2w_{01} & a_{22} \\ a_{03} & 0 & a_{23} \end{vmatrix}, \tag{5.33b}$$

$$A_{22} = \frac{1}{\Delta_2} \begin{vmatrix} a_{01} & a_{11} & w_{10} \\ a_{02} & a_{12} & -2w_{01} \\ a_{03} & a_{13} & 0 \end{vmatrix}, \tag{5.33c}$$

for $n = 2$, where the determinant Δ_2 is given by

$$\Delta_2 = \begin{vmatrix} a_{01} & a_{11} & a_{21} \\ a_{02} & a_{12} & a_{22} \\ a_{03} & a_{13} & a_{23} \end{vmatrix}, \tag{5.34}$$

and the coefficients a_{ij} are given by (2.117), (2.118) and (2.119).

V. 2.2. Calculation of Lift, Wave Drag and Moment Coefficients

Applying the above general formulae established in (§ V. 1.3), we find successively:

Lift coefficient. For the component wing of order n,

$$C_l^{(n)} = \frac{4}{(n+1)SU_\infty} \sum_{q=0}^{n} A_{n,q}(\cot\delta)^q \sum_{r=0}^{q} (-1)^r \frac{q!}{r!(q-r)!} \mathcal{J}_{n+1-r} \tag{5.35}$$

and for $n = 1, 2$, respectively

$$C_l^{(1)} = \frac{2}{SU_\infty}[A_{10}\mathcal{J}_2 + A_{11}\cot\delta(\mathcal{J}_2 - \mathcal{J}_1)], \tag{5.36a}$$

$$C_l^{(2)} = \frac{4}{3SU_\infty}[A_{20}\mathcal{J}_3 + A_{21}\cot\delta(\mathcal{J}_3 - \mathcal{J}_2) + A_{22}\cot^2\delta(\mathcal{J}_3 - 2\mathcal{J}_2 + \mathcal{J}_1)]. \tag{5.36b}$$

The total lift coefficient C_l of the complex wing is given by the sum of the above coefficients.

Wave drag coefficient. For general case,

$$C_{dm}^{(n)} = -\frac{4}{(m+n)SU_\infty^2} \sum_{k=0}^{m-1} \sum_{q=0}^{n} w_{m-k-1,k}$$
$$\times A_{n,q}(\cot\delta)^{k+q} \sum_{r=0}^{k+q} (-1)^r \frac{(k+q)!}{r!(k+q-r)!} \mathcal{J}_{m+n-r} \tag{5.37}$$

and for the particular cases $n = 1, 2$ and $m = 1, 2$,

$$C_{d1}^{(1)} = -\frac{w_{00}}{U_\infty} C_l^{(1)}, \tag{5.38a}$$

$$C_{d1}^{(2)} = -\frac{w_{00}}{U_\infty} C_l^{(2)}, \tag{5.38b}$$

$$C_{d2}^{(1)} = -\frac{4}{3SU_\infty}[w_{10}A_{10}\mathcal{J}_3 + (w_{10}A_{11} + w_{01}A_{10})(\mathcal{J}_3 - \mathcal{J}_2)\cot\delta$$
$$+ w_{01}A_{11}(\mathcal{J}_3 - 2\mathcal{J}_2 + \mathcal{J}_1)\cot^2\delta], \tag{5.38c}$$

$$C_{d2}^{(2)} = -\frac{1}{SU_\infty^2}\left[w_{10}A_{20}\mathcal{J}_4+(w_{10}A_{21}+w_{01}A_{20})(\mathcal{J}_4-\mathcal{J}_3)\cot\delta\right.$$
$$+(w_{10}A_{22}+w_{01}A_{21})(\mathcal{J}_4-2\mathcal{J}_3+\mathcal{J}_2)\cot^2\delta$$
$$\left.+w_{01}A_{22}(\mathcal{J}_4-3\mathcal{J}_3+3\mathcal{J}_2-\mathcal{J}_1)\cot^3\delta.\right. \tag{5.38d}$$

For the complex wing, the total wave drag coefficient is the sum of these coefficients.

Rolling moment coefficient. Analogously, we have for the general formula,

$$\mu_1^{(n)} = \frac{4}{(n+2)ShU_\infty}\sum_{q=0}^{n} A_{n,q}(\cot\delta)^{q+1}\sum_{r=0}^{q+1}(-1)^r\frac{(q+1)!}{r!(q+1-r)!}\mathcal{J}_{n+2-r}, \tag{5.39}$$

and for $n = 1, 2$, respectively

$$\mu_1^{(1)} = \frac{4}{3ShU_\infty}\left[A_{10}(\mathcal{J}_3-\mathcal{J}_2)\cot\delta+A_{11}(\mathcal{J}_3-2\mathcal{J}_2+\mathcal{J}_1)\cot^2\delta\right], \tag{5.40a}$$

$$\mu_1^{(2)} = \frac{1}{ShU_\infty}\left[A_{20}(\mathcal{J}_4-\mathcal{J}_3)\cot\delta+A_{21}(\mathcal{J}_4-2\mathcal{J}_3+\mathcal{J}_1)\cot^2\delta\right.$$
$$\left.+A_{22}(\mathcal{J}_4-3\mathcal{J}_3+3\mathcal{J}_2-\mathcal{J}_1)\cot^3\delta\right]. \tag{5.40b}$$

The sum of these two coefficients yields the total coefficient μ_1.

Pitching moment coefficient. The general formula is

$$\mu_2^{(n)} = -\frac{4}{(n+2)ShU_\infty}\sum_{q=0}^{n} A_{n,q}(\cot\delta)^{q}\sum_{r=0}^{q}(-1)^r\frac{q!}{r!(q-r)!}\mathcal{J}_{n+2-r}, \tag{5.41}$$

which gives for $n = 1, 2$

$$\mu_2^{(1)} = -\frac{4}{3ShU_\infty}\left[A_{10}\mathcal{J}_3+A_{11}(\mathcal{J}_3-\mathcal{J}_2)\cot\delta\right], \tag{5.42a}$$

$$\mu_2^{(2)} = -\frac{1}{ShU_\infty}\left[A_{20}\mathcal{J}_4+A_{21}(\mathcal{J}_4-\mathcal{J}_3)\cot\delta\right.$$
$$\left.+A_{22}(\mathcal{J}_4-2\mathcal{J}_3+\mathcal{J}_2)\cot^2\delta\right]. \tag{5.42b}$$

The total coefficient μ_2 is the sum of these two coefficients.

Yawing moment coefficient. The general expression for $\mu_{3m}^{(n)}$ is

$$\mu_{3m}^{(n)} = \frac{4}{(m+n+1)ShU_\infty^2}\sum_{k=0}^{m-1}\sum_{q=0}^{n} w_{m-k-1,k}$$
$$\times A_{n,q}(\cot\delta)^{k+q+1}\sum_{r=0}^{k+q+1}(-1)^r\frac{(k+q+1)!}{r!(k+q+1-r)!}\mathcal{J}_{m+n+1-r} \tag{5.43}$$

which yields the following formulae for $n = 1, 2$ and $m = 1, 2$,

$$\mu_{31}^{(1)} = \frac{w_{00}}{U_\infty}\mu_1^{(1)}, \tag{5.44a}$$

$$\mu_{31}^{(2)} = \frac{w_{00}}{U_\infty}\mu_1^{(2)}, \tag{5.44b}$$

$$\mu_{32}^{(1)} = \frac{1}{ShU_\infty^2}[w_{10}A_{10}(\mathcal{J}_4-\mathcal{J}_3)\cot\delta+(w_{10}A_{11}+w_{01}A_{10})(\mathcal{J}_4-2\mathcal{J}_3+\mathcal{J}_2)$$
$$\times\cot^2\delta+w_{01}A_{11}(\mathcal{J}_4-3\mathcal{J}_3+3\mathcal{J}_2-\mathcal{J}_1)\cot^3\delta], \tag{5.44c}$$

$$\mu_{32}^{(2)} = \frac{4}{5ShU_\infty^2}[w_{10}A_{20}(\mathcal{J}_5-\mathcal{J}_4)\cot\delta+(w_{10}A_{21}+w_{01}A_{20})(\mathcal{J}_5-2\mathcal{J}_4+\mathcal{J}_3)$$
$$\times\cot^2\delta+(w_{10}A_{22}+w_{01}A_{21})(\mathcal{J}_5-3\mathcal{J}_4+3\mathcal{J}_3-\mathcal{J}_2)\cot^3\delta$$
$$+w_{01}A_{22}(\mathcal{J}_5-4\mathcal{J}_4+6\mathcal{J}_3-4\mathcal{J}_2+\mathcal{J}_1)\cot^4\delta]. \tag{5.44d}$$

For the complex wing, the total coefficient is the sum of the above respective coefficients.

In all the above expressions of the lift, wave drag and moment coefficients, we have denoted by $\mathcal{J}_t$ the following general expression

$$\mathcal{J}_t = \frac{\pi}{2^{2(t-1)}}\sum_{p=0}^{t-1}\frac{(2p)![2(t-1-p)]!}{(p!)^2[(t-1-p)!]^2}\frac{1}{(1-l_1\tan\delta)^{t-p-1/2}(1+l_2\tan\delta)^{p+1/2}} \tag{5.45}$$

obtained by applying the residue method indicated above (§ V. 1.3).

The values of $\mathcal{J}_t$ which appear in the above expressions for the complex wing of order $n = 1, 2$, are given in Appendix II.1.

V. 2.3. Calculation of Suction Forces and Corresponding Moments

The suction forces have a complete effect: thrust force, lateral force and yawing moment.

Thrust force. The general formulae referring to the two leading edges OC_1 and OC_2 are given by (5.27), where a_1 and a_2 are the projections of the edges OC_1 and OC_2 upon Ox_1 (5.29).

Particularizing for $n = 1, 2$, we obtain successively:

$$\sigma_{11}^{(1)} = -\frac{\pi\sqrt{1-B^2l_1^2a_1^2}}{l_1(l_1+l_2)SU_\infty^2}(A_{10}+A_{11}l_1)^2, \tag{5.46a}$$

$$\sigma_{11}^{(2)} = \sigma_{12}^{(1)} = -\frac{2\pi\sqrt{1-B^2l_1^2}\,a_1^3}{3l_1(l_1+l_2)SU_\infty^2}(A_{10}+A_{11}l_1)(A_{20}+A_{21}l_1+A_{22}l_1^2), \quad (5.46b)$$

$$\sigma_{12}^{(2)} = -\frac{\pi\sqrt{1-B^2l_1^2}\,a_1^4}{2l_1(l_1+l_2)SU_\infty^2}(A_{20}+A_{21}l_1+A_{22}l_1^2)^2, \quad (5.46c)$$

for the leading edge OC_1, and

$$\sigma_{21}^{(1)} = -\frac{\pi\sqrt{1-B^2l_2^2}\,a_2^2}{l_2(l_1+l_2)SU_\infty^2}(A_{10}-A_{11}l_2)^2, \quad (5.47a)$$

$$\sigma_{21}^{(2)} = \sigma_{22}^{(1)} = -\frac{2\pi\sqrt{1-B^2l_2^2}\,a_2^3}{3l_2(l_1+l_2)SU_\infty^2}(A_{10}-A_{11}l_2)(A_{20}-A_{21}l_2+A_{22}l_2^2), \quad (5.47b)$$

$$\sigma_{22}^{(2)} = -\frac{\pi\sqrt{1-B^2l_2^2}\,a_2^4}{2l_2(l_1+l_2)SU_\infty^2}(A_{20}-A_{21}l_2+A_{22}l_2^2)^2 \quad (5.47c)$$

for the leading edge OC_2.

Further, by summing the above coefficients, we obtain σ_1 and σ_2, which refer to the two leading edges.

Lateral force. The coefficients of the lateral suction force are deduced from the above ones, by applying the simple formulae (3.109) and (3.110).

Yawing moment. On the basis of formulae (5.30), the yawing moment coefficients may be expressed as a function of the axial suction coefficients which are defined by (5.46) and (5.47).

The sum of these coefficients yields the total coefficients corresponding to each edge,

$$\gamma_1 = \gamma_{11}^{(1)}+2\gamma_{11}^{(2)}+\gamma_{12}^{(2)}, \quad \gamma_2 = \gamma_{21}^{(1)}+2\gamma_{21}^{(2)}+\gamma_{22}^{(2)}, \quad (5.48)$$

and respectively to both edges,

$$\gamma = \gamma_1+\gamma_2. \quad (5.49)$$

V. 2.4. Important Note on the Effect of Yaw

It should be first noticed that the angle of yaw (δ) cannot exceed a limit which can be practically considered about the value of 20°. Under such a hypothesis, the preponderant term in the yaw effect is that which contains the first power of $\tan\delta$, the other powers constituting secondary terms, out of which some might even be neglected, mainly for small values of δ.

Under such conditions, even from a brief examination of the formulae established above, it may be readily seen that the effect of the yaw manifests itself very strongly as regards the *rolling moment coefficient* which contains the terms beginning from $\tan\delta$ and which is hence equal to zero in the initial position, that is, it is cancelled at the same time with δ. Likewise, the *yawing moment coefficient* starts from the value zero in the initial position and increases as a function of $\tan\delta$, beginning with the first power of $\tan\delta$, as may also be seen from formulae (5.44). The same remarks also apply to the *suction yawing moment and lateral coefficient*.

As regards the remaining aerodynamic coefficients, the yaw effect manifests itself starting with terms in $(\tan\delta)^2$, and is thus negligible for the small values of δ, particularly if we take account of the fact that the aerodynamic coefficients C_l, C_d, μ_2 have already important values in their initial position.

In order to make the above statements self-evident (although they might in fact have been deduced intuitively) we have completely treated above the case of the yawed wing with both leading edges subsonic.

For the sake of simplicity, in the following cases we propose to compute only the *rolling and yawing moments coefficients* and possibly the *suction effect* for the subsonic edge of the yawed wing with one subsonic and one supersonic edge.

V. 3. Case of Yawed Thin Delta Wing having One Subsonic and One Supersonic Edge

V. 3.1. Determination of Axial Velocities

The downwashes will be deduced from the same wing as before, for which a natural symmetry of incidence has been considered. In this case, taking into account the wing position (Fig. 5.2, Mach angle μ_2), the axial velocity function is given by relation (2.84), into which we are now introducing double indices:

$$\mathcal{U}_{n-1} = \sum_{q=0}^{n-1} A_{n,q}\, x^q \sqrt{\frac{1+Bx}{B(l_1-x)}} + \sum_{q=0}^{n-1} H_{n,q}\, x^q \cos^{-1}\sqrt{\frac{(l_1+l_2)(1+Bx)}{(1+Bl_1)(l_2+x)}}. \tag{5.50}$$

With respect to the downwash coefficients, which are the same as in the preceding case (5.14), (5.15), the constants of the two component wings ($n = 1, 2$) are as follows:

$$A_{10} = -\frac{2w_{00}}{\pi B}\frac{Bl_1}{1+Bl_1}\sqrt{\frac{B(l_1+l_2)}{1+Bl_2}}, \tag{5.51a}$$

$$H_{10} = -\frac{2w_{00}}{\pi B}\frac{Bl_2}{\sqrt{B^2l_2^2-1}}, \tag{5.51b}$$

for $n = 1$, and

$$A_{20} = -\frac{2l_1}{3\pi(1+Bl_1)^2(B^2l_2^2-1)}\sqrt{\frac{B(l_1+l_2)}{1+Bl_2}}[(3B^3l_1l_2^2+5B^2l_2^2+2B^2l_1l_2$$
$$-2Bl_1+4Bl_2-6)w_{10}-(2l_1-l_2+4Bl_2^2+Bl_1l_2)w_{01}], \tag{5.52a}$$

$$A_{21} = \frac{2}{3\pi(1+Bl_1)^2(B^2l_2^2-1)^{3/2}}\sqrt{\frac{B(l_1+l_2)}{1+Bl_2}}[(B^3l_1^2l_2+B^3l_1l_2^2-B^2l_1^2+2B^2l_1l_2$$
$$+3B^2l_2^2+3Bl_2-3)w_{10}-(3B^3l_1^2l_2^2-B^2l_1^2l_2$$
$$+2B^2l_1l_2^2-2Bl_1^2+3Bl_2^2+Bl_1l_2)w_{01}], \tag{5.52b}$$

$$H_{20} = -\frac{2l_2}{\pi(B^2l_2^2-1)^{3/2}}[(B^2l_2^2-2)w_{10}+l_2w_{01}], \tag{5.52c}$$

$$H_{21} = \frac{2}{\pi(B^2l_2^2-1)^{3/2}}[w_{10}-B^2l_2^3w_{01}], \tag{5.52d}$$

for $n = 2$.

V. 3.2. Rolling and Yawing Moments Coefficients

As noted before (§ V. 2.4), we shall calculate only the rolling and yawing moments coefficients.

The same operations as in the preceding case are to be performed; taking into account the new expression for $\mathcal{U}_{n-1}$ and the formulae established in (§ V. 1.3), we obtain successively:

Rolling moment coefficient. For the nth order component wing, we have

$$\mu_1^{(n)} = \frac{4}{(n+2)ShU_\infty}\sum_{q=0}^{n-1}\sum_{r=0}^{q+1}(-1)^r\frac{(q+1)!}{r!(q+1-r)!}(\cot\,\delta)^{q+1}$$
$$\times\left\{A_{n,q}\left[\left(\frac{1}{B}+\cot\,\delta\right)\mathcal{J}'_{n+2-r}-\cot\,\delta\,\mathcal{J}'_{n+1-r}\right]+H_{n,q}\,\Lambda_{n+2-r}\right\}, \tag{5.53}$$

and for $n = 1, 2$,

$$\mu_1^{(1)} = \frac{4}{3ShU_\infty}\left[\frac{1}{B}A_{10}(\mathcal{J}'_3-\mathcal{J}'_2)\cot\,\delta+A_{10}(\mathcal{J}'_3-2\mathcal{J}'_2+\mathcal{J}'_1)\cot^2\,\delta\right.$$
$$\left.+H_{10}(\Lambda_3-\Lambda_2)\cot\,\delta\right], \tag{5.54a}$$

$$\mu_1^{(2)} = \frac{1}{ShU_\infty}\left[\frac{1}{B}A_{20}(\mathcal{J}'_4-\mathcal{J}'_3)\cot\delta+\left(A_{20}+\frac{1}{B}A_{21}\right)(\mathcal{J}'_4-2\mathcal{J}'_3+\mathcal{J}'_2)\cot^2\delta\right.$$
$$+A_{21}(\mathcal{J}'_4-3\mathcal{J}'_3+3\mathcal{J}'_2-\mathcal{J}'_1)\cot^3\,\delta+H_{20}(\Lambda_4-\Lambda_3)\cot\,\delta$$
$$\left.+H_{21}(\Lambda_4-2\Lambda_3+\Lambda_2)\cot^2\,\delta\right]. \tag{5.54b}$$

For the complex wing ($n = 1, 2$), the total coefficient is equal to their sum.

Yawing moment coefficient. For the general case we obtain

$$\mu_{3m}^{(n)} = \frac{4}{(m+n+1)ShU_\infty^2} \sum_{k=0}^{m-1} \sum_{q=0}^{n-1} w_{m-k-1,\,k} (\cot\delta)^{k+q+1}$$

$$\times \sum_{r=0}^{k+q+1} (-1)^r \frac{(k+q+1)!}{r!(k+q+1-r)!} \left\{ A_{n,\,q} \left[\left(\frac{1}{B} + \cot\delta \right) \mathcal{J}'_{m+n+1-r} \right.\right.$$

$$\left.\left. -\cot\delta \mathcal{J}'_{m+n-r} \right] + H_{n,\,q} \Lambda_{m+n+1-r} \right\}, \tag{5.55}$$

which applied to $n = 1, 2$ and $m = 1, 2$, yields the following:

$$\mu_{31}^{(1)} = \frac{w_{00}}{U_\infty} \mu_1^{(1)}, \tag{5.56a}$$

$$\mu_{31}^{(2)} = \frac{w_{00}}{U_\infty} \mu_1^{(2)}, \tag{5.56b}$$

$$\mu_{32}^{(1)} = \frac{1}{ShU_\infty^2} \left\{ A_{10} \left[\frac{1}{B} w_{10}(\mathcal{J}'_4 - \mathcal{J}'_3) \cot\delta + \left(w_{10} \right.\right.\right.$$

$$\left. + \frac{1}{B} w_{01} \right) (\mathcal{J}'_4 - 2\mathcal{J}'_3 + \mathcal{J}'_2) \cot^2\delta + w_{01}(\mathcal{J}'_4 - 3\mathcal{J}'_3 + 3\mathcal{J}'_2 - \mathcal{J}'_1) \cot^3\delta \Big]$$

$$\left. + H_{10}[w_{10}(\Lambda_4 - \Lambda_3) \cot\delta + w_{01}(\Lambda_4 - 2\Lambda_3 + \Lambda_2) \cot^2\delta] \right\}, \tag{5.56c}$$

$$\mu_{32}^{(2)} = \frac{4}{5ShU_\infty^2} \left\{ \frac{1}{B} w_{10} A_{20} (\mathcal{J}'_5 - \mathcal{J}'_4) \cot\delta + \left[w_{10} \left(A_{20} + \frac{1}{B} A_{21} \right) \right.\right.$$

$$\left. + \frac{1}{B} w_{01} A_{20} \right] (\mathcal{J}'_5 - 2\mathcal{J}'_4 + \mathcal{J}'_3) \cot^2\delta + \left[w_{10} A_{21} + w_{01} \left(A_{20} + \frac{1}{B} A_{21} \right) \right]$$

$$\times (\mathcal{J}'_5 + 3\mathcal{J}'_4 + 3\mathcal{J}'_3 - \mathcal{J}'_2) \cot^3\delta + w_{01} A_{21} (\mathcal{J}'_5 - 4\mathcal{J}'_4 + 6\mathcal{J}'_3 - 4\mathcal{J}'_2 + \mathcal{J}'_1) \cot^4\delta$$

$$+ w_{10} H_{20} (\Lambda_5 - \Lambda_4) \cot\delta + (w_{10} H_{21} + w_{01} H_{20})(\Lambda_5 - 2\Lambda_4 + \Lambda_3) \cot^2\delta$$

$$\left. + w_{01} H_{21} (\Lambda_5 - 3\Lambda_4 + 3\Lambda_3 - \Lambda_2) \cot^3\delta \right\}. \tag{5.56d}$$

The total yawing moment coefficient is of course the sum of the above four coefficients.

In the above expressions for the moment coefficients, we have denoted by $\mathcal{J}'_t$ and Λ_ν the following expressions:

$$\mathcal{J}'_t = \frac{\pi}{2^{2(t-1)}} \sum_{p=0}^{t-1} \frac{(2p)![2(t-1-p)]!}{(p!)^2[(t-1-p)!]^2} \frac{1}{(1-l_1 \tan \delta)^{t-p-1/2}\left(1+\frac{\tan \delta}{B}\right)^{p+1/2}}, \tag{5.57a}$$

which can be deduced from (5.45), by putting $l_2 = 1/B$, and

$$\Lambda_\nu = \frac{1}{2(\nu-1)} \sqrt{(l_1+l_2)\left(l_2-\frac{1}{B}\right)} \sum_{t=1}^{\nu-1} \frac{1}{(1+l_2 \tan \delta)^{\nu-t}} \mathcal{J}'_t. \tag{5.58}$$

The values of $\mathcal{J}'_t$ and Λ_ν which appear in the above expressions for the complex wing considered ($n = 1, 2$) are given in Appendix II.2.

V. 3.3. Suction Forces and Corresponding Moments

The suction refers only to the subsonic leading edge OC_1. The coefficient $\sigma_{1m}^{(n)}$ has exactly the same form as in formula (5.28); applying this formula for $n = 1, 2$ and $m = 1, 2$, we find successively:

$$\sigma_{11}^{(1)} = -\frac{\pi a_1^2(1+Bl_1)\sqrt{1-B^2 l_1^2}}{Bl_1 SU_\infty^2} A_{10}^2, \tag{5.59a}$$

$$\sigma_{11}^{(2)} = \sigma_{12}^{(1)} = -\frac{2\pi a_1^3(1+Bl_1)\sqrt{1-B^2 l_1^2}}{3Bl_1 SU_\infty^2} A_{10}(A_{20}+A_{21}l_1), \tag{5.59b}$$

$$\sigma_{12}^{(2)} = -\frac{\pi a_1^4(1+Bl_1)\sqrt{1-B^2 l_1^2}}{2Bl_1 SU_\infty^2} (A_{20}+A_{21}l_1)^2 . \tag{5.59c}$$

Their sum yields the total suction coefficient σ_1.

The coefficients of the lateral suction force are deduced from the above, by applying the simple formulae (3.109) and (3.110).

There is no difficulty in calculating the yawing moment coefficient if the axial suction coefficient σ_1 is known.

Indeed, the relationship between $\gamma_{1m}^{(n)}$ and $\sigma_{1m}^{(n)}$ is given by (5.30a).

The sum of the coefficients $\gamma_{1m}^{(n)}$, where $m = 1, 2$, $n = 1, 2$, yields the total coefficient γ_1.

V. 4. Yawed Thin Delta Wing having both Edges Supersonic

V. 4.1. Expression of the Axial Velocity

The downwashes on the wing retain the same form and are deduced from those of the delta wing which, under our assumption, has a symmetrical distribution of incidences in its initial position ($\delta = 0$).

In such a case, by applying formula (2.90), the axial velocity function $\mathcal{U}_{n-1}$ is of the form

$$\mathcal{U}_{n-1} = \sum_{q=0}^{n-1} K_{n,q}\, x^q \cos^{-1}\sqrt{\frac{(1+Bl_1)(1-Bx)}{2B(l_1-x)}} + \sum_{q=0}^{n-1} H_{n,q}\, x^q \cos^{-1}\sqrt{\frac{(1+Bl_2)(1+Bx)}{2B(l_2+x)}} + \sum_{q=0}^{n-2} D_{n,q} x^q \sqrt{\frac{1-B^2x^2}{B^2}}\,. \tag{5.60}$$

In this expression, the constants for the two first wings of order $n = 1, 2$, which depend on the constants of the downwashes (w_{00}, w_{10}, w_{01}) indicated above, (5.14, 5.15), are given by the following relations:

$$K_{10} = -\frac{2w_{00}}{\pi B}\frac{Bl_1}{\sqrt{B^2l_1^2-1}}, \tag{5.61a}$$

$$H_{10} = -\frac{2w_{00}}{\pi B}\frac{Bl_2}{\sqrt{B^2l_2^2-1}}, \tag{5.61b}$$

for $n = 1$, and

$$K_{20} = -\frac{2l_1}{\pi(B^2l_1^2-1)^{3/2}}\left[(B^2l_1^2-2)w_{10}-l_1w_{01}\right], \tag{5.62a}$$

$$K_{21} = -\frac{2}{\pi(B^2l_1^2+1)^{3/2}}\left[w_{10}+B^2l_1^3w_{01}\right], \tag{5.62b}$$

$$H_{20} = -\frac{2l_2}{\pi(B^2l_2^2-1)^{3/2}}\left[(B^2l_2^2-2)w_{10}+l_2w_{01}\right], \tag{5.62c}$$

$$H_{21} = \frac{2}{\pi(B^2l_2^2-1)^{3/2}}\left[w_{10}-B^2l_2^3w_{01}\right], \tag{5.62d}$$

$$D_{20} = -\frac{1}{\pi}\left[\left(\frac{Bl_1}{B^2l_1^2-1}+\frac{Bl_2}{(B^2l_2^2-1)}\right)w_{10}+\left(\frac{Bl_1^2}{B^2l_1^2-1}-\frac{Bl_2^2}{B^2l_2^2-1}\right)w_{01}\right], \tag{5.62e}$$

for $n = 2$.

V. 4.2. Rolling and Yawing Moment Coefficients

We give below the expressions for the rolling and yawing moment coefficients as functions of the above constants, by applying the same general formulae established previously (§ V. 1.3).

Rolling moment coefficient. Analogously with the preceding examples, we have for the general case,

$$\mu_1^{(n)} = \frac{4}{(n+2)ShU_\infty}\left\{\sum_{q=0}^{n-1}\sum_{r=0}^{q+1}(-1)^r \frac{(q+1)!}{r!(q+1-r)!}(\cot\delta)^{q+1}[K_{n,q}\Lambda'_{n+2-r}\right.$$

$$+H_{n,q}\Lambda''_{n+2-r}] + \sum_{q=0}^{n-2} D_{n,q}(\cot\delta)^{q+3}\sum_{r=0}^{q+1}\frac{(-1)^r}{n+1-r}\frac{(q+1)!}{r!(q+1-r)!}$$

$$\left.\times[\mathcal{J}''_{n+1-r}-\mathcal{J}''_{n-r}]\right\}, \tag{5.63}$$

and for $n = 1, 2$, respectively

$$\mu_1^{(1)} = \frac{4}{3ShU_\infty}[K_{10}(\Lambda'_3-\Lambda'_2)\cot\delta + H_{10}(\Lambda''_3-\Lambda''_2)\cot\delta], \tag{5.64a}$$

$$\mu_1^{(2)} = \frac{1}{ShU_\infty}\Big[K_{20}(\Lambda'_4-\Lambda'_3)\cot\delta + K_{21}(\Lambda'_4-2\Lambda'_3+\Lambda'_2)\cot^2\delta$$

$$+H_{20}(\Lambda''_4-\Lambda''_3)\cot\delta + H_{21}(\Lambda''_4-2\Lambda''_3+\Lambda''_2)\cot^2\delta +$$

$$+D_{20}\frac{\cot^3\delta}{6}(2\mathcal{J}''_3-5\mathcal{J}''_2+3\mathcal{J}''_1)\Big]. \tag{5.64b}$$

For the complex wing made up of component wings of order $n = 1, 2$, the total coefficient μ_1 is given by the sum of the above two coefficients.

Yawing moment coefficient. For the general case we have

$$\mu_{3m}^{(n)} = \frac{4}{(m+n+1)ShU_\infty^2}\sum_{k=0}^{m-1} w_{m-k-1,k}\left\{\sum_{q=0}^{n-1}\sum_{r=0}^{k+q+1}(-1)^r\frac{(k+q+1)!}{r!(k+q+1-r)!}\right.$$

$$\times(\cot\delta)^{k+q+1}[K_{n,q}\Lambda'_{m+n+1-r}+H_{n,q}\Lambda''_{m+n+1-r}]+\sum_{q=0}^{n-2}D_{n,q}$$

$$\left.\times(\cot\delta)^{k+q+3}\sum_{r=0}^{k+q+1}\frac{(-1)^r}{m+n+r}\frac{(k+q+1)!}{r!(k+q+1-r)!}[\mathcal{J}''_{m+n-r}-\mathcal{J}''_{m+n-1-r}]\right\} \tag{5.65}$$

which, applied to $n = 1, 2$ and $m = 1, 2$, yields

$$\mu_{31}^{(1)} = \frac{w_{00}}{U_\infty}\, \mu_1^{(1)}, \tag{5.66a}$$

$$\mu_{31}^{(2)} = \frac{w_{00}}{U_\infty}\, \mu_1^{(2)}, \tag{5.66b}$$

$$\mu_{32}^{(1)} = \frac{1}{ShU_\infty^2} \{K_{10}[w_{10}(\Lambda_4' - \Lambda_3') \cot \delta + w_{01}(\Lambda_4' - 2\Lambda_3' + \Lambda_2') \cot^2 \delta]$$

$$+ H_{10}[w_{10}(\Lambda_4'' - \Lambda_3'') \cot \delta + w_{01}(\Lambda_4'' - 2\Lambda_3'' + \Lambda_2'') \cot^2 \delta]\}, \tag{5.66c}$$

$$\mu_{32}^{(2)} = \frac{4}{5ShU_\infty^2} \Big[w_{10}K_{20}(\Lambda_5' - \Lambda_4') \cot \delta + (w_{10}K_{21} + w_{01}K_{20})(\Lambda_5' - 2\Lambda_4' + \Lambda_3')$$

$$\times \cot^2 \delta + w_{01}K_{21}(\Lambda_5' - 3\Lambda_4' + 3\Lambda_3' - \Lambda_2') \cot^3 \delta + w_{10}H_{20}(\Lambda_5'' - \Lambda_4'') \cot \delta$$

$$+ (w_{10}H_{21} + w_{01}H_{20})(\Lambda_5'' - 2\Lambda_4'' + \Lambda_3'') \cot^2 \delta + w_{01}H_{21}(\Lambda_5'' - 3\Lambda_4''$$

$$+ 3\Lambda_3'' - \Lambda_2'') \cot^3 \delta + D_{20} \frac{\cot^3 \delta}{12} (3\mathcal{J}_4'' - 7\mathcal{J}_3'' + 4\mathcal{J}_2'')$$

$$+ D_{20} \frac{\cot^4 \delta}{12} (3\mathcal{J}_4'' - 11\mathcal{J}_3'' + 14\mathcal{J}_2'' - 6\mathcal{J}_1'') \Big]. \tag{5.66d}$$

The total coefficient μ_3 is given by the sum of the above coefficients, which refer to the complex wing considered above ($n = 1, 2$).

In the above formulae, we have denoted by Λ_ν', Λ_ν'' and $\mathcal{J}_t''$ the following expressions:

$$\Lambda_\nu' = \frac{1}{2(\nu - 1)} \sqrt{l_1^2 - \frac{1}{B^2}} \sum_{t=1}^{\nu-1} \frac{1}{(1 - l_1 \tan \delta)^{\nu - t}} \mathcal{J}_t'', \tag{5.67a}$$

$$\Lambda_\nu'' = \frac{1}{2(\nu - 1)} \sqrt{l_2^2 - \frac{1}{B^2}} \sum_{t=1}^{\nu-1} \frac{1}{(1 + l_2 \tan \delta)^{\nu - t}} \mathcal{J}_t'', \tag{5.67b}$$

$$\mathcal{J}_t'' = \frac{\pi}{2^{2(t-1)}} \sum_{p=0}^{t-1} \frac{(2p)![2(t-1-p)]!}{(p!)^2[(t-1-p)!]^2}$$

$$\times \frac{1}{\left(1 - \dfrac{\tan \delta}{B}\right)^{t-p-1/2} \left(1 + \dfrac{\tan \delta}{B}\right)^{p+1/2}}. \tag{5.67c}$$

This latter formula is deduced from $\mathcal{J}_t$ (5.45), by putting $l_1 = 1/B$, $l_2 = 1/B$.

For $t = 1, 2, 3, 4$ and $\nu = 1, 2, 3, 4, 5$ we give the corresponding values in Appendix II.3.

V. 5. Yawed Delta Wing of Symmetrical Thickness

V. 5.1. Preliminary Considerations

We consider the case of subsonic edges since the obtained results may be also extended to supersonic edges.

We also assume that the initial delta wing has a natural and forced symmetry of slope, which implies that the wing should have a central ridge in its initial position.

This ridge is situated at the point $y = -s$ with respect to the system Ox_1x_2 (Fig. 5.1), where s is given by the following equation

$$s = \tan \delta. \tag{5.68}$$

The general expression for the axial velocity function $\mathcal{U}_{n-1} = u_{n-1} + iu'_{n-1}$ is

$$\begin{aligned}\mathcal{U}_{n-1} = & \sum_{q=0}^{n-1} C_{n,q} x^q \cosh^{-1} \sqrt{\frac{(1+Bl_1)(1-Bx)}{2B(l_1-x)}} \\ & + \sum_{q=0}^{n-1} G_{n,q} x^q \cosh^{-1} \sqrt{\frac{(1+Bl_2)(1+Bx)}{2B(l_2+x)}} \\ & + \sum_{q=0}^{n-1} F_{n,q} x^q \cosh^{-1} \sqrt{\frac{(1+Bs)(1+Bx)}{2B(s+x)}} + \sum_{q=0}^{n-2} D_{n,q} x^q \sqrt{\frac{1-B^2x^2}{B^2}}.\end{aligned} \tag{5.69}$$

This formula is also valid for the case when one leading edge is supersonic, the other being subsonic, or when both edges are supersonic. In these cases, it is more convenient to apply the following relations for each *supersonic edge* or *ridge*,

$$C_{n,q} \cosh^{-1} \beta = K_{n,q} \cos^{-1} \beta, \quad G_{n,q} \cosh^{-1} \beta = H_{n,q} \cos^{-1} \beta, \tag{5.70a}$$

$$F_{n,q} \cosh^{-1} \beta = E_{n,q} \cos^{-1} \beta. \tag{5.70b}$$

For the sake of simplicity, we shall treat the simple case of a complex wing of symmetrical thickness reduced to two component wings of order $n = 1$ and $n = 2$, defined by equation (5.16).

For determining the corresponding constants, we take into account (5.19) and (5.20) and write for the coefficients of the downwashes on the wing,

$$w_{00}^{(1)} = U_\infty t(b \cos \delta - h \sin \delta), \quad w_{00}^{(2)} = U_\infty t(b \cos \delta + h \sin \delta), \tag{5.71a}$$

$$w_{10}^{(1)} = U_\infty t\left[\sin 2\delta - \frac{b}{h}(1+\cos 2\delta)\right],$$

$$w_{10}^{(2)} = -U_\infty t\left[\sin 2\delta + \frac{b}{h}(1+\cos 2\delta)\right], \tag{5.71b}$$

$$w_{01}^{(1)} = U_\infty t\left[\cos 2\delta + \frac{b}{h}\sin 2\delta\right], \quad w_{01}^{(2)} = -U_\infty t\left[\cos 2\delta - \frac{b}{h}\sin 2\delta\right], \tag{5.71c}$$

where the superscript (1) refers to the right-hand side of the ridge OC and the superscript (2) to the left-hand side of the ridge OC, respectively.

As functions of these values, the constants $C_{n,q}$, $G_{n,q}$, $F_{n,q}$ and $D_{n,q}$ are given by the following relations corresponding to the component wings of order $n = 1, 2$:

$$C_{10} = -\frac{2w_{00}^{(1)}}{\pi B}\frac{Bl_1}{\sqrt{1-B^2l_1^2}}, \tag{5.72a}$$

$$G_{10} = -\frac{2w_{00}^{(2)}}{\pi B}\frac{Bl_2}{\sqrt{1-B^2l_2^2}}, \tag{5.72b}$$

$$F_{10} = -\frac{2(w_{00}^{(1)}-w_{00}^{(2)})}{\pi B}\frac{Bs}{\sqrt{1-B^2s^2}}, \tag{5.72c}$$

for $n = 1$, and

$$C_{20} = -\frac{2l_1}{\pi(1-B^2l_1^2)^{3/2}}[(2-B^2l_1^2)w_{10}^{(1)} + l_1 w_{01}^{(1)}], \tag{5.73a}$$

$$C_{21} = \frac{2}{\pi(1-B^2l_1^2)^{3/2}}[w_{10}^{(1)} + B^2l_1^3 w_{01}^{(1)}], \tag{5.73b}$$

$$G_{20} = -\frac{2l_2}{\pi(1-B^2l_2^2)^{3/2}}[(2-B^2l_2^2)w_{10}^{(2)} - l_2 w_{01}^{(2)}], \tag{5.73c}$$

$$G_{21} = -\frac{2}{\pi(1-B^2l_2^2)^{3/2}}[w_{10}^{(2)} - B^2l_2^3 w_{01}^{(2)}], \tag{5.73d}$$

$$F_{20} = -\frac{2s}{\pi(1-B^2s^2)^{3/2}}[(2-B^2s^2)(w_{10}^{(1)}-w_{10}^{(2)}) - s(w_{01}^{(1)}-w_{01}^{(2)})], \tag{5.73e}$$

$$F_{21} = -\frac{2}{\pi(1-B^2s^2)^{3/2}}[w_{10}^{(1)} - w_{10}^{(2)} - B^2s^3(w_{01}^{(1)}-w_{01}^{(2)})], \tag{5.73f}$$

$$D_{20} = \frac{1}{\pi}\left[\frac{Bl_1}{1-B^2l_1^2}w_{10}^{(1)} + \frac{Bl_2}{1-B^2l_2^2}w_{10}^{(2)} + \frac{Bs}{1-B^2s^2}(w_{10}^{(1)}-w_{10}^{(2)}) + \frac{Bl_1^2}{1-B^2l_1^2}w_{01}^{(1)} - \frac{Bl_2^2}{1-B^2l_2^2}w_{01}^{(2)} - \frac{Bs^2}{1-B^2s^2}(w_{01}^{(1)}-w_{01}^{(2)})\right], \tag{5.73g}$$

for $n = 2$.

V. 5.2. Yawing Moment Coefficient

As above, we shall give only the expression for the yawing moment coefficient for the wing of symmetrical thickness, since this is the case in which the effect of the yaw is the most important. In this case, by applying formula (3.60), we obtain

$$\mu_{3m}^{(n)} = \frac{4}{(m+n+1)ShU_\infty^2}\left[\int_{-l_2}^{-s}\left(\sum_{k=0}^{m-1} w_{m-k-1,\,k}^{(2)}\,y^k\right)\frac{u_{n-1}\,y\,dy}{(1-y\tan\delta)^{m+n+1}}\right.$$

$$\left.+\int_{-s}^{l_1}\left(\sum_{k=0}^{m-1} w_{m-k-1,\,k}^{(1)}\,y^k\right)\frac{u_{n-1}\,y\,dy}{(1-y\tan\delta)^{m+n+1}}\right] = \frac{4}{(m+n+1)ShU_\infty^2}$$

$$\times\,\mathrm{Re}\left\{\sum_{k=0}^{m-1}\sum_{q=0}^{n-1}\sum_{r=0}^{k+q+1}(-1)^r\frac{(k+q+1)!}{r!(k+q+1-r)!}(\cot\delta)^{k+q+1}\left[w_{m-k-1,\,k}^{(1)}\right.\right.$$

$$\times\left(C_{n,q}\,\psi_{m+n+1-r}^{(1)}+G_{n,q}\,\Gamma_{m+n+1-r}^{(1)}+F_{n,q}\,\Phi_{m+n+1-r}^{(1)}\right)$$

$$\left.+\,w_{m-k-1,\,k}^{(2)}\left(C_{n,q}\,\psi_{m+n+1-r}^{(2)}+G_{n,q}\,\Gamma_{m+n+1-r}^{(2)}+F_{n,q}\,\Phi_{m+n+1-r}^{(2)}\right)\right]$$

$$+\sum_{k=0}^{m-1}\sum_{q=0}^{n-2}D_{n,q}(\cot\delta)^{k+q+3}\sum_{r=0}^{k+q+1}\frac{(-1)^r}{m+n-r}\,\frac{(k+q+1)!}{r!(k+q+1-r)!}$$

$$\times B\left[w_{m-k-1,\,k}^{(1)}\left(\frac{\tan\delta}{B^2}\,\frac{\sqrt{1-B^2l_1^2}}{(1-l_1\tan\delta)^{m+n-r}}-\frac{\tan\delta}{B^2}\,\frac{\sqrt{1-B^2s^2}}{(1+s\tan\delta)^{m+n-r}}\right.\right.$$

$$\left.+\,\mathcal{J}_{m+n-r}^{(1)}-\mathcal{J}_{m+n-1-r}^{(1)}\right)+w_{m-k-1,\,k}^{(2)}\left(\frac{\tan\delta}{B^2}\,\frac{\sqrt{1-B^2s^2}}{(1+s\tan\delta)^{m+n-r}}\right.$$

$$\left.\left.\left.-\frac{\tan\delta}{B^2}\,\frac{\sqrt{1-B^2l_2^2}}{(1+l_2\tan\delta)^{m+n-r}}+\mathcal{J}_{m+n-r}^{(2)}-\mathcal{J}_{m+n-1-r}^{(2)}\right)\right]\right\}, \tag{5.74}$$

where the general expressions for the integrals $\mathcal{J}_t^{(\lambda)}$, $\psi_t^{(\lambda)}$, $\Gamma_t^{(\lambda)}$, $\Phi_t^{(\lambda)}$ ($\lambda = 1, 2$) are given in Appendix II.4. The same Appendix also includes the explicit values of the above integrals for $t = 1, 2, 3, 4, 5$, which appear in the expression for the yawing moment coefficient for the complex wing consisting of component wings of order $n = 1, 2$ ($m = 1, 2$).

V. 5.3. Case of One or Both Supersonic Edges

We have shown above that the expression for $\mathcal{U}_{n-1}$ (5.69) is also valid for the case when one or both edges are supersonic, possibly performing the formal changes indicated in (5.70). In this case, however, the constants $K_{n,q}$, $H_{n,q}$, $E_{n,q}$ are deduced from (5.72), (5.73), by applying the following relations:

$$\sqrt{B^2l_1^2-1}K_{n,q} = \sqrt{1-B^2l_1^2}C_{n,q}, \quad \sqrt{B^2l_2^2-1}H_{n,q} = \sqrt{1-B^2l_2^2}G_{n,q}, \tag{5.75a}$$

$$\sqrt{B^2s^2-1}E_{n,q} = \sqrt{1-B^2s^2}F_{n,q}, \quad \left(s > \frac{1}{B}\right). \tag{5.75b}$$

Analogously, the expression for the yawing moment coefficient $\mu_{3m}^{(n)}$ is the same as in the case of subsonic edges (5.74) with the remark that only the real part should be taken, considering at the same time relations (5.70) and (5.75).

V. 6. Delta Wing with Incidence or Slope Varying According to a Homogeneous Function

V. 6.1. Preliminary Considerations

In all the preceding problems, it has been assumed that the incidence α of the thin wing or the slope γ of the wing of symmetrical thickness and, consequently, the downwash w on the wing equal to $-\alpha U_\infty$ or γU_∞ varies according to a function represented by a superposition of homogeneous polynomials of various orders.

For the single or component wing of order n, the corresponding downwash varies according to a *homogeneous polynomial* of order $(n-1)$. In the following considerations we shall treat the case when this variation takes place according to a *homogeneous function* of order $(n-1)$. In order to simplify the presentation, simple concrete cases will be treated.

In this sense, we consider any wing whose slope γ for the thick wing or incidence α for the thin wing, varies as a function of y only, hence according to a zero-order homogeneous function; the downwash w follows the same variation which, as is known, is a *conical* one.

We shall assume from the beginning that this variation is symmetrical, with natural or forced symmetry; accordingly, we have the following corresponding representation of the downwash on the wing ($w = \gamma U_\infty$ or $w = -\alpha U_\infty$):

$$w = \sum_{k=0}^{k} w_{2k}\, y^{2k}, \tag{5.76a}$$

$$w = \pm \sum_{k=0}^{k} w_{2k+1}\, y^{2k+1}, \tag{5.76b}$$

in the latter expression the plus sign being valid for $y > 0$, and the minus sign for $y < 0$ (Fig. 5.3).

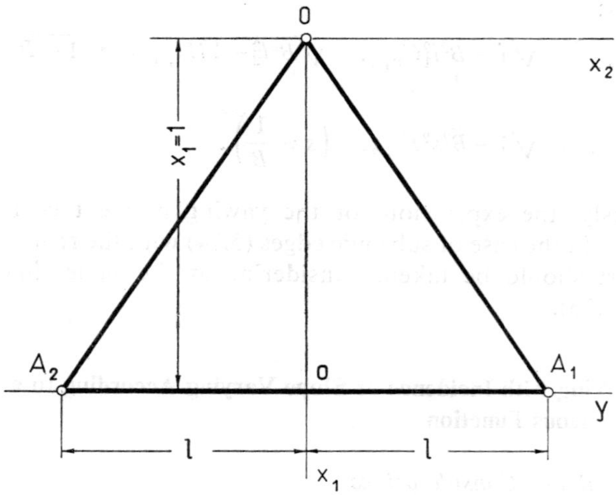

FIG. 5.3

By integration with respect to x_1 we also deduce the equation of the wing surface.

Let us consider the following cases:

$$w = w_0\left(1 \pm a\frac{y}{l}\right), \qquad Z = \frac{w_0}{U_\infty}x_1\left[1 \mp \frac{y}{l}\left(1 + a\ln\frac{y}{l}\right)\right], \tag{5.77a}$$

$$w = w_0\left(1 + \frac{y^2}{l^2}\right), \qquad Z = \frac{w_0}{U_\infty}x_1\left(1 - \frac{y^2}{l^2}\right), \tag{5.77b}$$

$$w = w_0\left(1 \pm 2\frac{y^3}{l^3}\right), \qquad Z = \frac{w_0}{U_\infty}x_1\left(1 \mp \frac{y^3}{l^3}\right), \tag{5.77c}$$

$$w = w_0\left[1 + (m-1)(\pm 1)^m\frac{y^m}{l^m}\right], \quad Z = \frac{w_0}{U_\infty}x_1\left[1 - (\pm 1)^m\frac{y^m}{l^m}\right]. \tag{5.77d}$$

All these surfaces are open at the trailing edge. The last four surfaces may be closed if they are combined with homogeneous *polynomial* wings of any order, which formed subject of Chapter IV.

Thus for example by adding a certain convenient term in x_1^m to the above downwashes (5.77), we obtain successively:

$$w = w_0\left(1 + \frac{y^2}{l^2}\right) - 2w_0x_1, \qquad Z = \frac{w_0}{U_\infty}x_1(1 - x_1)\left(1 - \frac{y^2}{l^2}\right), \tag{5.78a}$$

$$w = w_0\left(1 \pm 2\frac{y^3}{l^3}\right) - 3w_0 x_1^2, \qquad Z = \frac{w_0}{U_\infty}x_1(1-x_1^2)\left(1 \mp \frac{y^3}{l^3}\right), \tag{5.78b}$$

$$w = w_0\left[1+(m-1)(\pm 1)^m\frac{y^m}{l^m}\right] - mw_0x_1^{m-1},$$

$$Z = \frac{w_0}{U_\infty}x_1(1-x_1^{m-1})\left[1-(\pm 1)^m\frac{y^m}{l^m}\right], \tag{5.78c}$$

or other general forms,

$$w = w_0x_1^{n-1}\left[n+(m-n)(\pm 1)^m\frac{y^m}{l^m}\right] - mw_0x_1^{m-1},$$

$$Z = \frac{w_0}{U_\infty}x_1^n(1-x_1^{m-n})\left[1-(\pm 1)^m\frac{y^m}{l^m}\right]. \tag{5.79}$$

In the first case (5.78a), a second order homogenous wing (w being given in a polynomial form) is added, in the second case (5.78b), a third-order homogeneous wing is added, and in the last case (5.78c), a general homogeneous wing of order m is added. Thus, by these superpositions, closed complex wings are formed.

For thin wings, this is not necessary since both cases (5.77) and (5.78) may exist independently. In some cases, wings of symmetrical thickness may be open at the trailing edge (at high supersonic speeds this fact is generally less significant) as in case (5.77); for many reasons, however, they are generally closed, as are the cases (5.78). Since homogeneous (polynomial) wings have been comprehensively treated in all previous examples, we shall try to give below some applications pertaining only to the wings with slope or incidence varying according to a zero-order homogeneous function.

V. 6.2. *Wing of Symmetrical Thickness*

Linear variation of the slope. At the beginning we shall treat the simplest case, that when the wing surface has a forced symmetry, as follows from (5.77 a) where a is an arbitrary coefficient, it should be reminded that the plus sign is valid for $y > 0$, and the minus sign for $y < 0$.

Further, taking into consideration (2.75), (2.76) and (2.79), in which l_1 and l_2 will be replaced by $1/B$—since the wing is of symmetrical thickness—

the axial disturbance velocity is given by the following expression:

$$u = \operatorname{Re} C\left(\cosh^{-1}\sqrt{\frac{(1+Bl)(1-Bx)}{2B(l-x)}}+\cosh^{-1}\sqrt{\frac{(1+Bl)(1+Bx)}{2B(l+x)}}\right)$$
$$+\operatorname{Re}\frac{2}{\pi}\int_0^l\left[\left(\frac{dw}{dy}\right)_{y=s}\cosh^{-1}\sqrt{\frac{(1-Bx)(1+Bs)}{2B(s-x)}}\right.$$
$$\left.-\left(\frac{dw}{dy}\right)_{y=-s}\cosh^{-1}\sqrt{\frac{(1+Bx)(1+Bs)}{2B(s+x)}}\right]\frac{s\,ds}{\sqrt{1-B^2s^2}}\,. \quad (5.80)$$

The constant C is determined as indicated previously (§ II. 6); proceeding in this way we find

$$C = -\frac{2}{\pi}\frac{l}{\sqrt{1-B^2l^2}}(w)_{y=l} = -\frac{2w_0}{\pi}\frac{(1+a)l}{\sqrt{1-B^2l^2}}\,. \quad (5.81)$$

Further, noticing that

$$\left(\frac{dw}{dy}\right)_{y=s} = -\left(\frac{dw}{dy}\right)_{y=-s} = \frac{a}{l}w_0, \quad (5.82)$$

we may easily evaluate the above integrals, thus obtaining for the axial velocity $u = \operatorname{Re}\mathcal{U}$ the following formula

$$\mathcal{U} = -\frac{2}{\pi}\frac{w_0 l}{\sqrt{1-B^2l^2}}\left(1+\frac{a}{B^2l^2}\right)\left[\cosh^{-1}\sqrt{\frac{(1+Bl)(1-Bx)}{2B(l-x)}}\right.$$
$$\left.+\cosh^{-1}\sqrt{\frac{(1+Bl)(1+Bx)}{2B(l+x)}}\right]+\frac{2aw_0 l}{\pi B^2l^2}\left[2\cosh^{-1}\sqrt{\frac{1+Bx}{2Bx}}\right.$$
$$\left.-\sqrt{1-B^2x^2}\left(\ln\sqrt{\frac{l^2-x^2}{x^2}}+\frac{i\pi}{2}\right)\right]. \quad (5.83)$$

This expression is valid both for the case of subsonic edges and for that of supersonic edges. In the latter case, we observe that the terms in the first square bracket may be replaced according to the relation

$$\frac{\cosh^{-1}\beta}{\sqrt{1-B^2l^2}} = \frac{\cos^{-1}\beta}{\sqrt{B^2l^2-1}}, \quad (5.84)$$

the imaginary sign due to the radical $\sqrt{1-B^2l^2}$ thus disappearing.

Drag coefficient. Formula (3.38a), which reduces in our case to a simpler form and represents, at the same time, the drag coefficient, will be

$$C_d = -\frac{4}{lU_\infty^2}\int_0^l wu\,dy. \quad (5.85)$$

Taking into account (5.77a) and (5.83) we may perform the above integration which, although it is a simple quadrature, entails laborious operations; finally for the case of subsonic leading edges ($l < 1/B$), we find

$$C_d = \frac{8w_0^2}{\pi B U_\infty} \operatorname{Re}\left\{\left(1+\frac{a}{B^2l^2}\right)\frac{2Bl}{\sqrt{1-B^2l^2}}\cosh^{-1}\frac{1+Bl}{2\sqrt{Bl}}+\sin^{-1}Bl\right.$$
$$-\frac{a}{Bl}\left(2+a-\frac{a+B^2l^2}{\sqrt{1-B^2l^2}}\right)\cosh^{-1}\sqrt{\frac{1+Bl}{2Bl}}+\frac{3+2a}{6}\frac{a}{Bl}(1-\sqrt{1-B^2l^2})$$
$$-\frac{a^2}{3B^3l^3}\left[\ln\frac{2}{1+\sqrt{1-B^2l^2}}+(1-B^2l^2)^{3/2}\ln\frac{2\sqrt{1-B^2l^2}}{1+\sqrt{1-B^2l^2}}\right]$$
$$\left.+\frac{i\pi}{4}\frac{a}{B^2l^2}\left[\sin^{-1}Bl+Bl\sqrt{1-B^2l^2}+\frac{2a}{3Bl}(1-(1-B^2l^2)^{3/2})\right]+\frac{a}{2B^2l^2}I\right\}. \tag{5.86}$$

For the case of supersonic leading edges, setting $l > 1/B$ in the above expression and taking only the real part, we likewise have

$$C_d = \frac{8w_0^2}{\pi B U_\infty^2}\left\{\frac{\pi}{2}+\frac{a}{Bl}\frac{a+B^2l^2}{\sqrt{B^2l^2-1}}\cos^{-1}\sqrt{\frac{1+Bl}{2Bl}}+\frac{3+2a}{6}\frac{a}{Bl}\right.$$
$$-\frac{a^2}{3B^3l^3}\left[\ln\frac{2}{Bl}+(B^2l^2-1)^{3/2}\left(\cosh^{-1}\frac{1}{Bl}+\frac{\pi}{2}\right)\right]$$
$$-\frac{a}{B^2l^2}\frac{\pi}{4}\left[\cosh^{-1}Bl-Bl\sqrt{B^2l^2-1}-\frac{2a}{3Bl}(B^2l^2-1)^{3/2}\right]$$
$$\left.+\frac{a}{2B^2l^2}I\right\}. \tag{5.87}$$

In the two formulae for C_d, I denotes the following definite integral:

$$I = B\int_0^l \sqrt{1-B^2y^2}\ln\frac{l^2-y^2}{y^2}\,dy, \tag{5.88}$$

which is explicitly given below for the two cases $l < 1/B$ and $l > 1/B$, certain simplifying assumptions being introduced in the former case:

$$I \approx 2Bl(1-\tfrac{1}{6}B^2l^2-\tfrac{1}{40}B^4l^4)\ln 2Bl+\tfrac{1}{3}B^3l^3(1+\tfrac{1}{5}B^2l^2), \quad \left(l < \frac{0{\cdot}95}{B}\right), \tag{5.89a}$$

$$I = \frac{\pi}{2}(B^2l^2+\cosh^{-1}Bl-Bl\sqrt{B^2l^2-1}), \quad \left(l > \frac{1}{B}\right). \tag{5.89b}$$

Variation of slope according to a cubic parabola. The wing surface is that of a cone having arcs of a cubic parabola as directrix, as indicated in the example considered above (5.77c), which may be solved as simply as the first case. Indeed, performing the necessary substitutions and calculations, we finally obtain

$$\mathcal{U} = -\frac{2}{\pi}\frac{lw_0}{\sqrt{1-B^2l^2}}\left[1+\frac{2(2-B^2l^2)}{B^4l^4}\right]\left[\cosh^{-1}\sqrt{\frac{(1+Bl)(1-Bx)}{2B(l-x)}}\right.$$
$$\left.+\cosh^{-1}\sqrt{\frac{(1+Bl)(1+Bx)}{2B(l+x)}}\right]+\frac{4lw_0}{\pi B^4l^4}\left\{4\cosh^{-1}\sqrt{\frac{1+Bx}{2Bx}}\right.$$
$$\left.-\sqrt{1-B^2x^2}\left[(2+B^2x^2)\left(\ln\sqrt{\frac{l^2-x^2}{x^2}}+\frac{i\pi}{2}\right)+\tfrac{1}{2}B^2l^2\right]\right\}. \qquad (5.90)$$

which, making eventually the usual substitution (5.84), is also valid for supersonic edges ($l > 1/B$).

Slope variation according to a parabola of the 5th order. If the surface of the conical wing has a directrix in the shape of parabolic arcs of order 5, given by formula (5.77d) with $m = 5$, then, proceeding as above, we obtain

$$\mathcal{U} = -\frac{2}{\pi}\frac{lw_0}{\sqrt{1-B^2l^2}}\left(1+\tfrac{4}{3}\frac{8-4B^2l^2-B^4l^4}{B^6l^6}\right)\left[\cosh^{-1}\sqrt{\frac{(1+Bl)(1-Bx)}{2B(l-x)}}\right.$$
$$\left.+\cosh^{-1}\sqrt{\frac{(1+Bl)(1+Bx)}{2B(l+x)}}\right]+\frac{8lw_0}{3\pi B^6l^6}\left\{16\cosh^{-1}\sqrt{\frac{1+Bx}{2Bx}}\right.$$
$$-\sqrt{1-B^2x^2}\left[(8+4B^2x^2+3B^4x^4)\left(\ln\sqrt{\frac{l^2-x^2}{x^2}}+\frac{i\pi}{2}\right)\right.$$
$$\left.\left.+\tfrac{1}{4}B^2l^2(8+3B^2l^2+6B^2x^2)\right]\right\}. \qquad (5.91)$$

This expression is of course valid for supersonic edges too.

Slope variation according to even powers of y. As may be seen from (5.83), (5.90) and (5.91), the expressions for $\mathcal{U}$ are simple when the variation takes place according to odd powers of y. For even powers, however, notwithstanding all approximations which might be introduced, we are confronted with great calculation difficulties and the expression for $\mathcal{U}$ becomes complicated and lengthy in form.

It is simpler in such a case to replace the square variation

$$w = w_0\left(1+\alpha\frac{y^2}{l^2}+\beta\frac{y^4}{l^4}\right), \qquad (5.92)$$

by another variation of odd powers

$$w = w_0\left(1 \pm a\frac{y}{l} \pm b\frac{y^3}{l^3} \pm c\frac{y^5}{l^5}\right), \tag{5.93}$$

which is very close to the former one. Thus, for instance, for a parabolic variation of the slope, we may write

$$w = w_0\left(1+\frac{y^2}{l^2}\right) \approx w_0\left(1 \pm 0{\cdot}23\frac{y}{l} \pm 1{\cdot}1\frac{y^3}{l^3} \mp 0{\cdot}33\frac{y^5}{l^5}\right), \tag{5.94}$$

which represents a very good approximation, since the two representations practically coincide (Fig. 5.4).

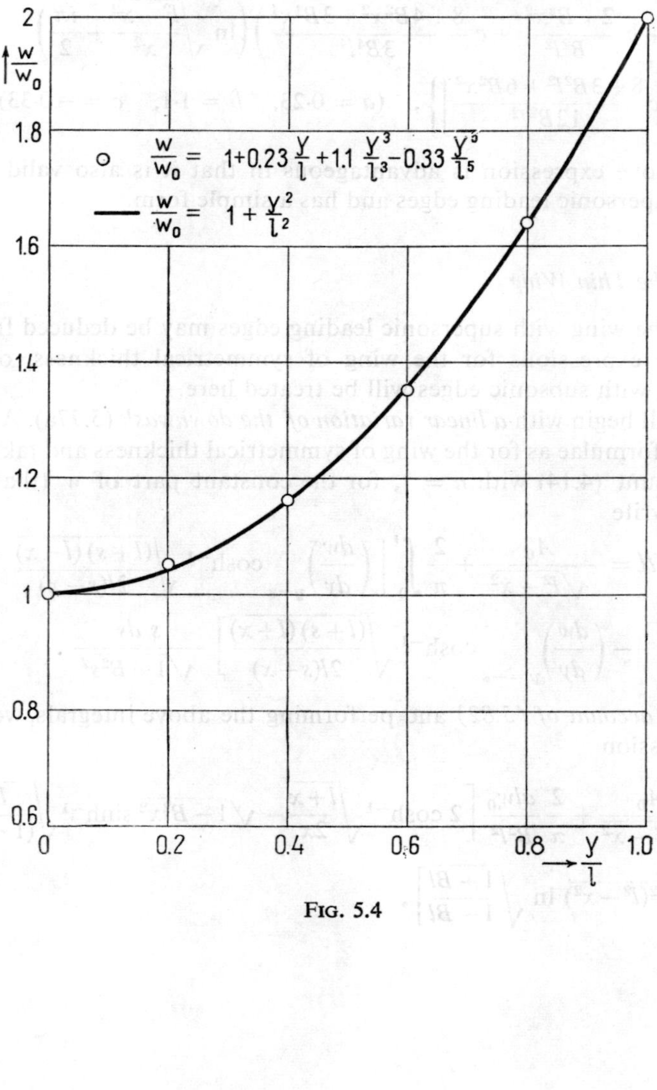

FIG. 5.4

For determining the axial velocity $u = \text{Re}\,\mathcal{U}$, taking into account (5.83), (5.90), (5.91) as well as (5.77a), (5.77c), and (5.77d) with $m = 5$, we may write finally

$$\mathcal{U} = -\frac{2}{\pi}\frac{lw_0}{\sqrt{1-B^2l^2}}\left(1+\frac{a}{B^2l^2}+b\frac{2-B^2l^2}{B^4l^4}+c\frac{8-4B^2l^2-B^4l^4}{3B^6l^6}\right)$$
$$\times\left[\cosh^{-1}\sqrt{\frac{(1+Bl)(1-Bx)}{2B(l-x)}}+\cosh^{-1}\sqrt{\frac{(1+Bl)(1+Bx)}{2B(l+x)}}\right]$$
$$+\frac{2}{\pi}\frac{lw_0}{B^2l^2}\left\{2\left(a+\frac{2b}{B^2l^2}+\frac{8c}{3B^4l^4}\right)\cosh^{-1}\sqrt{\frac{1+Bx}{2Bx}}-\sqrt{1-B^2x^2}\right.$$
$$\times\left[\left(a+b\,\frac{2+B^2x^2}{B^2l^2}+c\,\frac{8+4B^2x^2+3B^4x^4}{3B^4l^4}\right)\left(\ln\sqrt{\frac{l^2-x^2}{x^2}}+\frac{i\pi}{2}\right)\right.$$
$$\left.\left.+\frac{b}{2}+c\,\frac{8+3B^2l^2+6B^2x^2}{12B^2l^2}\right]\right\}, \quad (a = 0{\cdot}23, \quad b = 1{\cdot}1, \quad c = -0{\cdot}33). \quad (5.95)$$

The above expression is advantageous in that it is also valid for the case of supersonic leading edges and has a simple form.

V. 6.3. The Thin Wing

Since the wing with supersonic leading edges may be deduced from the preceding expressions for the wing of symmetrical thickness, only the thin wing with subsonic edges will be treated here.

We shall begin with *a linear variation of the downwash* (5.77a). Applying the same formulae as for the wing of symmetrical thickness and taking also into account (4.14) with $n = 1$, for the constant part of w (that is w_0) we may write

$$\mathcal{U} = \frac{A_0}{\sqrt{l^2-x^2}}+\frac{2}{\pi}\int_0^l\left[\left(\frac{dw}{dy}\right)_{y=s}\cosh^{-1}\sqrt{\frac{(l+s)(l-x)}{2l(s-x)}}\right.$$
$$\left.-\left(\frac{dw}{dy}\right)_{y=-s}\cosh^{-1}\sqrt{\frac{(l+s)(l+x)}{2l(s+x)}}\right]\frac{s\,ds}{\sqrt{1-B^2s^2}}\,. \quad (5.96)$$

Taking account of (5.82) and performing the above integrals, we obtain the expression

$$\mathcal{U} = \frac{A_0}{\sqrt{l^2-x^2}}+\frac{2}{\pi}\frac{alw_0}{B^2l^2}\left[2\cosh^{-1}\sqrt{\frac{l+x}{2x}}-\sqrt{1-B^2x^2}\sinh^{-1}\sqrt{\frac{l^2-x^2}{(1-B^2l^2)x^2}}\right.$$
$$\left.+\sqrt{B^2(l^2-x^2)}\ln\sqrt{\frac{1+Bl}{1-Bl}}\right], \quad (5.97)$$

in which the constant A_0 has the value

$$A_0 = -\frac{w_0 l^2}{E(k)}\left\{1+\frac{2a}{\pi}\left[\frac{\pi}{2}+K(k)(1-E(k'))\right.\right.$$
$$\left.\left.+(K(k)-E(k))\left(K(k')-\frac{1}{Bl}\ln\sqrt{\frac{1+Bl}{1-Bl}}\right)\right]\right\}, \tag{5.98}$$

deduced from the application of the compatibility relation in the form (1.159), where $K(k)$, $K(k')$ and $E(k)$, $E(k')$ are complete elliptic integrals of the first and second kind, having the modulus

$$k = \sqrt{1-B^2l^2}, \quad k' = Bl. \tag{5.99}$$

It may be easily seen that for $l > 1/B$, u has the same expression as in the case of the wing of symmetrical thickness.

For calculating the lift and drag coefficients, we use relations (3.26) and (3.38a), respectively, μ_2 being given by relation (4.10), where $n = 1$. We have thus

$$C_l = \frac{4}{lU_\infty}\int_0^l u\,dy = \frac{2\pi A_0}{lU_\infty}+\frac{alw_0}{B^2l^2U_\infty}\left[2-\frac{1-B^2l^2}{Bl}\ln\frac{1+Bl}{1-Bl}\right], \tag{5.100a}$$

$$C_d = -\frac{4}{lU_\infty^2}\int_0^l wu\,dy = -\frac{w_0}{U_\infty}C_l$$
$$-\frac{4aw_0}{U_\infty}\left\{\frac{A_0}{lU_\infty}+\frac{alw_0}{3\pi B^2l^2U_\infty}\left[1-\frac{1}{B^2l^2}\ln\frac{1}{1-B^2l^2}+2Bl\tanh^{-1}Bl\right]\right\}. \tag{5.100b}$$

If the edges are supersonic, then we introduce into (5.100 a) the expression of $u = \mathrm{Re}\,\mathcal{U}$ corresponding to the thick wing (5.83) and find

$$C_l = -\frac{4w_0}{BU_\infty}\left(1+\frac{a}{2}\right). \tag{5.101}$$

Further, it should be noticed that for $l > 1/B$ the coefficient C_d has the same expression as for the case of the wing of symmetrical thickness (5.87).

Variation of incidence according to a cubic parabola. Proceeding as above for the case in which the downwash varies according to a cubic parabola (5.77c), we find in a similar manner

$$\mathcal{U} = \frac{A_0'}{\sqrt{l^2-x^2}}+\frac{4lw_0}{\pi B^4l^4}\left\{4\cosh^{-1}\sqrt{\frac{l+x}{2x}}\right.$$
$$-\sqrt{1-B^2x^2}(2+B^2x^2)\sinh^{-1}\sqrt{\frac{l^2-x^2}{(1-B^2l^2)x^2}}$$
$$\left.-\sqrt{B^2(l^2-x^2)}\left[\tfrac{1}{2}Bl-\left(\frac{3+B^2l^2}{2}+B^2x^2\right)\ln\sqrt{\frac{1+Bl}{1-Bl}}\right]\right\}. \tag{5.102}$$

Where A_0' may be determined in the same way as for the linear variation.

Variation of incidence according to a parabola of the 5th order. In the case of the downwash variation given by (5.77d), with $m = 5$, proceeding in a similar manner we obtain

$$\mathcal{U} = \frac{A_0''}{\sqrt{l^2-x^2}} + \frac{8lw_0}{3\pi B^6 l^6}\left\{16 \cosh^{-1}\sqrt{\frac{l+x}{2x}}\right.$$
$$-\sqrt{1-B^2x^2}\,(8+4B^2x^2+3B^4x^4)\sinh^{-1}\sqrt{\frac{l^2-x^2}{(1-B^2l^2)x^2}}$$
$$-\tfrac{1}{8}\sqrt{B^2(l^2-x^2)}\Big[Bl(13+9B^2l^2+12B^2x^2)$$
$$\left.-(45+10B^2l^2+9B^4l^4+(20+12B^2l^2)B^2x^2+24B^4x^4)\ln\sqrt{\frac{1+Bl}{1-Bl}}\Big]\right\}. \tag{5.103}$$

Parabolic variation of incidence. As in the case corresponding to a wing of symmetrical thickness, we put the expression for the downwash in the form (5.93) or (5.94) and, taking into account the expressions for the axial velocities $u = \mathrm{Re}\,\mathcal{U}$ established above, we obtain

$$\mathcal{U} = \frac{A}{\sqrt{l^2-x^2}} + \frac{2lw_0}{\pi B^2 l^2}\left\{2\left(a+\frac{2b}{B^2l^2}+\frac{8c}{3B^4l^4}\right)\cosh^{-1}\sqrt{\frac{l+x}{2x}}\right.$$
$$-\sqrt{1-B^2x^2}\left(a+b\frac{2+B^2x^2}{B^2l^2}+c\frac{8+4B^2x^2+3B^4x^4}{3B^4l^4}\right)$$
$$\times\sinh^{-1}\sqrt{\frac{l^2-x^2}{(1-B^2l^2)x^2}}-\sqrt{B^2(l^2-x^2)}\Big[\frac{b}{2Bl}$$
$$+c\,\frac{13+9B^2l^2+12B^2x^2}{24B^3l^3}-\Big(a+b\,\frac{3+B^2l^2+2B^2x^2}{2B^2l^2}$$
$$+c\,\frac{45+10B^2l^2+9B^4l^4+(20+12B^2l^2)B^2x^2+24B^4x^4}{24B^4l^4}\Big)$$
$$\left.\times\ln\sqrt{\frac{1+Bl}{1-Bl}}\Big]\right\}, \tag{5.104}$$

where, as seen above,

$$a = 0{\cdot}23, \quad b = 1{\cdot}1, \quad c = -0{\cdot}33. \tag{5.105}$$

Note. The examples treated in this section (§ V. 6) refer to a particular class of wings whose surfaces are cones. Since such surfaces are employed in certain applications, we thought it useful to give the calculation method for determining the axial velocities and hence the pressures, etc.

CHAPTER VI

TRAPEZOIDAL AND RECTANGULAR WINGS

IN THIS chapter we shall study trapezoidal and rectangular wings. As in the two preceding chapters, we are going to treat a series of current applications of these wings employed in supersonic aerodynamics.

VI. 1. Generalities on Trapezoidal Wings

VI. 1.1. Preliminary Considerations

Let us consider a wing $O_1A_1A_2O_2$, of trapezoidal form, with leading edge $O_1\,O_2$ of length $2b$, trailing edge $A_1\,A_2$ of length $2(b+l_1)$, and chord equal to unity (Fig. 6.1).

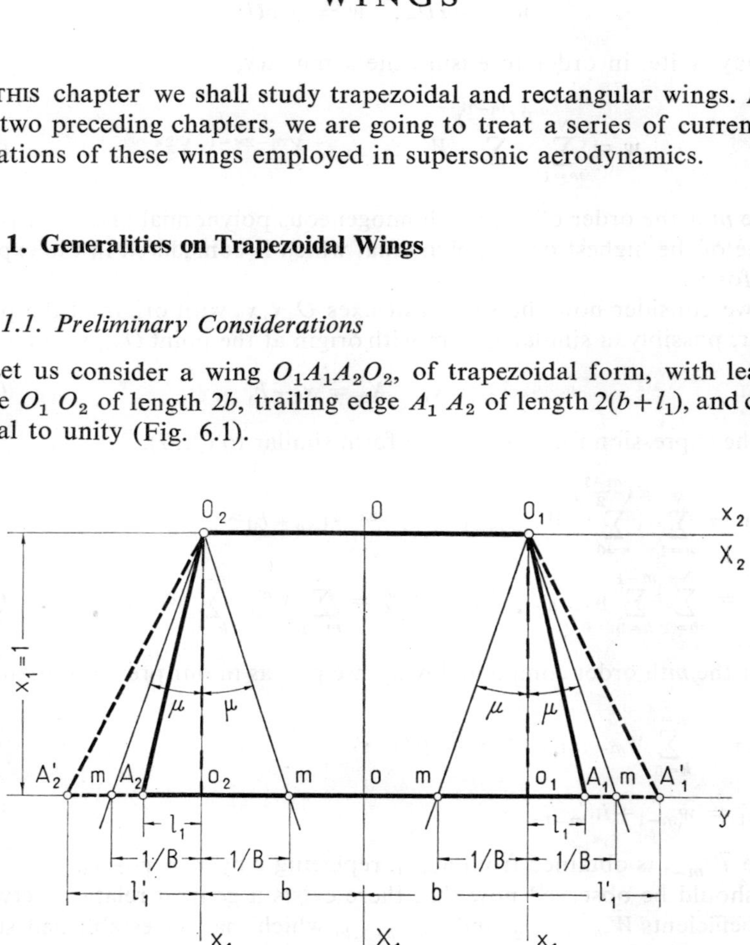

FIG. 6.1

We assume that the incidence α of a thin wing or the slope γ of a wing of symmetrical thickness varies according to a law represented in the form of a superposition of homogeneous polynomials of various orders, which ensures a *natural symmetry* with respect to the OX_1 axis. These polynomials contain X_2 in even powers.

Thus, taking into account the relations indicated previously (2.11), which refer to the downwash on a thin wing or on a wing of symmetrical thickness, that is

$$w = -\alpha U_\infty, \quad w = \pm\gamma U_\infty \tag{6.1}$$

we may write, in order to ensure the symmetry,

$$w = \sum_{m=1}^{N} \sum_{\varkappa=0}^{E\left(\frac{m-1}{2}\right)} W_{m-2\varkappa-1,\,2\varkappa}\, X_1^{m-2\varkappa-1}\, X_2^{2\varkappa}\,, \tag{6.2}$$

where m is the order of a single homogeneous polynomial and $N-1$ is the degree of the highest order polynomial which is considered in the expression for w.

If we consider now the system of axes $O_1x_1x_2$ with origin at the point O_1 (or, possibly, a similar system with origin at the point O_2), we may put

$$X_1 = x_1, \quad X_2 = x_2 + b, \tag{6.3}$$

and the expression for w assumes a form similar to (2.17),

$$\begin{aligned} w &= \sum_{m=1}^{N} \sum_{\varkappa=0}^{E\left(\frac{m-1}{2}\right)} W_{m-2\varkappa-1,\,2\varkappa}\, x_1^{m-2\varkappa-1}(x_2+b)^{2\varkappa} \\ &= \sum_{m=1}^{N} \sum_{k=0}^{m-1} w_{m-k-1,\,k}\, x_1^{m-k-1}\, x_2^{k} = \sum_{m=1}^{N} x_1^{m-1} \sum_{k=0}^{m-1} w_{m-k-1,\,k}\, y^{k}. \end{aligned} \tag{6.4}$$

For the mth order component wing, we put, as in our previous notation,

$$w_{m-1} = \sum_{k=0}^{m-1} w_{m-k-1,\,k}\, y^k = \operatorname{Re}\mathcal{W}_{m-1}\,, \tag{6.5a}$$

$$\mathcal{W}_{m-1} = w_{m-1} + i w'_{m-1}\,, \tag{6.5b}$$

where $\mathcal{W}_{m-1}$ is obtained from w_{m-1}, replacing y by $\mathfrak{x} = y + i\mathfrak{z}$.

It should be observed now that there exists a general relation between the coefficients $W_{m-2\varkappa-1,\,2\varkappa}$ and $w_{m-k-1,\,k}$, which may be established starting from relation (6.4).

For simplifying the general problem, we shall proceed from the fact that the calculations become extremely laborious for $N > 4$, and hence, we consider the highest order $N = 4$, since in this manner alone are we able to close the wing on all four edges, *a necessary condition in the case of a wing of symmetrical thickness*. In such a case, the equation of the wing may be written in the form

$$Z = tX_1(X_1-1)[(l_1X_1+b)^2-X_2^2], \tag{6.6}$$

where t is a constant proportional to the relative maximum thickness. The greatest order for w is $N-1 = 3$.

It should be mentioned that the minimum order $N = 4$, which was necessary for a wing of symmetrical thickness, is not necessary for thin wings, since these may exist for any N, as long as the symmetry conditions are obeyed.

It should be noticed, however, that in the case of a wing of symmetrical thickness, we may satisfy the closure condition even for a smaller N, for example $N = 2$. Indeed, considering the following two equations for the surface of the right-hand half of the wing,

$$Z' = t'x_1(1-x_1), \quad Z'' = t''(1-x_1)\left(x_1-\frac{x_2}{l}\right), \tag{6.7}$$

where the former refers to the left side of the Ox_1 axis (the rectangular part), and the latter to the right side of the Ox_1 axis (the triangular part), the wing closes on all sides (Fig. 6.1) and the highest order of the conical motion is $N = 2$. In such a case, however, we have to consider one ridge on each axis O_1x_1 and O_2x_1, which will be taken into account in the study of the wing of symmetrical thickness, in order to treat more easily certain applications.

In every case, at the points O_1 and O_2 we have two high order conical motions which are identical, but symmetrical with respect to the Oo axis and independent from each other. We shall further assume that the two Mach lines, O_1m and O_2m, issued from the two vertices, do not intersect on the wing, so that the effects of the two Mach cones superpose on no portion of the wing.

Under these conditions, it is sufficient to study one of the halves of the trapezoidal wing, OO_1A_1o or OO_2A_2o.

The case when the two cones with their vertices at O_1 and O_2 intersect on the wing will be treated in a more comprehensive form in our next chapter.

VI. 1.2. Determination of Axial Velocities

In order to find the motion corresponding to the cone with vertex at O_1 (Fig. 6.2), we start from an asymmetrical triangular wing with one supersonic edge OA_2; as regards the other leading edge, this may be either subsonic (O_1A_1, $l_1 < 1/B$) or supersonic (O_1A_1', $l_1' > 1/B$).

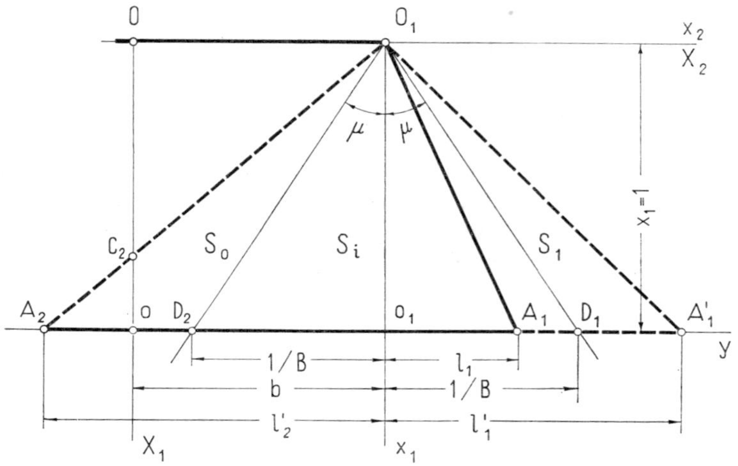

FIG. 6.2

TRIANGULAR THIN WING. The axial velocity u (or the analytic function $\mathcal{U}_{n-1} = u_{n-1} + iu'_{n-1}$) for the case of the subsonic lateral leading edge O_1A_1 of the thin wing will be, by applying formula (2.84),

$$\mathcal{U}_{n-1} = \sum_{q=0}^{n-1} \left[A_{nq} x^q \sqrt{\frac{1+Bx}{B(l_1-x)}} + H_{nq} x^q \cos^{-1} \sqrt{\frac{(l_1+l_2)(1+Bx)}{(1+Bl_1)(l_2+x)}} \right]. \tag{6.8}$$

For the case of the supersonic leading edge OA_1', by applying formula (2.89), we have

$$\mathcal{U}_{n-1} = \sum_{q=0}^{n-1} K_{nq} x^q \cos^{-1} \sqrt{\frac{(1+Bl_1)(1-Bx)}{2B(l_1-x)}} + \sum_{q=0}^{n-1} H_{nq} x^q \cos^{-1} \sqrt{\frac{(1+Bl_2)(1+Bx)}{2B(l_2+x)}} + \sum_{q=0}^{n-2} D_{nq} x^q \sqrt{\frac{1-B^2x^2}{B^2}}. \tag{6.9}$$

TRIANGULAR THICK WING. This last expression is also valid for the wing of symmetrical thickness, in both positions of the leading edge (O_1A_1 or O_1A_1'), introducing eventually also a formal change

$$K_{nq}\cos^{-1}\beta = C_{nq}\cosh^{-1}\beta, \tag{6.10}$$

in order to have real constants in all cases. On the other hand, taking into account the above remark in connection with the closure conditions of the wing with the degree N as small as possible, we also add a *central ridge*, in which case, applying (2.59), we may write:

$$\begin{aligned}\mathcal{U}_{n-1} &= \sum_{q=0}^{n-1} C_{nq}\,x^q \cosh^{-1}\sqrt{\frac{(1+Bl_1)(1-Bx)}{2B(l_1-x)}} \\ &+ \sum_{q=0}^{n-1} H_{nq}\,x^q \cos^{-1}\sqrt{\frac{(1+Bl_2)(1+Bx)}{2B(l_2+x)}} \\ &+ \sum_{q=0}^{n-1} Q_{nq}\,x^q \cosh^{-1}\sqrt{\frac{1}{B^2x^2}} + \sum_{q=0}^{n-2} D_{nq}x^q\sqrt{\frac{1-B^2x^2}{B^2}}\,. \end{aligned} \tag{6.11}$$

TRAPEZOIDAL WING. Now, making l_2 tend to infinity, we obtain the expression for $\mathcal{U}_{n-1}$ for the trapezoidal wing, in all the three cases indicated above:

$$\mathcal{U}_{n-1} = \sum_{q=0}^{n-1} A_{nq}x^q\sqrt{\frac{1+Bx}{B(l_1-x)}} + \sum_{q=0}^{n-1} H_{nq}x^q\cos^{-1}\sqrt{\frac{1+Bx}{1+Bl_1}}, \tag{6.12a}$$

$$\begin{aligned}\mathcal{U}_{n-1} &= \sum_{q=0}^{n-1} K_{nq}\,x^q\cos^{-1}\sqrt{\frac{(1+Bl_1)(1-Bx)}{2B(l_1-x)}} + \sum_{q=0}^{n-1} H_{nq}x^q\cos^{-1}\sqrt{\frac{1+Bx}{2}} \\ &+ \sum_{q=0}^{n-2} D_{nq}x^q\sqrt{\frac{1-B^2x^2}{B^2}}\,, \end{aligned} \tag{6.12b}$$

$$\begin{aligned}\mathcal{U}_{n-1} &= \sum_{q=0}^{n-1} C_{nq}\,x^q\cosh^{-1}\sqrt{\frac{(1+Bl_1)(1-Bx)}{2B(l_1-x)}} + \\ &+ \sum_{q=0}^{n-1} H_{nq}x^q\cos^{-1}\sqrt{\frac{1+Bx}{2}} \\ &+ \sum_{q=0}^{n-1} Q_{nq}x^q\cosh^{-1}\sqrt{\frac{1}{B^2x^2}} + \sum_{q=0}^{n-2} D_{nq}x^q\sqrt{\frac{1-B^2x^2}{B^2}}\,. \end{aligned} \tag{6.12c}$$

VI. 1.3. Considerations on the Calculation of the Aerodynamic Characteristics

Generally, the integration of forces and moments is best performed over portions of the wing which are most convenient for the appropriate calculations. Thus, considering the half of the trapezium OO_1A_1O (or $OO_1A_1'O$), it is more convenient to perform the computation of the aerodynamic characteristics of the half wing, of surface S, by dividing it into three portions (Fig. 6.2): S_0, the left-hand surface, outside the Mach cone, S_i, the surface inside the Mach cone, and S_1, the right-hand surface, also outside the Mach cone if we consider the case of the supersonic leading edge OA_1'. Thus we have for the integral of forces and moments,

$$\int_S = \int_{S_0} + \int_{S_i} + \int_{S_1}, \quad S = \tfrac{1}{2}(2b+l_1). \tag{6.13}$$

On the surfaces S_0 and S_1, the axial velocity u and the downwash w are represented by polynomials in x_1, x_2, hence the integration is a simple elementary quadrature. Indeed, considering a term of the form $E_{pr}x_1^{p-1}x_2^{r-1}$ from the expression of the aerodynamic characteristics (E_{pr} = constant) acting at any point on the elementary surface $dx_1\,dx_2 = x_1\,dx_1\,dy$, we have:

(a) on the surface S_0,

$$\begin{aligned}\int_{S_0} E_{pr}x_1^{p-1}x_2^{r-1}dx_1\,dx_2 &= E_{pr}\int_0^1 x_1^{p+r-1}\,dx_1\int_{-b/x_1}^{-1/B} y^{r-1}\,dy\\ &= -\frac{(-1)^r}{r}\left[\frac{b^r}{p}-\frac{1}{(p+r)B^r}\right]E_{pr},\\ S_0 &= \tfrac{1}{2}\left(2b-\frac{1}{B}\right);\end{aligned} \tag{6.14}$$

(b) on the surface S_1, if we consider the supersonic leading edge OA_1',

$$\begin{aligned}\int_{S_1} E_{pr}x_1^{p-1}x_2^{r-1}\,dx_1\,dx_2 &= E_{pr}\int_0^1 x_1^{p+r-1}\,dx_1\int_{1/B}^{l_1} y^{r-1}\,dy\\ &= \frac{E_{pr}}{(p+r)r}\left[(l_1)^r-\left(\frac{1}{B}\right)^r\right],\\ S_1 &= \tfrac{1}{2}\left(l_1-\frac{1}{B}\right);\end{aligned} \tag{6.15}$$

(c) on the surface S_i, in the interior of the Mach cone, we apply the formulae previously established, in which, however, the integrals have other limits; thus,

—for the *lift* we apply (3.26),

$$C_l^{(in)} = \frac{4}{(n+1)S_i U_\infty} \int_{-1/B}^{a} u_{n-1}\, dy, \tag{6.16}$$

—for the *wave drag* we apply (3.38a),

$$C_{dm}^{(in)} = -\frac{4}{(m+n)S_i U_\infty^2} \int_{-1/B}^{a} \left(\sum_{k=0}^{m-1} w_{m-k-1,k}\, y^k\right) u_{n-1}\, dy, \tag{6.17}$$

—for the *pitching moment* we apply (3.53) and (3.55a),

$$\mu_2^{(in)} = -\frac{4}{(n+2)S_i h U_\infty} \int_{-1/B}^{a} u_{n-1}\, dy = -\frac{1}{h} \cdot \frac{(n+1)}{n+2} C_l^{(in)}, \tag{6.18}$$

where we may take $h = 1$.

In the above formulae the limit a will be different according as the lateral edge is subsonic or supersonic and, consequently, S_i also has different values:

$$a = l_1, \quad S_i = \tfrac{1}{2}\left(\frac{1}{B} + l_1\right), \tag{6.19a}$$

$$a = \frac{1}{B}, \quad S_i = \frac{1}{B}. \tag{6.19b}$$

Derivation of some Useful Formulae needed for Computing the Aerodynamic Coefficients. The above considerations give us indications on the manner of computing the aerodynamic characteristics. Obviously, we may employ any other method for integrating the forces and moments on the wing portions outside and inside the Mach cone.

As previously indicated, the integrals (6.13, 6.15) extended over the surfaces S_0 or S_1 are simple quadratures; there are no difficulties in performing any computation. But the integrals over the surface S_i need special treatment.

Thus, for example, out of the methods employed in this respect, we can apply the residue method, but in a different way from that used in preceding cases.

Indeed, considering the different expressions for $\mathcal{U}_{n-1}$ (6.12), the following forms of integrals are to be calculated in formulae (6.16, 6.17):

$$\int y^q \sqrt{\frac{1+By}{B(l_1-y)}}\, dy = \int \frac{y^q(1+By)\, dy}{\sqrt{B(l_1-y)\,(1+By)}}, \tag{6.20a}$$

$$\int y^q \cos^{-1} \sqrt{\frac{1+By}{1+Bl_1}}\, dy = \frac{y^{q+1}}{q+1} \cos^{-1} \sqrt{\frac{1+By}{1+Bl_1}} + \frac{B}{2(q+1)} \int \frac{y^{q+1}\, dy}{\sqrt{B(1+By)\,(l_1-y)}}, \tag{6.20b}$$

$$\int y^q \cos^{-1} \sqrt{\frac{(1+Bl_1)(1-By)}{2B(l_1-y)}}\, dy = \frac{y^{q+1}-l_1^{q+1}}{q+1} \cos^{-1} \sqrt{\frac{(1+Bl_1)(1-By)}{2B(l_1-y)}} + \frac{\sqrt{B^2 l_1^2 - 1}}{2(q+1)} \sum_{j=0}^{q} l_1^j \int \frac{y^{q-j}\, dy}{\sqrt{1-B^2y^2}}\,. \tag{6.20c}$$

$$\int y^q \sqrt{1-B^2y^2}\, dy = \int \frac{y^q(1-B^2y^2)\, dy}{\sqrt{1-B^2y^2}}\,, \tag{6.20d}$$

$$\int y^q \cosh^{-1} \sqrt{\frac{(1+Bl_1)(1-By)}{2B(l_1-y)}}\, dy = \frac{y^{q+1}-l_1^{q+1}}{q+1} \cosh^{-1} \sqrt{\frac{(1+Bl_1)(1-By)}{2B(l_1-y)}} + \frac{\sqrt{1-B^2 l_1^2}}{2(q+1)} \sum_{j=0}^{q} l_1^j \int \frac{y^{q-j}\, dy}{\sqrt{1-B^2y^2}}\,, \tag{6.20e}$$

$$\int y^q \cosh^{-1} \sqrt{\frac{1}{B^2y^2}}\, dy = \frac{y^{q+1}}{q+1} \cosh^{-1} \sqrt{\frac{1}{B^2y^2}} + \frac{1}{q+1} \int \frac{y^q\, dy}{\sqrt{1-B^2y^2}}\,. \tag{6.20f}$$

It is to be noted that, in each of the above expressions, the following kinds of integrals have to be calculated:

(a) *For the thin trapezoidal wing,*

$$I_t = \int_{-1/B}^{l_1} \frac{y^t\, dy}{\sqrt{B(l_1-y)(1+By)}} = -\tfrac{1}{2} \oint \frac{x^t\, dx}{\sqrt{B(l_1-x)(1+Bx)}}$$
$$= \frac{\pi}{B^{t+1}} \left\{ \frac{(2t)!\,[(Bl_1)^t + (-1)^t]}{2^{2t}(t!)^2(1+\delta_0^t)} + \sum_{\nu=1}^{t-1} \frac{(-1)^{t-\nu}(2\nu)!\,(2t-2\nu)!\,(Bl_1)^\nu}{2^{2t}[(t-\nu)!]^2(\nu!)^2} \right\}, \tag{6.21a}$$

$$I_t' = \int_{-1/B}^{1/B} \frac{y^t\, dy}{\sqrt{1-B^2y^2}} = -\tfrac{1}{2} \oint \frac{x^t\, dx}{\sqrt{1-B^2x^2}} = \frac{\pi}{B^{t+1}} \left\{ \frac{(2t!)[1+(-1)^t]}{2^{2t}(t!)^2(1+\delta_0^t)} + \sum_{\nu=1}^{t-1} \frac{(-1)^{t-\nu}(2\nu)!\,(2t-2\nu)!}{2^{2t}[(t-\nu)!]^2(\nu!)^2} \right\}, \tag{6.21b}$$

this second integral I_t' being deduced from the preceding one I_t, by setting $l_1 = 1/B$.

In the above expressions δ_i^j is Kronecker's symbol,

$$\delta_i^j = \begin{cases} 1, & \text{for} \quad i = j \\ 0, & \text{for} \quad i \neq j. \end{cases}$$

(b) *For the thick trapezoidal wing,* $\mathcal{J}_t(a)$ is given by the following expressions:

$$\mathcal{J}_{2k}(a) = \int_0^a \frac{y^{2k}\,dy}{\sqrt{1-B^2y^2}} = \frac{(2k)!}{2^{2k}(k!)^2B^{2k+1}}\left[\sin^{-1} Ba - \sqrt{1-B^2a^2}\sum_{p=0}^{k-1}\frac{2^{2p}(p!)^2(Ba)^{2p+1}}{(2p+1)!}\right], (t = 2k) \qquad (6.22a)$$

$$\mathcal{J}_{2k+1}(a) = \int_0^a \frac{y^{2k+1}\,dy}{\sqrt{1-B^2y^2}} = \frac{2^{2k}(k!)^2}{(2k+1)!B^{2(k+1)}}\left[1 - \sqrt{1-B^2a^2}\sum_{p=0}^{k}\frac{(2p)!(Ba)^{2p}}{2^{2p}(p!)^2}\right], (t = 2k+1) \qquad (6.22b)$$

$(a = 1/B;\ -1/B;\ l_1)$.

VI. 2. Thin Trapezoidal Wing with Subsonic Lateral Edges

VI. 2.1. Constants in the Expressions for the Axial Velocities

As above, we consider only one-half of the trapezoidal wing (Fig. 6.3). The expression for the axial velocity is deduced from (6.12a) and we rewrite it below:

$$\mathcal{U}_{n-1} = \sum_{q=0}^{n-1} A_{nq}\, x^q \sqrt{\frac{1+Bx}{B(l_1-x)}} + \sum_{q=0}^{n-1} H_{nq}\, x^q \cos^{-1}\sqrt{\frac{1+Bx}{1+Bl_1}}\,. \qquad (6.23)$$

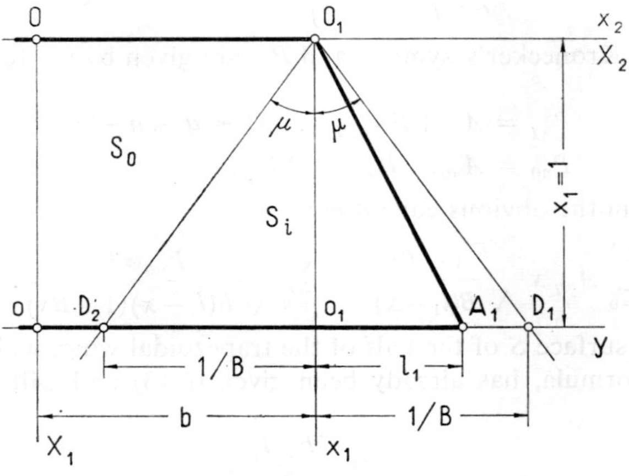

FIG. 6.3

The values of the constants for the two first motions $n = 1, 2$ are defined by the coefficients $w_{m-k-1,k}$ of the expression w_{m-1} (6.5a), namely: w_{00}, w_{10}, w_{01}.

Indeed, by means of the method for determining the constants presented previously (§ II. 5), we find

$$A_{10} = -\frac{2w_{00}}{\pi B}\frac{Bl_1}{1+Bl_1}, \quad H_{10} = -\frac{2w_{00}}{\pi B} \tag{6.24}$$

for $n = 1$;

$$A_{20} = -\frac{2Bl_1}{3\pi B(1+Bl_1)^2}[(5+3Bl_1)w_{10}-4Bw_{01}] \tag{6.25a}$$

$$A_{21} = \frac{2}{3\pi B(1+Bl_1)^2}[(3+Bl_1)Bw_{10}-(3+2Bl_1+3B^2l_1^2)w_{01}] \tag{6.25b}$$

$$H_{20} = -\frac{2w_{10}}{\pi B}, \quad H_{21} = -\frac{2w_{01}}{\pi B} \tag{6.25c}$$

for $n = 2$.

VI. 2.2. Lift Coefficient

Applying formulae (6.14) corresponding to the surface S_0 and (6.16) corresponding to the surface S_i, and taking into account (6.21), we find the general expression for the lift coefficient

$$C_l^{(n)} = \frac{8}{(n+1)(2b+l_1)U_\infty}\sum_{q=0}^{n}\left\{P_{nq}I_q+\frac{(1-\delta_q^n)H_{nq}}{2(q+1)}\left[BI_{q+1}\right.\right.$$
$$\left.\left.+\frac{(-1)^q\pi(n+1)b^{q+1}}{n-q}\right]\right\}, \tag{6.26}$$

where δ_q^n is Kronecker's symbol and P_{nq} are given by the following relations,

$$P_{nq} = A_{nq}+BA_{n,q-1}, \quad (1 \leqslant q \leqslant n-1), \tag{6.27a}$$
$$P_{n0} = A_{n0}, \quad P_{nn} = BA_{n,n-1}, \tag{6.27b}$$

deduced from the obvious equalities:

$$\sum_{q=0}^{n-1} A_{nq}\,x^q\sqrt{\frac{1+Bx}{B(l_1-x)}} = \sum_{q=0}^{n}\frac{P_{nq}\,x^q}{\sqrt{B(l_1-x)(1+Bx)}}. \tag{6.28}$$

The total surface S of the half of the trapezoidal wing, which occurs in the above formula, has already been given (6.13) and will be rewritten below:

$$S = \frac{2b+l_1}{2}.$$

Applying formula (6.26) for $n = 1, 2$ and taking at the same time account of (6.22), we obtain successively:

$$C_l^{(1)} = \frac{\pi}{B(2b+l_1)U_\infty}[2(1+Bl_1)A_{10}+(Bl_1+4Bb-1)H_{10}], \tag{6.29a}$$

$$C_l^{(2)} = \frac{\pi}{12B^2(2b+l_1)U_\infty}\{4(1+Bl_1)[4BA_{20}+(3Bl_1-1)A_{21}]$$
$$+8(Bl_1+3Bb-1)BH_{20}+(3B^2l_1^2-2Bl_1-24B^2b^2+3)H_{21}\}. \tag{6.29b}$$

The total lift coefficient of the complex wing, i.e. C_l, is equal to the sum of the above coefficients.

VI. 2.3. Pitching Moment Coefficient

The general formula for the pitching moment coefficient does not differ from that of the lift (6.26), except by the fact that n is replaced by $n+1$ when n appears in an arbitrary factor or as an exponent. This replacement *will be no longer made* for the summation limits, for the indices of Kronecker's symbol and for the indices of constants. We give below the coefficients $\mu_2^{(n)}$ for the general case:

$$\mu_2^{(n)} = -\frac{8}{(n+2)(2b+l_1)hU_\infty}\sum_{q=0}^{n}\left\{P_{nq}I_q+\frac{(1-\delta_q^n)H_{nq}}{2(q+1)}\left[BI_{q+1}\right.\right.$$
$$\left.\left.+\frac{(-1)^q\pi(n+2)b^{q+1}}{n-q+1}\right]\right\}, \tag{6.30}$$

where I_t and P_{nq} have the same significance as in the preceding formulae (6.21) and (6.27), respectively, h being a reference length.

Further, taking $h = 1$ and particularizing the above formula for $n = 1, 2$, we obtain successively:

$$\mu_2^{(1)} = -\frac{2\pi}{3B(2b+l_1)U_\infty}\{2(Bl_1+1)A_{10}+(Bl_1-1+3Bb)H_{10}\}, \tag{6.31a}$$

$$\mu_2^{(2)} = -\frac{\pi}{48B^2(2b+l_1)U_\infty}\{12(Bl_1+1)[4BA_{20}+(3Bl_1-1)A_{21}]$$
$$+8(3Bl_1-3+8Bb)BH_{20}+3(3B^2l_1^2-2Bl_1+3-16B^2b^2)H_{21},\} \tag{6.31b}$$

The sum of the above coefficients yields the total coefficient for the complex wing.

VI. 2.4. Wave Drag Coefficient

Applying (6.14) on the surface S_0 and (6.17) on S_1, and at the same time taking into account (6.21), we find the general formula for the wave drag coefficient:

$$C_{dm}^{(n)} = -\frac{8}{(m+n)(2b+l_1)U_\infty} \sum_{k=0}^{m-1} w_{m-k-1,k} \sum_{q=0}^{n} \Big\{ P_{nq} I_{k+q} + \frac{(1-\delta_q^n)H_{nq}}{2(k+q+1)} \Big[B I_{k+q+1} + \frac{(-1)^{k+q}\pi(m+n)b^{k+q+1}}{m+n-k-q-1} \Big] \Big\} \tag{6.32}$$

Particularizing for $m = 1, 2$ and $n = 1, 2$ and taking account of (6.21), we obtain:

$$C_{d1}^{(1)} = -\frac{\pi w_{00}}{(2b+l_1)BU_\infty^2} [2(Bl_1+1)A_{10} + (Bl_1-1+4Bb)H_{10}], \tag{6.33a}$$

$$\begin{aligned} C_{d2}^{(2)} = &-\frac{\pi}{48(2b+l_1)B^3U_\infty^2} \{Bw_{10}[12(Bl_1+1)(4BA_{20}+(3Bl_1-1)A_{21}) \\ &+8(3Bl_1-3+8Bb)BH_{20}+(3B^2l_1^2-2Bl_1+3-16B^2b^2)H_{21}] \\ &+w_{01}[6(1+Bl_1)(2(3Bl_1-1)BA_{20}+(5B^2l_1^2-2Bl_1+1)A_{21}) \\ &+3(3B^2l_1^2-2Bl_1+3-15B^2b^2)BH_{20} \\ &+((Bl_1-1)(5B^2l_1^2+2Bl_1+5)+64B^3b^3)H_{21}]\}, \end{aligned} \tag{6.33b}$$

$$\begin{aligned} C_{d1}^{(2)} = &-\frac{\pi w_{00}}{12(2b+l_1)B^2U_\infty^2} \{4(1+Bl_1)[4BA_{20}+(3Bl_1-1)A_{21}]+8(Bl_1-1 \\ &+3Bb)BH_{20}+(3B^2l_1^2-2Bl_1+3-24B^2b^2)H_{21}\}, \end{aligned} \tag{6.33c}$$

$$\begin{aligned} C_{d2}^{(1)} = &-\frac{\pi}{12(2b+l_1)B^2U_\infty^2} \{8Bw_{10}[2(1+Bl_1)A_{10}+(Bl_1-1+3Bb)H_{10}] \\ &+w_{01}[4(1+Bl_1)(3Bl_1-1)A_{10}+(3B^2l_1^2-2Bl_1+3-8B^2b^2)H_{10}]\}. \end{aligned} \tag{6.33d}$$

The sum of these coefficients is equal to the total coefficient for a complex wing constituted of the components $n = 1$ and $n = 2$.

VI. 2.5. Suction Force Coefficient

The intermediate suction coefficient due to the axial velocities of two conical motions of orders m and n is given by formula (3.103). In our case, considering the equation (6.13) for half of the trapezoidal wing surface, we obtain the following expression for the coefficient $\sigma_m^{(n)}$:

$$\begin{aligned} \sigma_m^{(n)} &= -\frac{4\pi\sqrt{1-B^2l_1^2}}{(2b+l_1)l_1U_\infty^2} \Big[\frac{u_{m-1}u_{n-1}}{m+n} (l_1-y) \Big]_{y\to l_1} \\ &= -\frac{4\pi\sqrt{1-B^2l_1^2}}{(2b+l_1)(m+n)l_1U_\infty^2} \frac{1+Bl_1}{B} \sum_{k=0}^{m-1} \sum_{q=0}^{n-1} A_{mk}A_{nq}l_1^{k+q}. \end{aligned} \tag{6.34}$$

Particularizing for $m = 1, 2$ and $n = 1, 2$, we obtain

$$\sigma_1^{(1)} = -\frac{2\pi(1+Bl_1)\sqrt{1-B^2l_1^2}}{U_\infty^2 Bl_1(2b+l_1)} A_{10}^2, \tag{6.35a}$$

$$\sigma_2^{(2)} = -\frac{\pi(1+Bl_1)\sqrt{1-B^2l_1^2}}{U_\infty^2 Bl_1(2b+l_1)} (A_{20}+A_{21}l_1)^2, \tag{6.35b}$$

$$\sigma_1^{(2)} = \sigma_2^{(1)} = -\frac{\pi(1+Bl_1)\sqrt{1-B^2l_1^2}}{U_\infty^2 Bl_1(2b+l_1)} A_{10}(A_{20}+A_{21}l_1). \tag{6.35c}$$

The sum of these coefficients represents the total suction coefficient of the trapezoidal wing.

VI. 3. Application to a Thin Rectangular Wing

VI. 3.1. Expression for the Axial Velocity and its Constants

If we put $l_1 = 0$ in formula (6.23), we obtain the following expression for the complex function $\mathcal{U}_{n-1}$, whose real part defines the axial velocity:

$$\mathcal{U}_{n-1} = \sum_{q=0}^{n-1} \left[A_{nq}\, x^q \sqrt{\frac{1+Bx}{-Bx}} + H_{nq}\, x^q \cos^{-1}\sqrt{1+Bx} \right], \quad A_{n0} = 0. \tag{6.36}$$

The constant A_{n0} must be equal to zero since, in the case of a rectangular wing, the lateral leading edge plays in the limit the part of a trailing edge (the trailing edge O_1A_1' passes in the limit into OA_1, Fig. 6.4); the axial

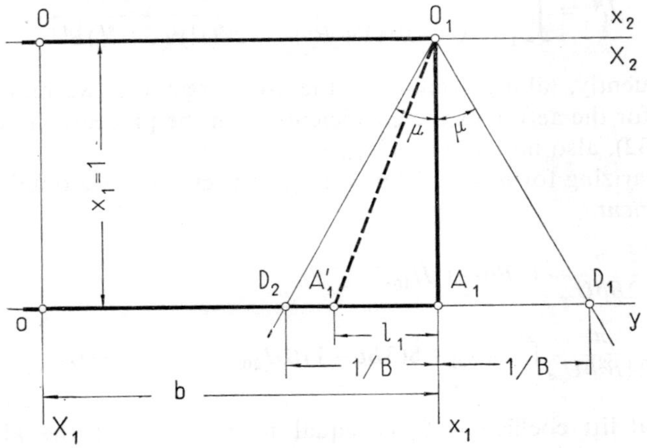

FIG. 6.4

disturbance velocity must be equal to zero along this edge, analogously with Jukowsky's hypothesis from the theory of aerodynamic profiles in incompressible flows. The other constants A_{nq} are determined from the continuity conditions of the axial velocity u when crossing the Mach cone (2.110), and the constants H_{nq}, which are not affected by the position of the subsonic lateral leading edge, have the same values as in the case of the trapezoidal wing.

The constants of the axial velocity for a rectangular wing have the following values for $n = 1, 2$:

$$A_{10} = 0, \quad H_{10} = -\frac{2w_{00}}{\pi B}, \tag{6.37}$$

for $n = 1$;

$$A_{20} = 0, \qquad A_{21} = \frac{2}{\pi B}[Bw_{10} - w_{01}], \tag{6.38a}$$

$$H_{20} = -\frac{2w_{10}}{\pi B}, \quad H_{21} = -\frac{2w_{01}}{\pi B}, \tag{6.38b}$$

for $n = 2$.

VI. 3.2. Aerodynamic Characteristics

The aerodynamic characteristics of a rectangular wing are obtained from those of trapezoidal wings by putting $l_1 = 0$ and replacing the constants of trapezoidal wings by those of rectangular wings given above. Thus, putting $l_1 = 0$ in (6.21), we find instead of I_t,

$$I_t^* = \int_{-1/B}^{0} \frac{y^t\, dy}{\sqrt{-By(1+By)}} = \frac{(-1)^t \pi (2t)!}{2^{2t}(t!)^2(1+\delta_0^t)B^{t+1}}. \tag{6.39}$$

Consequently, taking account of the above remarks, we find the same formulae for the aerodynamic coefficients as in the preceding cases (6.26), (6.30), (6.32), also noticing that $P_{n0} = A_{n0} = 0$.

Particularizing for $n = 1, 2$ ($m = 1, 2$, respectively), we obtain:

Lift coefficient,

$$C_l^{(1)} = \frac{\pi}{2BbU_\infty}(4Bb-1)H_{10}, \tag{6.40a}$$

$$C_l^{(2)} = \frac{\pi}{24B^2bU_\infty}[-4A_{21} + 8(3Bb-1)BH_{20} + 3(1-8B^2b^2)H_{21}]. \tag{6.40b}$$

The total lift coefficient C_l is equal to the sum of the above two coefficients.

Pitching moment coefficient,

$$\mu_2^{(1)} = -\frac{\pi}{3BbU_\infty}(3Bb-1)H_{10}, \tag{6.41a}$$

$$\mu_2^{(2)} = -\frac{\pi}{96B^2bU_\infty}[-12A_{21}+8(8Bb-3)BH_{20}+3(3-16B^2b^2)H_{21}]. \tag{6.41b}$$

The total pitching moment coefficient is, of course, equal to the sum of the above two coefficients.

Wave drag coefficient,

$$C_{d1}^{(1)} = -\frac{\pi w_{00}}{2BbU_\infty^2}(4Bb-1)H_{10} = -\frac{w_{00}}{U_\infty}C_l^{(1)}, \tag{6.42a}$$

$$C_{d2}^{(2)} = -\frac{\pi}{96B^3bU_\infty^2}\{Bw_{10}[-12A_{21}-8(3-8Bb)BH_{20}+3(3-16B^2b^2)H_{21}]$$
$$+w_{01}[6A_{21}+3(3-16B^2b^2)BH_{20}+(64B^3b^3-5)H_{21}]\}, \tag{6.42b}$$

$$C_{d1}^{(2)} = -\frac{\pi w_{00}}{24B^2bU_\infty^2}\{-4A_{21}-8(1-3Bb)BH_{20}+3(1-8B^2b^2)H_{21}\}, \tag{6.42c}$$

$$C_{d2}^{(1)} = -\frac{\pi H_{10}}{24B^2bU_\infty^2}\{8(3Bb-1)Bw_{10}+(3-8B^2b^2)w_{01}\}. \tag{6.42d}$$

VI. 4. Thin Trapezoidal Wing having Supersonic Lateral Leading Edges

VI. 4.1. Constants of Axial Velocities

The general expression for the axial velocity function for the nth order component wing, in the case of supersonic lateral edges (Fig. 6.5), is given by (6.12b) and is rewritten below:

$$\mathcal{U}_{n-1} = \sum_{q=0}^{n-1}\left(K_{nq}\,x^q\cos^{-1}\sqrt{\frac{(1+Bl_1)(1-Bx)}{2B(l_1-x)}}+H_{nq}\,x^q\cos^{-1}\sqrt{\frac{1+Bx}{2}}\right)$$
$$+\sum_{q=0}^{n-2}D_{nq}\,x^q\sqrt{\frac{1-B^2x^2}{B^2}}. \tag{6.43}$$

The expressions for the downwashes remain the same as in the preceding cases. As functions of the coefficients $w_{m-k-1,k}$ of the downwash (i.e. w_{00}, w_{10}, w_{01} for the two first motions $n = 1, 2$) we also define the constants of the two corresponding functions $\mathcal{U}_0$ and $\mathcal{U}_1$:

$$K_{10} = -\frac{2w_{00}}{\pi B}\frac{Bl_1}{\sqrt{B^2l_1^2-1}}, \quad H_{10} = -\frac{2w_{00}}{\pi B}, \tag{6.44}$$

for $n = 1$;

$$K_{20} = \frac{2l_1}{\pi(B^2l_1^2-1)^{3/2}}\left[(2-B^2l_1^2)w_{10}+l_1w_{01}\right], \tag{6.45a}$$

$$K_{21} = -\frac{2}{\pi(B^2l_1^2-1)^{3/2}}\left[w_{10}+B^2l_1^3w_{01}\right], \tag{6.45b}$$

$$H_{20} = -\frac{2w_{10}}{\pi B}, \quad H_{21} = -\frac{2w_{01}}{\pi B}, \tag{6.45c}$$

$$D_{20} = -\frac{w_{01}+Bl_1\cdot Bw_{10}}{\pi B(B^2l_1^2-1)}, \tag{6.45d}$$

for $n = 2$.

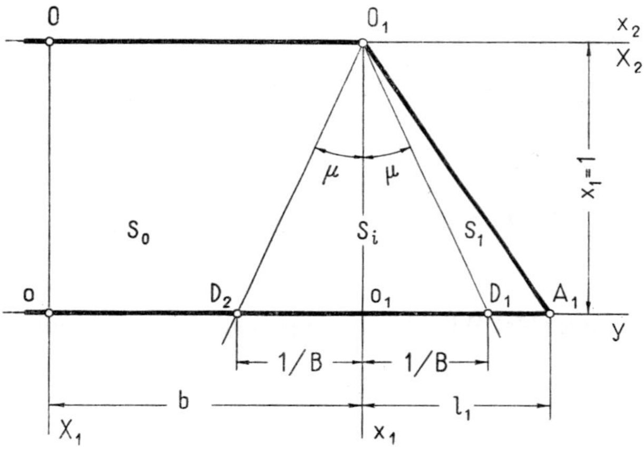

FIG. 6.5

VI. 4.2. Lift Coefficient

Applying formulae (6.14) and (6.15) corresponding to the two surfaces S_0 and S_1, and formula (6.16) for S_i, and taking further account of (6.21b), we find for the lift coefficient:

$$C_l^{(n)} = \frac{8}{(n+1)(2b+l_1)U_\infty}\sum_{q=0}^{n}\left\{P_{nq}I'_q + \frac{(1-\delta_q^n)K_{nq}\sqrt{B^2l_1^2-1}}{2(q+1)}\sum_{j=0}^{q} l_1^j\, I'_{q-j} + \frac{(1-\delta_q^n)H_{nq}}{2(q+1)}\left[BI'_{q+1}+\frac{(-1)^q\pi(n+1)b^{q+1}}{n-q}\right]\right\}. \tag{6.46}$$

In this formula P_{nq} is deduced from the relation

$$\sum_{q=0}^{n} \frac{P_{nq}x^q}{\sqrt{1-B^2x^2}} = \sum_{q=0}^{n-2} D_{nq}x^q \sqrt{\frac{1-B^2x^2}{B^2}} \tag{6.47}$$

from which we get:

$$P_{nq} = \frac{1}{B}(D_{nq} - B^2 D_{n,q-2}), \quad (2 \leq q \leq n-2) \tag{6.48a}$$

$$P_{n0} = \frac{1}{B} D_{n0}, \quad P_{n1} = \frac{1}{B} D_{n1}, \tag{6.48b}$$

$$P_{n,n-1} = -BD_{n,n-3}, \quad P_{nn} = -BD_{n,n-2}. \tag{6.48c}$$

Applying formula (6.46) for $n = 1, 2$, we obtain:

$$C_l^{(1)} = \frac{2\pi}{B(2b+l_1)U_\infty}[K_{10}\sqrt{B^2l_1^2-1} + 2BbH_{10}] = -\frac{4w_{00}}{BU_\infty}, \tag{6.49a}$$

$$C_l^{(2)} = \frac{\pi}{3B^2(2b+l_1)U_\infty}[4D_{20}+2B(2K_{20}+l_1K_{21})\sqrt{B^2l_1^2-1}$$
$$+6Bb\cdot BH_{20}+(1-6B^2b^2)H_{21}]. \tag{6.49b}$$

The sum of these two coefficients represents the total lift coefficient C_l.

VI. 4.3. *Pitching Moment Coefficient*

As in the preceding case, the pitching moment coefficient can be derived directly from the formula for the lift coefficient, by replacing n by $(n+1)$ when n appears as a factor or as an exponent (but not as an index either in Kronecker's symbol or in the summation indices).

Therefore, performing the replacements indicated above, we may write the following expression for the pitching moment coefficient:

$$\mu_2^{(n)} = -\frac{8}{(n+2)(2b+l_1)U_\infty} \sum_{q=0}^{n} \Bigg\{ P_{nq} I'_q$$
$$+ \frac{(1-\delta_q^n)K_{nq}\sqrt{B^2l_1^2-1}}{2(q+1)} \sum_{j=0}^{q} l_1^j I'_{q-j}$$
$$+ \frac{(1-\delta_q^n)H_{nq}}{2(q+1)} \left[BI'_{q+1} + \frac{(-1)^q\pi(n+2)b^{q+1}}{n-q+1} \right] \Bigg\}, \tag{6.50}$$

where I'_t has the same significance as in formula (6.21b).

Applying the above formula for $n = 1, 2$, we find successively:

$$\mu_2^{(1)} = -\frac{2\pi}{3B(2b+l_1)U_\infty}\left(2K_{10}\sqrt{B^2l_1^2-1}+3BbH_{10}\right), \tag{6.51a}$$

$$\mu_2^{(2)} = -\frac{\pi}{12B^2(2b+l_1)U_\infty}\Big[12D_{20}+6B(2K_{20}+l_1K_{21})\sqrt{B^2l_1^2-1}$$
$$+16Bb\cdot BH_{20}+3(1-4B^2b^2)H_{21}\Big], \tag{6.51b}$$

which yield the total pitching moment coefficient μ_2, by adding the above two coefficients.

VI. 4.4. *Wave Drag Coefficient*

Following exactly the same procedure as in the preceding cases, by applying formulae (6.14, 6.15) referring to the surfaces S_0, S_1 and formula (6.17) referring to S_i, we obtain the general expression for the drag coefficient of the nth order wing, due to the downwash of the mth order wing:

$$C_{dm}^{(n)} = -\frac{8}{(m+n)(2b+l_1)U_\infty^2}\sum_{k=0}^{m-1} w_{m-k-1,k}$$
$$\times\sum_{q=0}^{n}\left\{P_{nq}I'_{k+q}+\frac{(1-\delta_q^n)K_{nq}\sqrt{B^2l_1^2-1}}{2(k+q+1)}\sum_{j=0}^{k+q} l_1^j I'_{k+q-j}\right.$$
$$\left.+\frac{(1-\delta_q^n)H_{nq}}{2(k+q+1)}\left[BI'_{k+q+1}+\frac{(-1)^{k+q}\pi(m+n)b^{k+q+1}}{m+n-k-q-1}\right]\right\}, \tag{6.52}$$

where the integrals I'_t are given by (6.21b).

Although this formula seems to be long, it is, however, a general one, established once for all, and its application to low orders n and m is not difficult. Thus, for $n = 1, 2$ and $m = 1, 2$, we have:

$$C_{d1}^{(1)} = -\frac{2\pi w_{00}}{B(2b+l_1)U_\infty^2}\left[K_{10}\sqrt{B^2l_1^2-1}+2BbH_{10}\right] = -\frac{w_{00}}{U_\infty}C_l^{(1)}, \tag{6.53a}$$

$$C_{d2}^{(2)} = -\frac{\pi}{12B^3(2b+l_1)U_\infty^2}\Big\{Bw_{10}\big[12D_{20}+6B\sqrt{B^2l_1^2-1}\,(2K_{20}+l_1K_{21})$$
$$+16Bb\cdot BH_{20}+3(1-4B^2b^2)H_{21}\big]+w_{01}\big(2\sqrt{B^2l_1^2-1}\,[3Bl_1\cdot BK_{20}$$
$$+(1+2B^2l_1^2)K_{21}]+3(1-4B^2b^2)BH_{20}+16B^3b^3H_{21}\big)\Big\}, \tag{6.53b}$$

$$C_{d1}^{(2)} = -\frac{\pi w_{00}}{3B^2(2b+l_1)U_\infty^2}\{4D_{20}+2B(2K_{20}+l_1K_{21})\sqrt{B^2l_1^2-1}$$
$$+6Bb\cdot BH_{20}+(1-6B^2b^2)H_{21}\}, \tag{6.53c}$$
$$C_{d2}^{(1)} = -\frac{\pi}{3B^2(2b+l_1)U_\infty^2}\{2w_{10}B[2K_{10}\sqrt{B^2l_1^2-1}$$
$$+3BbH_{10}]+w_{01}[2Bl_1K_{10}\sqrt{B^2l_1^2-1}+(1-6B^2b^2)H_{10}]\}. \tag{6.53d}$$

The sum of the above four coefficients is equal to the total wave drag coefficient C_d.

Note. The same expressions as above are valid for a thick wing with supersonic lateral leading edges. We have to take account of all the closure conditions on the four edges of the wing if necessary, or of partial conditions if they are sufficient for practical applications.

VI. 5. Wing of Symmetrical Thickness

VI. 5.1. Considerations on the Axial Velocities and the Closure Conditions

For the sake of generality and having in view the closure on all sides of a trapezoidal wing of symmetrical thickness, we have given above the general formula for the axial velocity u, more precisely of the function $\mathcal{U}_{n-1}$ (6.12c), which is given below once again, in view of more detailed considerations:

$$\mathcal{U}_{n-1} = \sum_{q=0}^{n-1}\left(C_{nq}x^q\cosh^{-1}\sqrt{\frac{(1+Bl_1)(1-Bx)}{2B(l_1-x)}}\right.$$
$$\left.+Q_{nq}x^q\cosh^{-1}\sqrt{\frac{1}{B^2x^2}}+H_{nq}x^q\cos^{-1}\sqrt{\frac{1+Bx}{2}}\right)$$
$$+\sum_{q=0}^{n-2}D_{nq}x^q\sqrt{\frac{1-B^2x^2}{B^2}}. \tag{6.54}$$

On the basis of considerations of wing geometry and of the procedure for calculating the drag coefficient, we consider two cases:

(a) *Without a ridge on the O_1x_1 axis.* In such a case $Q_{nq} = 0$, of course, but the wing closes on its four sides only if $N = 4$.

Any other complex wing having $N < 4$ does not close on all the four sides of a trapezoidal wing.

If $N \leq 3$ ($n = 1, 2, 3$) the wing remains open at the trailing edge, a fact that diminishes the interest for practical utilization. Such a wing may possibly be used for the case of high supersonic flight, where the fact that the wing is open downstream presents little disadvantage.

We can consider, however, a rectangular wing open at the lateral edges and we may construct this complex wing by taking $n = 1, 2$, as shown above. Another very interesting case is that of the angular wedge wing, for which the motion is a conical one ($n = 1$):

$$\mathcal{U}_0 = C_{10} \cosh^{-1} \sqrt{\frac{(1+Bl_1)(1-Bx)}{2B(l_1-x)}} + H_{10} \cos^{-1} \sqrt{\frac{1+Bx}{2}} \,. \tag{6.55}$$

Taking into account (6.10) the constants C_{10} and H_{10} are deduced from (6.44).

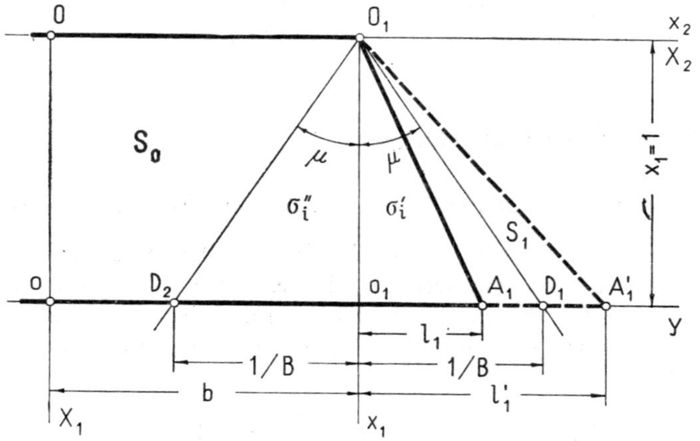

FIG. 6.6

(b) *With a ridge on the O_1x_1 axis.* This case is interesting because of the fact that we may obtain a closed wing even with two components only: $n = 1, 2$. Indeed, if the wing surface is given by the equations mentioned before (6.7),

$$Z' = t'x_1(1-x_1), \quad Z'' = t''(1-x_1)\left(x_1 - \frac{x_2}{l_2}\right),$$

where the former refers to the left side of the O_1x_1 axis (the rectangular part), and the latter to the right side of the axis (the triangular part), then the wing closes on all sides (Fig. 6.6); in such a case we also have a ridge on the O_1x_1 axis (obviously, another ridge O_2x_2 on the opposite side ensures the symmetry of the entire trapezoidal wing).

The downwashes are

$$w' = U_\infty \frac{\partial Z'}{\delta x_1} = U_\infty t' - 2U_\infty t' x_1, \tag{6.56a}$$

$$w'' = U_\infty \frac{\partial Z''}{\delta x_1} = U_\infty t'' - 2U_\infty t'' x_1 + U_\infty \frac{t''}{l_1} x_2, \tag{6.56b}$$

respectively, whence

$$w'_{00} = t' U_\infty, \quad w'_{10} = -2t' U_\infty, \quad w'_{01} = 0, \tag{6.57a}$$

$$w''_{00} = t'' U_\infty, \quad w''_{10} = -2t'' U_\infty, \quad w''_{01} = \frac{t''}{l_1} U_\infty. \tag{6.57b}$$

In this case, the expressions for the two functions $\mathscr{U}_0$ and $\mathscr{U}_1$ are obtained by putting n equal to 1 and 2 respectively, in formula (6.54).

The ridge $O_1 o_1$ is situated on the axis and we have previously established that the axial velocity u should be cancelled alongside this line, hence

$$Q_{n0} = 0, \quad Q_{10} = Q_{20} = 0. \tag{6.58}$$

The other constants in these formulae have the following expressions:

$$C_{10} = -\frac{2l_1 w'_{00}}{\pi\sqrt{1 - B^2 l_1^2}}, \quad H_{10} = -\frac{2w''_{00}}{\pi B}, \tag{6.59}$$

for $n = 1$;

$$C_{20} = -\frac{2l_1}{\pi(1 - B^2 l_1^2)^{3/2}} [(2 - B^2 l_1^2) w'_{10} + l_1 w'_{01}], \tag{6.60a}$$

$$C_{21} = \frac{2}{\pi(1 - B^2 l_1^2)^{3/2}} [w'_{10} + B^2 l_1^2 \cdot l_1 w'_{01}], \tag{6.60b}$$

$$H_{20} = -\frac{2w''_{10}}{\pi B}, \quad H_{21} = -\frac{2w''_{01}}{\pi B}, \tag{6.60c}$$

$$Q_{21} = \frac{1}{\pi}(w''_{10} - w'_{10}), \tag{6.60d}$$

$$D_{20} = \frac{w''_{01}}{\pi B} + \frac{Bl_1(w'_{10} + l_1 w'_{01})}{\pi(1 - B^2 l_1^2)}, \tag{6.60e}$$

for $n = 2$.

Note. Generally, the parabolic profiles obtained by the intersection of the surfaces Z'' and Z' with the plane $O_1x_1x_3$ may be different. In order that the surface should not include a sudden discontinuity in this plane (that is, in order to have the same profile), it is necessary that $t'' = t'$ in formulae (6.7).

VI. 5.2. Wave Drag Coefficient

In the case of a thin wing with subsonic or supersonic lateral edges and in the case of a wing of symmetrical thickness with supersonic lateral edges, we have employed the residue method for computing the drag coefficient.

In the following considerations we shall apply the method of direct integration in the special case of a wing of symmetrical thickness with subsonic or supersonic lateral leading edge, which may also be applied to all cases treated previously by means of the residue method.

In the following we shall give both the general expression for the drag coefficient of the entire semi-wing of total surface S, and the expressions for this coefficient on the various wing portions S_0, σ_i'', σ_i', S_1 (Fig. 6.6). The last portion S_1 appears only in the case of a wing with supersonic lateral leading edge.

We shall also consider that the semi-wing of symmetrical thickness is provided with a ridge on the O_1x_1 axis. We shall first determine the drag coefficient of the wing with subsonic lateral edges, which may possibly be provided with a central ridge.

In such a case, the disturbance downwash w on the wing has the following two forms:

$$x_1^{m-1}w''_{m-1} = x_1^{m-1}\sum_{k=0}^{m-1} w''_{m-k-1,\,k}\,y^k \tag{6.61}$$

on the surfaces σ_i'' and S_0, and

$$x_1^{m-1}w'_{m-1} = x_1^{m-1}\sum_{k=0}^{m-1} w'_{m-k-1,\,k}\,y^k \tag{6.62}$$

on the surface σ_i' and eventually S_1.

The axial disturbance velocity is derived from the real part of the formula (6.54). In the exterior of the Mach cone, on the portion S_0, this expression reduces to a polynomial form:

$$u^{(n-1)} = \frac{\pi}{2}\,x_1^{n-1}\sum_{q=0}^{n-1} H_{nq}\,y^q. \tag{6.63}$$

The following notation will be employed in the calculation of the wave drag coefficient:

$$\mathcal{J}_t(a) = \int_0^a \frac{y^t\,dy}{\sqrt{1-B^2y^2}}, \quad \left[t = 2k,\ 2k+1;\ a = l_1,\ \frac{1}{B},\ -\frac{1}{B}\right] \tag{6.64}$$

which is deduced from (6.22a) or (6.22b), according as t is even ($t = 2k$) or odd ($t = 2k+1$); also

$$\mathcal{G}_t = \mathcal{J}_t(l_1) - \mathcal{J}_t\left(-\frac{1}{B}\right). \tag{6.65}$$

Further, denoting by $C_{dm}^{(0n)}$, $C_{dm}^{\prime(in)}$, $C_{dm}^{\prime\prime(in)}$ the wave drag coefficients corresponding to the portions S_0, σ_i' and σ_i'', which are computed by considering formula (6.14) and twice formula (6.17)—once for σ_i' and the second time for σ_i''—we may write the wave drag coefficient $C_{dm}^{(n)}$ for the whole wing:

$$\begin{aligned}
C_{dm}^{(n)} &= \frac{1}{S}\left[S_0 C_{dm}^{(0n)} + \sigma_i' C_{dm}^{\prime(in)} + \sigma_i'' C_{dm}^{\prime\prime(in)}\right] \\
&= -\frac{8}{(m+n)(2b+l_1)U_\infty^2} \sum_{k=0}^{m-1}\sum_{q=0}^{n-1} \left\{ w'_{m-k-1,\,k}\left[\frac{C_{nq}}{2(k+q+1)}\right.\right. \\
&\quad \times \left(l_1^{k+q+1}\cosh^{-1}\frac{1}{Bl_1} + \sqrt{1-B^2l_1^2}\sum_{j=0}^{k+q} l_1^j \mathcal{J}_{k+q-j}(l_1)\right) \\
&\quad + \frac{H_{nq}}{2(k+q+1)}\left(2l_1^{k+q+1}\cos^{-1}\sqrt{\frac{1+Bl_1}{2}} + B\mathcal{J}_{k+q+1}(l_1)\right) \\
&\quad + \frac{Q_{nq}}{k+q+1}\left(l_1^{k+q+1}\cosh^{-1}\frac{1}{Bl_1} + \mathcal{J}_{k+q}(l_1)\right) + \frac{D_{nq}(1-\delta_q^{n-1})}{B(k+q+2)}\left(\mathcal{J}_{k+q}(l_1)\right. \\
&\quad \left.\left. + l_1^{k+q+1}\sqrt{1-B^2l_1^2}\right)\right] - w''_{m-k-1,\,k}\left[\frac{C_{nq}}{2(k+q+1)}\right. \\
&\quad \times \left(l_1^{k+q+1}\cosh^{-1}\frac{1}{Bl_1} + \sqrt{1-B^2l_1^2}\sum_{j=0}^{k+q} l_1^j \mathcal{J}_{k+q-j}\left(-\frac{1}{B}\right)\right) \\
&\quad + \frac{H_{nq}}{2(k+q+1)}\left(B\mathcal{J}_{k+q+1}\left(-\frac{1}{B}\right) + \frac{(-1)^{k+q+1}\pi(m+n)b^{k+q+1}}{m+n-k-q-1}\right) \\
&\quad \left.\left. + \frac{Q_{nq}}{k+q+1}\mathcal{J}_{k+q+1}\left(-\frac{1}{B}\right) + \frac{D_{nq}(1-\delta_q^{n-1})}{B(k+q+2)}\mathcal{J}_{k+q}\left(-\frac{1}{B}\right)\right]\right\}.
\end{aligned} \tag{6.66}$$

If the wing of symmetrical thickness with subsonic lateral leading edge has no central ridge ($w' = w'' = w$, $Q_{nq} = 0$), the expression (6.66) of the wave drag coefficient may be written in the simplified form:

$$C_{dm}^{(n)} = -\frac{8}{(m+n)(2b+l_1)U_\infty^2} \sum_{k=0}^{m-1} \sum_{q=0}^{n-1} w_{m-k-1,\,k} \left\{ \frac{C_{nq}\sqrt{1-B^2l_1^2}}{2(k+q+1)} \right.$$
$$\times \sum_{j=0}^{k+q} l_1^j \mathcal{J}_{k+q-j} + \frac{H_{nq}}{2(k+q+1)} \left[B\mathcal{J}_{k+q+1} \right.$$
$$\left. \left. + \frac{\pi(-1)^{k+q}(m+n)b^{k+q+1}}{m+n-k-q-1} \right] + \frac{D_{nq}(1-\delta_q^{n-1})(\mathcal{J}_{k+q} + l_1^{k+q+1}\sqrt{1-B^2l_1^2})}{B(k+q+2)} \right\}. \tag{6.67}$$

VI. 5.3. Wing of Symmetrical Thickness having both Leading Edges Supersonic

In this case we divide the surface of the half wing into four parts: S_0, σ', σ'' and S_1. As regards the drag coefficient, we retain the same notation $C_{dm}^{(0n)}$, $C_{dm}^{\prime(in)}$, $C_{dm}^{\prime\prime(in)}$ for the first three cases and denote by $C_{dm}^{(1,\,n)}$ the latter one.

In the same way as in the preceding case we may calculate each of these partial coefficients. Afterwards, by means of summation, we obtain the following expression for the wave drag coefficient of the whole wing:

$$C_{dm}^{(n)} = \frac{1}{S}\left[S_0 C_{dm}^{(0n)} + \sigma_i' C_{dm}^{\prime(in)} + \sigma_i'' C_{dm}^{\prime\prime(in)} + S_1 C_{dm}^{(1,\,n)}\right]$$
$$= -\frac{8}{(m+n)(2b+l_1)U_\infty^2} \sum_{k=0}^{m-1} \sum_{q=0}^{n-1} \left\{ w'_{m-k-1,\,k} \left[\frac{K_{nq}}{2(k+q+1)} \right. \right.$$
$$\times \left(l_1^{k+q+1} \cos^{-1}\frac{1}{Bl_1} + \sqrt{B^2l_1^2-1} \sum_{j=0}^{k+q} l_1^j \mathcal{J}_{k+q-j}\left(\frac{1}{B}\right) \right.$$
$$\left. + \frac{BH_{nq}}{2(k+q+1)} \mathcal{J}_{k+q+1}\left(\frac{1}{B}\right) + \frac{Q_{nq}}{k+q+1} \mathcal{J}_{k+q}\left(\frac{1}{B}\right) \right.$$
$$\left. + \frac{D_{nq}(1-\delta_q^{n-1})\mathcal{J}_{k+q+2}\left(\frac{1}{B}\right)}{B(k+q+2)} \right] - w''_{m-k-1,\,k} \left[\frac{K_{nq}}{2(k+q+1)} \left(l_1^{k+q+1} \right. \right.$$
$$\times \cos^{-1}\frac{1}{Bl_1} + \sqrt{B^2l_1^2-1} \sum_{j=0}^{k+q} l_1^j \mathcal{J}_{k+q+j}\left(-\frac{1}{B}\right)$$
$$+ \frac{H_{nq}}{2(k+q+1)} \left(B\mathcal{J}_{k+q+1}\left(-\frac{1}{B}\right) + \frac{(-1)^{k+q+1}(m+n)b^{k+q+1}}{m+n-k-q-1} \right)$$
$$\left. \left. + \frac{Q_{nq}}{k+q+1} \mathcal{J}_{k+q+1}\left(-\frac{1}{B}\right) + \frac{D_{nq}(1-\delta_q^{n-1})}{B(k+q+2)} \mathcal{J}_{k+q}\left(-\frac{1}{B}\right) \right] \right\}. \tag{6.68}$$

NOTE ON THE WING WITHOUT A RIDGE. If the wing has no ridge on the line O_1x_1, the downwash variation being given only by relation (6.5) over the entire wing ($w' = w'' = w$, $Q_{nq} = 0$), then we obtain the same expression for the drag coefficient as for the thin wing with supersonic edges (6.51). This result may also be obtained through direct integration, proceeding in the same way as for the preceding case with central ridge ($w' \neq w''$), where the coefficient $C_{dm}^{(n)}$ is given by (6.68).

VI. 5.4. Application to a Rectangular Wing of Symmetrical Thickness

In this case, the wing may be open at the lateral edges since these edges lie in the stream direction. On the other hand, the downwash is the same over the entire rectangular wing, the velocity w' being equal to zero. However, the ridge on the O_1x_1 axis remains. In this case, the axial disturbance velocity is obtained from relation (6.12c) in which we put $l_1 = 0$ and where the ridge contribution and that of the lateral edge (which now coincides with this ridge) are combined into one single term; further, employing the usual notation Q_{nq} for the central ridge, we have

$$\mathcal{U}_{n-1} = \sum_{q=0}^{n-1} \left(Q_{nq}\, x^q \cosh^{-1} \sqrt{\frac{1}{B^2x^2}} + H_{nq}\, x^q \cos^{-1} \sqrt{\frac{1+Bx}{2}} \right)$$

$$+ \sum_{q=0}^{n-2} D_{nq}\, x^q \sqrt{\frac{1-B^2x^2}{B^2}}, \quad (Q_{no} = 0), \tag{6.69}$$

because the axial velocity u is necessarily equal to zero on the axis (at the origin), as shown previously.

The constants in the expressions for $\mathcal{U}_{n-1}$ have the following values:

$$H_{10} = -\frac{2w_{00}}{\pi B}, \quad Q_{10} = 0 \tag{6.70}$$

for $n = 1$, and

$$Q_{20} = 0, \qquad Q_{21} = \frac{w_{10}}{\pi}, \quad D_{20} = \frac{w_{01}}{\pi B}, \tag{6.71a}$$

$$H_{20} = -\frac{2w_{10}}{\pi B}, \quad H_{21} = -\frac{2w_{01}}{\pi B}, \tag{6.71b}$$

for $n = 2$.

The expression for the drag coefficient is also given by formula (6.67) in which we have to put $l_1 = 0$. Particularizing this formula for $m = 1, 2$

and $n = 1, 2$, we obtain the following values for the partial drag coefficients:

$$C_{d1}^{(1)} = \frac{1-2\pi Bb}{BbU_\infty^2} w_{00}H_{10}, \tag{6.72a}$$

$$C_{d2}^{(2)} = -\frac{1}{144B^3b\,U_\infty^2}\{3Bw_{10}[-24Q_{21}+8BH_{20}(4\pi Bb-3)$$
$$+3\pi H_{21}(1-8B^2b^2)+12\pi D_{20}]+w_{01}[2\pi Q_{21}+9\pi BH_{20}(1-8B^2b^2)$$
$$+16H_{21}(6\pi B^3b^3-1)-48D_{20}]\}, \tag{6.72b}$$

$$C_{d1}^{(2)} = -\frac{w_{00}}{12B^2bU_\infty^2}\{8Q_{21}+4(2-3\pi Bb)BH_{20}+\pi(12B^2b^2-1)H_{21}-4\pi D_{20}\}, \tag{6.72c}$$

$$C_{d2}^{(1)} = -\frac{H_{10}}{12B^2bU_\infty^2}[4B(2-3\pi Bb)w_{10}+\pi(12B^2b^2-1)w_{01}]. \tag{6.72d}$$

The sum of these four coefficients represents the total wave drag coefficient C_d for the complex wing constituted of two components, $n = 1, 2$.

VI. 6. Thin Trapezoidal Wing having a Given Pressure Distribution (Indirect Problem)

VI. 6.1. Determination of Downwashes

Proceeding from the formula for the pressure coefficient under the assumption of small disturbances,

$$C_p = -\frac{2u}{U_\infty}, \tag{6.73}$$

the pressure distribution is given by the expression for the axial velocity u which, in its turn, is given by a superposition of homogeneous polynomials of order $n = 1, 2, \ldots (N-1)$:

$$u = \sum_{n=1}^{N} x_1^{n-1} \sum_{q=0}^{n-1} u_{n-q-1,\,q}\, y^q = \sum_{n=1}^{N} u_{n-1}, \tag{6.74}$$

where we have put

$$u_{n-1} = \frac{u^{(n-1)}}{x_1^{n-1}} = \sum_{q=0}^{n-1} u_{n-q-1,\,q}\, y^q, \tag{6.75}$$

corresponding to the nth order conical motion.

All considerations presented in Chapters II and IV on the indirect problem are applicable in just the same way for trapezoidal wings, with the only restriction that the coefficients $u_{n-q-1,\,q}$ should satisfy certain symmetry relations which are similar to those of the direct problem established for the general case (6.4).

Indeed, instead of the coefficients $w_{m-k-1,\,k}$ and $W_{m-2\varkappa-1,\,2\varkappa}$ from these relations, in our case we have $u_{n-q-1,\,q}$ and $U_{n-2\varkappa-1,\,2\varkappa}$, respectively.

In order to find the downwash in the nth order conical motion, we apply formula (2.155)—without an additional ridge, however—letting l_2 tend to infinity; thus, we find the downwash function $\mathcal{W}$,

$$\mathcal{W}_{n-1} = w_{n-1}+iw'_{n-1} = \sum_{q=0}^{n-1}\left(C_{nq}\,x^q\cosh^{-1}\sqrt{\frac{(1+Bl_1)(1-Bx)}{2B(l_1-x)}}\right.$$
$$\left.+H_{nq}\,x^q\cos^{-1}\sqrt{\frac{1+Bx}{2}}\right)+Q_n x^{n-1}\cosh^{-1}\sqrt{\frac{1}{B^2x^2}}$$
$$+\sum_{q=0}^{n-2}D_{nq}\,x^q\sqrt{\frac{1-B^2x^2}{B^2}}\,. \tag{6.76}$$

It should be noticed now that for $n = 1$ the downwash tends to infinity when crossing through the origin ($y = 0$), a fact that also leads to a surface which becomes infinite near the origin, as has been shown in our previous considerations (2.158). Hence, we do not consider $n = 1$.

Therefore, we shall deal with the cases $n = 2, 3$, for which the corresponding constants of $\mathcal{W}_{n-1}$ are given by the following relations:

$$C_{20} = -2\,\frac{B^2l_1u_{10}+u_{01}}{\pi\sqrt{1-B^2l_1^2}}\,,\qquad C_{21} = 2\,\frac{u_{10}+(2-B^2l_1^2)l_1u_{01}}{\pi l_1^2\sqrt{1-B^2l_1^2}}\,, \tag{6.77a}$$

$$H_{20} = -\frac{2Bu_{10}}{\pi}\,,\qquad H_{21} = -\frac{2Bu_{01}}{\pi}\,, \tag{6.77b}$$

$$Q_2 = -\frac{u_{10}+2l_1u_{01}}{\pi l_1^2}\,,\qquad D_{20} = -\frac{Bu_{10}}{\pi l_1}\,, \tag{6.77c}$$

for $n = 2$;

$$C_{30} = -\frac{(B^2l_1^2)l_1u_{02}+u_{11}+Bl_1(3-2B^2l_1^2)Bu_{20}}{\pi(1-B^2l_1^2)^{3/2}}\,, \tag{6.78a}$$

$$C_{31} = \frac{2[B^2u_{20}+(B^3l_1^3)Bu_{11}-(2-3B^2l_1^2)u_{02}]}{\pi(1-B^2l_1^2)^{3/2}}\,, \tag{6.78b}$$

$$C_{32} = \frac{(2-3B^2l_1^2)u_{20}+(3-4B^2l_1^2)l_1u_{11}+(6-9B^2l_1^2+2B^4l_1^4)l_1^2u_{02}}{\pi l_1^3(1-B^2l_1^2)^{3/2}}\,, \tag{6.78c}$$

$$H_{30} = -\frac{u_{02}-2Bu_{20}}{\pi B}, \quad H_{31} = \frac{2Bu_{11}}{\pi}, \quad H_{32} = \frac{2Bu_{02}}{\pi}, \tag{6.78d}$$

$$Q_3 = -\frac{2u_{20}+l_1(3+B^2l_1^2)u_{11}+6l_1^2u_{02}}{2\pi l_1^3}+\frac{B^2u_{11}}{2\pi}, \tag{6.78e}$$

$$D_{30} = -B\left[\frac{(1-2B^2l_1^2)u_{20}-(B^2l_1^2)l_1u_{11}-l_1^2u_{02}}{2\pi l_1(1-B^2l_1^2)}-\frac{u_{11}}{2\pi}\right], \tag{6.78f}$$

$$D_{31} = -B\left[\frac{(2+B^2l_1^2-2B^4l_1^4)u_{20}+(3-2B^2l_1^2)l_1u_{11}+l_1^2(2-B^2l_1^2)u_{02}}{2\pi l_1^2(1-B^2l_1^2)}\right.$$
$$\left.-\frac{2B^2u_{20}+u_{02}}{2\pi}\right], \tag{6.78g}$$

for $n = 3$.

VI. 6.2. Equation of the Wing Surface

After deducing w_{n-1} from (6.76), we apply formula (2.157) and find, with the addition of an arbitrary function of x_2, the following expression for the equation of the surface of the nth order wing:

$$Z_n(x_1, x_2) = -\frac{x_2^n}{U_\infty}\int\frac{w_{n-1}\,dy}{y^{n+1}}$$
$$= \frac{x_1^n}{U_\infty}\left\{\sum_{q=0}^{n-1}\frac{C_{nq}\,y^q}{n-q}\left[\left(1-\frac{y^{n-q}}{l_1^{n-q}}\right)\cosh^{-1}\sqrt{\frac{(1+Bl_1)(1-By)}{2B(l_1-y)}}\right.\right.$$
$$\left.+\frac{y^{n-q}\sqrt{1-B^2l_1^2}}{2l_1^{n-q}}\sum_{t=1}^{n-q}l_1^{t-1}\,\Gamma_t\right]+\sum_{q=0}^{n-1}\frac{H_{nq}\,y^q}{n-q}\left[\cos^{-1}\sqrt{\frac{1+By}{2}}\right.$$
$$\left.+\frac{By^{n-q}}{2}\,\Gamma_{n-q}\right]+Q_n\,y^{n-1}\left[\cosh^{-1}\sqrt{\frac{1}{B^2y^2}}-\sqrt{1-B^2y^2}\right]$$
$$\left.+\sum_{q=0}^{n-2}\frac{D_{nq}\,y^q}{B(n-q)}\left[\sqrt{1-B^2y^2}+B^2y^{n-q}\,\Gamma_{n-q-1}\right]\right\}, \tag{6.79}$$

where Γ_t has the following significance:

$$\Gamma_t = \Gamma_{2k+1} = \int\frac{dy}{y^{2k+1}\sqrt{1-B^2y^2}} = -\frac{(2k)!}{2^{2k}(k!)^2y^{2k}}$$
$$\times\left[(By)^{2k}\cosh^{-1}\sqrt{\frac{1}{B^2y^2}}+\sqrt{1-B^2y^2}\sum_{p=0}^{k-1}\frac{2^{2p}(p!)^2}{(2p+1)!}(By)^{2(k-p-1)}\right],$$
$$(t = 2k+1) \tag{6.80a}$$

$$\Gamma_t = \Gamma_{2(k+1)} = \int \frac{dy}{y^{2(k+1)}\sqrt{1-B^2y^2}} = -\frac{2^{2k}(k!)^2}{(2k+1)!\,y^{2k+1}}\sqrt{1-B^2y^2}$$
$$\times \sum_{p=0}^{k} \frac{(2p)!}{2^{2p}(p!)^2}(By)^{2(k-p)}, \quad [t = 2(k+1)]. \tag{6.80b}$$

As an example, we give below the expression for the component surface of order $n = 2$;

$$Z_2 = \frac{x_1^2}{U_\infty}\left\{\frac{C_{20}}{2}\left[\left(1-\frac{y^2}{l_1^2}\right)\cosh^{-1}\sqrt{\frac{(1+Bl_1)(1-By)}{2B(l_1-y)}}\right.\right.$$
$$\left.+\frac{\sqrt{1-B^2l_1^2}}{2l_1^2}\,y^2\cosh^{-1}\sqrt{\frac{1}{B^2y^2}}+\frac{y}{2l_1}\sqrt{(1-B^2l_1^2)(1-B^2y^2)}\right]$$
$$+C_{21}\,y\left[\left(1-\frac{y}{l_1}\right)\cosh^{-1}\sqrt{\frac{(1+Bl_1)(1-By)}{2B(l_1-y)}}\right.$$
$$\left.+\frac{\sqrt{1-B^2l_1^2}}{2l_1}\,y\cosh^{-1}\sqrt{\frac{1}{B^2y^2}}\right]+\frac{H_{20}}{2}\left[\cos^{-1}\sqrt{\frac{1+By}{2}}\right.$$
$$\left.+\frac{By}{2}\sqrt{1-B^2y^2}\right]+H_{21}\,y\left[\cos^{-1}\sqrt{\frac{1+By}{2}}-\frac{By}{2}\cosh^{-1}\sqrt{\frac{1}{B^2y^2}}\right]$$
$$+Q_2\,y\left[\cosh^{-1}\sqrt{\frac{1}{B^2y^2}}-\sqrt{1-B^2y^2}\right]$$
$$\left.+\frac{D_{20}}{2B}\left[\sqrt{1-B^2y^2}-B^2y^2\cosh^{-1}\sqrt{\frac{1}{B^2y^2}}\right]\right\}, \tag{6.81}$$

Note. If the lateral edge is supersonic, we take only the real part of the above expression.

Trapezoidal wing having a finite tangent at the lateral leading edge. In order to obtain a wing whose surface has a finite tangent along the edge $y = l_1$, the following condition should be satisfied:

$$C_{n0}+C_{n1}l_1+C_{n2}l_1^2+\ldots+C_{n,\,n-1}l_1^{n-1} = 0. \tag{6.82}$$

VI. 6.3. Application to a Thin Rectangular Wing having a Given Pressure Distribution

DOWNWASH. The lateral edge of the rectangular wing (Fig.6.4) becomes a central ridge. For this reason, a doublet also appears in the expression for the velocity, analogous to the general case (2.156).

In this case, the expression for the downwash function $\mathcal{W}_{n-1}$ becomes

$$\mathcal{W}_{n-1} = \sum_{q=0}^{n-1} \left(Q_{nq}\, x^q \cosh^{-1} \sqrt{\frac{1}{B^2x^2}} + H_{nq} x^q \cos^{-1} \sqrt{\frac{1+Bx}{2}} \right) + \sum_{q=-1}^{n-2} D_{nq}\, x^q \sqrt{\frac{1-B^2x^2}{B^2}}\,. \tag{6.83}$$

In order that Z_n be finite at the origin, one must have

$$Q_{n0} = D_{n,\,-1} = 0, \tag{6.84}$$

whence

$$u_{n-1,\,0} = u_{n-2,\,1} = 0, \tag{6.85}$$

which shows that such a surface does not exist for $n = 2$ since the coefficients u_{10} and u_{01}, which define the pressure distribution for $n = 2$, are equal to zero.

Hence, in order that the wing be geometrically compatible, we should have at least $n = 3$, for which the corresponding constants are given below:

$$Q_{31} = -\frac{4u_{02}}{\pi}, \qquad Q_{30} = Q_{32} = 0, \tag{6.86a}$$

$$H_{30} = -\frac{u_{02}}{\pi B}, \qquad H_{31} = 0, \qquad H_{32} = \frac{2Bu_{02}}{\pi}, \tag{6.86b}$$

$$D_{30} = D_{3,\,-1} = 0, \quad D_{31} = \frac{5Bu_{02}}{2\pi}. \tag{6.86c}$$

These values are obtained from formulae (6.78) in which we have put $l_1 = 0$ and $u_{20} = u_{11} = 0$, according to (6.85).

WING SURFACE. The disturbance downwash is deduced from formula (6.83), whence, by applying formula (2.157), we obtain the general equation of the wing surface:

$$Z_n = \frac{x_1^n}{U_\infty} \left\{ \sum_{q=1}^{n-1} \frac{Q_{nq}\, y^q}{n-q} \left[\cosh^{-1} \sqrt{\frac{1}{B^2y^2}} + y^{n-q}\, \Gamma_{n-q+1} \right] + \sum_{q=0}^{n-1} \frac{H_{nq}\, y^q}{n-q} \left[\cos^{-1} \sqrt{\frac{1+By}{2}} + \frac{B}{2} y^{n-q}\, \Gamma_{n-q} \right] + \sum_{q=0}^{n-2} \frac{D_{nq}\, y^q}{B(n-q)} \left[\sqrt{1-B^2y^2} + B^2 y^{n-q}\, \Gamma_{n-q-1} \right] \right\}, \tag{6.87}$$

where Γ_t is given by (6.80).

Applying this formula for $n = 3$, we obtain the following expression:

$$Z_3 = \frac{x_1^3}{U_\infty}\left\{\frac{Q_{31}y}{4}\left[(2-B^2y^2)\cosh^{-1}\sqrt{\frac{1}{B^2y^2}}-\sqrt{1-B^2y^2}\right]\right.$$
$$+\frac{H_{30}}{3}\left[\cos^{-1}\sqrt{\frac{1+By}{2}}-\frac{By}{4}\left(\sqrt{1-B^2y^2}+B^2y^2\cosh^{-1}\sqrt{\frac{1}{B^2y^2}}\right)\right]$$
$$+H_{32}\,y^2\left[\cos^{-1}\sqrt{\frac{1+By}{2}}-\frac{By}{2}\cosh^{-1}\sqrt{\frac{1}{B^2y^2}}\right]$$
$$\left.+\frac{D_{31}\,y}{2B}\left[\sqrt{1-B^2y^2}-B^2y^2\cosh^{-1}\sqrt{\frac{1}{B^2y^2}}\right]\right\} \tag{6.88}$$

VI. 7. Trapezoidal and Rectangular Wings of Symmetrical Thickness having a Given Pressure Distribution (Indirect Problem)

VI. 7.1. Trapezoidal Wing with Subsonic Lateral Leading Edge

DOWNWASH. For the nth order component wing, the function $\mathcal{W}_{n-1}$ referring to the downwash w_{n-1} in the case of a trapezoidal wing of symmetrical thickness with one subsonic edge, is deduced from formula (2.154) where we cancel the ridge ($G_q = 0$) and let $l_2 \to \infty$.

$$\mathcal{W}_{n-1} = \sum_{q=0}^{n-1}\left(A_{nq}\,x^q\sqrt{\frac{1+Bx}{B(l_1-x)}}+H_{nq}\,x^q\cos^{-1}\sqrt{\frac{1+Bx}{1+Bl_1}}\right)$$
$$+Q_n x^{n-1}\cosh^{-1}\sqrt{\frac{l_1(1+Bx)}{(1+Bl_1)x}}\,. \tag{6.89}$$

It is known that the expression for the downwash w contains an arbitrary constant (§ II. 8.3).

This constant is determined by imposing, for instance, the condition that the downwash w be finite on the subsonic lateral leading edge. This condition is expressed by the following relation:

$$A_{n0}+A_{n1}l_1+\ldots+A_{n,\,n-1}\,l_1^{n-1} = 0. \tag{6.90}$$

We avoid the conical motion ($n = 1$) which gives an infinite surface for $y = 0$.

Applying formula (6.90) for $n = 2$, we obtain the following values for the coefficients of the function $\mathcal{W}_1$:

$$A_{20} = -\frac{2Bu_{10}+u_{01}}{\pi}, \quad A_{21} = \frac{2Bu_{10}+u_{01}}{\pi l_1}, \tag{6.91a}$$

$$H_{20} = -\frac{2Bu_{10}}{\pi}, \quad H_{21} = -\frac{2Bu_{01}}{\pi}, \quad Q_2 = -\frac{2Bu_{10}+(1+3Bl_1)u_{01}}{\pi l_1\sqrt{Bl_1}}. \tag{6.91b}$$

WING SURFACE. Considering the expression for the downwash given in the form (6.89), we obtain the equation of the wing surface:

$$Z(x_1, x_2) = -\frac{x_1^n}{U_\infty}\left\{\sum_{q=0}^{n-1}\left[A_{nq}\, y^n(\Gamma'_{n-q+1}+B\Gamma'_{n-q}) - \frac{H_{nq}}{n-q}\left(\cos^{-1}\sqrt{\frac{1+By}{1+Bl_1}}+\frac{By^{n-q}}{2}\Gamma'_{n-q}\right)\right] + \frac{Q_n\, y^{n-1}}{2}\left(\frac{\sqrt{B(1+By)(l_1-y)}}{\sqrt{Bl_1}} - \frac{(Bl_1-1)y+2l_1}{l_1}\cosh^{-1}\sqrt{\frac{l_1(1+By)}{(1+Bl_1)y}}\right)\right\}, \tag{6.92}$$

where we have used the following notation:

$$\Gamma'_k = \int\frac{dy}{y^k\sqrt{B(l_1-y)(1+By)}} = -\frac{1}{(k-1)l_1}\left[\frac{\sqrt{B(1+By)(l_1-y)}}{By^{k-1}} + \frac{(2k-3)(Bl_1-1)}{2}\Gamma'_{k-1}-(k-2)B\Gamma'_{k-2}\right] \qquad (k \geqslant 2) \tag{6.93a}$$

$$\Gamma'_1 = -\frac{2}{\sqrt{Bl_1}}\cosh^{-1}\sqrt{\frac{l_1(1+By)}{(1+Bl_1)y}} \tag{6.93b}$$

As indicated above, the application of (6.92) concerns the n orders higher than 2 or equal to 2 ($n \geqslant 2$).

VI. 7.2. *Application to a Rectangular Wing*

Putting $l_1 = 0$ in the expression for the disturbance downwash from the case of a trapezoidal wing of symmetrical thickness (6.89), we obtain

$$\mathcal{W}_{n-1} = \sum_{q=0}^{n-1}\left(A_{nq}x^q\sqrt{\frac{1+Bx}{-Bx}}+H_{nq}\,x^q\cos^{-1}\sqrt{1+Bx}\right) \tag{6.94}$$

where the logarithmic terms vanish, because of the cancellation of l_1.

By imposing the condition that w be finite on the lateral leading edge, that is

$$A_{n0} = 0 \tag{6.95}$$

we find at the same time a relation between the coefficients u_{ij}.

The other constants may be determined uniquely. In the case of homogeneous motions of order $n = 2$, we have:

$$A_{20} = 0, \quad u_{01} = -2Bu_{10} \tag{6.96a}$$

$$H_{20} = -\frac{2Bu_{10}}{\pi}, \quad H_{21} = -\frac{2Bu_{01}}{\pi} = \frac{4Bu_{10}}{\pi}. \tag{6.96b}$$

WING SURFACE. Considering the expression for the downwash given by formula (6.94) and introducing this expression into (2.157), we obtain the wing equation:

$$Z_{(x_1x_2)} = -\frac{x_1^n}{U_\infty}\left\{\sum_{q=0}^{n-1} A_{nq}y^n(\Gamma''_{n-q+1}+B\Gamma''_{n-q}) - \sum_{q=0}^{n-1}\frac{H_{nq}y^q}{n-q}\left(\cos^{-1}\sqrt{1+By} + \frac{By^{n-q}}{2}\Gamma''_{n-q}\right)\right\} \tag{6.97}$$

in which we have denoted by Γ''_t the following integral:

$$\Gamma''_t = \int\frac{dy}{y^t\sqrt{-By(1+By)}} = \frac{2}{(2t-1)B}\left[\frac{-\sqrt{By(1+By)}}{y^t} - B^2(t-1)\Gamma''_{t-1}\right]$$
$$= -\sum_{\nu=0}^{k-1}\frac{(-1)^\nu 2^{2\nu+1}(2k-2\nu-2)![(k-1)!]^2B^{\nu-1}}{[(k-\nu-1)!]^2(2k-1)!x^{k-\nu}}\sqrt{-By(1+By)}. \tag{6.98}$$

Particularizing the formula (6.97) for $n = 2$, we obtain the following equation:

$$Z_2 = -\frac{x_1^2}{U_\infty}\left\{\frac{2(1+By)A_{21}}{3B}\sqrt{-By(1+By)} - \frac{H_{20}}{2}\left(\cos^{-1}\sqrt{\frac{1+By}{2}} - \frac{2By-1}{3}\sqrt{-By(1+By)}\right) - H_{21}\,y\left(\cos^{-1}\sqrt{\frac{1+By}{2}} + \sqrt{-By(1+By)}\right)\right\}. \tag{6.99}$$

CHAPTER VII

STUDY OF POLYGONAL WINGS

IN THIS chapter we are going to treat the polygonal, double-trapezoidal, swept-back, arrow-head wings, etc., having the usual shapes employed in supersonic aviation. We shall also establish some general formulae which may be easily applied to these forms of polygonal wings. Further, we shall deal with special wings having different shapes of the trailing edge.

VII. 1. Polygonal Wings of Symmetrical Thickness

VII. 1.1. Preliminary Considerations

Consider an ordinary form of polygonal wing defined in Fig. 7.1, from which we deduce certain current forms for practical applications.

The downwash w on the wing, which defines the wing from the geometrical standpoint as well as from the point of view of the boundary conditions, is generally represented, as in all cases previously examined, by a superposition of homogeneous polynomials of various orders, $n-1 = 0, 1, 2, \ldots (N-1)$.

According to the previous definition, the wing is a complex one, having an arbitrary number of components, among which also the component of the highest order $n = N$.

Such a wing of symmetrical thickness should close on all edges bordering it, at least for present velocities in supersonic aviation. The minimum number of component wings and the lowest number N of the highest order homogeneous motion result from the above condition.

It may be seen from Fig. 7.1 that the wing generates three high-order conical motions, issued from the vertices O, F and O_1—considering, of course, the effects of the half of the wing assumed to be of symmetrical planform, to which, in the case of interference, the motion corresponding to O_2, symmetrical to that corresponding to O_1, is added. It may be assumed, by means of the superposition system, that the actual wing is equivalent to a fictitious wing $OA_1'A_2'$, also of symmetrical thickness, having downwash w on the wing, to which the effect of the fictitious wing

FC_1C_2, having downwash $w^F = -w$ and also the effect of the fictitious wing $O_1A_1'C_1$, having downwash $w^I = -w$, are added. Consequently we have:

(a) The motion corresponding to the vertex O, which covers the triangular wing $OA_1'A_2$, is defined by the downwash w, whose general expres-

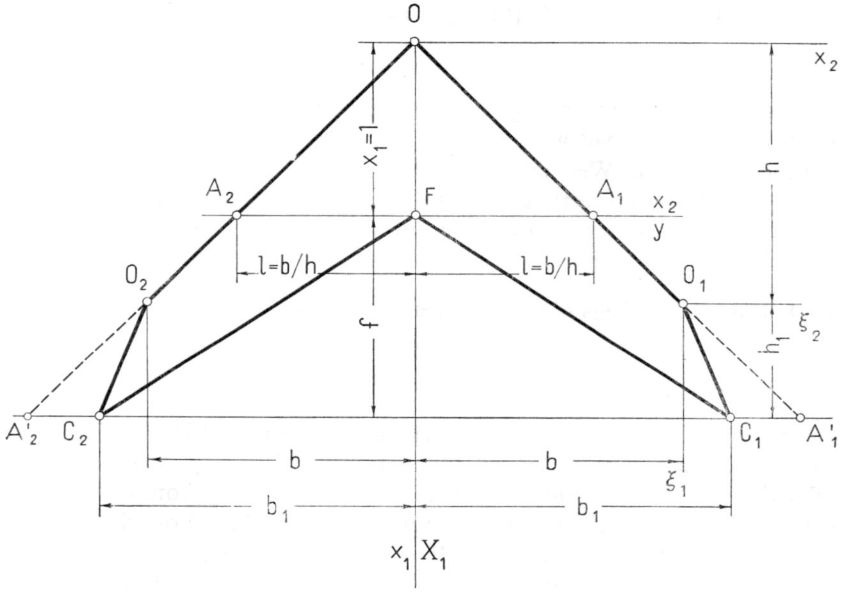

FIG. 7.1

sion, taking of course account of the natural and forced symmetry, is as follows:

$$w = \sum_{m=1}^{N} \sum_{k=0}^{m-1} (\pm 1)^k w_{m-k-1,\,k}\, x_1^{m-k-1} x_2^k$$
$$= \sum_{m=1}^{N} x_1^{m-1} \sum_{k=0}^{m-1} (\pm 1)^k w_{m-k-1,\,k}\, y^k, \tag{7.1}$$

where the plus sign (+) is valid for $x_2 > 0$, and the minus sign (−) for $x_2 < 0$.

(b) The motion corresponding to the vertex F is defined by the downwash $w^F = -w$, which is given as a function of the coordinates X_1, x_2 by

$$X_1 = x_1 - 1, \quad Y = \frac{x_2}{X_1} = \frac{x_2}{x_1 - 1}\,; \tag{7.2}$$

thus we may write

$$-w^F = w = \sum_{m=1}^{N}\sum_{k=0}^{m-1}(\pm 1)^k w_{m-k-1,\,k}(X_1+1)^{m-k-1}x_2^k$$

$$= \sum_{m=1}^{N}\sum_{k=0}^{m-1}(\pm 1)^k W_{m-k-1,\,k}X_1^{m-k-1}x_2^k$$

$$= \sum_{m=1}^{N}X_1^{m-1}\sum_{k=0}^{m-1}(\pm 1)^k W_{m-k-1,\,k}Y^k. \tag{7.3}$$

(c) The motion corresponding to the vertex O_1, defined by the downwash $w^I = -w$, which is given as a function of the coordinates ξ_1, ξ_2 by

$$\xi_1 = x_1-h, \quad \xi_2 = x_2-b, \quad \eta = \frac{\xi_2}{\xi_1} = \frac{x_2-b}{x_1-h}; \tag{7.4}$$

thus we may write

$$-w^I = w = \sum_{m=1}^{N}\sum_{k=0}^{m-1}w_{m-k-1,\,k}(\xi_1+h)^{m-k-1}(\xi_2+b)^k$$

$$= \sum_{m=1}^{N}\sum_{k=0}^{m-1}\beta_{m-k-1,\,k}\xi_1^{\,m-k-1}\xi_2^k = \sum_{m=1}^{N}\xi_1^{m-1}\sum_{k=0}^{m-1}\beta_{m-k-1,\,k}\eta^k. \tag{7.5}$$

All these motions represent a superposition of conical component motions of various orders, from $m = 1$ to $m = N$.

From the superposition of these motions we obtain the actual conditions on the wing:

> the velocity w on the surface of the wing proper;
> the velocity $w = 0$ on the areas FC_1C_2 and $O_1A_1'C_1$, as well as on the remaining part of the plane Ox_1x_2, outside the wing considered; this condition results from the symmetry of thickness.

Note. In the above presentation we have considered one-half of the wing, namely the right-hand one (Fig. 7.1), the left-hand one having exactly the same flow condition; on the other hand, we have assumed that the two flows do not interfere. If, however, the Mach cones having vertices at O_1 and O_2 (symmetrical to O_1), respectively, do interfere on the wing, then the effect of the motion corresponding to the vertex O_2 should be also added.

VII. 1.2. Definition of the Simplest Wings of Symmetrical Thickness closed at the Edges

As shown above, the wing of symmetrical thickness should be closed at the three edges OO_1, O_1C_1, FC_1.

The equation of the wing surface, which is assumed to be represented by a superposition of homogeneous polynomials of orders $n = 1,2, \ldots,$ N (2.16), should be cancelled on the three edges OO_1, O_1C_1, FC_1 and also on the three left-hand side edges symmetrical to the former ones, in order that the sides of the upper and lower surfaces (or the upper and lower sides) may intersect on these lines. On the other hand, since the wing axis is necessarily a ridge, this wing may be constituted of at least three components. Indeed, the upper or lower surface of the simplest closed wing of symmetrical thickness is represented by the equation

$$Z = t\left(x_1 \mp \frac{h}{b}x_2\right)\left(1 - x_1 \pm \frac{f}{b_1}x_2\right)[(b_1 - b)(x_1 - h) + h_1(b \mp x_2)]$$

$$= th_1\left(x_1 \mp \frac{x_2}{l}\right)\left(1 - x_1 \pm \frac{x_2}{L}\right)[l_1(x_1 - h) + b \mp x_2], \tag{7.6}$$

where h, f, b, b_1 are defined in Fig. 7.1, t is proportional to the wing thickness and l, L, l_1, h_1 are defined by the following relations:

$$l = \frac{b}{h}, \quad L = \frac{b_1}{f}, \quad l_1 = \frac{b_1 - b}{h_1}, \quad h_1 = 1 + f - h. \tag{7.7}$$

In equation (7.6) the upper signs refer to $x_2 > 0$, and the lower signs refer to $x_2 < 0$.

Therefore, the wing has three components ($n = 1, 2, 3$), as is easily seen from (7.6) and from its derivative with respect to x_1, which yields the slope.

It should be pointed out that although the complete calculation of the wave drag of such a wing involves no difficulties in principle, it does, however, require long and laborious computations.

In order to avoid this, we shall consider the case when the number of components may be reduced to two, that is, when the wing closes only on the leading edges OO_1, OO_2, and on the trailing edges FC_1, FC_2. In this case, the wing surface may be represented by the following equation deduced from (7.6):

$$Z = t\left(x_1 \mp \frac{h}{b}x_2\right)\left(1 - x_1 \pm \frac{f}{b}x_2\right) = t\left(x_1 \mp \frac{x_2}{l}\right)\left(1 - x_1 \pm \frac{x_2}{L}\right). \tag{7.8}$$

The downwash on the wing has the following expression:

$$w = \frac{\partial Z}{\partial x_1} U_\infty = w_{00} + (w_{10} x_1 \pm w_{01} x_2) = w_{00} + x_1 (w_{10} \pm w_{01} y), \quad (7.9a)$$

$$w_{00} = tU_\infty, \quad w_{10} = -2tU_\infty, \quad w_{01} = t\left(\frac{h}{b} + \frac{f}{b_1}\right) U_\infty. \quad (7.9b)$$

VII. 1.3. Definition of Axial Velocities

Let us put $m = n$ in the general expression for the downwash (7.1). For each $n = 1, 2 \ldots$, there is a corresponding motion.

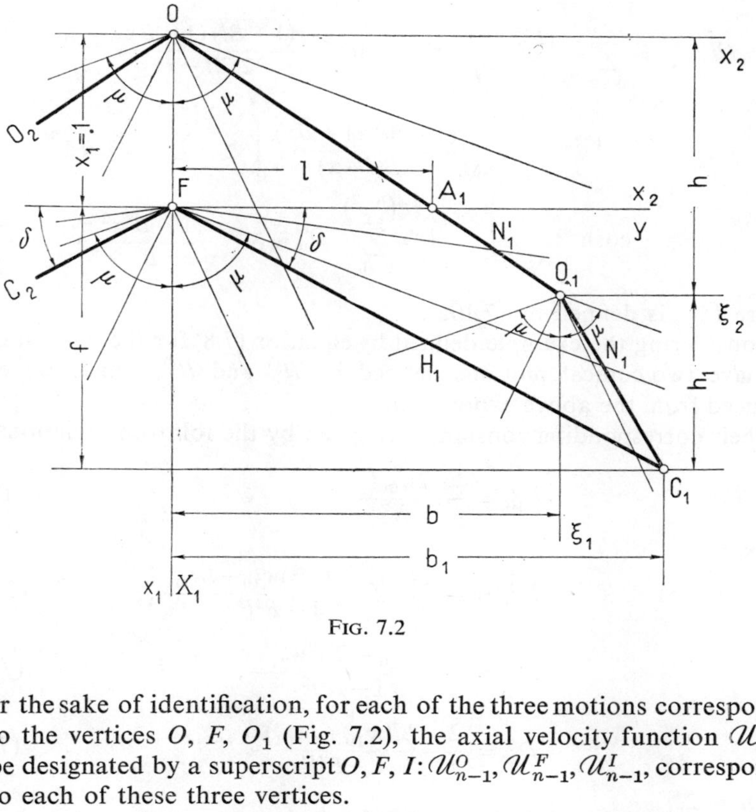

FIG. 7.2

For the sake of identification, for each of the three motions corresponding to the vertices O, F, O_1 (Fig. 7.2), the axial velocity function $\mathcal{U}_{n-1}$ will be designated by a superscript O, F, I: $\mathcal{U}^O_{n-1}, \mathcal{U}^F_{n-1}, \mathcal{U}^I_{n-1}$, corresponding to each of these three vertices.

Further, in order to have the same formulae for both subsonic and supersonic leading edges, we shall change the ordinary constants $C_{n,q}$,

$G_{n,q}$, used in the preceding chapters, to the following ones:

$$C^O_{nq} = \frac{\Gamma^O_{n,q}}{\sqrt{1-B^2l^2}}, \quad C^F_{n,q} = \frac{\Gamma^F_{n,q}}{\sqrt{1-B^2L^2}}, \tag{7.10a}$$

$$C^I_{n,q} = \frac{\Gamma^I_{n,q}}{\sqrt{1-B^2l^2}}, \quad G^I_{n,q} = \frac{\Theta^I_{n,q}}{\sqrt{1-B^2l_1^2}}. \tag{7.10b}$$

Under these conditions, we may write for the different motions corresponding to the three vertices:

THE VERTEX LYING AT O. The general axial velocity function is given by the following expression, already known from the study of delta wings with natural and forced symmetry (§§ (IV. 5.4, IV. 6.4):

$$\begin{aligned}\mathcal{U}^O_{n-1} &= \sum_{q=0}^{n-1} \frac{\Gamma^O_{n,q}}{\sqrt{1-B^2l^2}}\, x^q \left[\cosh^{-1}\sqrt{\frac{(1+Bl)(1-Bx)}{2B(l-x)}}\right. \\ &\left.+(-1)^q \cosh^{-1}\sqrt{\frac{(1+Bl)(1+Bx)}{2B(l+x)}}\right] + \sum_{q=1}^{E\left(\frac{n-1}{2}\right)} Q^O_{n,2q}\, x^{2q} \\ &\times \cosh^{-1}\sqrt{\frac{1}{B^2x^2}} + \sum_{q=0}^{E\left(\frac{n-2}{2}\right)} D^O_{n,2q}\, x^{2q}\sqrt{\frac{1-B^2x^2}{B^2}},\end{aligned} \tag{7.11}$$

where $\Gamma^O_{n,q}$ is defined by (7.10a).

Considering the example defined by equation (7.8) for the wing surface, we have two conical motions defined by $\mathcal{U}^O_0$ and $\mathcal{U}^O_1$, which are easily deduced from the above expression.

Their corresponding constants are given by the following relations:

$$\Gamma^O_{10} = -\frac{2w_{00}l}{\pi} \tag{7.12}$$

for $n = 1$;

$$\Gamma^O_{20} = -\frac{2l}{\pi}\,\frac{(2-B^2l^2)w_{10}+lw_{01}}{1-B^2l^2}, \tag{7.13a}$$

$$\Gamma^O_{21} = \frac{2}{\pi}\,\frac{w_{10}+B^2l^3w_{01}}{1-B^2l^2}, \tag{7.13b}$$

$$D^O_{20} = \frac{2}{\pi}\,\frac{Bl(w_{10}+lw_{01})}{1-B^2l^2} \tag{7.13c}$$

for $n = 2$.

Taking account of (7.10), the above constants are exactly the ones found for symmetrical triangular wings (delta wings) previously established

(§§ IV. 4 and IV. 5) where, however, the coefficients w_{ij} have the values given by (7.9b). In order to pass easily from the case of subsonic edges to that of supersonic edges, we take into consideration the known relation

$$\frac{\cosh^{-1}\beta}{\sqrt{1-B^1l^2}} = \frac{\cos^{-1}\beta}{\sqrt{B^2l^2-1}}. \tag{7.14}$$

The axial velocities of the general complex wing and of the wing made up of two components $n = 1, 2$ are respectively

$$u^O = \text{Re} \sum_{n=1}^{N} x_1^{n-1}\mathcal{U}_{n-1}^O, \quad u^O = \text{Re}\,(\mathcal{U}_0^O + x_1\mathcal{U}_0^O). \tag{7.15}$$

THE VERTEX LYING AT F. Let us first note that, if the Mach cone lies inside the angle with the vertex at F formed by the two trailing edges FC_1 and FC_2 ($\delta < \frac{1}{2}\pi - \mu$), then the motion corresponding to the vertex F has no effect on the wing; that is the reason why we take account only of the case when the trailing edges are subsonic ($L = b_1/f < 1/B$, Fig. 7.2).

The general expression for $\mathcal{U}_{n-1}^F$ has the same form as the preceding one where, however, l is replaced by L given by (7.7):

$$\begin{aligned} \mathcal{U}_{n-1}^F = \sum_{q=0}^{n-1} \frac{\Gamma_{n,q}^F}{\sqrt{1-B^2L^2}} X^q \Bigg[& \cosh^{-1}\sqrt{\frac{(1+BL)(1-BX)}{2B(L-X)}} \\ & + (-1)^q \cosh^{-1}\sqrt{\frac{(1+BL)(1+BX)}{2B(L+X)}}\Bigg] \\ & + \sum_{q=1}^{E\left(\frac{n-1}{2}\right)} Q_{n,2q}^F X^{2q} \cosh^{-1}\sqrt{\frac{1}{B^2X^2}} \\ & + \sum_{q=0}^{E\left(\frac{n-2}{2}\right)} D_{n,2q}^F X^{2q} \sqrt{\frac{1-B^2X^2}{B^2}}, \end{aligned} \tag{7.16}$$

where also $\Gamma_{n,q}^F$ is defined by (7.10a) and where

$$X = Y + iZ, \tag{7.17}$$

the abscissa Y being given by (7.2).

Considering the same example given by (7.8) and taking into account (7.9), as well as relation (7.3) we may write:

$$-w^F = w_{00} + w_{10}(X_1+1) \pm w_{01}x_2 = W_{00} + X_1(W_{10} \pm W_{01}Y), \tag{7.18}$$

where we have put

$$W_{00} = w_{00} + w_{10}, \quad W_{10} = w_{10}, \quad W_{01} = w_{01}. \tag{7.19}$$

There are also two conical motionsof order $n = 1, 2$ corresponding to the vertex F.

The total axial velocities u^F for the general complex wing and for the complex wing made up of the two components $n = 1,2$ are the following:

$$u^F = \mathrm{Re} \sum_{n=1}^{N} X_1^{n-1} \mathcal{U}_{n-1}^F, \quad u^F = \mathrm{Re}\,(\mathcal{U}_0^F + X_1 \mathcal{U}_1^F). \tag{7.20}$$

The constants of $\mathcal{U}_0^F$ and $\mathcal{U}_1^F$ are given by the following formulae:

$$\Gamma_{10}^F = \frac{2W_{00}L}{\pi} \tag{7.21}$$

for $n = 1$, and

$$\Gamma_{20}^F = \frac{2L}{\pi} \frac{(2-B^2L^2)W_{10}+LW_{01}}{1-B^2L^2}, \tag{7.22a}$$

$$\Gamma_{21}^F = -\frac{2}{\pi} \frac{W_{10}+B^2L^3W_{01}}{1-B^2L^2}, \tag{7.22b}$$

$$D_{20}^F = -\frac{2}{\pi} \frac{BL(W_{10}+LW_{01})}{1-B^2L^2} \tag{7.22c}$$

for $n = 2$; the coefficients W_{ij} are defined by (7.19).

THE VERTEX LYING AT O_1. The motion corresponding to this vertex exerts its effect at the wing extremity. Applying formula (2.91a) where s_1 is replaced by l—since the inclination of the leading edge $OO_1A'_1$ has not changed— and s'_2 by l_1 defined by relations (7.7), we find the following expression for $\mathcal{U}_{n-1}^I$:

$$\begin{aligned}\mathcal{U}_{n-1}^I = &\sum_{q=0}^{n-1} \frac{\Gamma_{n,q}^I}{\sqrt{1-B^2l^2}} \xi^q \cosh^{-1} \sqrt{\frac{(1+Bl)(1-B\xi)}{2B(l-\xi)}} \\ &+ \sum_{q=0}^{n-1} \frac{\Theta_{n,q}^I}{\sqrt{1-B^2l_1^2}} \xi^q \cosh^{-1} \sqrt{\frac{(1+Bl_1)(1-B\xi)}{2B(l_1-\xi)}} \\ &+ \sum_{q=0}^{n-2} D_{n,q}^I \xi^q \sqrt{\frac{1-B^2\xi^2}{B^2}},\end{aligned} \tag{7.23}$$

where $\Gamma_{n,q}^I$ and $\Theta_{n,q}^I$ are defined by (7.10b) and where we have written

$$\xi = \eta + i\zeta,$$

η being given by (7.4). Further, taking into account (7.9) as well as (7.5), we may write successively:

$$-w^I = w_{00}+w_{10}(\xi_1+h)+w_{01}(\xi_2+b) = \beta_{00}+\xi_1(\beta_{10}+\beta_{01}\eta), \tag{7.24}$$

where we have set

$$\beta_{00} = w_{00}+w_{10}h+w_{01}b, \quad \beta_{10} = w_{10}, \quad \beta_{01} = w_{01}. \tag{7.25}$$

Consequently, there are two conical motions of orders $n = 1, 2$, defined by $\mathcal{U}_0^I$ and $\mathcal{U}_1^I$, which are deduced from (7.19) and for which the constants are given function of β_{ij} by the relations:

$$\Gamma_{10}^I = \frac{2\beta_{00}l}{\pi}, \quad \Theta_{10}^I = -\frac{2\beta_{00}l_1}{\pi} \tag{7.26}$$

for $n = 1$;

$$\Gamma_{20}^I = \frac{2l}{\pi}\,\frac{(2-B^2l^2)\beta_{10}+l\beta_{01}}{1-B^2l^2}, \quad \Gamma_{21}^I = -\frac{2}{\pi}\,\frac{\beta_{10}+B^2l^3\beta_{01}}{1-B^2l^2}, \tag{7.27a}$$

$$\Theta_{20}^I = -\frac{2l_1}{\pi}\,\frac{(2-B^2l_1^2)\beta_{10}+l_1\beta_{01}}{1-B^2l^2}, \quad \Theta_{21}^I = \frac{2}{\pi}\,\frac{\beta_{10}+B^2l_1^3\beta_{01}}{1-B^2l^2}, \tag{7.27b}$$

$$D_{20}^I = -\frac{1}{\pi}\left[\left(\frac{Bl}{1-B^2l^2}-\frac{Bl_1}{1-B^2l_1^2}\right)\beta_{10}+\left(\frac{Bl^2}{1-B^2l^2}-\frac{Bl_1^2}{1-B^2l_1^2}\right)\beta_{01}\right] \tag{7.27c}$$

for $n = 2$.

If the leading edges are supersonic, the same expressions are valid, taking into consideration the relations similar to (7.14).

As in the preceding cases, we may write for the axial velocity of the complex wings,

$$u^I = \text{Re}\sum_{n=1}^{N}\xi_1^{n-1}\mathcal{U}_{n-1}^I, \quad u^I = \text{Re}\,(\mathcal{U}_0^I+\xi_1\mathcal{U}_1^I), \tag{7.28}$$

respectively.

Note. It may be easily seen now that the total velocity at any point on the wing is

$$u = u^0+u^F+u^I, \tag{7.29}$$

to which u^{II}, the axial velocity corresponding to the motion inside the Mach cone with vertex at O_2, equal and symmetrical to u^I, is also added, if the respective point falls in the interior of the Mach cone with vertex at O_2.

VII. 1.4. The Wave Drag by Considering the Interference between Various Motions round the Wing

Let us denote by S the surface of half a wing and by $d\sigma$ a surface element conveniently chosen as a function of each system of axes with the origin at O, F or O_1 (Fig. 7.2), such that we may write

$$d\sigma = dx_1\, dx_2 = x_1\, dx_1\, dy, \tag{7.30a}$$

$$d\sigma = dX_1\, dx_2 = X_1\, dX_1\, dY, \tag{7.30b}$$

$$d\sigma = d\xi_1\, d\xi_2 = \xi_1\, d\xi_1\, d\eta. \tag{7.30c}$$

The general formula for the drag coefficient is

$$C_d = \frac{1}{S}\int_S 2\frac{w}{U_\infty}\, C_p\, d\sigma = -\frac{4}{SU_\infty^2}\int_S uw\, d\sigma. \tag{7.31}$$

On the other hand, taking into account the total expression for the axial velocity u (7.29), this coefficient may be decomposed into three parts:

$$C_d = C_d^O + C_d^F + C_d^I. \tag{7.32}$$

Each term of this decomposition will be explicitly written by examining Fig. 7.2:

$$C_d^O = -\frac{4}{SU_\infty^2}\int_S u^O w\, d\sigma = \sum_{n=1}^{N}\sum_{m=1}^{N} C_{dm}^{On}, \tag{7.33a}$$

$$C_d^F = -\frac{4}{SU_\infty^2}\int_{FN_1C_1} u^F w\, d\sigma = \sum_{n=1}^{N}\sum_{m=1}^{N} C_{dm}^{Fn}, \tag{7.33b}$$

$$C_d^I = -\frac{4}{SU_\infty^2}\int_{O_1H_1C_1} u^I w\, d\sigma = \sum_{n=1}^{N}\sum_{m=1}^{N} C_{dm}^{In}, \tag{7.33c}$$

where the usual coefficient C_{dm}^n, which was defined and employed in all the preceding chapters, has now been marked with the superscripts adopted above, that is, C_{dm}^{On}, C_{dm}^{Fn}, C_{dm}^{In}. Employing further the general formulae determined previously, these latter coefficients will be given by the following relations:

$$C_{dm}^{On} = -\operatorname{Re}\frac{4}{SU_\infty^2}\iint_S x_1^{m+n-1}\, dx_1\left(\sum_{k=0}^{m-1} w_{m-k-1,k}\, y^k\right)\mathcal{U}_{n-1}^O\, dy, \tag{7.34a}$$

$$C_{dm}^{Fn} = -\operatorname{Re}\frac{4}{SU_\infty^2}\iint_{FN_1C_1} X_1^{m+n-1}\, dX_1\left(\sum_{k=0}^{m-1} W_{m-k-1,k}\, Y^k\right)\mathcal{U}_{n-1}^F\, dY, \tag{7.34b}$$

$$C_{dm}^{In} = -\operatorname{Re}\frac{4}{SU_\infty^2}\iint_{O_1H_1C_1} \xi_1^{m+n-1}\, d\xi_1\left(\sum_{k=0}^{m-1} \beta_{m-k-1,k}\, \eta^k\right)\mathcal{U}_{n-1}^I\, d\eta. \tag{7.34c}$$

If the Mach cone with origin at the vertex F intersects the leading edge OO_1 at N'_1, then the second integral (7.34b) will be performed on the surface $FN'_1O_1C_1$, instead of FN_1C_1 (Fig. 7.2).

These formulae are employed in the examples given below.

VII. 2. Application to a Double Trapezoidal Swept-back Wing

VII. 2.1. Preliminary Formulae

This wing is represented in Fig. 7.3 and is defined by equation (7.8); hence the downwash is given by (7.9) where $b_1 = b$.

The functions $\mathcal{U}^O_{n-1}$ and $\mathcal{U}^F_{n-1}$ are also given by (7.11) and (7.16); the function $\mathcal{U}^I_{n-1}$ which is deduced from (7.21) by setting $l_1 = 0$ and taking into account the conditions at the origin and the relations introduced previously (4.15), has the expression

$$\mathcal{U}^I_{n-1} = \sum_{q=0}^{n-1} \frac{\Gamma^I_{n,q}}{\sqrt{1-B^2l^2}}\, \xi^q \cosh^{-1} \sqrt{\frac{(1+Bl)(1-B\xi)}{2B(l-\xi)}} + \sum_{q=1}^{n-1} Q^I_{n,q}\, \xi^q \cosh^{-1} \sqrt{\frac{1}{B^2\xi^2}} + \sum_{q=0}^{n-2} D^I_{n,q}\, \xi^q \sqrt{\frac{1-B^2\xi^2}{B^2}}, \tag{7.35}$$

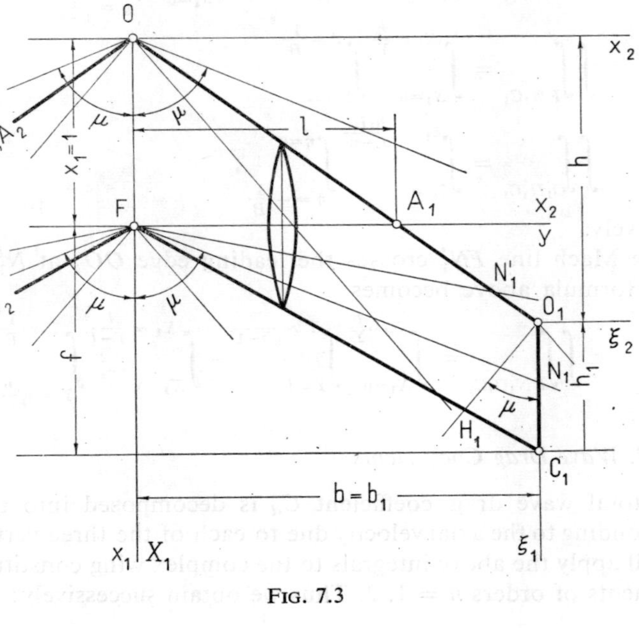

FIG. 7.3

where we have taken $Q^I_{n,0} = 0$, in order to avoid the discontinuity at the origin for u, which could not actually exist (§ II. 5).

For $n = 1, 2$, similarly to formulae (7.26), (7.27), the constants are written as follows:

$$\Gamma^I_{10} = \frac{2\beta_{00} l}{\pi}; \tag{7.36a}$$

$$\Gamma^I_{20} = \frac{2l}{\pi} \frac{(2-B^2l^2)\beta_{10}+l\beta_{01}}{1-B^2l^2}, \quad \Gamma^I_{21} = -\frac{2}{\pi} \frac{\beta_{10}+B^2l^3\beta_{01}}{1-B^2l^2}, \tag{7.36b}$$

$$Q^I_{21} = \frac{\beta_{10}}{\pi}, \qquad D^I_{20} = -\frac{Bl}{\pi} \frac{\beta_{10}+l\beta_{01}}{1-B^2l^2}. \tag{7.36c}$$

Further, we observe that the surface area of the half of the wing, upon which the integration is performed, is

$$S = \frac{(1+h_1)b}{2} = \frac{(2+f-h)b}{2}, \quad (b = hl = fL). \tag{7.37}$$

Noting that the surface covered by the Mach cone with the vertex at F is FN_1C_1, we obtain for the integrals in formulae (7.34),

$$\iint_S = \int_{x_1=0}^{x_1=\frac{L}{L-y}} \int_{y=0}^{y=\frac{b}{1+f}} + \int_{x_1=0}^{x_1=\frac{b}{y}} \int_{y=\frac{b}{1+f}}^{y=l}, \tag{7.38a}$$

$$\iint_{FN_1C_1} = \int_{X_1=0}^{X_1=\frac{b}{Y}} \int_{Y=L}^{Y=\frac{1}{B}}, \tag{7.38b}$$

$$\iint_{O_1H_1C_1} = \int_{\xi_1=0}^{\xi_1=\frac{h_1L}{L-\eta}} \int_{\eta=-\frac{1}{B}}^{\eta=0}, \tag{7.38c}$$

respectively.

If the Mach line FN_1' crosses the leading edge OO_1 at N_1', then the second formula above becomes

$$\iint_{FN_1'O_1C_1} = \int_{X_1=0}^{X_1=\frac{b}{Y}} \int_{Y=L}^{Y=\frac{b}{h-1}} + \int_{X_1=0}^{X_1=\frac{l}{Y-l}} \int_{Y=\frac{b}{h-1}}^{Y=\frac{1}{B}} \tag{7.39}$$

VII. 2.2. Wave Drag Coefficients

The total wave drag coefficient C_d is decomposed into three parts, corresponding to the axial velocity due to each of the three vertices (7.32). We shall apply the above integrals to the complex wing constituted of two components of orders $n = 1, 2$. Thus we obtain successively:

(a) As regards C_d^O, corresponding to the vertex σ:

$$C_{d1}^{O1} = -\frac{2w_{00}}{SU_\infty^2} \Gamma_{10}^O(E_{20}^O + b^2 H_{20}^O), \tag{7.40a}$$

$$C_{d1}^{O2} = -\frac{4w_{00}}{3SU_\infty^2} \{\Gamma_{20}^O(E_{30}^O + b^3 H_{30}^O) + \Gamma_{21}^O[L(E_{31}^O - E_{21}^O) + b^3 H_{21}^O] + D_{20}^O(P_3^O + b^3 R_3^O)\}, \tag{7.40b}$$

$$C_{d2}^{O1} = -\frac{4\Gamma_{10}^O}{3SU_\infty^2} \{w_{10}(E_{30}^O + b^3 H_{30}^O) + w_{01}[L(E_{30}^O - E_{20}^O) + b^3 H_{20}^O]\}, \tag{7.40c}$$

$$C_{d2}^{O2} = -\frac{w_{10}}{SU_\infty^2} \{\Gamma_{20}^O(E_{40}^O + b^4 H_{40}^O) + \Gamma_{21}^O[L(E_{41}^O - E_{31}^O) + b^4 H_{31}^O] + D_{20}^O(P_4^O + b^4 R_4^O)\} - \frac{w_{01}}{SU_\infty^2} \{\Gamma_{20}^O[L(E_{40}^O - E_{30}^O) + b^4 H_{30}^O] + \Gamma_{21}^O[L^2(E_{41}^O - 2E_{31}^O + E_{21}^O) + b^4 H_{21}^O] + D_{20}^O[L(P_4^O - P_3^O) + b^4 R_3^O]\}. \tag{7.40d}$$

The sum of these four coefficients represents C_d^O.

The above expressions are valid both for subsonic and supersonic edges.

In the above expressions we have denoted by $E_{t,q}^O$, $H_{t,q}^O$, P_t^O, R_t^O the following integrals:

$$E_{t,q}^O = \mathrm{Re}\,\frac{1}{\sqrt{1-B^2 l^2}} \int_0^{\frac{b}{1+f}} \frac{L^t}{(L-y)^t} \left[\cosh^{-1}\sqrt{\frac{(1+Bl)(1-By)}{2B(l-y)}} + (-1)^q \cosh^{-1}\sqrt{\frac{(1+Bl)(1+By)}{2B(l+y)}}\right] dy, \tag{7.41a}$$

$$H_{t,q}^O = \mathrm{Re}\,\frac{1}{\sqrt{1-B^2 l^2}} \int_{\frac{b}{1+f}}^{l} \frac{1}{y^t} \left[\cosh^{-1}\sqrt{\frac{(1+Bl)(1-By)}{2B(l-y)}} + (-1)^q \cosh^{-1}\sqrt{\frac{(1+Bl)(1+By)}{2B(l+y)}}\right] dy, \tag{7.41b}$$

$$P_t^O = \mathrm{Re} \int_0^{\frac{b}{1+f}} \frac{L^t}{(L-y)^t} \sqrt{\frac{1}{B^2} - y^2}\, dy,$$

$$R_t^O = \mathrm{Re} \int_{\frac{b}{1+f}}^{l} \frac{1}{y^t} \sqrt{\frac{1}{B^2} - y^2}\, dy. \tag{7.41c}$$

The values of these integrals corresponding to the wave drag coefficients indicated above (7.40) are given in Appendix III.1.

(b) As regards C_d^F, corresponding to the vertex F, by applying (7.34b) and (7.38b), (7.39):

$$C_{d1}^{F1} = -\frac{2W_{00}}{SU_\infty^2}\Gamma_{10}^F(b^2H_{20}^F+E_{20}^F), \tag{7.42a}$$

$$C_{d1}^{F2} = -\frac{4W_{00}}{3SU_\infty^2}\{\Gamma_{20}^F(b^3H_{30}^F+E_{30}^F)+\Gamma_{21}^F[b^3H_{21}^F+l(E_{31}^F+E_{21}^F)]$$
$$+D_{20}^F(b^3R_3^F+P_3^F)\}, \tag{7.42b}$$

$$C_{d2}^{F1} = -\frac{4\Gamma_{10}^F}{3SU_\infty^2}\{W_{10}(b^3H_{30}^F+E_{30}^F)+W_{01}[b^3H_{20}^F+l(E_{30}^F+E_{20}^F]\}, \tag{7.42c}$$

$$C_{d2}^{F2} = -\frac{W_{10}}{SU_\infty^2}\{\Gamma_{20}^F(b^4H_{40}^F+E_{40}^F)+\Gamma_{21}^F[b^4H_{31}^F+l(E_{41}^F+E_{31}^F)]$$
$$+D_{20}^F(b^4R_4^F+P_4^F)\}-\frac{W_{01}}{SU_\infty^2}\{\Gamma_{20}^F[b^4H_{30}^F+l(E_{40}^F+E_{30}^F)]+\Gamma_{21}^F[b^4H_{21}^F$$
$$+l^2(E_{41}^F+2E_{31}^F+E_{21}^F)]+D_{20}^F[b^4R_3^F+l(P_4^F+P_3^F)]\}. \tag{7.42d}$$

The total coefficient C_d^F is given by the sum of the above four coefficients, where we have employed the following integrals:

$$H_{t,q}^F = \mathrm{Re}\frac{1}{\sqrt{1-B^2L^2}}\int_L^a \frac{1}{Y^t}\left[\cosh^{-1}\sqrt{\frac{(1+BL)(1-BY)}{2B(L-Y)}}\right.$$
$$\left.+(-1)^q\cosh^{-1}\sqrt{\frac{(1+BL)(1+BY)}{2B(L+Y)}}\right]dY, \tag{7.43a}$$

$$E_{t,q}^F = \mathrm{Re}\frac{1}{\sqrt{1-B^2L^2}}\int_a^{1/B}\frac{l^t}{(Y-l)^t}\left[\cosh^{-1}\sqrt{\frac{(1+BL)(1-BY)}{2B(L-Y)}}\right.$$
$$\left.+(-1)^q\cosh^{-1}\sqrt{\frac{(1+BL)(1+BY)}{2B(L+Y)}}\right]dY, \tag{7.43b}$$

$$R_t^F = \mathrm{Re}\int_L^a \frac{1}{Y^t}\sqrt{\frac{1}{B^2}-Y^2}\,dY,\quad P_t^F = \mathrm{Re}\int_a^{1/B}\frac{l^t}{(Y-l)^t}\sqrt{\frac{1}{B^2}-Y^2}\,dY. \tag{7.43c}$$

The corresponding integrals included in the formulae of the wave drag coefficients (7.42) are given in Appendix III.2.

Note. The formulae determined above for the coefficients C_{dm}^{Fn} are valid both in the case when the Mach cone with vertex at F crosses the edge OO_1 at N_1', the integration limit being in such a case

$$a = \frac{b}{h-1}, \tag{7.44}$$

and in the case when the Mach cone with the vertex at F crosses the lateral edge O_1C_1 at N_1, for which the integration limit is

$$a = \frac{1}{B}. \tag{7.45}$$

In the latter case it is readily seen that

$$E^F_{t,q} = P^F_t = 0 \tag{7.46}$$

a fact which simplifies the expressions of C^{Fn}_{dm} very much.

(c) As regards the coefficient C^I_d, by applying relations (7.34c) and (7.38c), we obtain:

$$C^{I1}_{d1} = -\frac{2\beta_{00}h_1^2}{SU_\infty^2}\Gamma^I_{10}E^I_2, \tag{7.47a}$$

$$C^{I2}_{d1} = -\frac{4\beta_{00}h_1^3}{3SU_\infty^2}[\Gamma^I_{20}E^I_3+\Gamma^I_{21}L(E^I_3-E^I_2)+Q^I_{21}L(M^I_3-M^I_2)+D^I_{20}P^I_3], \tag{7.47b}$$

$$C^{I1}_{d2} = -\frac{4h_1^3\Gamma^I_{10}}{3SU_\infty^2}[\beta_{10}E^I_3+\beta_{01}L(E^I_3-E^I_2)], \tag{7.47c}$$

$$\begin{aligned} C^{I2}_{d2} = -\frac{h_1^4}{SU_\infty^2}\{&\beta_{10}[\Gamma^I_{20}E^I_4+\Gamma^I_{21}L(E^I_4-E^I_3)+Q^I_{21}L(M^I_4-M^I_3)+D^I_{20}P^I_4] \\ &+\beta_{01}[\Gamma^I_{20}L(E^I_4-E^I_3)+\Gamma^I_{21}L^2(E^I_4-2E^I_3+E^I_2)+Q^I_{21}L^2(M^I_4-2M^I_3+M^I_2) \\ &+D^I_{20}L(P^I_4-P^I_3)]\}. \end{aligned} \tag{7.47d}$$

The sum of these coefficients represents C^I_d, which is also valid for supersonic edges.

In the above formulae (7.47), we have employed the following integrals:

$$E^I_t = \mathrm{Re}\frac{1}{\sqrt{1-B^2l^2}}\int_{-1/B}^{0}\frac{L^t}{(L-\eta)^t}\cosh^{-1}\sqrt{\frac{(1+Bl)(1-B\eta)}{2B(l-\eta)}}\,d\eta, \tag{7.48a}$$

$$M^I_t = \mathrm{Re}\int_{-1/B}^{0}\frac{L^t}{(L-\eta)^t}\cosh^{-1}\sqrt{\frac{1}{B^2\eta^2}}\,d\eta, \tag{7.48b}$$

$$P^I_t = \mathrm{Re}\int_{-1/B}^{0}\frac{L^t}{(L-\eta)^t}\sqrt{\frac{1}{B^2}-\eta^2}\,d\eta, \tag{7.48c}$$

for which the particular values included in formulae (7.47) are given in Appendix III.3.

VII. 2.3. Case of Arrow-head Wing

If we set $h_1 = 0$ ($h = 1+f$) in Fig. 7.3, in which case O_1 falls at C_1, then we obtain the arrow-head wing (Fig. 7.4), where the motion corresponding to O_1 no longer exists. Because of this, the coefficients C_{dm}^{In} are equal to zero; as regards the coefficients C_{dm}^{Fn} of the formulae, the limit a in the

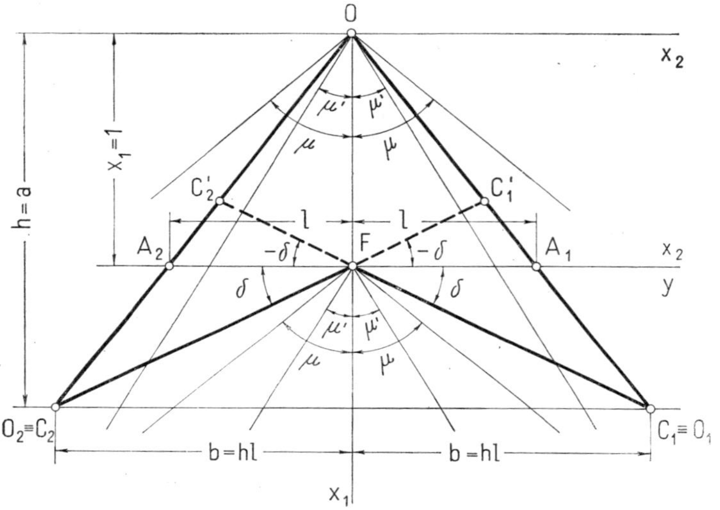

FIG. 7.4

respective integrals retains only the value

$$a = \frac{b}{h-1} = \frac{b}{f} = L, \tag{7.49}$$

a fact that implies

$$H_{t,q}^{F} = R_t^F = 0. \tag{7.50}$$

In a similar way, noting that one of the integration limits in C_{dm}^{On} (7.38a), (7.41) is

$$\frac{b}{1+f} = \frac{b}{h} = l, \tag{7.51}$$

it is readily seen that we have also

$$H_{t,q}^{O} = R_t^O = 0. \tag{7.52}$$

The drag coefficients may be determined by particularization from those of the double trapezoidal swept-back wing, by putting $h_1 = 0$. It is interesting however to compute the drag coefficient for this wing, under the assumption that the Mach cone with the vertex at F does not intersect the wing at all. In such a case, omitting the superscript O, the total drag coefficient will be given by relation

$$C_d = C_{d1}^1 + C_{d1}^2 + C_{d2}^1 + C_{d2}^2, \tag{7.53}$$

in which

$$C_{d1}^1 = -\frac{2w_{00}}{SU_\infty^2} \Gamma_{10}^O E_{20}, \tag{7.54a}$$

$$C_{d1}^2 = -\frac{4w_{00}}{3SU_\infty^2} [\Gamma_{20}^O E_{30} + \Gamma_{21}^O L(E_{31} - E_{21}) + D_{20}^O P_3], \tag{7.54b}$$

$$C_{d2}^1 = -\frac{4\Gamma_{10}^O}{3SU_\infty^2} [w_{10} E_{30} + w_{01} L(E_{30} - E_{20})], \tag{7.54c}$$

$$C_{d2}^2 = -\frac{1}{SU_\infty^2} \{w_{10}[\Gamma_{20}^O E_{40} + \Gamma_{21}^O L(E_{41} - E_{31}) + D_{20}^O P_4]$$
$$+ w_{01}[\Gamma_{20}^O L(E_{40} - E_{30}) + \Gamma_{21}^O L^2(E_{41} - 2E_{31} + E_{21})$$
$$+ D_{20}^O L(P_4 - P_3)]\}, \tag{7.54d}$$

where the integrals $E_{t,q}$ and P_t which are obtained from $E_{t,q}^O$ and P_t^O by means of the particularization mentioned above (7.51), are given in Appendix III.4.

If we note that we can write

$$\frac{f}{b} = \frac{1}{L} = \tan \delta, \tag{7.55}$$

then the above expressions are also valid for the case of diamond shaped wing, where δ is replaced now by $-\delta$ (Fig. 7.4).

VII. 3. Examples of Polygonal Wings of Symmetrical Thickness, the Interference between the Two Vertices O_1 and O_2 being taken into Consideration. Application to the Rectangular Wing

VII. 3.1. Preliminary Considerations

In the following we shall consider the case of wings of low aspect ratio, in order to emphasize the reciprocal influence between the motions corresponding to the vertices O_1 and O_2 (Fig. 7.5).

When the Mach cones with the vertices at O_1 and O_2 meet at the point M' or M'', the total axial velocity on the right hand side of the wing should also include the effect of the motion corresponding to the vertex O_2, that is

$$u = u^O + u^I + u^{II}. \tag{7.56}$$

As shown previously, the motion corresponding to the vertex O_2 is symmetrical to that corresponding to the vertex O_1 with respect to the axis Ox_1, and, accordingly, u^{II} is symmetrical to u^I with respect to the same axis. The drag coefficient which refers to the right-hand side of the wing should also include the effect of the motion corresponding to the vertex O_2 on the portion $M'H_2'F$ or on the portion $M''H_2''C_1F$; owing to

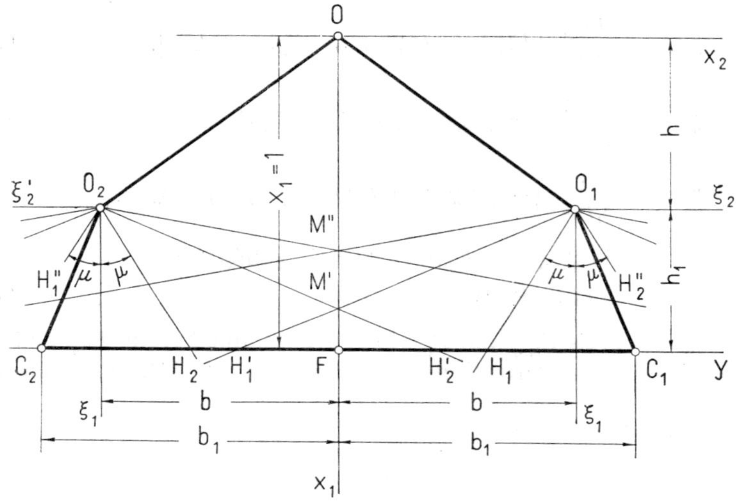

FIG. 7.5

the wing symmetry, this effect is equivalent to that of the motion corresponding to the vertex O_1 on the portions $M'H_1'F$ or $M''H_1''C_2F$, respectively. Accordingly, the total drag coefficient can be written for this case also as

$$C_d = C_d^O + C_d^I, \tag{7.57}$$

where C_d^O and C_d^I are given by relations (7.33a) and (7.33c); however, in the latter relations, the integration surface will be replaced by the surface $O_1H_1'C_1$ or $O_1H''C_2C_1$, respectively.

Proceeding in this manner, one may examine any polygonal wing of low aspect ratio. As an example, the case of the rectangular wing is treated below (Fig. 7.6).

This wing may be obtained from the polygonal wing (Fig. 7.1), by setting $h = f = 0$, $b = b_1$ ($l_1 = 0$, $l \to \infty$). It is defined by the equation of its surface,

$$Z = tx_1(1-x_1) \tag{7.58}$$

or by the resulting downwash,

$$w = U_\infty \frac{\partial Z}{\partial x_1} = tU_\infty - 2tU_\infty x_1 = w_{00} + w_{10}x_1, \tag{7.59}$$

this having the same expression as (7.9), but where

$$w_{00} = tU_\infty, \quad w_{10} = -2tU_\infty, \quad w_{01} = 0, \tag{7.60}$$

In this case, the motion corresponding to the vertex O reduces to a plane motion around the profile represented by (7.58).

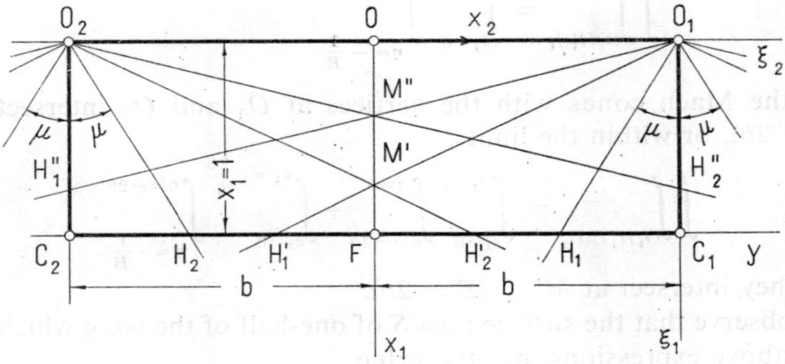

FIG. 7.6

The axial velocity u^0, at any point of the wing, results by applying formulae (1.68),

$$u^0 = -\frac{U_\infty}{B}\frac{\partial Z}{\partial x_1} = -\frac{tU_\infty(1-2x_1)}{B} = -\frac{w_{00}+w_{10}x_1}{B}, \tag{7.61}$$

which may be deduced from that established previously, (7.11) and (7.15), by making l tend to infinity ($l \to \infty$).

As regards the axial velocity u^I, given by relation (7.28), it may be obtained by making h, f and l_1 tend to zero, and l tend to infinity in expressions (7.23). We get thus

$$\mathcal{U}_0^I = \frac{2}{\pi}\frac{w_{00}}{B}\cos^{-1}\sqrt{\frac{1-B\xi}{2}}, \quad (\xi = \eta + i\zeta, \quad \eta = \frac{\xi_2}{\xi_1}, \quad w_{00} = tU_\infty), \tag{7.62a}$$

$$\mathcal{U}_1^I = \frac{2}{\pi}\frac{w_{10}}{B}\left[\cos^{-1}\sqrt{\frac{1-B\xi}{2}} + B\xi\cosh^{-1}\sqrt{\frac{1-B\xi}{-2B\xi}}\right], (w_{10} = -2tU_\infty). \tag{7.62b}$$

In the expression for the total velocity u (7.56), u^{II} is also contained; it is symmetrical to u^I with respect to the axis Ox_1.

VII. 3.2. Wave Drag Coefficient

The total drag coefficient is given by relation (7.57), in which C_d^O will be calculated by performing the integral of (7.33a) within the limits

$$\iint_S = \int_{x_1=0}^{x_1=1}\int_{x_2=0}^{x_2=b}. \tag{7.63}$$

The coefficient C_d^I, which is given by relation (7.33c) where $N = 2$, will be calculated by performing the integral of (7.34c) within the limits

$$\iint_{O_1H_1'C_1} = \int_{\xi_1=0}^{\xi_1=1}\int_{\eta=-\frac{1}{B}}^{\eta=0}, \tag{7.64}$$

when the Mach cones with the vertices at O_1 and O_2 intersect at M' $(1/B < 2b)$, or within the limits

$$\iint_{O_1H_1''C_2C_1} = \int_{\xi_1=0}^{\xi_1=1}\int_{\eta=-2b}^{\eta=0} + \int_{\xi_1=0}^{\xi_1=-\frac{2b}{\eta}}\int_{\eta=-\frac{1}{B}}^{\eta=-2b}, \tag{7.65}$$

when they intersect at M'' $(1/B > 2b)$.

We observe that the surface area S of one-half of the wing which occurs in the above expressions, has the value

$$S = b. \tag{7.66}$$

Thus, performing the above integrals, we find successively:

$$C_d^O = \frac{4b}{BSU_\infty^2}(w_{00}^2 + w_{00}w_{10} + \tfrac{1}{3}w_{10}^2), \tag{7.67a}$$

$$C_{d1}^{I1} = -\frac{4w_{00}^2}{\pi BSU_\infty^2}(\Phi_0 + 4b^2\Psi_2), \tag{7.67b}$$

$$C_{d1}^{I2} = -\frac{8w_{00}w_{10}}{3\pi BSU_\infty^2}[\Phi_0 + B\Lambda_1 - 8b^3(\Psi_3 + B\Theta_2)], \tag{7.67c}$$

$$C_{d2}^{I1} = -\frac{8w_{00}w_{10}}{3\pi BSU_\infty^2}(\Phi_0 - 8b^3\Psi_3), \tag{7.67d}$$

$$C_{d2}^{I2} = -\frac{2w_{10}^2}{\pi BSU_\infty^2}[\Phi_0 + B\Lambda_1 + 16b^4(\Psi_4 + B\Theta_3)], \tag{7.67e}$$

where we have written:

$$\Phi_0 = \int_{-a}^{0}\cos^{-1}\sqrt{\frac{1-B\eta}{2}}\,d\eta = a\cos^{-1}\sqrt{\frac{1+Ba}{2}} + \frac{1}{2B}(1-\sqrt{1-B^2a^2}), \tag{7.68a}$$

$$\Lambda_1 = \int_{-a}^{0} \eta \cosh^{-1} \sqrt{\frac{1-B\eta}{2B\eta}}\, d\eta = -\frac{a^2}{4} \cosh^{-1} \frac{1}{Ba}$$

$$-\frac{1}{4B^2}\left(1-\sqrt{1-B^2a^2}\,\right), \tag{7.68b}$$

$$\Psi_t = \int_{-1/B}^{-a} \frac{1}{\eta^t} \cos^{-1} \sqrt{\frac{1-B\eta}{2}}\, d\eta = -\frac{1}{t-1}\left[\frac{1}{(-a)^{t-1}} \cos^{-1} \sqrt{\frac{1+Ba}{2}}\right.$$

$$\left. -\frac{B}{2} I_{t-1}\right], \tag{7.68c}$$

$$\Theta_t = \int_{-1/B}^{-a} \frac{1}{\eta^t} \cosh^{-1} \sqrt{\frac{1-B\eta}{-2B\eta}}\, d\eta = -\frac{1}{2(t-1)}\left[\frac{1}{(-a)^{t-}}\right.$$

$$\left.\times \cosh^{-1} \frac{1}{Ba} + I_t\right]. \tag{7.68d}$$

The values of I_t for $t = 1, 2, 3$ are the following:

$$I_1 = \int_{-1/B}^{-a} \frac{d\eta}{\eta\sqrt{1-B^2\eta^2}} = -\cosh^{-1} \frac{1}{Ba}, \tag{7.69a}$$

$$I_2 = \int_{-1/B}^{-a} \frac{d\eta}{\eta^2\sqrt{1-B^2\eta^2}} = \frac{\sqrt{1-B^2\alpha^2}}{a}, \tag{7.69b}$$

$$I_3 = \int_{-1/B}^{-a} \frac{d\eta}{\eta^3\sqrt{1-B^2\eta^2}} = -\frac{1}{2a^2}\left(\sqrt{1-B^2a^2}\right.$$

$$\left. + B^2a^2 \cosh^{-1} \frac{1}{Ba}\right). \tag{7.69c}$$

In the above expressions,

$a = 2b$ if the Mach cones with the vertices at O_1 and O_2 intersect each other at M'' $(1/B > 2b)$,

$a = 1/B$ if the Mach cones with the vertices at O_1 and O_2 intersect each other at $\mathbf{M}'$ $(1/B < 2b)$.

In the latter case it is obvious that

$$\Psi_t = \Theta_t = I_t = 0. \tag{7.70}$$

VII. 4. Polygonal Wings of Symmetrical Thickness with Diamond-shaped Profile

VII. 4.1. Arrow-head Wing

Let us consider a symmetrical wing whose lower or upper surface consists, on each of its halves, of two planes tilted with respect to the plane Ox_1x_2 and intersecting on the lines GC_1 and GC_2, respectively (Fig. 7.7).

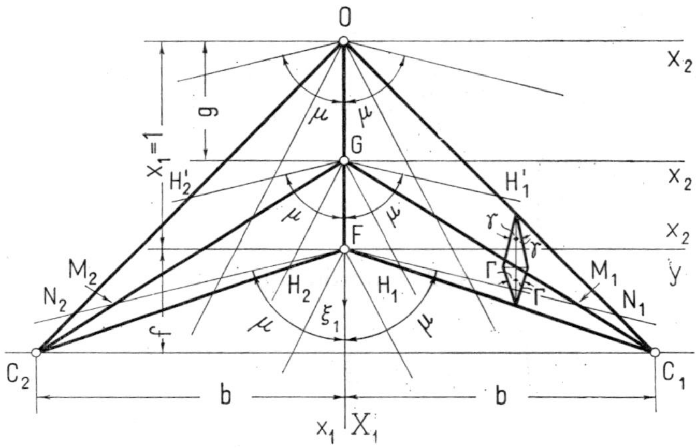

FIG. 7.7

If we denote by γ the angle made by the frontal plane OC_1GC_2 with the free stream direction, the rear plane GC_1FC_2 will make the angle

$$\Gamma = -\frac{g}{1-g}\gamma. \tag{7.71}$$

The downwash on the wing is

$$w = \gamma U_\infty \tag{7.72}$$

on the frontal portion OC_1GC_2, and

$$w = \Gamma U_\infty \tag{7.73}$$

on the rear portion GC_1FC_2.

THE MOTIONS AROUND THE WING. The motion around the right-hand half of the wing considered is equivalent to the superposition of the following motions which satisfy conditions (7.72) and (7.73) on the wing, as well as the condition $w = 0$ over the whole remaining plane outside the wing:

(a) The motion corresponding to the vertex O around the fictitious wing OC_1C_2, having the downwash $w = \gamma U_\infty$.
(b) The motion corresponding to the vertex G around the fictitious wing GC_1C_2, having the downwash $w^G = (\Gamma - \gamma)U_\infty$.
(c) The motion corresponding to the vertex F around the fictitious wing FC_1C_2, having the downwash $w^F = -\Gamma U_\infty$.

The determination of all these motions, whose superposition solves the problem of the wing considered, will be done by proceeding in the same way as in the case of polygonal wings of symmetrical thickness treated in (§ VII. 1).

In this sense, let us refer these motions to the reference systems Ox_1x_2, GX_1x_2 and $F\xi_1x_2$, respectively, systems related by

$$X_1 = x_1 - g, \quad Y = \frac{x_2}{X_1} = \frac{x_2}{x_1 - g}, \tag{7.74}$$

$$\xi_1 = x_1 - 1, \quad \eta = \frac{x_2}{\xi_1} = \frac{x_2}{x_1 - 1}. \tag{7.75}$$

Axial Velocities. Considering the way in which the three motions were defined above, one may readily deduce the expressions for the respective axial velocities. Thus:

(a) *The motion corresponding to the vertex O* is defined by the function $\mathcal{U}^O$, from which one may derive the axial velocity ($u^O = \text{Re}\, \mathcal{U}^O$),

$$\mathcal{U}^O = -\frac{2\gamma U_\infty l}{\pi\sqrt{1-B^2l^2}}\left[\cosh^{-1}\sqrt{\frac{(1+Bl)(1-Bx)}{2B(l-x)}} + \cosh^{-1}\sqrt{\frac{(1+Bl)(1+Bx)}{2B(l+x)}}\right] \tag{7.76}$$

where we have set

$$l = \frac{b}{1+f}, \quad x = y + i\mathfrak{z}. \tag{7.77}$$

(b) *The motion corresponding to the vertex G* is similarly defined by the function

$$\mathcal{U}^G = \frac{2(\gamma - \Gamma)U_\infty L}{\pi\sqrt{1-B^2L^2}}\left[\cosh^{-1}\sqrt{\frac{(1+BL)(1-BX)}{2B(L-X)}}\right.$$
$$\left. + \cosh^{-1}\sqrt{\frac{(1+BL)(1+BX)}{2B(L+X)}}\right], \tag{7.78}$$

where

$$L = \frac{b}{1+f-g}, \quad X = Y+iZ. \tag{7.79}$$

(c) *The motion corresponding to the vertex F* is likewise defined by the function

$$\mathcal{U}^F = \frac{2\Gamma U_\infty \lambda}{\pi\sqrt{1-B^2\lambda^2}}\left[\cosh^{-1}\sqrt{\frac{(1+B\lambda)(1-B\xi)}{2B(\lambda-\xi)}}\right.$$

$$\left.+\cosh^{-1}\sqrt{\frac{(1+B\lambda)(1+B\xi)}{2B(\lambda+\xi)}}\right], \tag{7.80}$$

where

$$\lambda = \frac{b}{f}, \qquad \xi = \eta+i\zeta. \tag{7.81}$$

These expressions are also valid when the edges OC_1, FC_1 or the ridge GC_1 are supersonic.

As regards the total axial disturbance velocity, this is given by the relation

$$u = u^O+u^G+u^F = \mathrm{Re}\,(\mathcal{U}^O+\mathcal{U}^G+\mathcal{U}^F). \tag{7.82}$$

WAVE DRAG COEFFICIENT. For determining the drag coefficient, we may use the known relation (7.31). Noticing, however, that in this expression u is given by relation (7.82) and w has on the wing two values defined by relations (7.72) and (7.73), we may decompose the total drag coefficient as follows:

$$C_d = C_d^{O1}+C_d^{O2}+C_d^{G1}+C_d^{G2}+C_d^{F1}+C_d^{F2}, \tag{7.83}$$

in which

$$C_d^{O1} = -\frac{4\Gamma}{SU_\infty}\iint_S x_1 u^O\, dx_1\, dy$$

$$= \mathrm{Re}\frac{4\gamma\Gamma l}{\pi}\frac{\lambda^2}{\lambda^2-l^2}\left\{\frac{l}{\sqrt{1-B^2l^2}}\left[2\cosh^{-1}\frac{1+Bl}{2\sqrt{Bl}}\right.\right.$$

$$\left.+\frac{l}{\lambda}\cosh^{-1}\frac{1}{Bl}\right]+\frac{\lambda}{\sqrt{1-B^2\lambda^2}}\left[2\cosh^{-1}\sqrt{\frac{(1+B\lambda)(1-Bl)}{2B(\lambda-l)}}\right.$$

$$\left.\left.-\cosh^{-1}\frac{1}{B\lambda}\right]\right\}, \tag{7.84a}$$

$$C_d^{O2} = -\frac{4(\gamma-\Gamma)}{SU_\infty}\iint_{OGC_1} x_1 u^O\, dx_1\, dy$$

$$= \mathrm{Re}\,\frac{4\gamma(\gamma-\Gamma)l}{\pi S}\,\frac{g^2L^2}{L^2-l^2}\left\{\frac{l}{\sqrt{1-B^2l^2}}\left[2\cosh^{-1}\frac{1+Bl}{2\sqrt{Bl}}\right.\right.$$

$$\left.+\frac{l}{L}\cosh^{-1}\frac{1}{Bl}\right]+\frac{L}{\sqrt{1-B^2L^2}}\left[2\cosh^{-1}\sqrt{\frac{(1+BL)(1-Bl)}{2B(L-l)}}\right.$$

$$\left.\left.-\cosh^{-1}\frac{1}{BL}\right]\right\}, \tag{7.84b}$$

$$C_l^{G1} = -\frac{4\Gamma}{SU_\infty}\iint_{GC_1F} X_1 u^G\, dX_1\, dY$$

$$= -\mathrm{Re}\,\frac{4(\gamma-\Gamma)\Gamma L}{\pi S}\,\frac{(1-g)^2\lambda^2}{\lambda^2-L^2}\left\{\frac{L}{\sqrt{1-B^2L^2}}\left[2\cosh^{-1}\frac{1+BL}{2\sqrt{BL}}\right.\right.$$

$$\left.+\frac{L}{\lambda}\cosh^{-1}\frac{1}{BL}\right]+\frac{\lambda}{\sqrt{1-B^2\lambda^2}}\left[2\cosh^{-1}\sqrt{\frac{(1+B\lambda)(1-BL)}{2B(\lambda-L)}}\right.$$

$$\left.\left.-\cosh^{-1}\frac{1}{B\lambda}\right]\right\}, \tag{7.84c}$$

$$C_d^{G2} = -\frac{4\gamma}{SU_\infty}\iint_{GH_1'C_1} X_1 u^G\, dX_1\, dY$$

$$= -\mathrm{Re}\,\frac{4\gamma(\gamma-\Gamma)L}{\pi S}\,\frac{g^2l^2}{L^2-l^2}\left[\frac{2L}{\sqrt{1-B^2L^2}}\cosh^{-1}\frac{1+BL}{2\sqrt{BL}}\right.$$

$$\left.+\frac{2l}{\sqrt{1-B^2l^2}}\cosh^{-1}\sqrt{\frac{(1-Bl)(1+BL)}{2B(L-l)}}\right], \tag{7.84d}$$

$$C_d^{F1} = -\frac{4\gamma}{SU_\infty}\iint_{FN_1C_1} \xi_1 u^F\, d\xi_1\, d\eta =$$

$$= -\mathrm{Re}\,\frac{4\gamma\Gamma\lambda}{\pi S}\,\frac{l^2}{\lambda^2-l^2}\left[\frac{2\lambda}{\sqrt{1-B^2\lambda^2}}\cosh^{-1}\frac{1+B\lambda}{2\sqrt{B\lambda}}\right.$$

$$\left.+\frac{2l}{\sqrt{1-B^2l^2}}\cosh^{-1}\sqrt{\frac{(1-Bl)(1+B\lambda)}{2B(\lambda-l)}}\right], \tag{7.84e}$$

$$C_d^{F2} = -\frac{4(\Gamma-\gamma)}{SU_\infty}\int\int_{FM_1C_1} \xi_1 u^F \, d\xi_1 \, d\eta =$$

$$= \mathrm{Re}\,\frac{4(\gamma-\Gamma)\Gamma\lambda}{\pi S}\,\frac{(1-g)^2L^2}{\lambda^2-L^2}\left[\frac{2\lambda}{\sqrt{1-B^2\lambda^2}}\cosh^{-1}\frac{1+B\lambda}{2\sqrt{B\lambda}}\right.$$

$$\left.+\frac{2L}{\sqrt{1-B^2L^2}}\cosh^{-1}\sqrt{\frac{(1-BL)(1+B\lambda)}{2B(\lambda-L)}}\right], \tag{7.84f}$$

where S is the surface area of half a wing

$$S = \frac{b}{2}. \tag{7.85}$$

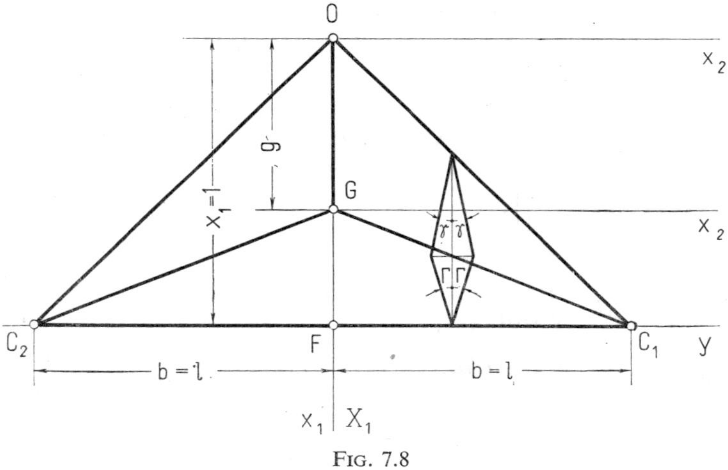

FIG. 7.8

The expressions determined above are also valid when $l > 1/B$, $L > 1/B$ or $\lambda > 1/B$; it should be mentioned, however, that if $\lambda > 1/B$ the coefficients C_d^{F1} and C_d^{F2} are zero, and if $\lambda > L > 1/B$ the coefficients C_d^{F1}, C_d^{F2} and C_d^{G2} are also zero.

The extraction of the real part from the above expressions requires particular attention. If we set $f = 0$ and $\lambda \to \infty$ in the above expressions, we obtain the wave drag coefficient of the delta wing with diamond-shaped profile (Fig. 7.8), for which $u^F = 0$ and consequently $C_d^{F1} = C_d^{F2} = 0$.

VII. 4.2. Rectangular Wing

THE MOTION AROUND THE WING. Let us consider the rectangular wing with diamond-shaped profile (Fig. 7.9); the downwashes on the frontal and rear surfaces of the wing are given by (7.72) and (7.73). Considering the case when the cones with their vertices at O_1, G_1 and those with their vertices at O_2 and G_2 do not intersect on the wing, the general motion

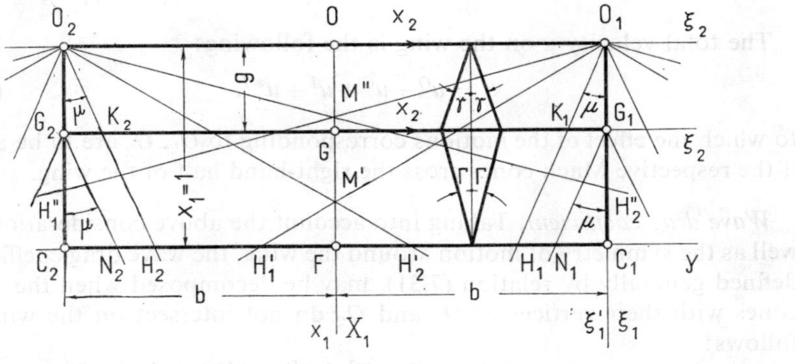

FIG. 7.9

around half a wing will consist of the superposition of the following motions:

(a) The plane motion corresponding to the plane $O_2O_1C_1C_2$ having the slope γ, the downwash being $w = \gamma U_\infty$.

(b) The plane motion corresponding to the plane $G_2G_1C_1C_2$ having the slope $(\Gamma-\gamma)$, the downwash being $w^G = -(\gamma-\Gamma)U_\infty$.

(c) The conical motion corresponding to the vertex O_1 which affects the fictitious wing O_1C_1FO motion defined by the downwash $w^I = -\gamma U_\infty$.

(d) The conical motion corresponding to the vertex G_1 which affects the plane G_1C_1FG, motion defined by the downwash $w_1^{*I} = -\Gamma U_\infty$.

If the Mach cones with their vertices at O_1, G_1, intersect on the wing with those having their vertices at O_2, G_2, then the effect of the conical motions corresponding to the vertices O_2, G_2, which are symmetrical to those corresponding to O_1, G_1, is to be added to the right-hand half of the wing.

AXIAL VELOCITIES. For the first two motions we have

$$u^O = -\frac{\gamma U_\infty}{B}, \quad u^G = \frac{(\gamma-\Gamma)U_\infty}{B}, \tag{7.86}$$

and for the remaining ones,

$$u^I = \mathrm{Re}\,\frac{2}{\pi}\frac{\gamma U_\infty}{B}\cos^{-1}\sqrt{\frac{1-B\xi}{2}}, \quad u^{*I} = \mathrm{Re}\,\frac{2}{\pi}\frac{\Gamma U_\infty}{B}\cos^{-1}\sqrt{\frac{1-B\xi^*}{2}}, \tag{7.87}$$

respectively, where (Fig. 7.9)

$$\xi = \eta + i\zeta, \quad \xi^* = \eta^* + i\zeta^*, \quad \eta = \frac{\xi_2}{\xi_1}, \quad \eta^* = \frac{\xi_2}{\xi_1^*} = \frac{\xi_2}{\xi_1 - g}. \tag{7.88}$$

The total velocity u on the wing is the following:

$$u = u^O + u^G + u^I + u^{*I}, \tag{7.89}$$

to which the effect of the motions corresponding to O_2, G_2 are to be added if the respective Mach cones cross the right-hand half of the wing.

Wave drag coefficient. Taking into account the above considerations, as well as the symmetry of motion around the wing, the wave drag coefficient, defined generally by relation (7.31), may be decomposed when the Mach cones with their vertices at O_1 and O_2 do not intersect on the wing, as follows:

$$C_d = C_d^{O1} + C_d^{O2} + C_d^G + C_d^{I1} + C_d^{I2} + C_d^{*I}. \tag{7.90}$$

The partial drag coefficients in the right-hand side have the following expressions:

$$C_d^{O1} = -\frac{4\gamma}{SU_\infty}\iint_{OO_1G_1G} u^O\,dx_1\,dx_2 = \frac{4gb\gamma^2}{BS}, \tag{7.91a}$$

$$C_d^{O2} = -\frac{4\Gamma}{SU_\infty}\iint_{GG_1C_1F} u^O\,dx_1\,dx_2 = \frac{4(1-g)b\gamma\Gamma}{BS}, \tag{7.91b}$$

$$C_d^{G} = -\frac{4\Gamma}{SU_\infty}\iint_{GG_1C_1F} u^G\,dx_1\,dx_2 = -\frac{4(1-g)b(\gamma-\Gamma)\Gamma}{BS}, \tag{7.91c}$$

$$C_d^{I1} = -\frac{4\Gamma}{SU_\infty}\iint_{O_1H_1C_1} \xi_1 u^I\,d\xi_1\,d\eta = -\frac{2\gamma\Gamma}{\pi B^2 S}, \tag{7.91d}$$

$$C_d^{I2} = -\frac{4(\gamma-\Gamma)}{SU_\infty}\iint_{O_1K_1G_1} \xi_1 u^I\,d\xi_1\,d\eta = -\frac{2g^2\gamma(\gamma-\Gamma)}{\pi B^2 S}, \tag{7.91e}$$

$$C_d^{*I} = -\frac{4\Gamma}{SU_\infty}\iint_{G_1N_1C_1} \xi_1^* u^{*I}\,d\xi_1^*\,d\eta^* = -\frac{2(1-g)^2\Gamma^2}{\pi B^2 S}, \tag{7.91f}$$

where, obviously, the surface area S of half a wing is

$$S = b. \tag{7.92}$$

It should be mentioned that the expressions established above are also valid when the Mach cones with their vertices at O_1 and O_2 intersect on the wing at M', and the intersection points H_1' and H_2' lie on the trailing edge ($b < 1/B < 2b$). In the remaining cases, when the Mach lines cross the lateral sides of the rectangle (the points H_1'' and H_2''), the corresponding integration surfaces are to be considered and the above results must consequently be amended.

VII. 5. Thin Polygonal Wings

VII. 5.1. Determination of the Different Motions around the Wing

In the previous subdivisions of this chapter we have been dealing exclusively with wings of symmetrical thickness. In the following we shall treat a few examples of thin polygonal wings.

Let us consider a thin polygonal wing (Fig. 7.10) defined by the downwash w, which is represented by a superposition of homogeneous polynomials of various orders (7.1).

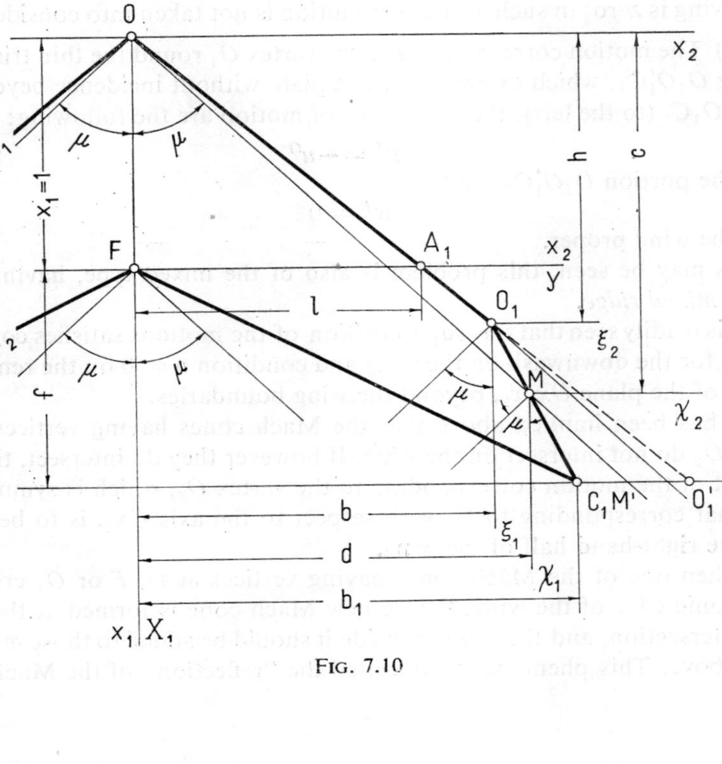

FIG. 7.10

Analogously with a wing of symmetrical thickness, the motion around half a wing (the right-hand side one) is equivalent to the following superposition of motions:

(a) The motion corresponding to the vertex O round the triangular wing $OO_1'O_2'$, defined by the velocity w (7.1). The axial velocity, which will be denoted by u^O, follows from the examples treated in Chapter IV.

(b) The motion corresponding to the vertex F around the fictitious triangular wing FC_1C_2, which extends as a flat plate without incidence beyond the Mach cone, under the assumption that this cone crosses the wing along an arbitrary line.

The following conditions are to be satisfied on the fictitious wing:

$$u^F = -u^O \tag{7.93}$$

on the portion FC_1C_2, and

$$w^F = 0 \tag{7.94}$$

on the wing proper. This motion corresponding to a triangular wing with supersonic edges, represents a mixed problem, where the lines FC_1 and FC_2 are *mixed ridges*.

If the Mach cone does not cross the wing, the effect of this motion upon the wing is zero; in such a case, the motion is not taken into consideration.

(c) The motion corresponding to the vertex O_1 round the thin triangular wing $O_1O_1'C_1$, which extends as a flat plate without incidence beyond the line O_1C_1 (to the left); the conditions of motion are the following:

$$u^I = -u^O \tag{7.95}$$

on the portion $O_1O_1'C_1$, and

$$w^I = 0 \tag{7.96}$$

on the wing proper.

As may be seen, this problem is also of the mixed type, having O_1C_1 as a *mixed ridge*.

It is readily seen that this superposition of the motions satisfies condition (7.1) for the downwash on the wing and condition $u = 0$ on the remaining part of the plane Ox_1x_2, beyond the wing boundaries.

It has been implied above that the Mach cones having vertices at O_1 and O_2 do not intersect on the wing. If however they do intersect, then the effect of the motion corresponding to the vertex O_2, which is symmetrical to that corresponding to O_1 with respect to the axis Ox_1, is to be added to the right-hand half of the wing.

When one of the Mach cones having vertices at O, F or O_1 crosses a subsonic edge of the wing, then a new Mach cone is formed at the point of intersection, and the motion inside it should be added to those mentioned above. This phenomenon is called the "reflection" of the Mach cone.

The problem becomes very complicated and laborious in the general case of the expression for w (7.1); hence, for the sake of simplicity, we propose to consider the case of the thin polygonal wing having a constant incidence and, consequently, a constant downwash:

$$\alpha = \alpha_0, \quad w = -\alpha_0 U_\infty = w_0 = \text{const.} \tag{7.97}$$

Under these conditions, we shall determine successively the axial velocities u for each of the motions defined above.

(a) *The motion corresponding to the vertex O.* Since we are dealing with a thin delta wing having the downwash w_0, by applying formulae (4.14) and (4.42) we may readily write for the axial velocity function $\mathcal{U}^O$ ($u^O = \text{Re}\,\mathcal{U}^O$)

$$\mathcal{U}^O = \frac{A_{10}^O}{\sqrt{l^2-x^2}}, \quad \left(A_{10}^O = -\frac{w_0 l^2}{E(k)}\right), \tag{7.98}$$

in the case of subsonic leading edges ($l < 1/B$), where $E(k)$ is the complete elliptical integral of the first kind having the modulus $k = \sqrt{1-B^2l^2}$, and

$$\mathcal{U}^O = K_{10}^O\left[\cos^{-1}\sqrt{\frac{(1+Bl)(1-Bx)}{2B(l-x)}}+\cos^{-1}\sqrt{\frac{(1+Bl)(1+Bx)}{2B(l+x)}}\right],$$

$$\left(K_{10}^O = -\frac{2w_0}{\pi B}\frac{Bl}{\sqrt{B^2l^2-1}}\right), \tag{7.99}$$

in the case of supersonic edges ($l > 1/B$).

(b) *The motion corresponding to the vertex F.* It should be again pointed out that this motion exerts its effect only if the edges FC_1 and FC_2 are subsonic ($L < 1/B$). In such a case, in order to find the axial velocity u^F on the wing, a mixed problem must be solved, in which the velocity $u^F = -u^O$ on the portion FC_1C_2 and $w^F = 0$ on the wing proper, are given. This is a mixed problem of a high-order conical motion, applied to a symmetrical wing with supersonic edges and two mixed *ridges*, FC_1 and FC_2, symmetrical with respect to the axis OX_1.

In order to solve this rather complex problem, we may apply the method of computation indicated in Chapters II and IV, referring to the mixed problem. For this purpose, $u^F = -u^O$ may be developed in an approximate manner under the form of a superposition of homogeneous polynomials of various orders as a function of X_1 and x_2.

For the sake of simplicity, however, we are going to treat a few examples in which the edges FC_1 and FC_2 are supersonic, so that the effect of the motion corresponding to F should be zero.

(c) *The motion corresponding to the vertex* O_1. This motion appears owing to the fact that the leading edge is kinked at the point O_1, being orientated in another direction.

The case when the leading edge OO_1 is subsonic presents great computational difficulties; hence, in the following we shall consider only the case when the leading edge OO_1 is supersonic, and the velocity $u = \mathrm{Re}\,\mathcal{U}^O$ (7.99) becomes

$$u^O = \frac{\pi}{2} K_{10}^O = -\frac{w_0 l}{\sqrt{B^2 l^2 - 1}} \tag{7.100}$$

on the supersonic portion of the wing.

In such a case, conditions (7.95) and (7.96) become

$$u^I = -u^O = \frac{w_0 l}{\sqrt{B^2 l^2 - 1}}, \quad w^I = 0, \tag{7.101}$$

respectively; therefore, we are again confronted with a simple mixed problem, of the same type as those treated earlier in Chapters II and IV. Thus we have for the axial velocity function $\mathcal{U}^I (u^I = \mathrm{Re}\,\mathcal{U}^I)$

$$\mathcal{U}^I = -K_{10}^O \cos^{-1} \sqrt{\frac{(1+Bl)(1-B\xi)}{2B(l-\xi)}} + K_{10}^I \cos^{-1} \sqrt{\frac{(1+Bl_1)(1-B\xi)}{2B(l_1-\xi)}}, \tag{7.102}$$

if $l_1 > 1/B$, where

$$K_{10}^O = -\frac{2w_0}{\pi B} \frac{Bl}{\sqrt{B^2 l^2 - 1}}, \quad K_{10}^I = -\frac{2w_0}{\pi B} \frac{Bl_1}{\sqrt{B^2 l_1^2 - 1}}; \tag{7.103}$$

$$\mathcal{U}^I = -K_{10}^O \cos^{-1} \sqrt{\frac{(1+Bl)(l_1-\xi)}{(1+Bl_1)(l-\xi)}} + A_{10}^I \sqrt{\frac{1+B\xi}{B(l_1-\xi)}} \tag{7.104}$$

if $l_1 < 1/B$, where the constant A_{10}^I is determined from the condition

$$\left[\frac{d\mathcal{U}^I}{d\xi}\right]_{\xi=0} = 0 \tag{7.105}$$

and hence

$$A_{10}^I = -\frac{2w_0}{\pi B} \frac{Bl_1}{1+Bl_1} \sqrt{\frac{B(l-l_1)}{Bl-1}}. \tag{7.106}$$

(d) *Case of Mach cone reflection.* As an example, let us consider the case when the Mach cone with its vertex at O intersects the edge O_1C_1 at the point M.

In this case, the condition $u^O + u^I = 0$ is not satisfied on the portion $MM'C_1$ since on this portion there still remains an axial velocity.

Hence, it follows that in order to satisfy that condition on the portion $MM'C_1$, the motion in the interior of the Mach cone with its vertex at M should take place around a thin wing having the left-hand edge supersonic and the right-hand edge sonic (at MM') and defined by the conditions

$$u^M = -(u^O + u^I) \tag{7.107}$$

on the portion $MM'C_1$ and

$$w^M = 0 \tag{7.108}$$

on the remaining part of the wing.

Thus we have to solve a new mixed problem having the *mixed separation ridge* MC_1.

A similar situation is also obtained when the Mach cone with its vertex at F crosses the edge O_1C_1 at an arbitrary point, from which a new Mach cone issues.

Since the effect of the conical motion corresponding to the vertex M and due to the reflection—which might also bring about other successive reflections—is very small and affects generally only a very small part of the wing surface, these effects may be neglected.

Note. The total velocity at any point on the right-hand half of the wing is the following:

$$u = u^O + u^F + u^I, \tag{7.109}$$

to which the axial velocities appearing becouse of the reflections of the Mach cones may eventually be added.

If the point considered lies at the same time in the interior of the Mach cone with vertex at O_2, then the velocity corresponding to this latter motion, symmetrical to that corresponding to the vertex O_1, should also be added to the above expression.

VII. 5.2. Determination of Aerodynamic Characteristics

Denoting by S the surface area of the half of the wing and by $d\sigma$ a surface element, we may write

$$C_l = \frac{4}{SU_\infty} \int_S u \, d\sigma \tag{7.110}$$

for the lift coefficient, and

$$C_d = -\frac{4}{SU_\infty^2} \int_S u w_0 \, d\sigma = -\frac{w_0}{U_\infty} C_l \tag{7.111}$$

for the drag coefficient.

As a matter of fact, we have to compute only the lift coefficient, which may be decomposed in the following manner, if relation (7.109) is taken into account,

$$C_l = C_l^O + C_l^F + C_l^I, \tag{7.112}$$

where

$$C_l^O = \frac{4}{SU_\infty}\iint_S x_1 u^O \, dx_1 \, dy, \qquad \left(y = \frac{x_2}{x_1}\right), \tag{7.113a}$$

$$C_l^F = \frac{4}{SU_\infty}\iint_{FN_1C_1} X_1 u^F \, dX_1 \, dY, \qquad \left(Y = \frac{x_2}{X_1}\right), \tag{7.113b}$$

$$C_l^I = \frac{4}{SU_\infty}\iint_{O_1H_1C_1} \xi_1 u^I \, d\xi_1 \, d\eta, \qquad \left(\eta = \frac{\xi_2}{\xi_1}\right). \tag{7.113c}$$

VII. 6. Some Examples of Thin Polygonal Wings

VII. 6.1. Double Trapezoidal Swept-back Wing

For the sake of simplicity, the positions of the Mach cones which lead to mixed problems with great calculation difficulties will be avoided.

Consequently, we shall consider only the case when the edges OO_1, OO_2, FC_1 and FC_2 are supersonic ($l > 1/B$ and $L > 1/B$), and when the Mach cone with vertex at O does not cross the lateral edges O_1C_1 and O_2C_2 (Fig. 7.11). In such a case it may be easily seen that relations (7.109)

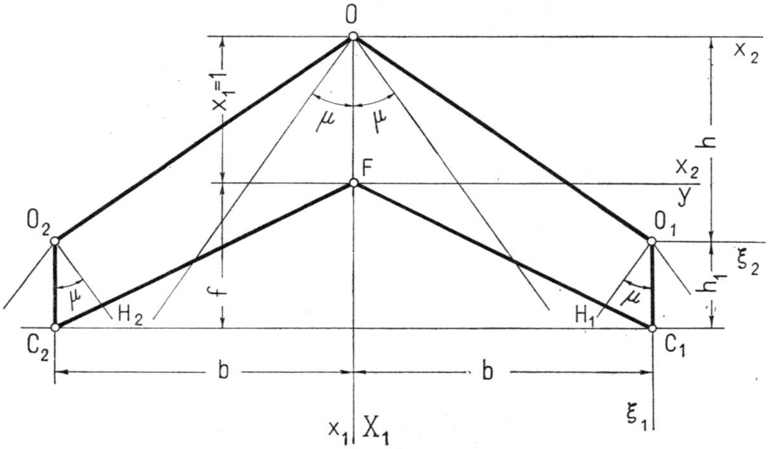

FIG. 7.11

and (7.112) reduce to

$$u = u^O + u^I = \operatorname{Re}(\mathcal{U}^O + \mathcal{U}^I), \quad C_l = C_l^O + C_l^I, \tag{7.114}$$

where $\mathcal{U}^O$ is given by (7.99) and $\mathcal{U}^I$ is deduced from (7.104) by setting $l_1 = 0$:

$$\mathcal{U}^I = -K_{10}^O \cos^{-1}\sqrt{\frac{(1+Bl)\xi}{\xi - l}}, \quad \left(K_{10}^O = -\frac{2w_0}{\pi B}\frac{Bl}{\sqrt{B^2l^2-1}}\right). \tag{7.115}$$

For determining the coefficients C_l^O and C_l^I, the respective integrals (7.113a) and (7.113c) are to be performed within the following limits:

$$\iint_S = \int_{x_1=0}^{x_1=\frac{L}{L-y}} \int_{y=0}^{y=\frac{b}{1+f}} + \int_{x_1=0}^{x_1=\frac{b}{y}} \int_{y=\frac{b}{1+f}}^{y=l}, \quad \left(L = \frac{b}{f}\right), \tag{7.116a}$$

$$\iint_{O_1H_1C_1} = \int_{\xi_1=0}^{\xi_1=\frac{h_1L}{L-\eta}} \int_{\eta=-\frac{1}{B}}^{\eta=0}. \tag{7.116b}$$

Performing the calculus operations, we obtain the following expressions:

$$\begin{aligned} C_l^O = \operatorname{Re}\frac{2K_{10}^O}{SU_\infty}\Bigg\{ & \frac{l^2L}{L^2-l^2}\cos^{-1}\frac{1}{Bl} \\ & + \frac{(1+f)l-b}{L-l}\left[b\left(\frac{L}{l}-1\right) - \frac{L^2}{(1+f)L-b}\right]\cos^{-1}\sqrt{\frac{(1+Bl)(1+f-Bb)}{2B[(1+f)l-b]}} \\ & + \frac{(1+f)l+b}{L+l}\left[b\left(\frac{L}{l}+1\right) + \frac{L^2}{(1+f)L-b}\right]\cos^{-1}\sqrt{\frac{(1+Bl)(1+f+Bb)}{2B[(1+f)l+b]}} \\ & + \frac{L^3}{L^2-l^2}\frac{\sqrt{B^2l^2-1}}{\sqrt{B^2L^2-1}}\left[2\cos^{-1}\sqrt{\frac{(1+BL)(1+f-Bb)}{2B[(1+f)L-b]}} - \cos^{-1}\frac{1}{BL}\right]\Bigg\}, \end{aligned} \tag{7.117a}$$

$$C_l^I = -K_{10}^O\frac{\pi h_1^2}{SU_\infty}\frac{L}{L-l}\left[L\sqrt{\frac{l(Bl+1)}{L(BL+1)}} - l\right], \tag{7.117b}$$

where the area S of the half wing will be given by relation (7.37).

As regards the wave drag coefficient C_d^O and C_d^I, it is easy to see that they are given by the following:

$$C_d^O = \alpha_0 C_l^O = -\frac{w_0}{U_\infty}C_l^O, \quad C_d^I = \alpha_0 C_l^I = -\frac{w_0}{U_\infty}C_l^I, \tag{7.118}$$

α_0 being the incidence of the wing.

VII. 6.2. Pentagonal Wing

We consider only the case when the edge OO_1 is supersonic, and the Mach cone with its vertex at O does not cross the lateral edge O_1C_1 (Fig. 7.12). In this case, the axial disturbance velocity and the total lift coefficient are given by relations (7.114) in which $\mathcal{U}^O$ has the expression (7.99) and $\mathcal{U}^I$ the expression (7.115).

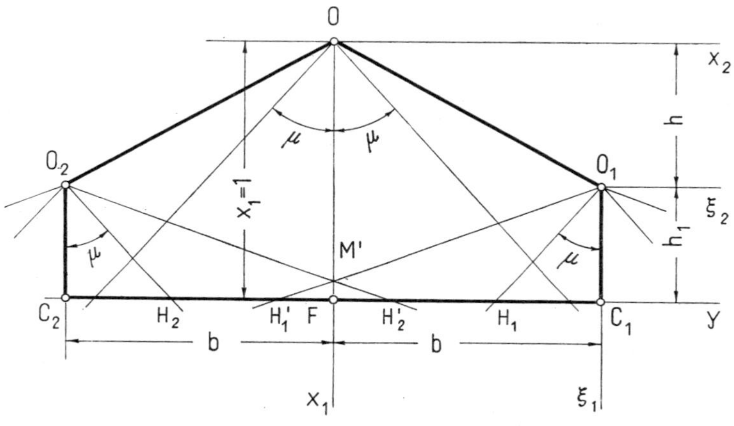

FIG. 7.12

For determining the coefficients C_l^O and C_l^I, the respective integrals (7.113a) and (7.113c) will be performed within the following limits:

$$\iint_S = \int_{x_1=0}^{x_1=1}\int_{y=0}^{y=b} + \int_{x_1=0}^{x_1=\frac{b}{y}}\int_{y=b}^{y=l}, \quad (b = hl), \tag{7.119a}$$

$$\iint_{O_1H_1C_1} = \int_{\xi_1=0}^{\xi_1=h_1}\int_{\eta=-\frac{1}{B}}^{\eta=0}, \quad (h_1 = 1-h). \tag{7.119b}$$

It should be noted that the above relations are also valid when the Mach cones with their vertices at O_1 and O_2 intersect on the wing, without, however, intersecting on the lateral edges O_1C_1 and O_2C_2 ($b < h_1/B < 2b$).

Performing the calculus operations, we obtain the following expressions for both cases:

$$C_l^O = \text{Re}\,\frac{2K_{10}^O}{lSU_\infty}\left[-(l-b)^2\cos^{-1}\sqrt{\frac{(1+Bl)(1-Bb)}{2B(l-b)}}\right.$$

$$\left.+(l+b)^2\cos^{-1}\sqrt{\frac{(1+Bl)(1+Bb)}{2B(l+b)}}+\frac{\sqrt{B^2l^2-1}}{B}\sin^{-1}Bb\right], \tag{7.120a}$$

$$C_l^I = -K_{10}^O\frac{\pi h_1^2}{SU_\infty}\left[\sqrt{l\left(\frac{1}{B}+l\right)}-l\right]. \tag{7.120b}$$

The surface area S of the half of the wing occurring in the above formulae is given by the relation

$$S = \frac{(2-h)b}{2} = \frac{(1+h_1)b}{2}. \tag{7.121}$$

VII. 6.3. Rectangular Wing

We consider only the case when the Mach cones with their vertices at O_1 and O_2 intersect on the wing without, however, intersecting on the lateral edges O_1C_1 and O_2C_2 ($b < 1/B < 2b$, Fig. 7.13).

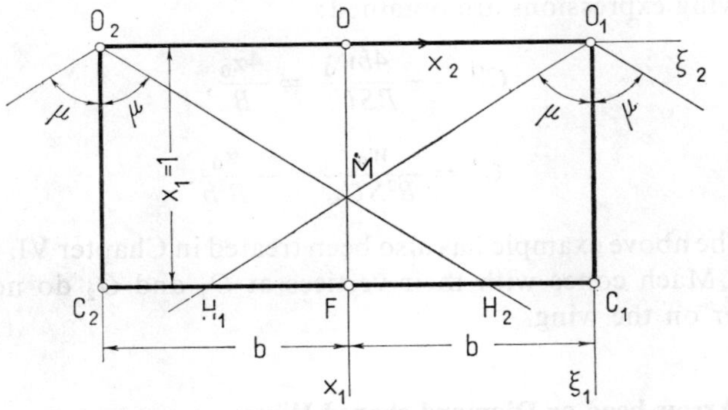

FIG. 7.13

In this case it is obvious that the total axial disturbance velocity at any point on the wing is given by the relation

$$u = u^O + u^I + u^{II}, \tag{7.122}$$

where, as in the case of the wing of symmetrical thickness, u^O is part of a plane motion and has the expression

$$u^O = -\frac{w_0}{B}, \tag{7.123}$$

and u^I is deduced from the expression established above (7.104), by making l tend to infinity ($l \to \infty$) and l_1 tend to zero ($l_1 = 0$),

$$u^I = \operatorname{Re} \frac{2}{\pi} \frac{w_0}{B} \cos^{-1} \sqrt{-B\xi} \,. \tag{7.124}$$

As regards the velocity u^{II}, this is symmetrical to u^I with respect to the Ox_1 axis.

Taking account of this fact, the lift coefficient is given in such a case by the relation

$$C_l = C_l^O + C_l^I, \tag{7.125}$$

where C_l^O and C_l^I are obtained by performing the integrals from relations (7. 113a) and (7.113c) within the limits

$$\iint_S = \int_{x_1=0}^{x_1=1} \int_{x_2=0}^{x_2=b}, \tag{7.126a}$$

$$\iint_{O_1H_1C_1} = \int_{\xi_1=0}^{\xi_1=1} \int_{\eta=-\frac{1}{B}}^{\eta=0}. \tag{7.126b}$$

Performing the calculus operations and taking account of relation (7.69), the following expressions are obtained:

$$C_l^O = -\frac{4bw_0}{BSU_\infty} = \frac{4\alpha_0}{B}, \tag{7.127a}$$

$$C_l^I = \frac{w_0}{B^2SU_\infty} = -\frac{\alpha_0}{B^2b}. \tag{7.127b}$$

Note. The above example has also been treated in Chapter VI, for the case when the Mach cones with their vertices at O_1 and O_2 do not intersect each other on the wing.

VII. 7. Arrow-head or Diamond-shaped Wings

VII. 7.1. General Introductory Considerations

Let us consider an arrow-head wing OC_1FC_2, or a diamond-shaped wing $OC_1'FC_2'$, which may have either subsonic or supersonic edges (Fig. 7.4). This wing may be deduced from that represented in Fig. 7.11, by setting $h_1 = 0$.

In this case, the motion corresponding to the vertex O_1 is omitted and there remain only the motions corresponding to the vertices O and F. Nevertheless, for the sake of simplicity, we shall assume that the Mach cone with its vertex at F does not intersect the wing at any point ($\delta < \frac{1}{2}\pi - \mu$).

In such a case, only the axial velocity function $\mathcal{U}^{0}_{n-1}$, which will be denoted simply by $\mathcal{U}_{n-1}$, is to be considered.

What is now of interest to us as regards this type of wing is the calculation of the aerodynamic characteristics, i.e. lift, drag and moment coefficients, in the general case, which yields interesting general formulae.

For determining the aerodynamic characteristics, we apply formulae (3.26), (3.38b), (3.53) where the integrals will be restricted to half the wing, due to the symmetrical planform, and thus obtain the following formulae, which will be of further use:

$$C_l^{(n)} = \frac{8}{(n+1)SU_\infty}\int_0^l \frac{u_{n-1}\,dy}{(1-y\tan\delta)^{n+1}}, \quad \left(C_l = \sum_{n=1}^{N} C_l^{(n)}\right) \tag{7.128}$$

for the lift coefficient,

$$C_{dm}^{(n)} = -\frac{8}{(m+n)SU_\infty^2}\int_0^l \frac{w_{m-1}u_{n-1}\,dy}{(1-y\tan\delta)^{m+n}}, \quad \left(C_d = \sum_{m=1}^{N}\sum_{n=1}^{N} C_{dm}^{(n)}\right) \tag{7.129}$$

for the wave drag coefficient, and

$$\mu_2^{(n)} = -\frac{8}{(n+2)ShU_\infty}\int_0^l \frac{u_{n-1}\,dy}{(1-y\tan\delta)^{n+2}}, \quad \left(\mu_2 = \sum_{n=1}^{N} \mu_2^{(n)}\right) \tag{7.130}$$

for the pitching moment coefficient, where

$$S = \frac{l}{1-l\tan\delta}, \quad h = a = \frac{1}{1-l\tan\delta}. \tag{7.131}$$

VII. 7.2. Case of Thin Wing with Subsonic Leading Edges

By considering the symmetry of incidence, the downwash and the axial velocity function $\mathcal{U}_{n-1}$ have the expressions

$$w^{(n-1)} = x_1^{n-1}w_{n-1} = x_1^{n-1}\sum_{k=0}^{n-1}(\pm 1)^k w_{n-k-1,\,k}\,y^k, \tag{7.132a}$$

$$\mathcal{U}_{n-1} = \sum_{q=0}^{E\left(\frac{n}{2}\right)} \frac{A_{n,\,2q}\,x^{2q}}{\sqrt{l^2-x^2}} + \sum_{q=1}^{E\left(\frac{n-1}{2}\right)} Q_{n,\,2q}\,x^{2q}\cosh^{-1}\sqrt{\frac{l^2}{x^2}}, \tag{7.132b}$$

where the constants are given in (§ IV. 2.1). As usual, $E(t)$ is the greatest integer contained in $t\left(t=\frac{n}{2}, \frac{n-1}{2}\right)$.

Calculating the aerodynamic characteristics we obtain the following expressions for the general case:

Lift coefficient

$$C_l^{(n)} = \frac{8}{(n+1)SU_\infty}\left[\sum_{q=0}^{E\left(\frac{n}{2}\right)} A_{n,2q}(\cot\delta)^{2q}\sum_{r=0}^{2q}(-1)^r\frac{(2q)!}{r!(2q-r)!}I_{n+1-r}\right.$$

$$\left.+\sum_{q=1}^{E\left(\frac{n-1}{2}\right)} Q_{n,2q}(\cot\delta)^{2q}\sum_{r=0}^{2q}(-1)^r\frac{(2q)!}{r!(2q-r)!}\,\frac{l}{n-r}\sum_{p=1}^{n-r}I_p\right]. \tag{7.133}$$

Wave drag coefficient

$$C_{dm}^{(n)} = -\frac{8}{(m+n)SU_\infty^2}\sum_{k=0}^{m-1} w_{m-k-1,k}$$

$$\times\left[\sum_{q=0}^{E\left(\frac{n}{2}\right)} A_{n,2q}\,(\cot\delta)^{k+2q}\sum_{r=0}^{k+2q}(-1)^r\frac{(k+2q)!}{r!(k+2q-r)}I_{m+n-r}\right.$$

$$+\sum_{q=1}^{E\left(\frac{n-1}{2}\right)} Q_{n,2q}\,(\cot\delta)^{k+2q}\sum_{r=0}^{k+2q}(-1)^r\frac{(k+2q)!}{r!(k+2q-r)!}$$

$$\left.\times\frac{l}{m+n-r-1}\sum_{p=1}^{m+n-r-1}I_p\right]. \tag{7.134}$$

Pitching moment coefficient

$$\mu_2^{(n)} = -\frac{8}{(n+2)ShU_\infty}\sum_{q=0}^{E\left(\frac{n}{2}\right)} A_{n,2q}\,(\cot\delta)^{2q}\sum_{r=0}^{2q}(-1)^r\frac{(2q)!}{r!(2q-r)!}I_{n+2-r}$$

$$\left.+\sum_{q=1}^{E\left(\frac{n-1}{2}\right)} Q_{n,2q}\,(\cot\delta)^{2q}\sum_{r=0}^{2q}(-1)^r\frac{(2q)!}{r!(2q-r)!}\,\frac{l}{n+1-r}\sum_{p=1}^{n+1-r}I_p\right]. \tag{7.135}$$

In the above expressions we denoted by I_p the following integrals:

$$I_p = \int_0^l \frac{dy}{(1-y\tan\delta)^p\sqrt{l^2-y^2}} = \frac{1}{(p-1)(1-l^2\tan^2\delta)}\Big[l\tan\delta$$

$$+(2p-3)I_{p-1}-(p-2)I_{p-2}\Big], \tag{7.136a}$$

$$I_1 = \int_0^l \frac{dy}{(1-y\tan\delta)\sqrt{l^2-y^2}} = \frac{1}{\sqrt{(1-l^2\tan^2\delta}}[\pi-\cos^{-1}(l\tan\delta)]. \tag{7.136b}$$

Suction coefficient. This coefficient is defined by expressions similar to those deduced for the delta wing with subsonic edges (§ IV. 2.4), in which account should be taken of the following two observations:

For the arrow-head wing the surface area S is given by relation (7.131).

The projection of the leading edge on the axis Ox_1, which has been denoted by a, is given by (7.131).

It follows that for the suction coefficients $\sigma_m^{*(n)}$ of the arrow-head wing, we may write the following relation:

$$\sigma_m^{*(n)} = \sigma_m^{(n)}\frac{1}{(1-l\tan\delta)^{m+n-1}}, \tag{7.137}$$

$\sigma_m^{(n)}$ being the suction coefficients of the delta wing having $h = 1$.

Note. The expressions indicated above are also valid for the diamond-shaped wing, δ being replaced by $-\delta$.

VII.7.3. Thin Wing with Supersonic Leading Edges

In the case of symmetry of incidence, considering the same expression (7.132a) for the downwash, similarly to (4.42), the axial velocity function $\mathcal{U}_{n-1}$ is the following:

$$\mathcal{U}_{n-1} = \sum_{q=0}^{n-1}\frac{\Gamma_{n,q}}{\sqrt{B^2l^2-1}}x^q\Bigg[\cos^{-1}\sqrt{\frac{(1+Bl)(1-Bx)}{2B(l-x)}}$$

$$+(-1)^q\cos^{-1}\sqrt{\frac{(1+Bl)(1+Bx)}{2B(l+x)}}\Bigg]$$

$$+\sum_{q=1}^{E\left(\frac{n-1}{2}\right)} Q_{n,2q}x^{2q}\cosh^{-1}\sqrt{\frac{1}{B^2x^2}}+\sum_{q=0}^{E\left(\frac{n-2}{2}\right)} D_{n,2q}x^{2q}\sqrt{\frac{1-B^2x^2}{B^2}}, \tag{7.138}$$

where, similarly to (7.10), we have employed the notation

$$\Gamma_{n,q} = K_{n,q}\sqrt{B^2 l^2 - 1} \tag{7.139}$$

and where the constants $K_{n,q}$ are given in § IV. 4.1.

For the general case, the aerodynamic characteristics are the following:

Lift coefficient

$$C_l^{(n)} = \frac{8}{(n+1)SU_\infty}\left\{\sum_{q=0}^{n-1}\Gamma_{n,q}(\cot\delta)^q\sum_{r=0}^{q}(-1)^r\frac{q!}{r!(q-r)!}E_{n+1-r,q}\right.$$

$$+\sum_{q=0}^{E\left(\frac{n-1}{2}\right)}\mathcal{Q}_{n,2q}(\cot\delta)^{2q}\sum_{r=0}^{2q}(-1)^r\frac{(2q)!}{r!(2q-r)!}\frac{1}{n-r}\sum_{p=1}^{n-r}\mathcal{J}_p$$

$$+\sum_{q=0}^{E\left(\frac{n-2}{2}\right)}D_{n,2q}(\cot\delta)^{2q}\sum_{r=0}^{2q}(-1)^r\frac{(2q!)}{r!(2q-r)!}\frac{1}{(n-r)B}$$

$$\left.\times\left[-\cot\delta+B^2\cot^2\delta(\mathcal{J}_{n-r}-\mathcal{J}_{n-r-1})\right]\right\}. \tag{7.140}$$

Wave drag coefficient

$$C_{dm}^{(n)} = -\frac{8}{(m+n)SU_\infty^2}\sum_{k=0}^{m-1}w_{m-k-1,k}\left\{\sum_{q=0}^{n-1}\Gamma_{n,q}(\cot\delta)^{k+q}\sum_{r=0}^{k+q}(-1)^r\right.$$

$$\times\frac{(k+q)!}{r!(k+q-r)!}E_{m+n-r,q}+\sum_{q=0}^{E\left(\frac{n-1}{2}\right)}\mathcal{Q}_{n,2q}(\cot\delta)^{k+2q}\sum_{r=0}^{k+2q}(-1)^r$$

$$\times\frac{(k+2q)!}{r!(k+2q-r)!}\frac{1}{m+n-r-1}\sum_{p=1}^{m+n-r-1}\mathcal{J}_p$$

$$+\sum_{q=0}^{E\left(\frac{n-2}{2}\right)}D_{n,2q}(\cot\delta)^{k+2q}\sum_{r=0}^{k+2q}(-1)^r\frac{(k+2q)!}{r!(k+2q-r)!}$$

$$\times\frac{1}{(m+n-r-1)B}\left[-\cot\delta+B^2\cot^2\delta(\mathcal{J}_{m+n-r-1}-\mathcal{J}_{m+n-r-2})\right]. \tag{7.141}$$

Pitching moment coefficient

$$\mu_2^{(n)} = -\frac{8}{(n+2)ShU_\infty}\left\{\sum_{q=0}^{n-1}\Gamma_{n,q}(\cot\delta)^q\sum_{r=0}^{q}(-1)^r\frac{q!}{r!(q-r)!}E_{n+2-r,q}\right.$$

$$+\sum_{q=0}^{E\left(\frac{n-1}{2}\right)}Q_{n,2q}(\cot\delta)^{2q}\sum_{r=0}^{2q}(-1)^r\frac{(2q)!}{r!(2q-r)!}\frac{1}{n+1-r}\sum_{p=1}^{n+1-r}\mathcal{J}_p$$

$$+\sum_{q=0}^{E\left(\frac{n-2}{2}\right)}D_{n,2q}(\cot\delta)^{2q}\sum_{r=0}^{2q}(-1)^r\frac{(2q)!}{r!(2q-r)!}$$

$$\left.\times\frac{1}{(n+1-r)B}\left[-\cot\delta+B^2\cot^2\delta(\mathcal{J}_{n+1-r}-\mathcal{J}_{n-r})\right]\right\}. \qquad (7.142)$$

In the above expressions we designated by $E_{t,q}$ and $\mathcal{J}_p$ the following integrals:

$$E_{t,q} = \operatorname{Re}\frac{1}{\sqrt{B^2l^2-1}}\int_0^l\frac{1}{(1-y\tan\delta)^t}\left[\cos^{-1}\sqrt{\frac{(1+Bl)(1-By)}{2B(l-y)}}\right.$$

$$\left.+(-1)^q\cos^{-1}\sqrt{\frac{(1+Bl)(1+By)}{2B(l+y)}}\right]dy$$

$$= \operatorname{Re}\frac{\cot\delta}{2(t-1)\sqrt{B^2l^2-1}}\left\{\left[\frac{(1+l\tan\delta)^{t-1}+(-1)^q(1-l\tan\delta)^{t-1}}{(1-l^2\tan^2\delta)^{t-1}}\right.\right.$$

$$\left.-1-(-1)^q\cos^{-1}\frac{1}{Bl}+(-1)^q\frac{2[(1+l\tan\delta)^{t-1}-(1-l\tan\delta)^{t-1}]}{(1-l^2\tan^2\delta)^{t-1}}\right.$$

$$\left.\times\cos^{-1}\frac{1+Bl}{2\sqrt{Bl}}\right\}+\frac{1}{2(t-1)}\sum_{p=1}^{t-1}\frac{(1-l\tan\delta)^{t-p}+(1+l\tan\delta)^{t-p}}{(1-l^2\tan^2\delta)^{t-p}}\mathcal{J}_p, \qquad (7.143a)$$

$$\mathcal{J}_p = \operatorname{Re}\int_0^l\frac{dy}{(1-y\tan\delta)^p\sqrt{1-B^2y^2}} = \operatorname{Re}\frac{B^2}{B^2-\tan^2\delta}\left[\frac{\tan\delta}{(p-1)B^2}\right.$$

$$\left.\times\left(1-\frac{\sqrt{1-B^2l^2}}{(1-l\tan\delta)^{p-1}}\right)+\frac{2p-3}{p-1}\mathcal{J}_{p-1}-\frac{p-2}{p-1}\mathcal{J}_{p-2}\right], \qquad (7.143b)$$

$$\mathcal{J}_1 = \operatorname{Re}\int_0^l\frac{dy}{(1-y\tan\delta)\sqrt{1-B^2y^2}} = \operatorname{Re}\frac{1}{\sqrt{B^2-\tan^2\delta}}$$

$$\times\left[2\cos^{-1}\sqrt{\frac{(\tan\delta+B)(1-Bl)}{2B(1-l\tan\delta)}}-\cos^{-1}\frac{\tan\delta}{B}\right]. \qquad (7.143c)$$

As remarked above, the aerodynamic coefficients are also valid in the case of the diamond-shaped wing, if δ is replaced by $-\delta$.

VII. 7.4. Case of Wings of Symmetrical Thickness

Let us similarly consider the case of symmetry of slope, for which the downwash on the upper part of the wing is given by the same expression (7.132a).

In the case of supersonic edges ($l > 1/B$), both the axial velocity u and the wave drag coefficients $C_{dm}^{(n)}$ will be given by the same expressions (7.138) and (7.141) as in the case of the thin wing with supersonic edges.

In the case of subsonic leading edges ($l < 1/B$), the axial velocity function $\mathcal{U}_{n-1} = u_{n-1} + iu'_{n-1}$ is given by the same expression (7.138), in which account should be taken of the known transformation relation

$$\frac{\cos^{-1}\beta}{\sqrt{B^2l^2-1}} = \frac{\cosh^{-1}\beta}{\sqrt{1-B^2l^2}}. \tag{7.144}$$

As regards the wave drag coefficient, we obtain the following expressions for the general case:

$$\begin{aligned} C_{dm}^{(n)} = &-\frac{8}{(m+n)SU_\infty^2}\sum_{k=0}^{m-1} w_{m-k-1,k}\left\{\sum_{q=0}^{n-1}\Gamma_{n,q}(\cot\delta)^{k+q}\sum_{r=0}^{k+q}(-1)^r\right.\\ &\times\frac{(k+q)!}{r!(k+q-r)!}E_{m+n-r,q} + \sum_{q=0}^{E\left(\frac{n-1}{2}\right)} Q_{n,2q}(\cot\delta)^{2q+k}\sum_{r=0}^{k+2q}(-1)^r\\ &\times\frac{(k+2q)!}{r!(k+2q-r)!}\frac{1}{m+n-r-1}\left[\left(\frac{\cot\delta}{(1-l\tan\delta)^{m+n-r-1}}-\cot\delta\right)\right.\\ &\left.\times\cosh^{-1}\frac{1}{Bl} + \sum_{p=1}^{m+n-r-1}\mathcal{J}_p\right] + \sum_{q=0}^{E\left(\frac{n-2}{2}\right)} D_{n,2q}(\cot\delta)^{k+2q}\sum_{r=0}^{k+2q}(-1)^r\\ &\times\frac{(k+2q)!}{r!(k+2q-r)!}\frac{1}{(m+n-r-1)B}\left[\left(\frac{\sqrt{1-B^2l^2}}{(1-l\tan\delta)^{m+n-r-1}}-1\right)\cot\delta\right.\\ &\left.\left.+B^2\cot^2\delta(\mathcal{J}_{m+n-r-1}-\mathcal{J}_{m+n-r-2})\right]\right\}, \end{aligned} \tag{7.145}$$

where $E_{t,q}$ and $\mathcal{J}_p$ are given by the same relations (7.143); in extracting the real part from these expressions, account should be taken of relation (7.144).

As in the case of the thin wing, these expressions are also valid for the diamond-shaped wing, if δ is replaced by $-\delta$.

VII. 8. Thin Angular Wing with Swept-back or Arrow-head-shaped Enclosing Line

VII. 8.1. Preliminary Considerations

Let us consider an enclosing line of the trailing edge given by (3.21) which, for the sake of simplicity, will be restricted to the following two forms:

$$x_1 = P(y) = 1+a_2y^2, \tag{7.146a}$$

$$x_1 = P(y) = 1+a_2y^2+a_4y^4; \tag{7.146b}$$

the variation of the coefficients a_2 and a_4 yields a very great variety of arrow-head wings (7.146a) or swept-back wings (7.146b), as shown in Fig. 7.14 and Fig. 7.15.

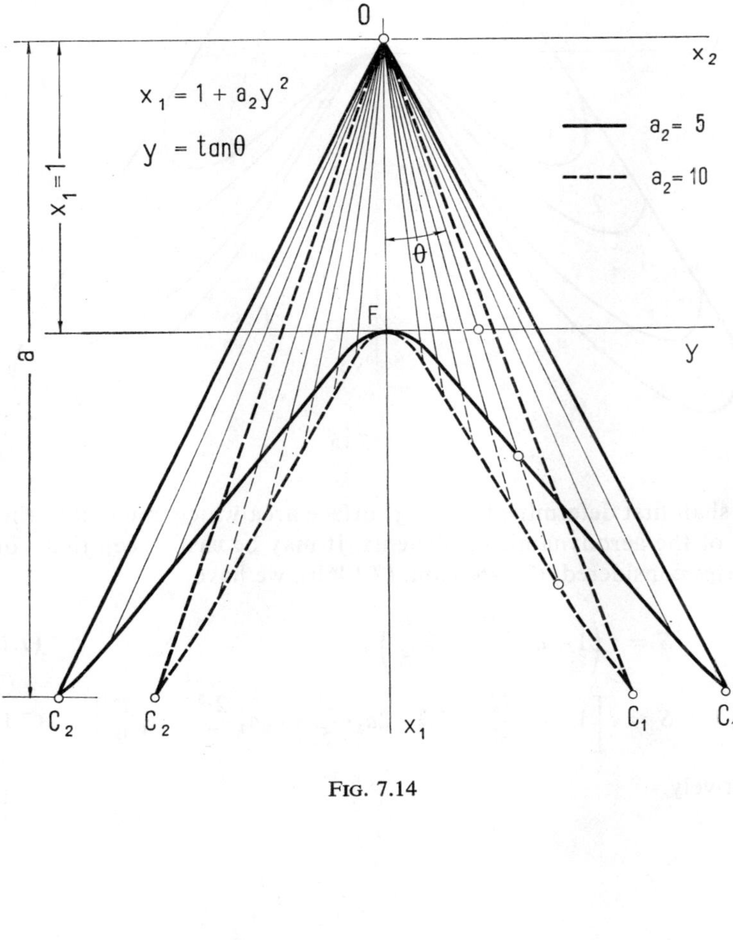

FIG. 7.14

These forms are very close to ordinary wings and may replace swept-back wings and arrow-head wings for which, as shown above, the trailing edge consists of two straight lines. The advantage of the wings indicated in Fig. 7.14 and Fig. 7.15 is very important since the enclosing line of the trailing edge can have a very great variety of forms.

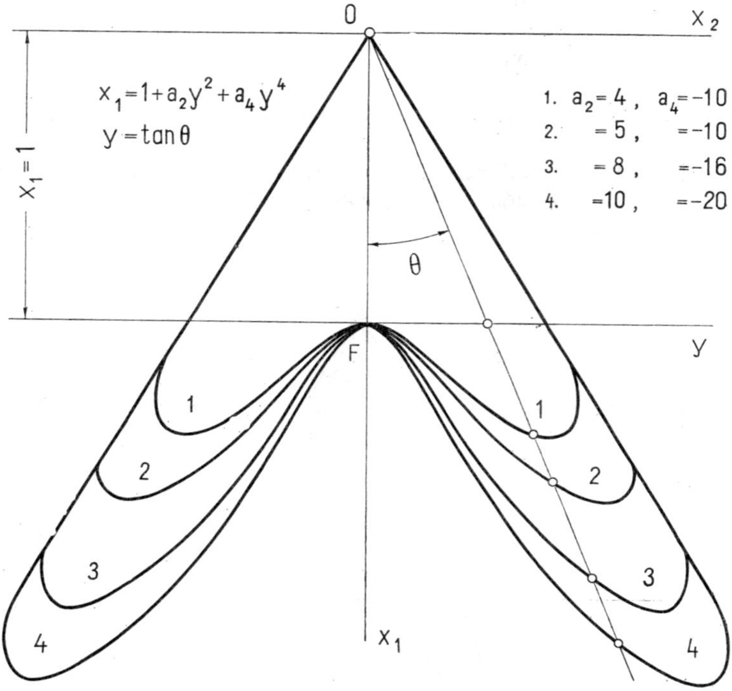

FIG. 7.15

We shall first determine the wing surface area which enters into the formulae of the aerodynamic coefficients. It may be easily seen that, for the two cases considered, (7.146a) and (7.146b), we have

$$S = l\left(1+a_2\frac{2l^2}{3}+a_2^2\frac{l^4}{5}\right), \qquad (7.147a)$$

$$S = l\left[1+a_2\frac{2l^2}{3}+(a_2^2+2a_4)\frac{l^4}{5}+a_2a_4\frac{2l^6}{7}+a_4^2\frac{l^8}{9}\right], \qquad (7.147b)$$

respectively.

For the sake of simplicity, we shall further assume that any Mach line issuing from the trailing edge does not interfere on the wing or has a very small interference, so that the effects of this interference may be neglected.

VII. 8.2. Arrow-head-shaped Enclosing Line: $P(y) = 1+a_2y^2$

For the sake of simplicity, the natural symmetry with respect to Ox_1 will be considered first, that is, the natural symmetrical downwashes

$$\frac{w^{(m-1)}}{x_1^{m-1}} = w_{m-1} = \sum_{\varkappa=0}^{E\left(\frac{m-1}{2}\right)} w_{m-2\varkappa-1,\,2\varkappa}\, y^{2\varkappa}. \tag{7.148}$$

In this case, the axial velocity function will be given by the same expressions (7.132b) and (7.138) for subsonic and supersonic leading edges respectively, but in which it will be assumed that $Q_{n,\,2q} = 0$ and $w_{n-2\varkappa-2,\,2\varkappa+1} = 0$.

In addition, we shall consider the ease of natural antisymmetrical incidence, where the downwashes are defined by the relation,

$$\frac{w^{(m-1)}}{x_1^{m-1}} = w_{m-1} = \sum_{\varkappa=0}^{E\left(\frac{m-2}{2}\right)} w_{m-2\varkappa-2,\,2\varkappa+1}\, y^{2\varkappa+1}, \tag{7.149}$$

and where the axial disturbance velocities are given by the following expressions for the function $\mathcal{U}_{n-1}$, deduced from (4.31) and (4.59):

$$\mathcal{U}_{n-1} = \sum_{q=0}^{E\left(\frac{n-1}{2}\right)} \frac{A_{n,\,2q+1}\, x^{2q+1}}{\sqrt{l^2-x^2}}, \quad \left(l < \frac{1}{B}\right), \tag{7.150a}$$

$$\mathcal{U}_{n-1} = \sum_{q=0}^{n-1} K_{n,\,q}\, x^q \left[\cos^{-1}\sqrt{\frac{(1+Bl)(1-Bx)}{2B(l-x)}} - (-1)^q \cos^{-1}\sqrt{\frac{(1+Bl)(1+Bx)}{2B(l+x)}}\right.$$
$$\left. + \sum_{q=0}^{E\left(\frac{n-3}{2}\right)} D_{n,\,2q+1}\, x^{2q+1}\sqrt{\frac{1-B^2x^2}{B^2}}\right], \quad \left(l > \frac{1}{B}\right), \tag{7.150b}$$

for a thin wing with subsonic or supersonic leading edges, respectively. The constants which appear in the above expressions have the same values as those determined in §§ IV. 3.1 and IV. 6.1, but with $w_{n-2\varkappa-1,\,2\varkappa} = 0$.

In the case of *symmetry of incidence*, the lift, wave drag and pitching moment coefficients are to be calculated; in the case of *antisymmetry of incidence*, only the drag and rolling moment coefficients are to be calculated. Although the computations are very simple, the operations are rather laborious and hence simpler examples will be given, considering the cases $n = 1, 2$ alone.

Natural Symmetry of Incidence

Lift. Applying formula (3.32) and taking into account (7.132b), where $Q_{n,\,2q} = 0$, we may write for the case of subsonic edges,

$$C_l^{(1)} = \frac{2\pi A_{10}}{SU_\infty}(1+a_2l^2+\tfrac{3}{8}a_2^2l^4), \tag{7.151a}$$

$$C_l^{(2)} = \frac{4\pi}{3SU_\infty}\Big[A_{20}(1+\tfrac{3}{2}a_2l^2+\tfrac{9}{8}a_2^2l^4+\tfrac{5}{16}a_2^3l^6)$$

$$+\frac{l^2}{2}A_{22}(1+\tfrac{9}{4}a_2l^2+\tfrac{15}{8}a_2^2l^4+\tfrac{35}{64}a_2^3l^6)\Big], \tag{7.151b}$$

and for supersonic edges, taking into account (7.138), we have similarly

$$C_l^{(1)} = \frac{4}{SU_\infty}K_{10}[H_{00}+2a_2H_{22}+a_2^2H_{44}], \tag{7.152a}$$

$$C_l^{(2)} = \frac{8}{3SU_\infty}\Big\{K_{20}[H_{00}+3a_2H_{22}+3a_2^2H_{44}+a_2^3H_{66}]$$

$$+K_{21}[H_{10}+3a_2H_{32}+3a_2^2H_{54}+a_2^3H_{76}]$$

$$+D_{20}\frac{\pi}{4B^2}\Big[1+\frac{3a_2l^2}{4B^2l^2}+\frac{3a_2^2l^4}{8B^4l^4}+\frac{5a_2^3l^6}{64B^6l^6}\Big]\Big\}, \tag{7.152b}$$

where the expressions for the integrals $H_{t,\,k}$ are given in Appendix I.3.

If the two wings $n = 1$ and $n = 2$ constitute a complex wing, then the total coefficient C_l will be the sum of the two coefficients given in (7.151) and (7.152).

Wave drag. Applying formula (3.42c) for $m = 1, 2$ and $n = 1, 2$, and taking account of (7.148) and (7.132), which refer to a wing with subsonic edges, we obtain

$$C_{d1}^{(1)} = -\frac{w_{00}}{U_\infty}C_l^{(1)}, \quad C_{d1}^{(2)} = -\frac{w_{00}}{U_\infty}C_l^{(2)}, \tag{7.153a}$$

$$C_{d2}^{(1)} = -\frac{4\pi}{3SU_\infty^2}w_{10}A_{10}(1+\tfrac{3}{2}a_2l^2+\tfrac{9}{8}a_2^2l^4+\tfrac{5}{16}a_2^3l^6), \tag{7.153b}$$

$$C_{d2}^{(2)} = -\frac{\pi w_{10}}{SU_\infty^2}\Big[A_{20}(1+2a_2l^2+\tfrac{9}{4}a_2^2l^4+\tfrac{5}{4}a_2^3l^6+\tfrac{35}{128}a_2^4)$$

$$+\frac{l}{2}A_{22}(1+3a_2l^2+\tfrac{15}{4}a_2^2l^4+\tfrac{35}{16}a_2^3l^6+\tfrac{63}{128}a_2^4l^8)\Big]. \tag{7.153c}$$

Similarly, taking account of (7.148) and (7.138), which refer to the supersonic edges, we have

$$C_{d1}^{(1)} = -\frac{w_{00}}{U_\infty} C_l^{(1)}, \quad C_{d1}^{(2)} = -\frac{w_{00}}{U_\infty} C_l^{(2)}, \tag{7.154a}$$

$$C_{d2}^{(1)} = -\frac{8w_{10}}{3SU_\infty^2} K_{10}[H_{00}+3a_2H_{22}+3a_2^2H_{44}+a_2^3H_{66}], \tag{7.154b}$$

$$C_{d2}^{(2)} = -\frac{2w_{10}}{SU_\infty^2}\Big[K_{20}(H_{00}+4a_2H_{22}+6a_2^2H_{44}+4a_2^3H_{66}+a_2^4H_{88})$$
$$+K_{21}(H_{10}+4a_2H_{32}+6a_2^2H_{54}+4a_2^3H_{76}+a_2^4H_{98})$$
$$+D_{20}\frac{\pi}{4B^2}\left(1+\frac{a_2l^2}{B^2l^2}+\frac{3a_2^2l^4}{4B^4l^4}+\frac{5a_2^3l^6}{16B^6l^6}+\frac{7a_2^4l^8}{128\,B^8l^8}\right)\Big], \tag{7.154c}$$

where the values of $H_{t,\,k}$ are given in **Appendix I.3.**

The sum of the coefficients (7.153) or (7.154) yields the total coefficient of the complex wing.

Pitching moment. Applying formula (3.55c) where, according to (7.146a) and Fig. 7.14, we put

$$h = a = 1+a_2l^2, \tag{7.155}$$

and taking account of (7.132) and (7.138), where $Q_{n,\,2q} = 0$, we deduce successively,

$$\mu_2^{(1)} = \frac{1}{h}\frac{U_\infty}{w_{10}} C_{d2}^{(1)}, \quad \mu_2^{(2)} = \frac{1}{h}\frac{U_\infty}{w_{10}} C_{d2}^{(2)}, \tag{7.156}$$

where $C_{d2}^{(1)}$ and $C_{d2}^{(2)}$ are given by (7.153b, c) for a wing with subsonic edges and by (7.154b, c) for a wing with supersonic edges, respectively.

Natural Antisymmetry of Incidence

Wave drag. For the wave drag coefficient of a wing with natural antisymmetry of incidence, taking into account (7.149) and (7.150) and considering only the case $n = 2$ (for $n = 1$ there is no natural antisymmetry), we have successively:

$$C_{d2}^{(2)} = -\frac{\pi l^2}{2SU_\infty^2} w_{01}A_{21}(1+3a_2l^2+\tfrac{15}{4}a_2^2l^4+\tfrac{35}{16}a_2^3l^6+\tfrac{63}{128}a_2^4l^8), \tag{7.157a}$$

$$C_{d2}^{(2)} = -\frac{2w_{01}}{SU_\infty^2}[K_{20}(H_{10}+4a_2H_{32}+6a_2^2H_{54}+4a_2^3H_{76}+a_2^4H_{98})$$
$$+K_{21}(H_{20}+4a_2H_{42}+6a_2^2H_{64}+4a_2^3H_{86}+a_2^4H_{10,\,8})], \tag{7.157b}$$

where relation (7.157a) corresponds to a wing with subsonic leading edges, and (7.157b) to a wing with supersonic leading edges, and where the values of $H_{t,k}$ are given in Appendix I.3.

Rolling moment. For calculating the rolling moment, formula (3.47) is to be applied, in which $\mathcal{U}_{n-1}$ is replaced by (7.150a) for a wing with subsonic leading edges and by (7.150b) for a wing with supersonic leading edges, respectively; in both cases $P(y)$ is defined by (7.146a).

In the case considered ($n = 2$), the rolling moment coefficient may be expressed as a function of the drag coefficient, as is readily seen from the formulae quoted above:

$$\mu_1^{(2)} = -\frac{1}{a}\frac{U_\infty}{w_{01}} C_{d2}^{(2)}, \tag{7.158}$$

$C_{d2}^{(2)}$ being given by (7.157a) or (7.157b).

Note on the suction force. The suction coefficient is given by the general relations (3.104) or (3.105), where a is the projection of the edges OC_1 and OC_2 upon Ox_1, defined by formula (7.155).

If $\sigma_m^{(n)}$ is the "intermediate" coefficient for an actual delta wing, then, for the arrow-head wing this coefficient is

$$\sigma_m^{*(n)} = a^{m+n}\,\sigma_m^{(n)}. \tag{7.159}$$

Comparison with an actual arrow-head wing. Let us consider the wing with subsonic leading edges of Fig. 7.16, where the enclosing line is defined by

$$x_1 = P(y) = 1+4y^2, \quad (a_2 = 4) \tag{7.160}$$

and where $l = 0{\cdot}5$, hence

$$S = 0{\cdot}93. \tag{7.161}$$

For the case of the conical motion ($n = 1$), by applying formulae (7.151a), (7.153a) and (7.156), one obtains the following values for the aerodynamic coefficients of the wing with symmetry of incidence:

$$C_l^{(1)} = 15{\cdot}95\frac{A_{10}}{U_\infty}, \quad C_{d1}^{(1)} = -15{\cdot}95\frac{w_{00}A_{10}}{U_\infty^2}, \quad \mu_2^{(1)} = -8{\cdot}85\frac{A_{10}}{U_\infty}. \tag{7.162}$$

If we consider now the arrow-head wing $OC_1O_1'C_2$ (Fig. 6.16) which has

$$l = 0{\cdot}5, \quad \tan\delta = 1{\cdot}09, \quad S' = 0{\cdot}92 \tag{7.163}$$

by applying formulae (7.133), (7.134) and (7.135), respectively, the following values are obtained for the aerodynamic coefficients,

$$C_l^{(1)} = 16\frac{A_{10}}{U_\infty}, \quad C_{d1}^{(1)} = -16\frac{w_{00}A_{10}}{U_\infty^2}, \quad \mu_2^{(1)} = -9\frac{A_{10}}{U_\infty}. \tag{7.164}$$

By comparing these results to those obtained for the wing whose enclosing curve is given by (7.160), it is readily seen that the differences are of order 1 per cent, so that the actual arrow-head wings may be quite well replaced by wings whose enclosing line is defined by (7.146a); in this latter case, the aerodynamic coefficients are more easily calculated.

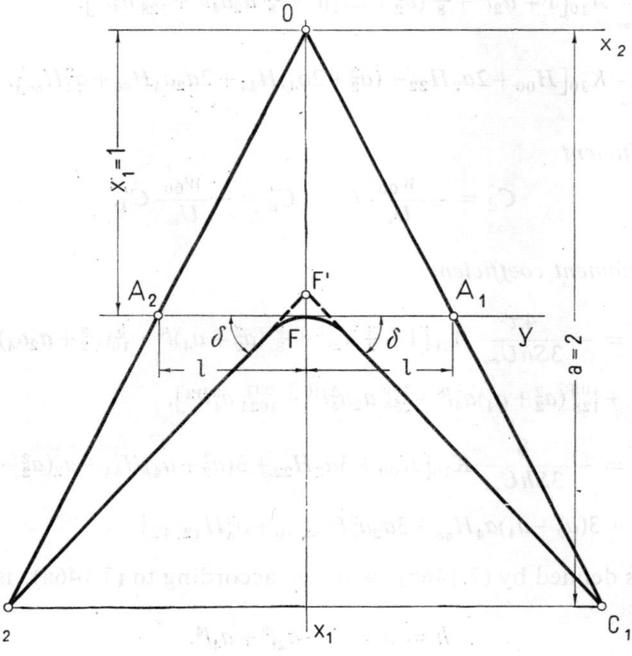

FIG. 7.16

VII. 8.3. Swept-back Wing whose Enclosing Line is of the Form: $P(y) = 1+a_2y^2+a_2y^4$

We consider only the conical motion ($n = 1$). In this case, (7.132a), (7.132b) and (7.138) are to be written under the following forms:

$$w_0 = w_{00}, \tag{7.165}$$

$$\mathcal{U}_0 = \frac{A_{10}}{\sqrt{l^2-x^2}}, \tag{7.166}$$

$$\mathcal{U}'_0 = K_{10}\left(\cos^{-1}\sqrt{\frac{(1+Bl)(1-Bx)}{2B(l-x)}}+\cos^{-1}\sqrt{\frac{(1+Bl)(1+Bx)}{2B(l+x)}}\right), \tag{7.167}$$

where the superscript (′) refers to the supersonic edges.

Introducing the above expressions into the formulae of the aerodynamic coefficients, we find successively:

Lift coefficient

$$C_l = \frac{2\pi}{SU_\infty} A_{10}[1+a_2l^2+\tfrac{3}{8}(a_2^2+2a_4)l^4+\tfrac{5}{8}a_2a_4l^6+\tfrac{35}{128}a_4^2l^8], \tag{7.168a}$$

$$C_l' = \frac{4}{SU_\infty} K_{10}[H_{00}+2a_2H_{22}+(a_2^2+2a_4)H_{44}+2a_2a_4H_{66}+a_4^2H_{88}]. \tag{7.168b}$$

Drag coefficient

$$C_d = -\frac{w_{00}}{U_\infty} C_l, \quad C_d' = -\frac{w_{00}}{U_\infty} C_l'. \tag{7.169}$$

Pitching moment coefficient

$$\mu_2 = -\frac{4\pi}{3ShU_\infty} A_{10}[1+\tfrac{3}{2}a_2l^2+\tfrac{9}{8}(a_2^2+a_4)l^4+\tfrac{5}{16}(a_2^3+a_2a_4)l^6$$
$$+\tfrac{105}{128}(a_2^2+a_4)a_4l^8+\tfrac{189}{256}a_2a_4^2l^{10}+\tfrac{231}{1024}a_4^3l^{12}], \tag{7.170a}$$

$$\mu_2' = -\frac{8}{3ShU_\infty} K_{10}[H_{00}+3a_2H_{22}+3(a_2^2+a_4)H_{44}+a_2(a_2^2+6a_4)H_{66}$$
$$+3(a_2^2+a_4)a_4H_{88}+3a_2a_4^2H_{10,\,10}+a_4^3H_{12,\,12}], \tag{7.170b}$$

where S is defined by (7.146b), while h, according to (7.146a), is given by formula

$$h = a = 1+a_2l^2+a_4l^4. \tag{7.171}$$

In the above formulae, the values of $H_{t,\,k}$ are given in Appendix I.3.

Note on the suction force. As in the preceding case, when the leading edges are subsonic, the suction coefficient $\sigma_m^{*(n)}$ is given by the same expressions (7.159) where, however, a will have the expression (7.171).

CHAPTER VIII

CRUCIFORM WINGS

IN THIS chapter we intend to treat a wing system fitted with a normal plate, constituting thus a complex lifting system which will be termed a *cruciform wing*, having numerous applications to the cruciform tails of rockets, to the lifting system of aeroplanes or to their tails.

VIII. 1. Theory of Cruciform Wing

VIII. 1.1. Definition of the Problem

Consider a cruciform wing referred to a system of three-rectangular axes $Ox_1x_2x_3$, where the Ox_1 axis is parallel to the free stream direction. The origin of the reference system lies at the true vertex O of the cruciform wing which consists of a triangular wing and a plate, also triangular, lying normally to the former and having equal arms (Fig. 8.1). Suppose that the

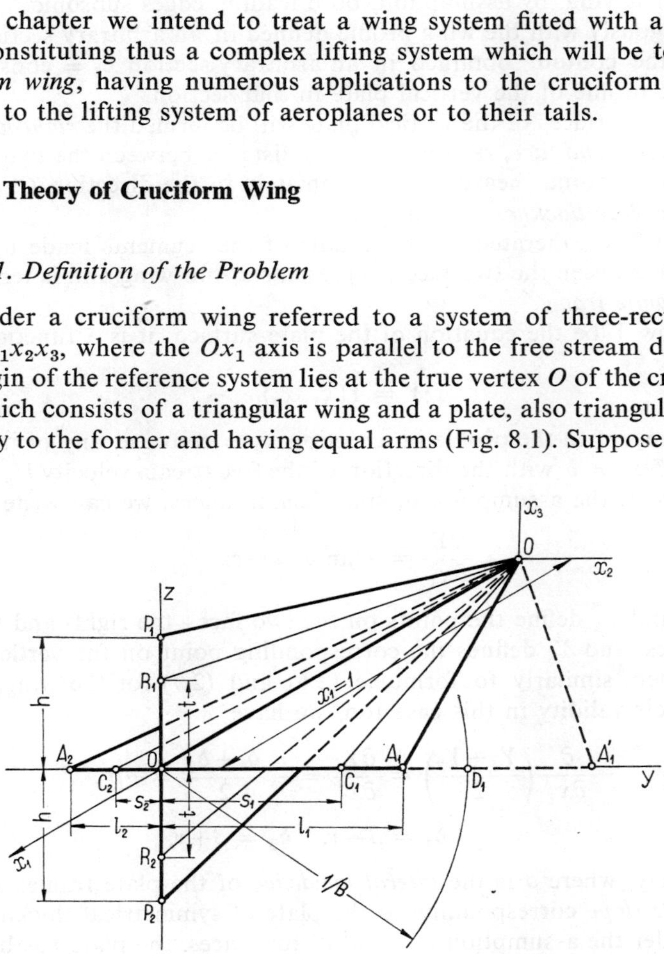

FIG. 8.1

wing incidence or slope, as well as the incidence or slope of the normal plate, are small enough that the wing can be considered to lie in the Ox_1x_2 plane, and the plate in the Ox_1x_3 plane. In the plane $x_1 = 1$, where we have defined the system of axes Oyz, the segment A_2A_1 is the wing trace —its extremities having the abscissae $y = -l_2$ and $y = l_1$, respectively— while the segment P_2P_1 is the plate trace, symmetrical with respect to the wing trace and normal to it, having the total height $2h$.

In connection with the geometry of the cruciform wing, all considerations presented in § II. 1 are valid here too, both for the wing and the plate, the latter having, by assumption, both leading edges subsonic.

Analogously with the wing profile defined in an arbitrary section $x_2 =$ const., the contour obtained in an arbitrary section $x_3 =$ const. represents the profile of the vertical plate in that section.

The two surfaces of the vertical plate will be termed the *right-hand face* and the *left-hand face*, respectively; the distance between the two faces at an arbitrary point, measured approximately in the direction Ox_2, represents the *plate thickness* at that point.

The surface generated by the middle of the segments made up of the distances between the two faces, represents a thin wing and is termed the *vertical plate frame*.

Let now Y be the equation of the plate surface; it is a function of x_1 and x_3,

$$Y = Y(x_1, x_3). \tag{8.1}$$

The tangent to the profile in a section $x_3 =$ const. at any of its points has a *deflexion* δ with the direction of the free stream velocity U_∞; taking into account the assumption of small disturbances, we can write

$$\frac{\partial Y}{\partial x_1} = -\tan\delta \approx -\delta. \tag{8.2}$$

If Y_1 and Y_2 define the points on the two faces, the right- and the left-hand ones, and Y_s defines the corresponding point on the vertical plate frame, then, similarly to formulae (2.6) and (2.7) for the wing, which retain their validity in this case too, we have

$$\frac{\partial}{\partial x_1}\left(\frac{Y_1+Y_2}{2}\right) = \frac{\partial Y_s}{\partial x_1} = -\frac{\delta_1+\delta_2}{2} = -\beta\,, \tag{8.3a}$$

$$\delta_1 = \beta - \nu, \quad \delta_2 = \beta + \nu, \tag{8.3b}$$

respectively, where β is the *lateral incidence* of the plate frame, and ν is the *profile slope* corresponding to the plate of symmetrical thickness; indeed, under the assumption of small disturbances, the plate has been decomposed into a thin plate (frame) and a plate of symmetrical thickness.

Hence, as is seen, the incidence β and the slope ν replace α and γ corresponding to the wing. Thus for example, instead of the downwash w referring to the wing, we have the *lateral velocity* v referring to the plate, the latter velocity being given by the general relation

$$v = -\delta U_\infty, \tag{8.4}$$

which becomes, for the thin plate or the plate of symmetrical thickness,

$$v = -\beta U_\infty, \quad v = \pm \nu U_\infty, \tag{8.5}$$

respectively.

These formulae are similar to (2.10) and (2.11) which refer to the wing. Likewise, denoting by the subscript 1 all that refers to the right-hand face and by the subscript 2 all that refers to the left-hand face, we can write

$$v_1 = v_2 = -\beta U_\infty, \quad u_1 = -u_2 \tag{8.6}$$

on the thin plate, and

$$v_1 = -v_2 = \nu U_\infty, \quad u_1 = u_2 \tag{8.7}$$

on the plate of symmetrical thickness. The last condition of (8.6) and (8.7) is valid only in the case when we have antisymmetry on the wing.

Similarly with Z which defines the wing equation given by (2.16), we have for the plate equation too, a superposition of homogeneous polynomials of order $n = 1, 2, \ldots, N$, in x_1, x_3,

$$Y = \sum_{n=1}^{N} \sum_{q=0}^{n} G_{n-q,q}\, x_1^{n-q} x_3^q = \sum_{n=1}^{N} x_1^n \sum_{q=0}^{n} G_{n-q,q}\, z^q, \quad \left(z = \frac{x_3}{x_1}\right), \tag{8.8}$$

whence

$$v = -\delta U_\infty = \frac{\partial Y}{\partial x_1} U_\infty = \sum_{n=1}^{N} \sum_{q=0}^{n-1} v_{n-q-1,q}\, x_1^{n-q-1} x_3^q$$

$$= \sum_{n=1}^{N} x_1^{n-1} \sum_{q=0}^{n-1} v_{n-q-1,q}\, z^q, \quad \left(z = \frac{x_3}{x_1}\right). \tag{8.9}$$

In the case of the isolated cruciform wing of order n, we have for the lateral velocity on the plate,

$$v = \sum_{q=0}^{n-1} v_{n-q-1,q}\, x_1^{n-q-1} x_3^q = x_1^{n-1} \sum_{q=0}^{n-1} v_{n-q-1,q}\, z^q, \quad \left(z = \frac{x_3}{x_1}\right), \tag{8.10}$$

analogously with (2.18) on the wing, which will be rewritten in the following form:

$$w = \sum_{q=0}^{n-1} w_{n-q-1,q}\, x_1^{n-q-1} x_2^q = x_1^{n-1} \sum_{q=0}^{n-1} w_{n-q-1,q}\, y^q, \quad \left(y = \frac{x_2}{x_1}\right). \tag{8.11}$$

The above relations regarding the vertical plate, together with those regarding the wing, define the data concerning the whole cruciform wing system, which will be used for finding the axial velocity u on the wing and plate.

This is the direct problem. If, however, in the nth order conical motion, the pressures or axial velocities u (to which they are proportional) are given on the wing and plate under the forms

$$\frac{u}{x_1^{n-1}} = \sum_{q=0}^{n-1} u_{n-q-1,\,q}\, y^q, \tag{8.12a}$$

$$\frac{u}{x_1^{n-1}} = \sum_{q=0}^{n-1} u_{n-q-1,\,q}\, z^q, \tag{8.12b}$$

respectively, and we are looking for w on the wing and v on the plate, that is, the equation of the respective surfaces,

$$Z = \frac{1}{U_\infty}\int w\, dx_1 + f_2(x_2), \quad Y = \frac{1}{U_\infty}\int v\, dx_1 + f_3(x_3), \tag{8.13}$$

where $f_2(x_2)$ and $f_3(x_3)$ are arbitrary functions of x_2 and x_3, respectively, then we are dealing with an *indirect problem*, a term that has been also employed for the single wing.

The mixed problem of the cruciform wing consists of the following cases:

(a) The pressures on the wing (or the axial velocities u) and the geometrical shape of the plate (or the lateral velocity v) are given, and the wing surface and the pressures on the plate (i.e. u) are required.

(b) The geometrical shape of the wing and the pressures on the plate are given, and the plate surface and pressure distribution on the wing are required.

VIII. 1.2. Representation of the Motion by Complex Variable Functions

Let us now return to the considerations presented in § I. 8.1, where we denoted by $\Phi(x_1, x_2, x_3)$ the nth order homogeneous function which reresents the velocity potential round the cruciform wing, and by $\Phi_{n-q-r,\,q,\,r}$ the derivative of order $n-q-r$ with respect to x_1, of order q with respect to x_2 and of order r with respect to x_3; this derivative is a zero-order homogeneous function. Further, we can write the following expressions for the disturbance velocities u, v, w:

$$u = \Phi_{100}, \quad v = \Phi_{010}, \quad w = \Phi_{001}, \tag{8.14}$$

which are homogeneous functions of order $(n-1)$.

In such a case, by applying formulae (1.169), we can write for points on the wing ($z = 0$):

$$\frac{(n-1)!}{x_1^{n-1}} u = \sum_{q=0}^{n-1} C_{n-1}^q y^q \Phi_{n-q,q,0}, \tag{8.15a}$$

$$\frac{(n-1)!}{x_1^{n-1}} w = \sum_{q=0}^{n-1} C_{n-1}^q y^q \Phi_{n-q-1,q,1}, \tag{8.15b}$$

respectively; and by applying (1.170), we likewise have for points on the plate ($y = 0$):

$$\frac{(n-1)!}{x_1^{n-1}} u = \sum_{r=0}^{n-1} C_{n-1}^r z^r \Phi_{n-r,0,r}, \tag{8.16a}$$

$$\frac{(n-1)!}{x_1^{n-1}} v = \sum_{r=0}^{n-1} C_{n-1}^r z^r \Phi_{n-r-1,1,r}. \tag{8.16b}$$

These two pairs of relations will be of use for the study of the cruciform wing in its various applications.

To this end, we shall first recall Busemann's geometrical transformation (1.138) which turns the plane Oyz into an auxiliary plane $O\eta\mathfrak{z}$ (Fig. 1.2) by the aid of a simple relation between the vector radii r in the physical plane and ϱ in the auxiliary plane, of the form

$$Br = \frac{2\varrho}{1+\varrho^2}, \quad \omega = \omega;$$

then, by means of a conformal transformation (1.145),

$$Bx = \frac{2\xi}{1+\xi^2}, \quad \xi = \eta + i\zeta, \quad x = y + i\mathfrak{z}, \tag{8.17}$$

we obtain the auxiliary plane x, which has the configuration shown in Fig. 8.2b, where the wing trace is reproduced to its actual dimension. However, as regards the plate trace, this undergoes an elongation, which is deduced from the above relations put in the following form:

$$B\mathfrak{z} = \frac{2\zeta}{1-\zeta^2}, \quad Bz = \frac{2\zeta}{1+\zeta^2}, \tag{8.18a}$$

$$\mathfrak{z} = \frac{z}{\sqrt{1-B^2z^2}}, \quad z = \frac{\mathfrak{z}}{\sqrt{1+B^2\mathfrak{z}^2}}, \tag{8.18b}$$

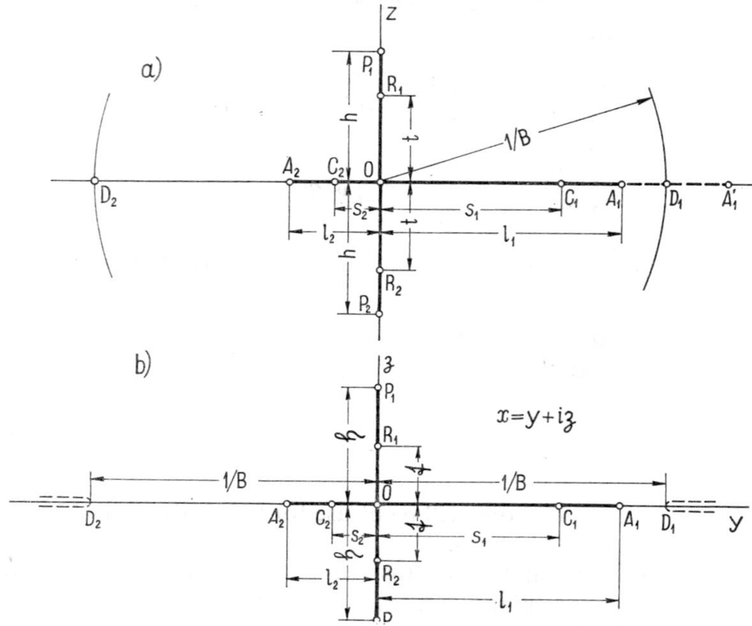

FIG. 8.2

whence we deduce the height $\mathfrak{h}$ in the auxiliary plane, as a function of the actual height h,

$$\mathfrak{h} = \frac{h}{\sqrt{1-B^2h^2}}\,. \tag{8.19}$$

Let us further consider the analytical function $F_{n-q-r,\,q,\,r}$ defined by (1.171),

$$F_{n-q-r,\,q,\,r} = \Phi_{n-q-r,\,q,\,r}+i\Phi'_{n-q-r,\,q,\,r}\,, \tag{8.20}$$

where $\Phi'_{n-q-r,\,q,\,r}$ is the harmonic function associated with $\Phi_{n-q-r,\,q,\,r}$. Noting that we can write, for the wing and plate, respectively,

$$y\Phi_{n-q,\,q,\,0} = \mathrm{Re}\;\mathfrak{x}F_{n-q,\,q,\,0}\,, \tag{8.21a}$$

$$z\Phi_{n-r,\,0,\,r} = \frac{\mathfrak{z}}{\sqrt{1+B^2\mathfrak{z}^2}}\,\Phi_{n-r,\,0,\,r} = \mathrm{Re}\,\frac{-i\mathfrak{x}}{\sqrt{1-B^2\mathfrak{x}^2}}\,F_{n-r,0,r}\,, \tag{8.21b}$$

relations (8.15) and (8.16) can be written, for points on the wing ($z = 0$), in the forms

$$\frac{(n-1)!}{x_1^{n-1}} u = \operatorname{Re} \sum_{q=0}^{n-1} C_{n-1}^q x^q F_{n-q, q, 0}, \tag{8.22a}$$

$$\frac{(n-1)!}{x_1^{n-1}} w = \operatorname{Re} \sum_{q=0}^{n-1} C_{n-1}^q x^q F_{n-q-1, q, 1}, \tag{8.22b}$$

which reproduce in fact relations (1.72) and, for points on the plate ($y = 0$), respectively, in the forms

$$\frac{(n-1)!}{x_1^{n-1}} u = \operatorname{Re} \sum_{r=0}^{n-1} C_{n-1}^r \left(\frac{-ix}{\sqrt{1-B^2x^2}} \right)^r F_{n-r, 0, r}, \tag{8.23a}$$

$$\frac{(n-1)!}{x_1^{n-1}} v = \operatorname{Re} \sum_{r=0}^{n-1} C_{n-1}^r \left(\frac{-ix}{\sqrt{1-B^2x^2}} \right)^r F_{n-r-1, 1, r}, \tag{8.23b}$$

which likewise reproduce relations (1.173). The functions $F_{n-q-r, q, r}$ are connected between themselves by the compatibility relations (1.180),

$$dF_{n00} = (-x)^q \left(\frac{ix}{\sqrt{1-B^2x^2}} \right)^r dF_{n-q-r, q, r}, \quad (F_{n00} = \Phi_{n00} + i\Phi'_{n00}), \tag{8.24}$$

which show that the terms in the right-hand side are equal to each other, irrespective of the values of q and r, within the limits of the inequality

$$0 \leqslant q+r \leqslant n. \tag{8.25}$$

VIII. 2. Determination of the Axial Velocities u on the Cruciform Wing

VIII. 2.1. Reduction of the Problem to a Hydrodynamic Analogy

By applying the compatibility relations (8.24), we might replace the functions $F_{n-q, q, 0}$ and $F_{n-r, 0, r}$, which enter into the expressions of u in (8.22a) and (8.23a), and obtain respectively

$$\frac{(n-1)!}{x_1^{n-1}} u = \operatorname{Re} \sum_{q=0}^{n-1} (-1)^q C_{n-1}^q x^q \int_{\text{Mach circle}}^{\text{wing}} \frac{dF_{n00}}{x^q}, \tag{8.26}$$

for points on the wing and, respectively,

$$\frac{(n-1)!}{x_1^{n-1}} u = \operatorname{Re} \sum_{r=0}^{n-1} (-1)^r C_{n-1}^r \left(\frac{x}{\sqrt{1-B^2x^2}} \right)^r \int_{\text{Mach circle}}^{\text{plate}} \left(\frac{\sqrt{1-B^2x^2}}{x} \right)^r dF_{n00}, \tag{8.27}$$

for points on the plate. The constants C^q_{n-1} or C^r_{n-1} represent the customary notations for the combinations.

From the above expressions it may be observed that the axial velocities u may be easily determined by simple quadratures if the function F_{n00} is known, for which we shall establish below some interesting properties that will allow us to determine it.

In this connection, considering the expressions for w given by (8.11) and for v given by (8.10), in which, however, q will be replaced by r, and comparing them with (8.15b) and (8.16b), respectively, we find the following relations:

$$C^q_{n-1}\Phi_{n-q-1,q,1} = (n-1)!w_{n-q-1,q}, \quad (d\Phi_{n-q-1,q,1} = 0) \quad (8.28)$$

valid at points on the wing ($z = 0$) and, respectively,

$$C^r_{n-1}\Phi_{n-r-1,1,r} = (n-1)!v_{n-r-1,r}, \quad (d\Phi_{n-r-1,1,r} = 0) \quad (8.29)$$

valid at points on the plate ($y = 0$).

Let us now observe that if in the compatibility relation (8.24), we set $r = 1$ for points on the wing ($z = 0$) and $q = 1$ for points on the plate ($y = 0$), there results the following condition on the wing and plate:

$$d\Phi'_{n00} = 0, \quad \Phi'_{n00} = \text{const.} \quad (8.30)$$

Thus we can assume that $F_{n00} = \Phi'_{n00}+i\Phi'_{n00}$ would represent the "potential" of a fictitious incompressible motion, where the traces of the wing and plate are "stream lines"; in such a case, as in the case of the simple wing, we may apply the *method of the hydrodynamic analogy* in order to find the expression for F_{n00}, which thus plays a particular part.

We have seen above that we can consider the trace of the cruciform wing as a plane contour, around which there takes place a fictitious motion defined by the "complex potential" F_{n00}. For determining this potential, we shall first consider the axial velocity u, whose behaviour is similar to that for the simple wing, viz.:

(a) If the edges of the wing and plate are subsonic, then at the corresponding points A_1 ($y = l_1$), A_2($y = -l_2$) on the wing, and P_1, P_2 ($z = \pm h$) on the plate (Fig. 8.2a), the axial velocity u tends towards infinity like the following expressions:

$$\frac{1}{\sqrt{l_1-y}}, \quad \frac{1}{\sqrt{l_2+y}}, \quad \frac{1}{\sqrt{h\pm z}}. \quad (8.31)$$

Hence it follows that the analytic function F_{n00}, whose real part is deduced from the derivative of u of order $(n-1)$, will contain in the auxiliary plane, as in the case of the simple wing, the following singularities referring to the leading edges A_1, A_2 and P_1, P_2, all edges being assumed subsonic (Fig. 8.2 b):

$$\frac{1}{(l_j \mp x)^{1/2}}, \quad \frac{1}{(l_j \mp x)^{3/2}}, \quad \ldots, \quad \frac{1}{(l_j \mp x)^{n-1/2}}, \tag{8.32a}$$

$$(-,\ j = 1; \quad +,\ j = 2), \tag{8.32b}$$

$$\frac{1}{(x \pm i\mathfrak{h})^{1/2}}, \quad \frac{1}{(x \pm i\mathfrak{h})^{3/2}}, \ldots, \quad \frac{1}{(x \pm i\mathfrak{h})^{n-1/2}}. \tag{8.32c}$$

Likewise, on the ridges situated at $C_i(y = s_i)$ on the wing and at R_i $(z = \pm t)$ on the plate (Fig. 8.2, $i = 1, 2$), the velocity u tends to infinity as

$$\ln |s_1 - y|, \quad \ln |s_2 + y|, \quad \ln |t - z|, \quad \ln |t + z|. \tag{8.33}$$

It follows that the function F_{n00} contains the following singularities in the x plane:

$$\ln (s_j \mp x), \quad \frac{1}{s_j \mp x}, \ldots, \quad \frac{1}{(s_j \mp x)^{n-1}}, \tag{8.34a}$$

$$(-,\ j = 1; \quad +,\ j = 2), \tag{8.34b}$$

$$\ln (x \mp it), \quad \frac{1}{x \mp it}, \ldots, \quad \frac{1}{(x \mp it)^{n-1}}, \tag{8.34c}$$

where t is the ordinate of the point $R_j(z = \pm t)$ in the x plane,

$$\mathrm{t} = \frac{t}{\sqrt{1 - B^2 t^2}}. \tag{8.35}$$

Hence, the problem reduces to the determination of a fictitious potential F_{n00}, which assumes the traces of the wing and plate to be "stream lines" and meanwhile contains the above singularities, also satisfying the other boundary conditions.

This problem admits a very simple solution, employing the *hydrodynamic analogy* which reduces the problem to that of a plane incompressible motion.

In this connection, we shall use another auxiliary X plane, given by the transformation function

$$X^2 = x^2 + \mathfrak{h}^2, \tag{8.36}$$

which leads to the configuration shown in Fig. 8.3. From this transformation we obtain the following relations which, in this new auxiliary plane, define the various magnitudes and coordinates in the physical plane:

$$L_i = \sqrt{l_i^2+\mathfrak{h}^2}, \quad S_i = \sqrt{s_i^2+\mathfrak{h}^2}, \quad T = \sqrt{\mathfrak{h}^2-\mathfrak{t}^2}, \quad (i = 1, 2), \tag{8.37a}$$

$$\frac{1}{\mathcal{B}} = \sqrt{\frac{1}{B^2}+\mathfrak{h}^2}\,. \tag{8.37b}$$

The points P_1 and P_2 fall at the origin itself.

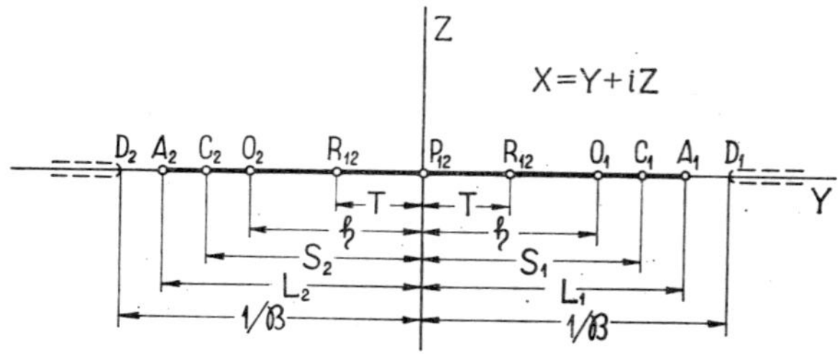

FIG. 8.3

Thus we have reduced the problem to a simple wing, a fact that considerably simplifies the finding of a solution, since the method to be employed is exactly the same as for the isolated simple wing, with the only remark that account should be taken of the singularities mentioned above.

The transformation (8.36) does not alter the character of these singularities, except of those lying at the origin and corresponding to the subsonic leading edges of the plate, which become "doublets" in the X plane in the hydrodynamic sense. It should be remarked that the singularities of the subsonic leading edges situated at A_1, A_2 and P_1, P_2, given by (8.32) and (8.34) in the plane $x = y+i\mathfrak{z}$, can be rewritten in the following form in the new plane $X = Y+iZ$:

$$\frac{1}{(L_j\mp X)^{1/2}}, \quad \frac{1}{(L_j\mp X)^{3/2}}, \quad \ldots, \quad \frac{1}{(L_j\mp X)^{n-1/2}}, \tag{8.38a}$$

$$(-, j = 1; \quad +, j = 2), \tag{8.38b}$$

$$\frac{1}{X}, \quad \frac{1}{X^3}, \quad \ldots, \quad \frac{1}{X^{2n-1}}\,. \tag{8.38c}$$

The singularities of the ridges situated at C_1, C_2, R_1 and R_2 can be written, respectively, as:

$$\ln (S_j \mp X), \quad \frac{1}{S_j \mp X}, \ldots, \quad \frac{1}{(S \mp X)^{n-1}}, \tag{8.39a}$$

$$(-, j = 1; \; +, j = 2), \tag{8.39b}$$

$$\ln (T \mp X), \quad \frac{1}{T \mp X}, \ldots, \quad \frac{1}{(T \mp X)^{n-1}}. \tag{8.39c}$$

VIII. 2.2. Expression of the Contribution of the Subsonic Edges and of the Ridges in Constructing the Function F_{n00}

CONTRIBUTION OF THE SUBSONIC EDGES OF THE THIN WING. As in the case of the isolated wing (2.51, 2.55), the edges A_1 ($Y = L_1$) and A_2 ($Y = -L_2$) give the following contributions, for each separate edge or for both edges considered together:

$$F_{n00}^{(1)} = \sum_{q=0}^{n-1} \frac{b_q X^q}{(L_1 - X)^{n-1}} \sqrt{\frac{L_2 + X}{L_1 - X}}, \tag{8.40a}$$

$$F_{n00}^{(2)} = \sum_{q=0}^{n-1} \frac{c_q X^q}{(L_2 + X)^{n-1}} \sqrt{\frac{L_1 - X}{L_2 + X}}, \tag{8.40b}$$

$$F_{n00} = \frac{a_0 + a_1 X + \ldots + a_{2n-1} X^{2n-1}}{(L_1 - X)^{n-1}(L_2 + X)^{n-1}\sqrt{(L_1 - X)(L_2 + X)}}. \tag{8.40c}$$

If in the contribution of the subsonic edge A_1, the other leading edge A_2 is supersonic, L_2 will be replaced by $1/\mathcal{B}$ in (8.40a); and if in the contribution of the subsonic edge A_2, the other leading edge A_1 is supersonic, L_1 will be replaced by $1/\mathcal{B}$. In such cases we have:

$$F_{n00}^{(1)} = \sum_{q=0}^{n-1} \frac{b_q X^q}{(L_1 - X)^{n-1}} \sqrt{\frac{1 + \mathcal{B}X}{\mathcal{B}(L_1 - X)}}, \tag{8.41a}$$

$$F_{n00}^{(2)} = \sum_{q=0}^{n-1} \frac{c_q X^q}{(L_2 + X)^{n-1}} \sqrt{\frac{1 - \mathcal{B}X}{\mathcal{B}(L_2 + X)}}. \tag{8.41b}$$

CONTRIBUTION OF THE SUBSONIC EDGES OF THE PLATE. In this case it should be pointed out that the lateral velocities v (8.10) may be different on the upper and lower portions of the plate; however, as may be readily seen, we can decompose this distribution of lateral velocities into two parts: a *symmetrical* one with respect to the real axis OY, and the other *antisymmetrical* with respect to the same axis.

The symmetrical distribution. For the part with symmetrical distribution, the resulting motion is also symmetrical. In such a case, the plate behaves like a simple wing, since the abscissae line upon which the wing trace is situated is a "stream line" for F_{n00} and hence the boundary conditions on the wing are fulfilled. The flow around the plate may therefore be easily determined, by employing the results already obtained in the previous chapters. The influence of this flow, both on calculating the pressures on the plate itself—which is generating this motion—and on calculating the pressures on the wing, i.e. the interference between the two planes, is expressed by an additional axial velocity u^*, which is "induced" on the wing by the symmetrical flow around the plate.

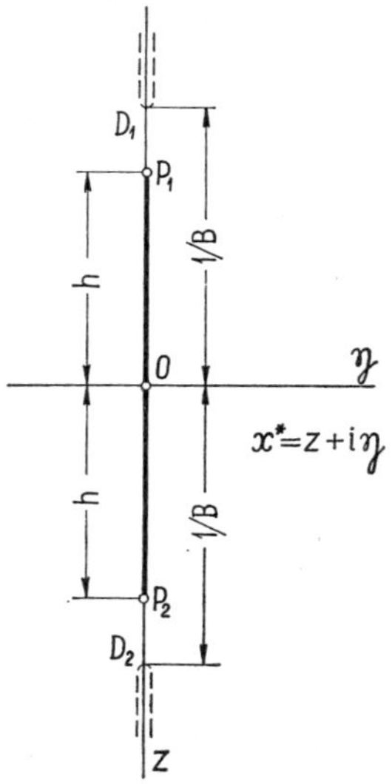

FIG. 8.4

As concerns, however, the influence of the wing upon the plate, this influence is included in the general solution of the flow around the cruciform wing, where this time, however, the plate reduces to a simple vertical plane without incidence.

In order to find now the axial velocity u^* on the isolated plate, we give its general expression in an auxiliary plane $x^* = z + i\mathfrak{y}$ (Fig. 8.4) which retains exactly the plate trace, inverting only the axes and coordinates (y into z, $\mathfrak{z}$ into $\mathfrak{y}$).

Analogously with (4.14) and (4.42), in this new plane x^*, the axial velocity functions are

$$\mathcal{U}^*_{n-1} = \sum_{q=0}^{E\left(\frac{n}{2}\right)} \frac{A^*_{n,2q}\, x^{*q}}{\sqrt{h^2 - x^{*2}}} \sum_{q=1}^{E\left(\frac{n-1}{2}\right)} Q^*_{n,2q}\, x^{*2q} \cosh^{-1} \sqrt{\frac{h^2}{x^{*2}}}, \quad (8.42a)$$

$$\mathcal{U}^*_{n-1} = \sum_{q=0}^{n-1} K^*_{n,q}\, x^{*q}\Bigg(\cos^{-1}\sqrt{\frac{(1+Bh)(1-Bx^*)}{2B(h-x^*)}}$$

$$+(-1)^q \cos^{-1}\sqrt{\frac{(1+Bh)(1+Bx^*)}{2B(h+x^*)}}\Bigg) + \sum_{q=0}^{E\left(\frac{n-2}{2}\right)} D^*_{n,\,2q}\, x^{*2q}\sqrt{\frac{1-B^2x^{*2}}{B^2}}$$

$$+\sum_{q=1}^{E\left(\frac{n-1}{2}\right)} Q^*_{n,\,2q}\, x^{*2q}\cosh^{-1}\sqrt{\frac{1}{B^2x^{*2}}}\,, \tag{8.42b}$$

respectively, the former for subsonic leading edges and the latter for supersonic leading edges, where $E(t)$ is the greatest integer contained in t.

The constants $A^*_{n,\,2q}$, $Q^*_{n,\,2q}$ are given in § IV. 2.1, and $K^*_{n,q}$, $Q^*_{n,\,2q}$ and $D^*_{n,\,2q}$ in § IV. 4.1; l will be replaced by h, and w_{ij} by v_{ij} in the expressions for these constants.

In order to find u^* on the wing, which is of use for calculating the pressures at its points, we shall apply a formula similar to (8.23a),

$$\frac{(n-1)!}{x_1^{n-1}}u^* = \operatorname{Re}\sum_{r=0}^{n-1}(-1)^{n+r-1}C_n^r\left(\frac{x^*}{\sqrt{1-B^2x^{*2}}}\right)^r$$

$$\times\int_{\text{Mach circle}}^{\text{wing trace}} x^{*n-1}\left(\frac{\sqrt{1-B^2x^{*2}}}{x^*}\right)^r dF^*_{1,\,n-1,\,0}\,, \tag{8.43}$$

where $F^*_{n-1,\,0}$ stands for the contribution of the two subsonic or supersonic leading edges of the plate,

$$F^*_{1,\,n-1,\,0} = \frac{d^{n-1}}{dx^{*n-1}}(\mathcal{U}^*_{n-1})\,. \tag{8.44}$$

Between the ordinate in the x^* plane and the abscissa y of the wing in the x plane, there exists the same relation as between $\mathfrak{z}$ and z (8.18):

$$\eta = \frac{y}{\sqrt{1-B^2y^2}}\,,\quad y = \frac{\eta}{\sqrt{1+B^2\eta^2}}\,. \tag{8.45}$$

The antisymmetrical distribution. For the antisymmetrical distribution of the lateral velocity v, an antisymmetrical motion is correspondingly obtained. This motion is given by the *multipoles* (8.38c) situated at the origin, having the abscissa axis as common axis and equal but of opposite signs with respect to this axis. The line A_2A_1 remains a "stream line"; this condition helps us find the contribution of these singularities.

For this purpose, it is more convenient to employ a new transformation similar to that employed for the simple wing (2.54),

$$\chi^2 = \frac{L_2+X}{L_1-X}\,, \tag{8.46}$$

with the aid of which the configuration of Fig. 8.5 is obtained.

In this plane, the singularities are situated at the points P_1 and P_2, having the abscissae

$$\chi_1 = \pm\sqrt{\frac{L_2}{L_1}}, \tag{8.47}$$

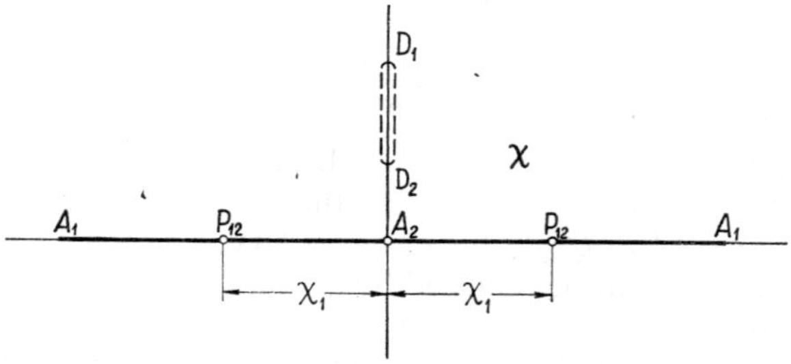

FIG. 8.5

which correspond to the points $x = \pm i\mathfrak{h}$ in the x plane and which appear in the χ plane in the form

$$\frac{1}{\chi\mp\chi_1}, \quad \frac{1}{(\chi\mp\chi_1)^3}, \quad \ldots, \quad \frac{1}{(\chi\mp\chi_1)^{2n-1}}. \tag{8.48}$$

All these multipoles have their axis on the abscissae line A_2A_1, which therefore retains its character of "stream line" and, two by two, they are equal and of the same sense. It is obvious that their contribution will be defined by the function

$$\begin{aligned} F_{n00}^{(p)} &= \sum_{k=1}^{n} e_{2k-1}\left[\frac{1}{(\chi-\chi_1)^{2k-1}}+\frac{1}{(\chi+\chi_1)^{2k-1}}\right] \\ &= \frac{\chi}{(\chi^2-\chi_1^2)^{2n-1}}\sum_{k=2}^{2n} \varepsilon_{2(2n-k)}\,\chi^{2(2n-k)}, \end{aligned} \tag{8.49}$$

which, in the X plane, becomes

$$F_{n00}^{(p)} = \sum_{q=0}^{2(n-1)} \frac{\alpha_q X^q}{X^{2n-1}}\sqrt{(L_1-X)(L_2+X)}. \tag{8.50}$$

When either the edge A_1 or A_2, or both of them, become supersonic, we replace L_1, or L_2, or L_1 and L_2, by $1/\mathcal{B}$ in the above formulae, a fact that results from the transformation formula (8.46); thus we get

$$F_{n00}^{(p)} = \sum_{q=0}^{2(n-1)} \frac{\alpha_q X^q}{X^{2n-1}}\sqrt{\frac{(1-\mathcal{B}X)(L_2+X)}{\mathcal{B}}}, \tag{8.51a}$$

$$F_{n00}^{(p)} = \sum_{q=0}^{2(n-1)} \frac{\alpha_q X^q}{X^{2n-1}}\sqrt{\frac{(L_1-X)(1+\mathcal{B}X)}{\mathcal{B}}}, \tag{8.51b}$$

$$F_{n00}^{(p)} = \sum_{q=0}^{2(n-1)} \frac{\alpha_q X^q}{X^{2n-1}} \sqrt{\frac{1-\mathscr{B}^2X^2}{\mathscr{B}^2}}, \tag{8.51c}$$

respectively.

CONTRIBUTION OF RIDGES. For the ridges situated at C_1 and C_2 ($Y = S_1$, $Y = -S_2$, Fig. 8.3), we obtain formulae similar to those for the simple wing in the x plane (2.63, 2.69a), replacing, however, x by X, l_1 by L_1, l_2 by L_2, s_1 by S_1, s_2 by S_2; thus we have

$$F_{n00}^{(1)} = C_w \cosh^{-1} \sqrt{\frac{(L_2+S_1)(L_1-X)}{(L_1+L_2)(S_1-X)}} + \sum_{q=0}^{n-2} \frac{c_q X^q}{(S_1-X)^{n-1}} \sqrt{(L_1-X)(L_2+X)}, \tag{8.52a}$$

$$F_{n00}^{(2)} = G_w \cosh^{-1} \sqrt{\frac{(L_1+S_2)(L_2+X)}{(L_1+L_2)(S_2+X)}} + \sum_{q=0}^{n-2} \frac{g_q X^q}{(S_2+X)^{n-1}} \sqrt{(L_1-X)(L_2+X)}. \tag{8.52b}$$

If one or both leading edges are supersonic, then for the respective edges we set $1/\mathscr{B}$ instead of L_1 or L_2.

Thus we have successively:

$$F_{n00}^{(1)} = c_w \cosh^{-1} \sqrt{\frac{(L_2+S_1)(1-\mathscr{B}X)}{(1+\mathscr{B}L_2)(S_1-X)}} + \frac{1}{(S_1-X)^{n-1}} \sqrt{\frac{(1-\mathscr{B}X)(L_2+X)}{\mathscr{B}}} \sum_{q=0}^{n-2} c_q X^q, \tag{8.53a}$$

$$F_{n00}^{(1)} = c_w \cosh^{-1} \sqrt{\frac{(1+\mathscr{B}S_1)(L_1-X)}{(1+\mathscr{B}L_1)(S_1-X)}} + \frac{1}{(S_1-X)^{n-1}} \sqrt{\frac{(L_1-X)(1+\mathscr{B}X)}{\mathscr{B}}} \sum_{q=0}^{n-2} c_q X^q, \tag{8.53b}$$

$$F_{n00}^{(1)} = c_w \cosh^{-1} \sqrt{\frac{(1+\mathscr{B}S_1)(1-\mathscr{B}X)}{2\mathscr{B}(S_1-X)}} + \frac{1}{(S_1-X)^{n-1}} \sqrt{\frac{1-\mathscr{B}^2X^2}{\mathscr{B}^2}} \sum_{q=0}^{n-2} c_q X^q, \tag{8.53c}$$

as well as three other similar expressions for $F_{n00}^{(2)}$, where the same replacements as above are to be performed.

For the ridges situated on the plate ($Y = \pm T$), the formulae are identical to the above ones, considering antisymmetrical v,

$$F_{n00}^{(p)} = C_p \cosh^{-1} \sqrt{\frac{(L_2+T)(L_1-X)}{(L_1+L_2)(T-X)}} + \sum_{q=0}^{n-2} \frac{c_q X^q}{(T-X)^{n-1}} \sqrt{(L_1-X)(L_2+X)}, \tag{8.54a}$$

$$F_{n00}^{(p)} = G_p \cosh^{-1} \sqrt{\frac{(L_1+T)(L_2+X)}{(L_1+L_2)(T+X)}}$$
$$+ \sum_{q=0}^{n-2} \frac{g_q X^q}{(T+X)^{n-1}} \sqrt{(L_1-X)(L_2+X)}. \tag{8.54b}$$

Likewise, when one or both leading edges of the wing are supersonic, we observe the above formula, by setting $1/\mathcal{B}$ instead of L_1 or L_2.

VIII. 2.3. Determination of the Function F_{n00} in some General Cases

Taking into consideration the contributions of the ridges and leading edges given in the above expressions, we shall try to construct the function F_{n00} in some concrete cases that will be useful for applications.

CRUCIFORM WING WHOLLY CONTAINED IN THE MACH CONE. Combining (8.40c) and (8.50), which represent the contributions of the four leading edges (A_1, A_2, P_1, P_2), we obtain a simple expression,

$$F_{n00} = \frac{A_0+A_1X+ \ldots +A_{4n-2}X^{4n-2}}{X^{2n-1}(L_1-X)^{n-1}(L_2+X)^{n-1}\sqrt{(L_1-X)(L_2+X)}}. \tag{8.55}$$

If the wing is of symmetrical planform ($L_1 = L_2 = L$), taking into account the fact that we have considered only the antisymmetry of incidence, we find

$$F_{n00} = \frac{A_0+A_2X^2+ \ldots +A_{4n-2}X^{4n-2}}{X^{2n-1}(L^2-X^2)^{n-1}\sqrt{L^2-X^2}}. \tag{8.56}$$

If one of the wing edges coincides with the axis of the cruciform wing then, at the corresponding point ($X = L_1 = \mathfrak{h}$ corresponding to $x = l_1 = 0$ or $X = -L_2 = -\mathfrak{h}$ corresponding to $x = -l_2 = 0$), the velocity is equal to zero. In such a case, formula (8.55) assumes a form similar to that for the isolated wing; thus, according as $l_1 = 0$ or $l_2 = 0$, we have respectively

$$F_{n00} = \frac{A_0+A_1X+ \ldots +A_{3n-2}X^{3n-2}}{X^{2n-1}(L_2+X)^{n-1}} \sqrt{\frac{\mathfrak{h}-X}{L_2+X}}, \tag{8.57a}$$

$$F_{n00} = \frac{A_0+A_1X+ \ldots +A_{3n-2}X^{3n-2}}{X^{2n-1}(L_1-X)^{n-1}} \sqrt{\frac{\mathfrak{h}+X}{L_1-X}}. \tag{8.57b}$$

CRUCIFORM WING WITH ONE SUBSONIC EDGE (A_2) AND THE OTHER SUPERSONIC, ALSO HAVING A RIDGE SITUATED AT C_1 ($Y = S_1$). Combining (8.41b) with (8.51) and (8.52a), where L_1 will be everywhere replaced by $1/\mathcal{B}$, and performing the required additions, we obtain

$$F_{n00} = C \cosh^{-1} \sqrt{\frac{(L_2+S_1)(1-\mathcal{B}X)}{(1+\mathcal{B}L_2)(S_1-X)}} + \frac{A_0+A_1X+\ldots+A_{4n-3}X^{4n-3}}{X^{2n-1}(L_2+X)^{n-1}(S_1-X)^{n-1}} \sqrt{\frac{1-\mathcal{B}X}{\mathcal{B}(L_2+X)}}. \quad (8.58)$$

If $s_1 > 1/B$ and, at the same time, s_1 is assumed to be the trace of the supersonic edge of the cruciform wing (OC_1 coincides with OA_1), then the above formula holds valid.

CRUCIFORM WING HAVING BOTH LEADING EDGES SUPERSONIC AND TWO RIDGES OC_1 ($Y = S_1$), OC_2 ($Y = -S_2$). In this case, we combine (8.51) with (8.52) by setting $L_1 = L_2 = 1/\mathcal{B}$ and, by addition, we obtain

$$F_{n00} = C_n \cosh^{-1} \sqrt{\frac{(1+\mathcal{B}S_1)(1-\mathcal{B}X)}{2\mathcal{B}(S_1-X)}} + G_n \cosh^{-1} \sqrt{\frac{(1+\mathcal{B}S_2)(1+\mathcal{B}X)}{2\mathcal{B}(S_2+X)}} + \frac{A_0+A_1X+\ldots+A_{4(n-1)}X^{4(n-1)}}{X^{2n-1}(S_1-X)^{n-1}(S_2+X)^{n-1}} \sqrt{\frac{1-\mathcal{B}^2X^2}{\mathcal{B}^2}}, \quad (8.59)$$

where S_1 and S_2 are deduced from the second formula (8.37a), where s_i is replaced by s_1, s_2, respectively.

If $s_1 > 1/B$, $s_2 > 1/B$ then s_1 and s_2 may be considered as the traces of the supersonic edges of the wing (OC_1 coincides with OA_1, and OC_2 with OA_2).

The above expressions for F_{n00} will be of use for studying some practical cases of the cruciform wing.

VIII. 3. Application to the Cruciform Wing having Subsonic Leading Edges and Antisymmetrical Incidences, in the True Conical Motion ($n = 1$)

VIII. 3.1. Determination of Constants in the Expression for the Axial Velocity

We shall consider the wing to have equal arms ($l_1 = l_2 = l$, $L_1 = L_2 = L$) and the incidence to be $+\alpha$ on the right-hand portion and $-\alpha$ on the left-hand one (Fig. 8.6).

Likewise, the plate incidence is $+\beta$ on the upper part (at the upper surface of the wing) and $-\beta$ on the lower part (at the lower surface of the wing). We shall put $n = 1$ in the general expression for F_{n00} given by (8.56); taking also account of (8.36), we can write successively:

$$F_{100} = \mathcal{U} = \frac{A_0+A_2X^2}{X\sqrt{L^2-X^2}} = \frac{A_{10}+A_{12}x^2}{\sqrt{(l^2-x^2)(\mathfrak{h}^2+x^2)}}, \quad (8.60)$$

where we have employed the current notations, previously used for the constants of motion ($A_{n,2q}$), in the representations in the x plane, denoting by n the order of the conical motion and by $2q$ the exponent of x. It is seen that for $\mathfrak{h} = 0$ (and $A_{10} = 0$), one obtains the antisymmetrical motion (with forced antisymmetry) for the simple delta wing.

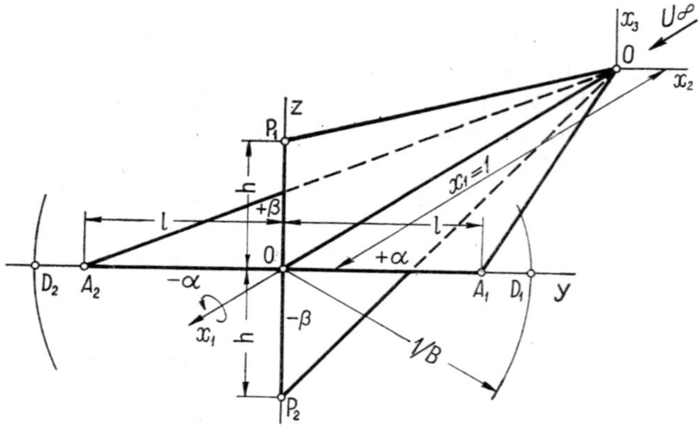

FIG. 8.6

For determining the constants A_{10} and A_{12} by employing the compatibility relations (8.24), we require that we have the downwash $w = -\alpha U_\infty$ on the right-hand part of the wing,

$$w = -\alpha U_\infty = \text{Re} \int_{\text{right-hand wing}}^{\text{Mach circle}} \frac{i\sqrt{1-B^2x^2}}{x}\, d\mathcal{U}, \tag{8.61}$$

and the lateral velocity $v = -\beta U_\infty$ on the upper part of the plate,

$$v = -\beta U_\infty = \text{Re} \int_{\text{upper plate}}^{\text{Mach circle}} \frac{1}{x}\, d\mathcal{U}. \tag{8.62}$$

It may be readily seen that by changing the limits, *Mach circle—left-hand wing* or *Mach circle—lower plate*, one automatically obtains $w = \alpha U_\infty$ or $v = \beta U_\infty$.

Let us now return to the integral (8.61); we can write successively (Fig. 8.7),

$$\alpha U_\infty = -\text{Re}\left[\frac{i\sqrt{1-B^2x^2}}{x}\,\mathcal{U}\Big|_{\text{right-hand wing}}^{\text{Mach circle}} - i\int_{\text{right-hand wing}}^{\text{Mach circle}} \mathcal{U}\, d\left(\frac{\sqrt{1-B^2x^2}}{x}\right)\right]$$

$$= \text{Re}\int_{AA_1A'} \frac{A_{10}+A_{12}x^2}{\sqrt{(x^2-l^2)(x^2+\mathfrak{h}^2)}}\,\frac{dx}{x^2\sqrt{1-B^2x^2}}. \tag{8.63}$$

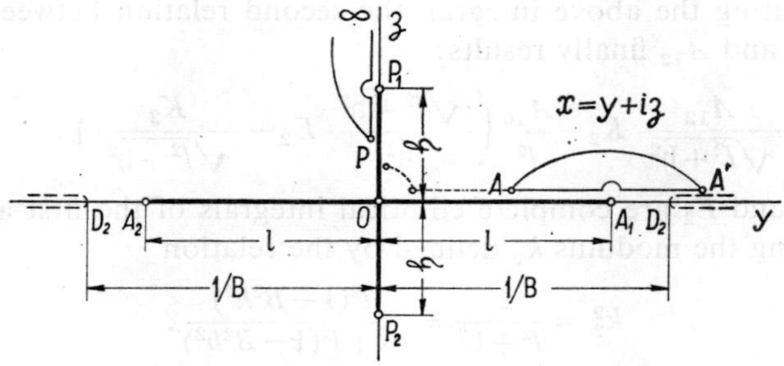

FIG. 8.7

Proceeding in the same way as in (§ II. 6), similarly to formula (2.114a), we can write

$$\alpha U_\infty = \frac{1}{B}\int_l^{1/B} \frac{(A_{10}+A_{12}y^2)\,dy}{y^2\sqrt{(y^2-l^2)(y^2+\mathfrak{h}^2)(1/B^2-y^2)}}\,. \tag{8.64}$$

Further employing the transformation

$$y^2 = \frac{l^2}{1-(1-B^2l^2)\sin^2\varphi}\,, \tag{8.65}$$

and performing the computations, a first relation between the constants A_{10} and A_{12} is obtained,

$$\alpha U_\infty = \frac{A_{12}}{\sqrt{l^2+\mathfrak{h}^2}}K_1 + \frac{A_{10}}{\mathfrak{h}^2}\left[\frac{\sqrt{l^2+\mathfrak{h}^2}}{l^2}E_1 - \frac{K_1}{\sqrt{l^2+\mathfrak{h}^2}}\right], \tag{8.66}$$

where K_1 and E_1 are complete elliptical integrals of the first and second kind, having the modulus k_1 defined by the relation

$$k_1^2 = \frac{\mathfrak{h}^2(1-B^2l^2)}{l^2+\mathfrak{h}^2} = \frac{h^2(1-B^2l^2)}{l^2+h^2(1-B^2l^2)}\,. \tag{8.67}$$

The second formula (8.62) may be written successively as

$$\beta U_\infty = -\operatorname{Re}\int_{\text{upper plate}}^{\text{Mach circle}} \frac{d\mathcal{U}}{x} = -\operatorname{Re}\left[\frac{\mathcal{U}}{x}\Big|_{\text{upper plate}}^{\text{Mach circle}} + \int_{\text{upper plate}}^{\text{Mach circle}} \frac{\mathcal{U}dx}{x^2}\right]$$

$$= \operatorname{Im}\int_{PP_1\infty} \frac{A_{10}+A_{12}x^2}{\sqrt{(x^2-l^2)(x^2+\mathfrak{h}^2)}}\frac{dx}{x^2} = \int_{\mathfrak{h}}^{\infty} \frac{A_{12}\mathfrak{z}^2-A_{10}}{\sqrt{(\mathfrak{z}^2+l^2)(\mathfrak{z}^2-\mathfrak{h}^2)}}\frac{d\mathfrak{z}}{\mathfrak{z}^2}\,. \tag{8.68}$$

Further employing the transformation

$$\mathfrak{z}^2 = \frac{\mathfrak{h}^2}{\cos^2\varphi} = \frac{h^2}{(1-B^2h^2)\cos^2\varphi}\,, \tag{8.69}$$

and evaluating the above integral, the second relation between the constants A_{10} and A_{12} finally results:

$$-\beta U_\infty = \frac{A_{12}}{\sqrt{l^2+\mathfrak{h}^2}} K_2 - \frac{A_{10}}{l^2}\left(\frac{\sqrt{l^2+\mathfrak{h}^2}}{\mathfrak{h}^2} E_2 - \frac{K_2}{\sqrt{l^2+\mathfrak{h}^2}}\right), \tag{8.70}$$

where K_2 and E_2 are complete elliptical integrals of the first and second kind, having the modulus k_2 defined by the relation

$$k_2^2 = \frac{l^2}{l^2+\mathfrak{h}^2} = \frac{l^2(1-B^2h^2)}{h^2+l^2(1-B^2h^2)}. \tag{8.71}$$

We obtain A_{10} and A_{12} from the two equations (8.65) and (8.70):

$$\frac{A_{10}}{U_\infty} = \frac{l^2\mathfrak{h}^2(\alpha K_2+\beta K_1)}{\sqrt{l^2+\mathfrak{h}^2}(K_1E_2+K_2E_1-K_1K_2)}, \tag{8.72a}$$

$$\frac{A_{12}}{U_\infty} = \frac{-\beta[(l^2+\mathfrak{h}^2)E_1-l^2K_1]+\alpha[(l^2+\mathfrak{h}^2)E_2-\mathfrak{h}^2K_2]}{\sqrt{l^2+\mathfrak{h}^2}(K_1E_2+K_2E_1-K_1K_2)}. \tag{8.72b}$$

In order to find the velocity on the wing we put $x = y$ in expression (8.60),

$$u_w = \frac{A_{10}+A_{12}y^2}{\sqrt{(l^2-y^2)(\mathfrak{h}^2+y^2)}}; \tag{8.73}$$

for the velocity on the plate we put $x = i\mathfrak{z}$, where $\mathfrak{z}$ will be replaced by its expression as a function of z (8.18b),

$$u_p = \frac{A_{10}-A_{12}\mathfrak{z}^2}{\sqrt{(l^2+\mathfrak{z}^2)(\mathfrak{h}^2-\mathfrak{z}^2)}} = \frac{A'_{10}+A'_{12}z^2}{\sqrt{(h^2-z^2)(\lambda^2+z^2)}}, \tag{8.74}$$

where the constants A'_{10} and A'_{12} are given by the following relations:

$$A'_{10} = A_{10}\sqrt{\frac{1-B^2h^2}{1-B^2l^2}}, \quad A'_{12} = -(A_{12}+B^2A_{10})\sqrt{\frac{1-B^2h^2}{1-B^2l^2}}, \tag{8.75}$$

while λ is obtained as a function of l, by a relation similar to (8.19),

$$\lambda = \frac{l}{\sqrt{1-B^2l^2}}. \tag{8.76}$$

VIII. 3.2. Calculation of Aerodynamic Coefficients

WAVE DRAG COEFFICIENT. Denoting by C_{dw} the wave drag coefficient of the wing, by applying formula (3.38a) and integrating over half a wing, we can write

$$C_{dw} = \frac{4\alpha}{lU_\infty}\int_0^l u\,dy = \frac{4\alpha}{l\sqrt{l^2+\mathfrak{h}^2}U_\infty}[A_{10}K_2+A_{12}((l^2+\mathfrak{h}^2)E_2-\mathfrak{h}^2K_2)], \tag{8.77}$$

where we have considered half of the wing surface area, viz. $\frac{1}{2}l$, as reference surface.

Likewise, denoting by C_{dp} the drag coefficient of the plate, taking account of (8.74) and integrating over half a plate, we can write

$$C_{dp} = \frac{4\beta}{hU_\infty}\int_0^h u_p\,dz = \frac{4\beta}{hU_\infty\sqrt{h^2+\lambda^2}}\,[A'_{10}K_1+A'_{12}((h^2+\lambda^2)E_1-\lambda^2K_1)], \tag{8.78}$$

where we have considered half of the plate surface area, viz. $\frac{1}{2}h$, as reference surface.

If the cruciform wing has all its arms equal ($l = h$) and the same antisymmetrical incidence ($-\beta = \alpha$), we shall have successively:

$$A_{10} = 0, \quad A_{12} = \frac{\alpha U_\infty l}{K(k)}\sqrt{\frac{2-B^2l^2}{1-B^2l^2}}, \quad k^2 = \frac{1-B^2l^2}{2-B^2l^2}, \tag{8.79a}$$

$$C_d = C_{dw} = C_{dp} = \frac{4\alpha^2 l}{K(k)}\,\frac{(2-B^2l^2)E(k)-K(k)}{1-B^2l^2}. \tag{8.79b}$$

Suction Force Coefficient. We shall proceed from the expression for $\mathcal{U}$ and apply formula (3.104) for calculating the suction force. Thus we can write the following expression for the two subsonic leading edges of the wing:

$$\sigma_w = -\frac{\pi\sqrt{1-B^2l^2}}{lS_wU_\infty^2}\left[\frac{(A_{10}+A_{12}y^2)^2}{(l^2-y^2)(\mathfrak{h}^2+y^2)}\,(l-y)\right]_{y\to l}$$

$$= -\frac{\pi\sqrt{1-B^2l^2}}{l^3(l^2+\mathfrak{h}^2)U_\infty^2}\,(A_{10}+A_{12}l^2)^2; \tag{8.80}$$

and for the subsonic leading edges of the plate, by taking account of (8.74), we can likewise write

$$\sigma_p = -\frac{\pi\sqrt{1-B^2h^2}}{hS_pU_\infty^2}\left[\frac{(A'_{10}+A'_{12}z^2)^2}{(h^2-z^2)(\lambda^2+z^2)}\,(h-z)\right]_{z\to h}$$

$$= -\frac{\pi\sqrt{1-B^2h^2}}{h^3(h^2+\lambda^2)U_\infty^2}\,(A'_{10}+A'_{12}h^2)^2. \tag{8.81}$$

If $l = h$, $\alpha = -\beta$ then, for the entire wing or the entire plate, we have

$$\sigma_w = \sigma_p = -\frac{\pi\alpha^2}{K(k)^2}\,l\sqrt{1-B^2l^2}. \tag{8.82}$$

Rolling Moment Coefficient. Denoting by μ_w the rolling moment coefficient of half a wing, applying formula (3.45) and designating by H a reference length that may even be considered equal to unity ($H = 1$), we an write successively

$$\tfrac{1}{2}\, lH\mu_w = \frac{4}{3U_\infty}\int_0^l u_w y\,dy$$
$$= \frac{2}{3U_\infty}\left[A_{12}l\mathfrak{h} + \left(A_{12}\frac{l^2-\mathfrak{h}^2}{2}+A_{10}\right)\cos^{-1}\frac{\mathfrak{h}^2-l^2}{\mathfrak{h}^2+l^2}\right]. \tag{8.83}$$

Likewise, denoting by μ_p the rolling moment coefficient of half the plate, we can write

$$\tfrac{1}{2}\, hH\mu_p = \frac{-4}{3U_\infty}\int_0^h u_p z\,dz$$
$$= \frac{-2}{3U_\infty}\left[A'_{12}h\lambda + \left(A'_{12}\frac{h^2-\lambda^2}{2}+A'_{10}\right)\cos^{-1}\frac{\lambda^2-h^2}{\lambda^2+h^2}\right]. \tag{8.84}$$

If $l = h$, $\alpha = -\beta$, then, taking also account of (8.79a), expression (8.83) becomes

$$H\mu = \tfrac{4}{3}\frac{\alpha l}{K}\sqrt{2-B^2l^2}\left[\frac{1}{1-B^2l^2}-\frac{B^2l^2}{2(1-B^2l^2)^{3/2}}\cos^{-1}\frac{B^2l^2}{2-B^2l^2}\right]. \tag{8.85}$$

VIII. 4. Cruciform Wing with Antisymmetrical Incidences in the Conical Motion of Order $n = 2$

VIII. 4.1. Determination of Constants in the Function F_{200}

Let us consider that the variation of the velocity components w and v on the wing and plate is

$$w = w_{01}x_2 = x_1 w_{01} y, \tag{8.86a}$$

$$v = v_{01}x_3 = x_1 v_{01} z, \tag{8.86b}$$

respectively.

In such a case, formula (8.26) which yields the expression of u on the wing, may be written in the form

$$\frac{u}{x_1} = \mathrm{Re}\left(F_{200} - x\int_{\text{Mach circle}}^{\text{wing}}\frac{dF_{200}}{x} = \mathrm{Re}\, x\int_{\text{Mach circle}}^{\text{wing}} F_{200}\, d\left(\frac{1}{x}\right)\right). \tag{8.87}$$

In the expression of F_{200} we have the contribution of the leading edges of the plate and wing, such that, by employing formula (8.56) (which also takes account of the antisymmetry), we can write the following expression:

$$F_{200} = \frac{A_0+A_2X^2+A_4X^4+A_6X^6}{X^3(X^2-L^2)\sqrt{L^2-X^2}}. \tag{8.88}$$

Returning to the x plane, we find the expression

$$F_{200} = i\,\frac{a_0+a_2x^2+a_4x^4+a_6x^6}{(x^2+\mathfrak{h}^2)(x^2-l^2)\sqrt{(x^2-l^2)(x^2+\mathfrak{h}^2)}}. \tag{8.89}$$

In applying relation (8.87) we have to perform the following integral:

$$\frac{u}{x_1} = \operatorname{Im} x\int_{\text{Mach circle}}^{\text{wing}} \frac{a_0+a_2x^2+a_4x^4+a_6x^6}{x^2(x^2+\mathfrak{h}^2)(x^2-l^2)\sqrt{(x^2-l^2)(x^2+\mathfrak{h}^2)}}\,dx. \tag{8.90}$$

We separate the rational expression in x^2 into simple fractions and integrate, starting from any point on the wing in the first quadrant, up to any point on the Mach circle. By performing the computations, we finally obtain the axial velocity u on the wing:

$$u_{1w} = \frac{u}{x_1} = \frac{A_{20}+A_{22}y^2}{\sqrt{(l^2-y^2)(\mathfrak{h}^2+y^2)}} + C_{21}y(F(\varphi,k)-K(k))+C_{22}y(E(\varphi,k)-E(k)), \tag{8.91}$$

where A_{20}, A_{22}, C_{21}, C_{22} are constants, $K(k)$ and $E(k)$ are complete elliptical integrals of the first and second kind, while $F(\varphi, k)$ and $E(\varphi, k)$ are the same elliptical, although incomplete, integrals, where we have to use the following:

$$F(\varphi,k)-K(k) = F(\varphi',k),\quad \varphi' = \sin^{-1}\sqrt{\frac{l^2-y^2}{l^2}},\quad \left(0 \leqslant \varphi' \leqslant \frac{\pi}{2}\right), \tag{8.92a}$$

$$k = \frac{l}{\sqrt{l^2+\mathfrak{h}^2}},\quad \varphi = \sin^{-1}\sqrt{\frac{y^2(l^2+\mathfrak{h}^2)}{l^2(y^2+\mathfrak{h}^2)}},\quad \left(0 \leqslant \varphi \leqslant \frac{\pi}{2}\right). \tag{8.92b}$$

For determining the constants, we impose the following conditions deduced from relations (8.28) and (8.29):

$$\Phi_{101} = w_{10} = 0,\quad \Phi_{011} = w_{01} \tag{8.93}$$

on the right-hand portion of the wing and, respectively,

$$\Phi_{110} = v_{10} = 0,\quad \Phi_{011} = v_{01} \tag{8.94}$$

on the upper portion of the plate.

Further, formula (8.91) may be written in the form

$$\Phi_{100} = u = x_1 u_1 = x_1 \operatorname{Re} \mathcal{U}_1, \quad (\mathcal{U}_1 = u_1 + u_1'), \tag{8.95}$$

whence

$$\Phi_{200} = \frac{\partial \Phi_{100}}{\partial x_1} = u_1 + x_1 \frac{\partial u_1}{\partial x_1} = u_1 - y\frac{du_1}{dy}, \tag{8.96a}$$

$$F_{200} = \mathcal{U}_1 - x\frac{d\mathcal{U}_1}{dx}, \qquad \frac{dF_{200}}{dx} = -x\frac{d^2\mathcal{U}_1}{dx^2}. \tag{8.96b}$$

Now we have also to determine the constants A_{20}, A_{22}, C_{21}, C_{22} in the expression for u_1. For this purpose we apply the compatibility relations and take (8.93) and (8.94) into consideration; further examining the limits of integration on Fig. 8.7, we can write

$$\operatorname{Im}\int_{\text{Mach circle}}^{\text{wing}} \frac{\sqrt{1-B^2x^2}}{x}\, dF_{200} = \operatorname{Im}\int_{\text{Mach circle}}^{\text{wing}} F_{200}\, d\left(\frac{\sqrt{1-B^2x^2}}{x}\right) = 0, \tag{8.97a}$$

$$-\operatorname{Im}\int_{\text{Mach circle}}^{\text{wing}} \frac{\sqrt{1-B^2x^2}}{x^2}\, dF_{200}$$
$$= \operatorname{Im}\int_{\text{Mach circle}}^{\text{wing}} F_{200}\, d\left(\frac{\sqrt{1-B^2x^2}}{x^2}\right) = w_{01}, \tag{8.97b}$$

$$\operatorname{Re}\int_{\text{Mach circle}}^{\text{plate}} \frac{1}{x}\, d\,F_{200} = \operatorname{Re}\left(\frac{d\mathcal{U}_1}{dx}\right)_{x=0} = 0, \tag{8.97c}$$

$$-\operatorname{Im}\int_{\text{Mach circle}}^{\text{plate}} \frac{\sqrt{1-B^2x^2}}{x^2}\, dF_{200}$$
$$= -\operatorname{Im}\int_{\text{Mach circle}}^{\text{wing}} - \operatorname{Im}\int_{\text{wing}}^{\text{plate}} = w_{01} - \operatorname{Im}\oint_{\gamma} \sqrt{\frac{1-B^2x^2}{x^2}}\, dF_{200} = v_{01}, \tag{8.97d}$$

where γ represents the quarter of a circle around the origin (Fig. 8.7).

Replacing dF_{200} or F_{200} in the above integrals and proceeding in the same way as in the preceding case in order to perform them, we obtain the required constants in the form of the solutions of the system of equations:

$$A_{20}I_{10} + A_{22}I_{12} + C_{21}J_{11} + C_{22}J_{12} = 0, \tag{8.98a}$$

$$A_{20}I_{20} + A_{22}I_{22} + C_{21}J_{21} + C_{22}J_{22} = w_{01}, \tag{8.98b}$$

$$C_{21}K(k)+C_{22}E(k) = 0, \quad \left(k = \frac{l}{\sqrt{l^2+\mathfrak{h}^2}}\right), \tag{8.98c}$$

$$(l^2-\mathfrak{h}^2)A_{20}-2l^2\mathfrak{h}^2A_{22}-2l^2\mathfrak{h}^2\sqrt{l^2+\mathfrak{h}^2}\,(C_{21}+C_{22}) = \frac{2}{\pi}\,l^3\mathfrak{h}^3(w_{01}-v_{01}). \tag{8.98d}$$

In the above equations we have introduced the following notations:

$$I_{10} = B^2[B^2(l^2+h^2-B^2l^2h^2)E_1-K_1], \tag{8.99a}$$

$$I_{12} = B^2[(l^2+h^2-B^2l^2h^2)E_1-l^2K_1], \tag{8.99b}$$

$$J_{11} = (1-B^2l^2)\sqrt{l^2+\mathfrak{h}^2}\,K_1, \tag{8.99c}$$

$$J_{12} = (1-B^2l^2)\sqrt{l^2+\mathfrak{h}^2}\,[K_1-(1-B^2h^2)E_1], \tag{8.99d}$$

$$I_{20} = \frac{\sqrt{1-B^2h^2}}{l^4\mathfrak{h}^4(1-B^2l^2)}\{l^2[(1-B^2l^2)(2\mathfrak{h}^2-l^2+B^2\mathfrak{h}^4)+B^2\mathfrak{h}^4)]K_2 -l^2\mathfrak{h}^2[3(1-B^2l^2)(1+B^2\mathfrak{h}^2)+B^4l^2\mathfrak{h}^2]E_2 +(1-B^2l^2)(l^4-\mathfrak{h}^4)\Pi(\varrho_2, k_2), \tag{8.99e}$$

$$I_{22} = \frac{\sqrt{1-B^2h^2}}{l^2\mathfrak{h}^2(1-B^2l^2)}\{l^2[2(1-B^2l^2)+B^2\mathfrak{h}^2]K_2 -B^2l^2\mathfrak{h}^2E_2-2(1-B^2l^2)(l^2+\mathfrak{h}^2)\Pi(\varrho_2, k_2)\}, \tag{8.99f}$$

$$J_{21} = \frac{\sqrt{1-B^2h^2}}{l^2\mathfrak{h}^2}\sqrt{l^2+\mathfrak{h}^2}\,[l^2(2+B^2\mathfrak{h}^2)K_2-2(l^2+\mathfrak{h}^2)\Pi(\varrho_2, k_2)], \tag{8.99g}$$

$$J_{22} = \frac{\sqrt{1-B^2h^2}}{l^2\mathfrak{h}^2\sqrt{l^2+\mathfrak{h}^2}}[2l^2(l^2+\mathfrak{h}^2)K_2+l^2\mathfrak{h}^2(2+B^2\mathfrak{h}^2)E_2 -2(l^2+\mathfrak{h}^2)^2\Pi(\varrho_2, k_2)], \tag{8.99h}$$

where K_i, E_i, $\Pi(\varrho_2, k_2)$ represent complete elliptical integrals of the first, second and third kind, having the modulus k_i and parameter ϱ_2 given by the following relations:

$$k_1 = h\sqrt{\frac{1-B^2l^2}{l^2+h^2-B^2l^2h^2}} \qquad k_2 = \sqrt{(1-B^2l^2)(1-B^2h^2)}, \tag{8.100a}$$

$$\varrho_2 = \frac{\mathfrak{h}^2(1-B^2l^2)}{l^2(1+B^2\mathfrak{h}^2)} = \frac{h^2}{l^2}(1-B^2l^2). \tag{8.100b}$$

VIII. 4.2. Calculation of Aerodynamic Characteristics

WAVE DRAG COEFFICIENT. By applying formula (3.38a) and integrating over half a wing, we obtain for the drag coefficient,

$$C_{dw} = -\frac{2w_{01}}{lU_\infty^2}\int_0^l u_{1w}y\,dy$$

$$= -\frac{w_{01}}{lU_\infty^2}\left\{A_{22}l\mathfrak{h}+\left(A_{22}\frac{l^2-\mathfrak{h}^2}{2}+A_{20}\right)\cos^{-1}\frac{\mathfrak{h}^2-l^2}{\mathfrak{h}^2+l^2}\right.$$

$$+\frac{1}{3\sqrt{l^2+\mathfrak{h}^2}}\left[(2C_{22}\mathfrak{h}^2-C_{21}(l^2+\mathfrak{h}^2))\left(l\mathfrak{h}+\frac{l^2-\mathfrak{h}^2}{2}\cos^{-1}\frac{\mathfrak{h}^2-l^2}{\mathfrak{h}^2+l^2}\right)\right.$$

$$\left.\left.-2C_{22}\mathfrak{h}^2l^2\cos^{-1}\frac{\mathfrak{h}^2-l^2}{\mathfrak{h}^2+l^2}\right]\right\}. \tag{8.101}$$

Likewise, we find the same expression as above for the drag coefficient of the plate, where, however, l will be replaced by h, $\mathfrak{h}$ by λ, w_{01} by $-v_{01}$; this substitution should also take place in the expression of the constants in formulae (8.98), which will be denoted now by A'_{20}, A'_{22}, C'_{21}, C'_{22}, in order to be distinguished from the former ones.

For the same reason, we shall denote by K'_i, E'_i, $\Pi(\varrho'_2, k'_2)$ the complete elliptical integrals of the first, second and third kind, and by k'_i, ϱ'_2 the respective modulus and parameter ($i = 1, 2$). Performing these interchanges in formulae (8.99, 8.100) and introducing the results obtained into (8.98), we get the values of the required constants.

Note. If the coefficients of the velocities w and v are $w_{01} = -v_{01}$, the above expressions are obviously simpler. This case corresponds to a rotation ω of the cruciform wing about its axis, where we have $w_{01} = -v_{01} = -\omega$, and therefore the constants that result from this condition.

Thus, for instance, if a rocket provided with a cruciform wing is in uniform rotation, then the forces and moments calculated above are exerted on its planes (wing and plate).

If, in addition, there occurs a deformation of the planes, such that the resulting velocities be equal to those of expressions (8.86), then the two motions are simply added and we are confronted with the same problem that we have been treating above where however the coefficients of the velocities are $w'_{01} = w_{01}-\omega$, $v'_{01} = v'_{01}+\omega$.

SUCTION COEFFICIENT. Applying formula (3.104) and taking account of the axial velocity expression (8.91), we can write for the suction coefficient of both the subsonic leading edges of the wing:

$$\sigma_w = -\frac{\pi\sqrt{1-B^2l^2}}{2lS_wU_\infty^2}\left[u_{1w}^2(l-y)\right]_{y\to l} = -\frac{\pi\sqrt{1-B^2l^2}}{2l^3(l^2+\mathfrak{h}^2)U_\infty}(A_{20}+A_{22}l^2)^2, \tag{8.102}$$

where S_w is the half of the wing surface ($S_w = \frac{1}{2}l$).

A similar formula is obtained for the plate, where we have considered the plate surface as reference length ($S_p = h$):

$$\sigma_p = -\frac{\pi\sqrt{1-B^2h^2}}{2hS_pU_\infty^2}\left[u_{1p}^2(h-z)\right]_{z\to h} = -\frac{\pi\sqrt{1-B^2h^2}}{2h^3(h^2+\lambda^2)}(A'_{20}+A'_{22}h^2)^2, \tag{8.103}$$

and where the constants A'_{20} and A'_{22} are the ones obtained by inverting the system, as mentioned above.

If $l = h$, $w_{01} = -v_{01}$, we obtain

$$\sigma = -\frac{\pi(1-B^2l^2)^{3/2}}{2l^5(2-B^2l^2)U_\infty^2}(A_{20}^*+A_{22}^*l^2)^2, \tag{8.104}$$

where the constants A_{20}^* and A_{22}^* are the ones corresponding to this particular case, obtained by putting $h = l$, $w_{01} = -v_{01}$ in all formulae that yield I_{ij} and I_{ij} (8.99).

ROLLING MOMENT COEFFICIENT. As in the preceding case, denoting by H a reference length and applying the formula for the wing rolling moment coefficient (3.45), we note that an expression similar to the drag coefficient is obtained:

$$H\mu_w = \frac{2}{lU_\infty}\int_0^l u_{1w}y\,dy = -\frac{U_\infty}{w_{01}}C_{dw}. \tag{8.105}$$

For the plate we obtain analogously

$$H\mu_p = \frac{U_\infty}{v_{01}}C_{dp}, \tag{8.106}$$

obviously taking account of the fact that the reference surface areas are those of the wing or plate, respectively.

VIII. 5. Thin Cruciform Wing with Antisymmetrical Incidences having Subsonic Plate Edges and Supersonic Wing Edges

VIII. 5.1. Determination of the Axial Velocity Function

Since the wing divides the space into two independent parts, we are going to treat the upper part, for which we shall find the axial velocity u on the wing and plate. On the lower part the results will be equal and of opposite sense.

We shall treat the motion of order $n = 1$.

The solution to this problem is given by (8.58), where we put $n = 1$ and, at the same time, $S_1 = S_2 = L$:

$$F_{100} = \mathcal{U} = \frac{A_0}{X}\sqrt{\frac{1-\mathcal{B}^2X^2}{B^2}} + K_{10}\left(\cos^{-1}\sqrt{\frac{(1+\mathcal{B}L)(1-\mathcal{B}X)}{2\mathcal{B}(L-X)}} - \cos^{-1}\sqrt{\frac{(1+\mathcal{B}L)(1+\mathcal{B}X)}{2\mathcal{B}(L+X)}}\right), \tag{8.107}$$

where we have replaced $C_1\cosh^{-1}$ by the equivalent expression $K_1\cos^{-1}$.

Evaluating the difference in the bracketed terms and passing to the x plane, we can finally write

$$F_{100} = \mathcal{U} = A_{10}\sqrt{\frac{1-B^2x^2}{\mathfrak{h}^2+x^2}} + K_{10}\cos^{-1}\sqrt{\frac{(l+\mathfrak{h}^2)(1-B^2x^2)}{(1+B^2\mathfrak{h}^2)(l^2-x^2)}}, \tag{8.108}$$

where the constant K_{10} is identical with the one corresponding to the wing, since the leading edge constitutes a jump, whence, according to the theory established previously (4.43), the following value results:

$$K_{10} = \frac{2\alpha U_\infty}{\pi B}\frac{Bl}{\sqrt{B^2l^2-1}}. \tag{8.109}$$

The constant A_{10} will be determined in the usual way, by putting

$$-\beta U_\infty = \text{Re}\int_{\text{upper plate}}^{\text{Mach circle}} \frac{1}{x}\,d\mathcal{U}, \tag{8.110}$$

whence, after evaluating this integral,

$$A_{10} = h^2U_\infty \frac{2\alpha[(l^2+\mathfrak{h}^2)K(k)-\mathfrak{h}^2\Pi(\varrho, k)]+\pi\beta l\sqrt{(l^2+\mathfrak{h}^2)(1+B^2\mathfrak{h}^2)}}{\pi l\sqrt{l^2+{}^2\mathfrak{h}}E(k)}, \tag{8.111}$$

where $K(k), E(k), \Pi(\varrho, k)$ are complete elliptical integrals of the first, second and third kind, having the modulus k and the parameter ϱ, given by the relations

$$k = \sqrt{1-B^2h^2}, \quad \varrho = -\frac{l^2}{l^2+\mathfrak{h}^2} = -\frac{l^2(1-B^2h^2)}{l^2+h^2-B^2l^2h^2}. \qquad (8.112)$$

VIII. 5.2. *Aerodynamic Characteristics*

WAVE DRAG COEFFICIENT. Integrating over half a wing, taking hence $S_w = \frac{1}{2}l$ as reference surface area, we find the following expression for the respective drag coefficient:

$$C_{dw} = \frac{4\alpha}{lU_\infty}\int_0^l u\,dy$$

$$= \frac{4\alpha}{lU_\infty}\left[A_{10}(K(k)-E(k))+K_{10}h^2\Pi(\varrho, k)\sqrt{\frac{B^2l^2-1}{l^2+h^2-B^2l^2h^2}}\right], \qquad (8.113)$$

where $K(k), E(k), \Pi(\varrho, k)$ are the elliptical integrals of the first, second and third kind, having as modulus and parameter k and ϱ, respectively, given by expressions (8.112).

For calculating the plate drag coefficient, we shall first perform a change of variable in the formula for the axial velocity, replacing $\mathfrak{z}$ by its expression in z previously given (8.18b); indeed, putting $x = i\mathfrak{z}$ in $\mathcal{U}$ (8.108), we have

$$u = \mathcal{U}_{x=i\mathfrak{z}} = A_{10}\sqrt{\frac{1+B^2\mathfrak{z}^2}{\mathfrak{h}^2-\mathfrak{z}^2}}+K_{10}\cos^{-1}\sqrt{\frac{(l^2+\mathfrak{h}^2)(1+B^2\mathfrak{z}^2)}{(1+B^2\mathfrak{h}^2)(l^2+\mathfrak{z}^2)}}$$

$$= \frac{A_{10}\sqrt{1-B^2h^2}}{\sqrt{h^2-z^2}}+K_{10}\cos^{-1}\sqrt{\frac{\lambda^2-h^2}{\lambda^2-z^2}}, \quad \left(\lambda = \frac{l}{\sqrt{B^2l^2-1}}\right). \qquad (8.114)$$

Applying now the usual formula (3.38a) and integrating over half a plate (having the surface area) $S_p = \frac{1}{2}h$), we find successively

$$C_{dp} = \frac{4\beta}{hU_\infty}\int_0^h u\,dz$$

$$= \frac{2\pi\beta}{hU_\infty}\left[A_{10}\sqrt{1-B^2h^2}-K_{10}(\lambda+\sqrt{\lambda^2-h^2})\right], \qquad (8.115)$$

where λ is the same as above.

SUCTION COEFFICIENT. Applying the general formula (3.104) to the subsonic leading edges of the plate, and taking into account the expression for the axial velocity (8.108), we obtain the following formula:

$$\sigma_p = -\frac{\pi\sqrt{1-B^2h^2}}{hS_pU_\infty^2}\,[u^2(h-z)]_{z\to h} = -\frac{\pi(1-B^2h^2)^{3/2}}{h^3U_\infty^2}\,A_{10}^2. \tag{8.116}$$

ROLLING MOMENT COEFFICIENT. Let us take H as reference length and integrate over half a wing, by applying formula (3.45); we can write

$$\begin{aligned} H\mu_w &= \frac{8}{3lU_\infty}\int_0^l uy\,dy \\ &= -\frac{2}{3BlU_\infty}\left\{A_{10}\left[2B\mathfrak{h}-(B^2\mathfrak{h}^2+1)\cos^{-1}\frac{B^2\mathfrak{h}^2-1}{B^2\mathfrak{h}^2+1}\right]\right. \\ &\quad -K_{10}\left[Bl^2\left(\pi+2\cos^{-1}\frac{\mathfrak{h}}{l}\sqrt{\frac{B^2l^2-1}{B^2\mathfrak{h}^2+1}}\right)\right. \\ &\quad \left.\left.+\sqrt{(l^2+\mathfrak{h}^2)(B^2l^2-1)}\cos^{-1}\frac{B^2\mathfrak{h}^2-1}{B^2\mathfrak{h}^2+1}\right)\right]. \end{aligned} \tag{8.117}$$

For the moment coefficient corresponding to half of the plate we take account of the expression for u as a function of z given above (8.114), and by taking H as reference length we can write

$$-H\mu_p = \frac{8}{3hU_\infty}\int_0^h uz\,dz$$

$$= \frac{4}{3hU_\infty}\left[K_{10}\lambda^2\cos^{-1}\sqrt{\frac{\lambda^2-h^2}{\lambda^2}}+h\left(2A_{10}\sqrt{1-B^2h^2}-K_{10}\sqrt{\lambda^2-h^2}\right)\right]. \tag{8.118}$$

Note. If the entire wing has the same incidence, then the plus sign will be put in the bracket of the right-hand side of expression (8.107), in order to ensure the symmetry of the motion. In such a case, however, the motions round a symmetrical wing with supersonic edges and round a plate with subsonic edges may be treated separately, as in § VIII. 2.2.

VIII. 5.3. Case of a Wing of Symmetrical Thickness

Let it be assumed that the wing, with subsonic edges (of symmetrical planform), is of symmetrical thickness, and that the thin plate has an incidence β^*, the same on both sides, and an incidence β, antisymmetrical on the two sides. The general motion about this system is decomposed into three separate motions:

1. A motion around the delta wing of symmetrical thickness, having slope γ.
2. A motion around the plate, due to the incidence β^* and symmetrical with respect to the wing axis, which will be given in the x^* plane (Fig. 8.4).
3. A motion similar to (8.60), having the antisymmetrical incidences: $\alpha = -\alpha = 0$ on the wing and $\beta, -\beta$ on the plate.

We can write respectively:

$$\mathcal{U}' = C_{10}\left[\cosh^{-1}\sqrt{\frac{(1+Bl)(1-Bx)}{2B(l-x)}}+\cosh^{-1}\sqrt{\frac{(1+Bl)(1+Bx)}{2B(l+x)}}\right], \tag{8.119a}$$

$$\mathcal{U}^* = \frac{A_{10}^*}{\sqrt{h^2-x^{*2}}}, \tag{8.119b}$$

$$\mathcal{U}'' = \frac{A_{10}+A_{12}x^2}{\sqrt{(l^2-x^2)(\mathfrak{h}^2+x^2)}}, \tag{8.119c}$$

where the constant C_{10} follows from (4.53, 4.43), making it real, however, and the constant A_{10}^* results from (4.16), replacing l by h and α by β^*:

$$C_{10} = -\frac{2\gamma U_\infty}{\pi B}\frac{Bl}{\sqrt{1-B^2l^2}}, \quad A_{10}^* = -\frac{\beta^* U_\infty h^2}{E(k)}, \quad (k = \sqrt{1-B^2h^2}). \tag{8.120}$$

As far as the coefficients A_{10} and A_{12} are concerned, they are deduced from (8.72), by making $\alpha = 0$:

$$\frac{A_{10}}{\beta U_\infty} = \frac{l^2\mathfrak{h}^2K_1}{\sqrt{l^2+\mathfrak{h}^2}(K_1E_2+K_2E_1-K_1K_2)},$$
$$\frac{A_{12}}{\beta U_\infty} = -\frac{(l^2+\mathfrak{h}^2)E_1-l^2K_1}{\sqrt{l^2+\mathfrak{h}^2}(K_1E_2+K_2E_1-K_1K_2)}. \tag{8.121}$$

The calculation of pressures presents no difficulty: besides the direct pressures, the reciprocal effects of the three motions will also be added on the wing or plate.

If the wing of symmetrical thickness has unequal arms, then (8.119a) will be written as

$$\mathcal{U}' = C_{10}\cosh^{-1}\sqrt{\frac{(1+\mathcal{B}L_1)(1-\mathcal{B}X)}{2\mathcal{B}(L_1-X)}}+G_{10}\cosh^{-1}\sqrt{\frac{(1+\mathcal{B}L_2)(1+\mathcal{B}X)}{2\mathcal{B}(L_2+X)}}, \tag{8.122}$$

L_1 and L_2 being the leading edges in the X plane, C_{10} and G_{10} being given by the following relations which are deduced by taking the semi-residue around the points $x = l_1$ and $x = -l_2$ (corresponding to $X = L_1$ and $X = -L_2$):

$$C_{10} = -\frac{2\gamma U_\infty}{\pi B}\frac{Bl_1}{\sqrt{B^2 l_1^2 - 1}}, \quad G_{10} = -\frac{2\gamma U_\infty}{\pi B}\frac{Bl_2}{\sqrt{B^2 l_2^2 - 1}}. \tag{8.123}$$

For the second motion $\mathcal{U}^*$ remains the same as in (8.119b), and instead of the third motion (8.119c) we have a general expression for $\mathcal{U}''$,

$$\mathcal{U}'' = \frac{A_0 + A_1 X + A_2 X^2}{X\sqrt{(L_1 - X)(L_2 + X)}}, \tag{8.124}$$

where the constants are to be determined.

The calculation of drag or eventually of the yawing moment involves no difficulty.

Note: When both the wing and plate are of symmetrical thickness, then the following positions occur:

(a) When the wing planform is symmetrical, the wing and plate are treated separately and the effect due to the reciprocal influence is added.

(b) When the wing is asymmetrical, the wing and the plane (thin) frame of the plate are treated separately, as in the above example, while, on the other hand, the plate of symmetrical thickness is also treated separately and then the effect of the reciprocal influence is added.

(c) When the upper and lower sides of the plate of symmetrical thickness have equal incidences of opposite sense, the thin cruciform wing made up of the plane frames and the cruciform wing of symmetrical thickness are treated separately, the effect of the reciprocal action being added.

VIII. 5.4. Determination of the Function F_{n00} in the Case of Conical Flow of Order 2

In the case when the wing has subsonic edges, the problem has been completely treated (§ VIII. 4).

In the case of supersonic edges, we shall apply formula (8.58), putting $S_1 = S_2 = L$, and replace $C_{20} \cosh^{-1}$ by $K_{20} \cos^{-1}$; thus we get:

$$F_{200} = K_{20} \left[\cos^{-1} \sqrt{\frac{(1+\mathcal{B}L)(1-\mathcal{B}X)}{2\mathcal{B}(L-X)}} - \cos^{-1} \sqrt{\frac{(1+\mathcal{B}L)(1+\mathcal{B}X)}{2\mathcal{B}(L+X)}} \right] + \frac{A_0+A_2X^2+A_4X^4}{X^3(L^2-X^2)} \sqrt{1-\mathcal{B}^2X^2}, \tag{8.125}$$

which, after performing the subtraction in the bracket, becomes

$$F_{200} = K_{20} \cos^{-1} \sqrt{\frac{(l^2+\mathfrak{h}^2)(1-B^2x^2)}{(1+B^2\mathfrak{h}^2)(l^2-x^2)}} + \frac{a_0+a_2x^2+a_4x^4}{(x^2+\mathfrak{h}^2)(x^2-l^2)} \sqrt{\frac{1-B^2x^2}{B^2(x^2+\mathfrak{h}^2)}}, \tag{8.126}$$

in the x plane.

The axial velocities, u_w on the wing and u_p on the plate are deduced from the application of formulae (8.26): and (8.27):

$$\frac{u_w}{x_1} = \operatorname{Re} x \int_{\text{Mach circle}}^{\text{right-hand wing}} F_{200}\, d\left(\frac{1}{x}\right), \tag{8.127a}$$

$$\frac{u_p}{x_1} = \operatorname{Re} \frac{x}{\sqrt{1-B^2x^2}} \int_{\text{Mach circle}}^{\text{upper plate}} F_{200}\, d\left(\frac{\sqrt{1-B^2x^2}}{x}\right). \tag{8.127b}$$

Evaluating the integral in the first relation, we obtain the following expression for the velocity on the wing:

$$u_{1w} = \frac{u_w}{x_1} = K_{20} \cos^{-1} \sqrt{\frac{(l^2+\mathfrak{h}^2)(1-B^2y^2)}{(1+B^2\mathfrak{h}^2)(l^2-y^2)}} + A_{20} \sqrt{\frac{1-B^2y^2}{\mathfrak{h}^2+y^2}} + C_{21}y[F(\varphi, k)-K(k)] + C_{22}y[E(\varphi, k)-E(k)] - \frac{\pi}{2} B^2lK_{20}y + Gy[\Pi(\varphi, \varrho, k)-\Pi(\varrho, k)], \tag{8.128}$$

where $F(\varphi, k)$, $E(\varphi, k)$, $K(k)$, $E(k)$, $\Pi(\varphi, \varrho, k)$, $\Pi(\varrho, k)$ are the incomplete and complete elliptical integrals having the modulus k, the parameter ϱ and the argument φ; in this connection, we have to use the following relations:

$$F(\varphi, k)-K(k) = F(\varphi', k), \quad \Pi(\varphi, \varrho, k)-\Pi(\varrho, k) = -\frac{l^2}{\mathfrak{h}^2} \left[F(\varphi', k) + \frac{B^2(l^2+\mathfrak{h}^2)}{1-B^2l^2} \Pi(\varphi', \varrho', k) \right], \tag{8.129a}$$

$$k = \sqrt{1-B^2h^2}, \quad \varrho = -\frac{l^2+h^2-B^2l^2h^2}{l^2},$$

$$\varphi = \sin^{-1}\sqrt{\frac{(1+B^2\mathfrak{h}^2)y^2}{\mathfrak{h}^2+y^2}}. \tag{8.129b}$$

$$\varrho' = \frac{1}{B^2l^2-1}, \quad \varphi' = \sin^{-1}\sqrt{1-B^2y^2}. \tag{8.129c}$$

For determining the constants, proceeding in the same way as above, we apply the compatibility relations and find, respectively:

$$K_{20} = \frac{2w_{01}}{\pi}\frac{l^2}{(B^2l^2-1)^{3/2}}, \tag{8.130a}$$

$$A_{20} = \frac{\mathfrak{h}^2\sqrt{1+B^2\mathfrak{h}^2}}{N}\{2w_{01}B^2l^2\mathfrak{h}^2\Pi(\varrho,k)-[w_{01}(l^2+\mathfrak{h}^2)$$
$$+(w_{01}+v_{01})\mathfrak{h}(B^2l^2-1)\sqrt{l^2+\mathfrak{h}^2}][(1+B^2\mathfrak{h}^2)E(k)-B^2\mathfrak{h}^2K(k)]\}, \tag{8.130b}$$

$$C_{21} = \frac{B^2\mathfrak{h}^2(1+B^2\mathfrak{h}^2)}{N}\{w_{01}B^2l^2\mathfrak{h}^2\Pi(\varrho,k)-[w_{01}(l^2+\mathfrak{h}^2)$$
$$+(w_{01}+v_{01})\mathfrak{h}(B^2l^2-1)\sqrt{l^2+\mathfrak{h}^2}]E(k)$$
$$+w_{01}l^2[(B^2\mathfrak{h}^2-1)E(k)-B^2\mathfrak{h}^2K(k)]\}, \tag{8.130c}$$

$$C_{22} = -\frac{(1+B^2\mathfrak{h}^2)}{N}\{w_{01}B^2l^2\mathfrak{h}^2(B^2\mathfrak{h}^2-1)\Pi(\varrho,\ k)-[w_{01}(l^2+\mathfrak{h}^2)$$
$$+(w_{01}+v_{01})\pi(B^2l^2-1)\sqrt{l^2+\mathfrak{h}^2}]E(k)\}, \tag{8.130d}$$

$$G = -\frac{B^2\mathfrak{h}^2\sqrt{B^2l^2-1}}{\sqrt{(l^2+\mathfrak{h}^2)(1+B^2\mathfrak{h}^2)}}K_{20}, \tag{8.130e}$$

$$N = \tfrac{1}{2}\pi(B^2l^2-1)(1+B^2\mathfrak{h}^2)^{3/2}\sqrt{l^2+\mathfrak{h}^2}\,[(B^2\mathfrak{h}^2-1)E(k)$$
$$-B^2\mathfrak{h}^2K(k)], \tag{8.130f}$$

$$k = \sqrt{1-B^2h^2}, \quad \varrho = -\frac{l^2}{l^2+\mathfrak{h}^2} = -\frac{l^2(1-B^2h^2)}{l^2+h^2-B^2l^2h^2}, \tag{8.130g}$$

where $K(k)$, $E\Pi(k)$, (ϱ, k) are the complete elliptical integrals having the modulus k and the parameter ϱ given by (8.130g).

In order to find the velocity u_p on the plate, we start from relation (8.127b), into which we introduce the expression for F_{200} from (8.126); evaluating the integral along the axis of ordinates ($y = 0$), we obtain an integral in $\mathfrak{z}$ alone, which will be replaced by its expression in z from relation (8.18b). Thus we find successively:

$$u_{1p} = \frac{u_p}{x_1} = \operatorname{Re} \frac{x}{\sqrt{1-B^2x^2}} \int_{\text{Mach circle}}^{\text{upper plate}} F_{200}\, d\left(\frac{\sqrt{1-B^2x^2}}{x}\right)$$
$$= \frac{\mathfrak{z}}{\sqrt{1+B^2\mathfrak{z}^2}} \int_{\infty}^{\text{upper plate}} F_{200}(\mathfrak{z})\, d\left(\frac{\sqrt{1+B^2\mathfrak{z}^2}}{\mathfrak{z}}\right)$$
$$= z \int_{\infty}^{\text{upper plate}} F_{200}(z)\, d\left(\frac{1}{z}\right) = K_{20} \cos^{-1} \sqrt{\frac{\lambda^2-h^2}{\lambda^2-z^2}}$$
$$+ \left[A_{20} - \left(B^2A_{20} + C_{22}\sqrt{1-B^2h^2}\right) z^2\right] \sqrt{\frac{1-B^2h^2}{h^2-z^2}}, \tag{8.131}$$

where the constants K_{20}, A_{20}, C_{20} are those of the wing, given by (8.130).

Aerodynamic Characteristics. As in the preceding cases, denoting by C_{dw} and C_{dp} the *drag coefficients* of the wing and plate, we can find respectively:

—on the wing,

$$C_{dw} = -\frac{2w_{01}}{lU_\infty^2} \int_0^l u_{1w} y\, dy$$
$$= -\frac{w_{01}}{3B^3lU_\infty^2} \Bigg\{ K_{20}B^3l^2(3-2B^2l^2) \left(\frac{\pi}{2} + \cos^{-1} \frac{\mathfrak{h}}{l} \sqrt{\frac{B^2l^2-1}{B^2\mathfrak{h}^2+1}}\right)$$
$$- \left[C_{21} + \frac{G}{2\mathfrak{h}^2} \left((3-2B^2l^2)(l^2+\mathfrak{h}^2) - 2l^2\right)\right] \sqrt{1+B^2\mathfrak{h}^2} \cos^{-1} \frac{B^2\mathfrak{h}^2-1}{B^2\mathfrak{h}^2+1}$$
$$- \left[3A_{20}B^2 + \left(C_{21} - 2C_{22}\frac{B^2\mathfrak{h}^2}{B^2\mathfrak{h}^2+1} - G\frac{l^2}{\mathfrak{h}^2}\right) \sqrt{B^2\mathfrak{h}^2+1}\right]$$
$$\times \left(B\mathfrak{h} - \frac{B^2\mathfrak{h}^2+1}{2} \cos^{-1} \frac{B^2\mathfrak{h}^2-1}{B^2\mathfrak{h}^2+1}\right)$$
$$- 2(B^3l^3-1)\left[C_{21}K(k) + C_{22}E(k) + G\Pi(\varrho,k)\right]\Bigg\}, \tag{8.132}$$

where the constant G is given by (8.130e), and the modulus k and parameter ϱ of the complete elliptical integrals $K(k)$, $E(k)$ and $\Pi(\varrho, k)$ are given by the following relations:

$$k = \sqrt{1-B^2h^2}, \quad \varrho = -\frac{l^2+h^2-B^2l^2h^2}{l^2}. \tag{8.133}$$

—on the plate,

$$C_{dp} = -\frac{2v_{01}}{hU_\infty^2}\int_0^h u_{1p}z\,dz = = -\frac{2v_{01}}{hU_\infty^2}\Bigg[\frac{1}{2}K_{20}\lambda^2\cos^{-1}\sqrt{\frac{\lambda^2-h^2}{\lambda^2}}$$

$$+h\Bigg\{\sqrt{1-B^2h^2}\left[A_{20}-\tfrac{2}{3}\left(B^2A_{20}+C_{22}\sqrt{1-B^2h^2}\right)h^2\right]$$

$$-\tfrac{1}{2}K_{20}\sqrt{\lambda^2-h^2}\Bigg\}\Bigg], \quad \left(\lambda = \frac{l}{\sqrt{B^2l^2-1}}\right). \tag{8.134}$$

For the *rolling moment coefficients*, μ_w and μ_p, we shall find the same simple expressions as in the preceding similar cases (8.105, 8.106):

$$H\mu_w = \frac{2}{lU_\infty}\int_0^l u_{1w}y\,dy = -\frac{U_\infty}{w_{01}}C_{dw}, \tag{8.135a}$$

$$H\mu_p = -\frac{2}{hU_\infty}\int_0^h u_{1p}z\,dz = \frac{U_\infty}{v_{01}}C_{dp}. \tag{8.135b}$$

For the *suction coefficient* of the plate, we shall likewise find the following expression:

$$\sigma_p = -\frac{\pi\sqrt{1-B^2h^2}}{2hS_pU_\infty^2}\left[u_{1p}^2(h-z)\right]_{z\to h}$$

$$= -\frac{\pi(1-B^2h^2)^{3/2}}{2h^3U_\infty^2}\left[A_{20}-\left(B^2A_{20}+C_{22}\sqrt{1-B^2h^2}\right)h^2\right]^2. \tag{8.136}$$

VIII. 6. Wing System with Three Arms

VIII. 6.1. System with Three Arms having all Edges Subsonic

If we set $l_2 = 0$ in Fig. 8.1, the plate becomes marginal and the edges OA_2 no longer exist, the axial velocity being zero at this point.

The resulting system represents an ensemble of three wings or an ordinary tail, having the plate as horizontal plane and the wing as vertical plane (Fig. 8.8).

For simplicity, we shall consider the true conical motion and apply formula (8.57b) by setting $n = 1$; thus, in the X plane defined by (8.36), we get

$$F_{100} = \mathcal{U} = \frac{A_{10}+A_{11}X}{X}\sqrt{\frac{X+\mathfrak{h}}{L_1-X}}, \tag{8.137}$$

where $\mathfrak{h}$ and L_1 are given by (8.19) and (8.37a) respectively.

The above expression corresponds to a plate with antisymmetrical incidences, β on the upper side and $-\beta$ on the lower side.

For determining the constants, we follow the same way as in the preceding cases, applying the compatibility relations and integrating in the manner indicated in Fig. 8.9:

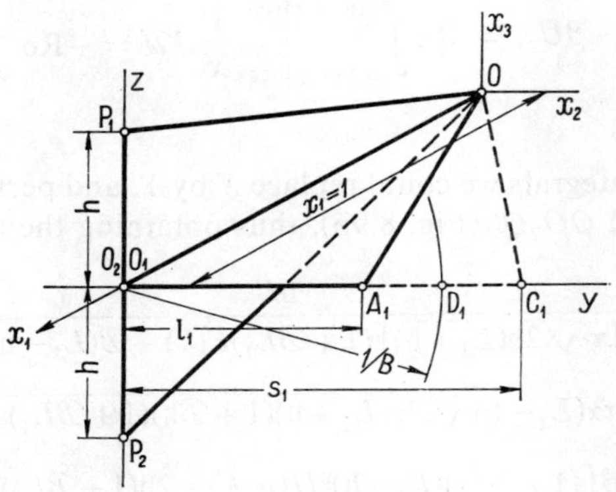

FIG. 8.8

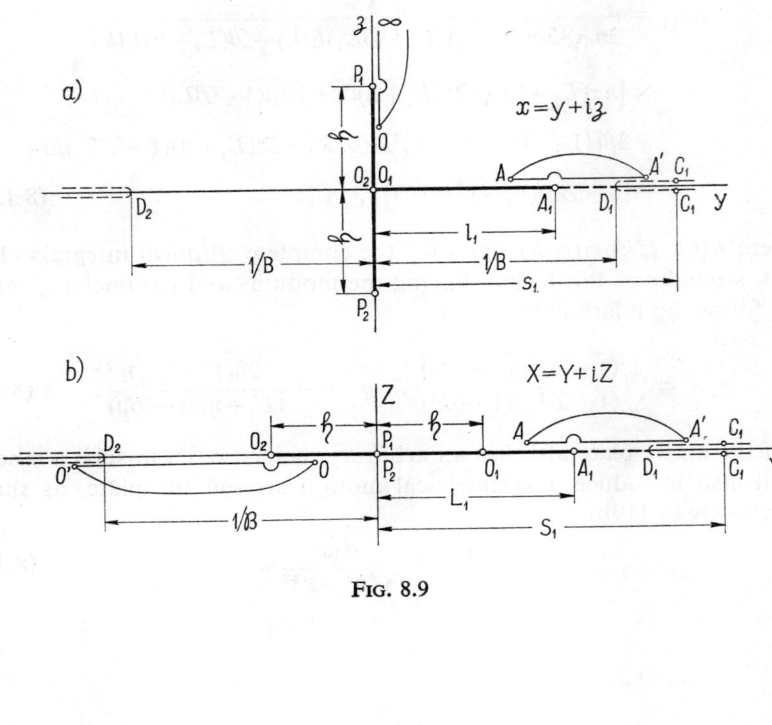

FIG. 8.9

$$w = -\alpha U_\infty = \operatorname{Re} \int_{\text{right wing}}^{\text{Mach circle}} \frac{i\sqrt{1-B^2x^2}}{x}\, d\mathcal{U}$$

$$= -\operatorname{Im} \frac{B}{\mathcal{B}} \int_{AA_1A'} \sqrt{\frac{1-\mathcal{B}^2X^2}{X^2-\mathfrak{h}^2}}\, d\mathcal{U}, \tag{8.138a}$$

$$v = -\beta U_\infty = \operatorname{Re} \int_{\text{upper plate}}^{\text{Mach circle}} \frac{1}{x}\, d\mathcal{U} = -\operatorname{Re} \int_{OO_2O'} \frac{d\mathcal{U}}{\sqrt{X^2-\mathfrak{h}^2}}\,. \tag{8.138b}$$

In the above integrals we could replace X by Y, and perform them on the paths AA_1A' and OO_2O' (Fig. 8.9b), thus obtaining the following values:

$$A_{10} = \frac{\mathfrak{h}U_\infty}{2\pi\sqrt{2\mathfrak{h}(L_1+\mathfrak{h})}\,[(1+\mathcal{B}L_1)E(k)-\mathcal{B}(L_1-\mathfrak{h})K(k)]}$$

$$\times\{\pi\alpha(L_1-\mathfrak{h})\sqrt{2\mathfrak{h}(L_1+\mathfrak{h})(1+\mathcal{B}\mathfrak{h})(1+\mathcal{B}L_1)}$$

$$+2\beta[(1+\mathcal{B}L_1)(L_1-\mathfrak{h})^2\Pi(\varrho, k)+2\mathfrak{h}(1+\mathcal{B}L_1)(L_1+\mathfrak{h})E(k)$$

$$-(1+3\mathcal{B}\mathfrak{h})(L_1^2-\mathfrak{h}^2)K(k)]\}, \tag{8.139a}$$

$$A_{11} = \frac{U_\infty}{2\pi\sqrt{2\mathfrak{h}(L_1+\mathfrak{h})}\,[(1+\mathcal{B}L_1)E(k)-\mathcal{B}(L_1-\mathfrak{h})K(k)}$$

$$\times\{\pi\alpha(L_1-\mathfrak{h})\sqrt{2\mathfrak{h}(L_1+\mathfrak{h})(1+\mathcal{B}\mathfrak{h})(1+\mathcal{B}L_1)}$$

$$+2\beta[(1+\mathcal{B}L_1)(L_1-\mathfrak{h})^2\Pi(\varrho, k)-2\mathfrak{h}(L_1+\mathfrak{h})(1+\mathcal{B}L_1)E(k)$$

$$-(1-\mathcal{B}\mathfrak{h})(L_1^2-\mathfrak{h}^2)K(k)]\}, \tag{8.139b}$$

where $K(k)$, $E(k)$, $\pi(\varrho, k)$ stand for the complete elliptical integrals of the first, second and third kind, having the modulus and parameter given by the following relations:

$$k = \sqrt{\frac{(1-\mathcal{B}L_1)(1+\mathcal{B}\mathfrak{h})}{(1+\mathcal{B}L_1)(1-\mathcal{B}\mathfrak{h})}}, \quad \varrho = -\frac{2\mathfrak{h}(1-\mathcal{B}L_1)}{(L_1+\mathfrak{h})(1-\mathcal{B}\mathfrak{h})}\,. \tag{8.140}$$

If the entire plate also has an arbitrary incidence β^*, in such a case we shall also introduce a symmetrical motion around the plate, as shown previously (8.119b),

$$\mathcal{U}^* = \frac{A_{10}^*}{\sqrt{h^2-x^{*2}}}, \tag{8.141}$$

where, examining Fig. 8.4, we have written

$$x^* = z + i\mathfrak{y}, \quad \left(\mathfrak{y} = \frac{y}{\sqrt{1-B^2\ ^2}}, \quad z = 0\right), \tag{8.142}$$

that is, inverting the axes and co-ordinates, changing y into z and $\mathfrak{z}$ into $\mathfrak{y}$. The constant A_{10}^* is given by (8.120).

VIII.6.2. Aerodynamic Characteristics

The pressure coefficient on the wing and plate will be given by both motions (8.130) and (8.134). The coefficient of "*side-force*" on the plate is made up of the coefficient C_{yp} due to the first motion (8.130), equal and of opposite sense on the two sides of the plate, and of the coefficient C_{yp}^* due to the motion (8.141), given by an expression similar to (4.23a):

$$C_{yp}^* = \frac{2\pi}{hU_\infty} A_{10}^*. \tag{8.143}$$

Thus, on the upper and lower sides we have the side-force coefficients

$$C_{yp}^{(u)} = C_{yp}^* + C_{yp}, \quad C_{yp}^{(l)} = C_{yp}^* - C_{yp}, \tag{8.144}$$

respectively.

It may be readily seen that the plate drag coefficient, differing on the two sides, has a total value given by the following expression:

$$C_{dp} = \beta^* C_{yp}^* + \beta C_{yp}. \tag{8.145}$$

Let us now compute C_{yp} on the upper half of the plate; for this purpose, by evaluating the integral along the abscissae axis in the X plane, we obtain successively:

$$\tfrac{1}{2} hC_{yp} = \frac{1}{U_\infty} \int_{O_1P_1O_2} u\, dz. \tag{8.146}$$

Let us evaluate this integral along the abscissae axis in the X plane. From relation (8.36), taking also account of (8.18b), the following relations result successively for points on the plate:

$$X^2 = x^2 + \mathfrak{h}^2 = -\mathfrak{z}^2 + \mathfrak{h}^2 = -\frac{z^2}{1-B^2z^2} + \mathfrak{h}^2 = Y^2, \tag{8.147a}$$

$$z = \sqrt{1-\mathcal{B}^2\mathfrak{h}^2}\sqrt{\frac{\mathfrak{h}^2 - Y^2}{1-\mathcal{B}^2Y^2}},$$

$$dz = -(1-\mathcal{B}^2\mathfrak{h}^2)^{3/2} \frac{Y\, dY}{(1-\mathcal{B}^2Y^2)\sqrt{(1-\mathcal{B}^2Y^2)(\mathfrak{h}^2-Y^2)}}. \tag{8.147b}$$

In such a case the integral above becomes

$$\tfrac{1}{2}hC_{yp} = \frac{(1-\mathcal{B}^2\mathfrak{h}^2)^{3/2}}{U_\infty}\int_{-\mathfrak{h}}^{\mathfrak{h}}\frac{uY\,dY}{(1-\mathcal{B}^2Y^2)\sqrt{(1-\mathcal{B}^2Y^2)(\mathfrak{h}^2-Y^2)}}$$

$$= \frac{(1-\mathcal{B}\mathfrak{h})\sqrt{1+\mathcal{B}\mathfrak{h}}}{U_\infty(1-\mathcal{B}L_1)\sqrt{1+\mathcal{B}L_1}}\Big\{A_{10}[(1+\mathcal{B}^2L_1\mathfrak{h})(E(\varphi_0,k)-E(k))$$

$$+(\mathcal{B}(L_1-\mathfrak{h})-2)(F(\varphi_0,k)-K(k))]-A_{11}(L_1+\mathfrak{h})[F(\varphi_0,k)$$

$$-E(\varphi_0,k)-K(k)+E(k)]$$

$$+(A_{10}\mathcal{B}+A_{11})(1-\mathcal{B}L_1)\sqrt{\frac{2\mathfrak{h}(L_1+\mathfrak{h})(1+\mathcal{B}\mathfrak{h})}{1+\mathcal{B}L_1}}\Big\}, \tag{8.148}$$

where the complete and incomplete elliptical integrals $K(k)$, $E(k)$, $F(\varphi_0, k)$, $E(\varphi_0, k)$ have the following modulus and argument:

$$k = \sqrt{\frac{(1+\mathcal{B}\mathfrak{h})(1-\mathcal{B}L_1)}{(1-\mathcal{B}\mathfrak{h})(1+\mathcal{B}L_1)}},\quad \varphi_0 = \sin^{-1}\frac{1-\mathcal{B}\mathfrak{h}}{1+\mathcal{B}\mathfrak{h}} = \sin^{-1}\frac{1-Bh}{1+Bh}. \tag{8.149}$$

In order to find the lift on the wing, we perform the following transformation deduced from relation (8.36):

$$X^2 = x^2+\mathfrak{h}^2 = y^2+\mathfrak{h}^2 = Y^2,\quad dy = \frac{Y\,dY}{\sqrt{Y^2-\mathfrak{h}^2}}. \tag{8.150}$$

Noticing further that the motion (8.141) has no effect on the wing lift, since u^* is the same on the two faces, and taking account of (8.150), we shall write successively

$$\tfrac{1}{2}l_1C_{lw} = \frac{2}{U_\infty}\int_0^{l_1}u\,dy = \frac{2}{U_\infty}\int_{\mathfrak{h}}^{L_1}\frac{uY\,dY}{\sqrt{Y^2-\mathfrak{h}^2}}$$

$$= \frac{\pi}{U_\infty}[2A_{10}+A_{11}(L_1+\mathfrak{h})]; \tag{8.151}$$

the respective drag coefficient is

$$C_{dw} = \alpha C_{lw}. \tag{8.152}$$

Further, the rolling moment coefficient on the wing will be given by the expression

$$\tfrac{1}{2}l_1H\mu_w = \frac{4}{3U_\infty}\int_{\mathfrak{h}}^{L_1}uY\,dY = \frac{4}{3U_\infty}\int_0^{l_1}uy\,dy$$

$$= \frac{1}{6U_\infty}\Big\{2\sqrt{2\mathfrak{h}(L_1-\mathfrak{h})}\,[4A_{10}+3A_{11}(L_1+\mathfrak{h})]$$

$$+(L_1+\mathfrak{h})[4A_{10}+A_{11}(3L_1-\mathfrak{h})]\cos^{-1}\frac{3\mathfrak{h}-L_1}{\mathfrak{h}+L_1}\Big\}. \tag{8.153}$$

The rolling moment coefficient due to the plate will be

$$\tfrac{1}{2}\,hH\mu_p = \frac{1}{6\mathcal{B}^2U_\infty}\left(\frac{1-\mathcal{B}^2\mathfrak{h}^2}{1-\mathcal{B}^2L_1^2}\right)^{3/2}$$

$$\times\Bigg\{\mathcal{B}^2\sqrt{\frac{2\mathfrak{h}(L_1-\mathfrak{h})(1-\mathcal{B}^2L_1^2)}{1-\mathcal{B}^2\mathfrak{h}^2}}\,[2A_{11}(L_1+\mathfrak{h})+A_{10}(2+\mathcal{B}^2L_1\mathfrak{h})$$

$$-(1-\mathcal{B}L_1)\sqrt{(1+\mathcal{B}\mathfrak{h})(1-\mathcal{B}L_1)}\,[A_{11}(\mathfrak{h}\mathcal{B}-5\mathcal{B}L_1-6)$$

$$-A_{10}\mathcal{B}(\mathcal{B}\mathfrak{h}-\mathcal{B}L_1-2)]\cos^{-1}\sqrt{\frac{(L_1-\mathfrak{h})\,(1-\mathcal{B}\mathfrak{h})}{(L_1+\mathfrak{h})\,(1+\mathcal{B}\mathfrak{h})}}$$

$$-(1+\mathcal{B}L_1)\sqrt{(1-\mathcal{B}\mathfrak{h})\,(1+\mathcal{B}L_1)}\,[A_{11}(\mathcal{B}\mathfrak{h}-5\mathcal{B}L_1+6)$$

$$+A_{10}\mathcal{B}(\mathcal{B}\mathfrak{h}-\mathcal{B}L_1+2)]\cos^{-1}\sqrt{\frac{(L_1-\mathfrak{h})\,(1+\mathcal{B}\mathfrak{h})}{(L_1+\mathfrak{h})\,(1-\mathcal{B}\mathfrak{h})}}\Bigg\}. \tag{8.154}$$

Further, the suction coefficients for the wing and plate will be given by the following expressions:

$$\sigma_w = -\frac{\pi\sqrt{1-B^2l^2}}{U_\infty^2S_wl}\left[\frac{(A_{10}+A_{11}Y)^2}{Y^2}\,\frac{Y+\mathfrak{h}}{L_1-Y}\,(L_1-Y)\right]_{Y\to L_1}$$

$$= -\frac{2\pi}{L_1^2U_\infty}\sqrt{\frac{(L_1+\mathfrak{h})\,(1-\mathcal{B}^2L_1^2)}{(L_1-\mathfrak{h})\,(1-\mathcal{B}^2\mathfrak{h}^2)}}\,(A_{10}+A_{11}L_1)^2, \tag{8.155a}$$

$$\sigma_p = -\frac{\pi\sqrt{1-B^2h^2}}{U_\infty^2S_ph}\left[\frac{(A_{10}+A_{11}\sqrt{\mathfrak{h}^2-\mathfrak{z}^2}}{\mathfrak{h}+\mathfrak{z}}\,\frac{\sqrt{\mathfrak{h}^2-\mathfrak{z}^2}+\mathfrak{h}}{L_1-\sqrt{\mathfrak{h}^2-\mathfrak{z}^2}}\right]_{\mathfrak{z}\to\mathfrak{h}}$$

$$= -\frac{\pi A_{10}^2}{h^2U_\infty^2\sqrt{l^2+h^2-B^2l^2h^2}}, \tag{8.155b}$$

respectively.

VIII. 6.3. System with Three Arms having a Supersonic Edge

If the leading edge of the wing lies at OC_1 (Fig. 8.8), hence being supersonic, then the contribution of the subsonic edges of the plate will be given by a formula similar to (8.51a) in which $L_2 = \mathfrak{h}$ and $n = 1$. On the other hand, the contribution of the supersonic edge will be given by a formula similar to (8.52a) in which $L_1 = 1/\mathcal{B}$; we shall note further that the ridge S_1 from the same formula lies along the supersonic leading edge OC_1, so that, according to the customary notation, we can put $S_1 = L_1$ and $L_2 = \mathfrak{h}$ (Fig. 8.10). Finally we can write

$$F_{100} = \mathcal{U} = A_{10}\,\frac{\sqrt{(X+\mathfrak{h})\,(1-\mathcal{B}X)}}{X\sqrt{\mathcal{B}}} + K_{10}\cos^{-1}\sqrt{\frac{(L_1+\mathfrak{h})\,(1-\mathcal{B}X)}{(1+\mathcal{B}\mathfrak{h})\,(L_1-X)}}, \tag{8.156}$$

where the constant K_{10} is the same as for the simple wing having the incidence α:

$$K_{10} = \frac{2\alpha U_\infty}{\pi B} \frac{Bl_1}{\sqrt{B^2 l_1^2 - 1}} . \tag{8.157}$$

As in the preceding case, $\mathcal{U}^*$ (8.141) is to be added to this motion.

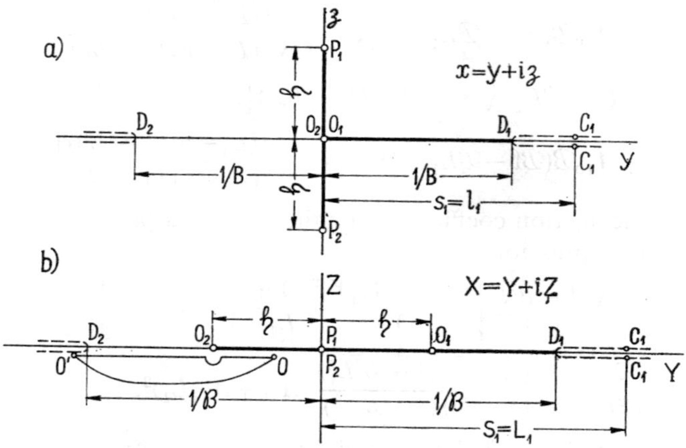

FIG. 8.10

For determining A_{10}, we write the same integral as above (8.138b), which will be performed along the path OO_2O' (Fig. 8.10):

$$v = -\beta U_\infty = \text{Re} \int_{\text{upper plate}}^{\text{Mach circle}} \frac{1}{x}\, d\mathcal{U} = -\text{Re} \int_{OO_2O'} \frac{d\mathcal{U}}{\sqrt{X^2 - \mathfrak{h}^2}} . \tag{8.158}$$

Finally,

$$A_{10} = \frac{2\mathfrak{h}\sqrt{2\mathcal{B}\mathfrak{h}}\, U_\infty}{\pi\sqrt{1+\mathcal{B}\mathfrak{h}}} \left\{\beta + \alpha\left[\sqrt{\frac{(L_1 - \mathfrak{h})(1 - \mathcal{B}\mathfrak{h})}{2\mathfrak{h}(1+\mathcal{B}L_1)}} - \sqrt{\frac{1 - \mathcal{B}^2\mathfrak{h}^2}{\mathcal{B}^2 L_1^2 - 1}}\right]\right\} . \tag{8.159}$$

Lift and drag coefficients. Taking into account (8.150), we shall evaluate the drag integral along the abscissae axis in the X plane; we find successively

$$\tfrac{1}{2}\,l_1 C_{lw} = \frac{2}{U_\infty}\int_0^{l_1} u\,dy = \frac{2}{U_\infty}\int_0^{1/B} u\,dy + \tfrac{1}{2}\left(l_1 - \frac{1}{B}\right)\frac{4\alpha}{B}\frac{Bl_1}{\sqrt{B^2 l_1^2 - 1}}$$

$$= \frac{\pi}{\mathcal{B}U_\infty}\{A_{10}(1-\mathcal{B}\mathfrak{h}) + K_{10}[\sqrt{1-\mathcal{B}^2\mathfrak{h}^2}$$

$$+ \sqrt{\mathcal{B}(L_1+\mathfrak{h})}\,(\sqrt{\mathcal{B}(L_1-\mathfrak{h})} + \sqrt{\mathcal{B}L_1 - 1})]\}$$

$$+ \frac{\pi K_{10}}{U_\infty}\left(l_1 - \frac{1}{\mathcal{B}}\right) = \frac{\pi}{\mathcal{B}U_\infty}[A_{10}(1-\mathcal{B}\mathfrak{h})$$

$$+ K_{10}\sqrt{\mathcal{B}(L_1+\mathfrak{h})}\,(2\sqrt{\mathcal{B}(L_1-\mathfrak{h})} + \sqrt{\mathcal{B}L_1 - 1})] \tag{8.160}$$

for the lift coefficient, and

$$C_{dw} = \alpha C_{lw} \tag{8.161}$$

for the wave drag coefficient, respectively.

For half of a plate, taking account of (8.147b), we likewise find

$$\tfrac{1}{2}\,hC_{yp} = \frac{1}{U_\infty}\int_{O_1P_1O_2} u\,dz$$

$$= \frac{(1-\mathcal{B}^2\mathfrak{h}^2)^{3/2}}{U_\infty}\int_{-\mathfrak{h}}^{\mathfrak{h}} \frac{uY\,dY}{(1-\mathcal{B}^2Y^2)\sqrt{(1-\mathcal{B}^2Y^2)(\mathfrak{h}^2-Y^2)}}$$

$$= \frac{1}{BU_\infty}\left\{A_{10}\sqrt{2\mathcal{B}\mathfrak{h}(1-\mathcal{B}\mathfrak{h})}\right.$$

$$- K_{10}\mathcal{B}(1-\mathcal{B}^2\mathfrak{h}^2)\sqrt{\frac{(L_1^2-\mathfrak{h}^2)(1-\mathcal{B}^2\mathfrak{h}^2)}{\mathcal{B}^2L_1^2-1}}\cos^{-1}\sqrt{\frac{(L_1-\mathfrak{h})(1-\mathcal{B}\mathfrak{h})}{(L_1+\mathfrak{h})(1+\mathcal{B}\mathfrak{h})}}$$

$$- (1+\mathcal{B}\mathfrak{h})\,[\tfrac{1}{2}A_{10}\sqrt{2(1-\mathcal{B}\mathfrak{h})}$$

$$\left. - K_{10}(1-\mathcal{B}\mathfrak{h})^2\sqrt{\frac{\mathcal{B}(L+\mathfrak{h})(1+\mathcal{B}\mathfrak{h})}{2(\mathcal{B}L_1-1)}}\right]\cos^{-1}\frac{1-\mathcal{B}\mathfrak{h}}{1+\mathcal{B}\mathfrak{h}}\bigg\} \tag{8.162}$$

for the coefficient of "side-force" and

$$C_{dp} = \beta C_{yp} \tag{8.163}$$

for the wave drag coefficient, respectively.

Coefficients of the rolling moments. The coefficients μ_w and μ_p, on the wing and plate, respectively, are yielded by the following formulae:

$$\tfrac{1}{2}\, l_1 H\mu_w = \frac{4}{3U_\infty}\int_0^{l_1} uy\, dy = \frac{4}{3U_\infty}\int_{\mathfrak{h}}^{L_1} uY\, dY$$

$$= \frac{1}{6\mathcal{B}^2 U_\infty}\Big\{\sqrt{2\mathcal{B}\mathfrak{h}(1-\mathcal{B}\mathfrak{h})}\,[A_{10}(1-3\mathcal{B}\mathfrak{h})$$

$$+K_{10}\sqrt{\mathcal{B}(L_1+\mathfrak{h})(\mathcal{B}L_1-1)}]$$

$$+2K_{10}\mathcal{B}^2(L^2-\mathfrak{h}^2)\left[\frac{\pi}{2}+\sin^{-1}\sqrt{\frac{(L+\mathfrak{h})\,(1-\mathcal{B}\mathfrak{h})}{(L-\mathfrak{h})\,(1+\mathcal{B}\mathfrak{h})}}\right]$$

$$[A_{10}(1+\mathcal{B}\mathfrak{h})^2+2K_{10}(1-\mathcal{B}\mathfrak{h}+2\mathcal{B}L)\sqrt{\mathcal{B}(L+\mathfrak{h})\,(\mathcal{B}L-1)}]$$

$$\times\sin^{-1}\sqrt{\frac{1-\mathcal{B}\mathfrak{h}}{1+\mathcal{B}\mathfrak{h}}}\Big\} \tag{8.164}$$

for the wing rolling moment coefficient, and

$$-\tfrac{1}{2}\, hH\mu_p = \frac{2}{3U_\infty}\int_{O_1P_1O_2} uz\, dz = \frac{2(1-\mathcal{B}^2\mathfrak{h}^2)}{3U_\infty}\int_{-\mathfrak{h}}^{\mathfrak{h}}\frac{uY\, dY}{1-\mathcal{B}^2Y^2}$$

$$= \frac{1}{6B^2U_\infty}\Big\{\tfrac{1}{2}(1+\mathcal{B}\mathfrak{h})\sqrt{2(1-\mathcal{B}\mathfrak{h})}\Big[A_{10}(5+\mathcal{B}\mathfrak{h})$$

$$-K_{10}\frac{\mathcal{B}L_1-1}{\mathcal{B}L_1+1}\sqrt{\frac{\mathcal{B}(L_1+\mathfrak{h})}{\mathcal{B}L_1-1}}\Big]\cos^{-1}\frac{1-\mathcal{B}\mathfrak{h}}{1+\mathcal{B}\mathfrak{h}}$$

$$-2K_{10}\frac{\mathcal{B}^2(L_1^2-\mathfrak{h}^2)}{\mathcal{B}^2L_1^2-1}\cos^{-1}\sqrt{\frac{(L_1+\mathfrak{h})\,(1-\mathcal{B}\mathfrak{h})}{(L_1-\mathfrak{h})\,(1+\mathcal{B}\mathfrak{h})}}$$

$$-\sqrt{2\mathcal{B}\mathfrak{h}(1-\mathcal{B}\mathfrak{h})}\Big[A_{10}(3+\mathcal{B}\mathfrak{h})-K_{10}\sqrt{\frac{\mathcal{B}(L_1+\mathfrak{h})}{\mathcal{B}L_1-1}}\Big]\Big\} \tag{8.165}$$

for the rolling moment coefficient of half the plate.

Suction coefficient. For the suction coefficient due to the entire plate we can write successively:

$$\sigma = -\frac{\pi\sqrt{1-\mathcal{B}^2\mathfrak{h}^2}}{U_\infty^2 S_p h}\,[u^2(h-z)]_{z\to h}$$

$$= -\frac{2\pi\sqrt{1-\mathcal{B}^2\mathfrak{h}^2}}{h^2U_\infty^2}\left[\frac{(\sqrt{\mathfrak{h}^2-\mathfrak{z}^2}+\mathfrak{h})(1-\mathcal{B}\sqrt{\mathfrak{h}^2-\mathfrak{z}^2})\,A_{10}^2}{\mathcal{B}(\mathfrak{h}^2-\mathfrak{z}^2)}(\mathfrak{h}-\mathfrak{z})\right]_{\mathfrak{z}\to\mathfrak{h}}$$

$$= -\frac{\pi A_{10}}{\mathcal{B}h^2U_\infty^2}\,. \tag{8.166}$$

VIII. 6.4. *Application to a Rectangular Wing fitted with End Plates*

We shall consider a plane rectangular wing (Fig. 8.11), having the total span equal to $2b$ and the chord equal to unity, and fitted with end plates whose height is $2h$. The wing incidence is $+\alpha$, and the plate incidence is $+\beta$ on the upper surface and $-\beta$ on the lower surface.

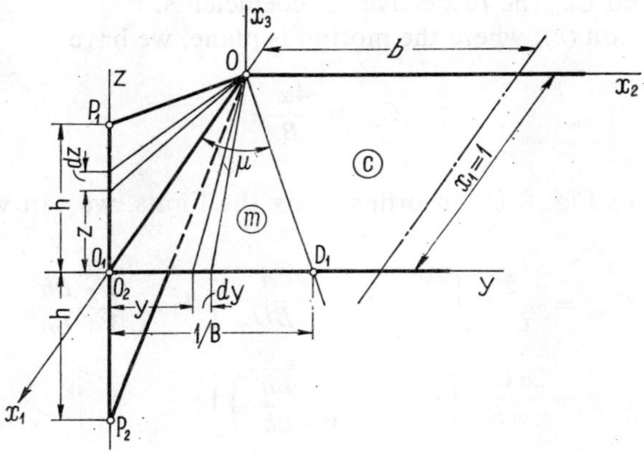

FIG. 8.11

The axial velocity is

$$u_c = \frac{\alpha U_\infty}{B} \tag{8.167}$$

on the wing, in the central region, outside the Mach cones (region C).

Inside the Mach cones (region m), the flow is a conical one.

In order to find the function $\mathcal{U} = u + iu'$ in the auxiliary plane X (Fig. 8.10), we apply formulae (8.156) and (8.157), where we make L_1 tend to infinity; thus we obtain

$$\mathcal{U} = \frac{A_{10}}{X}\sqrt{\frac{(X+\mathfrak{h})(1-\mathcal{B}X)}{\mathcal{B}}} + \frac{2\alpha U_\infty}{\pi\mathcal{B}}\cos^{-1}\sqrt{\frac{1-\mathcal{B}X}{1+\mathcal{B}\mathfrak{h}}}\,. \tag{8.168}$$

The constant A_{10} will be deduced from (8.159) by making L_1 also tend to infinity; thus we obtain

$$A_{10} = \frac{2hU_\infty}{\pi(1+Bh)}\left(\alpha + \beta\sqrt{\frac{2Bh}{1-Bh}}\right). \tag{8.169}$$

Aerodynamic Characteristics

Wing lift coefficient. We shall denote by S_c and S_m the surface areas of the wing regions (c) and (m),

$$S_c = b - \frac{1}{2B}, \quad S_m = \frac{1}{2B}, \tag{8.170}$$

and by C_{lc} and C_{lm} the respective lift coefficients.

For the region (c), where the motion is plane, we have

$$C_{lc} = \frac{4\alpha}{B}. \tag{8.171}$$

Referring to Fig. 8.11, in order to fix the limits, we can write for the region (m),

$$\frac{1}{2B} C_{lm} = \frac{2}{U_\infty} \int_0^{1/B} u\, dy = \frac{\pi}{BU_\infty} \left[A_{10} \sqrt{\frac{1-Bh}{1+Bh}} + \frac{2\alpha U_\infty}{\pi B} \left(1 - \tfrac{1}{2} \sqrt{\frac{1-Bh}{1+Bh}} \right) \right]. \tag{8.172}$$

The total lift coefficient C_l is given by the following relation:

$$C_{lw} = \left(1 - \frac{1}{2Bb}\right) C_{lc} + \frac{1}{2Bc} C_{lm}$$
$$= \frac{\pi}{Bb\sqrt{1-B^2h^2}\,U_\infty} \left[A_{10}(1-Bb) + \frac{\alpha U_\infty}{\pi B} \left(4Bb\sqrt{1-B^2h^2} - (1-Bh)\right) \right]. \tag{8.173}$$

When $h = 0$, we obtain the lift coefficient of a rectangular wing,

$$C_{lw} = \frac{4\alpha}{B} \left(1 - \frac{1}{4Bb}\right). \tag{8.174}$$

In all cases, the *drag coefficient* is

$$C_{dw} = \alpha C_{lw}. \tag{8.175}$$

In order to establish the influence of the plate, we note that its incidence β may be considered either positive or negative, as shown in Fig. 8.12.

The dotted lines show the position of the plates at zero incidence. In the case given in Fig. 8.12c and Fig. 8.12d, the influence of the incidence β is cancelled.

The "*side-force*" *coefficient* C_{yp} of half a plate is deduced from (8.162) by making L_1 (or l_1) tend to infinity; thus we get

$$\begin{aligned}\tfrac{1}{2}hC_{yp} &= \frac{1}{U_\infty}\int_{O_2P_1O_1} u\,dz = -\frac{1}{U_\infty}\int_{-\mathfrak{h}}^{\mathfrak{h}} u\,dz\\ &= \frac{\sqrt{1-Bh}}{BU_\infty}\Big[A_{10}\sqrt{2Bh}-\frac{2\alpha U_\infty}{\pi B}(1-B^2h^2)\sqrt{1+Bh}\cos^{-1}\sqrt{\frac{1-Bh}{1+Bh}}\\ &\quad -\frac{1}{\sqrt{2}}(1+Bh)\left(A_{10}-\frac{2\alpha U_\infty}{\pi B}(1-Bh)\sqrt{1-B^2h^2}\right)\cos^{-1}\frac{1-Bh}{1+Bh}\Big],\end{aligned} \tag{8.176}$$

and the corresponding *drag coefficient*,

$$C_{dp} = \beta C_{yp}. \tag{8.177}$$

The *plate rolling moment coefficient* will be given by the same expression (8.165), where, however, we make L_1 (or l_1) tend to infinity:

$$\begin{aligned}\tfrac{1}{2}hH\mu_p &= -\frac{2}{3U_\infty}\int_{O_2P_1O_1} uz\,dz = -\frac{2}{3U_\infty}\int_{-\mathfrak{h}}^{\mathfrak{h}} uz\,dz\\ &= \frac{1}{6B^2U_\infty}\Big\{\tfrac{1}{2}(1+Bh)\sqrt{2(1-Bh)}\Big[A_{10}(5+Bh)\\ &\quad -\frac{2\alpha U_\infty}{\pi B}\Big]\cos^{-1}\frac{1-Bh}{1+Bh}-\frac{4\alpha U_\infty}{\pi B}\cos^{-1}\sqrt{\frac{1-Bh}{1+Bh}}\\ &\quad -\sqrt{2Bh(1-Bh)}\Big[A_{10}(3+Bh)-\frac{2\alpha U_\infty}{\pi B}\Big]\Big\}.\end{aligned} \tag{8.178}$$

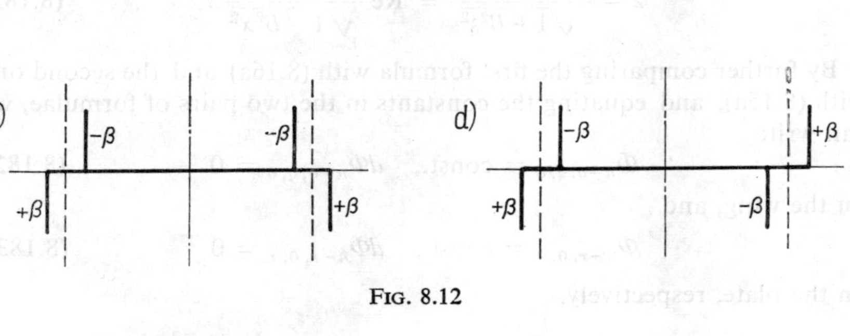

FIG. 8.12

The *suction coefficient* due to the subsonic edges of the plate is to be calculated by applying formula (8.166); in fact, we shall find exactly the same expression.

VIII. 7. Determination of the Wing and Plate Surfaces upon which a Given Pressure is exerted. Indirect Problem for the Cruciform Wing

VIII. 7.1. Reduction of the Problem to the Determination of a Characteristic Function $F_{n-2,1,1}$ which plays the Part of F_{n00} in the Direct Problem

If the pressure distribution on the wing and plate is given, and the equation of the respective surfaces or of the vertical velocity w on the wing and of the lateral one v on the plate is to be determined, then we are dealing with an indirect problem.

The pressure coefficient is proportional to the axial disturbance velocity u,

$$C_p = -\frac{2u}{U_\infty}, \tag{8.179}$$

so that it is sufficient that the distribution of axial velocities on the wing and plate be known; these velocities will be put in the following form in the case of the nth order motion:

$$u_w = \sum_{q=0}^{n-1} u_{n-q-1,\,q}\, x_1^{n-q-1} x_2^q = x_1^{n-1} \sum_{q=0}^{n-1} u_{n-q-1,\,q}\, y^q, \tag{8.180a}$$

$$u_p = \sum_{r=0}^{n-1} u_{n-r-1,\,r}\, x_1^{n-r-1} x_2^r = x_1^{n-1} \sum_{r=0}^{n-1} u_{n-r-1,\,r}\, z^r, \tag{8.180b}$$

where u_w is the axial velocity on the wing ($z=0$), and u_p is the one on the plate ($y = 0$). Let us note that z may be written as a function of $\mathfrak{z}$ defined previously (8.18b),

$$z = \frac{\mathfrak{z}}{\sqrt{1+B^2\mathfrak{z}^2}} = \mathrm{Re}\,\frac{-ix}{\sqrt{1-B^2x^2}}. \tag{8.181}$$

By further comparing the first formula with (8.16a) and the second one with (8.15a), and equating the constants in the two pairs of formulae, we can write

$$\Phi_{n-q,\,q,\,0} = \text{const.}, \quad d\Phi_{n-q,\,q,\,0} = 0 \tag{8.182}$$

on the wing, and

$$\Phi_{n-r,\,0,\,r} = \text{const.}, \quad d\Phi_{n-r,\,0,\,r} = 0 \tag{8.183}$$

on the plate, respectively.

The above results, when introduced into the compatibility relations (8.24) and now written in the forms

$$dF_{n-q-1,\,q,\,1} = -\frac{i\sqrt{1-B^2x^2}}{x}\,dF_{n-q,\,q,\,0}, \tag{8.184a}$$

$$dF_{n-r-1,\,1,\,r} = -\frac{1}{x}\,dF_{n-r,\,0,\,r}, \tag{8.184b}$$

give us the following relations, the first one on the wing and the second on the plate:

$$d\Phi'_{n-q-1,\,q,\,1} = 0, \quad d\Phi'_{n-r-1,\,1,\,r} = 0. \tag{8.185}$$

For the motions of order $n \geqq 2$, one of the functions $F_{n-q-r,\,q,\,r}$, viz. the function $F_{n-2,1,1}$, enjoys the property of having its imaginary part constant both on the plate and wing; indeed, by putting $q = 1$, $r = 1$ in the above relations, we get

$$\operatorname{Im} dF_{n-2,\,1,\,1} = d\Phi'_{n-2,\,1,\,1} = 0, \quad \Phi'_{n-2,\,1,\,1} = \text{const.}, \tag{8.186}$$

valid both on the wing and plate, a fact that makes the contour of the cruciform wing in the auxiliary plane x or X be a "streamline" of a fictitious motion having $F_{n-2,1,1}$ as a "potential". This fact will considerably facilitate the determination of the function $F_{n-2,1,1}$ by the usual means employed so far, and, consequently, the finding of the general solution, i.e. the finding of w on the wing and v on the plate.

Indeed, considering $q = r = 1$ in the compatibility equations (1.180), we can write

$$dF_{n-q-1,\,q,\,1} = \frac{dF_{n-2,\,1,\,1}}{(-x)^{q-1}}, \quad F_{n-q-1,\,q,\,1} = \int_{\text{Mach circle}}^{\text{wing}} \frac{dF_{n-2,\,1,\,1}}{(-x)^{q-1}}, \tag{8.187a}$$

$$dF_{n-r-1,\,1,\,r} = \left(\frac{\sqrt{1-B^2x^2}}{ix}\right)^{r-1} dF_{n-2,\,1,\,1},$$

$$F_{n-r-1,\,1,\,r} = \int_{\text{Mach circle}}^{\text{plate}} \left(\frac{\sqrt{1-B^2x^2}}{ix}\right)^{r-1} dF_{n-2,\,1,\,1}, \tag{8.187b}$$

respectively, whence the following expressions are obtained for w on the wing (8.22b) and v on the plate (8.23b),

$$\frac{(n-1)!}{x_1^{n-1}}\,w = \operatorname{Re}\sum_{q=0}^{n-1} C_{n-1}^{q}\,x^{q} \int_{\text{Mach circle}}^{\text{wing}} \frac{dF_{n-2,\,1,\,1}}{(-x)^{q-1}}, \tag{8.188a}$$

$$\frac{(n-1)!}{x_1^{n-1}}\,v = \operatorname{Re}\sum_{r=1}^{n-1} C_{n-1}^{r} \left(\frac{-ix}{\sqrt{1-B^2x^2}}\right)^{r}$$

$$\times \int_{\text{Mach circle}}^{\text{plate}} \left(\frac{\sqrt{1-B^2x^2}}{ix}\right)^{r-1} dF_{n-2,\,1,\,1}. \tag{8.188b}$$

Let us therefore notice that the problem of determining the components w and v, reduces for $n \geqq 2$ to finding the expression for the function $F_{n-2,1,1}$ analogous to the direct problem, where the determination of the function u on the wing and plate had been reduced to that of the function $F_{n,0,0}$ whose imaginary part is constant on the wing and plate; this property allowed us to determine it in a very simple manner. We shall follow the same method in order to determine $F_{n-2,1,1}$.

In this connection, we are going to find the boundary conditions that the function $F_{n-2,1,1}$ should satisfy, so as to be able to construct it in the same way; noticing, however, that there exists an equivalence between the boundary conditions of $F_{n-2,1,1}$ for certain indirect problems and $F_{n,0,0}$ for the equivalent direct problems, it is no longer necessary to determine the expression for the function $F_{n-2,1,1}$ directly; it is hence sufficient to know the equivalent function $F_{n,0,0}$ of the direct problem and use the respective formulae.

VIII. 7.2. Determination of the Characteristic Function $F_{n-2,1,1}$

In the direct problem it is considered that an arbitrary wing may be decomposed into two different wings: a thin wing and a wing of symmetrical thickness; in the indirect problem the wing may be decomposed into a lifting wing and a zero lift wing. We shall consider that an arbitrary plate is made up of the superposition of a lifting plate and a zero lift plate.

On a lifting wing or plate, the pressure distributions on the upper and lower faces are equal and of opposite sign; on a zero lift wing or plate, the pressure distribution is the same on the upper and lower faces. The lifting wing is always thin and the zero lift wing is always of symmetrical thickness, since the distributions of pressures on the plate are given in such a way as to have symmetry or antisymmetry with respect to the Ox_1x_2 plane.

Since the wing is not generally symmetrical or antisymmetrical with respect to the Ox_1x_3 plane, the lifting plate is not always a thin plate and the zero lift plate is not a plate of symmetrical thickness.

For determining $F_{n-2,1,1}$, we shall first study the behaviour of this function around the subsonic edges of the plate and wing, as well as around a ridge placed on the wing which separates two regions having different distributions of pressures.

Behaviour of $F_{n-2,1,1}$ *around a ridge placed on the wing or around the subsonic edge of a lifting wing*. We consider that at the point $y = s$ representing the ridge trace, $u_{n-q-1,q}$, which is proportional to $\Phi_{n-q,q,0}$, passes from the value $u^{(1)}_{n-q-1,q}$ to the value $u^{(2)}_{n-q-1,q}$; in the direct problem across a ridge, it is $w_{n-q-1,q}$, which is proportional to $\Phi_{n-q-1,q,1}$, that passes from the value $w^{(1)}_{n-q-1,q,1}$ to the value $w^{(2)}_{n-q-1,q}$ (Fig. 8.1).

It follows that $\Phi_{n-q,q,0} = \text{Re}F_{n-q,q,0}$ has the same behaviour at $y = s$ in the indirect problem as $\Phi_{n-q-1,q,1} = \text{Re}\,F_{n-q-1,q,p}$ in the direct problem. In this latter problem, the function $F_{n-q-1,q,1}$, as well as all the other functions $F_{p,q,r}$ $F_{n,0,0}$ included, contain, as we have seen, the following singularities around a ridge:

$$\ln(x-s), \quad \frac{1}{x-s}, \ldots, \frac{1}{(x-s)^{n-1}},$$

in the x plane, or

$$\ln(X-S), \quad \frac{1}{X-S}, \ldots, \frac{1}{(X-S)^{n-1}},$$

in the X plane.

The same thing will happen in the indirect problem for the function $F_{n-q,q,0}$, for any other function $F_{p,q,r}$, as well as for $F_{n-2,1,1}$ which will contain these singularities too.

The behaviour of $F_{n-2,1,1}$ around the subsonic edge of a zero lift wing which separates a region where $\Phi_{n-q-1,q,1} = \text{const.} = 0$ from another region where $\Phi_{n-q,q,0} = \text{const.} \neq 0$ in the Ox_1x_2 plane, is the same as that in the direct problem of the subsonic edge of a thin wing which separates a region where $\Phi_{n-q-1,q,1} = \text{const.} \neq 0$ from a region where $\Phi_{n-q,q,0} = \text{const.} = 0$. Hence it follows that in the indirect problem, u, v and w have the same behaviour around the subsonic edge of a zero lift wing as that in the direct problem around the subsonic edge of a thin wing. Consequently, at the point $y = l_1$ corresponding to the subsonic edge A_1, exactly like $F_{n,0,0}$ in the direct problem, the function $F_{n-2,1,1}$ should contain the singularities

$$(x-l_1)^{1/2-p}, \quad (p = 1, 2, \ldots, n)$$

in the x plane, or

$$(X-L_1)^{1/2-p}, \quad (p = 1, 2, \ldots, n)$$

in the X plane.

For the behaviour around the subsonic edges of the plate say P_1 and P_2, we shall consider two cases: the lifting wing and the zero lift wing. The velocities u, v, w have the same behaviour around the subsonic edge of a lifting wing or of a lifting plate; one may say the same thing about the behaviour around the subsonic edge of a zero lift wing or plate. Taking into account the above considerations on the behaviour around the edge of a wing, it follows that the function $F_{n-2,1,1}$ contains the singularities

$$(x \pm il_2)^{1/2-q}, \quad (q = 1, 2, \ldots, n)$$

in the x plane or

$$X^{1-2p}, \qquad (p = 1, 2, \ldots, n)$$

in the X plane, in the case of the zero lift plate; as well as the singularities

$$\ln (x \pm i\mathfrak{h}), \quad (x+i\mathfrak{h})^{-1}, \ldots, \quad (x \pm i\mathfrak{h})^{-(n-1)}$$

in the x plane or

$$\ln X, \quad X^{-2}, \ldots, \quad X^{-2(n-1)}$$

in the X plane, in the case of the lifting plate.

In addition, by examining formulae (8.184a) and (8.184b), we notice that in order that w and v shall be finite at the point $x = 0$ ($X = \pm \mathfrak{h}$), $F_{n-2,1,1}$ should not have the singularities corresponding to a ridge at that point. Consequently, the function $\Phi_{n,0,0}$ which is constant on the wing and plate (8.183), should not present any sudden change of value at $X = \pm \mathfrak{h}$.

Employing the compatibility relations written in the form

$$dF_{n\,0,0} = -\frac{ix^2}{\sqrt{1-B^2x^2}}\, dF_{n-2,\,1,\,1} = -i(X^2-\mathfrak{h}^2)\sqrt{\frac{1-\mathcal{B}^2\mathfrak{h}^2}{1-\mathcal{B}^2X^2}}\, dF_{n-2,\,1,\,1}, \tag{8.189}$$

it follows that since $F_{n,0,0}$ is analytic at $x = 0$ ($X = \pm \mathfrak{h}$), $F_{n-2,1,1}$ contains the singularities

$$\ln (X \pm \mathfrak{h}).$$

The contributions of the edges and ridges of the wing–plate system to $F_{n-2,1,1}$. We shall give below the expression for the function $F_{n-2,1,1}$, by pointing out the contribution of each edge of the wing–plate system and considering the equivalence existing between the functions $F_{n,0,0}$ *(direct problem)* and $F_{n-2,1,1}$ *(indirect problem)* and the results obtained in the direct problem.

Thus, for instance, referring to Fig. 8.3:

1. The contribution of the subsonic edge A_2 of a zero lift wing, by considering the edge $A_1(X = L_1)$ to be also subsonic, will be equivalent to (8.40b),

$$F^{(2)}_{n-2,\,1,\,1} = \frac{a_0+a_1X+\ldots+a_{n-1}X^{n-1}}{(X+L_2)^{n-1}}\sqrt{\frac{L_1-X}{X+L_2}}, \tag{8.190}$$

similarly, the contribution of the subsonic edge A_1, by considering the edge A_2 to be also subsonic, will be equivalent to (8.40a),

$$F^{(1)}_{n-2,\,1,\,1} = \frac{b_0+b_1X+\ldots+b_{n-1}X^{n-1}}{(L_1-X)^{n-1}}\sqrt{\frac{X+L_2}{L_1-X}}. \tag{8.191}$$

The formulae (8.190) and (8.191) are also valid in the case when A_1 in the former case and A_2 in the latter are supersonic, if L_1 and L_2, in (8.190) and (8.191), respectively, are replaced by $1/\mathcal{B}$. In the case of a zero lift wing with two subsonic edges, the effect of these edges is given by the following expression, corresponding to (8.40c),

$$F_{n-2,1,1} = \frac{c_0+c_1X+\ldots+C_{2n-1}X^{2n-1}}{(L_1-X)^{n-1}(L_2+X)^{n-1}\sqrt{(X+L_2)(L_1-X)}}. \tag{8.192}$$

2. The contribution of the subsonic edge of the lifting wing is given by the following expression, corresponding to (8.53c):

$$F^{(w)}_{n-2,1,1} = Q\cosh^{-1}\sqrt{\frac{(1+\mathcal{B}L_1)(1-\mathcal{B}X)}{2\mathcal{B}(L_1-X)}} + \frac{q_0+q_1X+\ldots+q_{n-2}X^{n-2}}{(L_1-X)^{n-1}}\sqrt{\frac{1-B^2X^2}{B^2}}. \tag{8.193}$$

3. The contribution of the subsonic edges of the zero lift plate is given by the following relation, equivalent to (8.50),

$$F^{(p)}_{n-2,1,1} = \frac{d_0+d_1X+\ldots+d_{2n-2}X^{2n-2}}{X^{2n-1}}\sqrt{(X+L_2)(L_1-X)}, \tag{8.194}$$

in the case when the wing, of zero lift too, has the subsonic edges A_2 and A_1. In the case when one or both edges are supersonic, then L_1 or L_2, or L_1 and L_2, respectively, are replaced by $1/\mathcal{B}$, as in the direct problem (8.51).

4. As regards the contribution of the subsonic edges of the lifting plate, we are now going to construct the respective expression by proceeding in the same way as that employed for the various contributions to $F_{n,0,0}$ in the direct problem.

Indeed, in the indirect problem, the edges of the lifting plate behave as the ridges do in the direct problem, the contributions of the latter being given by (8.54). If the points R_1, R_2 fall at P_1, P_2, respectively (Figs. 8.1 and 8.3), then $T = 0$; noting further that we have

$$\operatorname{Re}\cosh^{-1}\sqrt{\frac{L_2(X-L_1)}{(L_1+L_2)X}} = \operatorname{Re}\cosh^{-1}\sqrt{\frac{L_1(X+L_2)}{(L_1+L_2)X}}, \tag{8.195}$$

we shall add the two expressions (8.54) and obtain

$$F^{(p)}_{n-2,1,1} = Q\cosh^{-1}\sqrt{\frac{(X+L_2)L_1}{X(L_1+L_2)}} + \frac{a_0+a_1X+\ldots+a_{2n-3}X^{2n-3}}{X^{2(n-1)}}\sqrt{(X+L_2)(L_1-X)}, \tag{8.196}$$

in the case of the zero lift wing having the two edges A_1 and A_2 subsonic.

In the case when one or both edges are supersonic, L_1 or L_2, or L_1 and L_2, respectively, are replaced by $1/\mathcal{B}$.

In the case of the lifting wing, formula (8.196) holds its validity for this case too, replacing, however, L_1 and L_2 by $1/\mathcal{B}$.

5. We can write down the effect of the supersonic edge A_1 of a zero lift wing, A_2 being subsonic; this contribution is given by the formula equivalent to (8.53a) in the direct problem, where we put $S_1 = L_1$,

$$F^{(1)}_{n-2,1,1} = Q \cosh^{-1} \sqrt{\frac{(L_1+L_2)(1-\mathcal{B}X)}{(1+\mathcal{B}L_2)(L_1-X)}} + \frac{a_0+a_1X+\ldots+a_{n-2}X^{n-2}}{(X-L_1)^{n-1}} \sqrt{\frac{(X+L_2)(1-\mathcal{B}X)}{B}}. \tag{8.197}$$

Likewise, if the edge A_2 is supersonic and A_1 is subsonic, we shall have a similar formula,

$$F^{(2)}_{n-2,1,1} = Q \cosh^{-1} \sqrt{\frac{(L_1+L_2)(1+\mathcal{B}X)}{(1+\mathcal{B}L_1)(X+L_2)}} + \frac{a_0+a_1X+\ldots+a_{n-2}X^{n-2}}{(X+L_2)^{n-2}} \sqrt{\frac{(1+\mathcal{B}X)(L_1-X)}{\mathcal{B}}}. \tag{8.198}$$

If both edges of the zero lift wing are supersonic, the contribution of an edge, say A_2, has the expression

$$F^{(2)}_{n-2,1,1} = Q \cosh^{-1} \sqrt{\frac{(1+\mathcal{B}L_2)(1+\mathcal{B}X)}{2\mathcal{B}(X+L_2)}} + \frac{a_0+a_1X+\ldots+a_{n-2}X^{n-1}}{(X+L_2)^{n-1}} \sqrt{\frac{1-\mathcal{B}^2X^2}{\mathcal{B}^2}}. \tag{8.199}$$

The contribution of a supersonic edge of a lifting wing, say A_2, is given by this same formula.

6. The effects of the logarithmic singularities situated at X are given by the expressions

$$F^{(+)}_{n-2,1,1} = Q_+ \cosh^{-1} \sqrt{\frac{(L_1-\mathfrak{h})(X+L_2)}{(L_1+L_2)(X-\mathfrak{h})}}, \tag{8.200a}$$

$$F^{(-)}_{n-2,1,1} = Q_- \cosh^{-1} \sqrt{\frac{(L_1+\mathfrak{h})(X+L_2)}{(L_1+L_2)(X+\mathfrak{h})}}, \tag{8.200b}$$

in the case of a zero lift wing with subsonic edges. In the case when one or both edges are supersonic, L_1 or L_2, or L_1 and L_2, re y, are replaced by $1/\mathcal{B}$. Formulae (8.200) are valid in the case of a lifting wing too, replacing, however, L_1 and L_2 by $1/\mathcal{B}$.

VIII. 7.3. Application to the Cruciform Wing of Order $n = 2$

We are going to treat now a complete example, the lifting cruciform wing, in order to point out the manner in which one may apply the general results obtained above, by considering, for simplicity, $n = 2$.

We shall consider a lifting wing ($l_1 = l_2 = l$), fitted with a lifting plate ($-h<z<+h$), the wing–plate system being entirely included in the Mach cone. We assume that in the upper half-plane (on the wing upper surface)

$$u = \Phi_{200}x_1+\Phi_{110}x_2 = \pm ax_1+bx_2 \tag{8.201}$$

(the sign + on the left side of the plate and − on the right side of the plate), and on the plate

$$u = \Phi_{200}x_1+\Phi_{101}x_3 = \pm ax_1\pm cx_3 \tag{8.202}$$

(the sign + on the left side and − on the right side).

By employing formulae (8.196), (8.193) and (8.200), we can write the expression for the function F_{011}:

$$F_{011} = Q_0 \cosh^{-1}\sqrt{\frac{1+\mathcal{B}X}{2\mathcal{B}X}}+Q_1 \cosh^{-1}\sqrt{\frac{(1+\mathcal{B}L)(1-\mathcal{B}X)}{2\mathcal{B}(L-X)}}$$
$$+Q_2 \cosh^{-1}\sqrt{\frac{(1-\mathcal{B}L)(1+\mathcal{B}X)}{2\mathcal{B}(X+L)}}+Q_+ \cosh^{-1}\sqrt{\frac{(1+\mathcal{B}X)(1-\mathcal{B}\mathfrak{h})}{2\mathcal{B}(X-\mathfrak{h})}}$$
$$+Q_- \cosh^{-1}\sqrt{\frac{(1+\mathcal{B}X)(1+\mathcal{B}\mathfrak{h})}{2\mathcal{B}(X+\mathfrak{h})}}$$
$$+\left(\frac{A_0+A_1X}{X^2}+\frac{b_0}{X-L}+\frac{c_0}{X+L}\right)\sqrt{\frac{1-\mathcal{B}^2X^2}{\mathcal{B}^2}}. \tag{8.203}$$

Taking into account the fact that the function F_{011} should be symmetrical $[w(x_1,x_2) = -w(x_1,-x_2)]$, we must consider $A_1 = 0$, $b_0 = -c_0$, $Q_2 = Q_1$, $Q_+ = Q_- = Q$ and thus we obtain

$$F_{011} = \frac{Q_0}{2}\cosh^{-1}\frac{1}{\mathcal{B}X}+\frac{Q_1}{2}\cosh^{-1}\frac{2-\mathcal{B}(X^2+L^2)}{\mathcal{B}^2(X^2-L^2)}$$
$$+\frac{Q}{2}\cosh^{-1}\frac{2-\mathcal{B}^2(X^2+\mathfrak{h}^2)}{\mathcal{B}^2(X^2-\mathfrak{h}^2)}+\left(\frac{A_0}{X^2}+\frac{D_0}{X^2-L^2}\right)\sqrt{\frac{1-\mathcal{B}^2X^2}{\mathcal{B}^2}}, \tag{8.204}$$

or, by returning to the x plane,

$$F_{011} = \frac{Q_0}{2}\cosh^{-1}\sqrt{\frac{1+B^2\mathfrak{h}^2}{B^2(x^2+\mathfrak{h}^2)}} + \frac{Q_1}{2}\cosh^{-1}\frac{2-B^2(x^2+l^2)}{B^2(x^2-l^2)}$$
$$+\frac{Q}{2}\cosh^{-1}\frac{2-B^2x^2}{B^2x^2} + \left(\frac{A_0}{x^2+\mathfrak{h}^2}+\frac{D_0}{x^2-l^2}\right)\sqrt{\frac{1-B^2x^2}{1+B^2\mathfrak{h}^2}}. \tag{8.205}$$

For determining the constants Q_0, Q_1, Q, A_0, D_0 we proceed in the following manner:

1. At the point $X = -L$, on the real axis, Φ_{200} assumes the value $\Phi_{200} = 0$ on the left side and $\Phi_{200} = a$ on the right side. By employing the compatibility relations between the functions F_{200} and F_{011} and integrating round semicircle γ having its centre at $X = -L$, from a point $-L-\varepsilon$ up to a point $-L+\varepsilon$, we obtain

$$a = \operatorname{Re}\oint_{\gamma}\frac{-ix^2}{\sqrt{1-B^2x^2}}\,dF_{011}$$
$$= \frac{\pi\mathcal{B}}{2B}\left[\frac{Q_1 l^2}{\sqrt{1-\mathcal{B}^2L^2}} + D_0\frac{l^2\mathcal{B}^2+2(1-\mathcal{B}^2L^2)}{1-\mathcal{B}^2L^2}\right]. \tag{8.206}$$

At the same point $X = -L$, on the real axis, Φ_{110} assumes the value $\Phi_{110} = 0$ on the left and $\Phi_{110} = b$ on the right; by employing the compatibility relations between the functions F_{110} and F_{011} and integrating round the same semicircle γ, we have

$$b = \operatorname{Re}\oint_{\gamma}\frac{ix}{\sqrt{1-B^2x^2}}\,dF_{011} = \frac{\pi\mathcal{B}}{2B}\left[\frac{Q_1 l}{\sqrt{1-\mathcal{B}^2L^2}} + D_0\frac{1-\mathcal{B}^2\mathfrak{h}^2}{(1-\mathcal{B}^2L^2)l}\right]. \tag{8.207}$$

The two equations (8.206) and (8.207) have the following solutions:

$$Q_1 = \frac{2B[bl(2-2\mathcal{B}^2L^2+\mathcal{B}^2l^2)-a(1-\mathcal{B}^2\mathfrak{h}^2)]}{\pi\mathcal{B}l^2\sqrt{1-\mathcal{B}^2L^2}}, \quad D_0 = \frac{2B}{\pi\mathcal{B}}(a-bl). \tag{8.208}$$

2. At the point $X = 0$, on the real axis, Φ_{200} assumes the value $\Phi_{200} = +a$ on the left side and the value $\Phi_{200} = -a$ on the right side, and Φ_{101} assumes the value $\Phi_{101} = +c$ on the left and the value $\Phi_{101} = -c$ on the right.

By employing the compatibility relations between the functions F_{200} and F_{011}, F_{101} and F_{011}, and integrating round a semicircle γ having its centre at $X = 0$, from a point $-\varepsilon$ up to a point ε, we obtain

$$2a = \operatorname{Re}\oint_\gamma \frac{-ix^2}{\sqrt{1-B^2x^2}}\,dF_{011} = \frac{\pi\mathcal{B}}{2B}\,[Q_0\mathfrak{h}^2+2A_0(\mathfrak{h}^2\mathcal{B}^2-2)], \tag{8.209a}$$

$$2c = \operatorname{Re}\oint_\gamma -x\,dF_{011} = \frac{\pi}{2\mathfrak{h}}(-Q_0\mathfrak{h}^2+2A_0). \tag{8.209b}$$

We easily find

$$A_0 = -\frac{2B^2}{\pi\mathcal{B}^3}(aB+c\mathfrak{h}\mathcal{B}), \tag{8.210a}$$

$$Q_0 = -\frac{4}{\pi\mathcal{B}^3\mathfrak{h}^2}[a\mathcal{B}^2+c\mathfrak{h}(B^2+\mathcal{B}^2)\mathcal{B}]. \tag{8.210b}$$

3. From the expression for F_{011} there result the logarithmic singularities $\ln(X\pm 1/B)$ in F_{200} and F_{110}. In order to avoid this incompatibility we impose the condition

$$\left(\frac{dF_{011}}{dx}\right)_{x=\pm 1/\mathcal{B}} = 0, \tag{8.211}$$

and obtain two identical relations

$$\frac{Q}{\sqrt{1-\mathcal{B}^2\mathfrak{h}^2}}+\frac{Q_0}{2}+a_0\mathcal{B}^2+\frac{Q_1}{\sqrt{1-B^2L^2}}+\frac{D_0\mathcal{B}^2}{1-\mathcal{B}^2L^2} = 0. \tag{8.212}$$

Performing the computations we find

$$Q = \frac{2}{\pi}\left(a\,\frac{l^2+\mathfrak{h}^2+2B^2l^2\mathfrak{h}^2}{\mathfrak{h}^2l^2}-\frac{2b}{l}+\frac{2cB}{\mathfrak{h}\mathcal{B}}\right). \tag{8.213}$$

Having now the expression for F_{011}, by employing relations (8.188) we shall obtain the following expressions on the plate and wing:

$$\frac{v}{x_1} = \operatorname{Re}\left(\int_{\text{Mach cirle}}^{\text{plate}}\frac{ix}{\sqrt{1-B^2x^2}}\,dF_{011}-\frac{ix}{\sqrt{1-B^2x^2}}F_{011}\right)$$

$$= \operatorname{Re}\left[\frac{ib}{\pi}\cosh^{-1}\frac{x^2+l^2}{x^2-l^2}+\frac{cB+a\mathfrak{h}\mathcal{B}B^2}{\pi\mathcal{B}}\cosh^{-1}\frac{x^2-\mathfrak{h}^2}{x^2+\mathfrak{h}^2}\right.$$

$$-\frac{-ix}{\sqrt{1-B^2x^2}}\left(\frac{Q_0}{2}\cosh^{-1}\sqrt{\frac{1+B^2\mathfrak{h}^2}{B^2(x^2+\mathfrak{h}^2)}}+\frac{Q_1}{2}\cosh^{-1}\frac{2-B^2(x^2+l^2)}{(B^2(x^2-l^2)}\right.$$

$$\left.\left.+\frac{Q_2}{2}\cosh^{-1}\frac{2-B^2x^2}{B^2x^2}\right)\right], \tag{8.214a}$$

$$\frac{w}{x_1} = \mathrm{Re}\left(-\int_{\text{Mach circle}}^{\text{wing}} x\, dF_{011} + xF_{011}\right)$$

$$= \mathrm{Re}\frac{b-aB^2l}{\pi\sqrt{1-B^2l^2}}\cosh^{-1}\frac{l^2+x^2(1-2l^2B^2)}{l^2-x^2}$$

$$-\frac{ic}{\pi}\cosh^{-1}\frac{\mathfrak{h}^2-x^2(1+2\mathfrak{h}^2B)}{x^2+\mathfrak{h}^2}+x\left(\frac{Q_0}{2}\cosh^{-1}\sqrt{\frac{1+B^2\mathfrak{h}^2}{B^2(x^2+\mathfrak{h}^2)}}\right.$$

$$\left.+\frac{Q_1}{2}\cosh^{-1}\frac{2-B^2(x^2+l^2)}{B^2(l^2-x^2)}+\frac{Q_2}{2}\cosh^{-1}\frac{2-B^2x^2}{B^2x^2}\right)\Bigg]; \quad (8.214b)$$

the constants Q_1, Q_0, Q are given by relations (8.208), (8.210b) and (8.213), respectively.

It may be readily noticed from these same relations that, for given values a, b, c (8.201, 8.202), if we let $l \to 0$ or $\mathfrak{h} \to 0$, there result no infinitely high values for w and v.

That is the reason why, in order to avoid an incompatibility, we have imposed a restriction by considering the same value a.

VIII. 7.4. Indirect Problem of the Conical Flow of Order 1 ($n = 1$)

The method employed above is no longer valid for $n = 1$ since there is no function $F_{n-2,1,1}$ in such a case. Similarly, we are going to treat this problem directly. Thus, for instance, it is easy to find the expression for the function $\mathcal{U} = F_{100}$; this function assumes the wing, plate and Mach circle to be equipotential lines. As regards the behaviour of u around the edges of the wing and plate, it is obvious that this behaviour is the same as in the case $n \geq 2$.

For simplifying our exposition, we shall determine the expression for $\mathcal{U}$ in certain concrete applications.

1. We shall first study the problem of a lifting delta wing (thin wing) fitted with a plate, also lifting, the wing-plate system being entirely comprised in the Mach cone.

In order to avoid certain incompatibilities, we shall impose the following boundary conditions: in the upper half-plane we have on the wing $u = u_1$ on the left side of the plate and $u = -u_1$ on the right side of the plate, and on the plate $u = u_1$ on the left side and $u = -u_1$ on the right side.

In the lower half-plane u assumes equal values and of opposite sign with respect to those of the upper half-plane. On the axis outside the wing trace we have $u = 0$.

The function $\mathcal{U}$ satisfying all these conditions has the expression

$$\mathcal{U} = \frac{iu_1}{\pi}\ln(X^2-L^2) - \frac{2iu_1}{\pi}\ln X = \frac{iu_1}{\pi}\ln\frac{x^2+l^2}{x^2+\mathfrak{h}^2}. \qquad (8.215)$$

Employing the compatibility relations we can write

$$w = -\operatorname{Re} i\int_{\text{Mach circle}}^{\text{wing}} \frac{\sqrt{1-B^2x^2}}{x}\,d\mathcal{U} = -\frac{u_1 B}{\pi}\operatorname{Re}\left[\sqrt{\frac{1/B^2-l^2}{l}}\right.$$
$$\times\ln\frac{x+l}{x-l}\,\frac{1/B^2-lx+\sqrt{(1/B^2-l^2)(1/B^2-x^2)}}{1/B^2+lx+\sqrt{(1/B^2-l^2)(1/B^2-x^2)}}$$
$$\left.+i\frac{\sqrt{1/B^2+\mathfrak{h}^2}}{\mathfrak{h}}\ln\frac{x+i\mathfrak{h}}{x-i\mathfrak{h}}\,\frac{1/B^2-ix\mathfrak{h}+\sqrt{(1/B^2+\mathfrak{h}^2)(1/B^2-x^2)}}{1/B^2+ix\mathfrak{h}+\sqrt{(1/B^2+\mathfrak{h}^2)(1/B^2-x^2)}}\right], \qquad (8.216a)$$

$$v = -\operatorname{Re}\int_{\text{Mach circle}}^{\text{plate}} \frac{d\mathcal{U}}{x} = -\operatorname{Re}\frac{iu_1}{\pi}\left(\frac{1}{l}\ln\frac{x-l}{x+l}+\frac{i}{\mathfrak{h}}\ln\frac{x-i\mathfrak{h}}{x+i\mathfrak{h}}\right). \qquad (8.216b)$$

2. As regards the second example, we are going to determine the function $\mathcal{U}$ in the case of a zero lift wing (wing of symmetrical thickness) having one subsonic edge A_2 and the other, A_1, supersonic, fitted with a zero lift plate. We shall consider that $u = u_1$ on the wing and plate, while $w = 0$ on the axis Oy ($x = y+i\mathfrak{z}$) outside the trace of the wing, plate and Mach circle. As we have always done in the case of a supersonic edge, we shall first consider that at a point $A(y = l<1/B)$ on the wing there is a ridge across which u passes from the value $u = u_1$ on the left side, to the value $u = 0$ on the right side, and then we set $l = l_1$ in the final formulae.

Taking into account the boundary conditions, it follows that $du = 0$ *(equipotential lines)* in the X plane (Fig. 8.13a) on the OY axis on the segments $(-\infty D_2;\ A_2\infty)$, while $du' = 0$ *(streamline)* on D_2A_2.

For determining the function $\mathcal{U}$ in a simple manner, we shall perform a convenient conformal transformation

$$\chi^2 = \frac{\mathcal{B}(X+L_2)}{1+\mathcal{B}X}; \qquad (8.217)$$

in the χ plane (Fig. 8.13b), the real axis is an equipotential line ($du = 0$), while the imaginary axis is a streamline ($du' = 0$). The points χ_2, χ_p, χ_1 corresponding to the points $X = -L_2$, O, L_1 are

$$\chi_2 = 0, \quad \chi_p = \pm\sqrt{\mathcal{B}L_2}, \quad \chi_1 = \pm\sqrt{\frac{\mathcal{B}(L_1+L_2)}{1+\mathcal{B}L_1}}, \qquad (8.218)$$

respectively.

We can obtain the function $\mathcal{U}$ in the form of a potential of a fictitious motion due to a system of vortices and doublets conveniently situated in the χ plane,

$$\mathcal{U} = u+iu' = \frac{iM_2}{\chi}+iM_p\left(\frac{1}{\chi-\chi_p}-\frac{1}{\chi+\chi_p}\right)-\frac{iu_1}{\pi}\ln\frac{\chi-\chi_1}{\chi+\chi_1}. \tag{8.219}$$

It may be readily seen that the expression $\mathcal{U}$ satisfies the imposed conditions, the constants M_2 and M_p not being determined.

As in the preceding example, by employing the compatibility relations, we can obtain the expression for w on the wing and for v on the plate.

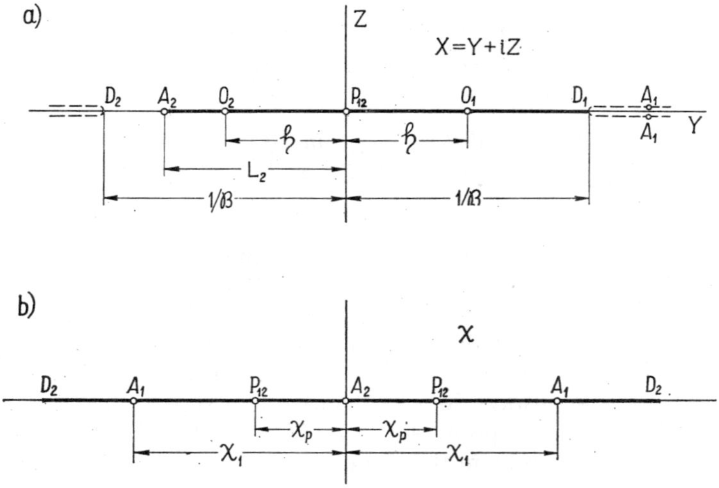

FIG. 8.13

VIII. 8. Mixed Problems of Cruciform Wings

The mixed problems we are going to treat are considered under two assumptions:

1. The pressures on the wing (or the axial velocity u) and the plate form (incidence or slope) are given; we have to determine the wing incidence or slope (for calculating its surface shape) and the pressures on the plate (or the axial velocity u).

2. The wing form (incidence or slope) and the pressures on the plate (or the axial velocity u) are given; we have to determine the plate incidence or slope (for calculating its surface shape) and the pressures on the wing (or the axial velocity u).

VIII. 8.1. Cruciform Wing having Given Pressures on the Wing and Velocity v Normal to the Plate

The pressures on the wing will be defined by the axial velocity u and the normal velocity on the plate by the lateral velocity v, both these velocities being given in the form of some homogeneous polynomials of order $n = 1$:

$$u = \sum_{q=0}^{n-1} u_{n-q-1,q}\, x_1^{n-q-1} x_2^q = x_1^{n-1} \sum_{q=0}^{n-1} u_{n-q-1.q}\, y^q, \qquad (8.220a)$$

$$v = \sum_{r=0}^{n-1} v_{n-r-1,r}\, x_1^{n-r-1} x_3^r = x_1^{n-1} \sum_{r=0}^{n-1} v_{n-r-1,r}\, z^q, \qquad (8.220b)$$

the former on the wing ($x_3 = 0$) and the latter on the plate ($x_2 = 0$).

The constants $u_{n-q-1,q}$, $v_{n-r-1,r}$ may assume different values on the wing and plate, respectively, separated by a line which passes through the wing vertex, constituting thus a separation *ridge*.

In the case of the thin wing, u assumes equal values of opposite sign on the upper and lower surfaces, while in the case of the wing of symmetrical thickness, u assumes equal values of the same sign; on the thin wing v assumes equal values of the same sign on both faces, while on the plate of symmetrical thickness v assumes equal values of opposite sign.

The problem we are confronted with consists of determining the component w on the wing, such that one might finally determine its shape, and the component u on the plate, useful for calculating the pressures.

Comparing formulae (8.220) with the real part of formulae (8.22) and (8.23), one obtains

$$\Phi_{n-q,q,0} = \text{const.}, \qquad d\Phi_{n-q,q,0} = 0, \qquad (8.221a)$$

$$\Phi_{n-r-1,1,r} = \text{const.}, \qquad d\Phi_{n-r-1,1,r} = 0, \qquad (8.221b)$$

the former on the wing and the latter on the plate.

Employing the compatibility relations, we obtain

$$\Phi'_{n-q-1,q,1} = 0, \qquad d\Phi'_{n-r,0,r} = 0, \qquad (8.222)$$

the former on the wing and the latter on the plate.

It may be easily seen that the function $F_{n-1,0,1}$ is the only $F_{n-q-r,q,r}$ function that assumes the wing and plate to be a streamline ($d\Phi'_{n-1,0,1} = 0$), in the sense of the hydrodynamic analogy, and which is to be found in the two unknown expressions (8.22b) and (8.23a).

Employing the compatibility relations again, we have

$$dF_{n-q-1,q,1} = \frac{dF_{n-1,0,1}}{(-x)^q}, \quad dF_{n-r,0,r} = \left(\frac{\sqrt{1-B^2x^2}}{ix}\right)^{r-1} dF_{n-1,0,1}, \qquad (8.223)$$

respectively, so that formulae (8.22b) and (8.23a) may be put in the form

$$\frac{(n-1)!}{x_1^{n-1}}\, w = \mathrm{Re} \sum_{q=0}^{n-1} C_{n-1}^{q}(-x)^q \int_{\text{Mach circle}}^{\text{wing}} \frac{dF_{n-1,0,1}}{x^q}, \tag{8.224a}$$

$$\frac{(n-1)!}{x_1^{n-1}}\, u = \mathrm{Re} \sum_{q=0}^{n-1} C_{n-1}^{r}\left(\frac{-ix}{\sqrt{1-B^2x^2}}\right)^r \int_{\text{Mach circle}}^{\text{plate}} \left(\frac{\sqrt{1-B^2x^2}}{ix}\right)^{r-1} dF_{n-1,0,1}. \tag{8.224b}$$

Hence we note that the problem of determining the components w and u reduces to finding the function $F_{n-1,0,1}$.

In the direct problem (§ VIII. 1), the determination of the function u on the wing and plate had been reduced to the determination of the function F_{n00} whose imaginary part is constant on the wing and plate; this property has allowed us to determine it in a very simple manner, by employing a hydrodynamic analogy.

In the following we shall give the boundary conditions that the function $F_{n-1,0,1}$ should satisfy, and meanwhile show the equivalence that should exist between these conditions and those of the function F_{n00} in the direct problem. Taking into account this equivalence, it is not necessary to determine the expression for the function $F_{n-1,0,1}$ directly; it is sufficient to know the equivalent function F_{n00} in the direct problem and consider the corresponding formulae.

We shall first study the behaviour of the function $F_{n-1,0,1}$ around the subsonic edges of the plate and wing, as well as around a ridge situated on the wing, which separates regions with different distributions of pressures.

For the behaviour around the subsonic edges of the plate, say P_1 and P_2, we shall consider two cases: a thin plate and a plate of symmetrical thickness. Since the problem is a direct one on the plate, u, v, w have the same behaviour at the plate edges in the mixed problem as in the direct problem. Taking into account the considerations stated for F_{n00} in the direct problem, it follows that $F_{n-1,0,1}$ should contain the singularities

$$\frac{1}{(x \pm i\mathfrak{h})^{q-1/2}}, \quad (q = 1, 2, \ldots, n)$$

in the x plane or

$$\frac{1}{X^{2p-1}}, \quad (p = 1, 2, \ldots, n)$$

in the X plane, in the case of the *thin* wing, and the singularities

$$\ln(x \pm i\mathfrak{h}), \quad \frac{1}{(x \pm i\mathfrak{h})^p}, \quad (p = 1, 2, \ldots, n-1)$$

in the x plane or

$$\ln X, \quad \frac{1}{X^{2p}}, \quad (p = 1, 2, \ldots, n-1)$$

in the X plane, in the case of the *thick* plate.

For the behaviour around a ridge that separates two regions on the wing with different pressure distributions, we assume that at the point $y = s$, $\Phi_{n-q,q,0}$ passes from the value $u^{(1)}_{n-q-1,q}$, to the value $u^{(2)}_{n-q-1,q}$. It follows that $F_{n-q,q,0}$, hence $F_{n-1,0,1}$ too, contain the singularities

$$\ln (x-l), \quad \frac{1}{(x-l)^p}, \quad (p = 1, 2, \ldots, n-1)$$

in the x plane or

$$\ln (X-L), \quad \frac{1}{(X-L)^p}, \quad (p = 1, 2, \ldots, n-1)$$

in the X plane, similarly to the function F_{n00} of the direct problem, in the case of a ridge that separates two regions on the wing with different incidence distributions.

In the mixed problem the subsonic edge of a *thin* wing plays the part of a ridge, since it separates a region where $\Phi_{n-q,q,0} = 0$ from another region where $\Phi_{n-q,0,0} = \text{const.} \neq 0$; as is known, in the direct problem the subsonic edge of a *thick* wing plays the part of a ridge.

The behaviour around the subsonic edge of a thick wing which separates a region where $\Phi_{n-q-1,q,1} = \text{const.} = 0$ from another region where $\Phi_{n-q,q,0} = \text{const.} \neq 0$, is the same as that in the direct problem around the subsonic edge of a *thin* wing which separates a region where $\Phi_{n-q-1,q,1} = \text{const.} \neq 0$ from a region where $\Phi_{n-q,q,0} = \text{const.} = 0$. It follows hence that in the mixed problem u, v and w have, around the subsonic edge of a *thick* wing, the same behaviour as in the direct problem around the subsonic edge of a *thin* wing. Consequently, at the point $y = l_1$, corresponding to the subsonic edge A_1, exactly like F_{n00} in the direct problem, the function $F_{n-1,0,1}$ should contain the singularities

$$\frac{1}{(x-l_1)^{p-1/2}}, \quad (p = 1, 2, \ldots, n)$$

in the x plane, or

$$\frac{1}{(X-L_1)^{p-1/2}}, \quad (p = 1, 2, \ldots, n)$$

in the X plane.

On the other hand, we notice from formulae (8.224) that it is necessary that w and u have finite values at the point $x = 0$ $(X = \pm\mathfrak{h})$, in order that $F_{n-1,1,0}$ should not have singularities at that point.

Employing the compatibility relations written in the form

$$dF_{n-1,\,0,\,1} = i\sqrt{1-B^2x^2}\; dF_{n-1,\,1,\,0}, \tag{8.225}$$

it follows that neither the function $F_{n-1,\,1,\,0}$ nor the function $F_{n-1,\,1,\,0}$ should contain singularities at the origin.

This function $F_{n-1,\,1,\,0}$ is to be found in (8.22), (8.23), and its real part, constant on the wing and plate, should have the same value; consequently, we must consider

$$u_{n-2,\,1} = v_{n-1,\,0}, \quad (n \geq 2) \tag{8.226}$$

in (8.220) in order that the function $F_{n-1,\,1,\,0}$ should not have logarithmic singularities at $x = 0$ ($X = \pm\mathfrak{h}$).

For $n = 1$, the component $v = \Phi_{010}$ should assume the same constant value on the wing and plate.

Taking into account the results obtained in the direct problem, we shall determine the contribution of each edge of the wing–plate system, and the expression for the function $F_{n-1,\,0,\,1}$ representing the sum of these contributions.

Thus, for example:

1. The contribution of the subsonic edge A_2 ($X = -L_2$) of a *thick* wing, by considering the edge A_1 ($X = L_1$) to be also subsonic, will be

$$F^{(2)}_{n-1,\,0,\,1} = \frac{a_0 + a_1X + \ldots + a_{n-2}X^{n-2}a_{n-1}X^{n-1}}{(X+L_2)^{n-1}}\sqrt{\frac{L_1-X}{X+L_2}}\,; \tag{8.227}$$

likewise, the contribution of the subsonic edge A_1, by considering the edge A_2 to be also subsonic, will be

$$F^{(1)}_{n-1,\,0,\,1} = \frac{b_0 + b_1X + \ldots + b_{n-2}X^{n-2} + b_{n-1}X^{n-1}}{(L_1-X)^{n-1}}\sqrt{\frac{X+L_2}{L_1-X}}\,. \tag{8.228}$$

2. The contribution of the subsonic edge A_1 ($X = L_1$) of a thin wing is given by the expression

$$F^{(w)}_{n-1,\,0,\,1} = Q\cosh^{-1}\sqrt{\frac{(1+\mathcal{B}L_1)(1-\mathcal{B}X)}{2\mathcal{B}(L_1-X)}} + \frac{+q_0 + q_1X + \ldots + q_{n-2}X^{n-2}}{(L_1-X)^{n-1}}\sqrt{\frac{1-\mathcal{B}^2X^2}{\mathcal{B}^2}}\,. \tag{8.229}$$

3. The contribution of the subsonic edges of the *thin* plate is given by the relation

$$F^{(p)}_{n-1,\,0,\,1} = \frac{d_0+d_1X+\ldots+d_{2n-2}X^{2n-2}}{X^{2n-1}}\sqrt{(X+L_2)(L_1-X)} \tag{8.230}$$

in the case of the *thick* wing with subsonic edges A_2 and A_1. In the case when one or both wing edges are supersonic, L_1 or L_2, or L_1 and L_2, respectively, are replaced by $1/\mathcal{B}$. In the case of the thin wing, the above formula is valid, obviously replacing L_1 and L_2 by $1/\mathcal{B}$.

4. The contribution of the subsonic edges of the *thick* plate is given by the formula

$$F^{(p)}_{n-1,\,0,\,1} = Q\cosh^{-1}\sqrt{\frac{(X+L_2)L_1}{X(L_1+L_2)}}$$
$$= \frac{a_0+a_1X+\ldots+a_{2n-3}X^{2n-3}}{X^{2n-2}}\sqrt{(X+L_2)(L_1-X)} \tag{8.231}$$

in the case of the *thick* wing having both the edges A_2 and A_1 subsonic. In the case when one or both edges are supersonic, then L_1 or L_2, or L_1 and L_2, respectively, are replaced by $1/\mathcal{B}$. In the case of the thin wing, the formula is valid, replacing L_1 and L_2 by $1/\mathcal{B}$.

5. Taking into account the results obtained in the direct problem, we can similarly write down the effect of a supersonic edge of a *thick* wing; this contribution is given by the formula

$$F^{(1)}_{n-1,\,0,\,1} = Q\cosh^{-1}\sqrt{\frac{(L_1+L_2)(1-\mathcal{B}X)}{(1+\mathcal{B}L_2)(L_1-X)}}$$
$$+\frac{a_0+a_1X+\ldots+a_{n-2}X^{n-2}}{(L_1-X)^{n-1}}\sqrt{\frac{(X+L_2)(1-\mathcal{B}X)}{\mathcal{B}}}, \tag{8.232}$$

considering the edge A_1 $(X = L_1)$ supersonic and A_2 $(X = -L_2)$ subsonic.

The formula

$$F^{(2)}_{n-1,\,0,\,1} = Q\cosh^{-1}\sqrt{\frac{(L_1+L_2)(1+\mathcal{B}X)}{(1+\mathcal{B}L_1)(L_2+X)}}$$
$$+\frac{a_0+a_1X+\ldots+a_{n-2}X^{n2}}{(X+L_2)^{n-1}}\sqrt{\frac{(L_1-X)(1+\mathcal{B}X)}{\mathcal{B}}} \tag{8.233}$$

represents the effect of the supersonic edge A_2, the edge A_1 being subsonic.

If both edges of the thick wing are supersonic, the contribution of an edge, say A_2 $(X=-L_2)$, has the expression (8.229) in which we put $L_1=-L_2$.

VIII. 8.2. Application to the True Conical Motion ($n = 1$)

Now we shall treat a complete example in order to point out the manner in which the general results obtained above may be applied, by considering, for simplicity, $n = 1$.

We shall consider a thin wing A_2A_1 fitted with a thin plate P_1P_2, the wing-plate system being entirely comprised in the interior of the Mach cone (Fig. 8.1).

For the moment we assume that on the wing upper surface $u = u_2$ on the left side of the plate, and $u = u_1$ on the right side of the plate, while $v = v_1$ on the plate.

Employing formulae (8.229) and (8.230) we obtain

$$F_{001} = \mathcal{W} = \frac{Q_1}{\pi}\cosh^{-1}\sqrt{\frac{(1+\mathcal{B}L_1)(1-\mathcal{B}X)}{2\mathcal{B}(L_1-X)}} + \frac{Q_2}{\pi}\cosh^{-1}\sqrt{\frac{(1+\mathcal{B}L_2)(1+\mathcal{B}X)}{2\mathcal{B}(X+L_2)}} + \frac{A_0}{X}\sqrt{\frac{1-\mathcal{B}^2X^2}{\mathcal{B}^2}}. \tag{8.234}$$

In order to determine the constants we act in the following way:

1. Taking into account the expression for $\mathcal{W}$, there result logarithmic singularities of the form $\ln(X \pm 1/\mathcal{B})$ in $\mathcal{U} = F_{100}$ and $\mathcal{V} = F_{010}$. In order to avoid this physical incompatibility, we impose $\left(\frac{d\mathcal{W}}{dX}\right)_{x=\pm 1/\mathcal{B}} = 0$ and obtain the following two relations:

$$Q_1\sqrt{\frac{1+\mathcal{B}L_1}{1-\mathcal{B}L_1}} + Q_2\sqrt{\frac{1-\mathcal{B}L_2}{1+\mathcal{B}L_2}} + 2\pi A_0 = 0, \tag{8.235a}$$

$$Q_1\sqrt{\frac{1-\mathcal{B}L_1}{1+\mathcal{B}L_1}} + Q_2\sqrt{\frac{1+\mathcal{B}L_2}{1-\mathcal{B}L_2}} - 2\pi A_0 = 0. \tag{8.235b}$$

2. At the point $X = L_1$, on the real axis, u passes from the value $u = 0$ of the right side to the value $u = u_1$ of the left side; employing the compatibility relation between the functions $\mathcal{U}$ and $\mathcal{W}$ and integrating round a semicircle γ with its centre at $X = L_1$ from a point $L_1 - \varepsilon$ up to a point $L_1 - \varepsilon$, we obtain

$$u_1 = \text{Re}\oint_\gamma \frac{ix}{\sqrt{1-B^2x^2}}\,d\mathcal{W} = \frac{Q_1 l_1}{2\sqrt{1-B^2l_1^2}}, \tag{8.236}$$

whence

$$Q_1 = \frac{2u_1\sqrt{1-B^2l_1^2}}{l_1}. \tag{8.237}$$

Using again the compatibility relation existing between $\mathcal{U}$ and $\mathcal{V}$, it is readily seen that at the point $X = L_1$, v passes from the value $v = 0$ to $v = -u_1/l_1$ on the part of the wing on the right-hand side of the plate.

We should therefore consider $v_1 = -u_1/l_1$ on the plate, in order that v shall have no jump at $X+\mathfrak{y}$.

For a given value of u_1, $v_1 (v_1 = -u_1/l_1)$ tends towards infinity if l_1 tends towards zero; hence a new restriction should be imposed, namely, for $v = v_1$ on the plate, one should consider $u = u_1 = -v_1 l_1$ on the wing portion on the right-hand side of the plate.

3. Since the plate is thin, one should have $v = v_1$ on both faces of the plate. Employing the compatibility relation between $\mathcal{U}$ and $\mathcal{W}$, it may be seen that v has no jump at $X = 0$.

We should likewise consider $v = v_1$ on the wing, on the left hand side of the plate, a necessary condition in order that v have no jump at $X = -\mathfrak{h}$.

We shall likewise impose that v should pass from $v = v_1$ to $v = 0$ at the point $X = -L_2$; thus we obtain

$$v_1 = \operatorname{Re} \oint_\gamma \frac{i d\mathcal{W}}{\sqrt{1-B^2x^2}} = \frac{Q_2}{2\sqrt{1-B^2 l_2^2}}, \tag{8.238}$$

and hence

$$Q_2 = 2v_1 \sqrt{1-B^2 l_2^2}. \tag{8.239}$$

In addition we deduce that u assumes the following value on the wing, on the left-hand side of the plate:

$$u_2 = \frac{Q_2 l_2}{2\sqrt{1-B^2 l^2}} = v_1 l_2. \tag{8.240}$$

We note that in the mixed problem that we are dealing with, we are obliged to renounce the generality of the problem, in order that we should not have physical incompatibilities.

Further, from any of the two relations obtained above, by replacing Q_1 and Q_2 by their respective values, it follows that

$$A_0 = \frac{v_1}{\pi} B(L_1+L_2). \tag{8.241}$$

Finally we can write

$$\mathcal{W} = -\frac{2v_1}{\pi}\Bigg[\sqrt{1-B^2 l_1^2}\cosh^{-1}\sqrt{\frac{(1+\mathcal{B}L_1)(1-\mathcal{B}X)}{2\mathcal{B}(L_1-X)}}$$

$$-\sqrt{1-B^2 l_2^2}\cosh^{-1}\sqrt{\frac{(1+\mathcal{B}L_2)(1+\mathcal{B}X)}{2\mathcal{B}(X+L_2)}}$$

$$+\frac{v_1}{\pi} B(L_1+L_2)\frac{1}{X}\sqrt{\frac{1-\mathcal{B}^2X^2}{\mathcal{B}^2}}\Bigg]. \tag{8.242}$$

Having the expression for $\mathcal{W}$, we get

$$\frac{d\mathcal{U}}{dX} = \frac{iv_1(L_1+L_2)[L_1L_2+(L_1-L_2)X]}{\pi(X-L_1)(X+L_2)X^2}\sqrt{X^2-\mathfrak{h}^2}, \tag{8.243}$$

whence, by means of an integration, we obtain the expression for $\mathcal{U}$,

$$\mathcal{U} = \frac{iv_1l_2}{\pi}\ln\frac{\mathfrak{h}^2+XL_2-\sqrt{(X^2-\mathfrak{h}^2)(L_2^2-\mathfrak{h}^2)}}{X+L_2}$$
$$+\frac{iv_1l_1}{\pi}\ln\frac{\mathfrak{h}^2-XL_1+\sqrt{(X^2-\mathfrak{h}^1)(L_1^2-\mathfrak{h}^2)}}{L_1-X}+i\frac{v_1(L_1+L_2)}{\pi}\frac{\sqrt{X^2-\mathfrak{h}^2}}{X}. \tag{8.244}$$

Making $\mathfrak{h}$ tend towards zero, we obtain $\mathcal{W}$ in the case of a thin wing having $u = v_1l_2$ on the segment A_2O and $u = -v_1l_1$ on the segment OA_1 (the indirect problem).

Setting $l_1 = l_2 = l$ and then $l \to 0$, we obtain $\mathcal{U}$ in the case of the isolated thin plate having $v = v_1$ on the upper surface and $v = v_1$ on the lower surface (the direct problem).

VIII. 8.3. *Cruciform Wing having Given Pressure on the Plate and Given Normal Velocity on the Wing*

In the following we are dealing with the mixed problem in which w on the wing and u on the plate are given, both velocities being put in the form of homogeneous polynomials of order $n-1$; thus we have on the wing and plate, respectively,

$$w = \sum_{q=0}^{n-1} w_{n-q-1,\,q}\,x_1^{n-q-1}\,x_2^q = x_1^{n-1}\sum_{q=0}^{n-1} w_{n-q-1,\,q}\,y^q, \tag{8.245a}$$

$$u = \sum_{r=0}^{n-1} u_{n-r-1,\,r}\,x_1^{n-r-1}\,x_3^r = x_1^{n-1}\sum_{r=0}^{n-1} u_{n-r-1,\,r}\,z^r. \tag{8.245b}$$

The constants $w_{n-q-1,\,q}$, $u_{n-r-1,\,r}$ may assume different values on the portions of the wing and plate, respectively.

The problem consists in determining the velocity u on the wing and the velocity v on the plate.

Taking into account the conditions indicated above, the following relations result from formulae (9.22) and (8.23),

$$d\Phi_{n-q-1,\,q,\,1} = 0, \quad d\Phi_{n-r,\,0,\,r} = 0, \tag{8.246}$$

the former on the wing and the latter on the plate.

Employing the compatibility relations we obtain

$$d\Phi'_{n-1,\,1,\,0} = 0. \tag{8.247}$$

Consequently, in this problem as in the preceding one, we find a function $F_{n-1,1,0}$ which admits the wing and plate to be streamlines; on the other hand, the same function is to be found in the required expressions of u on the wing and v on the plate, (8.22a) and (8.23b), respectively.

These formulae may be written in the following form:

$$\frac{(n-1)!}{x_1^{n-1}} u = \text{Re} \sum_{q=0}^{n-1} C_{n-1}^q x^q \int_{\text{Mach circle}}^{\text{wing}} \frac{dF_{n-1,1,0}}{(-x)^{q-1}}, \tag{8.248a}$$

$$\frac{(n-1)!}{x_1^{n-1}} v = \text{Re} \sum_{r=0}^{n-1} C_{n-1}^r \left(\frac{-ix}{\sqrt{1-B^2x^2}}\right)^r$$

$$\times \int_{\text{Mach circle}}^{\text{plate}} \left(\frac{\sqrt{1-B^2x^2}}{ix}\right)^r dF_{n-1,1,0}. \tag{8.248b}$$

Consequently, as is seen, the problem reduces to the determination of the function $F_{n-1,1,0}$.

The boundary problem which is to be set for $F_{n-1,1,0}$ is similar to the problem set for $F_{n-1,0,1}$ in the previous mixed problem.

The function $F_{n-1,1,0}$ is to be constructed in the same way as the function $F_{n-1,0,1}$ was in the preceding problem; namely, by taking account of the effect of each edge of the wing–plate system.

Since the problem is a direct one on the wing, the contribution of a wing edge is identical with the contribution of the same edge from the expression of the function F_{n00} in the direct problem. Thus, the contribution of a subsonic edge A_2 $(X = -L_2)$ of a thin wing is given by formula (8.227), considering the edge A_1 $(X = L_1)$ to be also subsonic. Formula (8.228) represents the effect of the subsonic edge A_1, the edge A_2 being also subsonic.

Formulae (8.232) and (8.233) represent the contribution of the supersonic edge A_1 of the thin wing, the edge A_2 being subsonic and, respectively, the contribution of the supersonic edge A_2, the edge A_1 being subsonic; the effect of the supersonic edge of a thin or thick wing, the other edge being also supersonic, is given for A_2 and A_1 by (8.232), (8.233), respectively, with $L_2 = 1/\mathcal{B}$ in the former and $L_1 = 1/\mathcal{B}$ in the latter formula. The contribution of a subsonic or supersonic edge of a thick wing is given by formula (8.229).

As regards the behaviour of the function $F_{n-1,1,0}$ around the plate edges, it should be remarked, taking into account the given boundary conditions, that in our problem, this behaviour around the edge of a lifting plate is analogous with the behaviour of the function F_{n00} around the edge of a thick wing in the direct problem; as regards the behaviour around the edge of a zero lift plate, this behaviour is analogous with that of the function F_{n00} around the edge of a thin wing in the direct problem. Thus, the effect of the edge of a lifting plate is given by formula (8.231) in the case of the *thin* wing with both edges A_2 and A_1 subsonic.

In the case when one or both edges are supersonic, L_1 or L_2, or L_1 and L_2, respectively, are replaced by $1/\mathcal{B}$. In the case of the thick wing, formula (8.221) is valid, obviously replacing L_1 and L_2 by $1/\mathcal{B}$.

The contribution of the subsonic edges of the zero lift wing is given by relation (8.230) in the case of the thick wing, obviously replacing L_1 and L_2 by $1/\mathcal{B}$.

Having the contributions of the edges of the wing and plate we can construct the function $F_{n-1,1,0}$ and then determine the axial velocity u on the wing and the velocity v normal to the plate.

CHAPTER IX

SIMPLE OR CRUCIFORM WING FITTED WITH A CONICAL BODY

In general, simple complex or cruciform wings, as defined in the previous chapters, do not exist by themselves; together with the other parts of a supersonic aircraft or of a rocket they form *complex systems* in which, from the aerodynamic standpoint, the various parts interfere with one another.

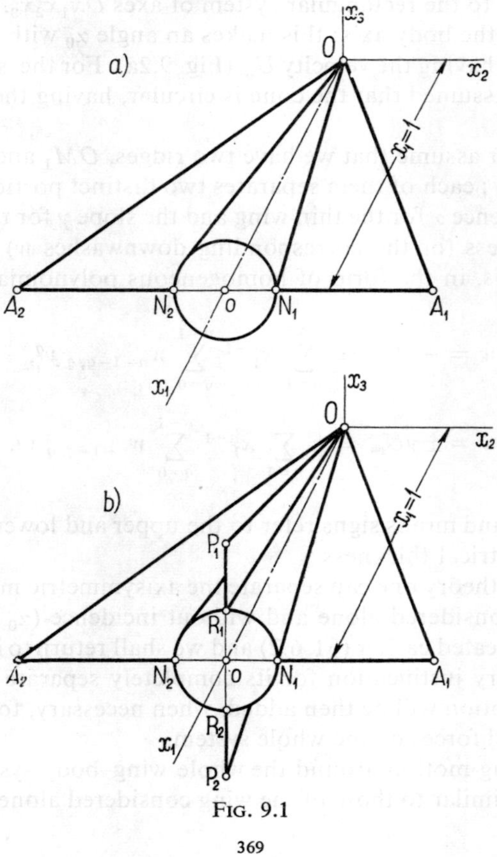

FIG. 9.1

The two principal parts, which are closeely relatd, are the wing and the body or the cruciform wing and any slender body, which together constitute the principal system. In the light of the results obtained for high-order conical flows and for solving this problem easily, we shall simplify it by considering a conical body having the same apex as the wing (Fig. 9.1).

This is in fact the object of the present chapter. By treating this schematized system, we shall be able to give rigorous solutions which could constitute useful indications for real practical cases.

IX. 1. Fundamental Preliminary Relations

IX. 1.1. Geometrical Definition

THE WING–CONICAL BODY SYSTEM. Let us consider a wing–conical body system referred to the rectangular system of axes $Ox_1x_2x_3$, having the axis Ox_1 parallel to the body axis; this makes an angle α_0 with the direction of the free stream having the velocity U_∞ (Fig. 9.2a). For the sake of simplicity, it will be assumed that the cone is circular, having the radius c in the section $x_1 = 1$.

Let us further assume that we have two ridges, OM_1 and OM_2, on each side of the body; each of them separates two distinct portions of the wing, where the incidence α for the thin wing and the slope γ for the wing of symmetrical thickness (or the corresponding downwashes w) vary according to different laws, in the form of homogeneous polynomials in x_1 and x_2 (§ II. 1.2):

$$w = -\alpha U_\infty = \sum_{n=1}^{N} x_1^{n-1} \sum_{q=0}^{n-1} w_{n-1-q,\,q}\, y^q, \tag{9.1a}$$

$$w = \pm\gamma U_\infty = \pm \sum_{n=1}^{N} x_1^{n-1} \sum_{q=0}^{n-1} w_{n-1-q,\,q}\, y^q, \tag{9.1b}$$

where the plus and minus signs refer to the upper and lower surfaces of the wing of symmetrical thickness.

In linearized theory one can separate the axisymmetric motion round the conical body considered alone and without incidence ($\alpha_0 = 0$). This motion has been treated earlier (§ I. 6.2) and we shall return to it below, by giving the necessary justification for its completely separate treatment. The effect of this motion will be then added, when necessary, to the calculation of pressures and forces on the whole system.

The remaining motion around the whole wing–body system may represent positions similar to those of the wing considered alone (isolated):

1. Thin wing with subsonic edges or with one subsonic and one supersonic edge, or with both edges supersonic, the body having an arbitrary incidence α_0.

2. Wing of symmetrical thickness with subsonic or supersonic edges, the body having no incidence ($\alpha_0 = 0$).

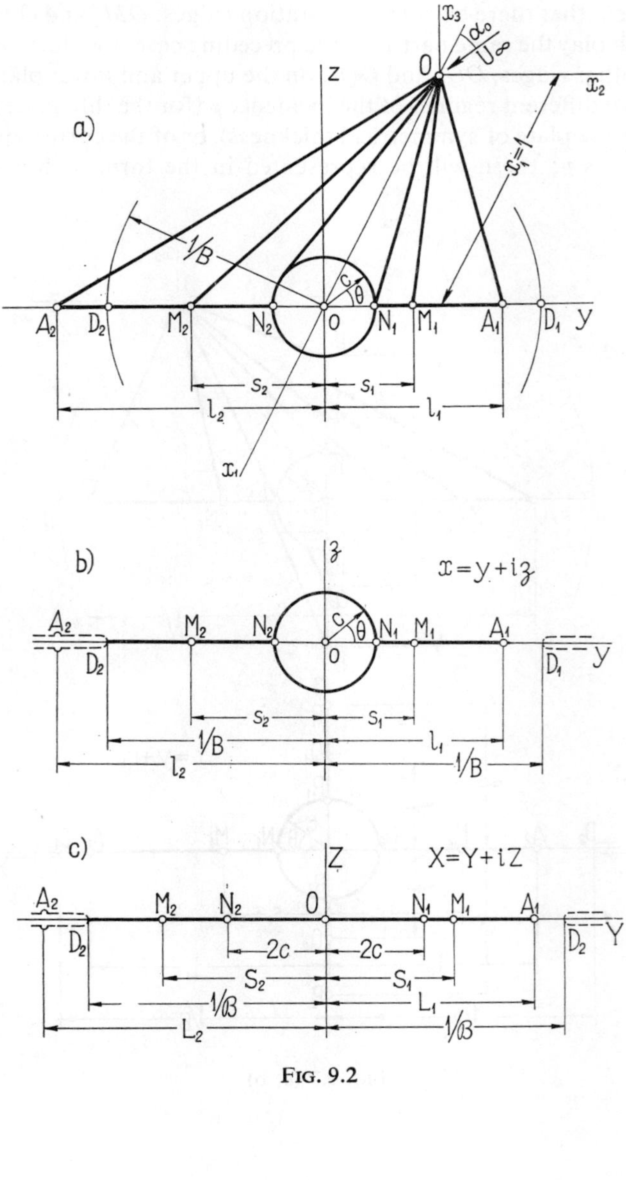

FIG. 9.2

THE CRUCIFORM WING–CONICAL BODY SYSTEM. This system differs from the preceding one only by the vertical plates R_1P_1 and R_2P_2 of equal height (Fig. 9.3a).

The body axis, however, may have, apart from the incidence α_0 in the vertical plane Ox_1x_3, the lateral incidence β_0 in the horizontal plane Ox_1x_2 (Fig. 9.4).

We assume that there exist two separation ridges, OM_1 and OM_2, on the wing, which play the same part as in the preceding case. In addition, we consider two other ridges, OQ_1 and OQ_2, on the upper and lower plates, which separate two different régimes of the incidence β (for the thin plate) or of the slope ν (for the plate of symmetrical thickness), or of the corresponding lateral velocities v; these will be represented in the form of homogeneous

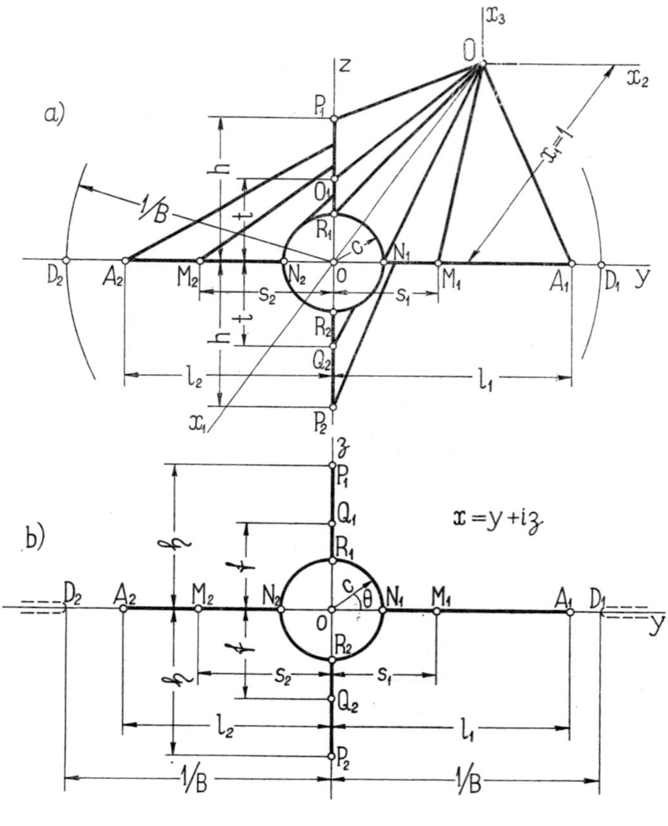

FIG. 9.3 (a, b)

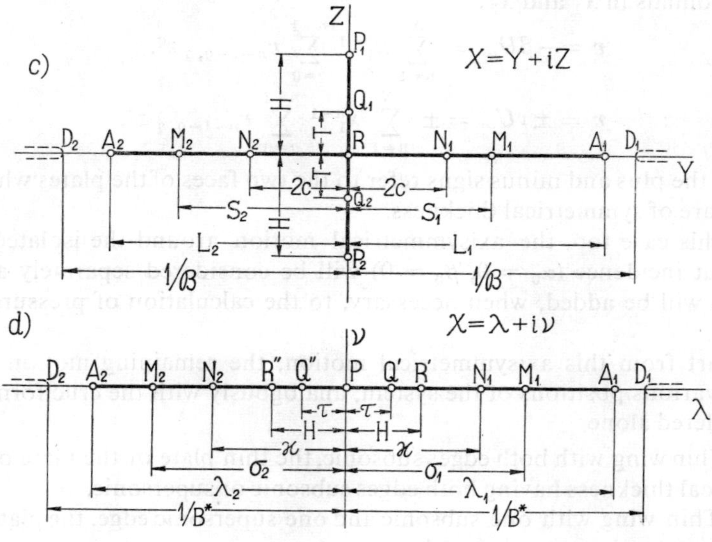

FIG. 9.3 (c, d)

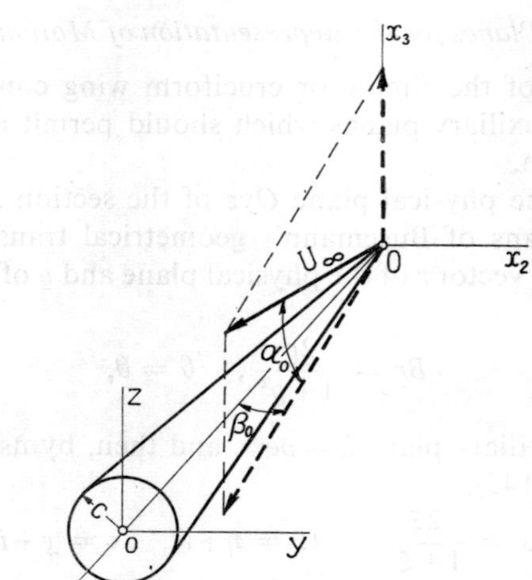

FIG. 9.4

polynomials in x_1 and x_3:

$$v = -\beta U_\infty = \sum_{n=1}^{N} x_1^{n-1} \sum_{q=0}^{n-1} v_{n-1-q,\,q}\, z^q, \tag{9.2a}$$

$$v = \pm \nu U_\infty = \pm \sum_{n=1}^{N} x_1^{n-1} \sum_{q=0}^{n-1} v_{n-1-q,\,q}\, z^q, \tag{9.2b}$$

where the plus and minus signs refer to the two faces of the plates when the latter are of symmetrical thickness.

In this case too, the axisymmetrical motion around the isolated cone without incidence ($\alpha_0 = 0$, $\beta_0 = 0$) will be considered separately and its effects will be added, when necessary, to the calculation of pressures and forces.

Apart from this axisymmetrical motion, the remaining motion represents various positions of the system, analogously with the cruciform wing considered alone:

1. Thin wing with both edges subsonic, the thin plate or the plate of symmetrical thickness having both edges subsonic or supersonic,
2. Thin wing with one subsonic and one supersonic edge, the plate having the same charcteristics as above.
3. Wing of symmetrical thickness. the plate having the same characteristics as above.

In the following we are going to treat certain cases of the motion around the body–wing system or body–cruciform wing system.

IX. 1.2. Auxiliary Planes for the Representation of Motion

As in the case of the simple or cruciform wing considered alone, we employ various auxiliary planes which should permit a convenient solution of the problem.

Starting from the physical plane Oyz of the section $x_1 = 1$ (Figs. 9.2a and 9.3a), by means of Busemann's geometrical transformation (1.138) between the radius vector r of the physical plane and ϱ of an auxiliary plane $O\eta\zeta$,

$$Br = \frac{2\varrho}{1+\varrho^2}, \qquad \theta = \theta, \tag{9.3}$$

we pass to the auxiliary plane $\xi = \varrho e^{i\theta}$, and then, by using the conformal transformation (1.145),

$$Bx = \frac{2\xi}{1+\xi^2}, \qquad (\xi = \eta + i\zeta, \quad x = y + i\mathfrak{z}) \tag{9.4}$$

we obtain the x plane having the configuration shown in Figs. 9.2b and 9.3b.

The wing trace and the horizontal diameter of the circle of radius c, as well as the whole abscissa axis, remain the same as for the physical plane; the ordinates, however, change. Thus, by setting $r = z$, $\varrho = \zeta$ into (9.3) and $\xi = i\zeta$ and $x = i\mathfrak{z}$ into (9.4), we obtain

$$Bz = \frac{2\zeta}{1+\zeta^2}, \quad B\mathfrak{z} = \frac{2\zeta}{1-\zeta^2}, \tag{9.5}$$

whence we deduce

$$\mathfrak{z} = \frac{z}{\sqrt{1-B^2z^2}}, \quad z = \frac{\mathfrak{z}}{\sqrt{1+B^2\mathfrak{z}^2}}. \tag{9.6}$$

It is seen that the circle radius in the vertical direction becomes k,

$$k = \frac{c}{\sqrt{1-B^2c^2}} \approx c(1+\tfrac{1}{2}B^2c^2) \approx c. \tag{9.7}$$

The circle is hence stretched in the vertical direction and becomes of approximately elliptical shape; but we have assumed above that the radius c is small enough with respect to the radius $1/B$ of the Mach cone, that is

$$B^2c^2 \ll 1. \tag{9.8}$$

Even if this distortion of the circle of radius c in the physical plane is more important, the boundary conditions in the auxiliary plane x may be considered on the circle of radius c.

Likewise, the plate height h, or any point $z = t$ on the plate, become respectively

$$\mathfrak{h} = \frac{h}{\sqrt{1+B^2h^2}}, \quad \mathfrak{t} = \frac{t}{\sqrt{1-B^2t^2}} \tag{9.9}$$

in the x plane.

Further we shall use a new auxiliary plane,

$$X = x+\frac{c^2}{x}, \quad (X = Y+iZ) \tag{9.10}$$

in which we obtain the configurations shown in Figs. 9.2c and 9.3c.

For the body–cruciform wing system, we shall perform yet another transforma tion of the form

$$\chi^2 = X^2+H^2, \tag{9.11}$$

where

$$H = \mathfrak{h}-\frac{c^2}{\mathfrak{h}}, \tag{9.12}$$

obtaining in this manner the configuration shown in Fig. 9.3d.

In the auxiliary plane X, the geometrical magnitudes from the physical plane Oyz (s_j, l_j, $1/B$, t) or from the auxiliary plane x (in which the abscissae remain unchanged) become

$$S_j = s_j + \frac{c^2}{s_j}, \quad L_j = l_j + \frac{c^2}{l_j}, \quad (j = 1, 2), \tag{9.13a}$$

$$\frac{1}{\mathcal{B}} = \frac{1+B^2c^2}{B}, \quad T = \mathrm{t} - \frac{c^2}{\mathrm{t}}, \tag{9.13b}$$

respectively.

In the auxiliary plane χ, we shall likewise write

$$\sigma_j = \sqrt{H^2+S_j^2}, \qquad \lambda_j = \sqrt{H^2+L_j^2}, \quad (j = 1, 2), \tag{9.14a}$$

$$\varkappa = \sqrt{H^2+4c^2}, \quad \frac{1}{B^*} = \sqrt{H^2+\frac{1}{\mathcal{B}^2}}, \quad \tau = \sqrt{H^2-T^2}. \tag{9.14b}$$

IX. 1.3. Flow Conditions on the Body, around the Ridges and the Leading Edges

BODY. The flow conditions on the body can be easily examined in the auxiliary x plane (Figs. 9.2b, 9.3b). We shall first note that, at an arbitrary point P, the direction cosines of the normal are

$$n_1 \approx -c, \quad n_2 \approx \cos\theta, \quad n_3 \approx \sin\theta, \tag{9.15}$$

respectively.

In the general case we assume that the body axis has the incidence α_0 in the plane Ox_1x_3 and β_0 in the plane (Fig. 9.4); the velocity components in the direction of the three axes at any point on the body are as follows:

$$u^* + U_\infty, \quad v^* + \beta_0 U_\infty, \quad w^* + \alpha_0 U_\infty, \tag{9.16}$$

where u^*, v^*, w^* are the disturbance velocities.

From the condition that the normal velocity on the body surface must be equal to zero, we deduce

$$(v^*+\beta_0 U_\infty)\cos\theta + (w^*+\alpha_0 U_\infty)\sin\theta = c(u^*+U_\infty) \approx cU_\infty. \tag{9.17}$$

In the axisymmetric flow around the cone, by setting $\alpha_0 = \beta_0 = 0$ in the above formulae and designating by u_c, v_c, w_c the velocities of this motion, we have similarly

$$v_c \cos\theta + w_c \sin\theta = c(u_c + U_\infty) \approx cU_\infty \tag{9.18}$$

at the same point on the body; further, if this axisymmetric motion is separated from the total flow around the system considered, as was already done above, we obtain the following condition for the residual velocities on the body, which have been customarily denoted by u, v, w:

$$(v+\beta_0 U_\infty)\cos\theta+(w+\alpha_0 U_\infty)\sin\theta \approx 0. \tag{9.19}$$

The downwashes w on the wing have been defined by (9.1), and the lateral velocities v on the plate by (9.2).

BEHAVIOUR OF THE DISTURBANCE VELOCITIES IN THE PROXIMITY OF THE LEADING EDGES AND THE SEPARATION RIDGES. This behaviour depends only on the local configuration of the subsonic leading edges (in the case of thin wings) or of the ridges. The principles set up for the wing in Chapter II and applied to the cruciform wing in Chapter VIII are also applicable to this problem in an identical manner; thus, for instance, in the auxiliary plane (X of Fig. 9.2c and χ of Fig. 9.3d), the behaviour of the analytic functions which define the disturbance velocities is exactly the same, since the conformal transformations (9.10) and (9.11) do not alter the character of this behaviour. Hence, instead of (2.35, 2.36) for the simple wing and (8.32) for the cruciform wing, we shall have the following behaviour for the subsonic edges of the wing given in Fig. 9.2c ($Y = L_1$, $Y=-L_2$),

$$\frac{1}{(L_1-X)^{1/2}},\ \frac{1}{(L_1-X)^{3/2}},\ \ldots,\ \frac{1}{(L_1-X)^{n-1/2}}, \tag{9.20a}$$

$$\frac{1}{(L_2+X)^{1/2}},\ \frac{1}{(L_2+X)^{3/2}},\ \ldots,\ \frac{1}{(L_2+X)^{n-1/2}}; \tag{9.20b}$$

and for the subsonic edges of the wing given in Fig. 9.3d ($\chi = \lambda_1$, $\chi = -\lambda_2$, $\chi = 0$) respectively

$$\frac{1}{(\lambda_1-\chi)^{1/2}},\ \frac{1}{(\lambda_1-\chi)^{3/2}},\ \ldots,\ \frac{1}{(\lambda_1-\chi)^{n-1/2}}, \tag{9.21a}$$

$$\frac{1}{(\lambda_2+\chi)^{1/2}},\ \frac{1}{(\lambda_2+\chi)^{3/2}},\ \ldots,\ \frac{1}{(\lambda_2+\chi)^{n-1/2}}, \tag{9.21b}$$

$$\frac{1}{\chi},\ \frac{1}{\chi^3},\ \ldots,\ \frac{1}{\chi^{2n-1}}. \tag{9.21c}$$

Likewise, analogously with (2.43), we shall have the following singularities on the wing ridges belonging to the first system (Fig. 9.2c, $Y = S_1$, $Y = -S_2$),

$$\ln(S_1-X),\ \frac{1}{S_1-X},\ \ldots,\ \frac{1}{(S_1-X)^{n-1}}, \tag{9.22a}$$

$$\ln(S_2+X),\ \frac{1}{S_2+X},\ \ldots,\ \frac{1}{(S_2+X)^{n-1}}, \tag{9.22b}$$

and, analogously with (8.34), the following singularities belonging to the second system (Fig. 9.3d, $\chi = \pm\tau$),

$$\ln(\tau \pm \chi), \quad \frac{1}{\tau \pm \chi}, \ldots, \quad \frac{1}{(\tau \pm \chi)^{n-1}}. \tag{9.23}$$

As will be seen below, these singularities are not sufficient for solving the problem.

THE CASE OF A NON-HOMOGENEOUS STREAM. In the case of a non-homogeneous stream of general velocity U_∞, in which the normal components, the vertical one w_∞ and the lateral one v_∞, may be developed into a sum of homogeneous polynomials of various orders, the problem reduces to the motion around a system lying in a flow of velocity U_∞, this system having, however, the incidences α and β, or the corresponding velocities w and v, given by the relations

$$w = -\alpha U_\infty = -w_\infty, \quad v = -\beta U_\infty = -v_\infty. \tag{9.24}$$

In such a case, however, it should be also considered that the body axis has a variable incidence α and β. On the other hand, we should take account of relations (9.24) when we are defining the velocity conditions on the body contour, a fact that rather complicates the problem.

IX. 2. Conical Flow ($n = 1$) around a Wing–Body System having Subsonic Leading Edges

IX. 2.1. Reduction of the Problem to a Wing considered alone (isolated) having Variable Incidence

As already mentioned above, we shall set aside, from the beginning, the axisymmetric motion around the cone, which was completely treated in the first chapter (§ I. 6.2); the remaining general motion will correspond to the following arrangements of the system:

(a) The body and the horizontal plane OM_1M_2 which crosses it, passing through its axis, and intersects it along the lines ON_1 and ON_2 have together the incidence α_0; whereas the lateral portions OM_1A_1 and OM_2A_2 have the incidences α_1 and α_2, both edges of the wing considered thin being subsonic.

(b) The body and the plane OM_1M_2 have the incidence α_0, the lateral portions OM_1A_1 and OM_2M_2 have the incidences α_1 and α_2, the wing being continued outside the portion OA_1A_2 and beyond the Mach cone by two flat plates without incidence.

(c) The body and the flat plate OM_1M_2 have zero incidence, the lateral portions OM_1A_1 and OM_2A_2 of the wing assumed to be of symmetrical thickness have the slopes γ_1 and γ_2.

Since all these cases may be deduced from the first scheme (a), we shall determine below the boundary conditions of this motion.

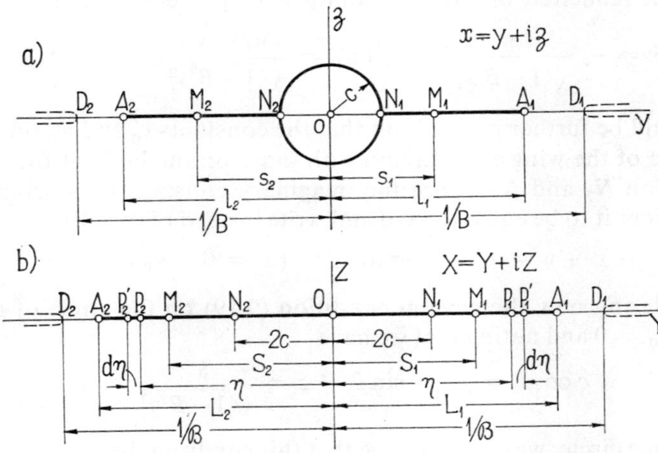

FIG. 9.5

From the compatibility relation (1.159), by taking also account of the assumption (9.8) which we apply at the points on the body ($x = ce^{i\theta}$ Fig. 9.5a) for $\mathcal{V} = v+iv'$, $\mathcal{W} = w+iw'$ we deduce the following relations,

$$-d(v+iv') = \frac{id(w+iw')}{\sqrt{1-B^2c^2e^{2i\theta}}} \approx id(w+iw') \tag{9.25a}$$

$$-dv' \approx dw, \quad v'+w = k_0 = \text{const.} \tag{9.25b}$$

At the points on any wing portion where the velocity is constant ($w_j = -\alpha_j U_\infty = \text{const.}$, $dw = 0$), we likewise find

$$dv' = dw = 0, \quad v' = v_j' = \text{const.} \quad w = w_j' = \text{const.}, \quad (j = 0, 1, 2). \tag{9.26}$$

In order to find the relation between v_j' and w_j on the wing, we proceed in a similar manner as for $\mathcal{U}$ (§ II.2.3) and developing the function $\mathcal{V}$ around the ridge M_j ($y = (-1)^{j+1}s_j$)—where we have the jumps $\Delta v' = (-1)^j(v_j'-v_0')$ and $\Delta w = (-1)^j(w_j-w_0)$—we find in fact an expression analogous to (2.38) and (2.39),

$$\mathcal{V}_{x\to(-1)^{j+1}s_j} = \frac{\Delta v'}{\pi} \ln (x+(-1)^j s_j) = \frac{\Delta v'}{\pi} \ln (\varepsilon\cdot e^{i\theta}), \tag{9.27}$$

where ε is the radius of a very small circle around the point $y = (-1)^{j+1}s_j$. Taking further into account the compatibility relation (1.159), and following the semicircle of radius ε from $\theta = 0$ up to $\theta = \pi$, we find

$$\Delta v' = -\frac{\Delta w}{\sqrt{1-B^2 s_j^2}}, \quad v_j' - v_0' = \frac{w_0 - w_j}{\sqrt{1-B^2 s_j^2}}, \quad (j = 1, 2). \tag{9.28}$$

It should be further pointed out that the constants v_0' and w_0 on the central plane of the wing are common with those on the body at the points of intersection N_1 and N_2. Since the imaginary constant v_0' is arbitrary, we can consider it to be equal to zero and write instead of (9.25b),

$$v' + w = w_0 = -\alpha_0 U_\infty, \quad (v' = 0, \quad k_0 = w_0). \tag{9.29}$$

Considering now the general condition (9.19) on the circle of radius c, putting $\beta_0 = 0$ and noting that we have

$$c \cos\theta = y, \quad c \sin\theta = z = \frac{\mathfrak{z}}{\sqrt{1-B^2\mathfrak{z}^2}} \approx \mathfrak{z} \tag{9.30}$$

on the same circle, we can easily see that this condition becomes

$$\text{Re}\,[x\mathcal{V}]_{\text{circle}} = 0. \tag{9.31}$$

On the other hand we have

$$\text{Im}\,[\mathcal{V}]_{N_j M_j} = 0, \quad (j = 1, 2) \tag{9.32}$$

on the segments N_1M_1 and N_2M_2, respectively (Fig. 9.5a), and

$$\text{Im}\,[\mathcal{V}]_{M_j A_j} = v_j' = \frac{w_0 - w_j}{\sqrt{1-B^2 s_j^2}} = \frac{(\alpha_j - \alpha_0)U_\infty}{\sqrt{1-B^2 s_j^2}} = C_j, \quad (j = 1, 2) \tag{9.33}$$

on the segments M_1A_1, M_2A_2.

On the real axis, outside the wing considered to be thin, beyond the subsonic leading edges, we shall have, analogously with the condition for $\mathcal{U}$ and $\mathcal{V}$ corresponding to the wing considered alone,

$$\text{Re}\,\mathcal{V} = v = 0. \tag{9.34}$$

Let us further consider a *fictitious* incompressible motion defined by a potential F_0 and write

$$\frac{dF_0}{dx} = \mathcal{V} = v - i(-v') = v - iv_{\mathfrak{z}}, \quad F_0 = \int \mathcal{V}\, dx; \tag{9.35}$$

one can thus see that the function F_0 would represent the true complex potential of a motion around the wing–body system, having the downwash

$$v_{\mathfrak{z}} = -v' \tag{9.36}$$

equal and of opposite sense with respect to the imaginary part of the function $\mathcal{V}$.

If F_0 is known, by applying the compatibility relation (1.159) we can write successively,

$$d\mathcal{U} = -x d\mathcal{V}, \tag{9.37a}$$

$$F_0 = \int \mathcal{U} dx = x\mathcal{V} + \int d\mathcal{U} = x\frac{dF_0}{dx} + \mathcal{U}. \tag{9.37b}$$

Let us now pass to the auxiliary plane $X = Y + iZ$ (9.10, Fig. 9.5b). In this plane, the complex velocity which derives from the potential F_0 may be written in the following forms:

$$\frac{dF_0}{dX} = \frac{dF_0}{dx}\,\frac{dx}{dX} = \frac{1}{2}\,\mathcal{V}\left(1 + \frac{X}{\sqrt{X^2 - 4c^2}}\right), \tag{9.38a}$$

$$\frac{dF_0}{dX} = \frac{dF_0}{dx}\,\frac{x}{x - \dfrac{c^2}{x}} = x\mathcal{V}\,\frac{1}{x - \dfrac{c^2}{x}}. \tag{9.38b}$$

Examining the conditions (9.31) and (9.32) in the X plane and taking account of (9.38), we note that we have

$$\text{Im}\left[\frac{dF_0}{dX}\right]_{M_2M_1} = 0 \tag{9.39}$$

on the entire M_2M_1 line.

On the other segments, M_1A_1 and M_2A_2, by taking account of (9.33) and (9.38a), we likewise have

$$\text{Im}\left[\frac{dF_0}{dX}\right]_{A_jM_j} = \frac{1}{2}\,C_j\left(1 + \frac{Y}{\sqrt{Y^2 - 4c^2}}\right), \quad (j = 1, 2). \tag{9.40}$$

On the remaining part of the real axis, beyond the subsonic leading edges of the wing, on taking account of (9.34) we obviously have

$$\text{Re}\,\frac{dF_0}{dX} = 0. \tag{9.41}$$

The above conditions must be completed with the singularities that result from the behaviours at the ridges and the leading edges (9.20, 9.22) as well as with those due from the relation (9.40).

In the absence of the body ($c = 0$), we have simply to satisfy the boundary conditions of a flat triangular wing, a problem which has been solved directly in the preceding chapters, without resorting to the above succession of operations.

By analogy with this problem for a flat wing, in the case of the wing-conical body system we shall consider that the motion in the X plane, defined above, corresponds to a conical motion about a wing having the trace $A_1A_2\,(-L_2 < X = Y < L_1)$ in this plane.

Let us write, analogously with (9.35) and (9.37b),

$$\mathcal{U}_a = u_a + iu'_a = F_0 - X\frac{dF_0}{dX}, \quad \mathcal{V}_a = v_a + iv'_a = \frac{dF_0}{dX} \tag{9.42}$$

for the complex functions corresponding to the two components of the velocities belonging to this motion which are connected by compatibility relations similar to (1.159),

$$d\mathcal{U}_a = -Xd\mathcal{V}_a = \frac{iXd\mathcal{W}_a}{\sqrt{1-\mathcal{B}^2X^2}}, \tag{9.43}$$

where $\mathcal{W}_a$ is the function of the third velocity component

$$\mathcal{W}_a = w_a + iw'_a \tag{9.44}$$

and, also, $\mathcal{B}$ is given by the following equation:

$$\frac{1}{\mathcal{B}} = \frac{1+B^2c^2}{B}. \tag{9.45}$$

The general formula for determining the function $\mathcal{U}$ (9.37b)—from which the axial velocity $u = \operatorname{Re}\mathcal{U}$ is derived—thus becomes

$$\mathcal{U} = F_0 - x\frac{dF_0}{dx} = F_0 - x\frac{dF_0}{dX}\frac{dX}{dx} = F_0 - \frac{dF_0}{dX}\left(X - \frac{2c^2}{x}\right)$$
$$= \mathcal{U}_a + \frac{2c^2}{x}\mathcal{V}_a. \tag{9.46}$$

The problem reduces now to the determination of the functions $\mathcal{U}_a$ and $\mathcal{V}_a$ of the true conical motion ($n = 1$) about the wing defined above in the X plane. *The functions $\mathcal{U}_a$ and $\mathcal{V}_a$ have the same behaviour near the ridges or the subsonic leading edges in the X plane as $\mathcal{U}$ and $\mathcal{V}$ on the single (isolated) wing in the x plane.*

Further, proceeding from the compatibility relation (9.43), we find as a function of a point $Y = \eta$ on the abscissa ($Z = 0$)

that
$$du'_a = -\eta dv'_a = \frac{\eta\, d\eta}{\sqrt{1-B^2\eta^2}}\left[\frac{dw_a}{dY}\right]_{Y=\eta}, \tag{9.47}$$

which indicates that the variation of the imaginary part of $\mathcal{U}_a$ or $\mathcal{V}_a$ given by (9.40) for each wing portion, M_1A_1 and M_2A_2, corresponds to a variation of incidence or slope on the fictitious wing in the X plane; as shown earlier for the simple wing (§ II.3.4), this variation corresponds to certain elementary ridges or, in the hydrodynamic sense, to a distribution of elementary sources, each of them bringing to $\mathcal{U}_a$ and $\mathcal{V}_a$ contributions similar to (2.79a,b):

$$d\mathcal{U}_a = -\eta d\mathcal{V}_a = \frac{2\eta}{\pi\sqrt{1-B^2\eta^2}} \cosh^{-1} \sqrt{\frac{(L_2+\eta)(L_1-X)}{(L_1+L_2)(\eta-X)}} \left[\frac{dw_a}{dy}\right]_{Y=\eta} d\eta$$

$$= -\frac{C_1\eta}{\pi} \cosh^{-1} \sqrt{\frac{(L_2+\eta)(L_1-X)}{(L_1+L_2)(\eta-X)}}\, d\left(\frac{\eta}{\sqrt{\eta^2-4c^2}}\right), \tag{9.48a}$$

$$d\mathcal{U}_a = \eta d\mathcal{V}_a = -\frac{2\eta}{\pi\sqrt{1-B^2\eta^2}} \cosh^{-1} \sqrt{\frac{(L_1+\eta)(L_2+X)}{(L_1+L_2)(\eta+X)}} \left[\frac{dw_a}{dy}\right]_{Y=-\eta}$$

$$= -\frac{C_2\eta}{\pi} \cosh^{-1} \sqrt{\frac{(L_1+\eta)(L_2+X)}{(L_1+L_2)(\eta+X)}}\, d\left(\frac{\eta}{\sqrt{\eta^2-4c^2}}\right), \tag{9.48b}$$

the former representing the contribution of an elementary ridge situated at the point $Y = \eta$ on the M_1A_1 portion, and the latter the contribution of a ridge situated at $Y = -\eta$ on the M_2A_2 portion of the wing (Fig. 9.5b).

Therefore, in conclusion, the study of the body–wing system reduces to the study of a wing considered alone having the same system of singularities as the real one and, in addition, a continuous variation of incidence or slope on the wing portions.

IX. 2.2. Thin Wing with Subsonic Leading Edges having the Same Incidence as the Body Axis

In such a case we have the condition (9.39) on the entire wing trace, which thus becomes a "stream line" for the function $\mathcal{V}_a$ (assimilated with a velocity potential) as well as for the function $\mathcal{U}_a$ derived from the relation (9.43). On the other hand, taking into consideration the behaviour of the velocities at the ends (9.20), we can write in a manner analogous to that for the simple wing (2.32),

$$\mathcal{U}_a = \frac{A_{u0}+A_{u1}X}{\sqrt{(L_1-X)(L_2+X)}}, \quad \mathcal{V}_a = \frac{A_{v0}+A_{v1}X}{\sqrt{(L_1-X)(L_2+X)}}, \tag{9.49}$$

whence, by applying (9.46) and returning to the x plane, we obtain

$$\mathcal{U} = \frac{A_{u0}+2c^2A_{v1}+A_{u1}x+(A_{u1}+2A_{v0})\dfrac{c^2}{x}+\dfrac{2c^4}{x^2}A_{v1}}{\sqrt{(l_1-x)(l_2+x)}\sqrt{\left(1-\dfrac{c^2}{l_1x}\right)\left(1+\dfrac{c^2}{l_2x}\right)}}. \tag{9.50}$$

From the compatibility relations (9.43) we obtain the connection between the constants A_{u0}, A_{u1}, A_{v0} and A_{v1},

$$A_{v0} = -\frac{L_1-L_2}{2}A_{v1} = A_{u1} = A_{u0}\frac{L_1-L_2}{2L_1L_2}. \tag{9.51}$$

It is to be mentioned that the expression (9.49) automatically satisfies the condition

$$[d\mathcal{U}_a]_{X=0} = 0, \tag{9.52}$$

which avoids $\mathcal{V}_a$ becoming infinite at the centre ($X = 0$, $x = ic$), as would result from (9.43).

Hence only a single constant is to be determined, namely A_{u0}.

In the case of wing symmetry ($l_1 = l_2 = l$, $A_{u1} = A_{v0} = 0$), on performing the necessary substitutions and also taking account of (9.51), we finally obtain:

$$\mathcal{U} = A_0\frac{1+\dfrac{c^4}{l^4}-\dfrac{2c^4}{l^2x^2}}{\sqrt{(l^2-x^2)\left(1-\dfrac{c^4}{l^2x^2}\right)}}, \quad \left(A_0 = \frac{A_{u0}}{\left(1+\dfrac{c^2}{l^2}\right)^2}\right), \tag{9.53}$$

which reduces to the expression obtained in the case of the triangular flat wing considered alone (4.3a), provided $c = 0$:

$$\mathcal{U}_p = \frac{A_{0p}}{\sqrt{l^2-x^2}}, \quad (A_{0p} = A_{10}). \tag{9.54}$$

Likewise, if it is only the conical body that remains alone ($l = c$), the expression (9.53) reduces to

$$\mathcal{U}_c = \frac{2iA_{0c}}{x}, \tag{9.55}$$

whence we deduce

$$u = \frac{2A_{0c}}{c}\sin\theta \tag{9.56}$$

for the axial velocity $u = \mathrm{Re}\,\mathcal{U}$ on the cone ($x = ce^{i\theta}$), thus finding again, under another notation, the expression corresponding to the lateral motion around a cone having the incidence α_0.

The constant A_{0c} may be determined either in the way indicated in (1.119) or in a way similar to the one which will be applied below for the wing–body system.

Indeed, returning to the expression of $\mathcal{U}$ (9.53) for determining the constant A_0, we shall follow the course indicated in (§ II. 5) and exemplified in (§ II. 6). Taking account of (2.107a), we find

$$w_0 = -\alpha_0 U_\infty = \operatorname{Im} \int_{1/B}^{y'} \frac{\sqrt{1-B^2x^2}}{x}\, d\mathcal{U}, \tag{9.57}$$

where α_0 is the incidence of the wing–conical body system and where y' is an arbitrary point on the wing.

For simplifying the computations, we shall consider that $c/l \lesssim 0{\cdot}95$ and obtain in such a case

$$A_0 = \frac{\alpha_0\, U_\infty\, l^2}{\left[\lambda + \dfrac{1}{2}\dfrac{c^4}{l^4} B^2 l^2 \dfrac{K(k)}{E(k)}\right] E(k)}, \tag{9.58}$$

where

$$\lambda = 1 - \frac{c^4}{l^4} B^2 l^2 + \frac{1}{3}\frac{c^8}{l^8} B^2 l^2 \left(\frac{1}{8} - B^2 l^2\right), \tag{9.59}$$

$K(k)$ and $E(k)$ being the complete elliptic integrals of the first and second kind, respectively, of modulus

$$k = \sqrt{1-B^2 l^2}\,. \tag{9.60}$$

Note. If the ratio c/l is sufficiently small ($c/l \lesssim 0{\cdot}5$) it may be seen from (9.58) and (9.59) that the terms c^4/l^4 and higher may be neglected, thus finding again for A_0 the same value as for the wing considered alone (4.16).

Calculation of Lift and Wave Drag. Let us denote by C_l the total lift coefficient of the wing–body system and by C_{lw}, C_{lf}, respectively, the lift coefficients of the wing and body, each of them referred to the total surface area of the whole system; we shall have

$$C_l = C_{lw} + C_{lf}\,. \tag{9.61}$$

In keeping with the considerations presented in (§ III. 2.1) and taking also into account the wing symmetry with respect to the Ox_1 axis, we can write

$$C_{lw} = \frac{4}{SU_\infty}\int_c^l u\, dy = \operatorname{Re} \frac{4}{SU_\infty}\int_c^l \mathcal{U}\, dx, \tag{9.62}$$

where, for the surface area S, we take the projection of the system on the plane Ox_1x_2, that is,

$$S = l. \tag{9.63}$$

In order to obtain a simple expression for C_{lw}, we carry out the required expansions, taking $c/l \lesssim 0{\cdot}95$ in this case too; thus we obtain

$$C_{lw} = \frac{4A_0}{lU_\infty}\left[\left(1+\frac{c^4}{l^4}\right)\left(\frac{\pi}{2}-\sin^{-1}\frac{c}{l}\right)-\sqrt{1-\frac{c^2}{l^2}}\left(\frac{3}{2}\frac{c^3}{l^3}+\frac{5}{24}\frac{c^5}{l^5}\right)\right], \tag{9.64}$$

in which the constant A_0 was determined as above (9.58).

As far as the lift coefficient of the body is concerned, it will be given by the relation

$$C_{lf} = -\frac{2}{S}\int_0^{\pi/2} C_p c \sin\theta \, d\theta = \frac{4}{SU_\infty}\int_0^c u \, dy. \tag{9.65}$$

Taking further account of (9.10), we have

$$X = Y = 2c\cos\theta = 2y, \quad (dY = 2dy); \tag{9.66}$$

substituting in expression (9.65), this becomes

$$C_{lf} = \mathrm{Re}\,\frac{2}{SU_\infty}\int_0^{2c} \mathcal{U}\, dX. \tag{9.67}$$

Performing the calculations we finally obtain

$$C_{lf} = \frac{A_0}{lU_\infty}\left[2\frac{c}{l}\left(1-\frac{c^2}{l^2}\right)+\left(1+\frac{c^2}{l^2}\right)^2\sin^{-1}\frac{2(c/l)}{1-(c^2/l^2)}\right]. \tag{9.68}$$

The sum of the two coefficients, C_{lw} given by (9.64) and C_{lf} given by (9.68), yields the total coefficient C_l.

It should be observed now that the wing and body axis both lie in the plane Ox_1x_2, consequently the total lift is normal to this plane which makes an incidence α_0 with the free stream direction. Since the total wave drag is the projection of the lift upon the stream direction, we have

$$C_{d\alpha} = \alpha_0 C_l = \alpha_0(C_{lw}+C_{lf}). \tag{9.69}$$

The drag coefficient due to the axisymmetrical flow around the isolated conical body will be also added to the above expression; that is

$$C_{d0} = \frac{\pi c^2}{S}C_p = \frac{\pi c^2}{l}C_p, \tag{9.70}$$

where C_p is given by relation (1.112) where $\tan\delta$ will be replaced by c:

$$C_p = 2\frac{C_1}{U_\infty}\cosh^{-1}\frac{1}{Bc} - \frac{C_1^2}{c^2 U_\infty^2}\left[1 - B^2c^2 + B^2c^2\left(\cosh^{-1}\frac{1}{Bc}\right)^2\right], \tag{9.71a}$$

$$C_1 = \frac{c^2 U_\infty^2}{\sqrt{1-B^2c^2} + c^2\cosh^{-1}\dfrac{1}{Bc}}. \tag{9.71b}$$

Hence, the total wave drag coefficient will be as follows:

$$C_d = C_{d\alpha} + C_{d0}. \tag{9.72}$$

IX. 2.3. The Case when the Body and the Wing with Subsonic Leading Edges have Different Indicences

Let us consider the incidence α for the wing and α_0 for the body axis. For simplicity, we shall consider from the beginning that we are dealing with the symmetrical wing (Fig. 9.6). In such a case, the contribution of the subsonic leading edges is given by the simple expression (9.53).

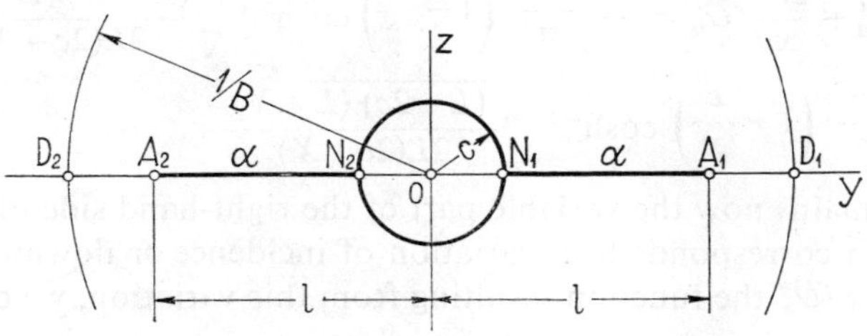

FIG. 9.6

In the X plane, the points M_1 and M_2 symmetrical with respect to the body axis ($Y = \pm S$) represent the traces of the separating ridges, both for $\mathcal{V}_a$ and $\mathcal{U}_a$, these two functions having the same behaviour.

The contributions of ridges to the construction of $\mathcal{U}_a$ are given by the same expressions as for the wing considered alone (isolated) (2.64, 2.69a), by performing the necessary substitutions, viz. $n = 1$, L instead of l, S instead of s. Designating by $\mathcal{U}_a'$ this contribution and taking account of the symmetry, we shall have

$$\mathcal{U}_a' = \frac{C_u}{\pi}\left[\cosh^{-1}\sqrt{\frac{(L+S)(L-X)}{2L(S-X)}} + \cosh^{-1}\sqrt{\frac{(L+S)(L+X)}{2L(S+X)}}\right]. \tag{9.73}$$

The function $\mathcal{V}'_a$ which has the same behaviour as $\mathcal{U}'_a$, being, however, antisymmetrical, will be written as

$$\mathcal{V}'_a = \frac{C}{\pi}\left[\cosh^{-1}\sqrt{\frac{(L+S)(L-X)}{2L(S-X)}} - \cosh^{-1}\sqrt{\frac{(L+S)(L+X)}{2L(S+X)}}\right]. \tag{9.74}$$

If the points M_1 and M_2 coincide with N_1 and N_2, the wing–body intersections, we put

$$S = 2c; \tag{9.75}$$

then, from the compatibility relation (9.43), we have

$$C_u = -2cC \tag{9.76}$$

where C is deduced from (9.33) by putting $s_1 = s_2 = c$ and $\alpha_1 = \alpha_2 = \alpha$:

$$C = C_1 = C_2 = \frac{(\alpha-\alpha_0)U_\infty}{\sqrt{1-B^2c^2}}. \tag{9.77}$$

It should be noted meanwhile that condition (9.52) is also automatically satisfied.

Further, the expression of $\mathcal{U}'$ follows:

$$\mathcal{U}' = \mathcal{U}'_a + \frac{2c^2}{x}\mathcal{V}'_a = -\frac{2cC}{\pi}\left(1-\frac{c}{x}\right)\cosh^{-1}\sqrt{\frac{(L+2c)(L-X)}{2L(2c-X)}}$$
$$-\frac{2cC}{\pi}\left(1+\frac{c}{x}\right)\cosh^{-1}\sqrt{\frac{(L+2c)(L+X)}{2L(2c+X)}}. \tag{9.78}$$

There remains now the variable part of the right-hand side of condition (9.40) which corresponds to a variation of incidence or downwash (9.47). Denoting by $\mathcal{V}^*_a$ the function resulting from this variation, we can write

$$\text{Im}\,[\mathcal{V}^*_a]_{N_jA_j} = \frac{C}{2}\frac{Y}{\sqrt{Y^2-4c^2}}, \quad (j = 1, 2), \tag{9.79a}$$

$$\text{Im}\,[\mathcal{V}^*_a]_{N_1N_2} = 0, \quad \text{Re}[\mathcal{V}^*_a]_{-\infty A_2, A_1\infty} = 0, \tag{9.79b}$$

which corresponds to a "potential" F^*_0 representing the fictitious motion around a horizontal plate of length $4c$, lying in a vertical stream and having the following expressions in the two planes X and x:

$$F^*_0 = i\frac{C}{2}\sqrt{X^2-4c^2} = i\frac{C}{2}\left(x-\frac{c^2}{x}\right). \tag{9.80}$$

Taking into account (9.37a), the corresponding contribution readily results:

$$\mathcal{U}^* = -x\frac{dF^*_0}{dx}+F^*_0 = -\frac{ic^2C}{x}. \tag{9.81}$$

Finally, by summing all these contributions (9.53), (9.78), (9.81), we obtain, after substituting for X and L in (9.78) as functions of x and l (9.10, 9.13),

$$\mathcal{U} = A_0 \frac{1+(c^4/l^4)-2(c^4/l^2x^2)}{\sqrt{(l^2-x^2)\{1-(c^4/l^2x^2)\}}}$$

$$-\frac{2cC}{\pi}\left(1-\frac{c}{x}\right)\cosh^{-1}\sqrt{\frac{(l+c)^2(l-x)(c^2-lx)}{2l(l^2+c^2)(x-c)^2}}$$

$$-\frac{2cC}{\pi}\left(1+\frac{c}{x}\right)\cosh^{-1}\sqrt{\frac{(l+c)^2(l+x)(c^2+lx)}{2l(l^2+c^2)(x+c)^2}}-\frac{ic^2C}{x}. \qquad (9.82)$$

If the incidence of the wing and body is the same ($\alpha = \alpha_0$), then $C = 0$ and it is only the first term that remains, which is identical with (9.53).

For determining the constant A_0 we shall use the compatibility relation and put

$$-\alpha U_\infty = \mathrm{Im}\int_{1/B}^{\text{wing}} \frac{\sqrt{1-B^2x^2}}{x}\, d\mathcal{U}. \qquad (9.83)$$

For simplifying the calculations we have considered $c/l \lesssim 0{\cdot}9$ in order to perform a convenient expansion and have found

$$A_0 = \frac{[\alpha U_\infty+(\alpha_0-\alpha)U_\infty(c/l)\Lambda]l^2}{\lambda E(k)+\frac{1}{2}(c^4/l^4)B^2l^2K(k)}, \qquad (9.84)$$

where $K(k)$ and $E(k)$ are complete elliptic integrals of the first and second kinds of modulus k, given by (9.60), where λ has the expression (9.59) and Λ is given by the following relation:

$$\Lambda = \frac{2}{\pi}\left(1-\frac{c^2}{l^2}\right)[\lambda_1 E(k)-\lambda_2 B^2l^2K(k)]$$

$$+\tfrac{1}{2}\frac{c}{l}\left[\sqrt{1-B^2l^2}-B^2l^2\cosh^{-1}\frac{1}{Bl}\right], \qquad (9.85)$$

in which

$$\lambda_1 = 1+\tfrac{1}{3}\frac{c^4}{l^4}(1-\tfrac{1}{2}B^2l^2)+\tfrac{1}{5}\frac{c^8}{l^8}\left(1-\tfrac{3}{40}B^2l^2-\frac{B^4l^4}{20}\right), \qquad (9.86a)$$

$$\lambda_2 = 1+\tfrac{1}{6}\frac{c^4}{l^4}+\tfrac{1}{10}\frac{c^8}{l^8}(1-\tfrac{5}{4}B^2l^2). \qquad (9.86b)$$

In order to form an idea of the pressure distribution on the wing and body, we have plotted in Fig. 9.7 diagrams of the variation of the disturbance axial velocities (which are proportional to the pressure coefficients) in the following cases:

1. The whole system has the same incidence ($\alpha_0 = \alpha$).
2. The body has zero incidence ($\alpha_0 = 0$), and the wing has the incidence α.
3. The body has the incidence α_0, and the wing has zero incidence ($\alpha = 0$).
4. The wing considered alone has the incidence α.

In the above we have taken the values

$$Bl = 0{\cdot}8, \quad \frac{c}{l} = 0{\cdot}5. \tag{9.87}$$

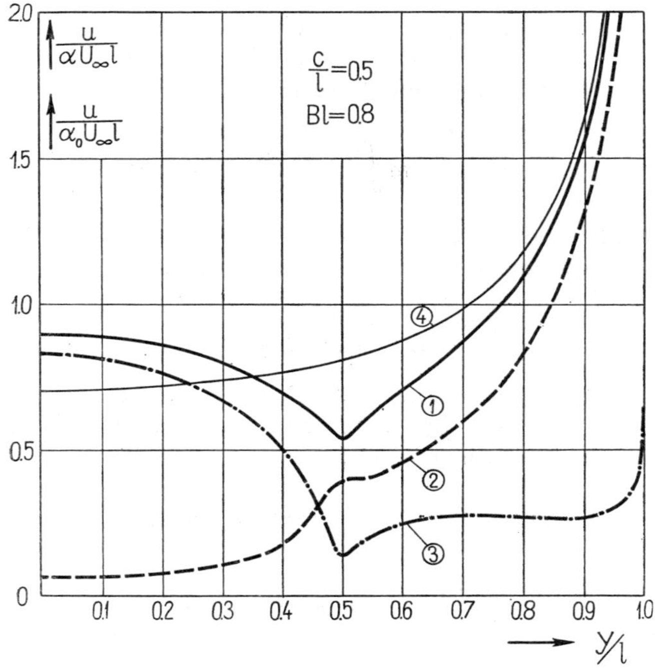

FIG. 9.7

IX. 2.4. The Contribution of Two Additional Symmetrical Ridges

If the thin wing of the preceding case also has two symmetrical ridges which make the transition from the incidence α to the incidence α' (Fig. 9.8)—this latter incidence may even be zero ($\alpha' = 0$)—then we have to add the effect of these two ridges to the expressions for $\mathcal{U}$ established above.

Let us assume that at M_1 and M_2 ($y = \pm s$) the wing incidence changes passing from α to α'; in such a case, the constant C_j defined by (9.33) will have the following value:

$$C_j = \frac{(\alpha'-\alpha)U_\infty}{\sqrt{1-B^2s^2}} = Q, \quad (j = 1, 2). \tag{9.88}$$

In keeping with the conditions given by (9.39) and (9.40), we have the jump

$$\operatorname{Im}\left[\Delta \frac{dF_0}{dX}\right]_{M_j} = \frac{Q}{2}\left(1+\frac{S}{\sqrt{S^2-4c^2}}\right), \quad (j = 1, 2) \tag{9.89}$$

at the ridges M_1 and M_2 ($Y = \pm S$) of the fictitious wing in the X plane.

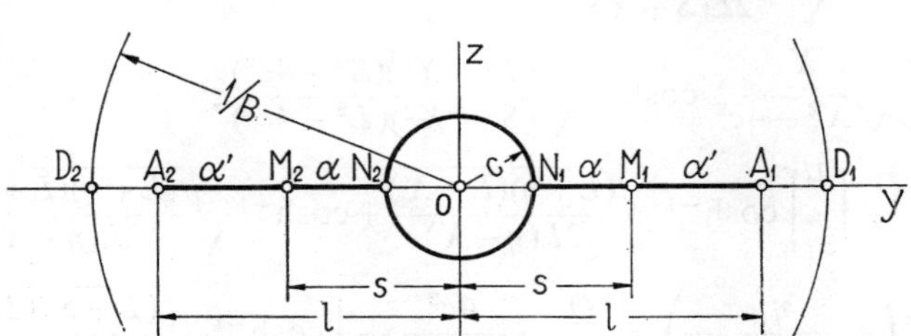

FIG. 9.8

These two ridges will have new contributions $\mathcal{U}''_a$ and $\mathcal{V}''_a$ which have expressions analogous to (9.73) and (9.74), where however the coefficients C_u and C are to be replaced now by Q_u and Q_v given by the following equations:

$$Q_u = -SQ_v = -QS\left(1+\frac{S}{\sqrt{S^2-4c^2}}\right). \tag{9.90}$$

Thus, similarly to (9.78), we finally obtain

$$\mathcal{U}'' = \mathcal{U}''_a + \frac{2c^2}{x}\mathcal{V}''_a = -\frac{Q}{\pi}\left(1+\frac{S}{\sqrt{S^2-4c^2}}\right)$$
$$\times\left[\left(S-\frac{2c^2}{x}\right)\cosh^{-1}\sqrt{\frac{(L+S)(L-X)}{2L(S-X)}}\right.$$
$$\left.+\left(S+\frac{2c^2}{x}\right)\cosh^{-1}\sqrt{\frac{(L+S)(L+X)}{2L(S+X)}}\right]. \tag{9.91}$$

Apart from the above contributions, account should be also taken of the incidence variation on the portions M_1A_1 and M_2A_2 of the fictitious wing, which corresponds to a continuous distribution of elementary ridges on those portions.

By applying formulae (9.48) where we put $L_1 = L_2 = L$ and $C_1 = C_2 = Q$, we obtain the following new contributions $\mathcal{V}_a^{**}$ and $\mathcal{U}_a^{**}$ due to the continuous distribution of elementary ridges on the portions M_1A_1 and M_2A_2:

$$\mathcal{V}_a^{**} = \frac{Q}{\pi}\int_S^L \left[\cosh^{-1}\sqrt{\frac{(L+\eta)(L-X)}{2L(\eta-X)}} - \cosh^{-1}\sqrt{\frac{(L+\eta)(L+X)}{2L(\eta+X)}}\right]$$

$$\times d\left(\frac{\eta}{\sqrt{\eta^2-4c^2}}\right) = -\frac{Q}{\pi}\frac{S}{\sqrt{S^2-4c^2}}\left[\cosh^{-1}\sqrt{\frac{(L+S)(L-X)}{2L(S-X)}}\right.$$

$$\left. - \cosh^{-1}\sqrt{\frac{(L+S)(L+X)}{2L(S+X)}}\right]$$

$$+\frac{Q}{\pi}\frac{X}{\sqrt{X^2-4c^2}}\cosh^{-1}\sqrt{\frac{(L^2-X^2)(S^2-4c^2)}{(S^2-X^2)(L^2-4c^2)}}, \tag{9.92a}$$

$$\mathcal{U}_a^{**} = -\frac{Q}{\pi}\int_S^L \left[\cosh^{-1}\sqrt{\frac{(L+\eta)(L-X)}{2L(\eta-X)}} + \cosh^{-1}\sqrt{\frac{(L+\eta)(L+X)}{2L(\eta+X)}}\right]$$

$$\times \eta d\left(\frac{\eta}{\sqrt{\eta^2-4c^2}}\right) = \frac{Q}{\pi}\frac{4c^2}{\sqrt{S^2-4c^2}}\left[\cosh^{-1}\sqrt{\frac{(L+S)(L-X)}{2L(S-X)}}\right.$$

$$\left. + \cosh^{-1}\sqrt{\frac{(L+S)(L+X)}{2L(S+X)}}\right]$$

$$-\frac{Q}{\pi}\frac{4c^2}{\sqrt{X^2-4c^2}}\cosh^{-1}\sqrt{\frac{(L^2-X^2)(S^2-4c^2)}{(S^2-X^2)(L^2-4c^2)}}. \tag{9.92b}$$

Finally, applying (9.46), we get

$$\mathcal{U}^{**} = \mathcal{U}_a^{**} + \frac{2c^2}{x}\mathcal{V}_a^{**} = \frac{Q}{\pi}\frac{2c^2}{\sqrt{S^2-4c^2}}$$

$$\times\left[\left(2-\frac{S}{x}\right)\cosh^{-1}\sqrt{\frac{(L+S)(L-X)}{2L(S-X)}}\right.$$

$$\left. + \left(2+\frac{S}{x}\right)\cosh^{-1}\sqrt{\frac{(L+S)(L+X)}{2L(S+X)}}\right]$$

$$-\frac{2}{\pi}\frac{Qc^2}{x}\cosh^{-1}\sqrt{\frac{(L^2-X^2)(S^2-4c^2)}{(S^2-X^2)(L^2-4c^2)}}. \tag{9.93}$$

Now, by summing up (9.91) and (9.93), we obtain the additional axial velocity function $\mathcal{U}_s$ due to the two new ridges M_1 and M_2 ($y = \pm s$) of the actual wing,

$$\mathcal{U}_s = \mathcal{U}'' + \mathcal{U}^{**} = -\frac{2Q}{\pi}\left(s - \frac{c^2}{x}\right)\cosh^{-1}\sqrt{\frac{(L+S)(L-X)}{2L(S-X)}}$$

$$-\frac{2Q}{\pi}\left(s + \frac{c^2}{x}\right)\cosh^{-1}\sqrt{\frac{(L+S)(L+X)}{2L(S+X)}}$$

$$-\frac{2}{\pi}\frac{Qc^2}{x}\cosh^{-1}\sqrt{\frac{(L^2-X^2)(S^2-4c^2)}{(S^2-X^2)(L^2-4c^2)}}, \tag{9.94}$$

which will be added to the previous expression of $\mathcal{U}$ (9.82), where no additional ridges are included.

IX. 3. Wing–Body System having a Thin Wing with Supersonic Edges or a Wedge-shaped Wing of Symmetrical Thickness (Conical Flow)

IX. 3.1. Case of a Thin Delta Wing with Supersonic Edges

For the beginning let us consider the case of a body–thin wing system, the body axis having the incidence α_0 and the wing having the incidence α on the portions N_1M_1 and N_2M_2 and zero incidence beyond the ridges M_1 and M_2 up to the leading edges which are supersonic, the wing being thus continued by two plane portions without incidence (Fig. 9.9).

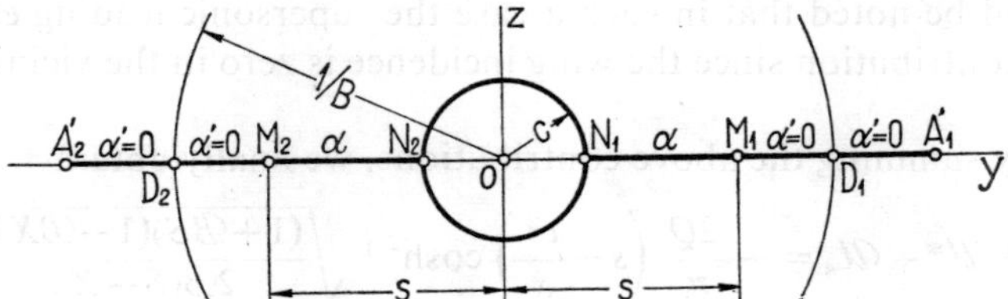

FIG. 9.9

In the construction of the axial velocity function $\mathcal{U}$ the following contributions occur, which will be expressed at will either in the plane X or x:

(a) The contributions of the ridges N_1 and N_2 ($Y = \pm 2c$) of the fictitious wing in the X plane, which will be given, by an expression similar to (9.78) where, however, L will be replaced by $1/\mathcal{B}$ since the edges are supersonic,

$$\mathcal{U}' = -\frac{2cC}{\pi}\left(1 - \frac{c}{x}\right)\cosh^{-1}\sqrt{\frac{(1+2\mathcal{B}c)(1-\mathcal{B}X)}{2\mathcal{B}(2c-X)}}$$

$$-\frac{2cC}{\pi}\left(1 + \frac{c}{x}\right)\cosh^{-1}\sqrt{\frac{(1+2\mathcal{B}c)(1+\mathcal{B}X)}{2\mathcal{B}(2c+X)}}, \tag{9.95}$$

the constant C being given by (9.77).

(b) The effect of the incidence variation on the fictitious wing in the X plane, due to the incidence jump from α_0 to the incidence α, whose contribution is given by (9.81),

$$\mathcal{U}^* = -\frac{ic^2C}{x}. \tag{9.96}$$

(c) The contribution of the additional ridges M_1 and M_2 ($y = \pm s$) of the actual wing, whose expression will be deduced from that established above (9.94) by setting $\alpha' = 0$ and replacing L by $1/B$ (since the wing edges are supersonic), that is

$$\begin{aligned}\mathcal{U}_s = &-\frac{2Q}{\pi}\left(s-\frac{c^2}{x}\right)\cosh^{-1}\sqrt{\frac{(1+\mathcal{B}S)(1-\mathcal{B}X)}{2\mathcal{B}(S-X)}}\\ &-\frac{2Q}{\pi}\left(s+\frac{c^2}{x}\right)\cosh^{-1}\sqrt{\frac{(1+\mathcal{B}S)(1+\mathcal{B}X)}{2\mathcal{B}(S+X)}}\\ &-\frac{2}{\pi}\frac{Qc^2}{x}\cosh^{-1}\sqrt{\frac{(1-\mathcal{B}^2X^2)(S^2-4c^2)}{(S^2-X^2)(1-4\mathcal{B}^2c^2)}},\end{aligned} \tag{9.97}$$

where Q has now the value

$$Q = -\frac{\alpha U_\infty}{\sqrt{1-B^2s^2}}. \tag{9.98}$$

It should be noted that in such a case the supersonic leading edges will make no contribution since the wing incidence is zero in the vicinity of the edge.

Now, by summing the above contributions, we finally obtain

$$\begin{aligned}\mathcal{U} = \mathcal{U}'+\mathcal{U}^*+\mathcal{U}_s = &-\frac{2Q}{\pi}\left(s-\frac{c^2}{x}\right)\cosh^{-1}\sqrt{\frac{(1+\mathcal{B}S)(1-\mathcal{B}X)}{2\mathcal{B}(S-X)}}\\ &-\frac{2Q}{\pi}\left(s+\frac{c^2}{x}\right)\cosh^{-1}\sqrt{\frac{(1+\mathcal{B}s)(1+\mathcal{B}X)}{2\mathcal{B}(S+X)}}-\frac{2}{\pi}\frac{Qc^2}{x}\\ &\times\cosh^{-1}\sqrt{\frac{(1-\mathcal{B}^2X^2)(S^2-4c^2)}{(S^2-X^2)(1-4\mathcal{B}^2c^2)}}-\frac{2cC}{\pi}\left(1-\frac{c}{x}\right)\cosh^{-1}\sqrt{\frac{(1+2\mathcal{B}c)(1-\mathcal{B}X)}{2\mathcal{B}(2c-X)}}\\ &-\frac{2cC}{\pi}\left(1+\frac{c}{x}\right)\cosh^{-1}\sqrt{\frac{(1+2\mathcal{B}c)(1+\mathcal{B}X)}{2\mathcal{B}(2c+X)}}-\frac{ic^2C}{x}.\end{aligned} \tag{9.99}$$

This expression also holds in the case when the ridges M_1 and M_2 become supersonic ($s > 1/B$, $S > 1/\mathcal{B}$), in which case we have to take into consideration a relation similar to (2.68) and (7.144),

$$\frac{\cosh^{-1}\Theta}{\sqrt{1-B^2s^2}} = \frac{\cos^{-1}\Theta}{\sqrt{B^2s^2-1}}. \tag{9.100}$$

Supposing now that the supersonic ridges M_1 and M_2 coincide with the edges A_1 and A_2 ($s = l$), we shall replace s by l and S by L in (9.99), obtaining thus the expression for the axial disturbance velocity in the case of a thin wing with supersonic edges having the incidence α (Fig. 9.10).

$$\begin{aligned}\mathcal{U} = {} & \frac{2K}{\pi}\left(l-\frac{c^2}{x}\right)\cos^{-1}\sqrt{\frac{(1+\mathcal{B}L)(1-\mathcal{B}X)}{2\mathcal{B}(L-X)}} \\ & +\frac{2K}{\pi}\left(l+\frac{c^2}{x}\right)\cos^{-1}\sqrt{\frac{(1+\mathcal{B}L)(1+\mathcal{B}X)}{2\mathcal{B}(L+X)}} \\ & +\frac{2}{\pi}\frac{Kc^2}{x}\cos^{-1}\sqrt{\frac{(1-\mathcal{B}^2X^2)(L^2-4c^2)}{(L^2-X^2)(1-4\mathcal{B}^2c^2)}} \\ & -\frac{2cC}{\pi}\left(1-\frac{c}{x}\right)\cosh^{-1}\sqrt{\frac{(1+2\mathcal{B}c)(1-\mathcal{B}X)}{2\mathcal{B}(2c-X)}} \\ & -\frac{2cC}{\pi}\left(1+\frac{c}{x}\right)\cosh^{-1}\sqrt{\frac{(1+2\mathcal{B}c)(1+\mathcal{B}X)}{2\mathcal{B}(2c+X)}}-\frac{ic^2C}{x},\end{aligned} \tag{9.101}$$

where

$$K = \frac{\alpha U_\infty}{\sqrt{B^2l^2-1}}, \qquad C = \frac{(\alpha-\alpha_0)U_\infty}{\sqrt{1-B^2c^2}}, \tag{9.102a}$$

$$X = x+\frac{c^2}{x}, \qquad L = l+\frac{c^2}{l}, \qquad \frac{1}{\mathcal{B}} = \frac{1+B^2c^2}{B}. \tag{9.102b}$$

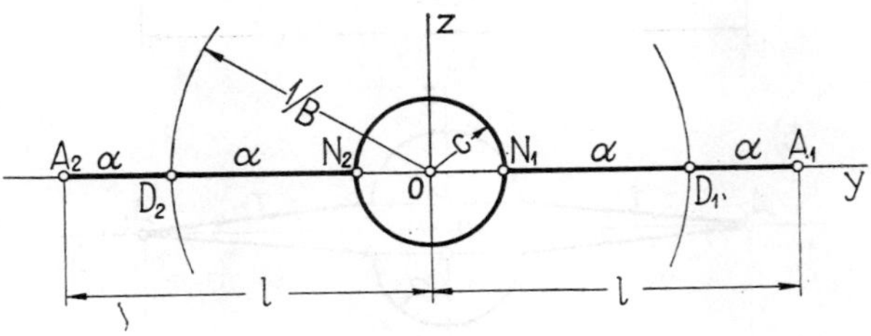

FIG. 9.10

IX. 3.2. Case of a Delta Wing of Symmetrical Thickness

Let us consider a delta wing of symmetrical thickness having the slope $\pm\gamma$ and fitted with a conical body without incidence (Fig. 9.11a).

Beyond the leading edges, as in the case of the single (isolated) wing of symmetrical thickness, the wing may be considered as being continued off the Mach cone by two plane portions without incidence. It may be easily seen now that, as regards the flow in the upper half-plane, the solution for this case can be deduced from (9.99), replacing s by l, S by L, α by γ and putting $\alpha_0 = 0$. In such a case, we obtain the following expression for $\mathcal{U}$,

$$\begin{aligned}\mathcal{U} = &\frac{2Q^*}{\pi}\left(l-\frac{c^2}{x}\right)\cosh^{-1}\sqrt{\frac{(1+\mathcal{B}L)(1-\mathcal{B}X)}{2\mathcal{B}(L-X)}}\\ &+\frac{2Q^*}{\pi}\left(l+\frac{c^2}{x}\right)\cosh^{-1}\sqrt{\frac{(1+\mathcal{B}L)(1+\mathcal{B}X)}{2\mathcal{B}(L+X)}}\\ &+\frac{2}{\pi}\frac{Q^*c^2}{x}\cosh^{-1}\sqrt{\frac{(1-\mathcal{B}^2X^2)(L^2-4c^2)}{(L^2-X^2)(1-4\mathcal{B}^2c^2)}}\\ &-\frac{2cC}{\pi}\left(1-\frac{c}{x}\right)\cosh^{-1}\sqrt{\frac{(1+2\mathcal{B}c)(1-\mathcal{B}X)}{2\mathcal{B}(2c-X)}}\\ &-\frac{2cC}{\pi}\left(1+\frac{c}{x}\right)\cosh^{-1}\sqrt{\frac{(1+2\mathcal{B}c)(1+\mathcal{B}X)}{2\mathcal{B}(2c+X)}}-\frac{ic^2C}{x},\end{aligned} \tag{9.103}$$

where

$$Q^* = -\frac{\gamma U_\infty}{\sqrt{1-B^2l^2}}, \quad C = -\frac{\gamma U_\infty}{\sqrt{1-B^2c^2}}. \tag{9.104}$$

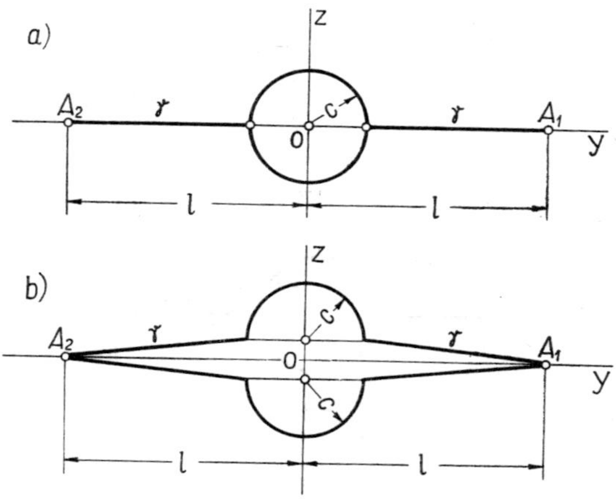

FIG. 9.11

When the wing edges are supersonic, the function $\mathcal{U}$ will have the same expression as in the case of a thin wing with supersonic edges (9.101), in which however the constant K now takes the value

$$K = -\frac{\gamma U_\infty}{\sqrt{B^2l^2-1}}, \tag{9.105}$$

while the constant C is given by (9.104).

Note. We have supposed above that the body is conical and without incidence. If, however, we suppose that the two halves of the conical body are inclined on both sides of the plane Ox_1, x_2 at the slope γ_0 (Fig. 9.11b), then the solution for this case will be obtained by replacing s by l, S by L, α by $-\gamma$ and α_0 by $-\gamma_0$ in (9.99) as well as in (9.77), (9.98).

IX. 3.3. Thin Trapezoidal Wing fitted with Two Marginal Conical Bodies

For simplifying the calculations we shall treat the case of subsonic lateral leading edges, considering for this purpose one half of the trapezoidal wing, symmetrical with respect to its central axis and having a *conical body* at its extremity (Fig. 9.12). It will be assumed that the two Mach cones issued from the vertices of the extremities do not intersect on the wing. Let now α_0

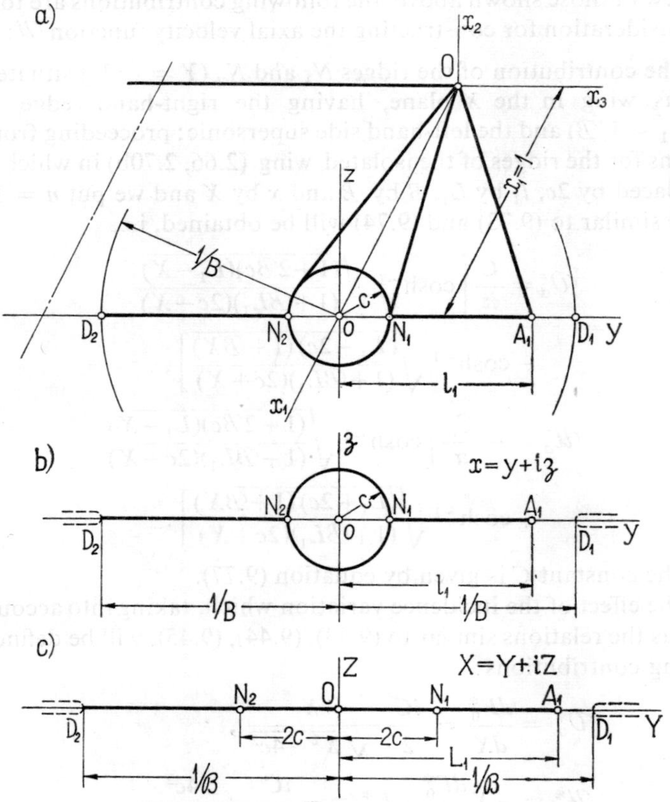

FIG. 9.12

be the incidence of the conical body axis and α the incidence throughout the wing span, both on the central portion, between the conical bodies, and on the lateral sides. In the region 1, outside the Mach cone, the motion is plane, the axial disturbance velocity being given by the equation

$$u_1 = \frac{\alpha U_\infty}{B}. \tag{9.106}$$

As regards the motion in the interior of the Mach cone, to which, in the planes x and X, there correspond the configurations given in Fig. 9.12, this motion is characterized by the same boundary conditions (9.31), (9.32), (9.33), (9.34) where, however, $\alpha_j = \alpha$, $C_j = C$ and where the abscissa of the point A_2 $(y = -l_2)$ is replaced by that of the point D_2 $(y = -1/B)$ since the lateral velocity is zero $(v = 0)$ in the plane motion outside the Mach cone.

In view of those shown above, the following contributions are to be taken into consideration for constructing the axial velocity function $\mathcal{U}$:

(a) The contribution of the ridges N_1 and N_2 $(Y = \pm 2c)$ situated on the fictitious wing in the X plane, having the right-hand edge subsonic $(Y = L_1 < 1/\mathcal{B})$ and the left-hand side supersonic; proceeding from the expressions for the ridges of the isolated wing (2.66, 2.70b) in which s_1 and s_2 are replaced by $2c$, l_1 by L_1, B by $\mathcal{B}$ and x by X and we put $n = 1$, contributions similar to (9.73) and (9.74) will be obtained, i.e.

$$\mathcal{V}'_a = \frac{C}{\pi}\left[\cosh^{-1}\sqrt{\frac{(1+2\mathcal{B}c)(L_1-X)}{(1+\mathcal{B}L_1)(2c-X)}} - \cosh^{-1}\sqrt{\frac{(L_1+2c)(1+\mathcal{B}X)}{(1+\mathcal{B}L_1)(2c+X)}}\right], \tag{9.107a}$$

$$\mathcal{U}'_a = -\frac{C_u}{\pi}\left[\cosh^{-1}\sqrt{\frac{(1+2\mathcal{B}c)(L_1-X)}{(1+\mathcal{B}L_1)(2c-X)}} + \cosh^{-1}\sqrt{\frac{(L_1+2c)(1+\mathcal{B}X)}{(1+\mathcal{B}L_1)(2c+X)}}\right], \tag{9.107b}$$

where the constant C is given by equation (9.77).

(b) The effect of the incidence variation which, taking into account (9.80) as well as the relations similar to (9.43), (9.44), (9.45), will be defined by the following contributions:

$$\mathcal{V}^*_a = \frac{dF^*_0}{dX} = \frac{iC}{2}\frac{X}{\sqrt{X^2-4c^2}}, \tag{9.108a}$$

$$\mathcal{U}^*_a = -X\frac{dF^*_0}{dX} + F^*_0 = -\frac{iC}{2}\frac{4c^2}{\sqrt{X^2-4c^2}}. \tag{9.108b}$$

(c) The contributions of the leading edges of the fictitious wing in the X plane, one of the edges being subsonic $A_1(Y = L_1)$ while the other tends to infinity ($L_2 \to \infty$); taking account of (2.84) in which we replace x by X, l_1 by L_1, B by $\mathcal{B}$, make l_2 tend to infinity ($l_2 \to \infty$) and put $n = 1$, we obtain the following expression for this contribution:

$$\mathcal{U}''_a = A_u \sqrt{\frac{1+\mathcal{B}X}{\mathcal{B}(L_1-X)}} + K \cos^{-1} \sqrt{\frac{1+\mathcal{B}X}{1+\mathcal{B}L_1}}, \tag{9.109a}$$

$$\mathcal{V}''_a = A \sqrt{\frac{1+\mathcal{B}X}{\mathcal{B}(L_1-X)}}. \tag{9.109b}$$

Note. Beyond the Mach cone, since the fictitious wing proceeds from a plane motion, Re $\mathcal{V}''_a = 0$; and hence, in this case, the contribution of the supersonic edge to the function $\mathcal{V}''_a$ is zero.

By adding all the above contributions we obtain the solution for the motion around the fictitious wing in the X plane,

$$\mathcal{U}_a = \mathcal{U}'_a + \mathcal{U}^*_a + \mathcal{U}''_a, \tag{9.110a}$$

$$\mathcal{V}_a = \mathcal{V}'_a + \mathcal{V}^*_a + \mathcal{V}''_a. \tag{9.110b}$$

Considering the relation (9.43) existing between $\mathcal{U}_a$ and $\mathcal{V}_a$, we obtain the following relations between the constants which occur in these functions:

$$C_u = -2cC, \quad A_u = -AL_1, \tag{9.111a}$$

$$K = A\left(L_1 + \frac{1}{\mathcal{B}}\right) + \frac{C}{\pi}\left[\sqrt{(L_1-2c)\left(\frac{1}{\mathcal{B}}+2c\right)} - \sqrt{(L_1+2c)\left(\frac{1}{\mathcal{B}}-2c\right)}\right]. \tag{9.111b}$$

Using these relations, the following expression will finally result for the function of the axial disturbance velocity

$$\mathcal{U} = -A\left(L_1 - \frac{2c^2}{x}\right)\sqrt{\frac{1+\mathcal{B}X}{\mathcal{B}(L_1-X)}} + K\cos^{-1}\sqrt{\frac{1+\mathcal{B}X}{1+\mathcal{B}L_1}}$$
$$-\frac{2cC}{\pi}\left(1-\frac{c}{x}\right)\cosh^{-1}\sqrt{\frac{(1+2\mathcal{B}c)(L_1-X)}{(1+\mathcal{B}L_1)(2c-X)}}$$
$$-\frac{2cC}{\pi}\left(1+\frac{c}{x}\right)\cosh^{-1}\sqrt{\frac{(L_1+2c)(1+\mathcal{B}X)}{(1+\mathcal{B}L_1)(2c+X)}} - \frac{ic^2C}{x}, \tag{9.112}$$

where X and $\mathcal{B}$ have the same significances as above and L_1 is given by the relation

$$L_1 = l_1 + \frac{c^2}{l_1}. \tag{9.113}$$

In the above relation a single constant has not been determined, namely K which represents a simple jump, being defined by the following expression:

$$K = \frac{2\alpha U_\infty}{\pi B}. \tag{9.114}$$

Introducing this value into (9.111b) and taking also account of (9.77), we can write for the other constants:

$$C = \frac{(\alpha-\alpha_0)U_\infty}{\sqrt{1-B^2c^2}}, \quad A = -\frac{\mathcal{B}}{1+\mathcal{B}L_1}\left[\frac{2\alpha U_\infty}{\pi B} - \frac{(\alpha-\alpha_0)U_\infty}{\pi\sqrt{(1-B^2c^2)B}}\right.$$
$$\left.\times\left(\sqrt{(L_1-2c)(1+2\mathcal{B}c)} - \sqrt{(L_1+2c)(1-2\mathcal{B}c)}\right)\right]. \tag{9.115}$$

When $\alpha = \alpha_0$, the expression for $\mathcal{U}$ is simplified and becomes

$$\mathcal{U} = \frac{2\alpha U_\infty}{\pi B}\left[\cos^{-1}\sqrt{\frac{1+\mathcal{B}X}{1+\mathcal{B}L_1}} + \frac{\mathcal{B}}{1+\mathcal{B}L_1}\left(L_1 - \frac{2c^2}{x}\right)\sqrt{\frac{1+\mathcal{B}X}{\mathcal{B}(L_1-X)}}\right]. \tag{9.116}$$

IX. 4. Considerations on High-order Conical Flows about the Wing–Body System

IX. 4.1. Preliminary Formulae

As in the preceding cases, we shall disregard the axisymmetrical motion about the isolated cone which has been treated as such in our first chapter (§ I. 6.2). Accordingly, we have to consider only the following cases which will be enumerated below with reference to Fig. 9.2a:

(a) The central thin wing OM_1M_2 crossing the body axis is parallel, just like the body axis, to the free stream U_∞, hence having its angle of incidence equal to zero ($\alpha_0 = 0$). The lateral sides of the thin wing, viz. OM_1A_1 and OM_2A_2, have angles of incidence α_1 and α_2, or the corresponding downwashes $w^{(1)} = -\alpha_1 U_\infty$ and $w^{(2)} = -\alpha_2 U_\infty$ which are given, under our assumption, by homogeneous polynomials of order $(n-1)$ of the form (9.1a),

$$w^{(j)} = -\alpha_j U_\infty = x_1^{n-1}\sum_{q=0}^{n-1} w_{n-q-1,\,q}^{(j)}\, y^q, \quad (j = 1, 2). \tag{9.117}$$

(b) The plane thin wing OM_1M_2 lies in the same position as in the preceding case, the lateral sides, however, OM_1A_1 and OM_2A_2, of symmetrical thickness, having the respective slopes $\pm\gamma_1$ and $\pm\gamma_2$, or the corresponding downwashes $w^{(1)} = \pm\gamma_1 U_\infty$ and $w^{(2)} = \pm\gamma_2 U_\infty$, which are given, under our assumption, by homogeneous polynomials of order $(n-1)$ of the form (9.1b),

$$w^{(j)} = \pm\gamma_j U_\infty = \pm x_1^{n-1} \sum_{q=0}^{n-1} w^{(j)}_{n-q-1,\,q}\, y^q, \quad (j = 1, 2). \tag{9.118}$$

(c) The downwashes on the entire wing being given as above (9.1a), the cone axis becomes deformed and the incidence α_0 of the axis with respect to the free stream direction varies according to an expression of the form

$$\alpha_0 = \alpha_{n-1,\,0}\, x_1^{n-1} = -\frac{w_{n-1,\,0}}{U_\infty}\, x_1^{n-1}. \tag{9.119}$$

Let us now determine the boundary conditions for the first case (a), from which one can deduce by different means all the other cases.

For this purpose, let us notice that on putting $\alpha_0 = \beta_0 = 0$ and observing that we have

$$\cos\theta = \frac{y}{c} = \frac{x_2}{x_1 c}, \quad \sin\theta = \frac{z}{c} = \frac{x_3}{x_1 c}, \tag{9.120}$$

respectively, on the circle of radius c, the general condition on the body (9.19) becomes in such a case

$$x_2 v + x_3 w = x_2 \Phi_{010} + x_3 \Phi_{001} = 0 \tag{9.121}$$

and, if it is differentiated $(n-1)$ times with respect to x_1, it becomes

$$x_2 \Phi_{n-1,\,1,\,0} + x_3 \Phi_{n-1,\,0,\,1} = x_1 (y\Phi_{n-1,\,1,\,0} + z\Phi_{n-1,\,0,\,1}) = 0. \tag{9.122}$$

It should be further noted that under the admitted hypothesis ($B^2c^2 \ll 1$), we may put $z = \mathfrak{z}$ and meanwhile write the compatibility relation round the body in the form

$$\begin{aligned} d\Phi_{n-1,\,1,\,0} + i\,d\Phi'_{n-1,\,1,\,0} &= dF_{n-1,\,1,\,0} = \frac{-i}{\sqrt{1-B^2x^2}}\, dF_{n-1,\,0,\,1} \\ &\approx -i\,d\Phi_{n-1,\,0,\,1} + d\Phi'_{n-1,\,0,\,1} \end{aligned} \tag{9.123}$$

whence

$$d\Phi_{n-1,\,0,\,1} = -d\Phi'_{n-1,\,1,\,0}, \quad \Phi_{n-1,\,1,\,0} = -\Phi'_{n-1,\,0,\,1}. \tag{9.124}$$

In this case, condition (9.122) becomes successively:

$$y\Phi_{n-1,\,1,\,0} + \mathfrak{z}\Phi_{n-1,\,0,\,1} = y\Phi_{n-1,\,1,\,0} - \mathfrak{z}\Phi'_{n-1,\,1,\,0} = \operatorname{Re}\,(xF_{n-1,\,1,\,0})_{\text{circle}} = 0. \tag{9.125}$$

On the other hand, on comparing relation (1.172c) to (9.117), it follows that in our case, on the wing portions M_1A_1 and M_2A_2,

$$\Phi_{n-q-1,q,1} = \operatorname{Re} F_{n-q-1,q,1} = \frac{(n-1)!}{C_{n-1}^q} w_{n-q-1,q}^{(j)}, \quad (j = 1, 2) \tag{9.126}$$

Thus, under the assumption that the incidence is zero ($w = 0$) on the central part of the wing–body system, we may write

$$\operatorname{Im} [F_{n-1,1,0}]_{N_jM_j} = 0, \ (j = 1, 2), \tag{9.127}$$

while on the lateral sides, on the wing and outside the wing,

$$\operatorname{Im} [F_{n-1,1,0}]_{M_jA_j} = C_j = \text{const.}, \quad (j = 1, 2), \tag{9.128a}$$

$$\operatorname{Re} [F_{n-1,1,0}]_{-l_2>y>l_1} = 0. \tag{9.128b}$$

Consider a "fictitious potential", $F_{n-1,0,0}$, defined in the following way:

$$\Phi_{n-1,0,0} = x_1 \operatorname{Re} F_{n-1,0,0} \tag{9.129a}$$

$$\Phi_{n-1,1,0} = \frac{\partial \Phi_{n-1,0,0}}{\partial x_2} = \operatorname{Re} \frac{\partial F_{n-1,0,0}}{\partial y}, \quad F_{n-1,1,0} = \frac{dF_{n-1,0,0}}{dx}. \tag{9.129b}$$

The function $F_{n-1,1,0}$ now plays the part of a complex velocity.
In the X plane, the "complex velocity" will be given by the relation

$$\frac{dF_{n-1,0,0}}{dX} = \frac{dF_{n-1,0,0}}{dx}\frac{dx}{dX} = F_{n-1,1,0}\frac{x}{x-c^2/x}$$

$$= F_{n-1,1,0}\tfrac{1}{2}\left(1+\frac{X}{\sqrt{X^2-4c^2}}\right) \tag{9.130}$$

which, on applying (9.125), becomes

$$\operatorname{Im}\left[\frac{dF_{n-1,0,0}}{dX}\right]_{\text{circle}} = \operatorname{Im}\left(\frac{xF_{n-1,1,0}}{2ic\sin\theta}\right)_{\text{circle}} = 0 \tag{9.131}$$

at the points corresponding to the circle; consequently, taking also account of (9.127),

$$\operatorname{Im}\left[\frac{dF_{n-1,0,0}}{dX}\right]_{M_2M_1} = 0 \tag{9.132}$$

on the entire central part M_2M_1.

On the two lateral sides, M_2A_2, M_1A_1, as well as outside the wing, we have, considering (9.13),

$$\operatorname{Im}\left[\frac{dF_{n-1,0,0}}{dX}\right]_{M_jA_j} = \tfrac{1}{2}\,C_j\left(1+\frac{Y}{\sqrt{Y^2-4c^2}}\right), \quad (j = 1, 2), \tag{9.133a}$$

$$\operatorname{Re}\left[\frac{dF_{n-1,0,0}}{dX}\right]_{-L_2>Y>L_1} = 0. \tag{9.133b}$$

As in the case $n = 1$ treated earlier (§ IX. 2.1), we assume that the motion in the X plane around the fictitious wing A_2A_1 corresponds to a conical motion of order n for which we have the compatibility relations analogous with (1.180) where, however, x will be replaced by X (9.10), B by $\mathcal{B}$ (9.13b), and the functions $F_{n-q-r,\,q.\,r}$ will be affected by a superscript (a) pointing out that we are dealing with the fictitious wing in the X plane. In such a case we can put

$$F^{(a)}_{n-1,1,0} = \frac{dF_{n-1,0,0}}{dX}, \tag{9.134}$$

whence

$$F_{n-1,1,0} = \frac{dF_{n-1,0,0}}{dX}\,\frac{dX}{dx} = \left(1-\frac{c^2}{x^2}\right)F^{(a)}_{n-1,1,0}. \tag{9.135}$$

In order to find $u = x_1^{n-1}u_{n-1} = x_1^{n-1}\operatorname{Re}\mathcal{U}_{n-1}$, we shall further apply Euler's formula. In this sense, we replace each function $F_{n-q,q,0}$, by its expression in terms of $F_{n-1,1,0}$ by applying the compatibility relations (1.180) put in the form

$$dF_{n-q,q,0} = (-x)^{1-q}\,dF_{n-1,1,0}. \tag{9.136}$$

Finally, the problem reduces to the determination of the function $F^{(a)}_{n-1,1,0}$ for the conical motion in the plane X, into which, besides the common singularities of the isolated wing, we shall introduce a continuous distribution of sources on M_1A_1 and M_2A_2, respectively, of a form similar to (9.48a),

$$dF^{(a)}_{n-1,1,0} = \frac{C_1}{\pi}\cosh^{-1}\sqrt{\frac{(L_2+\eta)(L_1-X)}{(L_1+L_2)(\eta-X)}}\,d\left(\frac{\eta}{\sqrt{\eta^2-4c^2}}\right), \tag{9.137a}$$

$$dF^{(a)}_{n-1,1,0} = \frac{C_2}{\pi}\cosh^{-1}\sqrt{\frac{(L_1+\eta)(L_2+X)}{(L_1+L_2)(\eta+X)}}\,d\left(\frac{\eta}{\sqrt{\eta^2-4c^2}}\right), \tag{9.137b}$$

the first expression relating to the contribution of an elementary ridge lying at the point $Y = \eta$ on the portion M_1A_1, and the latter relating to the contribution of a ridge lying at $Y = -\eta$ on the wing portion M_2A_2.

In conclusion, the study of the nth order conical flow about the wing–body system reduces to the corresponding study of an isolated wing having the boundary conditions (on the wing) completely determined.

IX. 4.2. Application to the Conical Flow of Order 2 ($n = 2$)

In the case $n = 2$, taking into account (2.103) and (9.129), the axial disturbance velocity may be written successively as

$$u = x_1 u_1 = x_1 \operatorname{Re} \mathcal{U}_1 = x_1 \operatorname{Re} F_{100}, \tag{9.138}$$

whence

$$F_{100} = \mathcal{U}_1, \quad F_{110} = \frac{d\mathcal{U}_1}{dx}, \quad F_{110}^{(a)} = \frac{d\mathcal{U}_1}{dX}. \tag{9.139}$$

On the other hand, taking account of (9.124), we can write for the points lying in the immediate neighbourhood of the body

$$F_{110} = \Phi_{110} + i\Phi'_{110} = \Phi_{110} - i\Phi_{101} = \frac{\partial \Phi_{100}}{\partial x_2} - i\frac{\partial \Phi_{100}}{\partial x_3} = \frac{d\mathcal{U}_1}{dx}, \tag{9.140}$$

whence it follows that $\mathcal{U}_1$ represents simply a real potential at the points lying around the body, which therefore has the same value in the planes x and X:

$$\mathcal{U}_1 = \mathcal{U}_{1a}. \tag{9.141}$$

One of the first applications of this result is the case when, on the entire wing–body system, we have a symmetrical variation of the form

$$w = w_{10} x_1, \quad (w_{10} = \Phi_{101} = \text{const.}). \tag{9.142}$$

In such a case, it is to be noted that the imaginary part of $\mathcal{U}_1$ ($\mathcal{U}_{1a}$) is zero on the actual and fictitious wings; from expression (9.140) we can see that it is also null on the body. Then, we can pass from the X plane, where $\mathcal{U}_{1a}$ is known, to the x plane, by applying the transformation (9.10):

$$\mathcal{U}_1 = \mathcal{U}_{1a} = \frac{A_{20} + A_{22}X^2}{\sqrt{L^2 - X^2}} = \frac{A_{20} + A_{22}(x + c^2/x)^2}{\sqrt{(l^2 - x^2)(1 - c^4/l^2x^2)}}. \tag{9.143}$$

In the general case, however, we assume that the central part, the body axis included, has zero incidence, while the lateral sides M_1A_1 and M_2A_2 have symmetrical incidences with respect to the wing axis, given in the common form

$$w = \alpha U_\infty = x_1(\alpha_{10} \pm \alpha_{01} y) U_\infty = x_1(w_{10} \pm w_{01} y), \tag{9.144}$$

where the (+) sign relates to the positive abscissae ($y > 0$) and the (−) sign to the negative ones ($y < 0$).

In this case, by applying (9.132) and (9.133), we have

$$\operatorname{Im}\left[\frac{d\mathcal{U}_{1a}}{dX}\right]_{M_2M_1} = \operatorname{Im}\left[F_{110}^{(a)}\right]_{M_2M_1} = 0, \tag{9.145a}$$

$$\operatorname{Im}\left[\frac{d\mathcal{U}_{1a}}{dX}\right]_{M_jA_j} = \operatorname{Im}\left[F_{110}^{(a)}\right]_{M_jA_j} = \frac{C}{2}\left(1+\frac{Y}{\sqrt{Y^2-4c^2}}\right), \tag{9.145b}$$

on the fictitious wing.

APPLICATION TO A THIN DELTA WING WITH SUBSONIC EDGES IN THE GENERAL CASE. The conditions above help us construct the function $\mathcal{U}_{1a}$ corresponding to the fictitious wing in the X plane, by summing the following contributions:

(a) The contribution of the subsonic leading edges A_1 and A_2 ($Y = \pm L$), which is deduced from (9.143):

$$\mathcal{U}'_{1a} = \frac{A'_{20}+A'_{22}X^2}{\sqrt{L^2-X^2}}. \tag{9.146}$$

(b) The contribution of the ridges M_1 and M_2 ($Y = \pm S$) which, taking account of (2.64), (2.69b) as well as of the wing symmetry, assumes the expression

$$\mathcal{U}''_{1a} = (C''_{20}+C''_{21}X)\cosh^{-1}\sqrt{\frac{(L+S)(L-X)}{2L(S-X)}} + (C''_{20}-C''_{21}X)\cosh^{-1}\sqrt{\frac{(L+S)(L+X)}{2L(S+X)}} + D''_{20}\sqrt{L^2-X^2}. \tag{9.147}$$

Further, considering relations (9.145), which will be applied on both sides of the ridge M_1 and in its immediate vicinity, we deduce the following relation between the constants C''_{21} and C,

$$C''_{21} = \frac{C}{\pi}\left(1+\frac{S}{\sqrt{S^2-4c^2}}\right). \tag{9.148}$$

(c) The effect of the incidence variation on the portions M_1A_1 and M_2A_2 of the fictitious wing in the X plane, corresponding to condition (9.145), which reduces to the effect of a continuous distribution of elementary ridges on those portions whose contribution to $F_{110}^{(a)}$ is to be obtained by

$$F_{110}^{*(a)} = \frac{C}{\pi}\int_S^L \left[\cosh^{-1}\sqrt{\frac{(L+\eta)(L-X)}{2L(\eta-X)}} - \cosh^{-1}\sqrt{\frac{(L+\eta)(L+X)}{2L(\eta+X)}}\right]$$
$$\times d\left(\frac{\eta}{\sqrt{\eta^2-4c^2}}\right) = -\frac{C}{\pi}\frac{S}{\sqrt{S^2-4c^2}}\left[\cosh^{-1}\sqrt{\frac{(L+S)(L-X)}{2L(S-X)}}\right.$$
$$\left.-\cosh^{-1}\sqrt{\frac{(L+S)(L+X)}{2L(S+X)}}\right]$$
$$+\frac{C}{\pi}\frac{X}{\sqrt{X^2-4c^2}}\cosh^{-1}\sqrt{\frac{(L^2-X^2)(S^2-4c^2)}{(S^2-X^2)(L^2-4c^2)}}. \tag{9.149}$$

As regards the contribution of this variation to $\mathcal{U}_{1a}$, by applying (9.139) we obtain

$$\mathcal{U}_{1a}^{*} = \int F_{11a}^{*(a)}\,dX = \frac{C}{\pi}\frac{4c^2-SX}{\sqrt{S^2-4c^2}}\cosh^{-1}\sqrt{\frac{(L+S)(L-X)}{2L(S-X)}}$$
$$+\frac{C}{\pi}\frac{4c^2+SX}{\sqrt{S^2-4c^2}}\cosh^{-1}\sqrt{\frac{(L+S)(L+X)}{2L(S+X)}}$$
$$+\frac{C}{\pi}\sqrt{X^2-4c^2}\cosh^{-1}\sqrt{\frac{(L^2-X^2)(S^2-4c^2)}{(S^2-X^2)(L^2-4c^2)}}. \tag{9.150}$$

By adding all the above contributions we finally obtain

$$\mathcal{U}_1 = \mathcal{U}_{1a} = \frac{A_{20}+A_{22}X^2}{\sqrt{L^2-X^2}} + (C_{20}+C_{21}X)\cosh^{-1}\sqrt{\frac{(L+S)(L-X)}{2L(S-X)}}$$
$$+(C_{20}-C_{21}X)\cosh^{-1}\sqrt{\frac{(L+S)(L+X)}{2L(S+X)}}$$
$$+C_{21}\sqrt{X^2-4c^2}\cosh^{-1}\sqrt{\frac{(L^2-X^2)(S^2-4c^2)}{(S^2-X^2)(L^2-4c^2)}}, \tag{9.151}$$

where we have used the notation:

$$A_{20} = A'_{20}+D''_{20}L^2, \quad A_{22} = A'_{22}-D''_{20}, \tag{9.152a}$$

$$C_{20} = C''_{20}+\frac{C}{\pi}\frac{4c^2}{\sqrt{S^2-4c^2}}, \quad C_{21} = \frac{C}{\pi}. \tag{9.152b}$$

For the determination of the new constants C_{20} and C_{21}, we perform the following integrals over a small semicircle σ, round the ridge $M_1(x = s)$:

$$w_{10} = \operatorname{Im} \oint \sqrt{1-B^2x^2}\, \frac{d^2\mathcal{U}_1}{dx^2}\, dx, \tag{9.153a}$$

$$w_{01} = -\operatorname{Im} \oint \frac{\sqrt{1-B^2x^2}}{x} \frac{d^2\mathcal{U}_1}{dx^2}\, dx \tag{9.153b}$$

Performing the calculations, we obtain the following values for C_{20} and C_{21}:

$$C_{20} = \frac{2s}{\pi} \frac{(2-B^2s^2)w_{10}+sw_{01}}{(1-B^2s^2)^{3/2}}, \tag{9.154a}$$

$$C_{21} = -\frac{1}{\pi} \frac{w_{10}+B^2s^3w_{01}}{(1-B^2s^2)^{3/2}}. \tag{9.154b}$$

As regards the constants A_{20} and A_{22}, they will be determined, as in the case of the isolated wing, from the conditions

$$w_{10} = \operatorname{Im} \int_{y'}^{1/B} \sqrt{1-B^2x^2}\, \frac{d^2\mathcal{U}_1}{dx^2}\, dx, \tag{9.155a}$$

$$w_{01} = -\operatorname{Im} \int_{y'}^{1/B} \frac{\sqrt{1-B^2x^2}}{x} \frac{d^2\mathcal{U}_1}{dx^2}\, dx, \tag{9.155b}$$

where y' is the abscissa of a point situated on the wing portion M_1A_1.

If we now suppose that on the entire wing we have the incidence α, or the corresponding downwash w given by (9.144), the body alone having zero incidence, then the expression for the axial disturbance velocity is to be deduced from (9.151), putting $S = 2c$, $s = c$, respectively, and thus obtaining

$$\mathcal{U}_1 = \frac{A_{20}+A_{22}X^2}{\sqrt{L^2-X^2}} + (C_{20}+C_{21}X)\cosh^{-1}\sqrt{\frac{(L+2c)(L-X)}{2L(2c-X)}}$$
$$+(C_{20}-C_{21}X)\cosh^{-1}\sqrt{\frac{(L+2c)(L+X)}{2L(2c+X)}} + \frac{i\pi}{2} C_{21}\left(x-\frac{c^2}{x}\right), \tag{9.156}$$

where

$$C_{20} = \frac{2c}{\pi} \frac{(2-B^2c^2)w_{10}+cw_{01}}{(1-B^2c^2)^{3/2}}, \quad C_{21} = -\frac{1}{\pi} \frac{w_{10}+B^2c^3w_{01}}{(1-B^2c^2)^{3/2}}. \tag{9.157}$$

IX. 4.3. Case of Delta Thin Wing with Supersonic Edges ($n = 2$)

It will be assumed from the beginning that the body axis has zero incidence with respect to the free stream U_∞, and that the wing has the incidence α, or the downwash w defined by (9.144), on the portions N_1M_1 and N_2M_2, and zero incidence beyond the ridges M_1 and M_2 up to the leading edges which are supersonic (Fig. 9.9). Under these conditions, the axial disturbance velocity function $\mathcal{U}_{1a}$ of the fictitious wing in the X plane will be made up of the following contributions:

(a) The contribution of the ridges M_1 and M_2 ($Y = \pm S$) of the fictitious wing in the X plane, whose expression is analogous to (9.147), where, however, L is to be replaced by $1/\mathcal{B}$,

$$\mathcal{U}'_{1a} = (C'_{20}+C'_{21}X)\cosh^{-1}\sqrt{\frac{(1+\mathcal{B}S)(1-\mathcal{B}X)}{2\mathcal{B}(S-X)}}$$
$$+(C'_{20}-C'_{21}X)\cosh^{-1}\sqrt{\frac{(1+\mathcal{B}S)(1+\mathcal{B}X)}{2\mathcal{B}(S+X)}}+D'_{20}\sqrt{1-\mathcal{B}^2X^2}\,. \tag{9.158}$$

Further we have to add the contribution of the incidence variation on the portions M_1D_1 and M_2D_2 of the fictitious wing in the X plane, whose expression is to be deduced from (9.150), on substituting L by $1/\mathcal{B}$ and taking account of (9.148),

$$\mathcal{U}^*_{1a} = C'_{21}\frac{4c^2-SX}{S+\sqrt{S^2-4c^2}}\cosh^{-1}\sqrt{\frac{(1+\mathcal{B}S)(1-\mathcal{B}X)}{2\mathcal{B}(S-X)}}$$
$$+C'_{21}\frac{4c^2+SX}{S+\sqrt{S^2-4c^2}}\cosh^{-1}\sqrt{\frac{(1+\mathcal{B}S)(1+\mathcal{B}X)}{2\mathcal{B}(S+X)}}$$
$$+C'_{21}\frac{\sqrt{S^2-4c^2}}{S+\sqrt{S^2-4c^2}}\sqrt{X^2-4c^2}\cosh^{-1}\sqrt{\frac{(1-\mathcal{B}^2X^2)(S^2-4c^2)}{(S^2-X^2)(1-4\mathcal{B}^2c^2)}}\,. \tag{9.159}$$

Now, by adding (9.158) and (9.159), we obtain the contribution of the two ridges M_1 and M_2 of the actual wing,

$$\mathcal{U}_1^{(M)} = \mathcal{U}'_{1a}+\mathcal{U}^*_{1a} = (Q_{20}+Q_{21}X)\cosh^{-1}\sqrt{\frac{(1+\mathcal{B}S)(1-\mathcal{B}X)}{2\mathcal{B}(S-X)}}$$
$$+(Q_{20}-Q_{21}X)\cosh^{-1}\sqrt{\frac{(1+\mathcal{B}S)(1+\mathcal{B}X)}{2\mathcal{B}(S+X)}}$$
$$+Q_{21}\sqrt{X^2-4c^2}\cosh^{-1}\sqrt{\frac{(1-\mathcal{B}^2X^2)(S^2-4c^2)}{(S^2-X^2)(1-4\mathcal{B}^2c^2)}}+D'_{20}\sqrt{1-\mathcal{B}^2X^2}, \tag{9.160}$$

where we have used the notations

$$Q_{20} = C'_{20}+C'_{21}\frac{4c^2}{S+\sqrt{S^2-4c^2}}, \quad Q_{21} = C'_{21}\frac{\sqrt{S^2-4c^2}}{S+\sqrt{S^2-4c^2}}. \qquad (9.161)$$

(b) The contribution of the intersections N_1 and N_2 ($y = \pm c$) of the wing and body, whose expression is to be deduced from (9.160) by making s tend to c and S to $2c$, thus obtaining

$$\mathcal{U}_1^{(N)} = (C_{20}+C_{21}X)\cosh^{-1}\sqrt{\frac{(1+2\mathcal{B}c)(1-\mathcal{B}X)}{2\mathcal{B}(2c-X)}}$$
$$+(C_{20}-C_{21}X)\cosh^{-1}\sqrt{\frac{(1+2\mathcal{B}c)(1+\mathcal{B}X)}{2\mathcal{B}(2c+X)}}$$
$$+\frac{i\pi}{2}C_{21}\sqrt{X^2-4c^2}+D''_{20}\sqrt{1-\mathcal{B}^2X^2}. \qquad (9.162)$$

On summing up the contributions (9.160) and (9.162), the following expression is obtained for $\mathcal{U}_1$:

$$\mathcal{U}_1 = \mathcal{U}_1^{(M)}+\mathcal{U}_1^{(N)} = (Q_{20}+Q_{21}X)\cosh^{-1}\sqrt{\frac{(1+\mathcal{B}S)(1-\mathcal{B}X)}{2\mathcal{B}(S-X)}}$$
$$+(Q_{20}-Q_{21}X)\cosh^{-1}\sqrt{\frac{(1+\mathcal{B}S)(1+\mathcal{B}X)}{2\mathcal{B}(S+X)}}$$
$$+Q_{21}\left(x-\frac{c^2}{x}\right)\cosh^{-1}\sqrt{\frac{(1-\mathcal{B}^2X^2)(S^2-4c^2)}{(S^2-X^2)(1-4\mathcal{B}^2c^2)}}$$
$$+(C_{20}+C_{21}X)\cosh^{-1}\sqrt{\frac{(1+2\mathcal{B}c)(1-\mathcal{B}X)}{2\mathcal{B}(2c-X)}}$$
$$+(C_{20}-C_{21}X)\cosh^{-1}\sqrt{\frac{(1+2\mathcal{B}c)(1+\mathcal{B}X)}{2\mathcal{B}(2c+X)}}$$
$$+\frac{i\pi}{2}C_{21}\left(x-\frac{c^2}{x}\right)+D_{20}\sqrt{1-\mathcal{B}^2X^2}, \qquad (9.163)$$

where we have written as before

$$X = x+\frac{c^2}{x}, \quad \frac{1}{\mathcal{B}} = \frac{1+B^2c^2}{B}, \quad S = s+\frac{c^2}{s}. \qquad (9.164)$$

The constants Q_{20}, Q_{21}, C_{20}, C_{21} are given by the following equation which are analogous to (9.154) and (9.157),

$$Q_{20} = -\frac{2s}{\pi}\frac{(2-B^2s^2)w_{10}+sw_{01}}{(1-B^2s^2)^{3/2}}, \quad Q_{21} = \frac{1}{\pi}\frac{w_{10}+B^2s^3w_{01}}{(1-B^2s^2)^{3/2}}, \tag{9.165a}$$

$$C_{20} = \frac{2c}{\pi}\frac{(2-B^2c^2)w_{10}+cw_{01}}{(1-B^2c^2)^{3/2}}, \quad C_{21} = -\frac{1}{\pi}\frac{w_{10}+B^2c^3w_{01}}{(1-B^2c^2)^{3/2}}. \tag{9.165b}$$

As regards the constant D_{20}, this will be determined, as in the case of the isolated wing, from the condition

$$\left[\sqrt{1-B^2x^2}\,\frac{d\mathcal{U}_1}{dx}\right]_{x=\pm 1/B} = 0. \tag{9.166}$$

Expression (9.163) retains its validity, taking (9.100) into account, also in the case when the ridges M_1 and M_2 become supersonic ($s > 1/B$, $S > 1/\mathcal{B}$).

Supposing now that the supersonic ridges M_1 and M_2 coincide with the edges A_1 and A_2 ($s = l$), the expression for the function $\mathcal{U}_1$ is to be obtained on replacing s by l and S by L in (9.163) and applying relation (9.100); thus we have

$$\begin{aligned}\mathcal{U}_1 &= (K_{20}+K_{21}X)\cos^{-1}\sqrt{\frac{(1+\mathcal{B}L)(1-\mathcal{B}X)}{2\mathcal{B}(L-X)}}\\ &+(K_{20}-K_{21}X)\cos^{-1}\sqrt{\frac{(1+\mathcal{B}L)(1+\mathcal{B}X)}{2\mathcal{B}(L+X)}}\\ &+K_{21}\left(x-\frac{c^2}{x}\right)\cos^{-1}\sqrt{\frac{(1-\mathcal{B}^2X^2)(L^2-4c^2)}{(L^2-X^2)(1-4\mathcal{B}^2c^2)}}\\ &+(C_{20}+C_{21}X)\cosh^{-1}\sqrt{\frac{(1+2\mathcal{B}c)(1-\mathcal{B}X)}{2\mathcal{B}(2c-X)}}\\ &+(C_{20}-C_{21}X)\cosh^{-1}\sqrt{\frac{(1+2\mathcal{B}c)(1+\mathcal{B}X)}{2\mathcal{B}(2c+X)}}\\ &+\frac{i\pi}{2}C_{21}\left(x-\frac{c^2}{x}\right)+D_{20}\sqrt{1-\mathcal{B}^2X^2},\end{aligned} \tag{9.167}$$

where the constants C_{20}, C_{21} are given by (9.165) and where

$$K_{20} = -\frac{2l}{\pi}\frac{(B^2l^2-2)w_{10}-lw_{01}}{(B^2l^2-1)^{3/2}},$$

$$K_{21} = \frac{-1}{\pi}\frac{w_{10}+B^2l^3w_{01}}{(B^2l^2-1)^{3/2}}, \quad L = l+\frac{c^2}{l}. \tag{9.168}$$

Note. When the body axis has the same incidence $\alpha = (w_{10}/U_\infty)x_1$ as the entire wing, then the contribution of the intersections N_1 and N_2 becomes zero; finally we obtain

$$\mathcal{U}_1 = (K_{20}+K_{21}X)\cos^{-1}\sqrt{\frac{(1+\mathcal{B}L)(1-\mathcal{B}X)}{2\mathcal{B}(L-X)}}$$
$$+(K_{20}-K_{21}X)\cos^{-1}\sqrt{\frac{(1+\mathcal{B}L)(1+\mathcal{B}X)}{2\mathcal{B}(L+X)}}$$
$$+K_{21}\left(x-\frac{c^2}{x}\right)\cos^{-1}\sqrt{\frac{(1-\mathcal{B}^2X^2)(L^2-4c^2)}{(L^2-X^2)(1-4\mathcal{B}^2c^2)}}+D_{20}\sqrt{1-\mathcal{B}^2X^2}, \tag{9.169}$$

where the constants K_{20}, K_{21} have the values given in (9.168) where $w_0 = 0$ while D_{20} is to be determined from condition (9.166).

IX. 4.4. Case of a Delta Wing of Symmetrical Thickness ($n = 2$)

Consider a conical body without incidence fitted with a delta wing of symmetrical thickness having the slope γ corresponding to the downwash on the wing upper surface

$$w = \gamma U_\infty = x_1(w_{10}\pm w_{01}y) = x_1(\gamma_{10}\pm\gamma_{01}y)U_\infty. \tag{9.170}$$

As in the case of the wing considered alone, we may consider in this case too that the wing is continued off the Mach cone by two flat portions with no incidence. In view of this fact, the solution of the flow in the upper half-plane will be deduced from (9.163) on substituting s by l, S by L (and α_{ij} by γ_{ij}). Thus we obtain

$$\mathcal{U}_1 = (Q_{20}+Q_{21}X)\cosh^{-1}\sqrt{\frac{(1+\mathcal{B}L)(1-\mathcal{B}X)}{2\mathcal{B}(L-X)}}$$
$$+(Q_{20}-Q_{21}X)\cosh^{-1}\sqrt{\frac{(1+\mathcal{B}L)(1+\mathcal{B}X)}{2\mathcal{B}(L+X)}}$$
$$+Q_{21}\left(x-\frac{c^2}{x}\right)\cosh^{-1}\sqrt{\frac{(1-\mathcal{B}^2X^2)(L^2-4c^2)}{(L^2-X^2)(1-4\mathcal{B}^2c^2)}}$$
$$+(C_{20}+C_{21}X)\cosh^{-1}\sqrt{\frac{(1+2\mathcal{B}c)(1-\mathcal{B}X)}{2\mathcal{B}(2c-X)}}$$
$$+(C_{20}-C_{21}X)\cosh^{-1}\sqrt{\frac{(1+2\mathcal{B}c)(1+\mathcal{B}X)}{2\mathcal{B}(2c+X)}}$$
$$+\frac{i\pi}{2}C_{21}\left(x-\frac{c^2}{x}\right)+D_{20}\sqrt{1-\mathcal{B}^2X^2}, \tag{9.171}$$

in which

$$Q_{20} = -\frac{2l}{\pi}\frac{(2-B^2l^2)\,w_{10}+l\,w_{01}}{(1-B^2l^2)^{3/2}}, \quad Q_{21} = \frac{1}{\pi}\frac{w_{10}+B^2l^3\,w_{01}}{(1-B^2l^2)^{3/2}}, \tag{9.172a}$$

$$C_{20} = \frac{2c}{\pi}\frac{(2-B^2c^2)w_{10}+cw_{01}}{(1-B^2c^2)^{3/2}}, \quad C_{21} = -\frac{1}{\pi}\frac{w_{10}+B^2c^3w_{01}}{(1-B^2c^2)^{3/2}}. \tag{9.172b}$$

As regards the constant D_{20}, it will be deduced similarly from condition (9.166).

When the wing edges are supersonic, the function $\mathcal{U}_1$ will assume the same expression (9.167) as in the case of the thin wing with supersonic edges.

IX. 5. Considerations on the Flow about a Cruciform Wing-Conical Body System

For simplifying our presentation, we shall deal below with the conical flow ($n = 1$) about a cruciform wing–conical body system.

Since our approach is made within the frame of the linearized theory, we shall separate from the beginning the flow about the system whose plate has its incidence or slope symmetrical with respect to the body axis, and whose wing is thin and without incidence, while the body axis has the incidence β_0 in the horizontal plane and zero incidence in the vertical plane; this system represents just the simple wing–body system treated above, but rotated through 90°.

We shall also separate the axisymmetrical flow about the isolated conical body having no incidence which was treated in our first chapter (§ I. 6.2).

After separating these motions, we have therefore to study the case of the system in which the wing has a determined incidence and slope, the body axis having the incidence α_0 in the vertical plane and no incidence in the horizontal plane, whereas the plate (which is considered to be thin) has antisymmetrical incidence.

In order to indicate the method of approach in this case, let us suppose that the wing central portion M_1M_2 together with the body axis has the incidence α_0, the wing lateral portions M_1A_1 and M_2A_2 have the angles of incidence α_1 and α_2, respectively, while the plate has the antisymmetrical lateral incidence $\pm\beta$, the (+) sign relating to the plate upper surface ($z > 0$) and the (−) sign to the lower surface ($z < 0$), (Fig. 9.13). Employing the same notations as above, the boundary conditions on the wing, body and real axis (outside the wing) will be given in the planes x (Fig. 9.13a) and X (Fig. 9.13b) by the same relations (9.31)–(9.34) and (9.39)–(9.41),

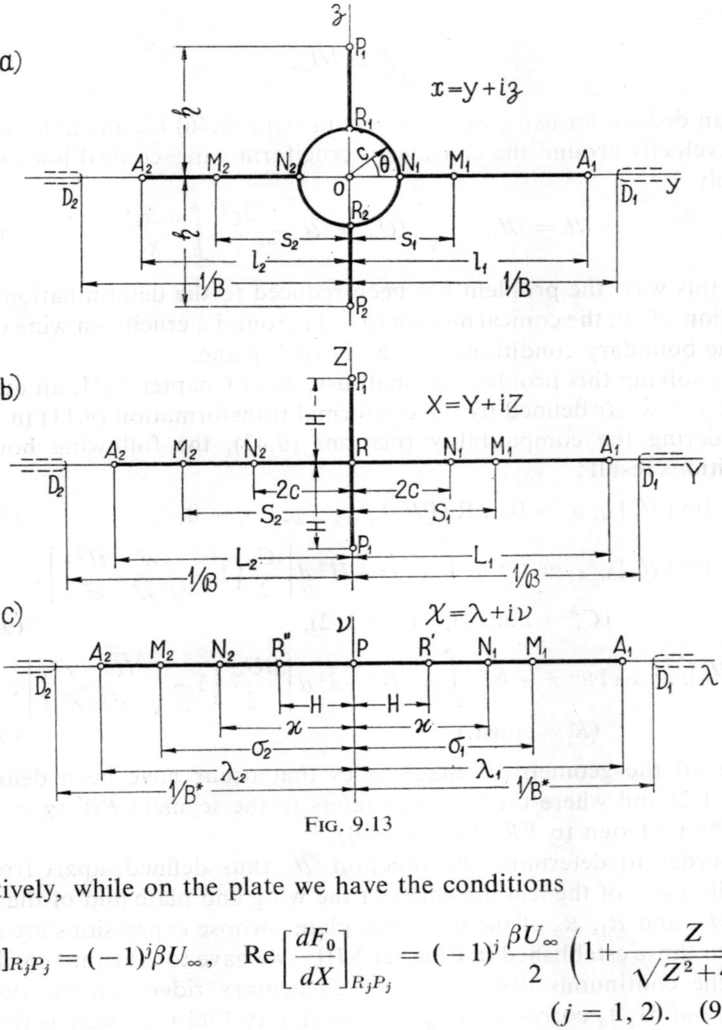

FIG. 9.13

respectively, while on the plate we have the conditions

$$\operatorname{Re}\,[\mathcal{V}]_{R_jP_j} = (-1)^j \beta U_\infty, \quad \operatorname{Re}\left[\frac{dF_0}{dX}\right]_{R_jP_j} = (-1)^j \frac{\beta U_\infty}{2}\left(1+\frac{Z}{\sqrt{Z^2+4c^2}}\right)$$

$$(j = 1, 2). \quad (9.173)$$

In this case too, by analogy with the cruciform wing problem, we can consider that the motion in the X plane defined above corresponds to a conical motion around a cruciform wing having the traces A_2A_1 $(-L_2 < X = Y < L_1)$ and P_2P_1 $(-iH < X = iZ < iH)$ in this plane.

Designating as above by $\mathcal{U}_a$, $\mathcal{V}_a$, $\mathcal{W}_a$ the complex functions corresponding to the three components of the disturbance velocity in this motion, connected between themselves by the compatibility relations (9.43), and putting

$$\frac{dF_0}{dX} = \mathcal{V}_a, \tag{9.174}$$

we can deduce an expression analogous with (9.46) for the axial disturbance velocity around the considered cruciform wing–conical body system, namely

$$\mathcal{U} = \mathcal{U}_a + \frac{2c^2}{x}\mathcal{V}_a = \mathcal{U}_a - \frac{2c^2}{x}\int \frac{d\mathcal{U}_a}{X}. \tag{9.175}$$

In this way, the problem has been reduced to the determination of the function $\mathcal{U}_a$ in the conical motion ($n = 1$) around a cruciform wing defined by the boundary conditions in the above X plane.

For solving this problem we shall use, as in Chapter VIII, an auxiliary plane $\chi = \lambda + i\nu$ defined by the conformal transformation (9.11) in which, considering the compatibility relations (9.43), the following boundary conditions result:

$$\mathrm{Im}\,[\mathcal{U}_a]_{M_2M_1} = 0, \quad \mathrm{Re}\,[\mathcal{U}_a]_{\lambda_1<\chi=\lambda<-\lambda_2} = 0, \tag{9.176a}$$

$$\mathrm{Im}\,[\mathcal{U}_a]_{M_jA_j} = C_j^* - \int \sqrt{\lambda^2 - H^2}\, d\left[\frac{C_j}{2}\left(1+\sqrt{\frac{\lambda^2-H^2}{\lambda^2-\varkappa^2}}\right)\right],$$

$$(C_j^* = \text{const.}), \quad (j = 1, 2), \tag{9.176b}$$

$$\mathrm{Im}\,[\mathcal{U}_a]_{PR'\text{ and }PR''} = \mp K \mp \int \sqrt{H^2-\lambda^2}\, d\left[\frac{\beta U_\infty}{2}\left(1+\sqrt{\frac{H^2-\lambda^2}{\varkappa^2-\lambda^2}}\right)\right],$$

$$(K = \text{const.}), \tag{9.176c}$$

where all the geometrical magnitudes that occur have been defined in (§ IX. 1.2) and where the $(-)$ sign refers to the segment PR' $(\chi = \lambda > 0)$ and the $(+)$ sign to PR'' $(\chi = \lambda < 0)$.

In order to determine the function $\mathcal{U}_a$ thus defined, apart from the contributions of the leading edges of the wing and plate and of the ridges M_1, M_2, and R_1, R_2 of the wing and plate—whose expressions are analogous to those established in Chapter VIII—we have to take into consideration the continuous distribution of elementary ridges on the portions M_1A_1 and M_2A_2 corresponding to condition (9.176b), as well as those on the portions PR' and PR'' corresponding to condition (9.176c); by applying the expressions established previously for the ridges of the wing and plate, we obtain the following contributions for these elementary ridges:

$$d\mathcal{U}_a = -\frac{C_1}{\pi}\sqrt{\eta^2-H^2}\cosh^{-1}\sqrt{\frac{(\lambda_2+\eta)(\lambda_1-\chi)}{(\lambda_1+\lambda_2)(\eta-\chi)}}\,\frac{d}{d\eta}\left(\sqrt{\frac{\eta^2-H^2}{\eta^2-\varkappa^2}}\right)d\eta,$$

$$\left(C_1 = \frac{(\alpha_1-\alpha_0)U_\infty}{\sqrt{1-B^2s_1^2}}\right); \tag{9.177a}$$

$$d\mathcal{U}_a = -\frac{C_2}{\pi}\sqrt{\eta^2-H^2}\cosh^{-1}\sqrt{\frac{(\lambda_1+\eta)(\lambda_2+\chi)}{(\lambda_1+\lambda_2)(\eta+\chi)}}\frac{d}{d\eta}\left(\sqrt{\frac{\eta^2-H^2}{\eta^2-\varkappa^2}}\right)d\eta,$$

$$\left(C_2 = \frac{(\alpha_2-\alpha_0)U_\infty}{\sqrt{1-B^2s_2^2}}\right), \tag{9.177b}$$

in the case of an elementary ridge situated at the point $\chi = \eta$ on the portion M_1A_1 and at the point $\chi = -\eta$ on the portion M_2A_2 respectively, and

$$d\mathcal{U}_a = -\frac{\beta U_\infty}{2}\sqrt{H^2-\eta^2}\cosh^{-1}\sqrt{\frac{(\lambda_2+\eta)(\lambda_1-\chi)}{(\lambda_1+\lambda_2)(\eta-\chi)}}\frac{d}{d\eta}\left(\sqrt{\frac{H^2-\eta^2}{\varkappa^2-\eta^2}}\right)d\eta, \tag{9.178a}$$

$$d\mathcal{U}_a = \frac{\beta U_\infty}{2}\sqrt{H^2-\eta^2}\cosh^{-1}\sqrt{\frac{(\lambda_1+\eta)(\lambda_2+\chi)}{(\lambda_1+\lambda_2)(\eta+\chi)}}\frac{d}{d\eta}\left(\sqrt{\frac{H^2-\eta^2}{\varkappa^2-\eta^2}}\right)d\eta, \tag{9.178b}$$

in the case of an elementary ridge situated at the point $\chi = \eta$ on the portion PR' and at the point $\chi = -\eta$ on the portion PR''.

Hence, in conclusion, the problem of the conical flow ($n = 1$) about a cruciform wing–conical body system has been reduced to the study of a fictitious cruciform wing having the same singularities as the actual one, having however in addition a continuous variation of incidence on the portions M_1A_1, M_2A_2 and R_1P_1, R_2P_2 corresponding to the conditions (9.40) and (9.173) in the X plane or (9.176b and c) in the χ plane.

In a similar way, the problem of the nth order conical flow about the cruciform wing–conical body system may be reduced to the study of the conical flow of the same order about a fictitious cruciform wing having, on certain portions, the incidence or slope distributed according to a homogeneous function of order $(n-1)$.

CHAPTER X

MISCELLANEOUS

IN THIS chapter we shall try to tackle several special problems which will complete the study of wings in supersonic flow. Thus we shall deal with the theory of ailerons, the theory of harmonic oscillations of wing systems, the study of wings with leading edge flaps, and the effect of flow separation at the subsonic leading edges of thin wings.

X. 1. Theory of Ailerons

X. 1.1. Approach to the Problem and Definition of Conical Motions about the Ailerons

Owing to their shape, position and way of approach, the ailerons may be included within the study of thin polygonal wings, such that the motions about them can be deduced from the results obtained in our earlier chapters.

Consider one half of a polygonal wing and let $O_1A_1C_2O_2$ be an aileron that can be rotated about the axis O_1O_2 which is called the *hinge* (Fig. 10.1). Apart from the *hinge* axis which must be straight, and from the line O_2C_2 which is generally parallel to the free stream direction, the aileron can have various shapes and different positions with respect to the wing (Fig. 10.2). Some of these positions present certain difficulties as regards the computation of the aerodynamic characteristics.

Therefore, for the sake of simplicity, we shall first consider a more convenient form, which has also the advantage of being a general one (Fig. 10.1). Thus we can establish the geometric and aerodynamic conditions of this form of aileron and then extend the results to other similar systems, if such be the case.

Assume first that the wing proper has no incidence and no aerodynamic effect on the aileron surface. In this case, the aileron surface lies in the field of the conical motions corresponding to O_1 and O_2 which are the origins of the reference systems $O_1x_1x_2$, $O_2\xi_1\xi_2$.

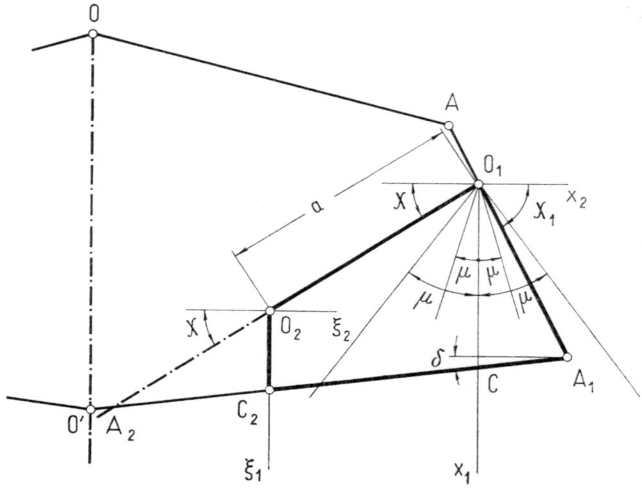

FIG. 10.1

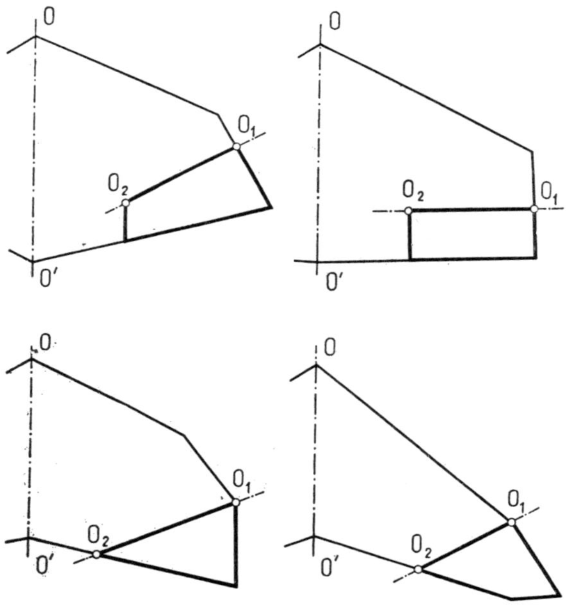

FIG. 10.2

In order to establish these motions, we shall denote by τ the dihedral angle made by the aileron with the wing plane, in its rotation about the hinge, the latter having in turn the angle χ with the axes O_1x_2 or $O_2\xi_2$. It can be easily seen that the *deflexion* of the aileron with respect to the free stream direction, which will be denoted by β, is given by the relation

$$\beta = \tau \cos \chi. \tag{10.1}$$

The downwash w is constant on the entire aileron,

$$w = -\beta U_\infty. \tag{10.2}$$

The two reference systems, $O_1x_1x_2$ and $O_2\xi_1\xi_2$, are connected through the relations

$$\xi_1 = x_1 - a \sin \chi, \quad \xi_2 = x_2 + a \cos \chi, \tag{10.3}$$

where a is the distance between O_1 and O_2; making use of the usual notation, we can write the following relations:

$$\frac{x_2}{x_1} = y, \quad \frac{\xi_2}{\xi_1} = \frac{x_2 + a \cos \chi}{x_1 - a \sin \chi} = \eta, \tag{10.4}$$

and define the complex variables x and ξ defined as follows:

$$x = y + i\mathfrak{z}, \quad \xi = \eta + i\zeta. \tag{10.5}$$

On the other hand, writing

$$l_1 = \cot \chi_1, \quad l = \cot \chi, \tag{10.6}$$

we shall establish the expressions for the functions $\mathcal{U}_0$ of the true conical motions ($n = 1$) corresponding to O_1 and O_2, from which we shall then deduce the axial disturbance velocity by adding them together.

In this sense, by applying formula (2.84) when the edge OA_1 is subsonic and formula (2.90) when OA_1 is supersonic, we can write for the motion corresponding to O_1:

$$\mathcal{U}_0^I = \frac{2\beta U_\infty}{\pi}\left[\frac{l_1}{1+Bl_1}\sqrt{\frac{B(l_1+l)}{1+Bl}}\sqrt{\frac{1+Bx}{B(l_1-x)}} + \frac{l}{\sqrt{B^2l^2-1}}\cos^{-1}\sqrt{\frac{(l_1+l)(1+Bx)}{(1+Bl_1)(l+x)}}\right], \quad (Bl_1 < 1), \tag{10.7a}$$

$$\mathcal{U}_0^I = \frac{2\beta U_\infty}{\pi}\left[\frac{l_1}{\sqrt{B^2l_1^2-1}}\cos^{-1}\sqrt{\frac{(1+Bl_1)(1-Bx)}{2B(l_1-x)}} + \frac{l}{\sqrt{B^2l^2-1}}\cos^{-1}\sqrt{\frac{(1+Bl)(1+Bx)}{2B(l+x)}}\right], \quad (Bl_1 > 1), \tag{10.7b}$$

where the constants in these expressions have been deduced from the cases studied previously.

It should be noted also that these expressions hold valid for both cases of the hinge position ($Bl < 1$, $Bl > 1$), by applying the known relation

$$\frac{\cos^{-1}\Theta}{\sqrt{B^2l^2-1}} = \frac{\cosh^{-1}\Theta}{\sqrt{1-B^2l^2}}, \tag{10.8}$$

in order that the coefficients of the second term in the right-hand side be real.

We shall further assume that the aileron stretches over the entire half of the wing and, consequently, the axis O_1O_2—which represents a ridge or a leading edge for the aileron—is prolonged beyond O_2; in that case, in order that the part $O'C_2O_2$ should return to its initial fixed position, it will be given a negative *deflexion* β. On the other hand, although O_2C_2 represents a central ridge, it has no effect on the conical motion of order $n = 1$, and consequently the function $\mathcal{U}_0^{II}$ will be given by the following simple expression which comes naturally from (7.35):

$$\mathcal{U}_0^{II} = -\frac{2\beta U_\infty l}{\pi\sqrt{B^2l^2-1}}\cos^{-1}\sqrt{\frac{(1+Bl)(1+B\xi)}{2B(l+\xi)}}\,; \tag{10.9}$$

this expression holds valid also in the case when the hinge axis is subsonic, making use in such a case of relation (10.8), in order to have real constants. The superscript II refers to the motion at the vertex O_2. At an arbitrary point, the total axial disturbance velocity will be given by the two motions corresponding to O_1 and O_2:

$$u_0 = \text{Re}\,(\mathcal{U}_0^I+\mathcal{U}_0^{II}), \tag{10.10}$$

to which we have to add the axial velocity u_w due to the motion around the wing, if the latter has an arbitrary incidence α.

X. 1.2. Calculation of Forces and Moments

Although the main role of the ailerons consists in establishing a rolling moment around the aircraft axis (which, in the present example, can be even considered as the axis of symmetry of the wing), it is, however, necessary to know the lift, drag and *hinge moment* of the aileron considered alone, whose surface area will be denoted by S_a, as well as its influence on the characteristics of the entire wing.

To this end, considering the inclination of the trailing edge to be zero, for simplifying our computations ($\delta = 0$, Fig. 10.3) and taking into account our earlier results (Chaps. III, VII), we shall calculate successively:

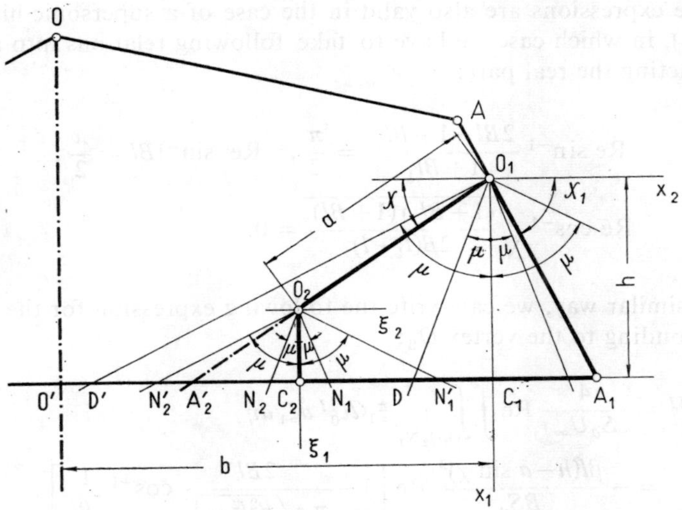

FIG. 10.3

Lift coefficient of the aileron. Designating by K_l^I the lift coefficient for the motion corresponding to the vertex O_1, we have

$$K_l^I = \frac{4}{S_a U_\infty} \operatorname{Re} \iint_{O_1 O_2 C_2 A_1} x_1 \mathcal{U}_0^I \, dx_1 \, dy. \tag{10.11}$$

Making the computations for the case of the subsonic hinge ($Bl < 1$), we obtain

$$K_l^I = \frac{\beta h^2}{B S_a} \sqrt{\frac{B(l_1+l)}{1+Bl}} \operatorname{Re}\left[(l_1+l)\left(1+\frac{2}{\pi}\sin^{-1}\frac{2Bl-1+Bl_1}{1+Bl_1}\right)\right.$$
$$\left.+\frac{4l_1}{\pi}\frac{\sqrt{B(l_1+l)(1-Bl)}}{1+Bl_1}\right] \tag{10.12}$$

when the edge O_1A_1 is subsonic, and

$$K_l^I = \frac{\beta h^2 (l_1+l)}{B S_a}\left[1+\frac{2}{\pi}\sin^{-1} Bl +\right.$$
$$\left.+\frac{4Bl_1}{\pi\sqrt{B^2 l_1^2-1}}\cos^{-1}\sqrt{\frac{(1+Bl_1)(1+Bl)}{2B(l_1+l)}}\right] \tag{10.13}$$

when the edge O_1A_1 is supersonic.

These expressions are also valid in the case of a supersonic hinge line ($Bl > 1$), in which case we have to take following relations into account in extracting the real part:

$$\operatorname{Re} \sin^{-1} \frac{2Bl-1+Bl_1}{1+Bl_1} = \frac{\pi}{2}, \quad \operatorname{Re} \sin^{-1} Bl = \frac{\pi}{2},$$

$$\operatorname{Re} \cos^{-1} \sqrt{\frac{(1+Bl_1)(1+Bl)}{2B(l_1+l)}} = 0. \tag{10.14}$$

In a similar way, we can write the following expression for the motion corresponding to the vertex O_2:

$$\begin{aligned} K_l^{II} &= \frac{4}{S_a U_\infty} \operatorname{Re} \iint_{O_2C_2N_1} \xi_1 \mathcal{U}_0^{II}\, d\xi_1\, d\eta \\ &= -\frac{\beta l(h-a \sin \chi)^2}{BS_a} \operatorname{Re}\left[1-\frac{2Bl}{\pi\sqrt{B^2l^2-1}} \cos^{-1}\frac{1}{Bl}\right], \end{aligned} \tag{10.15}$$

which is valid, taking account of relation (10.8), for both positions of the hinge ($Bl > 1$, $Bl < 1$).

On adding the expressions established above, we obtain the lift coefficient of the aileron due to the deflection β,

$$K_l = K_l^I + K_l^{II}. \tag{10.16}$$

If we denote by K_{lw} the lift coefficient on the aileron due to the initial flow about the wing, when the deflexion is zero ($\beta = 0$), we obtain the total lift coefficient of the aileron

$$K'_l = K_l + K_{lw}. \tag{10.17}$$

Total wave drag coefficient of the aileron. We shall denote by K'_d this coefficient which is deduced from the total lift coefficient,

$$K'_d = (\alpha+\beta)K'_l = (\alpha+\beta)(K_l+K_{lw}), \tag{10.18}$$

α being the incidence of the wing (assumed to be plane), and β the aileron deflexion.

Hinge moment. Denoting by μ_w the hinge moment coefficient due to the initial flow about the wing when the deflexion is zero ($\beta = 0$), we can write for the total hinge moment coefficient,

$$\mu = \mu^I + \mu^{II} + \mu_w, \tag{10.19}$$

where μ^{II} and μ^{I}, the coefficients corresponding to the motions at O_1 and O_2, are computed with the aid of the following relations:

$$\mu^{I} = \frac{4}{aS_a U_\infty} \operatorname{Re} \iint_{O_1O_2C_2A_1} x_1^2 \mathcal{U}_0^I (\cos\chi + y \sin\chi)\, dx_1\, dy, \qquad (10.20a)$$

$$\mu^{II} = \frac{4}{aS_a U_\infty} \operatorname{Re} \iint_{O_2C_2N_1} \xi_1^2 \mathcal{U}_0^{II} (\cos\chi + \eta \sin\chi)\, d\xi_1 d\eta, \qquad (10.20b)$$

where we have taken the hinge length a as reference length.

Influence of the aileron on the aerodynamic characteristics of the wing. This influence is shown in the increase of the lift and drag coefficients, ΔC_l and ΔC_d, and appearance of a rolling moment μ_1.

As regards the *increase of the lift coefficient* of one half of the wing of surface area S, it will be given by the relation

$$\Delta C_l = C_l^I + C_l^{II}, \qquad (10.21)$$

where, in the case of the supersonic hinge, C_l^I and C_l^{II} are computed with the aid of the relations

$$C_l^I = \frac{4}{SU_\infty} \operatorname{Re} \iint_{O_1A_1A_2'} x_1 \mathcal{U}_0^I\, dx_1\, dy, \qquad (10.22a)$$

$$C_l^{II} = \frac{4}{SU_\infty} \operatorname{Re} \iint_{O_2A_2'N_1} \xi_1 \mathcal{U}_0^{II}\, d\xi_1\, d\eta. \qquad (10.22b)$$

If the deflexion of the two ailerons is equal and of opposite sign, the total increase of the lift is zero. When the hinge O_1O_2 is subsonic, the surfaces of integration in the above relations will be substituted by O_1A_1D' and, respectively, $O_2N'_2N'_1$.

The increase of the drag coefficient of the wing will be given by the relation

$$\Delta C_d = (\alpha+\beta)\frac{S_a}{S} K_l + \alpha\left(\Delta C_l - \frac{S_a}{S} K_l\right). \qquad (10.23)$$

As for the *rolling moment* with respect to the axis OO' of the wing, due to the two ailerons on the right- and left-hand sides of the wing, its calculation involves no difficulty in principle, being, however, very laborious; therefore without making any significant error, we shall proceed in a simple way, considering the increase ΔC_l of one half of the wing and the distance d from the axis of the centroid of the whole surface influenced by the aileron deflexion in which the surface of the aileron and part of the wing surface are included.

In such a case, denoting by H_0 a reference length, we have

$$\mu_1 \approx \frac{d}{H_0} \Delta C_l. \tag{10.24}$$

X. 2. Low Frequency Harmonic Oscillations of Wings in Supersonic Flow

In the following, we shall consider only the oscillatory motion of wing systems about the position of zero incidence, to which a steady motion of the same kind as those studied in our previous chapters will generally be superposed.

X. 2.1. Preliminary Considerations

Under the assumption of small disturbances, the unsteady motion is defined by the following equation for the velocity potential which was established previously (1.35) and will be rewritten in the three-rectangular system $Ox_1x_2x_3$:

$$-B^2 \frac{\partial^2 \varphi}{\partial x_1^2} + \frac{\partial^2 \varphi}{\partial x_2^2} + \frac{\partial^2 \varphi}{\partial x_3^2} = \frac{1}{a_\infty^2} \left(\frac{\partial^2 \varphi}{\partial t^2} + 2U_\infty \frac{\partial^2 \varphi}{\partial x_1 \, \partial t} \right), \quad (B^2 = M_\infty^2 - 1), \tag{10.25}$$

where a_∞ is the speed of sound in the undisturbed stream, U_∞ is the velocity of the free stream, t represents the time and B has the usual significance. Let us assume that the motion is harmonic and its potential can be written in the following form [M–36]:

$$\varphi(x_1, x_2, x_3, t) = U_\infty e^{i(\omega t + kx_1)} \Phi(x_1, x_2, x_3), \tag{10.26}$$

where $\varphi(x_1, x_2, x_3)$, which is independent of time, is called the *reduced potential*, ω is the frequency of the periodic motion, and k is a coefficient which will be defined below. Introducing into (10.23) we obtain

$$-B^2 \frac{\partial^2 \Phi}{\partial x_1^2} + \frac{\partial^2 \Phi}{\partial x_2^2} + \frac{\partial^2 \Phi}{\partial x_3^2} = \lambda^2 \frac{1+B^2}{B^2} \Phi \approx 0, \tag{10.27}$$

the parameter λ and the coefficient k in (10.26) being given by the relations

$$\lambda = \frac{\omega}{U_\infty}, \quad k = -\lambda \frac{1+B^2}{B^2}. \tag{10.28}$$

Considering only harmonic oscillatory motions of low frequency, we shall assume that λ is very small (as well as k), so that the right-hand side of the above equation (10.27) can be neglected and considered equal to zero. Let us further assume that in the harmonic oscillatory motion, the equation of a point on the wing has the following variation in time,

$$Z = e^{i\omega t}P(x_1, x_2), \quad P(x_1, x_2) = \sum_{n=0}^{N} P_n(x_1, x_2), \tag{10.29}$$

where $P(x_1, x_2)$ represents in the general case a sum of homogeneous polynomials $P_n(x_1, x_2)$ of various orders in x_1, x_2. The downwash w of this point will be

$$w = \frac{\partial \varphi}{\partial x_3} \approx \frac{\partial Z}{\partial t} + U_\infty \frac{\partial Z}{\partial x_1} = U_\infty e^{i\omega t}\left(\frac{\partial P}{\partial x_1} + \frac{i\omega}{U_\infty} P\right). \tag{10.30}$$

By analogy with the notion of *reduced potential*, we shall likewise introduce the notion of *reduced downwash*, which will be denoted by β and defined on the wing as follows:

$$\beta = \frac{\partial \Phi}{\partial x_3} = \frac{1}{U_\infty} e^{-i(\omega t + kx_1)} \frac{\partial \Phi}{\partial x_3} = e^{-ikx_1}\left(\frac{\partial P}{\partial x_1} + i\lambda P\right)$$

$$\approx \frac{\partial P}{\partial x_1} + i\lambda\left(P + x_1 \frac{1+B^2}{B^2} \frac{\partial P}{\partial x_1}\right), \tag{10.31}$$

where, under our assumption that k is very small, we have written approximately

$$e^{\pm ikx_1} \approx 1 \pm ikx_1, \tag{10.32}$$

thus neglecting terms in k^2 and higher.

It can be seen from the expression for β that the reduced total potential Φ consists of two parts:

$$\Phi = \Phi' + i\lambda\Phi^* = \sum_{n=1}^{N} \Phi'_n + i\lambda \sum_{n=0}^{N} \Phi^*_{n+1} \tag{10.33}$$

where N is the highest order of the polynomial P.

The potential Φ'_n corresponds to the order $(n-1)$ from $\partial P/\partial x_1$, while Φ^*_{n+1} corresponds to the order n from the imaginary expression. Each of these potentials satisfies the general equation with partial derivatives established previously under the assumption of small disturbances (1.36),

$$-B^2 \frac{\partial^2 \Phi_m}{\partial x_1^2} + \frac{\partial^2 \Phi_m}{\partial x_2^2} + \frac{\partial^2 \Phi_m}{\partial x_3^2} = 0, \quad (m = n,\ n+1). \tag{10.34}$$

As a matter of fact, this equation can be deduced directly from (10.25) if the motion is steady, that is, independent of time.

CALCULATION OF PRESSURES. Proceeding from relation (1.51) we can write successively

$$C_p = -\frac{2}{U_\infty}\left(\frac{\partial \varphi}{\partial x_1}+\frac{1}{U_\infty}\frac{\partial \varphi}{\partial t}\right) = -2e^{i(\omega t+kx_1)}\left[\frac{\partial \Phi}{\partial x_1}+i(\lambda+k)\Phi\right]$$

$$= -2e^{i(\omega t+kx_1)}\left[\frac{\partial \Phi}{\partial x_1}-\frac{i\lambda}{B^2}\Phi\right]; \tag{10.35}$$

and further, if we substitute e^{ikx_1} by its expression in (10.32) and meanwhile neglect the second order terms, we obtain

$$C_p \approx -2e^{i\omega t}(1+ikx_1)\left(\frac{\partial \Phi}{\partial x_1}-\frac{i\lambda}{B^2}\Phi\right)$$

$$\approx -2e^{i\omega t}\left[\frac{\partial \Phi}{\partial x_1}-\frac{i\lambda}{B^2}\Phi+ikx_1\frac{\partial \Phi}{\partial x_1}\right]$$

$$\approx -2e^{i\omega t}\left[\frac{\partial \Phi'}{\partial x_1}-\frac{i\lambda}{B^2}\left(\Phi'-B^2\frac{\partial \Phi^*}{\partial x_1}+M_\infty^2\, x_1\frac{\partial \Phi'}{\partial x_1}\right)\right]. \tag{10.36}$$

Under the above form or under its more general form (10.35), the pressure coefficient C_p contains Φ' (given by 10.33) and the reduced axial velocity

$$q = \frac{\partial \Phi}{\partial x_1} = \frac{\partial \Phi'}{\partial x_1}+i\lambda\frac{\partial \Phi^*}{\partial x_1} = \sum_{n=1}^{N}\frac{\partial \Phi'_n}{\partial x_1}+i\lambda\sum_{n=0}^{N}\frac{\partial \Phi^*_{n+1}}{\partial x_1}. \tag{10.37}$$

As regards the reduced axial velocities, their expressions are similar to those established in our previous chapters, where however, in order to avoid a misunderstanding, we shall make use of the notations q and Q instead of u and $\mathcal{U}$:

$$\frac{\partial \Phi'_n}{\partial x_1} = x_1^{n-1}\, q_{n-1} = x_1^{n-1}\,\text{Re}\, Q_{n-1}, \tag{10.38a}$$

$$\frac{\partial \Phi^*_{n+1}}{\partial x_1} = x_1^n\, q_n^* = x_1^n\,\text{Re}\, Q_n^*. \tag{10.38b}$$

Therefore, the problem reduces to finding the functions Q_{n-1} and Q_n^*, a problem that has already been solved in our previous chapters, and to the determination of the potential Φ, which is in fact a very simple operation, as will be seen later.

X. 2.2. *Case of Translation and Rotation*

If the harmonic motion of the points on the wing proceeds from a general oscillating motion of the wing, then, under the assumption of small disturbances, this motion can be decomposed into three parts: a vertical translation ζ, a rolling rotation ψ about the axis Ox_1 and a pitching rotation θ about the axis Ox_2:

$$\zeta = \zeta_0 e^{i\omega t}, \quad \psi = \psi_0 e^{i\omega t}, \quad \theta = \theta_0 e^{i\omega t}. \tag{10.39}$$

One can eventually add a rotation about the axis Ox_3, but the effect of this rotation is less important and it is the rolling and yawing moments which are of particular interest.

In this case, expression (10.29) can be written as

$$Z = \zeta - x_1\theta + x_2\psi = e^{i\omega t}(\zeta_0 - x_1\theta_0 + x_2\psi_0), \tag{10.40}$$

and hence

$$P(x_1, x_2) = \zeta_0 - x_1\theta_0 + x_2\psi_0; \tag{10.41}$$

further, applying (10.31), one obtains

$$\beta = -\theta_0 + i\lambda\left[\zeta_0 - x_1\frac{2B^2+1}{B^2}\theta_0 + x_2\psi_0\right]. \tag{10.42}$$

The expression above for the reduced downwash corresponds to certain *equivalent fictitious conical motions* of order 1 or 2, defined respectively by the following values:

$$\beta_0 = \frac{\partial\Phi_1'}{\partial x_3} = -\theta_0, \quad \beta_0^* = \frac{\partial\Phi_1^*}{\partial x_3} = \zeta_0, \quad (n = 1), \tag{10.43a}$$

$$\beta_1^* = \frac{\partial\Phi^*}{\partial x_3} = x_1\beta_{10}^* + x_2\beta_{01}^* = -x_1\frac{2B^2+1}{B^2}\theta_0 + x_2\psi_0, \quad (n = 2) \tag{10.43b}$$

where the asterisk (*) refers to the motions represented by the reduced potential Φ^*, affected by the factor $i\lambda$ (10.33).

In an analogous way we can also obtain the corresponding *reduced axial velocities*, $q_{n-1} = \operatorname{Re} Q_{n-1}$, $q_{n-1}^* = \operatorname{Re} Q_{n-1}^*$, which can be now defined in terms of the reduced downwashes indicated above (10.43).

Further, referring to the formula for the pressure coefficient in its general form (10.36), we shall have to calculate also the reduced total potential

$$\Phi = \Phi_1' + i\lambda(\Phi_1^* + \Phi_2^*). \tag{10.44}$$

Thus we have for points on the wing ($z = 0$), denoting by v_0 the lateral reduced velocity,

$$\Phi_1' = x_1 q_0 + x_2 v_0 = x_1(q_0 + y v_0) = \operatorname{Re} x_1 x \int Q_0 \, d\left(\frac{1}{x}\right). \quad (10.45)$$

We have analogously

$$\Phi_1^* = \operatorname{Re} x_1 x \int Q_0^* \, d\left(\frac{1}{x}\right). \quad (10.46)$$

We also find in a way that will be developed later under its general form,

$$2\Phi_2^* = \operatorname{Re} (x_1 x)^2 \int Q_1^* \, d\left(\frac{1}{x^2}\right). \quad (10.47)$$

Since we have assumed that λ is very small ($\lambda \ll 1$) and have neglected in principle the terms in λ^2 and higher, the pressure coefficient C_p will be given by the following expression, deduced from (10.36):

$$C_p = -2e^{i\omega t}\left[\frac{\partial \Phi_1'}{\partial x_1} - \frac{i\lambda}{B^2}\left(\Phi_1' - B^2 \frac{\partial(\Phi_1^* + \Phi_2^*)}{\partial x_1} + M_\infty^2 x_1 \frac{\partial \Phi_1'}{\partial x_1}\right)\right]$$

$$= -2e^{i\omega t}\left[q_0 - \frac{i\lambda}{B^2}\left(\Phi_1' - B^2(q_0^* + x_1 q_1^*) + M_\infty^2 x_1 q_0\right)\right], \quad (10.48)$$

where, in the last term, we have taken account of (10.38).

Case of a Wing of Infinite Aspect Ratio. We shall restrict ourselves to a translation ζ and a rotation θ; the rotation ψ cannot be taken into consideration. The reduced velocity has the following expression,

$$\beta = -\theta_0 + i\lambda\left(\zeta_0 - \frac{2B^2+1}{B^2}\theta_0 x_1\right), \quad (10.49)$$

which represents the boundary condition of a fictitious profile having a parabolic contour. The reduced potentials Φ_1', Φ_1^* and Φ_2^* can be obtained readily:

$$\Phi_1' = \frac{\theta_0}{B}(x_1 \mp B x_3), \quad \Phi_1^* = -\frac{\zeta_0}{B}(x_1 \mp B x_3), \quad (10.50)$$

on the upper surface (the upper signs) and on the lower surface (the lower signs) respectively, and

$$\Phi_2^* = \frac{2B^2+1}{B^2}\,\frac{\theta_0}{2B}(x_1 \mp B x_3)^2, \quad (10.51)$$

also on the two surfaces. For points on the wing which are of interest, we put $x_3 = 0$. As regards the *reduced axial velocities*, we can obtain them by differentiating the above expressions, or in a simpler way as shown earlier (1.68a):

$$q_0 = \frac{\theta_0}{B}, \quad q_0^* = -\frac{\zeta_0}{B}, \quad x_1 q_1^* = \frac{2B^2+1}{B^2}\frac{\theta_0}{B}x_1. \tag{10.52}$$

Finally, applying formula (10.48), we obtain the pressure coefficient on the wing ($x_3 = 0$):

$$C_p = -\frac{2}{B}e^{i\omega t}\left[\theta_0 + i\lambda\left(\frac{B^2-1}{B^2}\theta_0 x_1 - \zeta_0\right)\right]. \tag{10.53}$$

By means of integration we obtain the lift coefficient and the pitching moment coefficient, with respect to the leading edge, viz.

$$C_l = -\frac{1}{c}\int_0^c 2C_p\,dx_1 = \frac{4}{B}e^{i\omega t}\left[\theta_0 + i\lambda\left(\frac{B^2-1}{2B^2}c\theta_0 - \zeta_0\right)\right], \tag{10.54a}$$

$$\mu_2 = -\frac{1}{c^2}\int_0^c 2C_p x_1\,dx_1 = \frac{2}{B}e^{i\omega t}\left[\theta_0 + i\lambda\left(\frac{2}{3}\frac{B^2-1}{B^2}\theta_0 c - \zeta_0\right)\right], \tag{10.54b}$$

where we have denoted by c the chord of the wing.

X. 2.3. *The Oscillatory Motion of a Delta Wing*

From the expression for the reduced velocity (10.43) it follows that the *equivalent conical motions* are of orders 1 and 2 (10.44). Depending on the wing position with respect to the Mach cone we have to examine two cases, according as the delta wing has subsonic or supersonic edges.

Delta Wing with Subsonic Edges. The formulae below, which are obtained by applying the expressions (4.14), (4.31), can be written for the functions Q_0, Q_0^* and Q_1^* corresponding to the reduced axial velocities (10.38)

$$Q_0 = \frac{\theta_0}{E(k)}\frac{l^2}{\sqrt{l^2-x^2}}, \quad Q_0^* = -\frac{\zeta_0}{E(k)}\frac{l^2}{\sqrt{l^2-x^2}}, \quad (k = \sqrt{1-B^2l^2}) \tag{10.55}$$

for the "conical motions" of order one, where $E(k)$ is the complete elliptic integral of the second kind, and

$$Q_1^* = \frac{A_{20}^* + A_{21}^* x + A_{22}^* x^2}{\sqrt{l^2 - x^2}}, \tag{10.56a}$$

$$A_{20}^* = 2l^2 \frac{2B^2+1}{B^2} \theta_0 \frac{1-B^2l^2}{(1-2B^2l^2)E(k)+B^2l^2K(k)}, \tag{10.56b}$$

$$A_{21}^* = -l^2\psi_0 \frac{1-B^2l^2}{(2-B^2l^2)E(k)-B^2l^2K(k)}, \tag{10.56c}$$

$$A_{22}^* = -\frac{2B^2+1}{B^2} \theta_0 \frac{1-B^2l^2}{(1-2B^2l^2)E(k)+B^2l^2K(k)} \tag{10.56d}$$

for the "conical motion" of order two, where $K(k)$ and $E(k)$ are complete elliptic integrals of the first and second kind, having the same modulus k (10.55).

For the potential Φ_1' we find

$$\frac{1}{x_1}\Phi_1' = \text{Re}\, \frac{\theta_0}{E(k)} \sqrt{l^2-x^2}. \tag{10.57}$$

We can easily obtain the pressure coefficient under its approximate form deduced from (10.48):

$$C_p = -2e^{i\omega t}\left[\frac{\theta_0 - i\lambda\zeta_0}{E(k)} \frac{l^2}{\sqrt{l^2-y^2}} + i\lambda x_1 \frac{D_0 + D_1 y + D_2 y^2}{\sqrt{l^2-y^2}}\right], \tag{10.58}$$

where we have written

$$D_0 = A_{20}^* - \frac{\theta_0 l^2}{E(k)}\left(1+\frac{2}{B^2}\right), \quad D_1 = A_{21}^*, \quad D_2 = A_{22}^* + \frac{\theta_0}{B^2E(k)}. \tag{10.59}$$

As far as the aerodynamic characteristics are concerned, as seen in the above, they can be deduced from those of the steady motion (Chap. IV), to which one should also add the terms which contain the *reduced potential.*

Thus, if we calculate the lift coefficient and the rolling and pitching moment coefficients, we obtain

$$C_l = 2\pi l e^{i\omega t}\left[\frac{\theta_0}{E(k)} + i\lambda\left(\frac{2}{3l^2} D_0 - \frac{\zeta_0}{E(k)} + \tfrac{1}{3} D_2\right)\right], \tag{10.60a}$$

$$H\mu_1 = i\lambda e^{i\omega t} \frac{\pi l}{2} D_1, \tag{10.60b}$$

$$H\mu_2 = -\pi l e^{i\omega t}\left[\tfrac{4}{3} \frac{\theta_0}{E(k)} + i\lambda\left(\frac{1}{l^2} D_0 - \tfrac{4}{3} \frac{\zeta_0}{E(k)} + \tfrac{1}{2} D_2\right)\right], \tag{10.60c}$$

where H is a reference length which can be taken as the wing root chord.

DELTA WING WITH SUPERSONIC EDGES. In an analogous way, taking into account the expression for the reduced downwash, we can write similarly to (4.42, 4.59)

$$Q_0 = \frac{2}{\pi} \frac{\theta_0 l}{\sqrt{B^2 l^2 - 1}} \left[\cos^{-1} \sqrt{\frac{(1+Bl)(1-Bx)}{2B(l-x)}} + \cos^{-1} \sqrt{\frac{(1+Bl)(1+Bx)}{2B(l+x)}} \right], \tag{10.61a}$$

$$Q_0^* = -\frac{2}{\pi} \frac{\zeta_0 l}{\sqrt{B^2 l^2 - 1}} \left[\cos^{-1} \sqrt{\frac{(1+Bl)(1-Bx)}{2B(l-x)}} + \cos^{-1} \sqrt{\frac{(1+Bl)(1+Bx)}{2B(l+x)}} \right] \tag{10.61b}$$

for the "conical motions" of order 1 , and

$$Q_1^* = \frac{2}{\pi(B^2l^2-1)^{3/2}} \left\{ \left[\theta_0 \frac{2B^2+1}{B^2} (l(B^2l^2-2)+x) + \psi_0 l^2 (1-B^2 lx) \right] \cos^{-1} \sqrt{\frac{(1+Bl)(1-Bx)}{2B(l-x)}} \right.$$
$$\left. + \left[\theta_0 \frac{2B^2+1}{B^2} (l(B^2l^2-2)-x) - \psi_0 l^2 (1+B^2 lx) \right] \cos^{-1} \sqrt{\frac{(1+Bl)(1+Bx)}{2B(l+x))}} \right\}$$
$$+ \frac{2Bl}{\pi(B^2l^2-1)} \theta_0 \frac{2B^2+1}{B^2} \sqrt{\frac{1-B^2x^2}{B^2}} \tag{10.62}$$

for the "conical motion" of order two.

As regards the reduced potential, we shall restrict ourselves, for the sake of simplicity, only to Φ_1,

$$\frac{1}{x_1} \Phi_1' = \frac{2\theta_0}{\pi \sqrt{B^2 l^2 - 1}} \left[(l-y) \cos^{-1} \sqrt{\frac{(1+Bl)(1-By)}{2B(l-y)}} + (l+y) \cos^{-1} \sqrt{\frac{(1+Bl)(1+By)}{2B(l+y)}} \right], \tag{10.63}$$

which is contained in the approximate formula for the pressure coefficient (10.48) where the terms in λ^2 and higher have been neglected.

In this case we have

$$C_p = -\frac{4l}{\pi\sqrt{B^2l^2-1}}\,e^{i\omega t}\left\{(\theta_0 - i\lambda\zeta_0)\left[\cos^{-1}\sqrt{\frac{(1+Bl)(1-By)}{2B(l-y)}}\right.\right.$$
$$\left.+\cos^{-1}\sqrt{\frac{(1+Bl)(1+By)}{2B(l+y)}}\right] - i\lambda x_1\left[\left(\theta_0\frac{3+l^2-B^2l^2-(2+l^2)y/l}{B^2l^2-1}\right.\right.$$
$$\left.-\psi_0 l\,\frac{1-B^2ly}{B^2l^2-1}\right)\cos^{-1}\sqrt{\frac{(1+Bl)(1-By)}{2B(l-y)}} + \left(\theta_0\frac{3+l^2-B^2l^2+(2+l^2)y/l}{B^2l^2-1}\right.$$
$$\left.\left.\left.+\psi_0 l\,\frac{1+B^2ly}{B^2l^2-1}\right)\cos^{-1}\sqrt{\frac{(1+Bl)(1+By)}{2B(l+y)}} - \theta_0\frac{2B^2+1}{B^2}\frac{\sqrt{1-B^2y^2}}{\sqrt{B^2l^2-1}}\right]\right\}. \tag{10.64}$$

For the lift coefficient we apply the similar formulae from the steady conical motions and obtain respectively

$$C_l = \frac{4}{B}\,e^{i\omega t}\left[\theta_0 + i\lambda\left(\theta_0\frac{2B^2-1}{3B^2} - \zeta_0\right)\right]. \tag{10.65}$$

For the moment coefficients we have analogously

$$H\mu_1 = -i\lambda\frac{2l^2}{3B}\,\psi_0 e^{i\omega t}, \tag{10.66a}$$

$$H\mu_2 = -\frac{1}{B}\,e^{i\omega t}\left[\tfrac{8}{3}\,\theta_0 + i\lambda\left(\theta_0\frac{2B^2-1}{B^2} - \tfrac{8}{3}\,\zeta_0\right)\right]. \tag{10.66b}$$

It should be recalled that H is a reference length which can be taken as the root chord of the delta wing.

X. 2.4. The Oscillatory Motion of a Cruciform Wing

In view of the fact that the cruciform delta wing with equal arms (Fig. 10.4) is of particular practical interest, we shall treat the oscillatory motion of such a wing, which can be decomposed into two translations along Ox_2 and Ox_3 and three rotations about the three axes of coordinates. It should, however, be noticed that the translations as well as the rotations about the axes Ox_2 and Ox_3 reduce to the preceding case of the isolated delta wing, since in the case of a translation along Ox_3 and rotation about the axis Ox_2, the plate does not influence the motion, while in the case of a translation along Ox_2 and rotation about the axis Ox_3, the wing plays no part. Hence, we have to study the *rotation of the cruciform wing* only *about the axis* Ox_1.

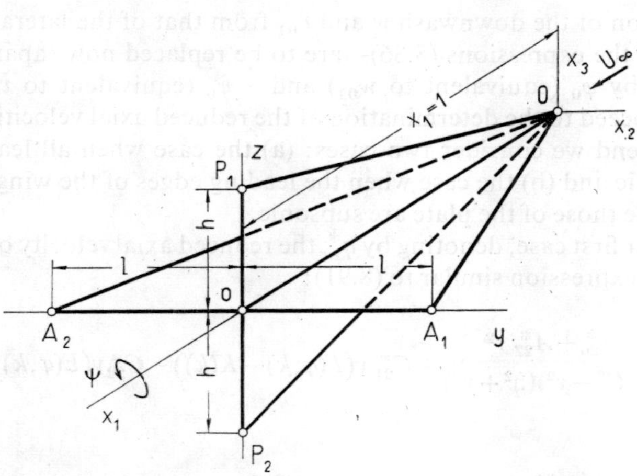

FIG. 10.4

In such a case we write, as above,

$$\psi = \psi_0 e^{i\omega} \tag{10.67}$$

whence, applying (10.42), we obtain the reduced downwash

$$\beta = \frac{\partial \Phi}{\partial x_3} = i\lambda\psi_0 x_2 \tag{10.68}$$

for points on the wing, and in a similar way, the lateral reduced velocity

$$v = \frac{\partial \Phi}{\partial x_2} = -i\lambda\psi_0 x_3 \tag{10.69}$$

for points on the plate.

The reduced potential is limited now to the expression

$$\Phi = i\lambda\Phi_2^*. \tag{10.70}$$

Taking into account the notation introduced above (10.38) and the expression (10.48) for the pressure coefficient, we can write in our case

$$C_p = -2e^{i\omega t}\, i\lambda \frac{\partial \Phi_2^*}{\partial x_1} = 2e^{i\omega t}\, i\lambda x_1 q_1^*, \quad (q_1^* = \operatorname{Re} Q_1^*). \tag{10.71}$$

Thus the problem reduces to an *equivalent fictitious conical motion* of order $n = 2$ about the cruciform wing, in which the coefficients—w_{01} from the variation of the downwash w and v_{01} from that of the lateral velocity v defined by the expressions (8.86)—are to be replaced now (apart from the factor $i\lambda$) by ψ_0 (equivalent to w_{01}) and $-\psi_0$ (equivalent to v_{01}). Let us further proceed to the determination of the reduced axial velocities.

To this end we consider two cases: (a) the case when all leading edges are subsonic and (b) the case when the leading edges of the wing are supersonic, while those of the plate are subsonic.

(a) In our first case, denoting by q^*_{1w} the reduced axial velocity on the wing, we find an expression similar to (8.91):

$$q^*_{1w} = \frac{A^*_{20}+A^*_{22}\,y^2}{\sqrt{(l^2-y^2)(\mathfrak{h}^2+y^2)}}+C^*_{21}\,y\big(F(\varphi,k)-K(k)\big)+C^*_{22}y\big(E(\varphi,k)-E(k)\big), \tag{10.72}$$

in which the notation has the same significance as in formula (8.91), except the constants which have been marked with an asterisk (*) as superscript but, nevertheless, will be determined in the same way as in § IV. 4.1, substituting, however, w_{01} by ψ_0 and v_{01} by $-\psi_0$.

For points on the plate we obtain analogous formulae through the interchange of the wing by the plate.

The results obtained in the steady motion of order $n = 2$ in conjunction with the rolling moment coefficient are also applicable in the case of the harmonic oscillatory motion, with the approximation of certain characteristic multiplying factors. Denoting by μ_{1w} the appropriate coefficient, we have for the wing

$$Hl\mu_{1w} = i\lambda e^{i\omega t}\Bigg\{A^*_{22}l\mathfrak{h}+\left(A^*_{22}\,\frac{l^2-\mathfrak{h}^2}{2}+A^*_{20}\right)\cos^{-1}\frac{\mathfrak{h}^2-l^2}{\mathfrak{h}^2+l^2}$$
$$-\frac{1}{3\sqrt{l^2+\mathfrak{h}^2}}\Bigg[C^*_{21}l\mathfrak{h}(l^2+\mathfrak{h}^2)-2C^*_{22}\mathfrak{h}^3l+\left(C^*_{21}\frac{l^4-\mathfrak{h}^4}{2}+2C^*_{22}(l^2+\mathfrak{h}^2)\mathfrak{h}^2\right)$$
$$\times\cos^{-1}\frac{\mathfrak{h}^2-l^2}{\mathfrak{h}^2+l^2}\Bigg]\Bigg\}, \tag{10.73}$$

while for the plate, the coefficient μ_{1p} is to be deduced from the expression above through the interchange of l by h and $\mathfrak{h}$ by $\lambda^* = \dfrac{l}{\sqrt{1-B^2l^2}}$ both in (10.73) and in the expression for the constants that are contained in (10.72).

(b) In our second case, the reduced velocities on the wing (q_{1w}) and plate (q^*_{1p}) will have expressions analogous to (8.128) and (8.131):

$$q_{1w}^* = K_{20}^* \cos^{-1} \sqrt{\frac{(l^2+\mathfrak{h}^2)(1-B^2y^2)}{(1+B^2\mathfrak{h}^2)(l^2-y^2)}} + A_{20}^* \sqrt{\frac{1-B^2y^2}{\mathfrak{h}^2+y^2}}$$
$$+C_{21}^* y[(F(\varphi,k)-K(k)]+C_{22}^* y[E(\varphi,k)-E(k)]-G^* y[\Pi(\varphi,\varrho,k)-\Pi(\varrho,k)]$$
$$-\frac{\pi B^2 l}{2} K_{20}^* y, \tag{10.74a}$$

$$q_{1p} = K_{20}^* \cos^{-1} \sqrt{\frac{\lambda^{*2}-h^2}{\lambda^{*2}-z^2}} + [A_{20}^* - (B^2 A_{20}^* + C_{22}^* \sqrt{1-B^2h^2})\, z^2] \sqrt{\frac{1-B^2h^2}{h^2-z^2}},$$
$$\lambda^* = \frac{l}{\sqrt{1-B^2l^2}}. \tag{10.74b}$$

The constants in these expressions are given by the same relations (8.130), in which, however, w_{01} is to be replaced by ψ_0 and v_{01} by $-\psi_0$.

As regards the rolling moment coefficient, we can write analogously to (8.135a) and (8.135b):

$$Hl\mu_{1w} = \frac{i\lambda}{3B^3} e^{i\omega t} \left\{ K_{20}^* B^3 l^2 (3-2B^2l^2) \left(\frac{\pi}{2} + \cos^{-1} \frac{\mathfrak{h}}{l} \sqrt{\frac{B^2l^2-1}{B^2\mathfrak{h}^2+1}} \right) \right.$$
$$- \left[C_{21}^* + \frac{G^*}{2\mathfrak{h}^2} ((3-2B^2l^2)(l^2+\mathfrak{h}^2)-2l^2) \right] \sqrt{1+B^2\mathfrak{h}^2} \cos^{-1} \frac{B^2\mathfrak{h}^2-1}{B^2\mathfrak{h}^2+1}$$
$$- \left[3A_{20}^* B^2 + \left(C_{21}^* - 2C_{22}^* \frac{B^2\mathfrak{h}^2}{B^2\mathfrak{h}^2+1} - G^* \frac{l^2}{\mathfrak{h}^2} \right) \sqrt{B^2\mathfrak{h}^2+1} \right]$$
$$\times \left(B\mathfrak{h} - \frac{B^2\mathfrak{h}^2+1}{2} \cos^{-1} \frac{B^2\mathfrak{h}^2-1}{B^2\mathfrak{h}^2+1} \right) - 2(B^3l^3-1)$$
$$\left. \times [C_{21}^* K(k) + C_{22}^* E(k) + G^* \Pi(\varrho, k)] \right\}, \tag{10.75a}$$

$$Hl\mu_{1p} = i\lambda e^{i\omega t} \left[K_{20}^* \lambda^{*2} \cos^{-1} \sqrt{\frac{\lambda^{*2}-h^2}{\lambda^{*2}}} + 2A_{20}^* h (1-\tfrac{2}{3} B^2h^2) \sqrt{1-B^2h^2} \right.$$
$$\left. - \tfrac{4}{3} C_{22}^* h^3 \sqrt{1-B^2h^2} - K_{20}^* h \sqrt{\lambda^{*2}-h^2} \right]. \tag{10.75b}$$

X. 2.5. Example of an Oscillatory Motion of a Wing–Body System

Examining Fig. 10.5 we can consider, as in the case of an isolated simple wing, a translation ζ, a rotation ψ about the axis Ox_1 and a rotation θ about the axis Ox_2 (10.39).

For the sake of simplicity, we shall consider only the translation ζ and the rotation θ which reduce the problem to solutions similar to those of

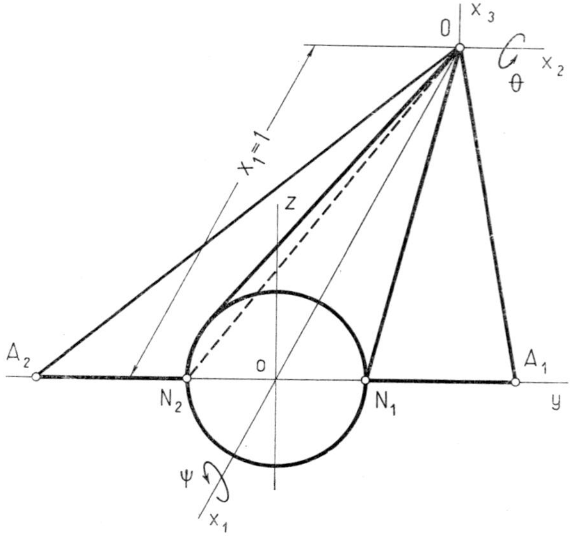

FIG. 10.5

steady conical motions of order 1 and 2, presenting therefore no difficulty in principle as regards their approach.

In order to further simplify the problem, we shall consider only a wing with subsonic edges having the same incidence as the body axis. In such a case, the solution can be deduced readily from the results obtained previously for a wing–body system in steady conical motions. Indeed, taking into account the results obtained in (§§ IX. 2, IX. 4) and proceeding in a way analogous to that pursued in the case of the oscillations of the isolated wing, the problem of the oscillatory motion of a wing–body system reduces to the study of the following steady reduced motions:

(a) Two "fictitious conical motions", in which the wing and the body have the same incidence, defined by the following expression for the axial reduced velocity

$$q = \text{Re}\, Q = \text{Re}\, A\, \frac{1 + c^4/l^4 - 2c^4/l^2x^2}{\sqrt{(l^2 - x^2)(1 - c^4/l^2x^2)}}, \tag{10.76}$$

which is similar to formula (9.53). The constant A is given by an expression analogous with (9.58),

$$A = \frac{-\beta l^2}{E(k)\left[1 - B^2c^2\,\frac{c^2}{l^2}\left(1 - \frac{1}{2}\,\frac{K(k)}{E(k)} - \frac{1}{24}\,\frac{c^4}{l^4}\right)\right]}, \tag{10.77}$$

where $K(k)$ and $E(k)$ are elliptic integrals of the first and second kind, having the modulus

$$k = \sqrt{1-B^2l^2}. \tag{10.78}$$

We have respectively, for the two fictitious motions defined by (10.76),

$$\beta = \beta_0 = \frac{\partial \Phi_1'}{\partial x_3} = -\theta_0, \quad q = q_0 = \frac{\partial \Phi_1'}{\partial x_1}, \quad A = A_0, \tag{10.79a}$$

$$\beta = \beta_0^* = \frac{\partial \Phi_1^*}{\partial x_3} = \zeta_0, \quad q = q_0^* = \frac{\partial \Phi_1^*}{\partial x_1}, \quad A = A_0^*. \tag{10.79b}$$

(b) A "fictitious conical motion" of order 2, in which the wing and the body axis have the same incidence corresponding to the reduced downwash

$$\beta_1^* = \frac{\partial \Phi_2^*}{\partial x_3} = -x_1 \frac{2B^2+1}{B^2}\theta_0, \tag{10.80}$$

which will be, however, first applied to a fictitious isolated wing in a plane $X = Y+iZ$.

Using the notation (10.38b), we shall put

$$\frac{\partial \Phi_2^*}{\partial x_1} = x_1 q_1^* = x_1 \operatorname{Re} Q_1^*, \tag{10.81}$$

where the function Q_1^* replaces $\mathcal{U}_1$ from the steady motions.

Further applying (2.57) for symmetrical incidences and putting

$$l_1 = l_2 = L \tag{10.82}$$

where $2L$ represents the span of the fictitious delta wing, as considered below, we may write

$$Q_1^* = \frac{A_{20}^* + A_{22}^* X^2}{\sqrt{L^2-X^2}}. \tag{10.83}$$

On the other hand, we have demonstrated (9.141) that this function is also valid in the plane x which results from the well-known transformation

$$X = x+\frac{c^2}{x}, \quad L = l+\frac{c^2}{l}, \tag{10.84}$$

that yields the wing–body system (Fig. 10.6), by assuming

$$B^2c^2 \ll 1. \tag{10.85}$$

In these conditions, we can write analogously with (9.143),

$$Q_1^* = \frac{A_{20}^* + A_{22}^*(x+c^2/x)^2}{\sqrt{(l^2-x^2)(1-c^4/l^2x^2)}}, \tag{10.86}$$

where the constants A_{20}^* and A_{22}^* will be determined in the same way that has been used so far (§ IX. 4.2) by putting

$$-\frac{2B^2+1}{B^2}\theta_0 = \text{Im}\int_{y'}^{1/B} \sqrt{1-B^2x^2}\,\frac{d^2Q_1^*}{dx^2}\,dx, \tag{10.87a}$$

$$0 = \text{Im}\int_{y'}^{1/B} \frac{\sqrt{1-B^2x^2}}{x}\,\frac{d^2Q_1^*}{dx^2}\,dx, \tag{10.87b}$$

y' being an arbitrary point on the wing. Obviously, the evaluation of these integrals is very laborious and we shall content ourselves with approximate values, taking the same expressions (10.56b, d) for the constants A_{20}^* and A_{22}^* as for the isolated wing without a body. This approximation can easily be checked in the case $n = 1$, noticing that, provided we have, as usual, $B^2c^2 \ll 1$, the constant from (10.77) becomes the same as in the case of the isolated wing (10.55).

Now we have to calculate the reduced potential Φ_1' which corresponds to the reduced downwash $\beta_0 = -\theta_0$.

A general method for calculating the potential of order n, based on Euler's theory on the homogeneous functions, will be given later.

In our particular case, we shall proceed from the potential given in (10.57) applied to an isolated wing in a plane $X = Y+iZ$ and write

$$\frac{1}{x_1}\Phi_1' = \text{Re}\,A'\sqrt{L^2-X^2}. \tag{10.88}$$

The imaginary part is zero on the wing ($Z = 0$) and one can apply the conformal transformation (10.84) from which the wing–body system results (Fig. 10.6). The value of the potential holds valid on the body too. Indeed,

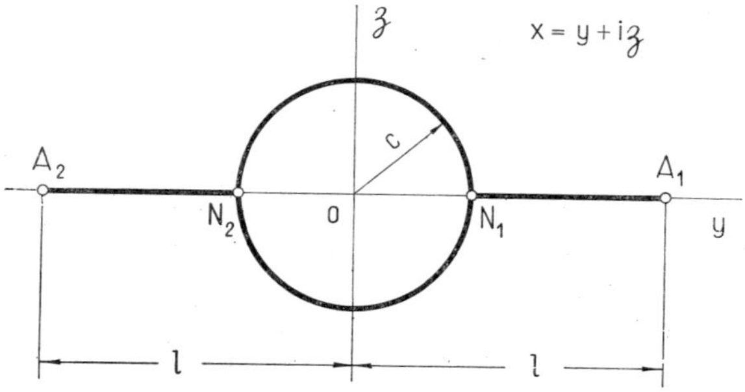

FIG. 10.6

in the immediate vicinity of the body, the motion has the character of an incompressible plane motion, while the "stream line" which also includes the body circle, remains a stream line both on the wing and body.

We have implicitly assumed that the downwash on the wing and body is zero, although, in reality, it is β_0. In order to take account of this fact, we have to add also the term $z\beta_0 = -z\theta_0$. Therefore, introducing the transformation (10.84) into (10.88), we obtain finally

$$\frac{1}{x_1}\Phi_1' = \operatorname{Re} A' \sqrt{(l^2-x^2)\left(1-\frac{c^4}{l^2x^2}\right)} - z\theta_0, \tag{10.89}$$

which is valid on the wing and in the vicinity of the body and where, by comparison with (10.77) and (10.79), the constant A' is given by the following relation

$$A' = \frac{A}{l^2} = \frac{A_0}{l^2}. \tag{10.90}$$

Having included these additional considerations, we can now write the pressure coefficient C_p by applying expression (10.48):

$$C_p = -2e^{i\omega t}\left[q_0 - \frac{i\lambda}{B^2}\left(\Phi_1' - B^2(q_0^* + xq_1^*) + x_1 M_\infty^2 q_0\right)\right]. \tag{10.91}$$

X. 3. Delta Wings with Subsonic Edges and Two Symmetrical Lateral Ridges

X. 3.1. Theory of Leading Edge Flaps

Consider a delta wing with subsonic edges and two symmetrical lateral ridges lying at points C_1 $(y = s)$ and C_2 $(y = -s)$. On applying formula (2.80) and taking account of the symmetry, we can write the following expression for the function $\mathcal{U}_{n-1}$ which defines the axial disturbance velocity:

$$\mathcal{U}_{n-1} = \frac{1}{\sqrt{l^2-x^2}} \sum_{q=0}^{E\left(\frac{n}{2}\right)} A_{n,2q}\, x^{2q} + \sum_{q=0}^{n-1} C_{n,q} x^q \left[\cosh^{-1}\sqrt{\frac{(l+s)(l-x)}{2l(s-x)}} + (-1)^q \cosh^{-1}\sqrt{\frac{(l+s)(l+x)}{2l(s+x)}}\right]. \tag{10.92}$$

The downwash on the wing is of the form

$$\frac{w}{x_1^{n-1}} = \sum_{q=0}^{n-1} w_{n-q-1,\,q}\, y^q, \tag{10.93}$$

but the value of the coefficients $w_{n-q-1,\,q}$ differs on the three regions indicated in Fig. 10.7.

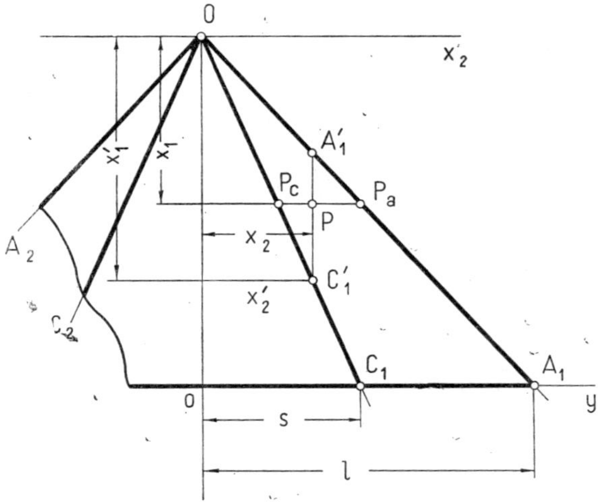

FIG. 10.7

To allow for this effect, we shall adopt a superscript notation in order to write $w_{n-q-1,\,q}$ in a corresponding manner:

$w^0_{n-q-1,\,q}$ for the central part;
$w'_{n-q-1,\,q}$ for the right-hand side, towards the leading edge OA_1;
$w''_{n-q-1,\,q}$ for the left-hand side, towards the leading edge OA_2.

Owing to the symmetry, q is even ($q = 2p$) or odd ($q = 2p+1$) in the coefficients above, as follows:

$$w^0_{n-2p,\,2p+1} = 0, \qquad w^0_{n-2p-1,\,2p} \neq 0, \tag{10.94a}$$

$$w'_{n-2p,\,2p+1} = -w''_{n-2p,\,2p+1}, \quad w'_{n-2p-1,\,2p} = w''_{n-2p-1,\,2p}. \tag{10.94b}$$

In order to determine the constants we shall follow the same procedure as in numerous previous cases, thus finding the various formulae which give these constants in terms of the values of the coefficients $w_{n-q-1,\,q}$.

As seen in expression (10.92), the axial disturbance velocity becomes infinite at the leading edges and we have to see if we can avoid this. The answer is in the affirmative and the condition is simple, viz.

$$\sum_{q=0}^{E\left(\frac{n}{2}\right)} A_{n,2q}\, l^{2q} = 0, \tag{10.95}$$

implying a certain connection between the coefficients of the downwashes. In this case, the first term in the right-hand part of expression (10.92) can be written in the form

$$\frac{1}{\sqrt{l^2-x^2}} \sum_{q=0}^{E\left(\frac{n}{2}\right)} A_{n,2q}\, x^{2q} = \sqrt{l^2-x^2} \sum_{q=0}^{E\left(\frac{n-2}{2}\right)} a_{n,2q}\, x^{2q}, \tag{10.96}$$

such that the constants may be determined in this case by starting from this form, which ensures a finite velocity at the leading edges from the beginning.

By avoiding the infinite velocity at the leading edges, one can avoid a characteristic pheonomenon, namely the appearance of vortices at the leading edges. The avoidance of these vortices represents an advantage, but the logarithmic singularity on the separation ridge OC_1 (OC_2, respectively) remains, its effect being borne, however, more easily in aeronautical constructions.

Application to the Leading Edge Flap in the Case ($n = 1$). We consider the case of the true conical motion ($n = 1$), for which we write

$$\mathcal{U}_0 = \frac{A_{10}}{\sqrt{l^2-x^2}} + C_{10}\left[\cosh^{-1}\sqrt{\frac{(l+s)(l-x)}{2l(s-x)}} + \cosh^{-1}\sqrt{\frac{(l+s)(l+x)}{2l(s+x)}}\right] \tag{10.97}$$

where the constants A_{10} and C_{10} will be given in terms of the incidence α_0 of the central part and α_1 of the lateral sides, or in terms of the corresponding downwashes:

$$w_0 = -\alpha_0 U_\infty, \quad w_1 = -\alpha_1 U_\infty. \tag{10.98}$$

Thus we find

$$-A_{10} = \frac{2B^2l^2\sqrt{l^2-s^2}\, s(w_1-w_0)\Pi(\varrho,k)}{\pi\sqrt{1-B^2s^2}\, E(k)} + \frac{w_0 l^2}{E(k)}, \tag{10.99a}$$

$$C_{10} = \frac{2s(w_1-w_0)}{\pi\sqrt{1-B^2s^2}}, \tag{10.99b}$$

where $E(k)$ is the complete elliptic integral of the second kind, and $\Pi(\varrho, k)$ is the elliptic integral of the third kind having the modulus k and parameter ϱ given by the expressions

$$k = \sqrt{1-B^2l^2}, \quad \varrho = B^2s^2-1. \tag{10.100}$$

In order to avoid the infinite velocity at the leading edges, we shall cancel the constant A_{10} and find the condition

$$\alpha_1 = \alpha_0\left[1-\frac{\pi\sqrt{1-B^2s^2}}{2B^2s\sqrt{l^2-s^2}\,\Pi(\varrho, k)}\right]. \tag{10.101}$$

Let us now calculate the lift coefficient; we find

$$C_l = -\frac{1}{U_\infty}\oint \mathcal{U}_0\, dx = 4\pi\alpha\,\frac{s\sqrt{l^2-s^2}}{2B^2sl\sqrt{l^2-s^2}\,\Pi(\varrho,k)-\pi(l-s)\sqrt{1-B^2s^2}}, \tag{10.102}$$

where we have denoted by α a *mean incidence*:

$$\alpha = \frac{s}{l}\alpha_0+\left(1-\frac{s}{l}\right)\alpha_1 = \alpha_0\left[1-\left(1-\frac{s}{l}\right)\frac{\pi\sqrt{1-B^2s^2}}{2B^2s\sqrt{l^2-s^2}\,\Pi(\varrho, k)}\right]. \tag{10.103}$$

It is of interest to note the fact that such a wing with the flaps deflected in a certain position so as to avoid the infinite velocity at the leading edges, can explain the overall effect of the detachment of the vortices from the leading edges, as done by Squire [S—24].

X. 3.2. Leading Edge Flaps in the (Case $n = 2$)

On applying formula (10.96), we can find for the axial velocity function

$$\mathcal{U}_1 = \frac{A_{20}+A_{22}x^2}{\sqrt{l^2-x^2}}+(C_{20}+C_{21}x)\cosh^{-1}\sqrt{\frac{(l+s)(l-x)}{2l(s-x)}}$$
$$+(C_{20}-C_{21}x)\cosh^{-1}\sqrt{\frac{(l+s)(l+x)}{2l(s+x)}}, \tag{10.104}$$

which corresponds to the downwashes in the three regions connected through the following relations:

$$w'_{10} = w''_{10}, \quad w'_{01} = -w''_{01}, \quad w^0_{10}, \quad w^0_{01} = 0. \tag{10.105}$$

In order to determine the constants, we shall apply the system (2.127) which, in this case, reduces to the following values of the constants:

$$C_{20} = \frac{2s}{\pi(1-B^2s^2)^{3/2}}\left[(2-B^2s^2)(w'_{10}-w^0_{10})+sw'_{01}\right], \tag{10.106a}$$

$$C_{21} = -\frac{2}{\pi(1-B^2s^2)^{3/2}}\left[(w'_{10}-w^0_{10})+B^2s^3w'_{01}\right]. \tag{10.106b}$$

As regards the constants A_{20}, A_{22}, on applying conditions (2.114a, c), the following system can be obtained in this case:

$$w'_{10} = (A_{20}+A_{22}l^2)\frac{B^2}{1-B^2l^2}[E(k)-K(k)]$$

$$+A_{22}E(k)-\frac{\sqrt{l^2-s^2}}{s^2}[C_{20}B^2s^2I+C_{21}s(I-K(k)], \quad (10.107a)$$

$$0 = A_{20}+2l^2A_{22}+\frac{l^2\sqrt{l^2-s^2}}{s^2}(C_{20}+2sC_{21}), \quad (10.107b)$$

in which we have used the notation

$$I = \frac{\pi}{2}\frac{s}{\sqrt{(l^2-s^2)(1-B^2s^2)}}-\frac{B^2s^2}{1-B^2s^2}\Pi(\varrho, k)+\frac{1}{1-B^2s^2}K(k), \quad (10.108)$$

and where $K(k), E(k), \Pi(\varrho, k)$ represent elliptic integrals of the first, second and third kind respectively, having the modulus ϱ and parameter k given by expressions (10.100).

In order to avoid the infinite velocity at the leading edges we have to put, applying (10.95),

$$A_{20}+A_{22}l^2 = 0; \quad (10.109)$$

in this case, the expression for $\mathcal{U}_1$ can be written now in the form

$$\mathcal{U}_1 = \frac{A_{20}}{l^2}\sqrt{l^2-x^2}+(C_{20}+C_{21}x)\cosh^{-1}\sqrt{\frac{(l+s)(l-x)}{2l(s-x)}}$$

$$+(C_{20}-C_{21}x)\cosh^{-1}\sqrt{\frac{(l+s)(l+x)}{2l(s+x)}}. \quad (10.110)$$

Working out the system constituted by equations (10.107) and (10.109), we obtain

$$A_{20} = \frac{2l^2\sqrt{l^2-s^2}}{\pi s(1-B^2s^2)^{3/2}}[B^2s^2(w^0_{10}-w'_{10})+s(1-2B^2s^2)w'_{01}]; \quad (10.111)$$

the downwash coefficients w^0_{10}, w'_{10}, w'_{01} satisfy the following condition:

$$\frac{\pi}{2}\frac{s(1-B^2s^2)^{3/2}}{\sqrt{l^2-s^2}}w^0_{10}+B^2s^2(w'_{10}-w^0_{10})[(1-B^2s^2)\Pi(\varrho, k)$$

$$+K(k)-E(k)]+sw'_{01}[(1-2B^2s^2)E(k)+B^2s^2K(k)] = 0, \quad (10.112)$$

which ensures a finite velocity at the leading edges.

Let us now calculate the lift coefficient, which will be given by the expression

$$C_l = \frac{4l\sqrt{l^2-s^2}}{3U_\infty s(1-B^2s^2)^{3/2}}\left\{\left[(3-2B^2s^2)\frac{s^2}{l^2}-B^2s^2\right](w'_{10}-w^0_{10})\right.$$
$$\left.+\left[1-2B^2s^2+(2-B^2s^2)\frac{s^2}{l^2}\right]sw'_{01}\right\}. \tag{10.113}$$

One of the coefficients w^0_{10}, w'_{10}, w'_{01} can be eliminated from this formula through equation (10.112), and the other two remain.

As in the preceding case, we can introduce here also a mean downwash at an arbitrary section x_1 = const.,

$$w_{10} = \frac{s}{l}w^0_{10}+\left(1-\frac{s}{l}\right)w'_{10}+\frac{1}{l}\int_s^l w'_{01}y\,dy$$
$$= \frac{s}{l}w^0_{10}+\left(1-\frac{s}{l}\right)w'_{10}+\left(1-\frac{s^2}{l^2}\right)\frac{l}{2}w'_{01}, \tag{10.114}$$

which corresponds to the mean incidence

$$\alpha = \alpha_{10}x_1 = -\frac{w_{10}}{U_\infty}x_1, \quad (w = w_{10}x_1), \tag{10.115}$$

in which we can likewise eliminate one of the coefficients on the right-hand side, as in the expression for the lift coefficient.

If we establish further one more condition, then we can express (10.113) in terms of the mean incidence or of the corresponding mean downwash (10.113).

To this end, we can consider two particular simple cases, defined through the downwash coefficients:

$$(1) \quad w'_{10} = w''_{10} \neq 0, \quad w'_{01} = -w''_{01} = 0, \tag{10.116a}$$

$$(2) \quad w'_{10} = w''_{10} = 0, \quad w'_{01} = -w''_{01} \neq 0. \tag{10.116b}$$

(1) In the first case we have the system

$$\frac{\pi}{2}\frac{s(1-B^2s^2)^{3/2}}{\sqrt{l^2-s^2}}w^0_{10}+B^2s^2(w'_{10}-w^0_{10})[(1-B^2s^2)\,\Pi(\varrho,k)+K(k)-E(k)] = 0, \tag{10.117a}$$

$$w_{10} = w^0_{10}+\left(1-\frac{s}{l}\right)(w'_{10}-w^0_{10}); \tag{10.117b}$$

having solved it, we can express the coefficients w_{10}^0, w_{10}' in terms of the mean coefficient w_{10}, i.e.

$$w_{10}^0 = w_{10}\frac{2B^2s^2\sqrt{l^2-s^2}\,[(1-B^2s^2)\Pi(\varrho, k)+K(k)-E(k)]}{2B^2s^2\sqrt{l^2-s^2}[(1-B^2s^2)\Pi(\varrho,k)+K(k)-E(k)]-\pi s(1-s/l)(1-B^2s^2)^{3/2}} \tag{10.118}$$

and

$$w_{10}' = w_{10} \times\left\{1-\frac{\pi s(s/l)(1-B^2s^2)^{3/2}}{2B^2s^2\sqrt{l^2-s^2}\,[(1-B^2s^2)\Pi(\varrho,k)+K(k)-E(k)]-\pi s(1-s/l)(1-B^2s^2)^{3/2}}\right\} \tag{10.119}$$

Introducing these values into (10.113) we obtain the espression for the lift coefficient in terms of the mean downwash,

$$C_l = -\frac{4lw_{10}}{3U_\infty} \times\frac{\pi\sqrt{l^2-s^2}[(3-2B^2s^2)s^2/l^2-B^2s^2]}{2B^2s^2\sqrt{l^2-s^2}\,[(1-B^2s^2)\Pi(\varrho, k)+K(k)-E(k)]+\pi s(1-s/l)(1-B^2s^2)^{3/2}}. \tag{10.120}$$

We can thus equate the wing with leading edge flaps to a fictitious wing of mean incidence defined by (10.115), having, however, a finite axial velocity at the leading edges.

(2) In the second case we have the system

$$\frac{\pi}{2}\frac{s(1-B^2s^2)^{3/2}}{\sqrt{l^2-s^2}}w_{10}^0+sw_{01}'[(1-2B^2s^2)E(k)+B^2s^2K(k)] = 0, \tag{10.121a}$$

$$w_{10} = \frac{s}{l}w_{10}^0+\left(1-\frac{s^2}{l^2}\right)\frac{l}{2}w_{01}', \tag{10.121b}$$

whence we obtain the following expressions for w_{10} and w_{01},

$$w_{10}^0 = w_{10}\frac{4l[(1-2B^2s^2)E(k)+B^2s^2K(k)]}{4s[(1-2B^2s^2)E(k)+B^2s^2K(k)]-\pi\sqrt{l^2-s^2}(1-B^2s^2)^{3/2}} \tag{10.122}$$

and

$$w_{01}' = -\frac{2w_{10}l}{\sqrt{l^2-s^2}}\frac{\pi(1-B^2s^2)^{3/2}}{4s[(1-2B^2s^2)E(k)+B^2s^2K(k)]-\pi\sqrt{l^2-s^2}(1-B^2s^2)^{3/2}}. \tag{10.123}$$

Substituting these values into (10.113), we obtain for the lift coefficient

$$C_l = -\frac{8w_{10}l^2}{3U_\infty s(1-B^2s^2)^{3/2}}$$
$$\times\left(\frac{2\sqrt{l^2-s^2}[(3-2B^2s^2)s^2/l^2-B^2s^2][(1-2B^2s^2)E(k)+B^2s^2K(k)]}{4s[(1-2B^2s^2)E(k)+B^2s^2K(k)]-\pi\sqrt{l^2-s^2}\,(1-B^2s^2)^{3/2}}\right.$$
$$\left.+\frac{s(1-B^2s^2)^{3/2}[1-2B^2s^2+(2-B^2s^2)s^2/l^2]}{4s[(1-2B^2s^2)E(k)+B^2s^2K(k)]-\pi\sqrt{l^2-s^2}(1-B^2s^2)^{3/2}}\right) \tag{10.124}$$

In such a case, we shall consider the ridge line ($y = s$) around which the incidence of each section of the leading edge flap varies as a function of x_2' (Fig. 10.7) according to the relation

$$w_{01}'x_2 = w_{01}'x_2' = w_{10}'x_1's. \tag{10.125}$$

At an arbitrary point P of coordinates x_1, x_2, the incidence or the corresponding downwash have just the same expression.

This case is interesting because of the fact that it can be associated with the plane wing ($n = 1$), thus constituting a complex wing made up of two components $n = 1$ and $n = 2$.

In this way we can ensure the condition of finite velocity at the leading edges only by means of a "deflexion" of the flap in any position of the wing (corresponding to the respective positions of the aircraft fitted with such a wing having a parabolic profile in each section).

Thus from the practical standpoint one can construct a complex wing which should have the property of avoiding the infinite velocity at the leading edges.

X. 4. Theory of a Thin Delta Wing with Flow Separation at the Leading Edges

X. 4.1. The Finding of an Equivalent Wing for explaining the Overall Effect of the Flow Detachment

As is known, in the case of a wing with subsonic leading edges the flow may detach from these edges, giving rise to very complicated motions which are characterized, in particular, by the appearance of a complex system of vortices in the space situated in the vicinity of the wing upper surface. A scheme of this phenomenon is represented in Fig. 10.8. A layer of vortices detaches from the wing leading edges and rolls up in the shape of a cornet, then turning into two nuclei of vortices where the major part of its vortical intensity is concentrated. The photographs of Fig. 10.9 illustrate

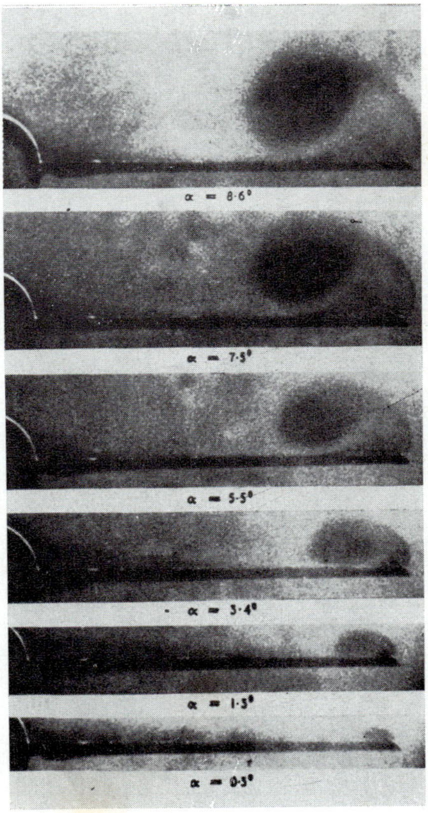

FIG. 10.9

the evolution of this phenomenon as a function of the incidence [S−25].†
Among the numerous authors who have dealt with the flow over a delta wing and given a general description of the vortex cornet are M. Roy and R. Legendre. As a matter of fact, this problem has been amply treated and discussed by many authors [K. Mangler, I. Smith, C. Brown, W. Michael, etc.], and very interesting results have been sometimes reached, particularly for slender wings, for which $Bl \ll 1$ [M−4]. When Bl is fairly large and even close to unity, the problem is difficult to treat; therefore, we shall try to approach it in an indirect way.

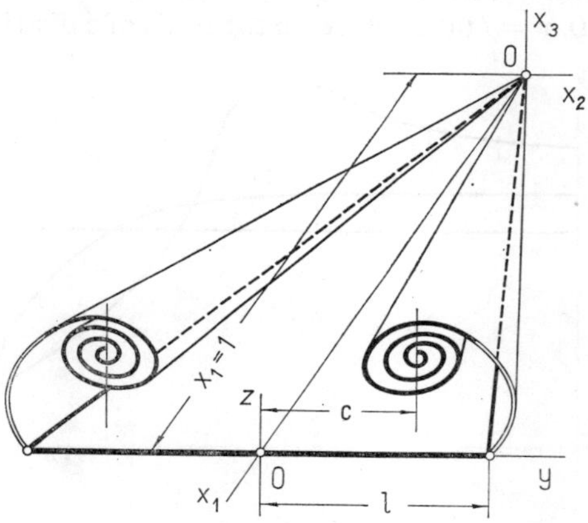

FIG. 10.8

Thus, we shall assume that the effect of the flow detachment at the wing leading edges and the appearance of vortices on the upper surface consists in producing a downwash field which alters the flow, thus leading to the avoidance of the infinite velocity at the leading edges. In reality, there also exist longitudinal disturbance velocities, but one may assume that these velocities could be replaced by a corresponding downwash distribution. To this effect, we shall consider a distribution of downwashes that differs on the two wing surfaces, the upper and the lower one: hence, a *conical wing* whose incidence varies according to a well-determined law.

Following a suggestion made by Küchemann (K−20] and Squire [S−24], we shall suppose that the real thin wing, having finite velocities at the leading edges owing to the effect of flow detachment, can be equated to such a fictitious thin wing having a convenient variation of incidences.

To this end, we shall assume that on the fictitious wing, the variable inci-

† The photographs are published with the permission of the controller of Her Britannic Majesty's Stationery Office.

dence α', the downwash w' respectively, vary along the lines U on the upper surface and L on the lower surface, having as mean line the straight line, which represents the incidence or the downwash of the real thin wing:

$$w = -\alpha U_\infty, \quad (w'_u = -\alpha'_u U_\infty, \quad w'_l = -\alpha'_l U_\infty, \quad w'_1 = -\alpha_1 U_\infty). \qquad (10.126)$$

The mean incidence or downwash follow from the relation below which can be deduced on examining Fig. 10.10:

$$\int_0^l w'_u \, dy + \int_0^l w'_l \, dy = 2wl. \qquad (10.127)$$

On the other hand, it should be noticed that the variation of velocities in the interval $y = 0$, $y = l$ takes place on the basis of a distribution of sources

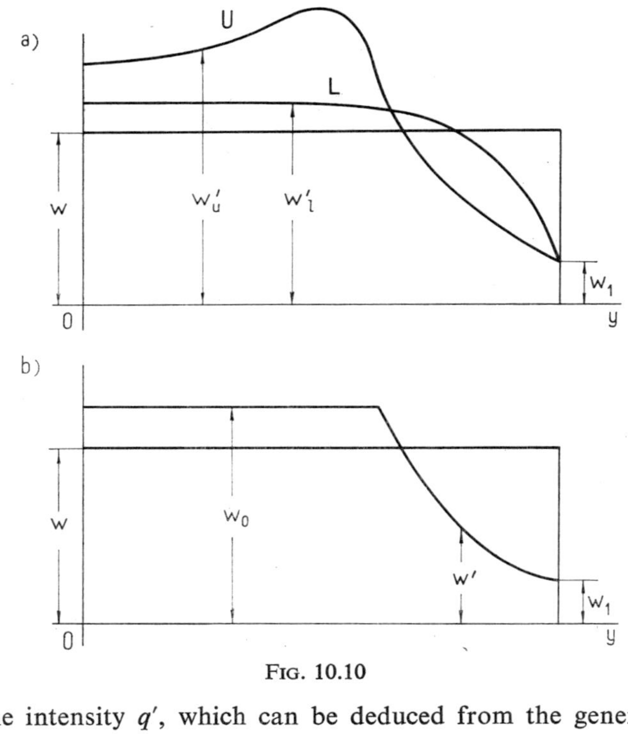

FIG. 10.10

having the intensity q', which can be deduced from the general theory (2.41):

$$q' = \frac{y}{\sqrt{1-B^2y^2}} \frac{dw'}{dy}. \qquad (10.128)$$

and will be introduced into (10.127) in order to obtain a relation between the constants of the variation of the velocities w' and w (or between the incidences α' and α).

Now we have to find the distribution of sources q', chosen conveniently in order to correspond to the distribution of the downwashes considered, by applying formula (10.128). For this purpose we shall make use of the methods presented in our previous chapters and decompose the fictitious wing into three components:

(1) A thin wing with a conveniently chosen variation of *incidence*, in order to represent, to a certain extent, the phenomena taking place on the upper surface.

(2) A wing of symmetrical thickness whose slope varies according to the same law as the incidence in (1). Owing to this fact we can avoid the disturbance axial velocity being equal and of opposite sign on the two surfaces of the wing, as would follow from assumption (1).

(3) A wing of symmetrical thickness whose slope, in contrast to the preceding one, can combine with it in order to obtain a zero mean slope. In this manner, the two thick wings behave together like a *mean* thin wing, as is the actual case.

The choice of the distribution of velocities or distribution of sources which should represent them, must be connected with the actual phenomenon and the indications given by experiments. The problem becomes very difficult to treat, and for this reason we shall try simplify to it, by choosing the simplest expressions with the aid of which we shall try to represent the actual phenomenon as far as possible. We shall proceed successively as follows.

X. 4.2. Determination of the Disturbance Axial Velocities

We shall consider in turn each of the three constituent wings defined above and determine by turns the disturbance axial velocities.

1. THIN WING. Let us assume that the downwash is constant and equal to w_0 from 0 to s and varies according to an arbitrary law from s to l where this velocity is w_1. We shall assume that this variation is given by a linear distribution of sources from s to l:

$$q' = q_0\left(1-\frac{y}{l}\right). \tag{10.129}$$

Thus, taking account of (10.128), we find

$$w_1 - w_0 = q_0 \int_s^l \left(1-\frac{y}{l}\right)\frac{\sqrt{1-B^2y^2}}{y}\,dy$$

$$= \frac{q_0}{2Bl}\left[Bl\left(\sqrt{1-B^2l^2}-\sqrt{1-B^2s^2}\right)-B(l-s)\sqrt{1-B^2s^2}\right.$$

$$+2Bl\left(\cosh^{-1}\frac{1}{Bs}-\cosh^{-1}\frac{1}{Bl}\right)$$

$$\left.+\sin^{-1}Bs-\sin^{-1}Bl\right], \tag{10.130}$$

from which we deduce q_0 in terms of w_0 and w_1.

Applying the formula established in § II. 3.4 and taking into account (10.128), (10.129), where y will be replaced by η, in order to avoid a confusion with y from $x = y+i\mathfrak{z}$, we find:

$$\mathcal{U}_l = \frac{A_{10}}{\sqrt{l^2-x^2}}+\frac{2}{\pi}q_0\int_s^l\left(1-\frac{\eta}{l}\right)\left[\cosh^{-1}\sqrt{\frac{(l+\eta)(l-x)}{2l(\eta-x)}}\right.$$

$$\left.+\cosh^{-1}\sqrt{\frac{(l+\eta)(l+x)}{2l(\eta+x)}}\right]d\eta = \frac{A_{10}}{\sqrt{l^2-x^2}}-\frac{q_0}{\pi}\left[(s-x)\left(1-\frac{s+x}{2l}\right)\right.$$

$$\times\cosh^{-1}\frac{l^2-sx}{l(s-x)}+(s+x)\left(1-\frac{s-x}{2l}\right)\cosh^{-1}\frac{l^2+sx}{l(s+x)}$$

$$\left.+\sqrt{l^2-x^2}\left(\sqrt{1-\frac{s^2}{l^2}}-2\cos^{-1}\frac{s}{l}\right)\right]. \tag{10.131}$$

The constant A_{10} may be determined in the usual way used above. In order to avoid some very laborious calculations, we shall make an approximation in determining the constant A_{10}, by considering a source situated at $y = s'$ whose intensity Q_0 is equal to the total intensity of the sources given by the relations:

$$s' = s+\frac{l-s}{3}, \quad Q_0 = \frac{q_0}{2l}(l-s)^2. \tag{10.132}$$

Proceeding in this manner and denoting, as usual, by $E(k)$ and $\Pi(\varrho, k)$ the complete elliptic integrals of the second and third kind, we obtain:

$$A_{10} = -\frac{w_0l^2}{E(k)}-\frac{2Q_0}{\pi}\frac{B^2l^2\sqrt{l^2-s'^2}\,\Pi(\varrho,k)}{E(k)}$$

$$= -\frac{w_0l^2}{E(k)}-\frac{2q_0}{3\pi E(k)}B^2l(l-s)^{5/2}\sqrt{2l+s}\;\Pi(\varrho,k), \tag{10.133a}$$

$$k = \sqrt{1-B^2l^2}, \quad \varrho = B^2s'^2-1 = \tfrac{1}{9}B^2(l+2s)^2-1. \tag{10.133b}$$

Now, imposing the condition that the velocity shall be finite at the leading edges, we can cancel the constant A_{10} and thus obtain a new relation between w_0 and q_0:

$$3\pi l w_0 + 2q_0 B^2 (l-s)^{5/2} \sqrt{2l+s}\ \Pi(\varrho, k) = 0. \qquad (10.134)$$

On the other hand, since the mean downwash is equal to w or the mean incidence to α, as shown above, we have

$$wl = w_0 s + \int_s^l w' dy = w_0 s + (w'y)\Big|_s^l - \int_s^l y dw'$$

$$= w_1 l - \frac{q_0}{2B}\Big[Bl\sqrt{1-B^2l^2} - Bs\sqrt{1-B^2s^2} + \sin^{-1}Bl - \sin^{-1}Bs$$

$$+ \frac{2}{3Bl}\left((1-B^2l^2)^{3/2} - (1-B^2s^2)^{3/2}\right)\Big]\cdot \qquad (10.135)$$

w_0, w_1 and q_0 can be obtained from relations (10.130, 10.134, 10.135) in terms of α, the incidence of the actual wing, or in terms of w ($w = -\alpha U_\infty$).

It should be noticed that, as a consequence of the cancellation of the constant A_{10}, the axial velocity function for the resultant fictitious "lifting wing" is

$$\mathcal{U}_l = -\frac{q_0}{\pi}\Big[(s-x)\left(1-\frac{s+x}{2l}\right)\cosh^{-1}\frac{l^2-sx}{l(s-x)}$$

$$+(s+x)\left(1-\frac{s-x}{2l}\right)\cosh^{-1}\frac{l^2+sx}{l(s+x)}$$

$$+\sqrt{l^2-x^2}\left(\sqrt{1-\frac{s^2}{l^2}} - 2\cos^{-1}\frac{s}{l}\right)\Big], \qquad (10.136)$$

in which the constant q_0 can be expressed in terms of α, the *real incidence of the wing*, as pointed out above.

2. Wing of Symmetrical Thickness having a Slope equal to the Incidence α of the Fictitious Wing. This component levels the pressure peak on the lower surface and thus introduces a different distribution of pressures on the two wing surfaces. On applying formula (5.80) which corresponds to a wing of symmetrical thickness with slope variation, defined by the same distribution of sources as (10.129), we can write the

following expression for the axial velocity function which will be denoted by $\mathcal{U}_t$:

$$\mathcal{U}_t = \frac{2}{\pi} q_0 \int^l \left(1-\frac{\eta}{l}\right) \left[\cosh^{-1} \sqrt{\frac{(1+B\eta)(1-Bx)}{2B(\eta-x)}}\right.$$

$$\left. +\cosh^{-1} \sqrt{\frac{(1+B\eta)(1+Bx)}{2B(\eta+x)}}\right] d\eta + L$$

$$= \frac{q_0}{\pi} \left[(l-x)\left(1-\frac{l+x}{2l}\right) \cosh^{-1} \frac{1-B^2lx}{B(l-x)} + (l+x)\left(1-\frac{l-x}{2l}\right)\right.$$

$$\times \cosh^{-1} \frac{1+B^2lx}{B(l+x)} - (s-x)\left(1-\frac{s+x}{2l}\right) \cosh^{-1} \frac{1-B^2sx}{B(s-x)}$$

$$-(s+x)\left(1-\frac{s-x}{2l}\right) \cosh^{-1} \frac{1+B^2sx}{B(s+x)} + \frac{2}{B}\sqrt{1-B^2x^2}$$

$$\left.\times\left(\sin^{-1} Bl - \sin^{1} Bs + \frac{\sqrt{1-B^2l^2}-\sqrt{1-B^2s^2}}{2Bl}\right)\right] + L, \qquad (10.137)$$

in which the term L is due to the leading edge having the downwash w_1 or the slope $\gamma_1 = \alpha_1$. In the expression above, y from the expression of q' (10.129) is replaced by η.

3. Wing of Symmetrical Thickness compensating for the Slope. Through the introduction of the effect mentioned in item (2), the wing has become a wing of symmetrical thickness in a fictitious sense, a fact that is contrary to reality. In order to compensate for this situation and restore the wing to a zero mean thickness, we shall introduce a new distribution of sources of any form which should correspond to the velocity $-w_1(-\gamma = \alpha_1)$ at the extremity. This is necessary in order to cancel the effect of the slope γ_1 pointed out at item (2), by providing a term $-L$ which cancels the preceding one. Thus the term that gives a velocity tending to infinity logarithmically at the leading edge is eliminated.

As to the variation of the slope of this new wing, this should reflect the distribution of pressures on the lower surface, which varies slowly, except near the edges.

For simplicity, we shall assume that the distribution law is given by the expression

$$q'' = k_0 y(1+k_1 y) = \frac{y}{\sqrt{1-B^2y^2}} \frac{dw''}{dy}, \qquad (10.138)$$

in which we shall determine one of the constants by putting

$$\int_0^l w''dy = (w''y)_0^l - \int_0^l y\,dw''$$

$$= -w_1 l - k_0 \int_0^l y(1+k_1 y)\sqrt{1-B^2y^2}\,dy$$

$$= -w_1 l - k_0 \left\{ \frac{1}{3B^2}[1-(1-B^2l^2)^{3/2}] + \frac{k_1}{8B^3}[Bl\sqrt{1-B^2l^2}\right.$$

$$\left. + \sin^{-1}Bl - 2Bl(1-B^2l^2)^{3/2}] \right\} = -wl, \tag{10.139}$$

because the mean total slope of both wings of symmetrical thickness must be zero.

On adding these two terms and the corresponding terms from (10.135), we obtain the following relation between q_0, k_0 and k_1:

$$\frac{q_0}{2}\left\{ Bl\sqrt{1-B^2l^2} - Bs\sqrt{1-B^2s^2} + \sin^{-1}Bl - \sin^{-1}Bs \right.$$

$$\left. + \frac{2}{3Bl}[(1-B^2l^2)^{3/2} - (1-B^2s^2)^{3/2}] \right\}$$

$$+ k_0 \left\{ \frac{1}{3B}[1-(1-B^2l^2)^{3/2}] \right.$$

$$\left. + \frac{k_1}{8B^2}[Bl\sqrt{1-B^2l^2} + \sin^{-1}Bl - 2Bl(1-B^2l^2)^{3/2}] \right\} = 0, \tag{10.140}$$

which causes the two components (2) and (3) to be reduced to a wing of zero mean slope. We shall now establish another relation necessary for determining the constants.

The axial velocity function which will be denoted by $\mathcal{U}_c$, will be given for this "compensating wing" by an expression analogous with (10.137), where we take (5.80) into consideration, replacing y by η:

$$\mathcal{U}_c = \frac{2}{\pi} k_0 \int_0^l \eta(1+k_1\eta) \left(\cosh^{-1} \sqrt{\frac{(1+B\eta)(1-Bx)}{2B(\eta-x)}} \right.$$

$$\left. + \cosh^{-1} \sqrt{\frac{(1+B\eta)(1+Bx)}{2B(\eta+x)}} \right) d\eta - L$$

$$= \frac{k_0}{2\pi B^2}\left[B^2(l^2-x^2)\left(\cosh^{-1}\frac{1-B^2lx}{B(l-x)}+\cosh^{-1}\frac{1+B^2lx}{B(l+x)}\right)\right.$$

$$\left.+2B^2x^2\cosh^{-1}\frac{1}{Bx}+2(1-\sqrt{1-B^2l^2})\sqrt{1-B^2x^2}\right]$$

$$+\frac{k_0k_1}{3\pi B^3}\left[B^3(l^3-x^3)\cosh^{-1}\frac{1-B^2lx}{B(l-x)}+B^3(l^3+x^3)\cosh^{-1}\frac{1+B^2lx}{B(l+x)}\right.$$

$$\left.+(2\sin^{-1}Bl B^2x^2+\sin^{-1}Bl-Bl\sqrt{1-B^2l^2})\sqrt{1-B^2x^2}\right]-L, \quad (10.141)$$

L being the same term as in (10.137), considered now to be of opposite sign.

On adding (10.136, 10.137, 10.141), we finally obtain the function of the axial total velocity:

$$\mathcal{U} = \mathcal{U}_l+\mathcal{U}_t+\mathcal{U}_c. \quad (10.142)$$

X.4.3. *Calculation of Pressure Distribution and Aerodynamic Characteristics*

The constant q_0 which is included in the expressions (10.136) and (10.137) of the axial velocity functions $\mathcal{U}_l$ and $\mathcal{U}_t$, can be deduced from equations (10.130), (10.134) and (10.135). Indeed, by eliminating w_0 and w_1 from these equations, we find q_0 as a function of the real incidence α of the wing:

$$\frac{q_0}{\alpha U_\infty} = Bl\left[\frac{2}{3\pi}B^3l^3\left(1-\frac{s}{l}\right)^{5/2}\sqrt{2+\frac{s}{l}}\,\Pi(\varrho, k)\right.$$

$$+Bl\left(1-\frac{s}{l}\right)\sqrt{1-B^2s^2}+\frac{1}{3Bl}\left((1-B^2l^2)^{3/2}-(1-B^2s^2)^{3/2}\right)$$

$$\left.+Bl\left(\cosh^{-1}\frac{1}{Bl}-\cosh^{-1}\frac{1}{Bs}\right)+\sin^{-1}Bl-\sin^{-1}Bs\right]^{-1}, \quad (10.143)$$

where $\Pi(\varrho, k)$ is the elliptical integral of the third kind in which the modulus k and the argument ϱ were defined above (10.133b). In order to make use of the usual computation table, it may be more convenient to express $\Pi(\varrho, k)$ as a function of the elliptical, complete and incomplete, integrals of the first and second kind, respectively $K(k)$ and $F(\varphi_0, k')$, $E(k)$ and $E(\varphi_0, k')$,

$$\Pi(\varrho, k) = K(k)+\frac{\sqrt{1-B^2s'^2}}{B^2s'\sqrt{l^2-s'^2}}\left[\frac{\pi}{2}-K(k)E(\varphi_0, k')\right.$$

$$\left.+\big(K(k)-E(k)\big)F(\varphi_0, k')\right], \quad (10.144)$$

where s' is given by (10.132), ϱ and k by (10.133b), and φ_0 and k' by the expressions below:

$$\varphi_0 = \sin^{-1}\frac{s'}{l}, \quad k' = Bl. \tag{10.145}$$

Let us further note that the other two constants k_0 and k_1 are included in the expression (10.141) of the velocity $\mathcal{U}_c$. For the sake of simplification k_1 may be given the following value,

$$k_1 = -\frac{1}{2l}, \tag{10.146}$$

with the aid of which one obtains a convenient pressure distribution for this compensatory wing of symmetrical thickness, that is, an almost constant pressure over the whole wing, except the area lying in the proximity of the leading edge. In such a case, introducing the above value of k_1 into equation (10.140), one obtains the expression of k_0,

$$k_0 = -24B^2 l q_0 \Big[Bl\sqrt{1-B^2l^2} - Bs\sqrt{1-B^2s^2} + \sin^{-1}Bl - \sin^{-1}Bs$$
$$+\frac{2}{3Bl}\left((1-B^2l^2)^{3/2}-(1-B^2s^2)\right)^{3/2}\Big]\left[2Bl(8-5(1-B^2l^2)^{3/2})\right.$$
$$\left.-3(\sin^{-1}Bl + Bl\sqrt{1-B^2l^2})\right]^{-1}. \tag{10.147}$$

Let us now pass to the calculation of the lift coefficient. To this purpose, in view of the fact that the wings of symmetrical thickness do not yield the lift, we shall consider only the pressure coefficient due to the "lifting thin wing",

$$C_p = -2\frac{u_l}{U_\infty} = -2\,\mathrm{Re}\,\frac{\mathcal{U}_l}{U_\infty}. \tag{10.148}$$

Introducing this value into the formula of the lift coefficient (3.38a), we obtain

$$C_l = \frac{4}{lU_\infty}\int_0^l u_l\,dy = \frac{2q_0 l}{U_\infty}\left[\cos^{-1}\frac{s}{l} - \frac{s}{l}\left(1-\frac{s^2}{l^2}\right)^{1/2} - \frac{2}{3}\left(1-\frac{s^2}{l^2}\right)^{3/2}\right]. \tag{10.149}$$

As concerns, however, the pressure distribution, we have to consider the axial total velocity (10.142) and put

$$C_p = -2\frac{u}{U_\infty} = -2\,\mathrm{Re}\,\frac{\mathcal{U}}{U_\infty}. \tag{10.150}$$

In the above expressions of the lift and pressure coefficients, it is only the parameter s/l, which is part of these expressions, that has to be determined. To this end we shall note that in the case in which the source distribution varies according to relation (10.129) considered as fundamental assumption, the absolute value of u becomes maximum around the abscissa $y = c$ given by the equation

$$\frac{c}{l} \approx \frac{s}{l} + \frac{1}{6}\left(1 - \frac{s}{l}\right), \quad \left(\frac{s}{l} = 1{\cdot}2\frac{c}{l} - 0{\cdot}2\right). \tag{10.151}$$

From the physical standpoint it is easy to assume that c is the very abscissa of the centre of the vortex nucleus schematically defined in Fig. 10.8.

We shall present below some considerations on the position of these vortex nuclei.

THE POSITION OF THE VORTEX NUCLEI CENTRE. The scheme plotted in Fig. 10.8 is simple; actually, the phenomenon is very complicated. The vortex cornet turns into a main nucleus which will be dealt with in a particular manner, and part of the vortex layer situated on the cornet turns into a small marginal nucleus (tip vortex) in a proportion of about 80 per cent and 20 per cent. At large angles of attack there also appears another inboard vortex between the main nucleus and the small marginal one, of opposite sign as compared with the former ones. A more accurate examination cannot be made owing to the complexity of this phenomenon; consequently, we shall consider the simple scheme given above (Fig. 10.8), meanwhile also assuming the existence of that small marginal area of vortices which has the effect of reducing the total intensity of the main nucleus whose centre has the abscissa $y = c$ on the wing, which we just want to determine in an approximate manner.

For the case $Bl \ll 1$, the rigorous treatment of the problem [M–4] shows that $c = l$ for $\alpha = 0$, and varies, according to a very complicated law, towards the interior; however, the distance c forecast by theory differs from the one found experimentally.

We shall therefore follow a simple reasoning for the general case, which should, however, lead us to a result close to reality. Thus, for instance, for very small incidences ($\alpha \to 0$), the concentrated vortices might be considered as being stuck to the upper surface of the wing, a fact that would lead, no matter how small their intensity would be, to very high local velocities which are incompatible with the very actual effects of the flow detachment. Hence, we have to assume that $c = l$ for the incidence $\alpha = 0$; as a matter of fact, this assumption is consistent with the theoretical results obtained for the slender wing. For larger incidences, we assume that the vortex-filaments—that would proceed approximately from the leading edges, being proportional to the wing width at these points—roll up an axis representing the line of the centroids, and thus bring the nucleus to an appropriate position.

We shall notice, however, that, as mentioned above, the intensity of the vortex nucleus is somewhat smaller than the total intensity of the vorticity, the rest being included in the thin layer that constitutes the cornet surface which generally yields the tip vortex. Therefore, as illustrated by experiments, this minimum distance could oscillate about the value

$$C \approx 0{\cdot}6l \div 0{\cdot}5l, \tag{10.152}$$

which can be deduced through a reasoning identical to the above one, on considering that the intensity of the vortex nucleus is somewhat smaller than the total intensity, as written above.

This position is also difficult to be established owing to the appearance of vortex nuclei of opposite sense at large angles of attack, as pointed out above.

Nevertheless, taking also account of the experimental data obtained by various authors which, as a matter of fact, are rather inconclusive, we can suggest an approximate law between these limit positions,

$$\frac{c}{l} \approx \frac{1}{1+1{\cdot}7(\alpha)^{1/2}}. \tag{10.153}$$

The value of c that has been thus determined, will be substituted in (10.152) in order to find s which is to be introduced into the formulae for the pressure and lift coefficients and in this way, we obtain their expressions in terms of α/l.

X. 4.4. Comparison with Experiments

We have determined above the manner of calculating the position c of the vortex nucleus centre, from which the parameter s is being then deduced (10.152). This parameter will be introduced into the formula of the lift coefficient C_l (10.149) which has been calculated for various values of Bl and for which we also have experimental results owing to the direct measurements carried out by A. O. Ormerod and A. B. Haines [S-24].

As one can see in the diagrams of Fig. 10.11, it may be asserted that there exists a fair agreement between theory and experiment.

Similarly, as regards the distribution of pressures on the wing, one can remark a fair agreement between theoretical predictions and the experimental results obtained by W. H. Michael [M-16] for various values of Bl. Indeed, in the diagrams plotted in Fig. 10.12 a, b, c, d the theoretical dis-

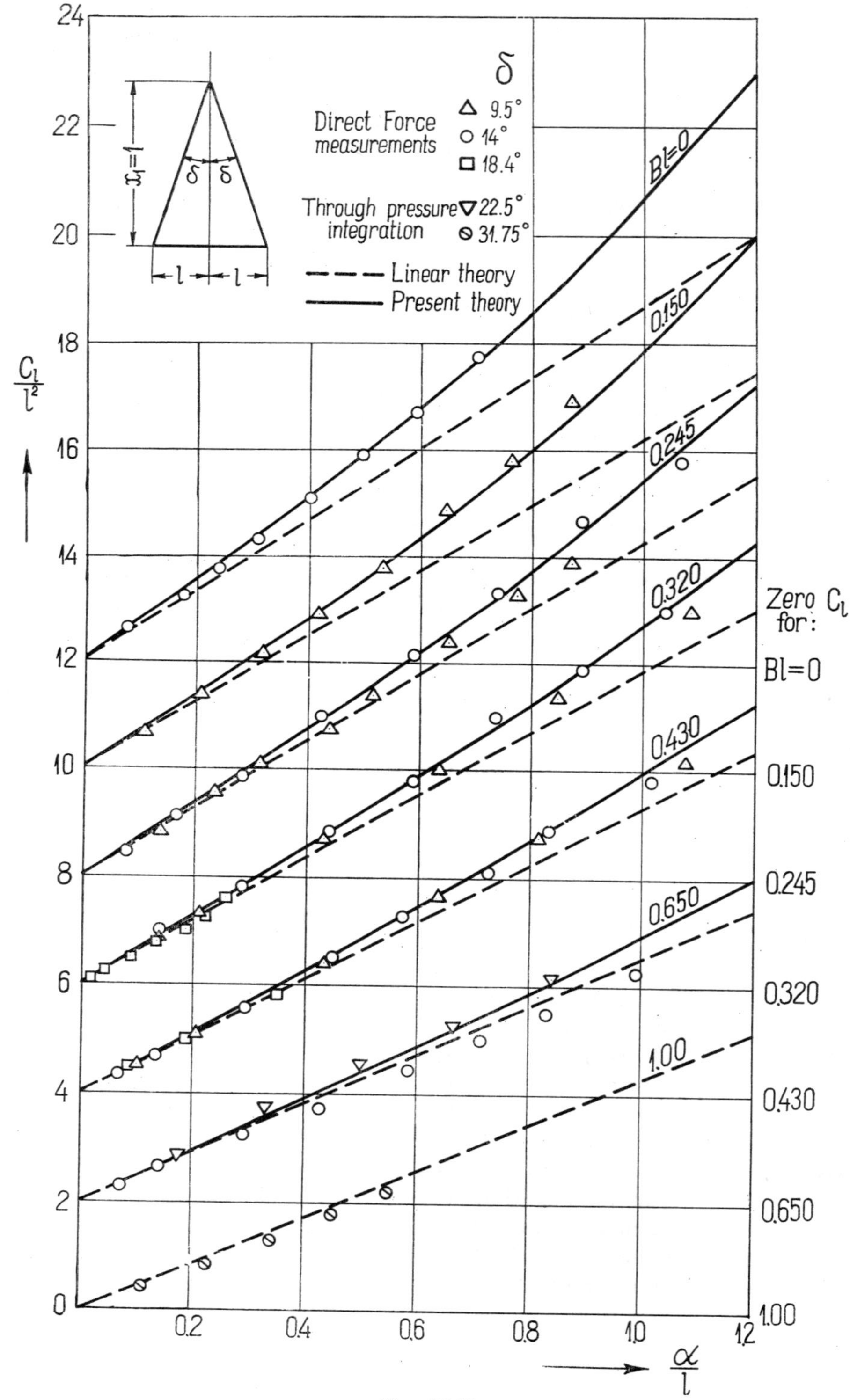

δ
Direct Force measurements
9.5°
14°
18.4°
Through pressure integration
22.5°
31.75°
Linear theory
Present theory
$x_1 = 1$
δ δ
l l
Bl=0
0.150
0.245
0.320
0.430
0.650
1.00
$\frac{C_l}{l^2}$
$\frac{\alpha}{l}$
Zero C_l for:
Bl=0
0.150
0.245
0.320
0.430
0.650
1.00
0
2
4
6
8
10
12
14
16
18
20
22
24
0.2
0.4
0.6
0.8
1.0
1.2

FIG. 10.11

tribution of the pressure coefficient follows closely the experimental values, not only quantitatively, but also qualitatively, a fact that testifies convincingly to the good choice of the theoretical course. To this end, making the maximum point of depressions correspond to the vortex centre posi-

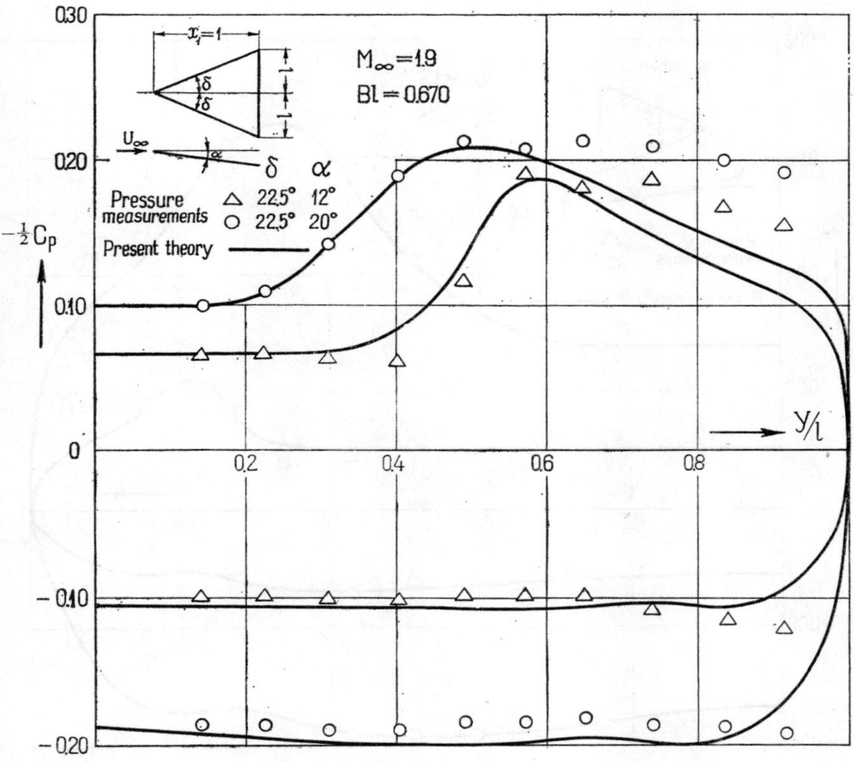

FIG. 10.12a

tion, where formula (10.154) does not give this maximum precisely, as is the case of Fig. 10.12d, for $Bl = 0{\cdot}45$, $\alpha = 4°$ and $\alpha/l = 0{\cdot}26$, and repeating the calculation by using the new experimental position of c, we obtain a theoretical distribution which is very close to the experimental one (Fig. 10.12, the dotted line), a fact which proves that the hypothesis of considering the position c of the nucleus centre to coincide with the point of maximum pressure on the wing corresponds to the physical phenomenon.

The main difficulty regarding this problem consists in the fact that the variation of c in terms of α or α/l cannot be easily deduced theoretically; and the experiments carried out so far in subsonic or supersonic flow are not conclusive; hence we have been content with the approximate formula (10.154).

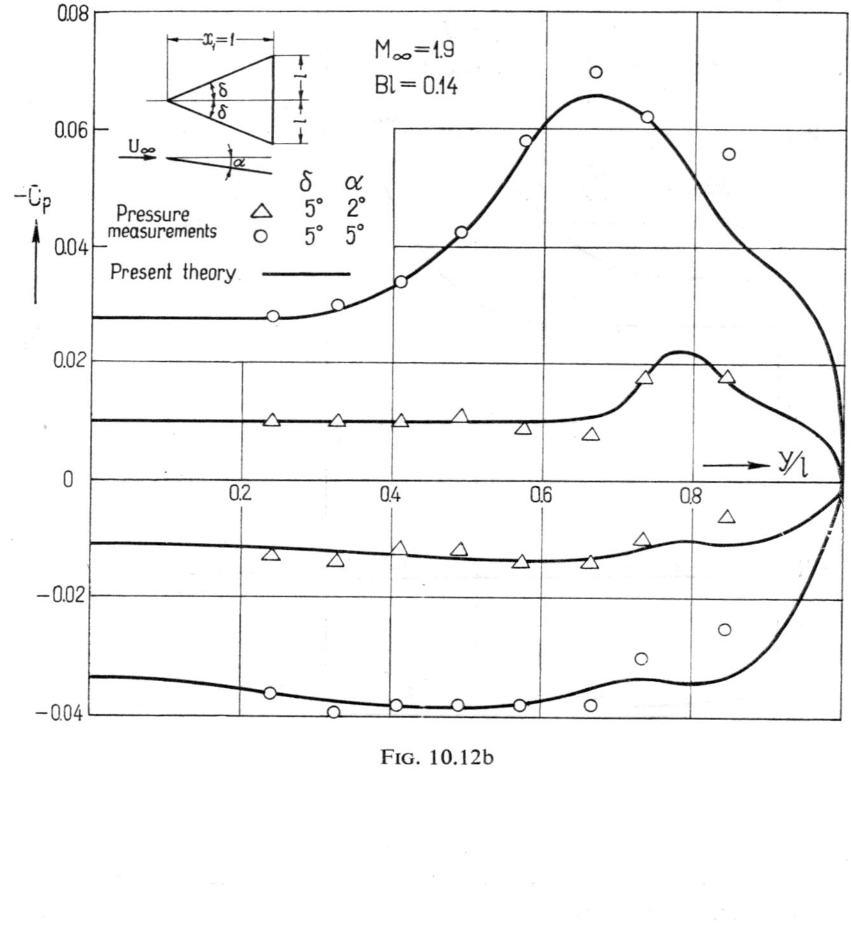

FIG. 10.12b

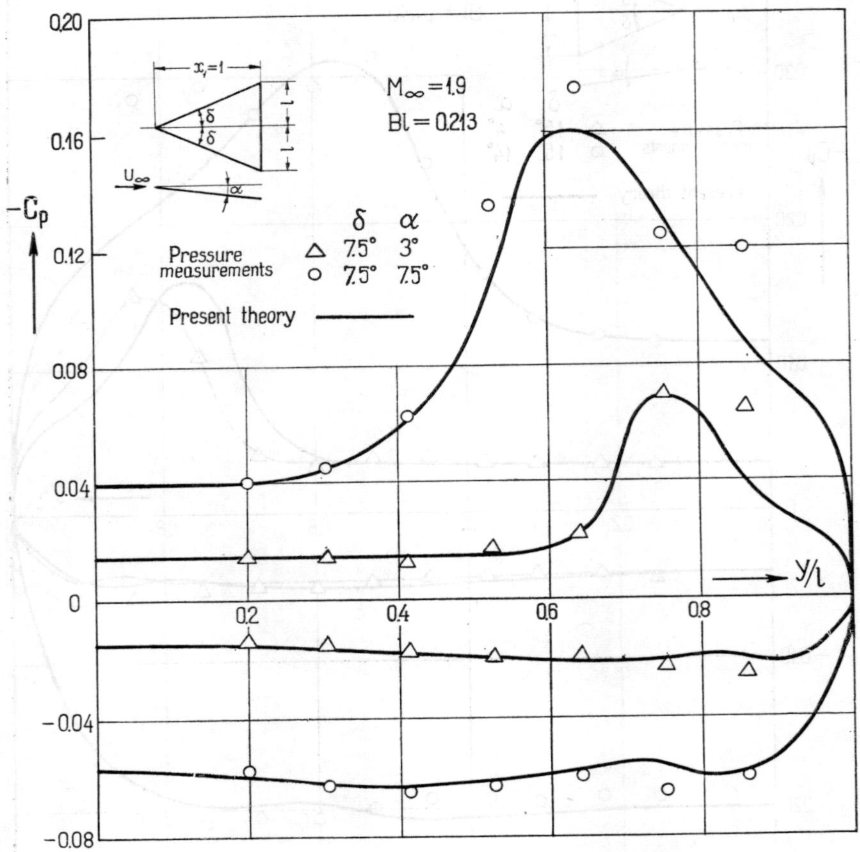
0.20
0.16
-Cp
0.12
0.08
0.04
0
-0.04
-0.08
0.2
0.4
0.6
0.8
y/l
$x_1=1$
δ
δ
l
l
U∞
α
M∞ = 1.9
Bl = 0.213
δ α
Pressure measurements
△ 7.5° 3°
○ 7.5° 7.5°
Present theory

FIG. 10.12c

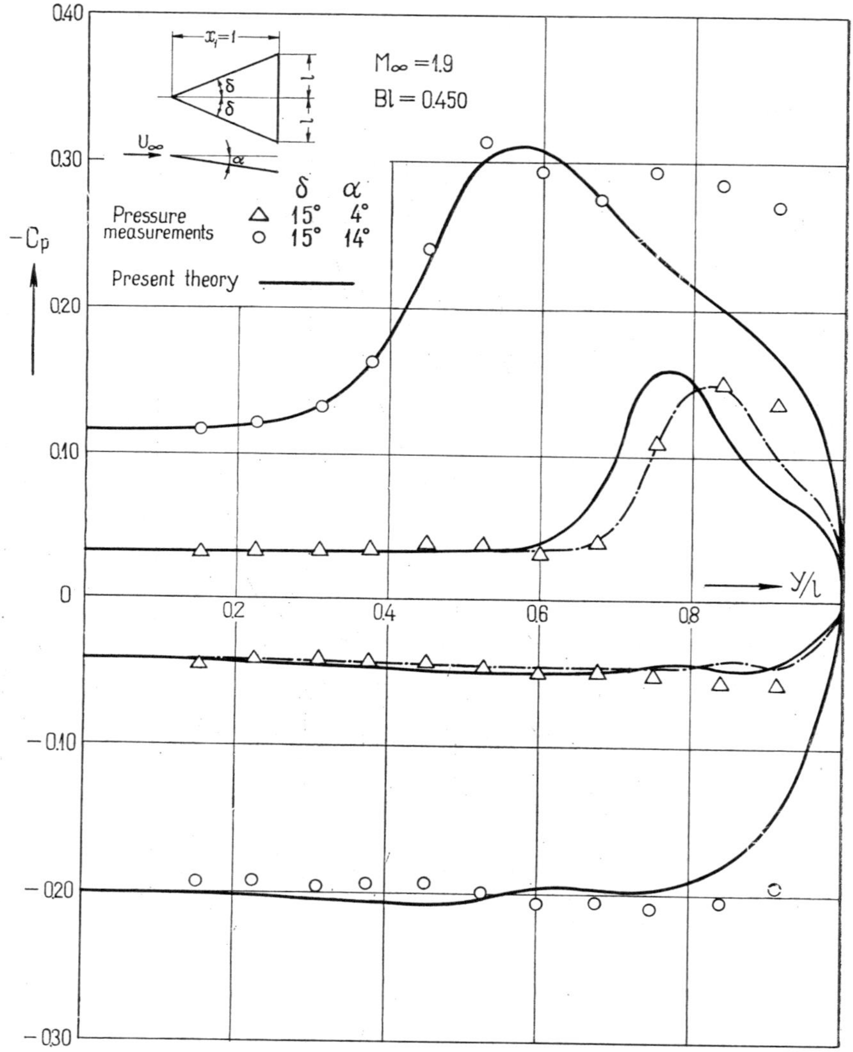

$M_\infty = 1.9$
$Bl = 0.450$
$x_i = 1$
δ
δ
1
1
U_∞
α
δ α
Pressure measurements
△ 15° 4°
○ 15° 14°
Present theory
$-C_p$
0.40
0.30
0.20
0.10
0
−0.10
−0.20
−0.30
0.2
0.4
0.6
0.8
y/l

FIG. 10.12d

CHAPTER XI

DETERMINATION OF PRESSURES AND AERODYNAMIC CHARACTERISTICS OF WINGS BEYOND THE ASSUMPTION OF SMALL DISTURBANCES, IN SUPERSONIC AND MODERATE HYPERSONIC FLOWS. EXPERIMENTAL VERIFICATIONS

IN THIS chapter we shall overstep the bounds of the small disturbances assumption and extend the calculation of pressures beyond this assumption, by establishing a more general formula which covers the supersonic and *moderate hypersonic* régimes for a wide range of Mach numbers and free stream deflexion.

On the other hand, in order to determine the degree of validity of the various formulae established in this way, a comparison between the theoretical values and the experimental results obtained in various researches will be made.

XI. 1. Determination of Pressure in Supersonic and Hypersonic Flows

XI. 1.1. Shock Wave Compression Formula valid for both Supersonic and Hypersonic Flows

We shall first establish the pressure formula in the case of oblique shock waves.

In this respect let τ be the angle of deflexion from the free stream, of velocity U_∞, and β the angle made by the plane shock wave with the direction of this stream (Fig. 11.1); behind this shock wave the velocity W has the deflexion τ. On both sides of the shock wave we have the components U_n, U_n' of the velocity normal to it and the components U_t, U_t' parallel to it, respectively, which will be written as a function of U_∞ and of the velocity components U and V beyond the shock wave:

$$U_n = U_\infty \sin\beta, \quad U'_n = U\sin\beta - V\cos\beta, \quad \frac{V}{U} = \tan\tau, \tag{11.1a}$$

$$U_t = U_\infty \cos\beta, \quad U'_t = U\cos\beta + V\sin\beta. \tag{11.1b}$$

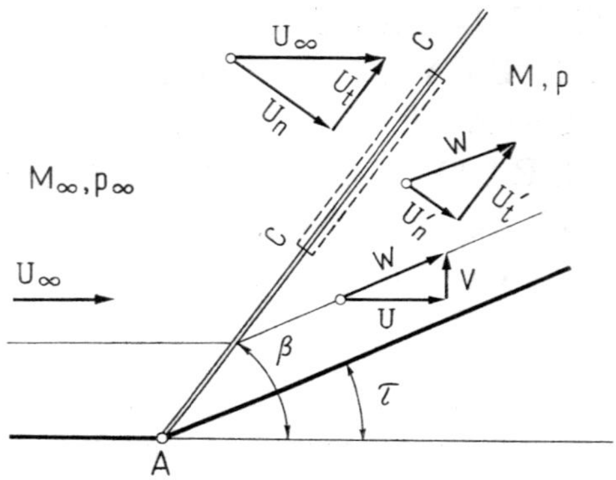

FIG. 11.1

We consider a control surface CC on the shock wave and apply the continuity equation, the momentum theorem as well as the pressure equation; we obtain successively:

$$\varrho_\infty U_\infty \sin\beta = \varrho(U\sin\beta - V\cos\beta), \tag{11.2a}$$

$$U_t = U'_t, \quad (U_\infty - U)\cos\beta = V\sin\beta, \quad \tan\beta = \frac{U_\infty - U}{V}, \tag{11.2b}$$

$$\begin{aligned} p - p_\infty &= \varrho_\infty U_\infty (U_n - U'_n) = \varrho_\infty U_\infty[(U_\infty - U)\sin^2\beta + V\sin\beta\cos\beta] \\ &= -\varrho_\infty U_\infty(U - U_\infty) = -\varrho[U(U_\infty - U) - V^2], \end{aligned} \tag{11.2c}$$

$$U_\infty^2 + \frac{2\gamma}{\gamma-1}\frac{p_\infty}{\varrho_\infty} = (U^2 + V^2) + \frac{2\gamma}{\gamma-1}\frac{p}{\varrho} = U_\infty^2 + \frac{2a_\infty^2}{\gamma-1}, \tag{11.2d}$$

$$U_n^2 + \frac{2\gamma}{\gamma-1}\frac{p_\infty}{\varrho_\infty} = U_n'^2 + \frac{2\gamma}{\gamma-1}\frac{p}{\varrho} = U_\infty^2 + \frac{2a_\infty^2}{\gamma-1} - U_t^2. \tag{11.2e}$$

From these equations, we obtain by simple operations the following formulae:

$$\frac{p-p_\infty}{p_\infty} = \frac{2\gamma}{\gamma+1}(M_\infty^2 \sin^2 \beta - 1), \tag{11.3a}$$

$$\frac{\varrho}{\varrho_\infty} = \frac{\gamma+1}{\gamma-1+\dfrac{2}{M_\infty^2 \sin^2 \beta}}, \tag{11.3b}$$

$$\left(\frac{\gamma-1}{2}+\frac{1}{M_\infty^2 \sin^2 \beta}\right)\left(\frac{\gamma-1}{2}+\frac{1}{M_\infty^2 \sin^2 (\beta-\tau)}\right) = \left(\frac{\gamma+1}{2}\right)^2, \tag{11.3c}$$

$$\tan(\beta-\tau) = \frac{2 \tan \beta}{\gamma+1}\left(\frac{1}{M_\infty^2 \sin^2 \beta}+\frac{\gamma-1}{2}\right), \tag{11.3d}$$

which will be used later on in this investigation.

From equation (11.2c) we get the pressure coefficient,

$$C_p = \frac{p-p_\infty}{\frac{1}{2}\varrho_\infty U_\infty^2} = -2\frac{U-U_\infty}{U_\infty}. \tag{11.4}$$

In the case of small perturbations, this formula reduces to that established previously (1.52) and used so far for calculating the pressures on the wing.

Further, we shall try to set C_p as a function of the stream deflexion τ. For this purpose on the one hand we write

$$\frac{V^2}{(U_\infty-U)^2} = \frac{\dfrac{2}{\gamma+1}(U_\infty^2-a_\infty^2)-U_\infty(U_\infty-U)}{\dfrac{2a_\infty^2}{\gamma+1}+U_\infty(U_\infty-U)}$$

$$= \frac{M_\infty^2-1-\dfrac{\gamma+1}{2}M_\infty^2\dfrac{(U_\infty-U)}{U_\infty}}{1+\dfrac{\gamma+1}{2}M_\infty^2\dfrac{(U_\infty-U)}{U_\infty}}, \tag{11.5}$$

and on the other hand, considering (11.1a), we write

$$\frac{V}{U_\infty} = \frac{U}{U_\infty}\tan\tau = \left(1-\frac{U_\infty-U}{U_\infty}\right)\tan\tau. \tag{11.6}$$

From these two equations we get the following relation between the pressure coefficient C_p and the deflexion τ:

$$\left(1+\frac{\gamma+1}{2}M_\infty^2\frac{C_p}{2}\right)^{1/2}\left(1-\frac{C_p}{2}\right)\tan\tau$$

$$= \frac{C_p}{2}\sqrt{M_\infty^2-1}\left(1-\frac{\gamma+1}{2}\frac{M_\infty^2}{M_\infty^2-1}\frac{C_p}{2}\right)^{1/2}. \tag{11.7}$$

Further, writing

$$m = \frac{M_\infty}{\sqrt{M_\infty^2-1}}, \quad \varkappa = \frac{1-\frac{1}{2}C_p}{(1-(\gamma+1/4)m^2C_p)^{1/2}}, \quad K' = m\varkappa M_\infty \tan\tau, \tag{11.8}$$

the above expression becomes

$$\frac{C_p}{m^2\varkappa^2\tan^2\tau} = \frac{\gamma+1}{2} \pm \sqrt{\left(\frac{\gamma+1}{2}\right)^2+\frac{4}{K'^2}} = \frac{\gamma+1}{2}+\frac{2}{K'}\sqrt{1+\left(\frac{\gamma+1}{4}\right)^2 K'^2}, \tag{11.9}$$

where the plus sign corresponds to the case of the shock wave proper ($\tau > 0$), and the minus sign would represent an *expansion* ($\tau < 0$) defined by the same equations (11.2).

Equation (11.9) cannot be used in the above form, since C_p is contained in $\varkappa$. For this reason, we shall try to see if the expression

$$Z = \varkappa^2 \tan^2\tau \tag{11.10}$$

could not be written as a function of τ only. In this respect, considering relation (11.9) with the plus (+) sign in the form

$$\frac{C_p}{m^2 Z} = \frac{\gamma+1}{2}+\sqrt{\left(\frac{\gamma+1}{2}\right)^2+\frac{4}{M_\infty^2 m^2 Z}} = P, \tag{11.11}$$

and having made the necessary substitutions, expression (11.10) takes the following form:

$$Z = \frac{(1-\frac{1}{2}C_p)^2}{1-\frac{(\gamma+1)m^2}{4}C_p}\tan^2\tau = \frac{\left(1-\frac{m^2PZ}{2}\right)^2}{1-\frac{(\gamma+1)m^2}{4}Pm^2Z}. \tag{11.12}$$

After performing several simple operations, we can write finally

$$Z = \frac{\sin^2\tau}{1-\left(1-\frac{m^2}{2}P\right)\sin^2\tau-\frac{1}{2}(1-R)\cos^2\tau} = \frac{\sin^2\tau}{1-\eta}, \tag{11.13}$$

where we have written

$$R = \sqrt{1-m^2P[m^2(\gamma+1)-2]\tan^2\tau}, \tag{11.14a}$$

$$\eta = \left(1-\frac{m^2}{2}P\right)\sin^2\tau+\frac{1}{2}(1-R)\cos^2\tau$$

$$= \left[1-\frac{m^2P}{2}\left(1-\frac{m^2(\gamma+1)-2}{1+R}\right)\right]\sin^2\tau. \tag{11.14b}$$

Let us now look for the value of η in the following cases defined approximately as follows:

1. $M_\infty \sin\tau \ll 1$ (supersonic flow under the assumption of small disturbances);
2. $M_\infty \sin\tau \sim 1$ (moderate hypersonic flow);
3. $M_\infty \sin\tau > 1$ (hypersonic flow) and in addition the limiting case, $M_\infty \sin\tau \to \infty$, (Newtonian flow (impact theory)).

For the last flow régime, we observe that the density ϱ behind the shock wave, increases substantially with respect to the density ϱ_∞ of the free stream; indeed, applying formula (11.3b), we have

$$\frac{\varrho}{\varrho_\infty} = \frac{\gamma+1}{\gamma-1+\dfrac{2}{M_\infty^2 \sin^2\beta}} \approx \frac{\gamma+1}{\gamma-1} \to \infty, \quad \gamma \to 1, \tag{11.15}$$

whence it follows that at such high régimes, the gas specific heats become equal. Introducing into (11.14) and (11.13) and observing that $m = 1$, we may write successively:

$$m = 1, \quad R = 1, \quad \eta \to 0, \quad Z = \sin^2\tau, \tag{11.16a}$$

$$C_p = 2\sin^2\tau; \tag{11.16b}$$

as is seen, we have thus obtained Newton's formula.

Let us now return to the supersonic régime and suppose first that we have

$$M_\infty m\sqrt{Z} = M_\infty m \frac{\sin\tau}{\sqrt{1-\eta}} = K' < \frac{4}{\gamma+1}; \quad (\sin\tau \ll 1). \tag{11.17}$$

In this case, we may expand the square root in (11.11) in the form

$$\frac{C_p(1-\eta)}{m^2\sin^2\tau} = \frac{\gamma+1}{2} + \frac{2}{K'} + \left(\frac{\gamma+1}{4}\right)^2 K' = P, \tag{11.18}$$

and further, successively deduce the following:

1. Under the assumption of small disturbances we have found previously that the pressure coefficient reduces to the linear relation (1.69),

$$C_p = -\frac{2}{B}\tau \approx -\frac{2}{B}\sin\tau, \tag{11.19}$$

whence it may seen that P reduces only to the term $2/K'$, and η, which would introduce secondary order terms, is considered as negligible.

2. In the case of *supersonic–hypersonic* flow, we find first

$$R = \sqrt{1-m^2P[m^2(\gamma+1)-2]\tan^2\tau} = 1-\frac{m^2P}{2}[m^2(\gamma+1)-2]\tan^2\tau$$

$$\approx 1-[m^2(\gamma+1)-2]\frac{m^2\tan^2\tau}{K'} \approx 1-[m^2(\gamma+1)-2]\frac{\sin\tau}{B\cos^2\tau}, \quad (11.20a)$$

$$1-R = \frac{m^2(\gamma+1)-2}{B}\frac{\sin\tau}{\cos^2\tau}, \quad (B = \sqrt{M_\infty^2-1}). \quad (11.20b)$$

Further, after performing the necessary substitutions into (11.14b) and neglecting the second-order terms, we have

$$\eta \approx \left[\frac{m^2(\gamma+1)}{2}-2\right]\frac{\sin\tau}{B}. \quad (11.21)$$

It should be noted now that the bracketed term vanishes for $M_\infty = 1{\cdot}6$, and for $M_\infty > 1{\cdot}6$ the coefficient of $\sin\tau$ has a small value ($\approx -0{\cdot}2$), so that η may be neglected. On the other hand, for smaller values of M_∞, the angle τ must be very small in order that the shock wave be attached at the vertex; in this case, however, as shown at item (1), we find (11.19) and hence we may put $\eta \approx 0$.

3. In hypersonic flow, considering

$$K' = M_\infty m\frac{\sin\tau}{\sqrt{1-\eta}} = M_\infty\frac{\sin\tau}{\sqrt{1-\eta}} > \frac{4}{\gamma+1}, \quad (M_\infty^2 \gg 1,\ m^2 \approx 1), \quad (11.22)$$

we perform the following expansion of the square root from (11.11);

$$P = \gamma+1+\frac{4}{\gamma+1}\frac{1}{K'^2}. \quad (11.23)$$

Introducing into R and observing that $M_\infty^2 \gg 1$ and $m^2 \approx 1$ for this régime, we find

$$\eta \approx \left(\frac{\gamma-1}{2}\right)^2\sin^2\tau-\frac{3-\gamma}{\gamma+1}\frac{1}{M_\infty^2} \approx 0 \quad (11.24)$$

since M_∞ is sufficiently large.

Accordingly, for all cases and for sufficiently large values of τ ($\tau \leqq 30°$) we may put $\eta = 0$ and hence write

$$K' \approx K = M_\infty m\sin\tau, \quad (11.25)$$

which will be termed the *unitary similarity parameter*, both for supersonic and hypersonic régimes, such that finally formula (11.11) becomes

$$\frac{C_p}{m^2\sin^2\tau} = \frac{\gamma+1}{2}\pm\sqrt{\left(\frac{\gamma+1}{2}\right)^2+\frac{4}{K^2}} = \frac{\gamma+1}{2}+\frac{2}{K}\sqrt{1+\left(\frac{\gamma+1}{4}\right)^2K^2}. \quad (11.26)$$

This formula is valid over a large interval, from the supersonic régime —proceeding first from the assumption of small disturbances ($K \ll 1$) and reaching then formula (11.19)—up to the high hypersonic régime which is obtained from the same formula, after replacing P by its expression given in (11.22),

$$C_p = \left(\frac{\gamma+1}{2} + \sqrt{\left(\frac{\gamma+1}{2}\right)^2 + \frac{4}{K^2}}\right) \sin^2 \tau \approx (\gamma+1) \sin^2 \tau + \frac{4}{\gamma+1} \frac{1}{M_\infty^2} . \tag{11.27}$$

Further we obtain Newton's formula if M_∞ tends to infinity and consequently γ tends to 1 (11.15).

Note. Formula (11.26) is only valid as long as the shock wave does not detach from the vertex of the wedge which causes the stream deflexion.

XI. 1.2. Compression-Expansion Unitary Formula for both Supersonic and Hypersonic Flows

Let us now consider the case in which the stream deflexion is negative ($\tau < 0$, Fig. 11.2): in such a case we have what is usually termed a Prandtl–Meyer expansion. The analysis of this expression is well known and for

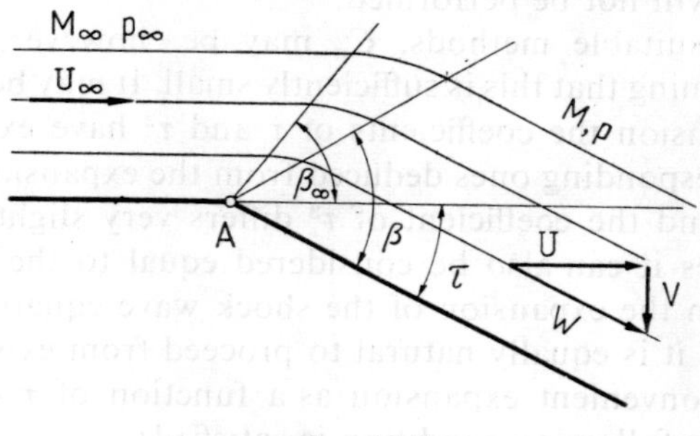

FIG. 11.2

for this reason we shall restrict ourselves to giving certain results useful in our subsequent considerations. Thus, by writing

$$\lambda^2 = \frac{\gamma-1}{\gamma+1}, \tag{11.28}$$

we obtain the following expression for the deflexion τ, as a function of the initial and final Mach numbers M_∞ and M:

$$\tau = \frac{1}{\lambda} [\tan^{-1} (\lambda \sqrt{M^2 - 1}) - \tan^{-1} (\lambda \sqrt{M_\infty^2 - 1})]$$
$$- [\tan^{-1} \sqrt{M^2 - 1} - \tan^{-1} \sqrt{M_\infty^2 - 1}]. \tag{11.29}$$

On the other hand, there exists the known relation between pressures and Mach numbers:

$$\frac{p}{p_\infty} = \frac{p}{p_0}\frac{p_0}{p_\infty} = \left[\frac{1+\frac{\gamma-1}{2}M_\infty^2}{1+\frac{\gamma-1}{2}M^2}\right]^{\frac{\gamma}{\gamma-1}} = \left[\frac{1+\lambda^2(M_\infty^2-1)}{1+\lambda^2(M^2-1)}\right]^{\frac{\gamma}{\gamma-1}} \tag{11.30}$$

whence, denoting by C'_{pe} the pressure coefficient due to expansion, we have

$$C'_{pe} = \frac{p-p_\infty}{p_\infty}\frac{2}{\gamma M_\infty^2} = \frac{2}{\gamma M_\infty^2}\left\{\left[\frac{1+\frac{\gamma-1}{2}M_\infty^2}{1+\frac{\gamma-1}{2}M^2}\right]^{\frac{\gamma}{\gamma-1}} - 1\right\}. \tag{11.31}$$

In principle, the expression for M can be obtained from (11.29) and introduced into (11.31), in order to obtain a relation between the pressure coefficient and the deflexion.

This operation is, however, extremely laborious and very complicated and therefore it will not be performed.

By means of suitable methods, C_p may be, however, expanded in powers of τ, assuming that this is sufficiently small. It may be demonstrated that in this expansion the coefficients of τ and τ^2 have exactly the same value as the corresponding ones deduced from the expansion of the shock wave equation, and the coefficient of τ^3 differs very slightly, so that for practical purposes it can also be considered equal to the corresponding one deduced from the expansion of the shock wave equation.

In such a case, it is equally natural to proceed from expression (11.26) and perform a convenient expansion as a function of τ or $\sin\tau$ or K, assuming that the following condition is satisfied:

$$|K| = M_\infty m\,|\sin\tau| < \frac{4}{\gamma+1}; \tag{11.32}$$

indeed, we obtain

$$C_p = m^2\sin^2\tau\left[\frac{\gamma+1}{2}+\frac{2}{K}+\left(\frac{\gamma+1}{4}\right)^2 K+\dots\right]$$

$$\approx \frac{2}{\sqrt{M_\infty^2-1}}\sin\tau+\frac{\gamma+1}{2}\frac{M_\infty^2}{M_\infty^2-1}\sin^2\tau+\left(\frac{\gamma+1}{4}\right)^2\frac{M_\infty^4}{(M_\infty^2-1)^{3/2}}\sin^3\tau. \tag{11.33}$$

Accordingly, this expression is valid both for compressions and expansions. However, it should be noted that this validity appears naturally from the exact solution of the problem, as has been established above in *closed form* (11.26), where the plus sign in the right-hand side stands for a compression and the minus sign for an expansion. Hence, it may be concluded that formula (11.33), valid both for compression and expansion, follows naturally from the two values of the general solution (11.26).

Accordingly, it may be further concluded that the two solutions of the pressure equation behind the shock waves represent almost rigorously the two phenomena of compression and expansion.

We should now examine to what extent the formula obtained above represent the Prandtl–Meyer expansion.

For angles τ ranging up to sufficiently large values ($\tau \leqq 30°$), but satisfying condition (11.32), expression (11.33) is valid for both phenomena, i.e. compression–expansion.

The general formula (11.26) is, however, also valid beyond the limit (11.32), that is, also for

$$|K| > \frac{4}{\gamma+1}\,. \tag{11.34}$$

In order to render this fact evident, this formula will be given another form; indeed, considering that at high Mach numbers ($M_\infty \gg 1$) we have

$$m^2 = \frac{M_\infty^2}{M_\infty^2 - 1} \approx 1, \tag{11.35}$$

the expansion coefficient obtained from the general formula (11.26) becomes

$$\begin{aligned} C_{pe} &= m^2 \sin^2 \tau \left(\frac{\gamma+1}{2} - \sqrt{\left(\frac{\gamma+1}{2}\right)^2 + \frac{4}{K^2}} \right) \\ &= -\frac{4}{(\gamma+1) M_\infty^2} \, \frac{2}{1 + \sqrt{1 + \left(\frac{4}{\gamma+1}\right)^2 \frac{1}{K^2}}}\,. \end{aligned} \tag{11.36}$$

We shall try to compare this last expression with the form of the Prandtl–Meyer expansion formula over another interval of K satisfying condition (11.34).

In this connection it should be noted that the introduction of the result of (11.29) into expression (11.31) for the coefficient C_{pe} so as to obtain a convenient formula in K is a cumbersome process; therefore, we shall restrict ourselves to showing that the two values of C_p (11.36) and (11.31) are very close for high Mach numbers ($M_\infty^2 \gg 1$). A brief examination shows in this case formula (11.31) becomes approximately

$$C'_{pe} \approx -\frac{2}{\gamma M_\infty^2}\,. \tag{11.37}$$

Hence, the following ratio

$$\frac{C'_{pe}}{C_{pe}} \approx \frac{\gamma+1}{4\gamma}\left(1+\sqrt{1+\left(\frac{4}{\gamma+1}\right)^2\frac{1}{K^2}}\right) \tag{11.38}$$

is obtained. Within the limits of the accepted approximations, this ratio is very close to unity, being actually equal to one around the value $K = 1{\cdot}8$ which fully corresponds to the case of hypersonic flow. Indeed, considering the incidence (or the wing slope) restricted to about 18°, $M_\infty = 6$ is correspondingly obtained.

This good agreement is confirmed in the graph of Fig. 11.3, where compression and expansion coefficient deduced from formula (11.26),

$$\frac{C_p}{m^2\sin^2\tau} = \frac{\gamma+1}{2}+\sqrt{\left(\frac{\gamma+1}{2}\right)^2+\frac{4}{K^2}} = P_+, \tag{11.39a}$$

$$\frac{C_p}{m^2\sin^2\tau} = \frac{\gamma+1}{2}-\sqrt{\left(\frac{\gamma+1^2}{2}\right)+\frac{4}{K^2}} = P_-, \tag{11.39b}$$

are compared with the corresponding values calculated according to shock wave and Prandtl–Meyer expansion tables. Further, if we take

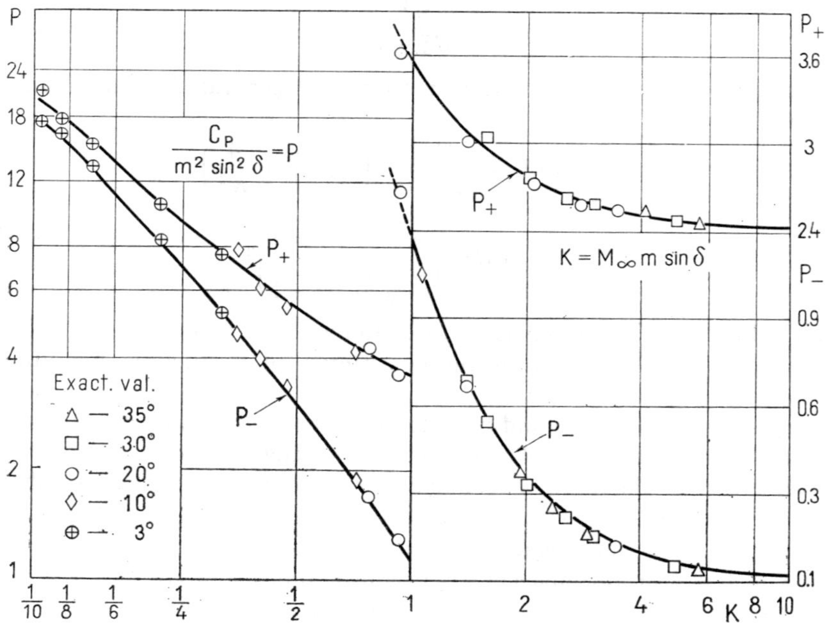

FIG. 11.3

account of the fact that at high K numbers the absolute value of the expansion coefficient tends to zero, and is negligible, then so much the more a slight difference between C_{pe} (11.36) and C'_{pe} (11.31) will be quite negligible.

In view of this fair agreement one may conclude, proceeding from formula (11.7) deduced from the theory of shock waves, that the stream expansion may be considered almost isentropic and hence depends only on the deflexion τ, a fact which is very convenient, as shown by the applications given below.

On the other hand, as regards the compression, since this is deduced from the shock wave phenomenon, it is perfectly valid *as long as the wave is attached to the wedge vertex.*

However, in contradistinction to an expansion, for a compression we must also consider the loss of total pressure behind the shock wave. As long as this loss of pressure is negligible—and this happens for τ angles up to 18° and Mach numbers ranging up to the values encountered in hypersonic flow—we may assume that the compression coefficient also depends only on the deflexion τ.

In this manner we obtain a *unitary compression–expansion formula valid over a wide supersonic–hypersonic range* which, written in various convenient forms, may be used for calculating the aerodynamic characteristics of wings in supersonic–hypersonic flow. We may thus write

$$\frac{C_p}{m^2 \sin^2 \tau} = \frac{\gamma+1}{2} + \frac{2}{K}\sqrt{1+\left(\frac{\gamma+1}{4}\right)^2 K^2}\,, \tag{11.40a}$$

$$C_p = \frac{\gamma+1}{2} m^2 \sin^2 \tau + \frac{2m \sin \tau}{M_\infty}\sqrt{1+\left(\frac{\gamma+1}{4}\right)^2 M_\infty^2 m^2 \sin^2 \tau}\,, \tag{11.40b}$$

$$M_\infty^2 C_p = \frac{\gamma+1}{2} K^2 + 2K\sqrt{1+\left(\frac{\gamma+1}{4}\right)^2 K^2}\,, \tag{11.40c}$$

$$C_p \approx \frac{2}{M_\infty} m \sin \tau + \frac{\gamma+1}{2}(m \sin \tau)^2 + \left(\frac{\gamma+1}{4}\right)^2 M_\infty (m \sin \tau)^3, \tag{11.40d}$$

$$M_\infty^2 C_p = 2\left(M_\infty^2 \frac{\sin \tau}{B}\right) + \frac{\gamma+1}{2}\left(M_\infty^2 \frac{\sin \tau}{B}\right)^2 + \left(\frac{\gamma+1}{4}\right)^2\left(M_\infty^2 \frac{\sin \tau}{B}\right)^3, \tag{11.40e}$$

which is in good agreement with the shock–expansion theory [L–34].

XI. 2. Aerodynamic Characteristics of Profiles

XI. 2.1. Straight Line Profiles (Double Wedge Profiles)

The application of the above formulae is simple and presents no difficulties for calculating the lift coefficient (C_l), the drag coefficient (C_d) and the pitching moment coefficient (μ_2).

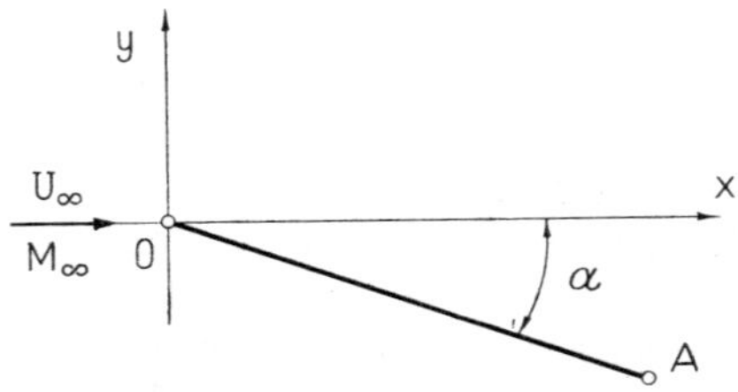

FIG. 11.4

THE FLAT PLATE (Fig. 11.4). Denoting by α the incidence, we obtain

$$C_l = 2m^2 \sin^2 \alpha \sqrt{\left(\frac{\gamma+1}{2}\right)^2 + \frac{4}{K^2}} \approx \frac{4\alpha}{\sqrt{M_\infty^2-1}}\left[1+\left(\frac{\gamma+1}{4}\right)^2 K^2\right]^{1/2},$$

$$(K = M_\infty m \sin \alpha \approx M_\infty m\alpha), \tag{11.41a}$$

$$C_d = C_l \tan\alpha \approx \alpha C_l \quad \mu_2 = \tfrac{1}{2}\frac{C_l}{\cos\alpha} \approx \tfrac{1}{2} C_l. \tag{11.41b}$$

THE PROFILE OF SYMMETRICAL THICKNESS (Fig. 11.5). Denoting by k the profile portion up to the maximum thickness which will be denoted by

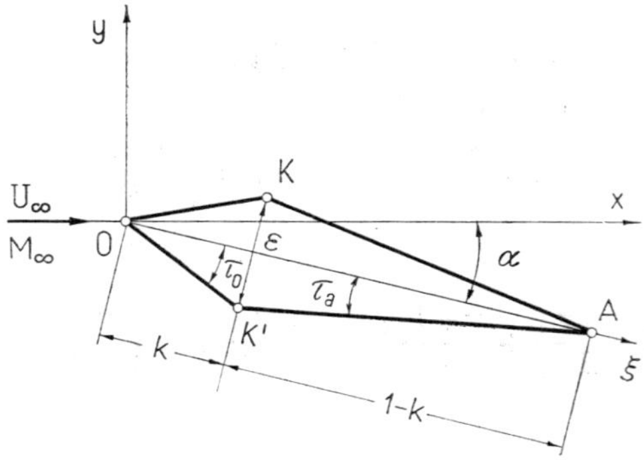

FIG. 11.5

ε, we may write

$$k\tau_0 = (1-k)\tau_a = \frac{\varepsilon}{2}. \tag{11.42}$$

Integrating over the separate portions, after some simplifications we obtain the following expressions for the aerodynamic characteristics:

$$C_l = [k(C_{pl}^{(o)} - C_{pu}^{(o)}) + (1-k)(C_{pl}^{(a)} - C_{pu}^{(a)})] \cos\alpha$$

$$\approx \frac{4\alpha}{\sqrt{M_\infty^2 - 1}} \left[1 + \tfrac{1}{2}\left(\frac{\gamma+1}{4}\right)^2 m^2 M_\infty^2 \left(\alpha^2 + \frac{3\varepsilon^2}{4k(1-k)}\right)\right], \tag{11.43a}$$

$$\mu_2 = \tfrac{1}{2}[k^2(C_{pl}^{(o)} - C_{pu}^{(o)}) + (1-k^2)(C_{pl}^{(a)} - C_{pu}^{(a)})]$$

$$\approx \frac{2\alpha}{\sqrt{M_\infty^2 - 1}} \left[1 + \tfrac{1}{2}\left(\frac{\gamma+1}{4}\right)^2 m^2 M_\infty^2 \left(\alpha^2 + \frac{3\varepsilon^2}{2(1-k)}\right)\right]$$

$$-\frac{\gamma+1}{2} m^2 \varepsilon\alpha, \tag{11.43b}$$

$$C_d = [k(C_{pl}^{(o)}\tau_{ol} + C_{pu}^{(o)}\tau_{ou}) + (1-k)(C_{pl}^{(a)}\tau_{al} + C_{pu}^{(a)}\tau_{au})] \cos\alpha$$

$$\approx \frac{4}{\sqrt{M_\infty^2 - 1}} \left[\alpha^2 + \frac{\varepsilon^2}{4k(1-k)} + \frac{1}{2}\left(\frac{\gamma+1}{4}\right)^2 m^2 M_\infty^2 \left(\alpha^4 + \frac{3\alpha^2\varepsilon^2}{2k(1-k)}\right.\right.$$

$$\left.\left. + \frac{\varepsilon^4(1-3k+3k^2)}{16k^3(1-k)^3}\right)\right] + \frac{(\gamma+1)m^2\varepsilon^2(1-2k)}{8k^2(1-k)^2}, \tag{11.43c}$$

where $C_{pl}^{(o)}$, $C_{pu}^{(o)}$, $C_{pl}^{(a)}$, and $C_{pu}^{(a)}$ are given by (11.40b), replacing τ respectively by

$$\tau_{ol} = \alpha + \tau_o, \quad \tau_{ou} = -\alpha + \tau_o, \quad \tau_{al} = \alpha - \tau_a, \quad \tau_{au} = -\alpha - \tau_a. \tag{11.43d}$$

For $k = {}^1/_2$ we obtain the symmetrical double wedge profile, with the following characteristics:

$$C_l = \frac{2\cos\alpha}{\sqrt{M_\infty^2 - 1}} \sum_{i=1}^{2} (-1)^i \sin(\varepsilon + \alpha_i) \left(\sqrt{1 + \left(\frac{\gamma+1}{4}\right)^2 m^2 M_\infty^2 \sin^2(\varepsilon + \alpha_i)}\right)$$

$$\approx \frac{4\alpha}{\sqrt{M_\infty^2 - 1}} \left[1 + \tfrac{1}{2}\left(\frac{\gamma+1}{4}\right)^2 m^2 M_\infty^2 (\alpha^2 + 3\varepsilon^2)\right], \tag{11.44a}$$

$$[\alpha_i = (-1)^i\alpha, \ i = 1,2],$$

$$\mu_2 = \frac{C_l}{2\cos\alpha} - \frac{\gamma+1}{8} m^2 [\sin^2(\varepsilon+\alpha) - \sin^2(\varepsilon-\alpha)] \approx \tfrac{1}{2} C_l - \frac{\gamma+1}{2} m^2 \varepsilon\alpha, \tag{11.44b}$$

$$C_d = \frac{2\cos\alpha}{\sqrt{M_\infty^2-1}} \sum_{i=0}^{2} \frac{\sin^2(\varepsilon+\alpha_i)}{\cos(\varepsilon+\alpha_i)} \sqrt{1+\left(\frac{\gamma+1}{4}\right)^2 m^2 M_\infty^2 \sin^2(\varepsilon+\alpha_i)}$$

$$\approx \frac{4}{\sqrt{M_\infty^2-1}}\left[\alpha^2+\varepsilon^2+\tfrac{1}{2}\left(\frac{\gamma+1}{4}\right)^2 m^2 M_\infty^2(\alpha^4+6\alpha^2\varepsilon^3+\varepsilon^4)\right] \approx C_{do}+\alpha C_l, \tag{11.44c}$$

where

$$C_{d0} = \frac{4\varepsilon^2}{\sqrt{M_\infty^2-1}}\left[1+\tfrac{1}{2}\left(\frac{\gamma+1}{4}\right)^2 m^2 M_\infty^2 \varepsilon^2\right] \tag{11.45}$$

is the drag coefficient for $\alpha = 0$.

THE PROFILE OF ANTISYMMETRICAL THICKNESS (Fig. 11.6). Writing

$$k\tau_a = (1-k)\tau_0 = \frac{\varepsilon}{2} \tag{11.46}$$

we obtain successively:

$$C_l = [(1-k)(C_{pl}^{(o)}-C_{pu}^{(a)})+k(C_{pl}^{(a)}-C_{pu}^{(o)})]\cos\alpha$$

$$\approx \frac{4}{\sqrt{M_\infty^2-1}}\left[\alpha+\tfrac{1}{2}\left(\frac{\gamma+1}{4}\right)^2 m^2 M_\infty^2\left(\alpha^3+\frac{3\alpha\varepsilon^2}{4k(1-k)}+\frac{\varepsilon^3(1-2k)}{8k^2(1-k)^2}\right)\right], \tag{11.47a}$$

$$\mu_2 = \tfrac{1}{2}[C_{pl}^{(a)}-C_{pu}^{(a)}+(1-k)^2(C_{pl}^{(o)}-C_{pl}^{(a)})-k^2(C_{pu}^{(o)}-C_{pu}^{(a)})]$$

$$\approx \tfrac{1}{2}C_l-\frac{\gamma+1}{2}m^2\varepsilon\left[\alpha-\frac{\varepsilon(1-2k)}{4k(1-k)}\right], \tag{11.47b}$$

$$C_d = [(1-k)(C_{pl}^{(o)}\tau_{ol}+C_{pu}^{(a)}\tau_{au})+k(C_{pl}^{(a)}\tau_{al}+C_{pu}^{(o)}\tau_{ou})]\cos\alpha$$

$$\approx \frac{4}{\sqrt{M_\infty^2-1}}\left[\alpha^2+\tfrac{1}{2}\left(\frac{\gamma+1}{2}\right)^2 m^2 M_\infty^2\left(\alpha^4+\frac{3\alpha^2\varepsilon^2}{2k(1-k)}-\frac{\alpha\varepsilon^3(1-k)}{2k^2(1-k)^2}\right.\right.$$

$$\left.\left.+\frac{\varepsilon^4(1-3k+3k^2)}{16k^3(1-k)^3}\right)\right], \tag{11.47c}$$

$$\tau_{ou} = -\alpha+\tau_a, \quad \tau_{au} = -\alpha-\tau_0, \quad \tau_{ol} = \alpha+\tau_0, \quad \tau_{al} = \alpha-\tau_a, \tag{11.47d}$$

where $C_{pu}^{(o)}$, $C_{pu}^{(a)}$, $C_{pl}^{(o)}$ and $C_{pl}^{(a)}$ are given by relation (11.40b), replacing τ respectively by (11.47d).

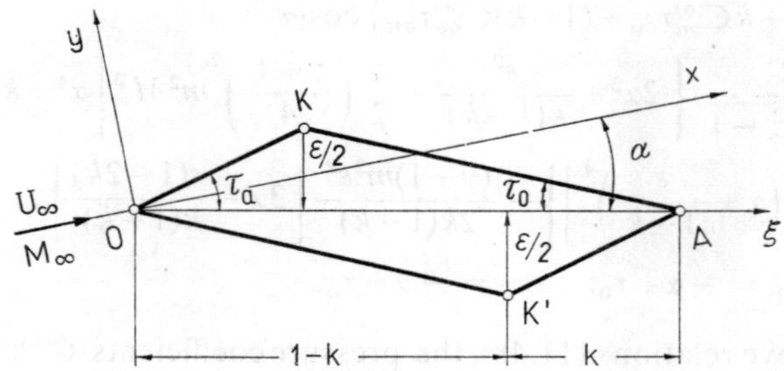

FIG. 11.6

THE PROFILE WITH FLAT LOWER SURFACE (Fig. 11.7). Using the same symbols as above and making the approximations which are usual for small incidences, we obtain:

$$C_l = [C_{pl}^{(o)} - kC_{pu}^{(o)} - (1-k)C_{pu}^{(a)}]\cos\alpha$$

$$\approx \frac{2}{\sqrt{M_\infty^2-1}}\left\{2\alpha + \tfrac{1}{2}\left(\frac{\gamma+1}{4}\right)^2 m^2 M_\infty^2 \left[\alpha^3 + k\left(\alpha - \frac{\varepsilon}{k}\right)^3 + (1-k)\left(\alpha + \frac{\varepsilon}{1-k}\right)^3\right]\right\} - \frac{\gamma+1}{2} m^2 \frac{\varepsilon^2}{k(1-k)}, \tag{11.48a}$$

$$\mu_2 = \tfrac{1}{2}[C_{pl}^{(o)} - k^2 C_{pu}^{(o)} - (1-k^2)C_{pu}^{(a)}] \approx \frac{1}{\sqrt{M_\infty^2-1}}\left\{2\alpha + \varepsilon + \tfrac{1}{2}\left(\frac{\gamma+1}{4}\right)^2 m^2 M_\infty^2 \left[\alpha^3 + k^2\left(\alpha - \frac{\varepsilon}{k}\right)^3 + (1-k^2)\left(\alpha + \frac{\varepsilon}{1-k}\right)^3\right]\right\} - \frac{\gamma+1}{2} m^2 \varepsilon \left(\alpha + \frac{\varepsilon}{1-k}\right), \tag{11.48b}$$

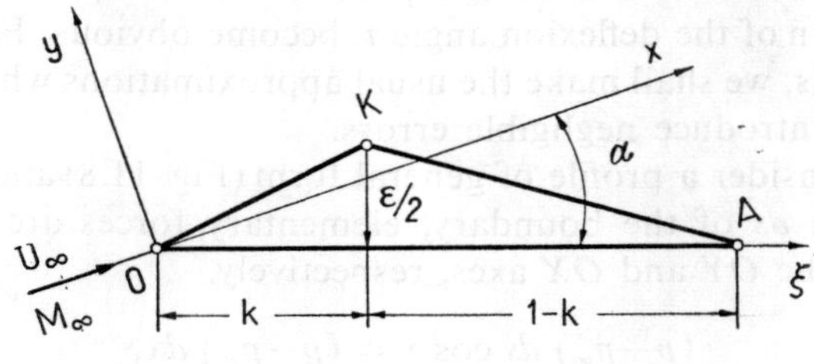

FIG. 11.7

$$C_d = \left[C_{pl}^{(o)}\alpha + kC_{pu}^{(o)}\tau_{ou} + (1-k)C_{pu}^{(a)}\tau_{au}\right]\cos\alpha$$

$$\approx \frac{2}{\sqrt{M_\infty^2-1}}\left\{2\alpha^2 + \frac{\varepsilon^2}{k(1-k)} + \frac{1}{2}\left(\frac{\gamma+1}{4}\right)^2 m^2 M_\infty^2\left[\alpha^4 + k\left(\alpha - \frac{\varepsilon}{k}\right)^4\right.\right.$$

$$\left.\left. + (1-k)\left(\alpha + \frac{\varepsilon}{1-k}\right)^4\right]\right\} - \frac{(\gamma+1)m^2\varepsilon^2}{2k(1-k)}\left[3\alpha - \frac{\varepsilon(1-2k)}{k(1-k)}\right], \quad (11.48c)$$

$$\tau_{ol} = \alpha, \quad \tau_{ou} = -\alpha + \tau_0, \quad \tau_{au} = -\alpha - \tau_a. \quad (11.48d)$$

In the above relations (11.48), the pressure coefficients $C_{pl}^{(o)}$, $C_{pu}^{(o)}$ and $C_{pu}^{(a)}$ are given by (11.40 b), replacing τ respectively by (11.48 d). The approximate expressions of the above formulae are valid for moderate values of M_∞ and α ($|K| < 1{\cdot}6$).

XI. 2.2. Cambered Profiles

In the following we shall establish formulae for calculating the aerodynamic characteristics of parabolic profiles. In such cases, as will be seen below, the advantages of applying the above formulae (11.40) which give

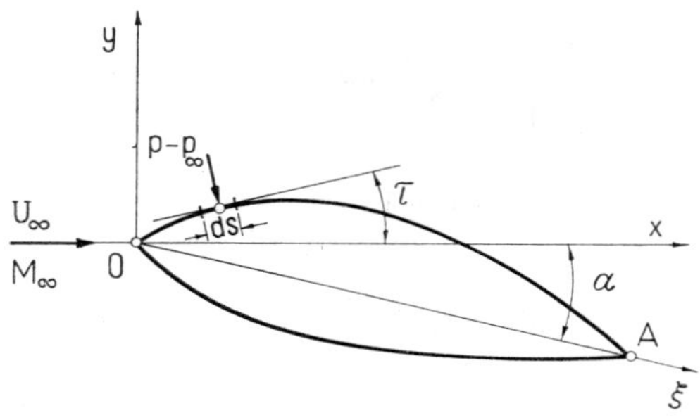

FIG. 11.8

the pressure distribution, both in expansion and compression, depending only on the sign of the deflexion angle τ, become obvious. For simplifying the calculations, we shall make the usual approximations which, under our assumptions, introduce negligible errors.

We shall consider a profile of general form (Fig. 11.8) and suppose that on an element ds of the boundary, elementary forces are acting in the directions of the OY and OX axes, respectively,

$$(p-p_\infty)\, ds\cos\tau = (p-p_\infty)\, dx, \quad (11.49a)$$

$$(p-p_\infty)\, ds\sin\tau = (p-p_\infty)\tan\tau\, dx. \quad (11.49b)$$

Adopting now the indices *u* and *l* for the upper and lower surfaces, respectively, introducing the pressure coefficient and taking the profile chord equal to unity ($\overline{OA} = 1$), we obtain successively:

the *lift coefficient*,

$$C_l = \int_0^{\cos\alpha} (C_{pl} - C_{pu})\, dx, \tag{11.50}$$

the *moment coefficient* with respect to the leading edge,

$$\mu_2 = \int_0^{\cos\alpha} (C_{pl} - C_{pu})(x\, dx + y\, dy) \approx \int_0^1 (C_{pl} - C_{pu}) x\, dx \tag{11.51}$$

and the *wave drag coefficient*,

$$C_d = \int_0^{\cos\alpha} (C_{pi} \tan \tau_l + C_{pu} \tan \tau_u) dx \approx \int_0^1 C_{pl}\tau_l + C_{pu}\tau_u)\, dx. \tag{11.52}$$

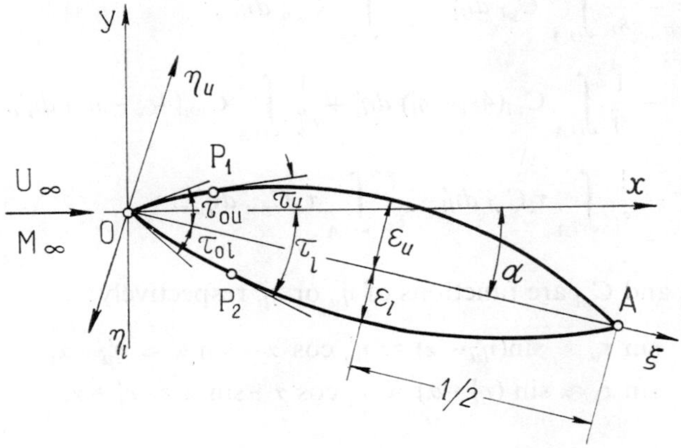

FIG. 11.9

For our subsequent calculations it is more advantageous to use the system $\xi O \eta$ rotated through an angle α with respect to the system XOY. In this system, considering as above the chord to be equal to unity, the equation of the parabolic profile contour may be written in the form

$$\eta_u = 4\varepsilon_u \xi(1-\xi), \quad \eta_l = 4\varepsilon_l \xi(1-\xi), \tag{11.53}$$

where ε_u and ε_l stand for the profile thickness with respect to the chord (Fig. 11.9).

It is to be noticed that *dx* and *xdx* may be replaced on the upper and lower surfaces, respectively, as functions of the tangent angular coefficient, as follows:

$$dx = -\frac{\cos\alpha}{8\varepsilon_u}(1+\eta'_u \tan\alpha)\, d\eta'_u \approx -\frac{d\eta'_u}{8\varepsilon_u}, \quad xdx = -\frac{4\varepsilon_u-\eta'_u}{64\varepsilon_u^2}\, d\eta',$$

$$\left(\eta'_u = \frac{d\eta_u}{d\xi}\right), \tag{11.54a}$$

$$dx = -\frac{\cos\alpha}{8\varepsilon_l}(1-\eta'_l \operatorname{tg}\alpha)\, d\eta'_l \approx -\frac{d\eta'_l}{8\varepsilon_l}, \quad xdx \approx -\frac{4\varepsilon_l-\eta'_l}{64\varepsilon_l^2}\, d\eta'_l,$$

$$\left(\eta'_l = \frac{d\eta_l}{d\xi}\right). \tag{11.54b}$$

In this way we obtain from equations (11.50), (11.51) and (11.52) expressions for calculating the aerodynamic characteristics of parabolic profiles

$$8C_l = -\frac{1}{\varepsilon_l}\int_{OA} C_{pl}\, d\eta'_l + \frac{1}{\varepsilon_u}\int_{OA} C_{pu}\, d\eta'_u, \tag{11.55a}$$

$$64\mu_2 = -\frac{1}{\varepsilon_l^2}\int_{OA} C_{pl}(4\varepsilon_l-\eta'_l)\, d\eta'_l + \frac{1}{\varepsilon_u^2}\int_{OA} C_{pu}(4\varepsilon_u-\eta'_u)\, d\eta'_u, \tag{11.55b}$$

$$8C_d = -\frac{1}{\varepsilon_l}\int_{OA} \tau_l C_{pl}\, d\eta'_l - \frac{1}{\varepsilon_u}\int_{OA} \tau_u C_{pu}\, d\eta'_u, \tag{11.55c}$$

where C_{pu} and C_{pl} are functions of η'_u or η'_l, respectively:

$$\sin\tau_u \approx \sin(\eta'_u-\alpha) \approx \eta'_u\cos\alpha - \sin\alpha \approx \eta'_u-\alpha, \tag{11.56a}$$

$$\sin\tau_l \approx \sin(\eta'_l+\alpha) \approx \eta'_l\cos\alpha + \sin\alpha \approx \eta'_l+\alpha, \tag{11.56b}$$

η'_u and η'_l being small.

DOUBLY PARABOLIC PROFILES (Fig. 11.9). Applying directly formulae (11.55), we obtain

$$C_l = \frac{\cos^2\alpha}{4\sqrt{M_\infty^2-1}}\left(\frac{1}{\varepsilon_l}H_{l1}+\frac{1}{\varepsilon_u}H_{u1}\right) - \tfrac{8}{3}(\gamma+1)m^2(\varepsilon_u^2-\varepsilon_l^2)\cos^3\alpha$$

$$\approx \frac{4\alpha}{\sqrt{M_\infty^2-1}}\left\{1+\tfrac{1}{2}\left(\frac{\gamma+1}{4}\right)^2 m^2M_\infty^2[\alpha^2+8(\varepsilon_u^2+\varepsilon_l^2)]\right\}$$

$$-\tfrac{8}{3}(\gamma+1)\, m^2(\varepsilon_u^2-\varepsilon_l^2), \tag{11.57a}$$

$$\mu_2 = \frac{\cos\alpha}{32\sqrt{M_\infty^2-1}}\left\{\frac{1}{\varepsilon_l^2}\left[H_{l2}+(4\varepsilon_l+\alpha)H_{l1}\right]+\frac{1}{\varepsilon_u^2}\left[-H_{u2}+(4\varepsilon_u-\alpha)H_{u1}\right]\right\}$$

$$-\tfrac{2}{3}(\gamma+1)m^2[2(\varepsilon_u^2-\varepsilon_l^2)+\alpha(\varepsilon_u+\varepsilon_l)] \approx \frac{2}{\sqrt{M_\infty^2-1}}\left\{\alpha+\tfrac{2}{3}(\varepsilon_u-\varepsilon_l)+\right.$$

$$\left.+\tfrac{1}{2}\left(\frac{\gamma+1}{4}\right)^2 m^2M_\infty^2\left[\alpha^3+2\alpha^2(\varepsilon_u-\varepsilon_l)+8\alpha(\varepsilon_u^2+\varepsilon_l^2)+\tfrac{32}{5}(\varepsilon_u^3-\varepsilon_l^3)\right]\right\}$$

$$-\tfrac{2}{3}(\gamma+1)m^2[2(\varepsilon_u^2-\varepsilon_l^2)+\alpha(\varepsilon_u+\varepsilon_l)], \tag{11.57b}$$

$$C_d = -\frac{\cos^2\alpha}{4\sqrt{M_\infty^2-1}}\left(\frac{1}{\varepsilon_u}H_{u2}+\frac{1}{\varepsilon_l}H_{l2}\right)-8(\gamma+1)m^2(\varepsilon_u^2-\varepsilon_l^2)\cos^3\alpha\tan\alpha$$

$$\approx \frac{4}{\sqrt{M_\infty^2-1}}\left\{\alpha^2+\tfrac{8}{3}(\varepsilon_u^2+\varepsilon_l^2)+\tfrac{1}{2}\left(\frac{\gamma+1}{4}\right)^2 m^2M_\infty^2\left[\alpha^4+16\alpha^2(\varepsilon_u^2+\varepsilon_l^2)\right.\right.$$

$$\left.\left.+\tfrac{128}{5}(\varepsilon_u^4+\varepsilon_l^4)\right]\right\}-8(\gamma+1)m^2(\varepsilon_u^2-\varepsilon_l^2)\alpha, \tag{11.57c}$$

where we have denoted, respectively:

$$H_{u1} = \frac{1}{3a^2}(R_{0u}^3-R_{au}^3), \tag{11.58a}$$

$$H_{u2} = -\frac{1}{8a^2}\left[(2R_{0u}^2-1)R_{0u}\alpha_{0u}+(2R_{au}^2-1)R_{au}\alpha_{au}+\frac{1}{a}\ln\frac{R_{0u}-a\alpha_{0u}}{R_{au}+a\alpha_{au}}\right], \tag{11.58b}$$

$$R_{0u} = \sqrt{1+a^2\alpha_{0u}^2}, \quad R_{au} = \sqrt{1+a^2\alpha_{au}^2},$$

$$\alpha_{0u} = \alpha+4\varepsilon_u, \quad \alpha_{au} = -\alpha+4\varepsilon_u, \quad a^2 = \left(\frac{\gamma+1}{4}\right)^2 m^2M_\infty^2\cos^2\alpha, \tag{11.58c}$$

$$H_{l1} = \frac{1}{3a^2}(R_{0l}^3-R_{al}^3), \tag{11.58d}$$

$$H_{l2} = -\frac{1}{8a^2}\left[(2R_{0l}^2-1)R_{0l}\alpha_{0l}+(2R_{al}^2-1)R_{al}\alpha_{al}+\frac{1}{a}\ln\frac{R_{0l}-a\alpha_{0l}}{R_{al}+a\alpha_{al}}\right], \tag{11.58e}$$

$$R_{0l} = \sqrt{1+a^2\alpha_{0l}^2}, \quad R_{al} = \sqrt{1+a^2\alpha_{al}^2}, \quad \alpha_{0l} = \alpha+4\varepsilon_l, \quad \alpha_{al} = -\alpha+4\varepsilon_l. \tag{11.58f}$$

Denoting by ε the profile total thickness ($\varepsilon = \varepsilon_u + \varepsilon_l$) and considering a symmetrical doubly parabolic profile ($\varepsilon_u = \varepsilon_l = \varepsilon/2$) we obtain

$$C_l = \frac{\cos^2\alpha}{\varepsilon\sqrt{M_\infty^2-1}} H_1 \approx \frac{4\alpha}{\sqrt{M_\infty^2-1}}\left[1+\tfrac{1}{2}\left(\frac{\gamma+1}{4}\right)^2 m^2 M_\infty^2(\alpha^2+4\varepsilon^2)\right], \tag{11.59a}$$

$$\mu_2 = \frac{C_l}{2\cos\alpha} - \tfrac{2}{3}(\gamma+1)m^2\varepsilon\tan\alpha \approx \tfrac{1}{2}C_l - \tfrac{2}{3}(\gamma+1)m^2\varepsilon\alpha, \tag{11.59b}$$

$$C_d = -\frac{\cos^2\alpha}{\varepsilon\sqrt{M_\infty^2-1}} H_2 \approx \frac{4}{\sqrt{M_\infty^2-1}}\left[\alpha^2+\tfrac{4}{3}\varepsilon^2+\tfrac{1}{2}\left(\frac{\gamma+1}{4}\right)^2 m^2 M_\infty^2 \right.$$
$$\left.\times(\alpha^4+8\alpha^2\varepsilon^2+\tfrac{16}{5}\varepsilon^4)\right] = C_{d0}+\alpha C_l + 2\left(\frac{\gamma+1}{2}\right)^2 m^2 M_\infty^2\varepsilon^2\alpha^2 \approx C_{d0}+\alpha C_l \tag{11.59c}$$

where H_1 and H_2 are given by (11.58a) and (11.58b), respectively, replacing ε_u by $\frac{1}{2}\varepsilon$, and C_{d0} is the wave drag coefficient for $\alpha = 0$.

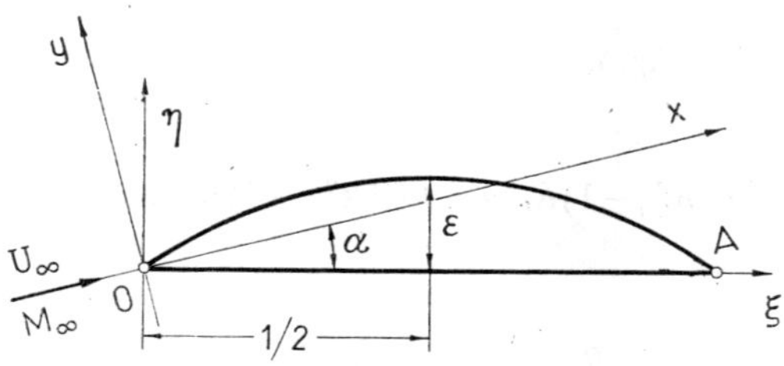

FIG. 11.10

PLANE PARABOLIC PROFILES (Fig. 11.10). The respective characteristics may be obtained from the doubly parabolic profiles by setting $\varepsilon_u = \varepsilon$ and $\varepsilon_l = 0$.

In this way, formulae (11.57) become respectively:

$$C_l = C_{pl}^{(0)}\cos\alpha + \frac{\cos^2\alpha}{4\varepsilon\sqrt{M_\infty^2-1}} H_{u1} - \tfrac{8}{3}(\gamma+1)m^2\varepsilon^2\cos^3\alpha$$
$$\approx \frac{4\alpha}{\sqrt{M_\infty^2-1}}\left[1+\tfrac{1}{2}\left(\frac{\gamma+1}{4}\right)^2 m^2 M_\infty^2(\alpha^2+8\varepsilon^2)\right] - \tfrac{8}{3}(\gamma+1)m^2\varepsilon^2, \tag{11.60a}$$

$$\mu_2 = \tfrac{1}{2}C_{pl}^{(0)} + \frac{\cos\alpha}{32\varepsilon^2\sqrt{M_\infty^2-1}}[-H_{u2}+(4\varepsilon-\alpha)H_{u1}]$$

$$-\tfrac{2}{3}(\gamma+1)m^2\varepsilon(\alpha+2\varepsilon)\cos^2\alpha \approx \frac{2}{\sqrt{M_\infty^2-1}}\Bigg[\alpha+\tfrac{2}{3}\varepsilon$$

$$+\tfrac{1}{2}\left(\frac{\gamma+1}{4}\right)^2 m^2M_\infty^2(\alpha^3+2\alpha^2\varepsilon+8\alpha\varepsilon^2+\tfrac{32}{5}\varepsilon^3)\Bigg]-\tfrac{2}{3}(\gamma+1)m^2\varepsilon(\alpha+2\varepsilon), \tag{11.60b}$$

$$C_d = C_{pl}^{(0)}\sin\alpha - \frac{\cos^2\alpha}{4\varepsilon\sqrt{M_\infty^2-1}}H_{u2} - 8(\gamma+1)m^2\varepsilon^2\cos^3\alpha\tan\alpha$$

$$\approx \frac{4}{\sqrt{M_\infty^2-1}}\left[\alpha^2+\tfrac{8}{3}\varepsilon^2+\tfrac{1}{2}\left(\frac{\gamma+1}{4}\right)^2 m^2M_\infty^2(\alpha^4+16\alpha^2\varepsilon^2+\tfrac{128}{5}\varepsilon^4)\right]$$

$$-8(\gamma+1)m^2\varepsilon^2\alpha, \tag{11.60c}$$

where $C_{pl}^{(0)}$ results from the relation (11.40b) for $\tau = \alpha$, and H_{u1}, H_{u2}, are given by (11.58a) and (11.58b), respectively, for $\varepsilon_u = \varepsilon$.

Note. The approximate expressions of the above formulae are valid for moderate Mach numbers and incidences ($|K| < 1{\cdot}6$).

XI. 2.3. Comparison with Experimental Data and Final Remarks

In Figs. 11.11, 11.12, 11.13, the pressure coefficient calculated with the aid of formulae (11.40) is compared with experimental data for various angles of incidence (α) and Mach numbers (M_∞). Both curved and double wedge profiles have been considered. One may observe very good agreement of the theoretical pressure distribution with experimental tests, except in a small area situated near the trailing edge of the profile. This disagreement is due to the effect of boundary layer separation, which causes an alteration of the pressure distribution transmitted on to the profile through the boundary layer, where the velocity is subsonic. The effect increases with increasing angle of attack and is more important in the case of an expansion. To show the nonlinear effects, the pressure curve to a first approximation, deduced from the assumption of small disturbances, has also been plotted. In Figs. 11.11 and 11.12, the experimental data have been taken from [F–15].

Figure 11.13 is a plot of the pressure coefficient C_p on a diamond-shaped profile versus the incidence angle α at a Mach number $M_\infty = 6{\cdot}9$, that is, in the high supersonic régime. One may conclude that for this régime, too, the agreement between theory and experimental given by data [L–23] is fairly good.

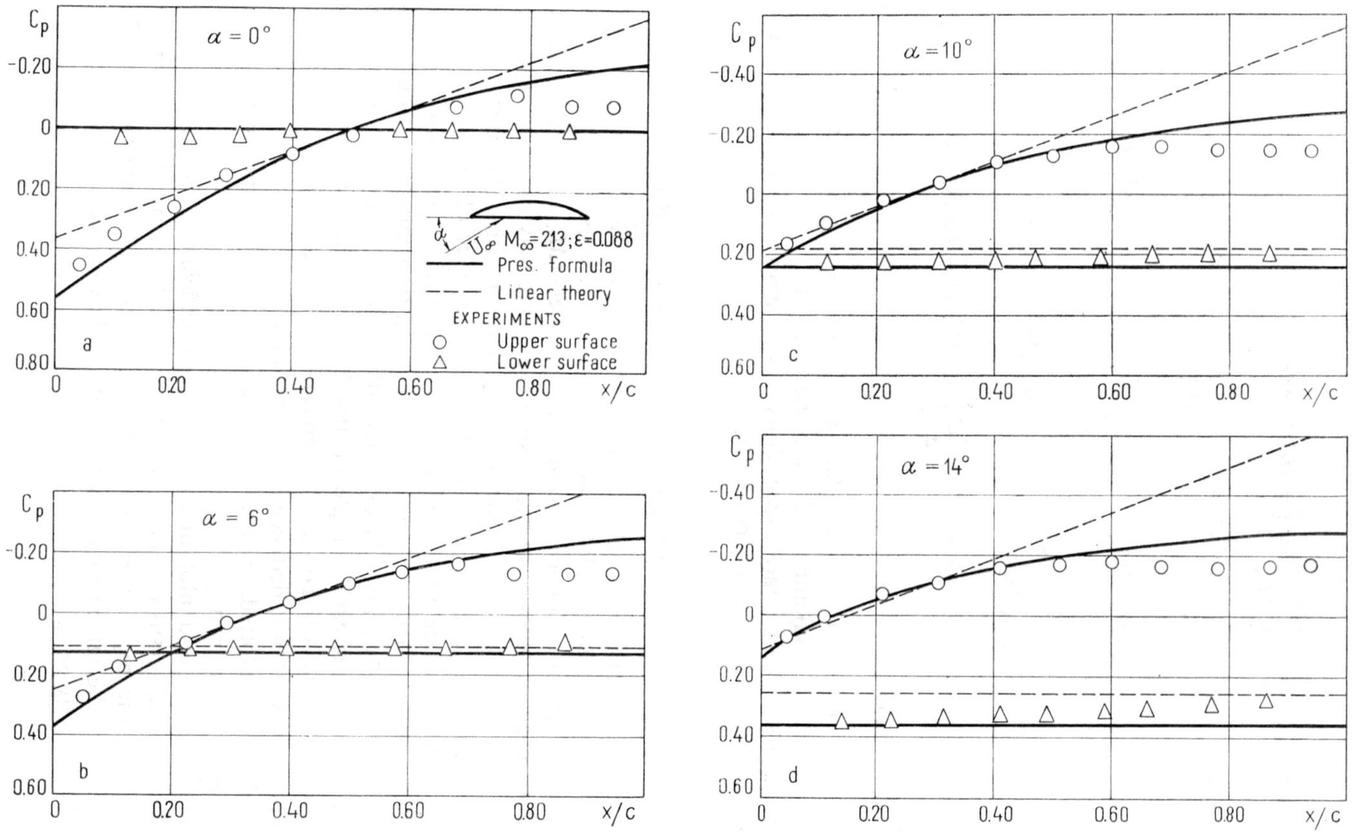

C_p
$\alpha = 0°$
a
$M_\infty = 2.13$; $\varepsilon = 0.088$
Pres. formula
Linear theory
EXPERIMENTS
Upper surface
Lower surface
$\alpha = 6°$
b
$\alpha = 10°$
c
$\alpha = 14°$
d
x/c

Fig. 11.11.

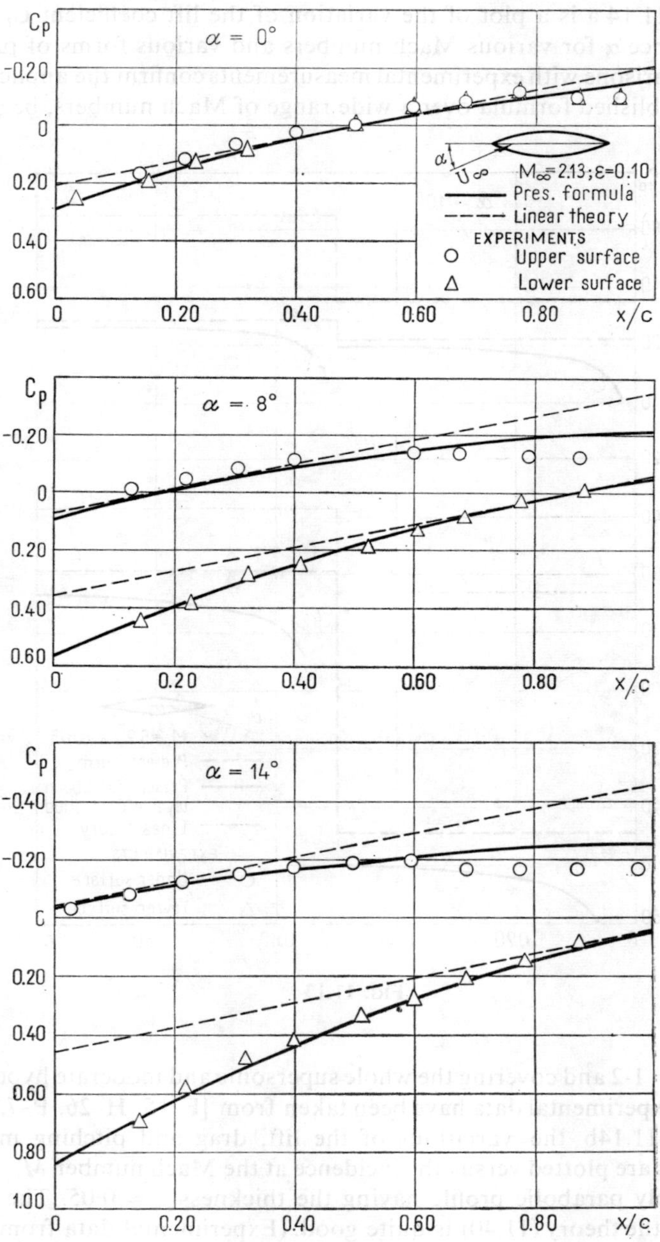
Cp
α = 0°
M∞=2.13; ε=0.10
Pres. formula
Linear theory
EXPERIMENTS
Upper surface
Lower surface
x/c
α = 8°
α = 14°

FIG. 11.12

Figure 11.14 a is a plot of the variation of the lift coefficient C_l versus the incidence α for various Mach numbers and various forms of profiles. The comparisons with experimental measurements confirm the applicability of the established formula over a wide range of Mach numbers, beginning

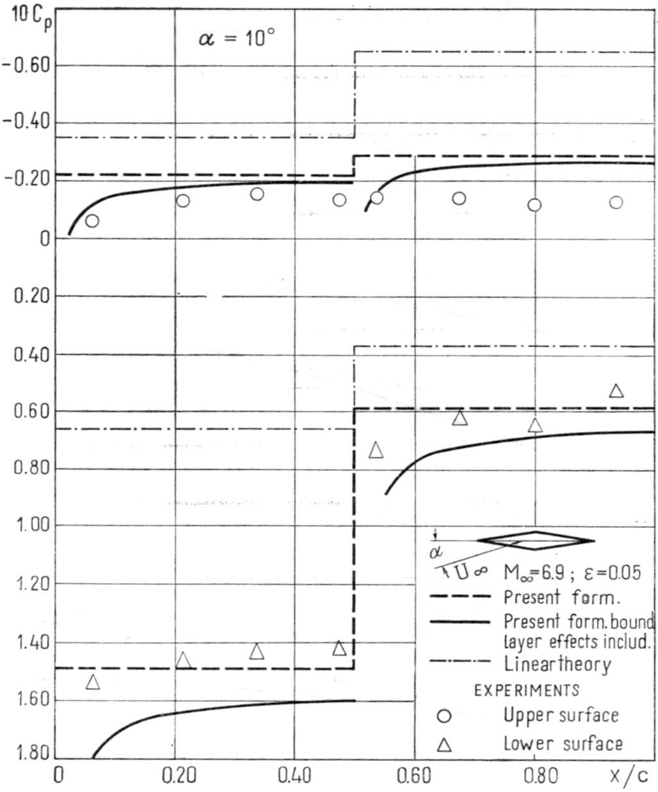

FIG. 11.13

with $M_\infty = 1{\cdot}2$ and covering the whole supersonic and moderate hypersonic régimes. Experimental data have been taken from [F–15, H–26, P–7, D–6].

In Fig. 11.14b the variations of the lift, drag and pitching moment coefficients are plotted versus the incidence at the Mach number $M_\infty = 6{\cdot}86$ for a doubly parabolic profile having the thickness $\varepsilon = 0{\cdot}05$. The agreement with the theory (11.40) is quite good. (Experimental data from [G–8, E–2].)

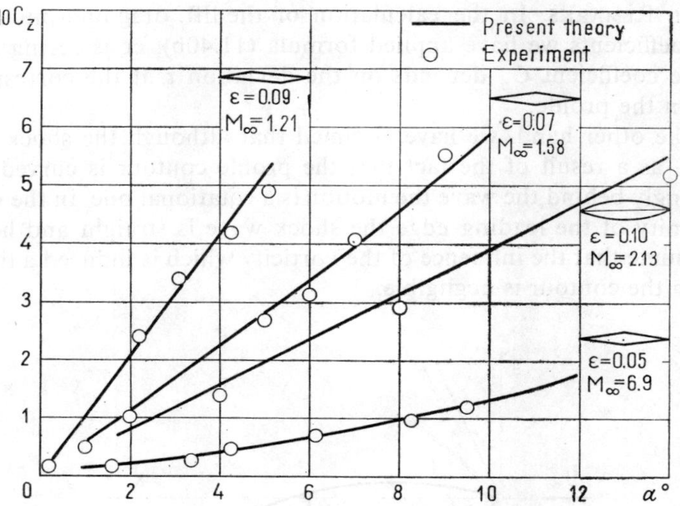
10 C_z
Present theory
Experiment
ε = 0.09
M∞ = 1.21
ε=0.07
M∞=1.58
ε = 0.10
M∞ = 2.13
ε=0.05
M∞=6.9
α°

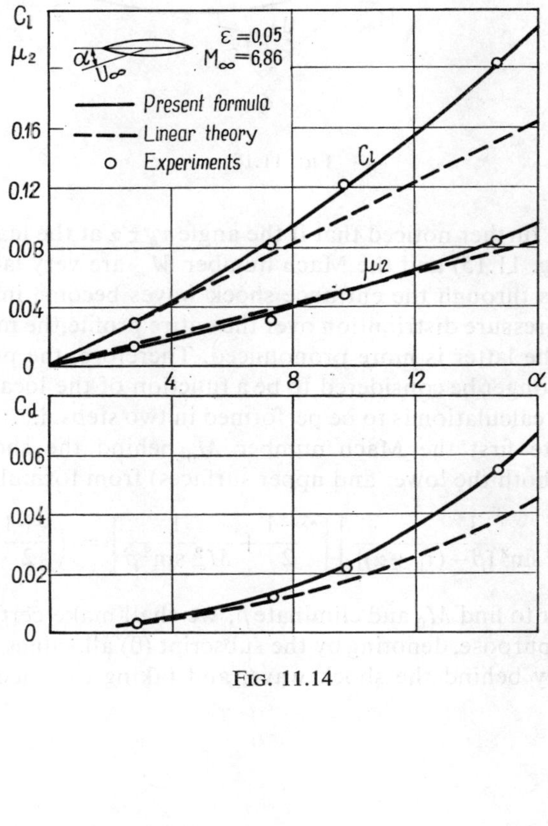
C_l
μ_2
ε =0,05
M∞=6,86
α
U∞
Present formula
Linear theory
Experiments
C_l
μ_2
α°
C_d

FIG. 11.14

FINAL REMARKS. In the calculation of the lift, drag and pitching moment coefficients we have applied formula (11.40b), considering that the pressure coefficient C_p depends on the deviation τ at the corresponding point on the profile.

On the other hand, we have assumed that although the shock wave is curved—as a result of the fact that the profile contour is curved — and accordingly behind the wave the motion is a rotational one, in the immediate vicinity of the leading edge the shock wave is straight and hence we may assume that the influence of the vorticity which is induced a little way off from the contour is negligible.

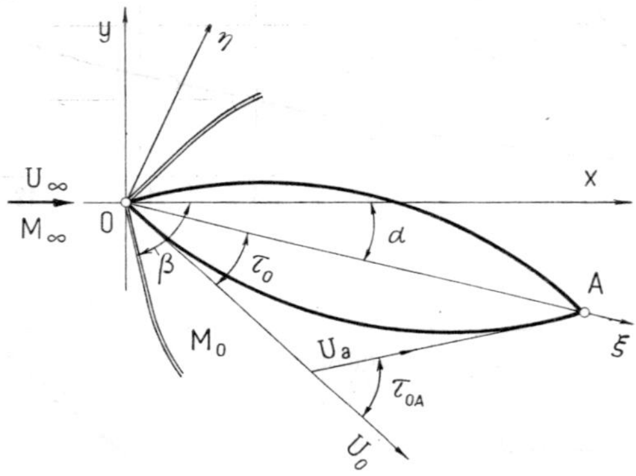

FIG. 11.15

It should be further noticed that if the angle $\tau_0 \pm \alpha$ at the leading edge of the profile (Fig. 11.15) and the Mach number M_∞ are very large, then the pressure losses through the entrance shock waves become important and influence the pressure distribution over the entire profile,the more so as the curvature of the latter is more pronounced. Therefore, the pressure coefficient can no longer be considered to be a function of the local deviation τ alone, and the calculation is to be performed in two steps, i.e.:

We calculate first the Mach number M_0 behind the shock wave (if necessary, on both the lower and upper surfaces) from formula (11.3c):

$$\left[\frac{\gamma-1}{2}+\frac{1}{M_0^2 \sin^2[\beta-(\tau_0 \pm \alpha)]}\right]\left[\frac{\gamma-1}{2}+\frac{1}{M_\infty^2 \sin^2 \beta}\right]=\left(\frac{\gamma+1}{2}\right)^2, \quad (11.61)$$

where, in order to find M_0 and eliminate β, we shall make certain substitutions. For this purpose, denoting by the subscript (0) all values at the vertex O, immediately behind the shock wave, and taking into account (11.3),

(11.11), (11.13), (11.14), we can write successively:

$$\frac{(1-\eta_0)C_{p0}}{m^2 \sin^2(\tau_0 \pm \alpha)} = \frac{\gamma+1}{2} + \sqrt{\left(\frac{\gamma+1}{2}\right)^2 + \frac{4}{K^2}} = P_0, \tag{11.62a}$$

$$\eta_0 = \left(1 - \frac{m^2}{2} P_0\right) \sin^2(\tau_0 \pm \alpha) + \tfrac{1}{2}(1-R_0)\cos^2(\tau_0 \pm \alpha), \tag{11.62b}$$

$$R_0 = \sqrt{1 - m^2 P_0 [m^2(\gamma+1) - 2] \tan^2(\tau_0 \pm \alpha)}, \tag{11.62c}$$

$$M_\infty^2 \sin^2 \beta = \frac{\gamma+1}{2\gamma} \frac{p - p_\infty}{p_\infty} + 1 = \frac{\gamma+1}{4} M_\infty^2 C_{p0} + 1, \tag{11.62d}$$

$$\sin^2[\beta - (\tau_0 \pm \alpha)] = \frac{[m^2 P_0 - 2(1-\eta_0)]^2 \tan^2(\tau_0 \pm \alpha)}{4(1-\eta_0)^2 + [m^2 P_0 - 2(1-\eta_0)]^2 \tan^2(\tau_0 \pm \alpha)}. \tag{11.62e}$$

With the aid of these formulae we may obtain very exact values of M_0, a fact which allows us to apply further, along the contour, the expansion coefficient C_{pe} which, as shown from the beginning, is valid over a wide range of the deflexion τ, i.e. from the velocity U_0 at the leading edge up to the velocity U_a at the trailing edge, the expansion angle being τ_{0A} (Fig. 11.15).

It is worth noting in this respect that $\sin \tau$ may be replaced by $\tan \tau$ in formulae (11.40), provided this is convenient in calculations (for integrating the pressures). Indeed, if we refer to the form (11.36) of C_{pe} and denote by C_{pe} the same expression, setting, however, $\tan \tau$ instead of $\sin \tau$, we obtain

$$\frac{C'_{pe}}{C_{pe}} = \frac{1 + \sqrt{1 + \left(\frac{4}{\gamma+1}\right)^2 \frac{1}{M_\infty^2 m^2 \sin^2 \tau}}}{1 + \sqrt{1 + \left(\frac{4}{\gamma+1}\right)^2 \frac{1}{M_\infty^2 m^2 \tan^2 \tau}}} \approx 1, \tag{11.63}$$

which justifies the approximation suggested above.

Finally, at an arbitrary point of the profile we have the following total pressure coefficient

$$C_p = C_{p0} + \frac{M_0^2}{M_\infty^2}\left(1 + \frac{\gamma}{2} M_\infty^2 C_{p0}\right) C_{pe}^*, \tag{11.64}$$

in which, however, C_{pe}^* must correspond to the conditions at the leading edge, where the expansion begins:

$$\frac{C_{pe}^*}{m_0^2 \sin^2(\tau_0 - \tau)} = \frac{\gamma+1}{2} - \sqrt{\left(\frac{\gamma+1}{2}\right)^2 + \frac{4}{K^2}}. \tag{11.65}$$

In this formula we have introduced K_0 and m_0:

$$K_0 = m_0 M_0 \sin(\tau_0 - \tau), \quad m_0^2 = \frac{M_0^2}{M_0^2 - 1}. \tag{11.66}$$

In the case of an expansion on the profile lower surface, beginning just from the leading edge (Fig. 11.16), the stream has an initial expansion $(\alpha - \tau_0')$, after which the expansion on the profile follows. In order to find

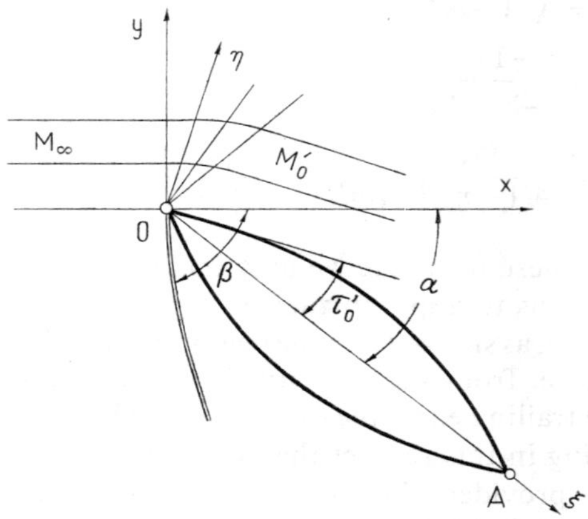

FIG. 11.16

M_0' at the point O, averaged according to the expansion, we apply equation (11.29):

$$(\alpha - \tau_0') = \frac{1}{\lambda}\left[\tan^{-1}(\lambda\sqrt{M_0'^2 - 1}) - \tan^{-1}(\lambda\sqrt{M_\infty^2 - 1})\right]$$

$$-\left[\tan^{-1}\sqrt{M_0'^2 - 1} - \tan^{-1}\sqrt{M_\infty^2 - 1}\right]. \tag{11.67}$$

The same technique could be developed further; but we restrict our considerations here since, for extending the applications to delta or trapezoidal wings, the pressure losses through the shock wave are neglected. We apply only formulae (11.40) in which the pressure coefficient depends on the stream deflexion alone.

XI. 3. Extension of the Unitary Formula for Compression Expansion to Wings of Finite Span

XI. 3.1. Effect of Yaw on the Pressures behind the Shock Wave. Application to Wings with Supersonic Leading Edges

Let us consider an infinite flat plate whose leading edge AA makes the angle χ with the normal to the free stream direction U_∞ (Fig. 11.17). Let us denote by τ_n the angle of the dihedron formed by the plate with the plane

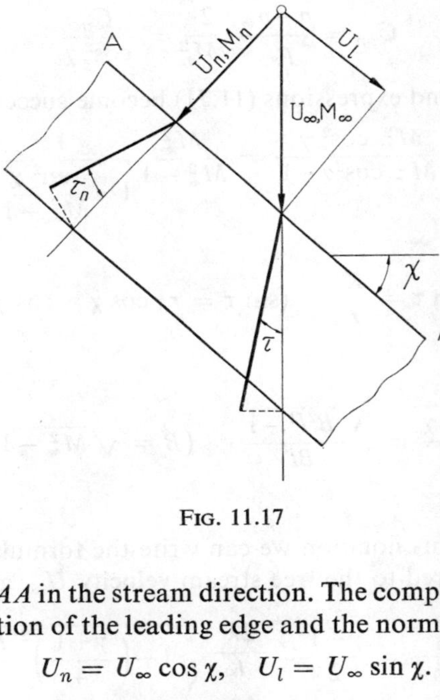

FIG. 11.17

through the edge AA in the stream direction. The components of the velocity U_∞ in the direction of the leading edge and the normal to it are

$$U_n = U_\infty \cos \chi, \quad U_l = U_\infty \sin \chi. \tag{11.68}$$

If the effects of friction are neglected, the lateral component U_l has no effect, and the normal component U_n, the only one which has an effect on the plate, is deflected through the angle τ_n related by the equation

$$\tau_n \cos \chi = \tau \tag{11.69}$$

with the *deflexion* τ, as defined previously (Fig. 11.17).

On the other hand, it is easy to see that the Mach number of the normal component is

$$M_n = \frac{U_n}{a_\infty} = \frac{U_\infty}{a_\infty} \cos \chi = M_\infty \cos \chi. \tag{11.70}$$

If we write

$$K_n = m_n M_n \sin \tau_n, \quad m_n = \frac{M_n}{\sqrt{M_n^2-1}}, \tag{11.71}$$

then the pressure coefficient C_{pn} with respect to the velocity U_n will be given by the same formulae established above, from which we apply (11.40a):

$$\frac{C_{pn}}{m_n^2 \sin^2 \tau_n} = \frac{\gamma+1}{2} + \frac{2}{K_n} \sqrt{1+\left(\frac{\gamma+1}{4}\right)^2 K_n^2}\,. \tag{11.72}$$

Let us note that on the one hand we have

$$C_{pn} = \frac{p-p_\infty}{p_\infty} \frac{2}{\gamma M_n^2} = \frac{C_p}{\cos^2 \chi} \tag{11.73}$$

and on the other hand expressions (11.71) become successively

$$m_n^2 = \frac{M_n^2}{M_n^2-1} = \frac{M_\infty^2 \cos^2 \chi}{M_\infty^2 \cos^2\chi - 1} = \frac{M_\infty^2}{M_\infty^2 - 1} \frac{1}{1-\dfrac{\tan^2 \chi}{M_\infty^2-1}} = \frac{m^2}{\lambda^2}, \tag{11.74a}$$

$$K_n = \frac{m}{\lambda} M_\infty \sin \tau = \frac{K}{\lambda}, \quad (\sin \tau = \tau_n \cos \chi \approx \cos \chi \sin \tau_n), \tag{11.74b}$$

where we have set

$$\lambda = \sqrt{1-\frac{\tan^2 \chi}{B^2}} = \frac{\sqrt{B^2 l^2 - 1}}{Bl}, \quad (B = \sqrt{M_\infty^2-1}\,,\ l = \cot \chi). \tag{11.75}$$

With the aid of this notation we can write the formula for the pressure coefficient C_p, referred to the free stream velocity U_∞, as a function of τ:

$$\frac{\lambda^2 C_p}{m^2 \sin^2 \tau} = \frac{\gamma+1}{2} + \frac{2\lambda}{K} \sqrt{1+\left(\frac{\gamma+1}{4}\right)^2 \frac{K^2}{\lambda^2}}\,. \tag{11.76}$$

We notice that everything takes place as if we were confronted with an *equivalent deflexion* θ with respect to the free stream velocity.

$$\sin \theta \approx \theta = \frac{\tau}{\lambda} \approx \frac{\sin \tau}{\lambda}, \tag{11.77}$$

with the aid of which we can define an equivalent new similarity parameter.

$$Q = M_\infty m \theta = M_\infty m \frac{\sin \tau}{\lambda} = \frac{K}{\lambda}, \tag{11.78}$$

such that the above formula becomes

$$\frac{C_p}{m^2 \sin^2 \theta} = \frac{\gamma+1}{2} + \frac{2}{Q}\sqrt{1+\left(\frac{\gamma+1}{4}\right)Q^2}\,. \qquad (11.79)$$

This formula may be also written as a function of the disturbance velocity u, corresponding to the first approximation. Indeed, as shown previously (1.84), this velocity may be written as

$$\frac{u}{U_\infty} = -\frac{Bl}{\sqrt{B^2l^2-1}}\frac{\sin\tau}{B} \approx -\frac{\theta}{B} \approx -\frac{m}{M_\infty}\theta, \qquad (11.80)$$

whence we deduce

$$-M_\infty^2 \frac{u}{U_\infty} = mM_\infty\,\theta = Q. \qquad (11.81)$$

Accordingly, replacing in (11.79), we find

$$M_\infty^2 C_p = \frac{\gamma+1}{2}\left(M_\infty^2 \frac{u}{U_\infty}\right)^2 - 2\left(M_\infty^2 \frac{u}{U_\infty}\right)\sqrt{1+\left(\frac{\gamma+1}{4}\right)^2\left(M_\infty^2 \frac{u}{U_\infty}\right)^2} \qquad (11.82)$$

in this closed form or – if

$$\left| M_\infty^2 \frac{u}{U_\infty} \right| < \frac{4}{\gamma+1},$$

in which case the expression under the square root sign can be conveniently expanded – in forms equivalent to (11.40c, d, e),

$$M_\infty^2 C_p = -2\left(M_\infty^2 \frac{u}{U_\infty}\right) + \frac{\gamma+1}{2}\left(M_\infty^2 \frac{u}{U_\infty}\right)^2 - \left(\frac{\gamma+1}{4}\right)^2\left(M_\infty^2 \frac{u}{U_\infty}\right)^3. \qquad (11.83a)$$

$$C_p = -2\frac{u}{U_\infty} + \frac{\gamma+1}{2}M_\infty^2\left(\frac{u}{U_\infty}\right)^2 - \left(\frac{\gamma+1}{4}\right)^2 M_\infty^4\left(\frac{u}{U_\infty}\right)^3. \qquad (11.83b)$$

It will be assumed that C_p is a function of u alone and that the losses of total head through the shock wave are negligible. These formulae will be used later for calculating the characteristics of delta wings. Indeed on a *delta wing with supersonic edges* having constant incidence (or constant slope if the wing is of symmetrical thickness), the above formulae are valid on the portion outside the Mach cone. The situation inside the Mach cone will be examined subsequently.

Let us extend the application of these formulae to a *delta wing of perfectly conical shape* with supersonic leading edges, whose incidence or slope varies according to a homogeneous function of zero order (Fig. 11.18). At point P on the generatrix PO which represents the ridge of an elementary dihedron, the stream undergoes the deflexion $d\tau^*$ and passes to P' on the infinitesimally close generatrix OP', while the *elementary equivalent deflexion* $d\theta^*$ will be given by the same formula (11.77),

$$d\theta^* = \frac{d\tau^*}{\lambda^*}, \quad \left[\lambda^* = \frac{\sqrt{B^2y^2-1}}{By}, \quad \tau^* = \tau_0^*\left(1+\frac{y^2}{l^2}\right)\right], \qquad (11.84)$$

where τ^* varies as a function of y, and λ^* obviously has its local value as a function of y.

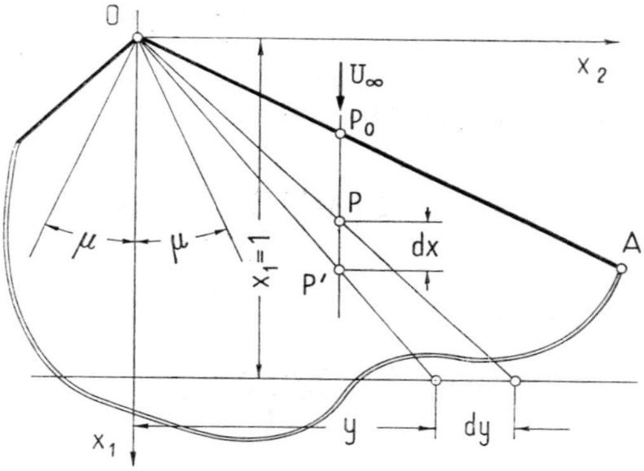

FIG. 11.18

The *equivalent total deflexion* follows easily by integrating from P_0 to P:

$$\theta^* = \theta_0^* + \int_{\tau_c}^{\tau^*} \frac{d\tau^*}{\lambda^*} = \tau_0^* - \int_y^l \frac{d\tau^*}{dy} \frac{By\,dy}{\sqrt{B^2y^2-1}}. \qquad (11.85)$$

On the other hand, applying expression (1.159) at point P, we obtain

$$\frac{du}{U_\infty} = -\frac{d\tau^*}{B}\frac{By}{\sqrt{B^2y^2-1}} = -\frac{m}{M_\infty}d\theta^*, \quad \frac{u}{U_\infty} = -\frac{m}{M_\infty}\theta^*, \quad (11.86)$$

whence we deduce that u is formed in the same manner as θ^*, hence formulae (11.82) and (11.83) are also perfectly valid for the conical wing with supersonic edges.

Let us now consider the case of a conical motion of order n, for which the axial velocity u can be easily deduced from the formulae established in our previous chapters. At the leading edge of the wing there occurs a transformation through the shock wave or expansion wave, for which formula (11.82) is locally valid, since one can assume that at the points lying on the edge line the conditions are similar to those which have been used for establishing this formula.

Owing to the slope variation along the leading edge, the wave will also have a variable intensity on the edge line and, consequently, the motion will be influenced by the Mach cones of that line.

If we assume that the losses through the shock wave at the leading edges are negligible, the transformations that are taking place behind this wave can be considered isentropic and hence formula (11.82) is valid at any point of the wing starting from the leading edge and the development (11.83) can be employed for computations, as long as we have

$$\left| M_\infty^2 \frac{u}{U_\infty} \right| < \frac{4}{\gamma+1} \,. \tag{11.87}$$

Let us notice further that the situation in the interior of the Mach cone issuing from the wing tip falls under the same considerations as above. As a matter of fact, the axial velocity does not vary too much as against its value on the Mach line, such that the transformation can be considered isentropic all the time and formula (11.82) can be valid in the interior of the Mach cone too.

We may conclude hence that the general formula (11.82), (11.83) respectively—valid for supersonic and moderate hypersonic flows and for both compressions and expansions—is applicable to any wing with supersonic edges, having a great advantage in that it depends only on the axial disturbance velocity u, accurately determined in most ordinary cases dealt with in the preceding examples.

Thus the results obtained in conjunction with the determination of u will be turned to good account by applying the above mentioned formula.

For this purpose, we have to see if this formula agrees with the experimental data.

XI. 3.2. Experimental Verifications of the Pressure Distribution on Wings with Supersonic Edges

As a comparison of the theoretical formulae obtained above with the results of experiments carried out by various authors, we give below several representative diagrams which show a good agreement between theory and experiments.

An important remark is to be made in connection with the *boundary layer* which, owing to its variable thickness, changes the local slope of the contour. For this reason, this slope variation is to be considered and added to the initial slope of the contour. It is to be noticed however, that instead of the boundary layer thickness proper, we consider only the *displacement thickness*, which is defined in terms of the thickness δ itself by the following expression:

$$\delta^* = \int_0^\delta \left(1 - \frac{\varrho_y u_y}{\varrho U}\right) dy \tag{11.88}$$

where u_y stands for the velocity in the boundary layer which varies from 0 up to δ, and U is the velocity at the boundary, unaffected by friction (Fig. 11.19).

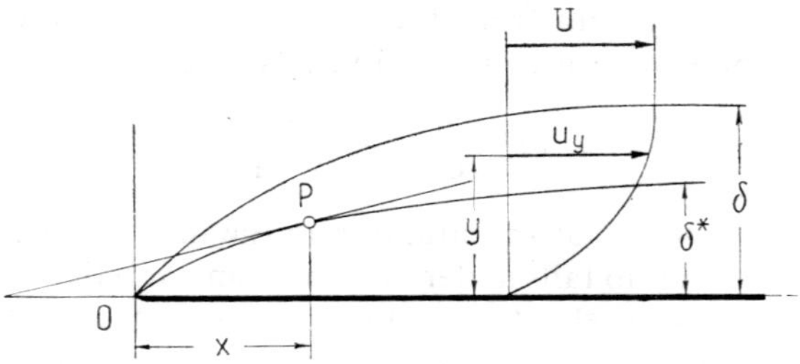

FIG. 11.19

It may be proved that on a flat plate parallel to the flow, δ^* is given in terms of the Mach number and Reynolds number

$$\delta^* = 1{\cdot}72(1+0{\cdot}277M^2)x\left(\frac{Ux}{\nu}\right)^{-1/2}, \tag{11.89}$$

where ν is the kinematic viscosity; an additional slope $\Delta\tau$ on the plate

$$\Delta\tau = \frac{d\delta^*}{dx} = 0{\cdot}86(1+0{\cdot}277M^2)\left(\frac{Ux}{\nu}\right)^{-1/2} \tag{11.90}$$

follows. For applications on the wing, M represents the Mach number after compression or expansion, far downstream, where the boundary layer thickness is almost constant (theoretically at infinity).

Though this scheme corresponds to a flat plate, this formula may also be applied with a good approximation to the actual wing. The results obtained in this way are in good agreement with those obtained by other, authors using different methods [*L*–11, *W*–14].

1. *Delta wing with constant slope.* $M_\infty = 7{\cdot}35$ (Fig. 11.20). We consider a wing of symmetrical thickness with leading edges swept at 45° and slope

equal to 5.4°, and reproduce the pressure measurements on various wing sections (Fig. 11.20) and various incidences carried out at a very high free stream Mach number $M_\infty = 7{\cdot}35$.

In the calculation of the theoretical pressures, the boundary layer effect will also be included since it adds a supplementary slope $\Delta\tau$ given by (11.90),

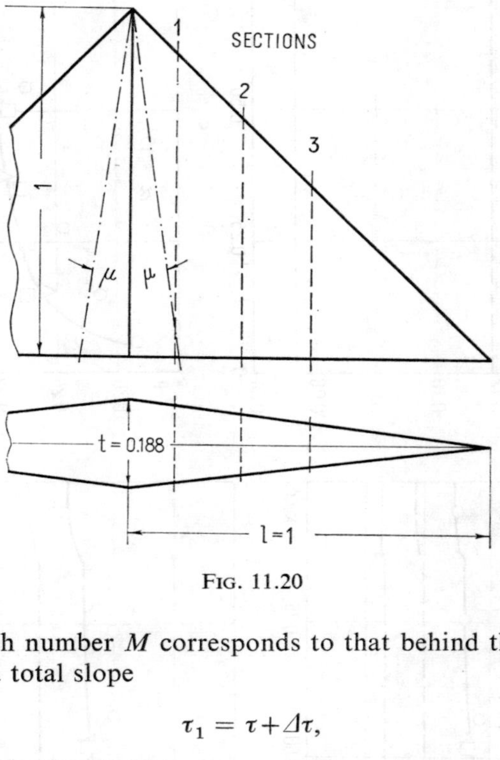

FIG. 11.20

where the Mach number M corresponds to that behind the shock wave; we have thus a total slope

$$\tau_1 = \tau + \Delta\tau, \tag{11.91}$$

which modifies the pressure coefficient substantially. This modified slope, introduced into formulae (11.40) or (11.83), gives a real variation of the pressure coefficient which is in very fair agreement with that given by experiments, as may be observed in Fig. 11.21, where the variation of the coefficient on the lower surface is plotted in accordance with the results obtained previously [W–14]. It is worth noting that these experiments also confirm the aspects inside the Mach cone.

In Fig. 11.22 similar diagrams are plotted for the upper surface, in accordance with the experiments carried out by the same author, indicating a good agreement for positive τ (compression, Fig. 11.22c) and an increasing disagreement when τ becomes negative and increases in absolute value (ex-

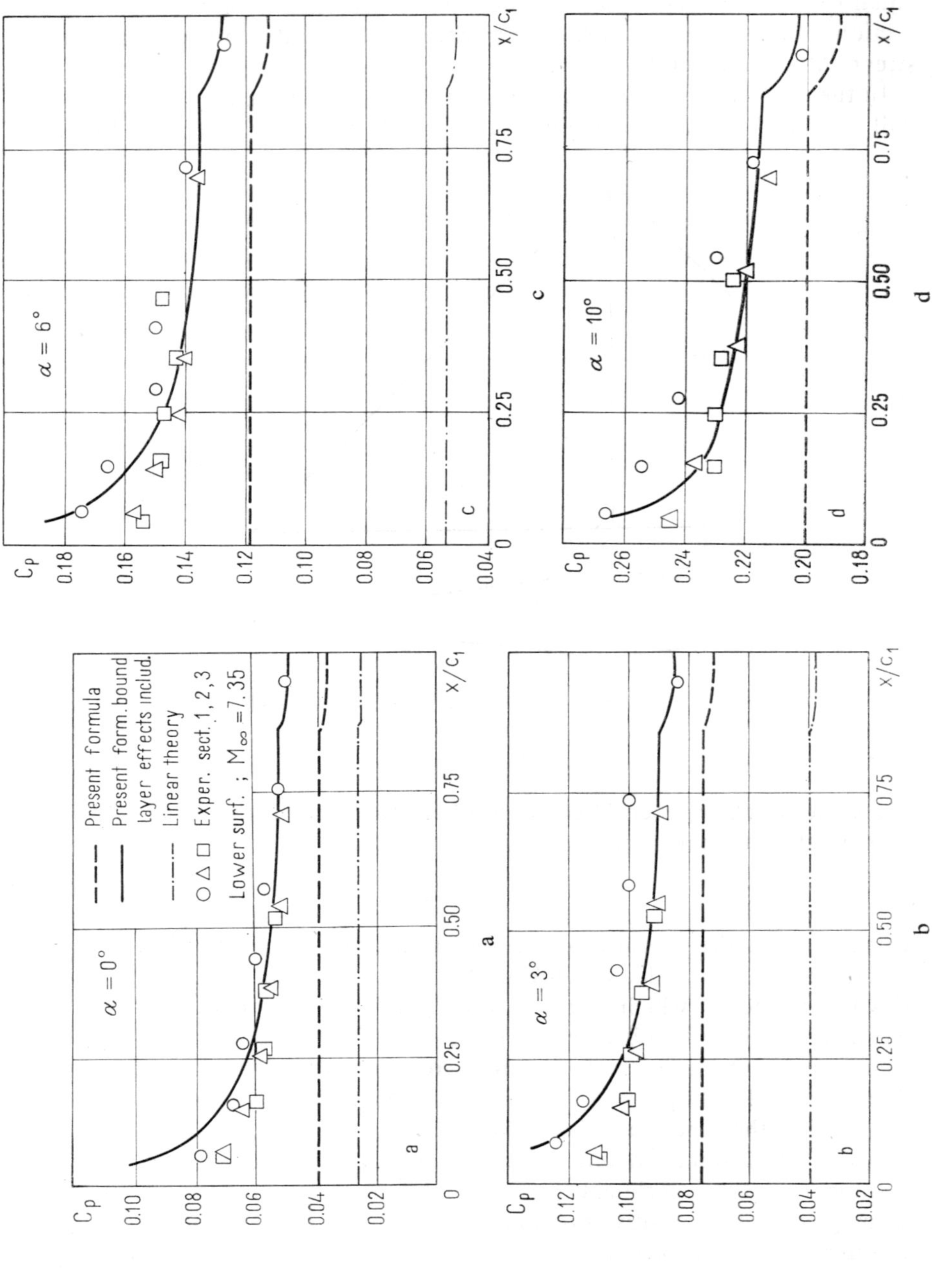
Present formula
Present form. bound layer effects includ.
Linear theory
Exper. sect. 1, 2, 3
Lower surf. ; M_∞ = 7.35
α = 0°
α = 3°
α = 6°
α = 10°
C_p
x/c_1
a
b
c
d

FIG. 11.21

pansion, Fig. 11.22 a). This is mainly due to the fact that the boundary layer becomes substantially thicker, as shown by the photographs taken by various experimenters [W – 14]. Practically, the expansion does not take place, even at sufficiently large negative angles.

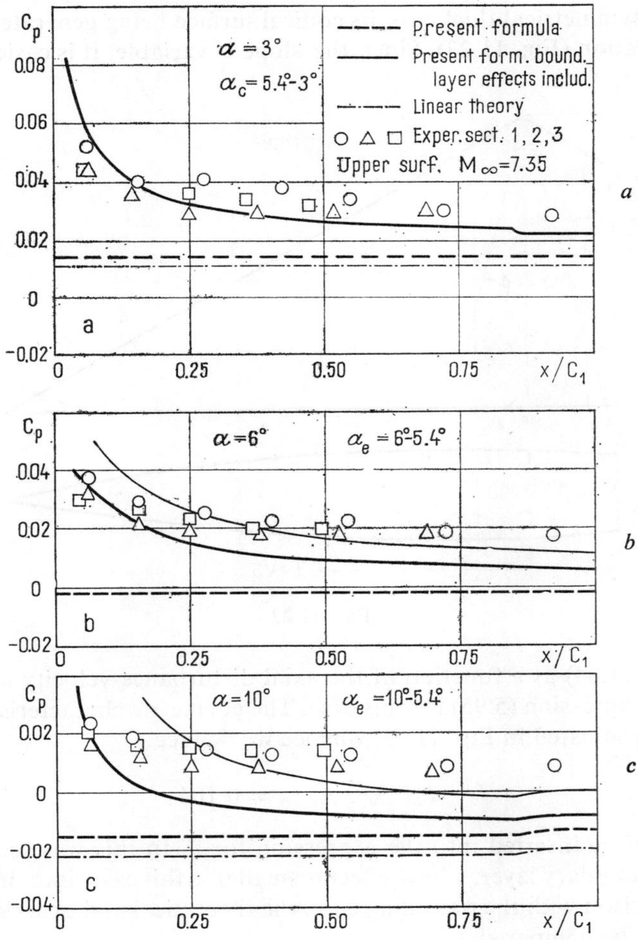

FIG. 11.22

This boundary layer thickening invalidates the additional slope given by (11.90), where M is the Mach number corresponding to an expansion; we consider therefore a thickness δ^* and an additional slope $d\delta^*/dx$ greater than those given by (11.89) and (11.90), respectively, and also in terms of the absolute value of the expansion angle τ. Thus, for instance, it is sufficient to consider the slope $\Delta\tau$ 1·75 times greater than that given by (11.90), in order to make the curve represented by the thin line in Fig. 11.22b agree reasonably well with experimental data.

2. *Conical delta wing with parabolic cross-section, $M_\infty = 4$.* This is a wing of symmetrical thickness, its conical surface being generated by a parabolic section (Fig. 11.23). Since the slope is variable, it is easier to apply

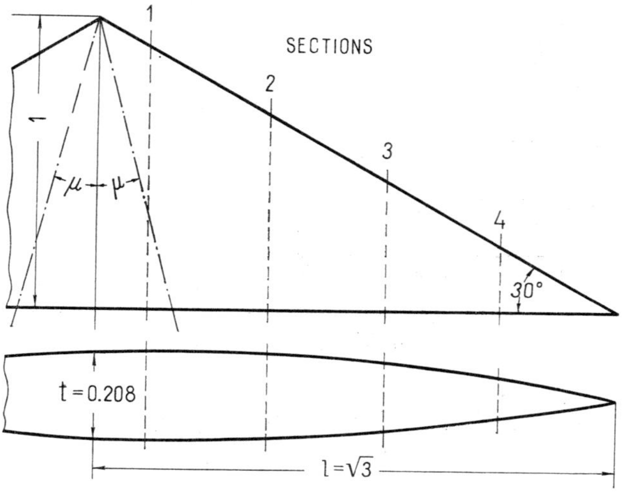

FIG. 11.23

formula (11.84) as a function of the axial disturbance velocity u, which is given by expression (5.95) for this case. The goemetric characteristics of the wing are indicated in Fig. 11.23, whence we deduce

$$l = \sqrt{3}, \quad \tau_0 = 0{\cdot}104 \tag{11.92}$$

which will be inserted into the expression for u. In this way, considering also the boundary layer, whose effect is smaller in this case since M_∞ is small in comparison with the preceding case, the theoretical and experimental results may be compared.

In order to establish in an approximate way the additional slope, we assume that in an arbitrary section of the wing parallel to U_∞, whose form (Fig. 11.24) is the parabola

$$Z = \tau_0 x_1 \left(1 - \frac{y^2}{l^2}\right), \quad x_2 = d = \text{const.}, \tag{11.93}$$

the boundary layer slope does not differ too much from the case of a plate having the deflexion β equal to the slope of the tangent to the parabola (11.93) at the point P situated at a quarter of the local slope c starting from the leading edge O. Thus one obtains the angle β:

$$\tan\beta = \tau_0\left[1+\frac{1}{\left(\frac{lc}{4d}+1\right)^2}\right]. \tag{11.94}$$

In the above expressions we have designated by d the distance of a section to the of plane symmetry of the wing.

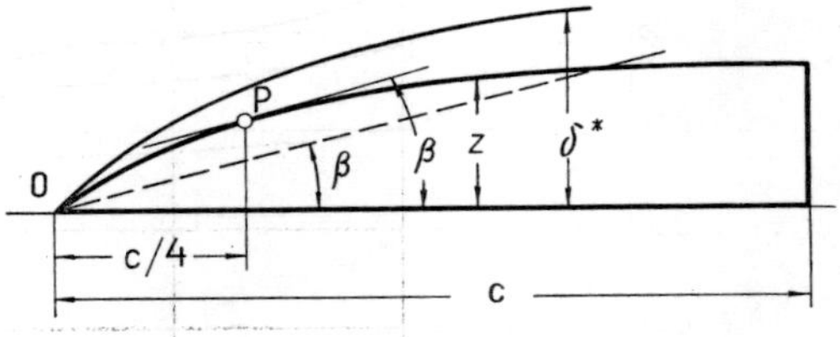

FIG. 11.24

Then we have to calculate the Mach number M, resulting from the compression with angle β, introduce it into (11.90) and find the additional slope $\Delta\tau$. Figure 11.25 represents the pressure coefficient C_p at various wing sections (Fig. 11.23), as evaluated by the experiments [L−11].

The agreement between theory and experiment is very good, with a small exception for points on the trailing edge.

Owing to the fact that this edge ends abruptly, the effect of the expansion which takes place immediately may also be transmitted upstream through the boundary layer.

This effect is more marked at section (1) near the axis (Figs. 11.23 and 11.25) which lies partly inside the Mach cone and where the penetration of the depression from the trailing edge is transmitted very far upstream; for this reason, there is also a great difference between theoretical and experimental values almost up to the middle of the wing.

3. *Delta wing with constant slope of symmetrical thickness* $M_\infty = 4$. We consider a wing having an apex semiangle $\approx 26{\cdot}5°$ ($l = 0{\cdot}5$), such that the Mach cone covers more than half of the wing surface. The measurements carried out at sections 1 and 2 of the lower surface, parallel to the span allow us to follow better the aspects related to the presence of the Mach cone. As is seen, the experiments [S−26] confirm the tendency of the pressure coefficient to decrease (in absolute value) on entering this cone,

both at positive incidences (Fig. 11.27) and negative ones (Fig. 11.28). The conical character of the motion is also confirmed (the measurements car-

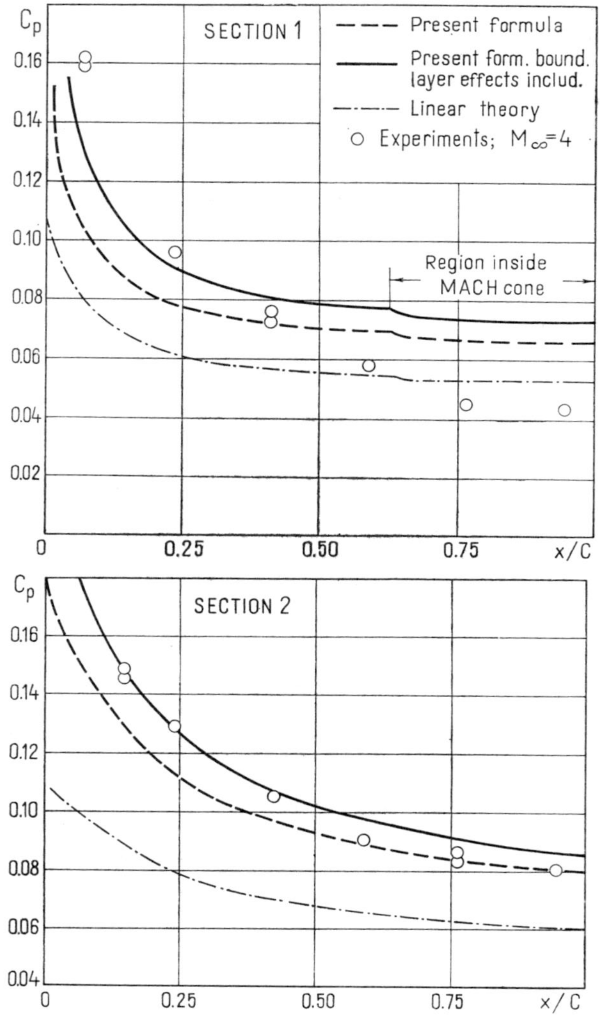

FIG.

ried out at different points of the two sections, but with the same y, practically coincide). In the case of larger positive incidences ($\alpha \geqslant 10°$), the shock waves are no longer attached to the wing leading edge and hence formula (11.26) can no longer be applicable. Indeed, calculations show that at $\alpha = 10°$ the maximum incidence at which the wave is still attached has been reached; on the other hand, the photographs taken by some experimenters [S–26] emphasize the fact that the shock waves detach a little earlier than stipulated by theory.

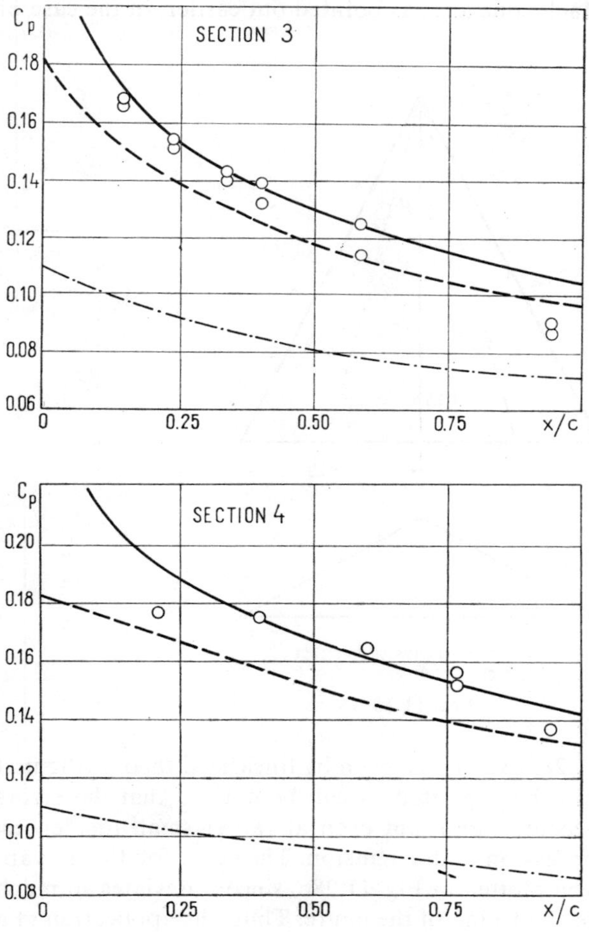

11.25

Consequently, we have not reproduced the experiments carried out by the same author [S–26] for angles $\alpha > 10°$, since they cannot be compared with the theoretical formulae established under conditions only applicable to an attached wave.

It can be noted further that, particularly in the case of positive incidences (Fig. 11.27), the experiment indicates greater values of the pressure coefficient inside the Mach cone than those stipulated by theory. A general tendency of the pressure increase is noticed at large negative incidences (Fig. 11.28c). These effects are due to the boundary layer which thickens very much inside the Mach cone and, as pointed out earlier, in the case of an expansion.

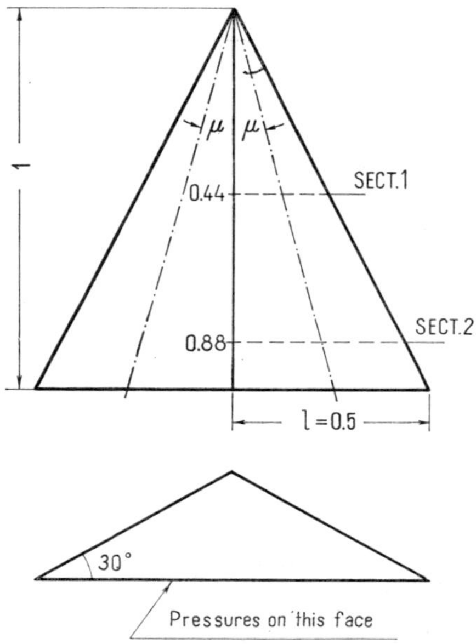

FIG. 11.26

In Figs. 11.27 and 11.28, the curves given by linearized theory (the first approximation) have also been plotted. It can be noticed that the errors given by this theory become significant even at rather small incidences, both in the case of compression and expansion. The curve for the first approximation has not been plotted in Fig. 11.28c, since it deviates so much that it would lie outside the frame of the figure. Thus, the application of a more exact formula for the pressure calculation appears once more to be thoroughly justified.

GENERAL NOTE. As shown above, the boundary layer causes a modification of the pressure distribution, particularly near the wing leading edge. This modification, however, extends over a small distance which does not

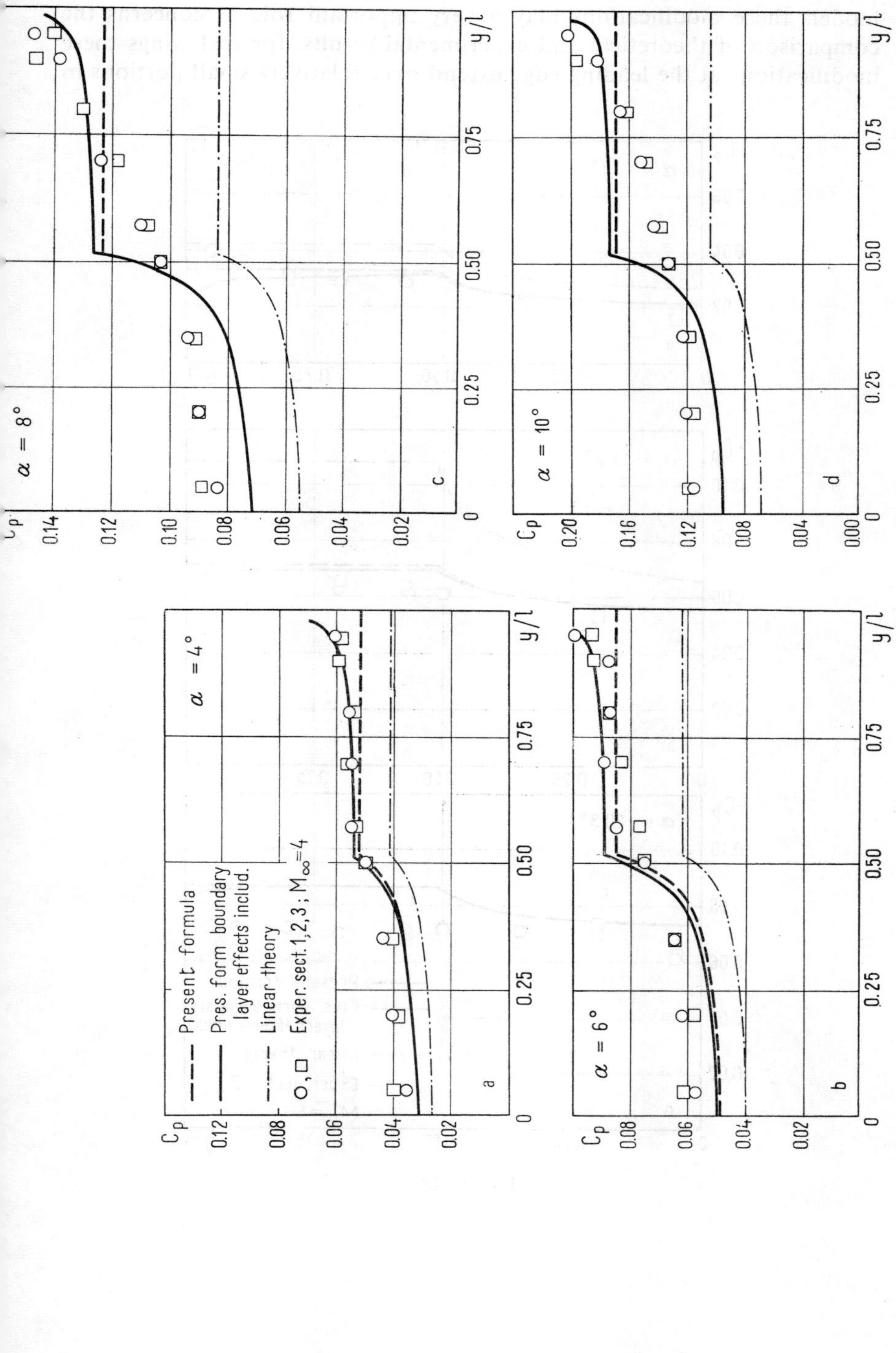
Present formula
Pres. form boundary layer effects includ.
Linear theory
Exper. sect. 1,2,3 ; $M_\infty = 4$
$\alpha = 4°$
$\alpha = 6°$
$\alpha = 8°$
$\alpha = 10°$
a
b
c
d
C_p
y/l

FIG. 11 27

depend on the wing dimensions. Consequently, if for small experimental models these modifications play a very important role as concerns the comparison of theoretical and experimental results, for real wings these modifications at the leading edge extend over relatively small portions in

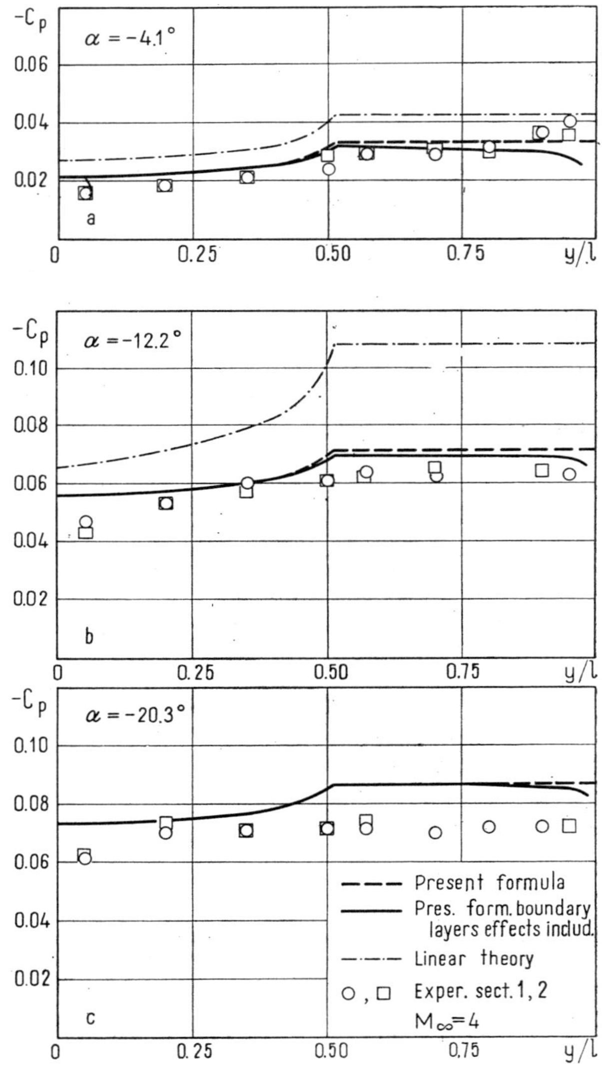

FIG. 11.28

comparison with the wing total surface and hence, expressed in percentages, the boundary layer effect is quite negligible, particularly at low Mach numbers when the effect itself is much reduced. Accordingly, for real wings of full size, this effect need not be considered.

XI. 4. Aerodynamic Characteristics of Delta and Rectangular Wings in Supersonic–Moderate Hypersonic Flow

XI. 4.1. General Formulae for Lift, Drag and Moments

Let us note from the beginning that formulae (11.83), which give the pressure coefficient, have the same "conical character" as the corresponding ones obtained under the assumption of small perturbations, since u has the same value; consequently, the methods of integration are similar to those previously established.

On the other hand, however, on account of the non-linear character of the formulae then thin wing cannot be separated from that of symmetrical thickness; for this reason, in the following approach we treat the real, thick and lifting wing.

Let us hence consider a thick wing (without being necessarily of symmetrical thickness) on which we define the limits of a surface element $dx_1 dx_2$, exactly as in our preceding calculations (3.23, Fig. 3.1):

$$dx_1\, dx_2 = x_1\, dx_1\, dy; \quad x_1\, dx_1\, dx_2 = x_1^2\, dx_1\, dy; \quad x_2\, dx_1\, dx_1^2 = x_1^2\, dx_1 y\, dy. \tag{11.95}$$

This element is subjected to the pressure $(p-p_\infty)$ whose coefficient C_p is given by (11.82) or, for calculations which refer to a continuous contour, by (11.83). This latter expression includes terms in u, u^2 and u^3; for calculating the aerodynamic coefficients, these terms are multiplied by the differential elements of (11.95) and by the terms of the downwash w, such that we find expressions of the form:

$$w = \sum_{m=1}^{N} x_1^{m-1} \sum_{q=0}^{m-1} W_{m-q-1,\,q}\, y^q; \quad u^k = \left(\sum_{n=1}^{N} x_1^{n-1} u_{n-1}\right)^k, \quad (k = 1, 2, 3), \tag{11.96a}$$

$$\left(\sum_{n=1}^{N} x_1^{n-1} u_{n-1}\right)^k x_1\, dx_1\, dy, \quad \left(\sum_{n=1}^{N} x_1^{n-1} u_{n-1}\right)^k x_1^2\, dx_1\, dy,$$

$$\left(\sum_{n=1}^{N} x_1^{n-1} u_{n-1}\right)^k x_1\, dx_1 y\, dy. \tag{11.96b}$$

$$\left(\sum_{n=1}^{N} x_1^{n-1}u_{n-1}\right)^k x_1\, dx_1\, dy \sum_{m=1}^{N} x_1^{m-1} \sum_{q=0}^{m-1} W_{m-q-1,\,q}\, y^q$$
$$= \left(\sum_{n=1}^{N} x_1^{n-1}u_{n-1}\right)^k \sum_{m=1}^{N} x_1^{m} \sum_{q=0}^{m-1} W_{m-q-1,\,q}\, y^q\, dx_1\, dy, \qquad (11.96c)$$

$$\left(\sum_{n=1}^{N} x_1^{n-1}u_{n-1}\right)^k x_2\, dx_1\, dx_2 \sum_{m=1}^{N} x_1^{m-1} \sum_{q=0}^{m-1} W_{m-q-1,\,q}\, y^q$$
$$= \left(\sum_{n=1}^{N} x_1^{n-1}u_{n-1}\right)^k \sum_{m=1}^{N} x_1^{m+1} \sum_{q=0}^{m-1} W_{m-q-1,\,q}\, y^{q+1}\, dx_1\, dy. \qquad (11.96d)$$

Further, denoting by C_{p1} and C_{p2} the pressure coefficients on the wing upper and lower surfaces, respectively, we can write successively:

The lift coefficient

$$C_l = \frac{1}{S}\int_S (C_{p2}-C_{p1})dx_1\, dx_2$$
$$= \frac{1}{S}\sum_{p=1}^{2}(-1)^p\left\{\sum_{k=1}^{3} N_k \int_0^1\int_{-l_2}^{l_1}\left(\sum_{n=1}^{N} x_1^{n-1}\frac{u_{n-1}}{U_\infty}\right)^k x_1\, dx_1\, dy;\right. \qquad (11.97)$$

The wave drag coefficient

$$C_d = \frac{1}{S}\int_S (C_{p2}-C_{p1})\,\frac{w}{U_\infty}\,dx_1\, dx_2$$
$$= \frac{1}{S}\sum_{p=1}^{2}(-1)^p\left\{\sum_{k=1}^{3} N_k\int_0^1\int_{-l_2}^{l_1}\left(\sum_{n=1}^{N} x_1^{n-1}\frac{u_{n-1}}{U_\infty}\right)^k\right.$$
$$\left.\times\left(\sum_{m=1}^{N} x_1^m \sum_{q=0}^{m-1}\frac{W_{m-q-1,\,q}}{U_\infty}\, y^q\right) dx_1 dy\right\}; \qquad (11.98)$$

The rolling moment coefficient

$$\mu_1 = \frac{1}{SH}\int_S (C_{p2}-C_{p1})\, x_2\, dx_1\, dx_2$$
$$= \frac{1}{SH}\sum_{p=1}^{2}(-1)^p\left\{\sum_{k=1}^{3} N_k\int_0^1\int_{-l_2}^{l_1}\left(\sum_{n=1}^{N} x_1^{n-1}\frac{u_{n-1}}{U_\infty}\right)^k x_1^2\, dx_1 y\, dy\right\}; \qquad (11.99)$$

The pitching moment coefficient

$$\mu_2 = \frac{1}{SH}\int_S (C_{p2}-C_{p1})\, x_1\, dx_1\, dx_2$$
$$= \frac{1}{SH}\sum_{p=1}^{2}(-1)^p\left\{\sum_{k=1}^{3} N_k \int_0^1 \int_{-l_2}^{l_1}\left(\sum_{n=1}^{N} x_1^{n-1}\frac{u_{n-1}}{U_\infty}\right)^k x_1^2\, dx_1\, dy\right\}; \tag{11.100}$$

The yawing moment coefficient

$$\mu_3 = \frac{1}{SH}\int_S (C_{p2}-C_{p1})\frac{w}{U_\infty} x_2\, dx_1\, dx_2 = \frac{1}{SH}\sum_{p=1}^{2}(-1)^p$$
$$\times\left\{\sum_{k=1}^{3} N_k \int_0^1 \int_{-l_2}^{l_1}\left(\sum_{n=1}^{N} x_1^{n-1}\frac{u_{n-1}}{U_\infty}\right)^k \left(\sum_{m=1}^{N} x_1^{m+1}\sum_{q=0}^{m-1}\frac{w_{m-q-1,q}}{U_\infty} y^q\right) y\, dx_1\, dy\right\}, \tag{11.101}$$

where $p = 1, 2$ corresponds to the upper and lower surfaces respectively, S is the wing area, H is a reference height and N_k are the coefficients of expression (11.84):

$$N_1 = -2,\quad N_2 = \frac{\gamma+1}{2}M_\infty^2,\quad N_3 = -\left(\frac{\gamma+1}{4}\right)^2 M_\infty^4. \tag{11.102}$$

Note. In the above integrals, the approximate value of $x_1 = \cos\alpha \approx 1$ was considered as the upper limit of x_1.

If, however, α and the exponent of x_1 have large values, mainly if the theoretical results are to be compared with the experimental ones, the exact value $x_1 = \cos\alpha$ should be introduced. In such a case, for the above integrals we write

$$\int_0^{\cos\alpha} x_1^r\, dx_1 = \frac{(\cos\alpha)^{r+1}}{r+1}, \tag{11.103}$$

r being an arbitrary whole number.

XI. 4.2. Application to a Delta Wing with Constant Slope

The wing is in the shape of a wedge with constant slope on both sides of the symmetry plane. On the wing axis at the trailing edge ($x_1 = 1$), the total thickness ε is equal to twice the slope, as it was defined previously (Fig. 11.29).

If the symmetry plane has the incidence α, then the stream deflexion is

$$\tau = \frac{\varepsilon}{2} \pm \alpha, \tag{11.104}$$

where the (+) sign corresponds to the lower surface and the (−) sign to the wing lower surface. On the portion outside the Mach cone, the axial disturbance velocity is constant and it is more convenient to use the exact closed formula (11.82).

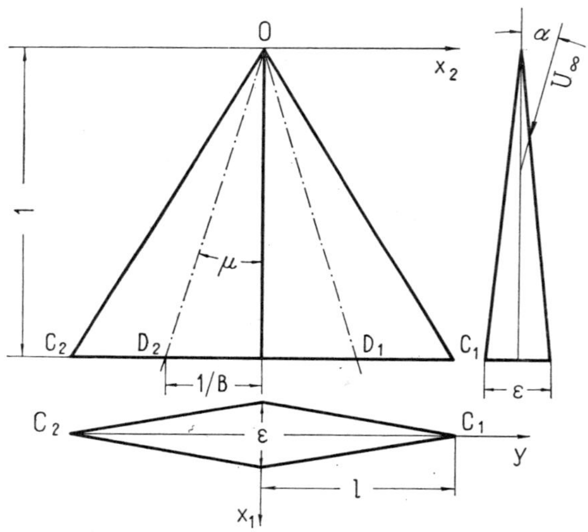

FIG. 11.29

Inside the Mach cone we use the expansion (11.83), taking for the axial disturbance velocity its expression previously determined (4.42), which we write in the following form:

$$\frac{u}{U_\infty} = \mathrm{Re}\left[-\frac{2\tau Bl}{\pi B\sqrt{B^2l^2-1}}\left(\cos^{-1}\sqrt{\frac{(1+Bl)(1-Bx)}{2B(l-x)}}\right.\right.$$
$$\left.\left.+\cos^{-1}\sqrt{\frac{(1+Bl)(1+Bx)}{2B(l+x)}}\right)\right] = \mathrm{Re}\left[-\frac{2\tau Bl}{\pi B\sqrt{B^2l^2-1}}\cos^{-1}\sqrt{\frac{1-B^2x^2}{B^2(l^2-x^2)}}\right]. \tag{11.105}$$

Performing the calculations, after some simplifications we obtain successively:

The lift coefficient is given by the following expression:

$$C_l = \frac{l}{\sqrt{B^2l^2-1}}[(\varepsilon+2\alpha)R_1-(\varepsilon-2\alpha)R_2]-\frac{4\alpha}{B}\left(\frac{Bl}{\sqrt{B^2l^2-1}}-1\right)$$
$$+\frac{(\gamma+1)M_\infty^2\alpha\varepsilon l^2}{B^2l^2-1}-\Delta C_l, \tag{11.106}$$

where we have written

$$R_1 = \sqrt{1+\left(\frac{\gamma+1}{8}\right)^2\frac{M_\infty^4(\varepsilon+2\alpha)^2l^2}{B^2l^2-1}},$$
$$R_2 = \sqrt{1+\left(\frac{\gamma+1}{8}\right)^2\frac{M_\infty^4(\varepsilon-2\alpha)^2l^2}{B^2l^2-1}}, \tag{11.107a}$$

$$\Delta C_l = \alpha(\gamma+1)M_\infty^2\left[\varepsilon a_2+\frac{\gamma+1}{32}M_\infty^2(4\alpha^2+3\varepsilon^2)a_3\right], \tag{11.107b}$$

$$a_k = \frac{2l^{k-1}}{45B(B^2l^2-1)^{k/2}}\left[19-3\left(\frac{2}{\pi}\cos^{-1}\frac{1}{Bl}\right)^k\right.$$
$$\left.-16\left(\frac{2}{\pi}\cos^{-1}\sqrt{\frac{3}{4B^2l^2-1}}\right)^k\right], \quad (k = 2, 3). \tag{11.107c}$$

Note. The term ΔC_l includes the effect of the pressure variation inside the Mach cone, corresponding to the secondary terms u^2, u^3.

This term is negligible in practice, both at small Mach numbers, in which case the effect of the secondary terms u^2, u^3 is unimportant, and at large Mach numbers, when the portion of the wing lying inside the Mach cone is small.

The wave drag coefficient will be deduced in a simpler way by applying the approximate formula established above for the plane motion. Thus we write

$$C_d = C_{d0}+\alpha C_l, \tag{11.108}$$

where C_l is given above (11.106) and C_{d0} is to be calculated for zero incidence ($\alpha = 0$). In this way, using the expansion (11.83) for the pressure coefficient, we find

$$C_{d0} = \frac{\varepsilon^2}{B}+\frac{\gamma+1}{8}\frac{M_\infty^2l^2\varepsilon^3}{B^2l^2-1}+\tfrac{1}{2}\left(\frac{\gamma+1}{8}\right)^2\frac{M_\infty^4l^3\varepsilon^4}{(B^2l^2-1)^{3/2}}-\Delta C_{d0}+C_{dl}, \tag{11.109}$$

where the correction term ΔC_{do}, which includes the phenomenon taking place inside the Mach cone corresponding to the secondary terms u^2, u^3 is given by the following approximate expression:

$$\Delta C_{do} = \frac{\varepsilon^3}{8}(\gamma+1)M_\infty^2\left(a_2+\frac{\gamma+1}{16}M_\infty^2\varepsilon a_3\right), \tag{11.110}$$

the coefficient a_k ($k = 2,3$) being given by (11.107). This term is negligible, as shown above. The term C_{dl} is due to the downstream deflexion where the stream expands suddenly. This depression is difficult to determine, but it may be written in the form

$$C_p = -\frac{2\nu}{\gamma M_\infty^2}, \tag{11.111}$$

where the coefficient ν is close to unity. Under these conditions we have

$$C_{dl} = \frac{2\varepsilon\nu}{\gamma M_\infty^2}. \tag{11.112}$$

XI. 4.3. A Doubly Conical Delta Wing

Let us consider a doubly conical wing, similar to that treated earlier (Chapter VII), having a relative maximum thickness ε at the vertex G (Fig. 11.30). On the portion OC_1GC_2 of the wing, the velocity u_0 is that corresponding to the conical motion at the vertex O and is given by the same expression as the preceding one (11.105). On the enclosing surface GC_1C_2, the axial velocity consists of u_0, to which $u_0^{(G)}$ is also added, due to the motion at the vertex G:

$$u_0^{(G)} = \text{Re}\left[-\frac{2\tau_G}{\pi B}\frac{2Bl}{\sqrt{4B^2l^2-1}}\cos^{-1}\sqrt{\frac{1-4B^2x^2}{4B^2(l^2-x^2)}}\right]. \tag{11.113}$$

The stream deflexions in the case of the first (τ_0) and second motions corresponding to the vertex G (τ_G) are

$$\tau_0 = \varepsilon\pm\alpha, \quad \tau_G = -2\varepsilon \tag{11.114}$$

respectively; the (+) sign is for the lower surface and the (−) sign for the upper surface.

For the wing portions situated outside the Mach cone, we use the closed formula (11.82) and inside the cone the development (11.83). We shall consider the contribution of each plane side upon which the pressure are constant, if we are not taking account of the Mach cone whose effect will be included, as before, in a separate term. On summing these contributions we obtain successively:

The lift coefficient has the expression

$$C_l = \frac{1}{\lambda B}\sum_{k=1}^{4}(-1)^{k-1}\alpha_k R_k - \frac{4\alpha}{B}\left(\frac{1}{\lambda}-1\right) - \frac{2(\gamma+1)M_\infty^2\alpha\varepsilon}{B^2\lambda^2}\left(1-\frac{\lambda}{\lambda'}\right) - \Delta C_l, \tag{11.115}$$

$$\alpha_1 = \varepsilon+\alpha, \quad \alpha_2 = \varepsilon-\alpha, \quad \alpha_3 = \varepsilon+\alpha-2\varepsilon\frac{\lambda}{\lambda'}, \tag{11.116a}$$

$$\alpha_4 = \varepsilon-\alpha-2\varepsilon\frac{\lambda}{\lambda'},$$

where

$$\lambda = \frac{\sqrt{B^2l^2-1}}{Bl}, \quad \lambda' = \frac{\sqrt{4B^2l^2-1}}{2Bl}; \quad R_k = \sqrt{1+\left(\frac{\gamma+1}{4}\right)^2\frac{M_\infty^4\alpha_k^2}{B^2\lambda^2}}, \quad (k = 1, 2, 3, 4). \tag{11.116b}$$

As for ΔC_l, its expression is rather complicated; however, it will be given below since it is to be used later on for a comparison between the theoretical and experimental values:

$$\Delta C_l = 2(\gamma+1)\left[a_2-\frac{2}{B^2\lambda\lambda'}(I_1+H_{11})\right]M_\infty^2\alpha\varepsilon + \tfrac{1}{2}\left(\frac{\gamma+1}{2}\right)^2 M_\infty^4\alpha\left\{(\alpha^2+3\varepsilon^2)a_3-\frac{12\varepsilon^2}{B^3\lambda^2\lambda'}\left[I_2+H_{21}-\frac{\lambda}{\lambda'}(I_1+H_{12})\right]\right\}, \tag{11.117}$$

where the coefficient a_k $(k = 2, 3)$ is given by (11.107), and I_p and H_{pr} have the following expressions:

$$I_p = \frac{20Bl-1}{32B^2l^2}\left[1-\left(\frac{2}{\pi}\cos^{-1}\frac{1}{Bl}\right)^p\right] + \frac{5(6Bl-1)}{144B^2l^2}\left[1-\left(\frac{2}{\pi}\cos^{-1}\sqrt{\frac{3}{4B^2l^2-1}}\right)^p\right], \quad (p = 1, 2, 3), \tag{11.118a}$$

$$H_{pr} = \frac{1}{4Bl}\left[1-\tfrac{1}{3}\left(\frac{2}{\pi}\cos^{-1}\frac{1}{Bl}\right)^p\right]\left[1-\left(\frac{2}{\pi}\cos^{-1}\frac{1}{Bl}\right)^p\left(\frac{2}{\pi}\cos^{-1}\frac{1}{2Bl}\right)^r\right], \quad (p, r = 1, 2). \tag{11.118b}$$

If the deflexions τ do not exceed certain limits, we may apply the development (11.83); in such a case, the expression of the lift coefficient becomes much simpler:

$$C_l = \frac{4\alpha}{B} + \frac{2(\gamma+1)M_\infty^2\alpha\varepsilon}{B^2\lambda^2}\left(1-\frac{\lambda}{\lambda'}\right)$$
$$+\tfrac{1}{2}\left(\frac{\gamma+1}{2}\right)^2 \frac{M_\infty^4\alpha}{B^3\lambda^3}\left[3\varepsilon^2+\alpha^2-6\varepsilon^2\frac{\lambda}{\lambda'}\left(1-\frac{\lambda}{\lambda'}\right)\right]-\Delta C_l \quad (11.119)$$

where λ and λ' are given by (11.116) and ΔC_l by (11.117). The latter term may be neglected since it results from the high order approximations, as shown previously.

The wave drag coefficient will be determined by means of the simplified method, considering firstly the drag coefficient at zero lift, to which we add the drag due to the lift:

$$C_d = C_{d0} + \alpha C_l \quad (11.120)$$

where C_l is given by the above formulae (11.115) or (11.119).

For calculating C_{d0} we use the development (11.83) of the pressure coefficient which, together with formula (11.98) of the drag coefficient where we set $\alpha = 0$, leads to

$$C_{d0} = \frac{8\varepsilon^2}{3\pi B\lambda}\left[\frac{2\lambda}{\lambda'}\left(\pi-\cos^{-1}\frac{1}{2Bl}\right)+\cos^{-1}\frac{1}{Bl}\right]+\frac{2(\gamma+1)M_\infty^2\varepsilon^3}{B^2\lambda\lambda'}\left(1-\frac{\lambda}{\lambda'}\right)$$
$$+\tfrac{1}{2}\left(\frac{\gamma+1}{2}\right)^2\frac{M_\infty^4\varepsilon^4}{B^3\lambda^2\lambda'}\left(3-\frac{6\lambda}{\lambda'}+\frac{4\lambda^2}{\lambda'^2}\right)-\Delta C_{d0}, \quad (11.121)$$

where

$$\Delta C_{d0} = \frac{(\gamma+1)M_\infty^2\varepsilon^3}{B^2\lambda^2}\left[B^2\lambda^2a_2-2(I_2+\mathcal{J}_2)+4\frac{\lambda}{\lambda'}(I_1+H_{11})\right]$$
$$+\tfrac{1}{2}\left(\frac{\gamma+1}{2}\right)^2\frac{M_\infty^4\varepsilon^4}{B^3\lambda^3}\left\{B^3\lambda^3a_3-2(I_3+\mathcal{J}_3)\right.$$
$$\left.+6\frac{\lambda}{\lambda'}\left[I_2+H_{21}-2\frac{\lambda}{\lambda'}(I_1+H_{12})\right]\right\}, \quad (11.122a)$$

$$\mathcal{J}_k = \frac{1}{4Bl}\left[1-\left(\frac{2}{\pi}\cos^{-1}\frac{1}{Bl}\right)^k\right], \quad (k = 2, 3), \quad (11.122b)$$

the coefficients a_k, I_p and H_{pr} being given by relations (11.107) and (11.118). The quantity ΔC_{d0} is negligible.

Experimental Verifications. In order to verify the theoretical results obtained above, Fig. 11.30 compares the theoretical variation of the lift coefficient and the experimental values, and Fig. 11.31 the variation of the drag coefficient and the experimental values obtained at $M_\infty = 6{\cdot}9$ [L–23].

For the drag coefficient, the effect of friction has also been added, which has been estimated at $C_f = 0{\cdot}0028$ by various investigators.

Since boundary layer phenomena have an important effect on small models, particularly at $M_\infty = 6{\cdot}9$, we have considered an approximate mean value of this effect, which was added particularly at angles where the effect is more marked.

Thus, from a comparative analysis of the theoretical and experimental results, very good agreement is observed even for large angles ($\alpha = 16°$), a

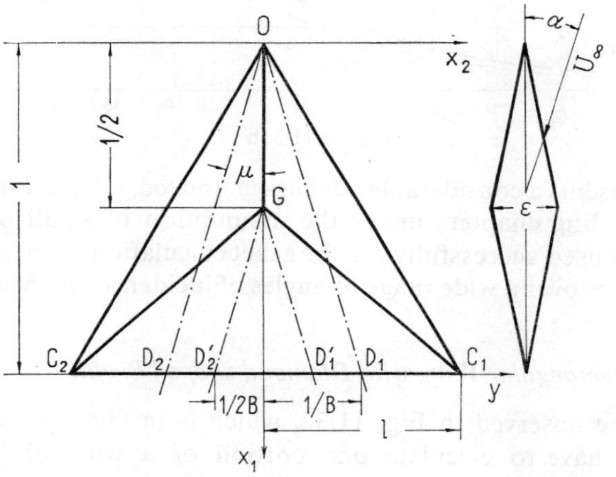

FIG. 11.30

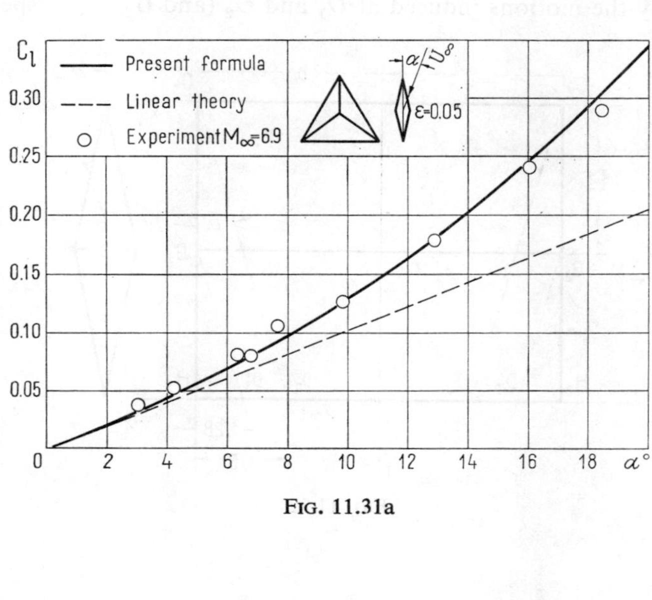

FIG. 11.31a

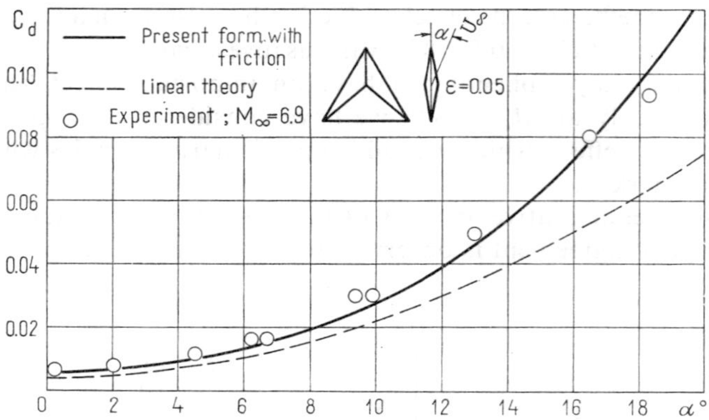

FIG. 11.31b

fact that presents a considerable advantage. Indeed, all the results obtained in the preceding chapters under the assumption of small perturbations may now be used successfully for the exact calculation of the aerodynamic characteristics over a wide range of angles of incidence and Mach numbers.

XI. 4.4. A Rectangular Wing with Diamond-shaped Profile

As may be observed in Fig. 11.32, which is in fact a reproduction of Fig. 7.9, we have to calculate one portion of a wing of infinite span $O_1D_1D_2O_2$ and two portions at each of the two ends of the wing O_1 and O_2 affected by the motions induced at O_1 and O_2 (and G_2, G_1, respectively).

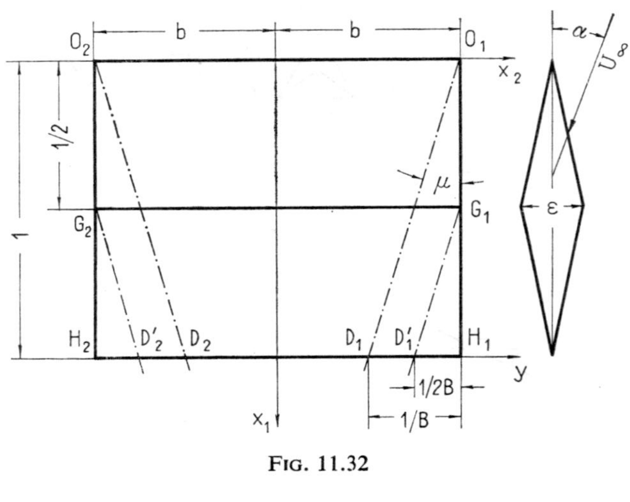

FIG. 11.32

These last portions are generally reduced, since we are dealing with rather high Mach numbers; in any case, the lines O_1D_1 and O_2D_2 do not cross on the wing.

For calculating the axial disturbance velocities, we may follow the same procedure as in (§ VII. 4.2). However, since on the central portion it is more accurate to use the closed formula (11.82), we shall use it and put

$$\gamma = \pm\varepsilon, \quad \Gamma = 2\varepsilon. \tag{11.123}$$

Further, taking into account the incidence α, we obtain

$$\tau_u = \gamma - \alpha, \quad \tau_l = \gamma + \alpha \tag{11.124}$$

on the upper and lower surfaces, respectively.

Finally, we may write successively:

The lift coefficient is given by the expression

$$C_l = \frac{2}{B}[(\varepsilon+\alpha)R_1 - (\varepsilon-\alpha)R_2] - \frac{\alpha}{bB^2} - \Delta C_l \tag{11.125}$$

where we have written

$$R_1 = \sqrt{1+\left(\frac{\gamma+1}{4}\right)^2 \frac{M_\infty^4(\varepsilon+\alpha)^2}{B^2}}, \quad R_2 = \sqrt{1+\left(\frac{\gamma+1}{4}\right)^2 \frac{M_\infty^4(\varepsilon-\alpha)^2}{B^2}}. \tag{11.126}$$

As regards ΔC_l, this term represents the effect of the Mach cones situated at the ends:

$$\Delta C_l = -\frac{9(\gamma+1)M_\infty^2\alpha\varepsilon}{50bB^3} + \left(\frac{\gamma+1}{4}\right)^2 \frac{M_\infty^4\alpha}{6B^4}\left(\tfrac{3}{4}\alpha^2 + \tfrac{11}{5}\varepsilon^2\right). \tag{11.127}$$

If instead of the closed formula (11.82) we use the expansion (11.83) for C_p, then the lift coefficient becomes

$$C_l = \frac{4\alpha}{B}\left(1 - \frac{1}{46B}\right) + \tfrac{1}{2}\left(\frac{\gamma+1}{2}\right)^2 \frac{M_\infty^4\alpha(\alpha^2+3\varepsilon^2)}{B^3} - \Delta C_l \tag{11.128}$$

where ΔC_l has the same significance as above (11.127).

The wave drag coefficient may be written in the simplified form

$$C_d = C_{d0} + \alpha C_l, \tag{11.129}$$

where C_{d0} is the drag coefficient at zero incidence, which may be calculated by applying formula (11.98),

$$C_{d0} = \frac{4\varepsilon^2}{B}\left[1 + \tfrac{1}{2}\left(\frac{\gamma+1}{4}\right)^2 \frac{M_\infty^4\varepsilon^2}{B^2}\right] - \left[\frac{7(\gamma+1)M_\infty^2\varepsilon^2}{40bB^2} + \tfrac{18}{31}\left(\frac{\gamma+1}{4}\right)^2 \frac{M_\infty^4\varepsilon^4}{bB^4}\right]. \tag{11.130}$$

EXPERIMENTAL VERIFICATIONS. The above calculations have been carried out in order to compare them with tests carried out on a similar model [L–23]. The rather high Mach number implies the application of the pressure coefficient *Cp* in *supersonic–hypersonic* flow.

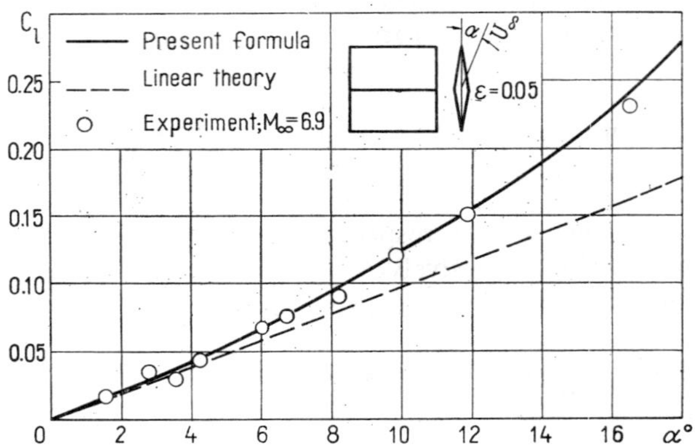

FIG. 11.33

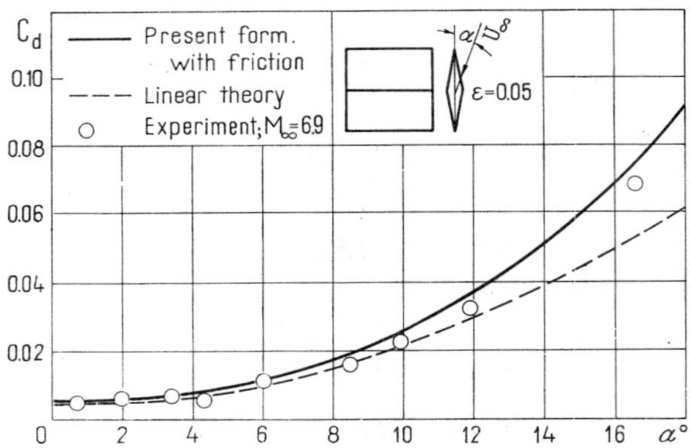

FIG. 11.34

From the curves of Fig. 11.33 for the lift coefficient and Fig. 11.34 for the drag coefficient, we can see a good agreement between theory and experimental data up to quite high angles of incidence ($\alpha = 16°$). The friction coefficient which has been estimated as being $C_f = 0{\cdot}0028$ has also been added to the drag coefficient. For small angles, a slightly increased value, estimated approximatively to be $C_{pl} \approx 0{\cdot}0010$, has also been taken, owing to the overpressures caused by the boundary layer.

XI. 4.5. Wing with Doubly Parabolic Profile

If the relative maximum thickness is also denoted by ε, the equation Z of a point P of the surface of a delta wing with parabolic profile is given by the following relation:

$$Z = 2\frac{\varepsilon}{l}(1-x_1)(x_1 l \mp x_2), \qquad (11.131)$$

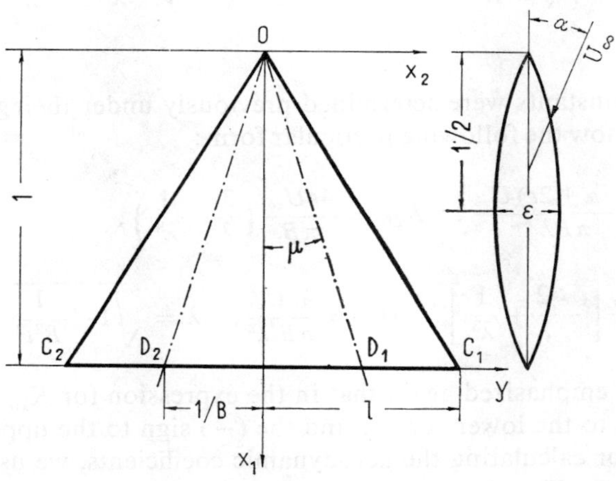

FIG. 11.35

where the $(-)$ sign refers to the right-hand leading edge and the $(+)$ sign to the left-hand leading edge (Fig. 11.35). If we also consider the incidence α, then the downwash on the upper wing surface is given by the equation

$$w = \left(-\alpha + \frac{\partial Z}{\partial x_1}\right) U_\infty = (-\alpha + 2\varepsilon)U_\infty - \frac{2\varepsilon}{l}(2x_1 l \mp x_2)U_\infty, \quad (11.132)$$

whence it can be seen that the wing consists of two components: the first, of order 1, having

$$w_{00} = (-\alpha \pm 2\varepsilon)U_\infty, \qquad (11.133)$$

where the $(+)$ sign refers to the upper surface and the $(-)$ sign to the lower surface; and a second one, of order $n = 2$, having

$$\frac{w}{x_1} = -\frac{2\varepsilon}{l}(2l \mp y)U_\infty, \quad \left(w_{10} = -4\varepsilon U_\infty, \quad w_{01} = \pm 2\frac{\varepsilon}{l}U_\infty\right), \qquad (11.134)$$

where the $(-)$ sign refers to the portion where the abscissae are positive and the $(+)$ sign to the portion where the abscissae are negative.

The functions $\mathcal{U}_0$ and $\mathcal{U}_1$, corresponding to the axial disturbance velocity for the first and second component, have the following expressions:

$$\mathcal{U}_0 = K_{10}\cos^{-1}\sqrt{\frac{1-B^2x^2}{B^2(l^2-x^2)}}, \tag{11.135a}$$

$$\mathcal{U}_1 = K_{20}\cos^{-1}\sqrt{\frac{1-B^2x^2}{B^2(l^2-x^2)}}+K_{21}x\cos^{-1}\sqrt{\frac{l^2(1-B^2x^2)}{l^2-x^2}}+D_{20}\sqrt{\frac{1-B^2x^2}{B^2}}, \tag{11.135b}$$

where the constants were determined previously under their general form, but assume now the following particular form:

$$K_{10} = -\frac{2(-\alpha\pm 2\varepsilon)\,U_\infty}{\pi B\lambda}, \quad K_{20} = \frac{4\varepsilon U_\infty}{\pi B}\left(\frac{3}{\lambda}-\frac{1}{\lambda^3}\right), \tag{11.136a}$$

$$K_{21} = \frac{2\varepsilon U_\infty}{\pi Bl}\left[-\frac{2}{\lambda}+\frac{1}{\lambda^3}\right], \quad D_{20} = \frac{4\varepsilon U_\infty}{\pi Bl\lambda^2}, \quad \lambda = \sqrt{1-\frac{1}{B^2l^2}}. \tag{11.136b}$$

It is to be emphasized again that in the expression for K_{10} the (+) sign corresponds to the lower surface and the (−) sign to the upper surface of the wing. For calculating the aerodynamic coefficients, we use the expansions (11.84) for C_p.

Further, applying formula (11.97) for the lift coefficient and formulae (11.98) for the wave drag total coefficient, we obtain successively:

For the *lift coefficient*,

$$C_l = \frac{4\alpha}{B}+\frac{2(\gamma+1)M_\infty^2(3-4\lambda^2)\alpha\varepsilon}{3B^2\lambda^4} + \frac{1}{2}\left(\frac{\gamma+1}{2}\right)^2\frac{M_\infty^4\alpha}{B^3\lambda^3}\left[\alpha^2+\varepsilon^2\left(10-\frac{11}{\lambda^2}+\frac{7}{2\lambda^4}\right)\right]-\Delta C_l, \tag{11.137}$$

where the correction term ΔC_l is given by the expression

$$\Delta C_l = \frac{4(\gamma+1)M_\infty^2}{3B^3\lambda^2 l}(3b_{20}-b_{11})\alpha\varepsilon + \left(\frac{\gamma+1}{4}\right)^2\frac{M_\infty^4\alpha}{B^4\lambda^3 l}\left[2(\alpha^2+12\varepsilon^2)b_{30}-\varepsilon^2(16b_{21}-3b_{12})\right], \tag{11.138}$$

and concerns the secondary terms u^2, u^3 inside the Mach cone, therefore negligible.

For the *drag coefficient* at zero incidence,

$$C_{d0} = \frac{8\varepsilon^2}{3B}\left[\left(1+\frac{1}{2\lambda^2}\right)-\left(\frac{3}{2}-\frac{1}{\lambda^2}\right)\frac{2}{\pi\lambda}\cos^{-1}\frac{1}{Bl}-\frac{1}{\pi Bl}\right]$$
$$+\frac{2(\gamma+1)M_\infty^2\varepsilon^3}{15B^2\lambda^2}\left(3-\frac{1}{\lambda^2}\right)\left(\frac{5}{\lambda^2}-6\right)$$
$$+\frac{2}{15}\left(\frac{\gamma+1}{4}\right)^2\frac{M_\infty^4\varepsilon^4}{B^3\lambda^3}\left(3-\frac{1}{\lambda^2}\right)\left(\frac{17}{\lambda^4}-\frac{48}{\lambda^2}+40\right)-\Delta C_{d0}, \qquad (11.139)$$

where ΔC_{d0} is given by the expression

$$\Delta C_{d0} = \frac{4\varepsilon}{Bl}\sum_{k=2}^{3}N_k\left(\frac{\varepsilon}{B\lambda}\right)^k\sum_{r=0}^{k}\frac{(-2)^{k-r}C_k^r}{(r+2)(r+3)}\left[(r+2)A_{k-r,r}-(r+1)b_{k-r,r}\right]. \qquad (11.140)$$

Like ΔC_l, ΔC_{d0} is negligible.

The coefficients N_k $(k = 2, 3)$ are given by the previous relation (11.102), while the coefficients $b_{k-r,r}$, $A_{k-r,r}$ have the following expressions:

$$b_{k-r,r} = \frac{2}{45}\left(\frac{B\lambda}{\varepsilon U_\infty}\right)^r\left[3\left(\frac{\pi}{2}K_{20}\right)^r+16\left(\frac{\pi}{2}K_{20}+\frac{\pi}{4B}K_{21}\right)^r\right.$$
$$\left.-3\left(\frac{2}{\pi}\cos^{-1}\frac{1}{Bl}\right)^{k-r}u_1^r(0)-16\left(\frac{2}{\pi}\cos^{-1}\sqrt{\frac{3}{4B^2l^2-1}}\right)^{k-r}u_1^r(1/2B)\right], \qquad (11.141a)$$

$$A_{k-r,r} = \frac{4}{9Bl}\left(\frac{B\lambda}{\varepsilon U_\infty}\right)^r\left[-\frac{3}{16}\left(\frac{\pi}{2}K_{20}\right)^r+\left(\frac{\pi}{2}K_{20}+\frac{\pi}{4B}K_{21}\right)^r\right.$$
$$\left.+\frac{3}{16}\left(\frac{2}{\pi}\cos^{-1}\frac{1}{Bl}\right)^{k-r}u_1^r(0)-\left(\frac{2}{\pi}\cos^{-1}\sqrt{\frac{3}{4B^2l^2-1}}\right)^{k-r}u_1^r(1/2B)\right],$$
$$(k = 2, 3;\ r = 0, 1, \ldots k), \qquad (11.141b)$$

$u_1(0)$ and $u_1(1/2B)$ being the real values of $\mathcal{U}_1$ (11.135b) for $x = 0$ and $x = 1/2B$, respectively.

The above calculations have been used for plotting the lift and drag coefficients for Mach numbers $M_\infty = \sqrt{10}$ and $M_\infty = \sqrt{37}$.

The results obtained under the assumption of small perturbations gradually diverge from the new improved formula as the incidence and the Mach number increase (Fig. 11.36).

For certain values of the incidence and of the Mach number, we can no longer restrict ourselves to linearized theory, and we must apply the more exact formula for the pressure coefficient.

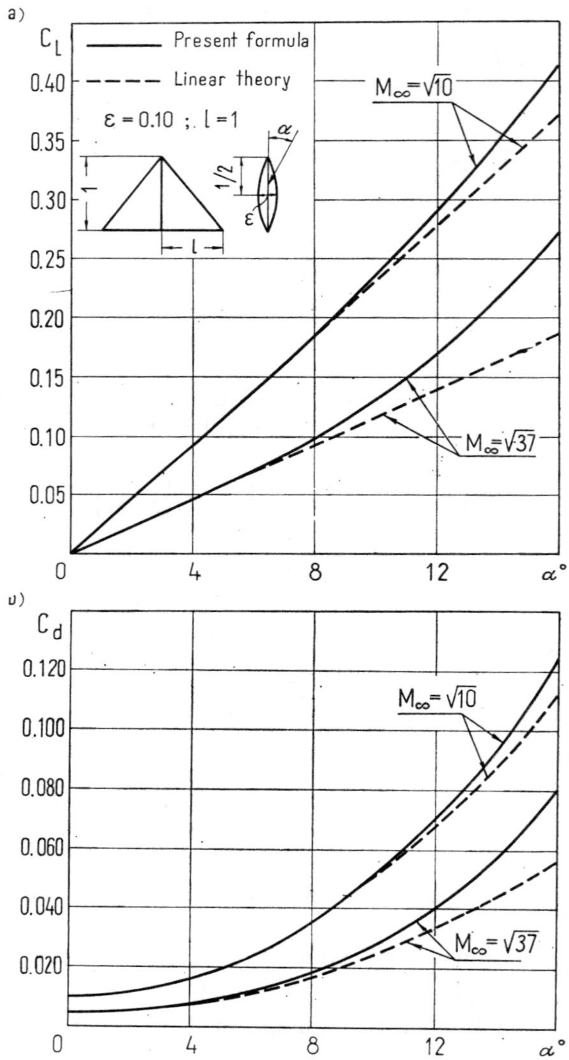
a)
C_L
Present formula
Linear theory
ε = 0.10 ; l = 1
1
l
1/2
α
ε
$M_\infty = \sqrt{10}$
$M_\infty = \sqrt{37}$
0.40
0.35
0.30
0.25
0.20
0.15
0.10
0.05
0
4
8
12
α°
b)
C_d
0.120
0.100
0.080
0.060
0.040
0.020
$M_\infty = \sqrt{10}$
$M_\infty = \sqrt{37}$
0
4
8
12
α°

FIG. 11.36

GENERAL CONCLUSION. Although the calculations are very laborious, they may be performed without intricate operations. Obviously, in view of the comparison made with the experimental results, it is of importance to obtain the accurate results above.

In fact, the numerical examples were given above in view of emphasizing this consistency and in order to turn to account the results obtained under the assumption of small disturbances by extending them over a much wider range including also the moderate hypersonic flow régime.

CHAPTER XII

SOME THEORETICAL CONSIDERATIONS ON WINGS WITH CURVED LEADING EDGES AND WING-FUSIFORM BODY SYSTEMS

THE present chapter will be devoted to the study of certain motions that are very close to conical ones, for which reason they will be termed *quasi-conical motions*, and a corresponding theory will be set forth for these motions. A general method for calculating the pressures and aerodynamic characteristics of wings with curved leading edges, instead of the straight leading edges studied so far in our previous chapters, has been elaborated on the basis of this theory. The same theory will be applied to other more complex systems, as for instance a wing fitted with a slender body, etc.

XII. 1. Definition of Quasi-conical Motions

XII. 1.1. Determination of the Potential of a Conical Motion of Arbitrary Order n

Let us consider a conical motion ($n = 1$) around an angular wing referred to the ordinary system of axes Ox_1x_2 (Fig. 12.1) and let u be the axial disturbance velocity and v the lateral velocity in the direction of the Ox_2 axis.

The velocity potential at *points lying on the wing* ($z = 0$) will be given by the known relation

$$\Phi_1 = x_1u+x_2v = x_1(u+yv) = x_1 \operatorname{Re}(\mathcal{U}+x\mathcal{V}), \tag{12.1}$$

where $\mathcal{U}$ and $\mathcal{V}$ have the customary significance:

$$\mathcal{U} = u+iu', \quad \mathcal{V} = v+iv', \tag{12.2}$$

being linked by the compatibility relation put in the following forms:

$$\mathcal{U} = -\int x\, d\mathcal{V} = -x\mathcal{V}+\int \mathcal{V}\, dx, \tag{12.3a}$$

$$\mathcal{V} = -\int \frac{1}{x}\, d\mathcal{U} = -\frac{1}{x}\mathcal{U}+\int \mathcal{U}\, d\left(\frac{1}{x}\right). \tag{12.3b}$$

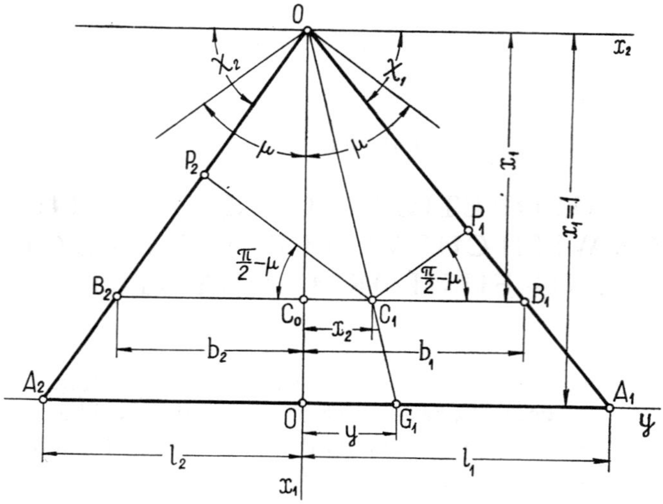

FIG. 12.1

Introducing any of these relations into (12.1), we obtain respectively

$$\frac{1}{x_1}\Phi_1 = \operatorname{Re} x \int \mathcal{U}\, d\left(\frac{1}{x}\right) = \operatorname{Re}\int \mathcal{V}\, dx. \tag{12.4}$$

We can find further a general formula for the n order conical motion. To this end, we shall write successively

$$\begin{aligned}
\frac{n!}{x_1^n}\Phi_n &= \operatorname{Re}\,(F_{n,0,0}+C_n^1 xF_{n-1,1,0}+\cdots+C_n^n x^n F_{0,n,0}) \\
&= \operatorname{Re}\, x^n \int d\left(\frac{F_{n,0,0}}{x^n}+C_n^1\frac{F_{n-1,1,0}}{x^{n-1}}+\cdots+C_n^n F_{0,n,0}\right) \\
&= \operatorname{Re}\, x^n \left\{\int \frac{1}{x^n}\,(dF_{n,0,0}+C_n^1 x\, dF_{n-1,1,0}+\cdots+C_n^n x^n\, dF_{0,n,0})\right. \\
&\quad + \int\left(F_{n,0,0}+C_n^1\frac{n-1}{n}xF_{n-1,1,0}+C_n^2\frac{n-2}{n}x^2F_{n-2,2,0}\right. \\
&\quad \left.\left. +\cdots+C_n^{n-1}\frac{1}{n}x^{n-1}F_{1,n-1,0}\right)d\left(\frac{1}{x^n}\right)\right\} \\
&= \operatorname{Re}\, x^n\left\{\int\frac{dF_{n,0,0}}{x^n}(1-C_n^1+C_n^2-C_n^3+\cdots+(-1)^nC_n^n)\right. \\
&\quad +(F_{n,0,0}+C_{n-1}^1 xF_{n-1,1,0}+C_{n-1}^2x^2F_{n-2,2,0} \\
&\quad \left. +\cdots+C_{n-1}^{n-1}x^{n-1}F_{1,n-1,0})\, d\left(\frac{1}{x^n}\right)\right\}.
\end{aligned} \tag{12.5}$$

If we notice that the last bracketed term is just $(n-1)!\,\mathcal{U}_{n-1}$, while in the first term the bracketed sum is equal to zero, we can write finally

$$\frac{n}{x_1^n}\Phi_n = \operatorname{Re} x^n \int \mathcal{U}_{n-1}\, d\left(\frac{1}{x^n}\right). \tag{12.6}$$

Following the same way, we find analogously

$$\frac{n!}{x_1^n}\Phi_n = \operatorname{Re}\left\{\int (dF_{n,0,0} + C_n^1 x\, dF_{n-1,1,0} + C_n^2 x^2\, dF_{n-2,2,0} + \ldots + x^n\, dF_{0,n,0})\right.$$
$$+ \int n (F_{n-1,1,0} + C_{n-1}^1\, xF_{n-2,2,0} + C_{n-2}^2\, x^2 F_{n-3,3,0}$$
$$\left. + \cdots + x^{n-1} F_{0,n,0})\, dx\right\}. \tag{12.7}$$

Applying the compatibility relations of the first bracket and noticing that the second term under the integral is just $(n-1)!\,\mathcal{V}_{n-1}$, we can write

$$\frac{n!}{x^n}\Phi_n = \operatorname{Re}\left\{\int (1 - C_n^1 + C_n^2 - \cdots + (-1)^n C_n^n)\, dF_{n,0,0} + n! \int \mathcal{V}_{n-1}\, dx\right\}$$
$$= \operatorname{Re} n! \int \mathcal{V}_{n-1}\, dx. \tag{12.8}$$

THE CASE WHEN THERE EXISTS A CENTRAL CONICAL BODY. If the wing is fitted with any conical body, $z \neq 0$ in its vicinity, and hence, the application of Euler's formula for determining the potential yields

$$\Phi_1 = x_1 u + x_2 v + x_3 w = x_1(u + yv + zw). \tag{12.9}$$

It has been shown previously (9.25) that in the case when the dimensions of the central body are small ($B^2c^2 \ll 1$) we have successively:

$$d\mathcal{V} = d(v + iv') = -\frac{id(w + iw')}{\sqrt{1 - B^2x^2}} \approx -id(w + iw') \tag{12.10}$$

whence we obtain

$$d(v' + w) = 0, \quad v' + w = w_0, \tag{12.11}$$

w_0 being the downwash corresponding to the incidence α of the wing–body system in the plane x, since $v' = 0$ on this system, as follows from the general formula for $\mathcal{V}$.

In this case, on setting $z \approx \mathfrak{z}$ for points lying around the body and taking also account of (12.4), the expression (12.9) for the potential becomes

$$\frac{1}{x_1}\Phi_1 = u+yv+z(w_0-v') \approx u+yv-zv'+zw_0$$

$$= \mathrm{Re}\,(\mathcal{U}+x\mathcal{V})+zw_0 = \mathrm{Re}\; x\int \mathcal{U}\, d\left(\frac{1}{x}\right)+zw_0$$

$$= \mathrm{Re}\int \mathcal{V}\, dx+zw_0 \qquad (12.12)$$

on the wing and on the body.

It should be, however, remarked that the term zw_0 plays no part in the calculation of the axial disturbance velocity, since this is to be determined by differentiating Φ_1 with respect to x_1. We have seen, however, that we have to take account of this term in the case of oscillations (10.89).

XII. 1.2. Examples of Applications

In view of the important role of the conical potential Φ in the theory of *quasi-conical* motions which will be presented below, we shall give several examples of applications, some of which will be used later on for solving certain interesting and practical problems in modern aviation.

(a) FLAT TRIANGULAR WING WITH SUBSONIC EDGES. The axial disturbance velocity function $\mathcal{U}$ for a triangular wing (Fig. 12.1) having a constant incidence α, can be deduced from (5.31):

$$\frac{1}{\alpha U_\infty}\mathcal{U} = \frac{2l_1l_2+(l_1-l_2)x}{2A\sqrt{(l_1-x)(l_2+x)}}\,; \qquad (12.13)$$

hence, the expression for the conical potential results readily

$$\frac{1}{\alpha U_\infty}\Phi_1 = \mathrm{Re}\,\frac{x_1\sqrt{(l_1-x)(l_2+x)}}{A} = \frac{\sqrt{(b_1-x_2)(b_2+x_2)}}{A}, \qquad (12.14a)$$

$$b_1 = x_1l_1, \quad b_2 = x_1l_2 \quad (l_1 = \text{const.}, \quad l_2 = \text{const.}) \qquad (12.14b)$$

where the coefficient A derives from the well-known constant given by (5.32).

(b) FLAT DELTA WING WITH SUBSONIC EDGES. On setting

$$l_1 = l_2 = l, \quad b = x_1 l \qquad (12.15)$$

in the above expressions, we find respectively

$$\frac{1}{\alpha U_\infty}\mathcal{U} = \frac{1}{E(k)} \cdot \frac{1}{\sqrt{l^2-x^2}}, \quad k = \sqrt{1-B^2l^2} \tag{12.16}$$

for the velocity function, and

$$\frac{1}{\alpha U_\infty}\Phi_1 = \mathrm{Re}\,\frac{x_1\sqrt{l^2-x^2}}{E(k)} = \frac{\sqrt{b^2-x_2^2}}{E(k)} \tag{12.17}$$

for the potential, where $E(k)$ is the complete elliptical integral of the second kind, having the modulus k.

(c) Reverse Example. Conical Motion about a Wing–Conical Body System. Consider now a potential of the form (12.14) but in an auxiliary plane X,

$$\frac{1}{\alpha U_\infty}\Phi_1 = \mathrm{Re}\,Qx_1\sqrt{L^2-X^2}. \tag{12.18}$$

Assume further that the auxiliary plane X proceeds from x the transformation:

$$X = x+\frac{c^2}{x} \quad \left(L = l+\frac{c^2}{l}\right). \tag{12.19}$$

By means of this transformation, the trace A_2A_1 of the fictitious wing in the plane X turns into a *wing–body system*, having the configuration given in Fig. 12.2.

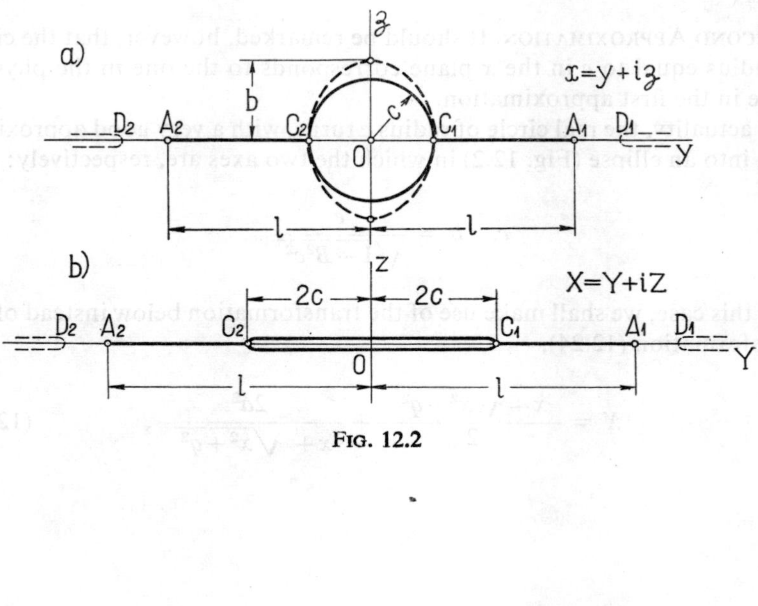

Fig. 12.2

The potential (12.18) is also valid in the plane x at points lying on the wing ($z = 0$) and it has been demonstrated earlier that is also valid for points lying on the body, provided that we have $B^2c^2 \ll 1$, an assumption that was made from the beginning.

In this case, however, the term zw_0 is also to be added to the above potential, thus obtaining

$$\frac{1}{\alpha U_\infty}\Phi_1 = x_1\left(\operatorname{Re} Q\sqrt{(l^2-x^2)\left(1-\frac{c^4}{l^2x^2}\right)}+zw_0\right)$$
$$= \operatorname{Re} Q\sqrt{(b^2-(xx_1)^2)\left(1-\frac{c^4x_1^4}{b^2(x_1x)^2}\right)}+x_3w_0. \quad (12.20)$$

We note that, as a matter of fact, this last term plays no part in the calculation of the axial velocity, which will be found by differentiating the above expression with respect to x_1. To this effect, we have inserted x_1 under the radical and, taking account of the fact that x_1x and c are independent in terms of x_1, we find easily

$$\frac{u}{\alpha U_\infty} = \frac{1}{\alpha U_\infty}\frac{\partial\Phi_1}{\partial x_1} = \operatorname{Re} Ql^2\frac{1+\dfrac{c^4}{l^4}-2\dfrac{c^4}{l^4}\dfrac{l^2}{x^2}}{\sqrt{(l^2-x^2)\left(1-\dfrac{c^4}{l^4}\dfrac{l^2}{x^2}\right)}}. \quad (12.21)$$

Thus we have refound the axial disturbance velocity established formerly (9.53) for the case of the thin wing with subsonic edges, where Q has the value of formula (9.58), divided, however, by $\alpha_0U_\infty l^2$.

SECOND APPROXIMATION. It should be remarked, however, that the circle of radius equal to c in the x plane corresponds to the one in the physical plane in the first approximation.

In actuality, the real circle of radius c turns, with a very good approximation, into an ellipse (Fig. 12.2) in which the two axes are, respectively:

$$c, \quad b = \frac{c}{\sqrt{1-B^2c^2}}.$$

In this case, we shall make use of the transformation below instead of the transformation (12.24),

$$X = \frac{x+\sqrt{x^2+q^2}}{2}+\frac{2a^2}{x+\sqrt{x^2+q^2}}, \quad (12.22)$$

where

$$a = \frac{1+\sqrt{1-B^2c^2}}{2\sqrt{1-B^2c^2}}\, c, \quad q^2 = \frac{B^2c^2}{1-B^2c^2}\, c^2.$$

Further we have

$$\frac{1}{\alpha U_\infty}\Phi_1 = \operatorname{Re} Qx_1\sqrt{L^2-X^2} = \operatorname{Re}\tfrac{1}{2}Qx_1$$

$$\times\sqrt{[(l+\sqrt{l^2+q^2})^2-(x+\sqrt{x^2+q^2})^2]\left[1-\frac{16a^4}{(l+\sqrt{l^2+q^2})^2(x+\sqrt{x^2+q^2})^2}\right]}, \tag{12.23a}$$

and consequently

$$\frac{u}{\alpha U_\infty} = \frac{1}{\alpha U_\infty}\frac{\partial\Phi_1}{\partial x_1} = \operatorname{Re} Q$$

$$\times\frac{(l+\sqrt{l^2+q^2})^2\left[1+\dfrac{16a^4}{(l+\sqrt{l^2+q^2})^4}\right]-\dfrac{x+\sqrt{x^2+q^2}}{\sqrt{x^2+q^2}}\left[\dfrac{16a^4}{(x+\sqrt{x^2+q^2})^2}+q^2\right]}{2\sqrt{(l+\sqrt{l^2+q^2})^2-(x+\sqrt{x^2+q^2})^2}\sqrt{1-\dfrac{16a^4}{(l+\sqrt{l^2+q^2})^2(x+\sqrt{x^2+q^2})^2}}} \tag{12.23b}$$

which reduces to (12.21) when

$$B^2c^2 \ll 1, \quad b \approx a \approx c, \quad q^2 \approx 0.$$

Note. As mentioned formerly, we have to add to the value of the axial velocity given in the above formulae, the one referring to the symmetrical axial motion around the revolution cone considered alone, corresponding to the wing without incidence (§ I. 6.2).

XII. 1.3. Quasi-conical Potential and Axial Velocity

Before entering into the essence of the problem, we shall recall that in certain problems of three-dimensional motion, as for instance the lateral motion of a slender body (Fig. 12.3), it is assumed that the motion potential in an arbitrary section x_1, as well as for a cylinder of circular section having the radius c, is given by the expression

$$\varphi = -w_\infty\left(r+\frac{c^2}{r}\right)\cos\omega \tag{12.24a}$$

whence

$$u = \frac{\partial \varphi}{\partial x_1} = -w_0 \frac{2c}{r} \cos \omega \frac{dc}{dx_1} \tag{12.24b}$$

for the axial velocity *u*.

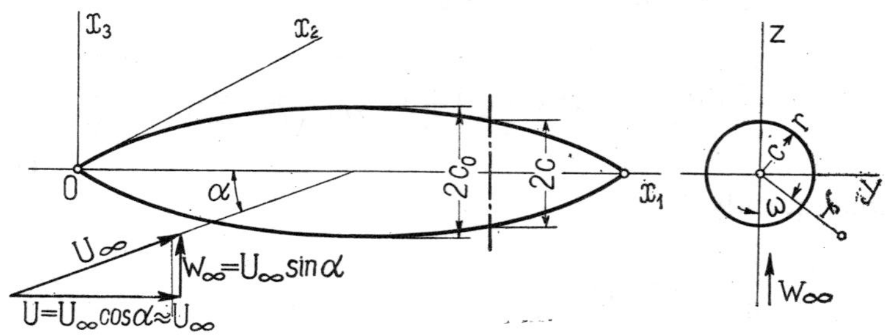

FIG. 12.3

This fact also results from the classical theory elaborated by Munk for the incompressible motions ($M_\infty \approx 0$) and by Jones for the compressible motions ($M_\infty^2 \gg 0$), which proceeds from the assumption that, in the case of slender bodies, the variation of the velocity u with respect to x_1 is sufficiently slow in order to enable us to neglect the term containing the second derivative with respect to x_1 of the potential equation having the form (1.36); thus one can write

$$\frac{\partial^2 \Phi}{\partial x_2^2} + \frac{\partial^2 \Phi}{\partial x_3^2} = (M_\infty^2 - 1) \frac{\partial^2 \Phi}{\partial x_1^2} = (M_\infty^2 - 1) \frac{\partial u}{\partial x_1} \approx 0 . \tag{12.25}$$

In such a manner, the problem is reduced to the determination of a *quasi-plane potential* which should satisfy the two-dimensional Laplace equation in any plane normal to x_1. It is in this way that the aerodynamic characteristics of slender wings, fitted with central bodies or not, have been deduced.

Let us now consider the change of variables

$$x_1' = x_1, \ y = \frac{x_2}{x_1}, \ z = \frac{x_3}{x_1}, \tag{12.26}$$

with the aid of which the potential and the disturbance velocities can be set as functions of x_1', y and z.

In view of this change of variables, the potential linearized equation can be written in the form

$$-B^2\frac{\partial^2\Phi}{\partial x_1^2}+\frac{\partial^2\Phi}{\partial x_2^2}+\frac{\partial^2\Phi}{\partial x_3^2}=-B^2\frac{\partial u}{\partial x_1}+\frac{\partial v}{\partial x_2}+\frac{\partial w}{\partial x_3}$$

$$=-B^2\frac{\partial u}{\partial x_1'}+\frac{B^2}{x_1'}\left(y\frac{\partial u}{\partial y}+z\frac{\partial u}{\partial z}\right)+\frac{1}{x_1'}\left(\frac{\partial v}{\partial y}+\frac{\partial w}{\partial z}\right)=0 \qquad (12.27a)$$

or

$$\frac{\partial v}{\partial y}+\frac{\partial w}{\partial z}+B^2\left(y\frac{\partial u}{\partial y}+z\frac{\partial u}{\partial z}\right)-B^2x_1'\frac{\partial u}{\partial x_1'}=0. \qquad (12.27b)$$

Let us examine this new form of the potential.

When the cross dimensions of the body we are dealing with are small and vary slowly with respect to x_1', the last two terms can be neglected and then we fall again upon Jones's approximation which reduces the potential of the real motion to a *quasi-plane potential.*

When the body has a conical shape, the disturbance velocities are constant on a radius vector (y = const., z = const.) and do not vary with x_1', and hence the last term of equation (12.27*b*) is identically zero and reduces to the equation of the conical motion potential

$$\frac{\partial v}{\partial y}+\frac{\partial w}{\partial z}+B^2\left(y\frac{\partial u}{\partial y}+z\frac{\partial u}{\partial z}\right)=0 \qquad (12.28)$$

which was treated in our previous chapters.

If the elongated body has a shape which varies slowly with respect to x_1', the disturbance velocities also vary slowly along a radius vector (and, hence, with respect to x_1', too) and thus, without imposing any restriction regarding the magnitude of the cross dimensions of the body with respect to the Mach cone, we can neglect the last term of equation (12.27b) and consequently obtain

$$\frac{\partial v}{\partial y}+\frac{\partial w}{\partial z}+B^2\left(y\frac{\partial u}{\partial y}+z\frac{\partial u}{\partial z}\right)=B^2x_1'\frac{\partial u}{\partial x_1'}\approx 0. \qquad (12.29)$$

In this manner, the real motion potential is reduced to a *quasi-conical potential* which satisfies, with a very good approximation, the motion conditions, and which can be deduced through the comparison with the conical motion.

As an example, we shall consider two cases: the wing having slightly curved edges and the delta wing–elongated body system.

XII. 1.4. The Wing with Slightly Curred Edges

Let us consider the *conical potential* for a triangular wing with subsonic edges (12.14) and assume that the curve of the leading edge varies according to an arbitrary function of x_1 (Fig. 12.4); in this case, one obtains a quasi-conical potential, which will be denoted by Φ_1^*, and the *quasi-conical velocity* u^* easily results from its differentiation:

$$\frac{u^*}{\alpha U_\infty} = \frac{1}{\alpha U_\infty}\frac{\partial \Phi_1^*}{\partial x_1} = \frac{b_2\dfrac{db_1}{dx_1}+b_1\dfrac{db_2}{dx_1}+x_2\left(\dfrac{db_1}{dx_1}-\dfrac{db_2}{dx_1}\right)}{2A^*\sqrt{(b_1-x_2)(b_2+x_2)}} - \frac{\sqrt{(b_1-x_2)(b_2+x_2)}}{A^{*2}}\,\frac{dA^*}{dx_1}. \tag{12.30}$$

In this expression we have employed again the initial coordinates x_1, x_2, x_3.

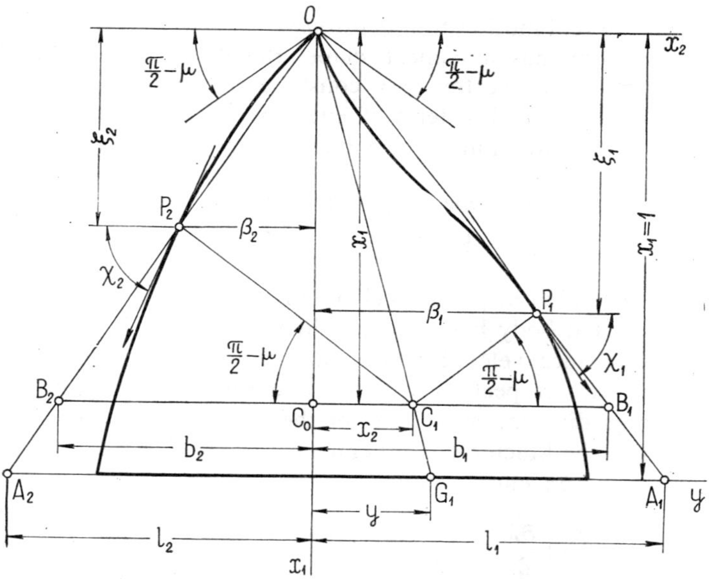

FIG. 12.4

When the leading edges are straight, the above formula reduces to the former one (12.13).

It is to be noticed that, at a point C_1 of coordinates x_1, x_2, the wing influence reduces to that of the portion delimited by the Mach lines C_1P_1 and C_1P_2 (Fig. 12.1), l_1 and l_2 remaining constant for every y. In the case of curved edges, the wing influence also reduces to that of the portion delimited by C_1P_1 and C_1P_2, but l_1 and l_2 as well as b_1 and b_2 represent now the magnitudes indicated in Fig. 12.4. Thus, we shall put, analogously with the triangular wing,

$$l_1 = \frac{b_1}{x_1} = \frac{\beta_1}{\xi_1}, \quad l_2 = \frac{b_2}{x_1} = \frac{\beta_2}{\xi_2}, \tag{12.31}$$

where ξ_1 and ξ_2 are given by the following relations,

$$\xi_1 + B\beta_1 = \xi_1(1+Bl_1) = x_1 + Bx_2, \tag{12.32a}$$

$$\xi_2 + B\beta_2 = \xi_2(1+Bl_2) = x_1 - Bx_2, \tag{12.32b}$$

β_1 and β_2 being, of course, functions of ξ_1 and ξ_2, which can be deduced from these two relations in terms of x_1.

From the above equations we further have:

$$b_1 = \frac{x_1}{\xi_1}\beta_1 = \frac{1+Bl_1}{1+By}\beta_1, \tag{12.33a}$$

$$b_2 = \frac{x_1}{\xi_2}\beta_2 = \frac{1+Bl_2}{1-By}\beta_2. \tag{12.33b}$$

For y tending to l_1, we can write with a good approximation,

$$\frac{db_1}{dx_1} \approx \frac{d\beta_1}{dx_1} \approx \frac{d\beta_1}{d\xi_1}, \tag{12.34}$$

a relation which is very accurate for $y = l_1$, when $\xi_1 = x_1$ too. On the other hand, one can see from (12.30) that, for $y = l_1$ $(x_2 = b_1)$, the influence of the term db_2/dx_1 is cancelled irrespective of its value. Therefore, the importance of this latter term is small even in the case when $y < l_1$.

Taking into account this remark and considering further that the coefficients of β_1 and β_2 from (12.33) vary only slightly with respect to x_1, we can assume them to remain almost constant. In this case we can write, in an approximate manner,

$$\frac{\xi_1}{x_1} \approx \text{const.}, \quad \frac{\xi_2}{x_1} \approx \text{const.}, \tag{12.35}$$

and deduce from (12.33), in the framework of this approximation:

$$\frac{db_1}{dx_1} \approx \frac{d\beta_1}{d\xi_1} = \lambda_1 = \cot\chi_1, \tag{12.36a}$$

$$\frac{db_2}{dx_1} \approx \frac{d\beta_2}{d\xi_2} = \lambda_2 = \cot\chi_2. \tag{12.36.b}$$

This latter relation, although very approximate, as well as the second equation of (12.35), are nevertheless acceptable, because of the small importance of the term λ_2 for the right-hand half of the wing.

With the aid of these simplifications and notations, formula (12.30) will be set finally in the following form:

$$\frac{u^*}{\alpha U_\infty} = \frac{l_2\lambda_1 + l_1\lambda_2 + (\lambda_1 - \lambda_2)y}{2A^*\sqrt{(l_1 - y)(l_2 + y)}} - \frac{x_1\sqrt{(l_1 - y)(l_2 + y)}}{A^{*2}} \frac{dA^*}{dx_1}. \tag{12.37}$$

If the wing is symmetrical, the above formulae remain the same and it is only the point situated at the middle of the wing ($y = 0$) that benefits by the property of symmetry, in which case we shall set

$$l_1 = l_2 = l_0, \quad \lambda_1 = \lambda_2 = \lambda_0. \tag{12.38}$$

Nevertheless, it should be remarked now that, in the case when the leading edges of a delta wing do not deviate too much from a mean straight line, taking also account of (12.36), one can write

$$l_1 \approx l_2 \approx l, \quad \lambda_1 \approx \lambda_2 \approx \lambda \tag{12.39}$$

in which case the above expression of u^* can be set in the following form:

$$\frac{u^*}{\alpha U_\infty} = \frac{l\lambda}{E(k)\sqrt{l^2 - y^2}} - \frac{x_1\sqrt{l^2 - y^2}}{E(k)^2} \frac{dE(k)}{dx_1}, \tag{12.40}$$

which is similar to the one given in the first approximation in our previous [(C−10)] paper and where $E(k)$ is the complete elliptical integral of the second kind, having the modulus k given by (12.16).

XII. 1.5. Determination of the Constant A^*

The determination of this constant constitutes a more difficult operation since there is no direct way for this operation, but an indirect one, by analogy and by considerations on the boundary conditions.

From Fig. 12.4 it may be seen that we can assimilate the real wing with a triangular wing OA_1A_2 passing through the points P_1 and P_2 and hence having l_1 and l_2 as parameters of the leading edges; or with a fictitious triangular wing characterized by the parameters λ_1 and λ_2 corresponding to the tangents at P_1 and P_2 to the leading edges. In fact, the values of the constant A, determined in these two ways, differ only slightly.

Therefore, taking into account the behaviour of the axial velocity at both leading edges, we shall imagine a triangular wing whose two leading edges assumed to be straight—should have λ_1 and λ_2 instead of l_1 and l_2 which appear in the expression of the axial velocity indicated previously (12.13).

In this case, we shall write, similarly to the latter formula,

$$\frac{u}{\alpha U_\infty} = \frac{2\lambda_1\lambda_2+(\lambda_1-\lambda_2)y}{2A\sqrt{(\lambda_1-y)(\lambda_2+y)}}. \tag{12.41}$$

At each of the leading edges, the behaviour of the axial velocity corresponds precisely to the local conditions, in keeping, as a matter of fact, with the fundamental assumptions in the conical motions. Indeed, when y tends to λ_1, we can write

$$\left(\frac{u}{\alpha U_\infty}\right)_{y\to\lambda_1} = \frac{\sqrt{\lambda_1(\lambda_1+\lambda_2)}}{2A\sqrt{1-y/\lambda_1}}, \tag{12.42}$$

which must be equivalent to the behaviour of the axial velocity deduced from (12.37), when y tends to l_1:

$$\left(\frac{u^*}{\alpha U_\infty}\right)_{y\to l_1} = \frac{\lambda_1\sqrt{l_1+l_2}}{2A^*\sqrt{l_1}\sqrt{1-y/l_1}}. \tag{12.43}$$

We obtain similar relations for the case when $y\to -l_2$.

From the above two expressions we deduce

$$A^* = \sqrt{\frac{\lambda_1(l_1+l_2)}{l_1(\lambda_1+\lambda_2)}}\,A \tag{12.44}$$

which is valid for the right-hand side of the wing and which will be introduced into (12.37). It should be noticed, however, that the above given expression of A^* is very complicated in order to be also introduced into the second term in the right-hand side of (12.37). However, apart from the fact that this term is of secondary order, it is zero for $y = l_1$ and negligible for the values of y in the proximity of the leading edge ($y \to l_1$); whereas, without making an important error on the whole, we can write approximately

$$\sqrt{\frac{\lambda_1(l_1+l_2)}{l_1(\lambda_1+\lambda_2)}} \approx \text{const.} \tag{12.45}$$

with respect to x_1.

In this case, the axial disturbance velocity will finally have the following expression:

$$\frac{u^*}{\alpha U_\infty} \sqrt{\frac{l_1(\lambda_1+\lambda_2)}{\lambda_1(l_1+l_2)}} \left\{\frac{l_2\lambda_1+l_1\lambda_2+(\lambda_1-\lambda_2)y}{2A\sqrt{(l_1-y)(l_2+y)}} - \frac{x_1\sqrt{(l_1-y)(l_2+y)}}{A^2}\frac{dA}{dx_1}\right\}. \tag{12.46a}$$

It remains now to establish the formula of the constant A which is, analogously with the results obtained in the theory of triangular wings (5.32),

$$A = \frac{(1+B\lambda_1)(1+B\lambda_2)E(k_1)-B(\lambda_1+\lambda_2)K(k_1)}{\sqrt{(1+B\lambda_1)(1+B\lambda_2)}}, \tag{12.46b}$$

where $K(k_1)$ and $E(k_1)$ are complete elliptical integrals of the first and second kind, having the modulus

$$k_1 = \sqrt{\frac{(1-B\lambda_1)(1-B\lambda_2)}{(1+B\lambda_1)(1+B\lambda_2)}}. \tag{12.47}$$

Let us further set

$$2\lambda = \lambda_1+\lambda_2 \tag{12.48}$$

and notice that we can write in an approximate manner

$$(1-B\lambda_1)(1-B\lambda_2) \approx (1-B\lambda)^2, \tag{12.49a}$$

$$(1+B\lambda_1)(1+B\lambda_2) \approx (1+B\lambda)^2, \tag{12.49b}$$

whence

$$k_1 \approx \frac{1-B\lambda}{1+B\lambda}. \tag{12.50}$$

Further, denoting by k and k', respectively,

$$k = \sqrt{1-B^2\lambda^2}, \qquad k' = \sqrt{1-k^2} = B\lambda, \tag{12.51}$$

and noticing that, taking account of these moduli, the functional relations between the complete elliptical integrals yield

$$K\left(\frac{1-B\lambda}{1+B\lambda}\right) = \frac{1+B\lambda}{2} K(k), \tag{12.52a}$$

$$E\left(\frac{1-B\lambda}{1+B\lambda}\right) = \frac{1}{1+B\lambda}\left(E(k)+B\lambda K(k)\right), \tag{12.52b}$$

respectively, we can write finally

$$A \approx E(k). \tag{12.53}$$

The value of $E(k)$ can be taken from the tables of the complete elliptical integrals. The derivative of $E(k)$ will be given by the expression

$$\frac{dE}{dx_1} = \frac{B^2\lambda\big(K(k)-E(k)\big)}{1-B^2\lambda^2}\frac{d\lambda}{dx_1} \tag{12.54}$$

where the complete elliptical integral of the first kind $K(k)$ will be also taken from the tables.

However, we can also calculate $E(k)$ and its derivative in an easier manner, by giving it a convenient development with respect to $B^2\lambda^2$:

$$E(k) = 1+0{\cdot}87B^2\lambda^2-0{\cdot}30B^4\lambda^4, \tag{12.55a}$$

$$\frac{dE(k)}{dx_1} = \frac{dE(k)}{d\lambda^2}\frac{d\lambda^2}{dx_1} = 2B^2\lambda(0{\cdot}87-0{\cdot}60B^2\lambda^2)\frac{d\lambda}{dx_1}. \tag{12.55b}$$

It remains now to determine the derivative of λ with respect to x_1, which is to be written in the following form:

$$\frac{d\lambda}{dx_1} = \frac{1}{2}\left(\frac{d\lambda_1}{dx_1} + \frac{d\lambda_2}{dx_1}\right). \tag{12.56}$$

Further, on taking account of (12.33) and (12.35), we can set approximately:

$$\frac{d\lambda_1}{dx_1} = \frac{d\lambda_1}{d\xi_1}\frac{d\xi_1}{dx_1} \approx \frac{d\lambda_1}{d\xi_1}\frac{1+By}{1+Bl_1}, \tag{12.57a}$$

$$\frac{d\lambda_2}{dx_1} = \frac{d\lambda_2}{d\xi_2}\frac{d\xi_2}{dx_1} \approx \frac{d\lambda_2}{d\xi_2}\frac{1-By}{1+Bl_2}. \tag{12.57b}$$

Consequently, we shall write

$$\frac{d\lambda}{dx_1} \approx \frac{1}{2}\left(\frac{d\lambda_1}{d\xi_1}\frac{1+By}{1+Bl_1} + \frac{d\lambda_2}{d\xi_2}\frac{1-By}{1+Bl_2}\right). \tag{12.58}$$

Finally, formula (12.46) of the axial velocity becomes

$$\frac{u^*}{\alpha U_\infty} = \sqrt{\frac{l_1(\lambda_1+\lambda_2)}{\lambda_1(l_1+l_2)}}\left\{\frac{l_2\lambda_1+l_1\lambda_2+(\lambda_1-\lambda_2)y}{2E(k)\sqrt{(l_1-y)(l_2+y)}}\right.$$
$$\left. - \frac{x_1\sqrt{(l_1-y)(l_2-y)}}{(E(k))^2}(0{\cdot}87-0{\cdot}60B^2\lambda^2)B^2\lambda\left(\frac{1+By}{1+Bl_1}\frac{d\lambda_1}{d\xi_1} + \frac{1-By}{1+Bl_2}\frac{d\lambda_2}{d\xi_2}\right)\right\}, \tag{12.59}$$

where λ is given by (12.48) and the modulus k of the complete integral of the second kind $E(k)$ is given by (12.51).

Note. The results obtained above refer to the classic theory of the thin wing in supersonic flow, under the assumption of small disturbances. This problem can be approached in the same manner by taking, however, account of the flow detachment at the subsonic leading edges. In this case, the problem is much more complicated.

XII .2. Wing with Central Fusiform Body

Let us consider a delta wing with a central fusiform body. Although this problem is very important, for the sake of simplicity we shall treat below only a simple case—that of the subsonic leading edges.

XII. 2.1. Delta Wing fitted with a Body of Variable Shape

Let us consider the conical potential (12.20) and assume that Q and c vary with respect to x_1:

$$Q = Q(x_1), \quad c = c(x_1). \tag{12.60}$$

In this case, the potential Φ_1 of (12.20) becomes the quasi-potential Φ_1^*, whose derivative is the following:

$$\frac{u^*}{\alpha U_\infty} = \frac{1}{\alpha U_\infty}\frac{\partial \Phi_1^*}{\partial x_1} = \mathrm{Re}\, Ql^2 \frac{1+\dfrac{c^4}{l^4}-2\dfrac{c^4}{l^2x^2}+2\dfrac{c^4}{l^2}\left(\dfrac{1}{l^2}-\dfrac{1}{x^2}\right)\dfrac{x_1}{c}\dfrac{dc}{dx_1}}{\sqrt{(l^2-x^2)\left(1-\dfrac{c^4}{l^2x^2}\right)}}$$

$$+\mathrm{Re}\sqrt{(l^2-x^2)\left(1-\frac{c^4}{l^2x^2}\right)}\, x_1\frac{dQ}{dx_1}. \tag{12.61}$$

For the constant Q we consider the value A_0 found out previously (9.58), but divided by $\alpha_0 U_\infty l^2$, reducing it, however, to a simpler form:

$$Q = \frac{A_0}{\alpha_0 U_\infty l^2} = \frac{1}{E(k)-B^2c^2\dfrac{c^2}{l^2}\left(E(k)-\frac{1}{2}K(k)\right)}, \quad (k = \sqrt{1-B^2l^2}), \tag{12.62}$$

$K(k)$ and $E(k)$ being the complete elliptical integral of the first and second kind, having the modulus k. We can further write

$$\frac{dQ}{dx_1} = \frac{4B^2c^2\dfrac{c^2}{l^2}\left(E(k)-\frac{1}{2}K(k)\right)}{\left[E(k)-B^2c^2\dfrac{c^2}{l^2}\left(E(k)-\frac{1}{2}K(k)\right)\right]^2}\frac{1}{c}\frac{dc}{dx_1}. \tag{12.63}$$

Introducing these two last expressions into (12.61), we find the axial disturbance velocity in an explicit form.

In order to illustrate the above method made use of so far, we shall treat a simple example, considering a polynomial variation of the shape of the fusiform body. Thus we shall put (Fig. 12.5)

$$f = x_1 c, \quad c = c_0(1+\nu_1 x_1+\nu_2 x_1^2); \tag{12.64}$$

hence

$$\frac{dc}{dx_1} = c_0(\nu_1+2\nu_2 x_1). \tag{12.65}$$

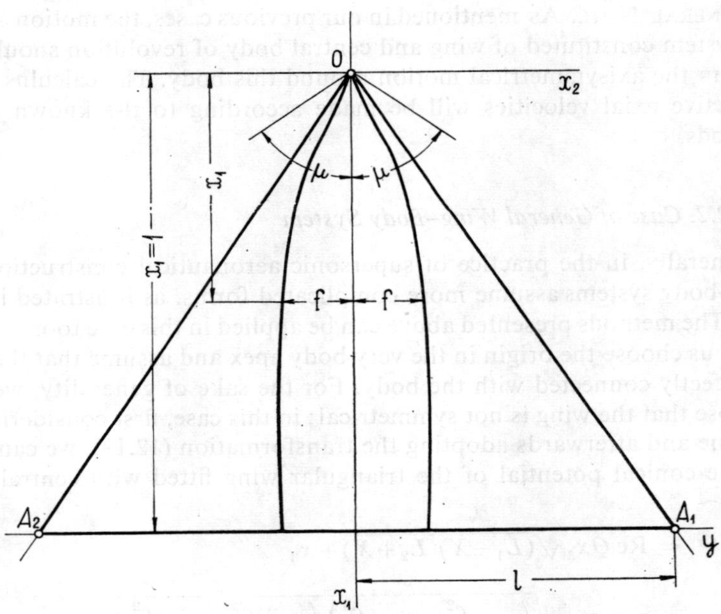

FIG. 12.5

Substituting into (12.61), we find the axial disturbance velocity in each section x_1:

$$\frac{u^*}{\alpha U_\infty} = \frac{l^2}{E(k) - B^2c^2\frac{c^2}{l^2}\left(E(k)-\frac{1}{2}K(k)\right)}$$

$$\times \frac{1+\frac{c^4}{l^4}2\frac{c^4}{l^2y^2}+2\frac{c^4}{l^2}\left(\frac{1}{l^2}-\frac{1}{y^2}\right)\frac{(\nu_1+2\nu_2x_1)x_1}{1+\nu_1x_1+\nu_2x_1^2}}{\sqrt{(l^2-y^2)\left(1-\frac{c^4}{l^2y^2}\right)}}$$

$$+\sqrt{(l^2-y^2)\left(1-\frac{c^4}{l^2y^2}\right)}$$

$$\times\frac{4B^2c^2\frac{c^2}{l^2}\left(E(k)-\frac{1}{2}K(k)\right)\frac{(\nu_1+2\nu_2x_1)x_1}{1+\nu_1x_1+\nu_2x_1^2}}{\left[E(k)-B^2c^2\frac{c^2}{l^2}\left(E(k)-\frac{1}{2}K(k)\right)\right]^2}. \qquad (12.66)$$

GENERAL NOTE. As mentioned in our previous cases, the motion around the system constituted of wing and central body of revolution should also contain the axisymmetrical motion around this body. The calculus of the respective axial velocities will be made according to the known classic methods.

XII. 2.2. Case of General Wing–Body System

Generally, in the practice of supersonic aeronautical construction, the wing–body systems assume more complicated forms, as illustrated in Fig. 12.6. The methods presented above can be applied in this case too.

Let us choose the origin in the very body apex and assume that the wing is perfectly connected with the body. For the sake of generality, we shall suppose that the wing is not symmetrical; in this case, first considering the X plane and afterwards adopting the transformation (12.19), we can write for the conical potential of the triangular wing fitted with central body

$$\frac{1}{\alpha U_\infty}\Phi_1 = \mathrm{Re}\, Qx_1\sqrt{(L_1-X)(L_2+X)}+x_3$$

$$= \mathrm{Re}\; Qx_1\sqrt{\left(l_1+\frac{c^2}{l_1}-x-\frac{c^2}{x}\right)\left(l_2+\frac{c^2}{l_2}+x+\frac{c^2}{x}\right)}+x_3, \quad (12.67)$$

where l_1 and l_2 are the coefficients of the straight leading edges of the triangular wing.

We shall further consider that the leading edges of the wing and the shape of the fuselage are curved and write

$$b_1 = x_1 l_1, \quad b_2 = x_1 l_2, \quad f = x_1 c; \quad (12.68)$$

in this case, the expression of the quasi-conical potential will be

$$\frac{1}{\alpha U_\infty}\Phi_1^* = \mathrm{Re}\; Q\sqrt{\left(b_1+\frac{f^2}{b_1}-x_1x-\frac{f^2}{x_1x}\right)\left(b_2+\frac{f^2}{b_2}+x_1x+\frac{f^2}{x_1x}\right)}+x_3. \quad (12.69)$$

The axial velocity is

$$\frac{u^*}{\alpha U_\infty} = \frac{1}{\alpha U_\infty}\frac{\partial\Phi_1^*}{\partial x_1} = \mathrm{Re}\; Q\frac{N}{2\sqrt{(l_1-x)(l_2+x)\left(1-\frac{c^2}{l_1x}\right)\left(1+\frac{c^2}{l_2x}\right)}}$$

$$+\mathrm{Re}\;\sqrt{(l_1-x)(l_2+x)\left(1-\frac{c^2}{l_1x}\right)\left(1+\frac{c^2}{l_2x}\right)}\,x_1\frac{dQ}{dx_1}, \quad (12.70)$$

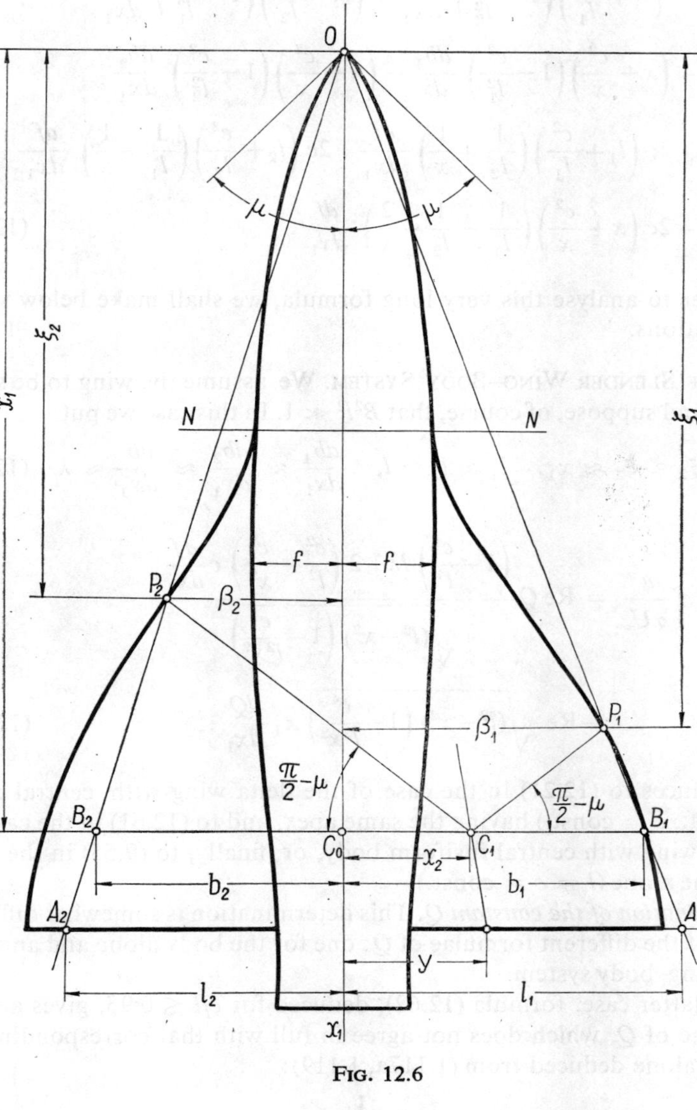
O
x_2
μ
μ
ξ_2
x_1
N
N
ξ_1
$x_1=1$
f
f
P_2
β_2
P_1
β_1
$\frac{\pi}{2}-\mu$
$\frac{\pi}{2}-\mu$
B_2
C_0
C_1
B_1
x_2
b_2
b_1
A_2
A_1
y
l_2
l_1
x_1

FIG. 12.6

where N has the following form:

$$N = \left(l_1+\frac{c^2}{l_1}\right)\left(1-\frac{c^2}{l_2^2}\right)\frac{db_2}{dx_1}+\left(l_2+\frac{c^2}{l_2}\right)\left(1-\frac{c^2}{l_1^2}\right)\frac{db_1}{dx_1}$$
$$+\left(x+\frac{c^2}{x}\right)\left(1-\frac{c^2}{l_1^2}\right)\frac{db_1}{dx}-\left(x+\frac{c^2}{x}\right)\left(1-\frac{c^2}{l_2^2}\right)\frac{db_2}{dx_1}$$
$$+2c\left(l_1+\frac{c^2}{l_1}\right)\left(\frac{1}{l_2}+\frac{1}{x}\right)\frac{df}{dx_1}+2c\left(l_2+\frac{c^2}{l_2}\right)\left(\frac{1}{l_1}-\frac{1}{x}\right)\frac{df}{dx_1}$$
$$+2c\left(x+\frac{c^2}{x}\right)\left(\frac{1}{l_1}-\frac{1}{l_2}-\frac{2}{x}\right)\frac{df}{dx_1}. \qquad (12.71)$$

In order to analyse this very long formula, we shall make below some simplifications.

CASE OF SLENDER WING–BODY SYSTEM. We assume the wing to be symmetrical and suppose, of course, that $B^2l^2 \ll 1$. In this case we put

$$\xi_2 \approx \xi_1 \approx x_1, \quad l_1 \approx l_2 \approx l, \quad \frac{db_1}{dx_1} \approx \frac{db_2}{dx_1} \approx \frac{db}{dx_1} \approx \lambda, \qquad (12.72)$$

hence

$$\frac{u^*}{\alpha U_\infty} = \operatorname{Re} Q \frac{\left(1-\frac{c^4}{l^4}\right) l\lambda + 2\left(\frac{c^2}{l^2}-\frac{c^2}{x^2}\right) c\,\frac{df}{dx_1}}{\sqrt{(l^2-x^2)\left(1-\frac{c^4}{l^2x^2}\right)}}$$
$$+\operatorname{Re}\sqrt{(l^2-x^2)\left(1-\frac{c^4}{l^2x^2}\right)}\, x_1 \frac{dQ}{dx_1}, \qquad (12.73)$$

which reduces to (12.21) in the case of the delta wing with central cone (l = const., c = const.) having the same apex, and to (12.61) in the case of the delta wing with central fusiform body, or, finally, to (9.55) in the case of the cone alone ($l = c$ = const.).

Determination of the constant Q. This determination is somewhat difficult because of the different formulae of Q: one for the body alone and another for the wing–body system.

In the latter case, formula (12.62), deduced for $c/l \lesssim 0{\cdot}95$, gives a very good value of Q, which does not agree in full with that corresponding to the body alone deduced from (1.117a, 1.119):

$$Q = \frac{1}{1+\frac{B^2c^2}{\sqrt{1-B^2c^2}}\cosh^{-1}\frac{1}{Bc}+2c^2}. \qquad (12.74)$$

In this case of the body alone, relation (12.73) becomes

$$\frac{u^*}{\alpha U_\infty} = \operatorname{Im}\left[\frac{2Qc}{x}\frac{df}{dx_1}+\left(x-\frac{c^2}{x}\right)x_1\frac{dQ}{dx_1}\right]. \tag{12.75a}$$

This formula can be used for the fore portion up to the connecting section *NN*.

For the wing–body mixed portion, we shall use formula (12.73) in which the constant Q will be given by (12.62). As concerns the zone which connects the wing with the body, comprised between $c/l = 1$ and $c/l = 0{\cdot}95$ and occupying a small portion, in the whole system, we can adopt an intermediate solution.

Therefore, we can write approximately:

$$\frac{dQ}{dx_1} = \frac{\dfrac{B^2c^2(3-2B^2c^2)}{(1-B^2c^2)^{3/2}} - \dfrac{B^2c^2}{1-B^2c^2}+4c^2}{\left[1+\dfrac{B^2c^2}{\sqrt{1-B^2c^2}}\cosh^{-1}\dfrac{1}{Bc}+2c^2\right]^2}\frac{1}{c}\frac{dc}{dx_1} \tag{12.75b}$$

for the body alone, and

$$\frac{dQ}{dx_1} \approx Q\frac{K(k)-E(k)}{E(k)}\frac{B^2l}{\sqrt{1-B^2l^2}}\frac{dl}{dx_1} \tag{12.76}$$

for the wing–body system.

Simplified case. Generally, the shape of the fusiform body varies slightly. Thus, we can assume that the body is a cylinder upon a large length. In this case we have

$$\frac{df}{dx_1} \approx 0, \quad c = \frac{f}{x_1} \tag{12.77}$$

which are to be introduced into (12.72).

GENERAL CASE. As for the wing alone, we shall put

$$\frac{db_1}{dx_1} \approx \frac{d\beta_1}{d\xi_1} = \lambda_1, \quad \frac{db_2}{dx_1} \approx \frac{d\beta_2}{d\xi_2} = \lambda_2 \tag{12.78}$$

in the expression of N (12.71). We shall assume further the same simplified form of the constant Q, as in the previous case, for small values of c^2/l^2. For very small values, we shall neglect the influence of the body, considering the same formula (12.44, 12.53) for the constant Q as in the case of the wing alone.

General Note. For an accurate consideration of the problem we have to take account of the flow separation at the leading edges, as we have done in the case of the wing alone in Chapter X. Besides, we must also take into consideration the vortices that detach from the body apex and stretch generally parallel to the body over its whole extent. The problem is, however, rather complicated and it is not so easy to approach it, even in the simpler case illustrated in Fig. 12.5.

APPENDIX I

THIS appendix refers to problems in Chapter IV.

A.I.1.

The coefficients $\Omega^{(k)}_{n,\,2q}$, Ω_n and $\Omega^{(k)}_{n,\,2q+1}$, Ω^*_n occurring in the expressions for the constants $A_{n,\,2q}$ (4.18b) and $A_{n,2q+1}$ (4.34a).

Wing with Symmetrical Incidences

$$\Omega_3 = (4-19B^2l^2+4B^4l^4)E(k)^2+8B^2l^2(B^2l^2+1)E(k)K(k)-5B^4l^4K(k)^2$$

$$\Omega^{(0)}_{30} = 2\pi[2(3-5B^2l^2+B^4l^4)E(k)-(3-5B^2l^2)B^2l^2K(k)]$$

$$\Omega^{(1)}_{30} = 6[(4+B^2l^2)E(k)^2-8B^2l^2E(k)K(k)+3B^4l^4K(k)^2]$$

$$\Omega^{(2)}_{30} = 2\pi[(1+B^2l^2)E(k)-2B^2l^2K(k)]$$

$$\Omega^{(0)}_{32} = 2\pi[(4-7B^2l^2+B^4l^4)E(k)+2(1-2B^2l^2)B^2l^2K(k)]$$

$$\Omega^{(1)}_{32} = 2(12-17B^2l^2+10B^4l^4)E(k)^2-8(2-B^2l^2+B^4l^4)B^2l^2E(k)K(k) \\ +2(1+2B^2l^2)B^4l^4K(k)^2$$

$$\Omega^{(2)}_{32} = -2\pi B^2l^2[2(2-B^2l^2)E(k)-(3-B^2l^2)B^2l^2K(k)]$$

$$\Omega_4 = (12-119B^2l^2+151B^4l^4-64B^6l^6+32B^8l^8)E(k)^2 \\ +2(24-5B^2l^2-15B^4l^4-16B^6l^6)B^2l^2E(k)K(k)-(27-31B^2l^2 \\ -8B^4l^4)B^4l^4K(k)^2$$

$$\Omega^{(0)}_{40} = 8\pi(1-B^2l^2)[(6-15B^2l^2+5B^4l^4-4B^6l^6)E(k)-(3-9B^2l^2-2B^4l^4) \\ \times B^2l^2K(k)]$$

$$\Omega^{(1)}_{40} = \tfrac{2}{3}[(72-14B^2l^2-91B^4l^4+13B^6l^6)E(k)^2-2(96-106B^2l^2-7B^4l^4 \\ -3B^6l^6)B^2l^2E(k)K(k)+(78-95B^2l^2-3B^4l^4)B^4l^4K(k)^2]$$

$$\Omega^{(2)}_{40} = \frac{8\pi}{3}(1-B^2l^2)[(3+7B^2l^2-2B^4l^4)E(k)-(9-B^2l^2)B^2l^2K(k)]$$

$$\Omega^{(3)}_{40} = 4[(4+11B^2l^2-11B^4l^4)E(k)^2-2(8-B^2l^2-3B^4l^4)B^2l^2E(k)K(k) \\ +(7-3B^2l^2)B^4l^4K(k)^2]$$

$$\Omega^{(0)}_{42} = -2\pi(1-B^2l^2)[(18-51B^2l^2+B^4l^4-8B^6l^6)E(k)-(9-45B^2l^2 \\ -4B^4l^4)B^2l^2K(k)]$$

$$\Omega_{42}^{(1)} = -\tfrac{2}{3}\left[(36+165B^2l^2-571B^4l^4+472B^6l^6-106B^8l^8)E(k)^2 - 2(108-171B^2l^2-17B^4l^4+112B^6l^6-36B^8l^8)B^2l^2E(k)K(k) + (99-201B^2l^2+136B^4l^4-36B^6l^6)B^4l^4K(k)^2\right]$$

$$\Omega_{42}^{(2)} = \frac{2\pi}{3}(1-B^2l^2)[(3-131B^2l^2+134B^4l^4-48B^6l^6)E(k)+(81-65B^2l^2 + 24B^4l^4)B^2l^2K(k)]$$

$$\Omega_{42}^{(3)} = -8B^2l^2[(19-34B^2l^2+19B^4l^4)E(k)^2-4(3-2B^2l^2-2B^4l^4 + 3B^6l^6)E(k)K(k)+2(3-4B^2l^2+3B^4l^4)B^2l^2K(k)^2]$$

$$\Omega_{44}^{(0)} = -2\pi(1-B^2l^2)B^2l^2[(3+7B^2l^2-2B^4l^4)E(k)-(9-B^2l^2)B^2l^2K(k)]$$

$$\Omega_{44}^{(1)} = -\frac{2B^2l^2}{3}[(153-202B^2l^2+37B^4l^4-4B^6l^6)E(k)^2-2(45-6B^2l^2 - 71B^4l^4+16B^6l^6)B^2l^2E(k)K(k)+(45-78B^2l^2+17B^4l^4)B^4l^4K(k)^2]$$

$$\Omega_{44}^{(2)} = -\frac{2\pi}{3}(1-B^2l^2)[(6-55B^2l^2-65B^4l^4-24B^6l^6)E(k)+(27-31B^2l^2 + 12B^4l^4)B^2l^2K(k)]$$

$$\Omega_{44}^{(3)} = -4B^2l^2[(11-11B^2l^2-4B^4l^4)E(k)^2-2(3+B^2l^2-8B^4l^4)B^2l^2E(k) \times K(k)+(3-7B^2l^2)B^4l^4K(k)^2]$$

Wing with Antisymmetrical Incidences

$$\Omega_3^* = 2(1-B^2l^2+B^4l^4)E(k)-(1+B^2l^2)B^2l^2K(k)$$

$$\Omega_{31}^{(0)} = \tfrac{4}{3}[(3-11B^2l^2+4B^4l^4)E(k)-(3-20B^2l^2+9B^4l^4)B^2l^2K(k)]$$

$$\Omega_{31}^{(1)} = 2\pi(1-B^2l^2)^2$$

$$\Omega_{31}^{(2)} = \tfrac{4}{3}[2(1-B^2l^2)E(k)-(1-3B^2l^2)B^2l^2K(k)]$$

$$\Omega_{33}^{(0)} = \tfrac{2}{3}B^2l^2[(9-2B^2l^2+B^4l^4)E(k)-4(3-B^2l^2)B^4l^4K(k)]$$

$$\Omega_{33}^{(1)} = -\pi(1-B^2l^2)^2$$

$$\Omega_{33}^{(2)} = \tfrac{4}{3}B^2l^2[(1+B^2l^2)E(k)-2B^2l^2K(k)]$$

$$\Omega_4^* = (32-64B^2l^2+151B^4l^4-119B^6l^6+12B^8l^8)E(k)^2-2(16+15B^2l^2 + 5B^4l^4-24B^6l^6)B^2l^2E(k)K(k)+(8+31B^2l^2-27B^4l^4)B^4l^4K(k)^2$$

$$\Omega_{41}^{(0)} = (64-608B^2l^2+554B^4l^4-145B^6l^6+51B^8l^8)E(k)^2-2(32-570B^2l^2 + 679B^4l^4-261B^6l^6+36B^8l^8)B^2l^2E(k)K(k)+(16-418B^2l^2 + 507B^4l^4-189B^6l^6)B^4l^4K(k)^2$$

$$\Omega_{41}^{(1)} = 2\pi(1-B^2l^2)[(24-65B^2l^2+55B^4l^4-6B^6l^6)E(k)-(12-31B^2l^2 + 27B^4l^4)B^2l^2K(k)]$$

$$\Omega_{41}^{(2)} = 2(32-128B^2l^2+83B^4l^4+9B^6l^6)E(k)^2-4(16-89B^2l^2 + 69B^4l^4)B^2l^2E(k)K(k)+2(8-57B^2l^2+45B^4l^4)B^4l^4K(k)^2$$

$$\Omega_{41}^{(3)} = 6\pi(1-B^2l^2)[(2-7B^2l^2-3B^4l^4)E(k)-(1-9B^2l^2)B^2l^2K(k)]$$

$$\Omega_{43}^{(0)} = B^2l^2[(352-256B^2l^2+87B^4l^4-125B^6l^6+26B^8l^8)E(k)^2-2(416 -463B^2l^2+169B^4l^4-44B^6l^6+6B^8l^8)B^2l^2E(k)K(k)+(328-369B^2l^2 +119B^4l^4+6B^6l^6)B^4l^4K(k)^2]$$

$$\Omega_{43}^{(1)} = -2\pi(1-B^2l^2)[(16-44B^2l^2+39B^4l^4-3B^6l^6)E(k)-(8-21B^2l^2 +21B^4l^4)B^2l^2K(k)]$$

$$\Omega_{43}^{(2)} = -\tfrac{2}{3}[(32-192B^2l^2+7B^4l^4+115B^6l^6+10B^8l^8)E(k)^2-2(16 -193B^2l^2+117B^4l^4+26B^6l^6+6B^8l^8)B^2l^2E(k)K(k)+(8-145B^2l^2 +103B^4l^4+6B^6l^6)B^4l^4K(k)^2]$$

$$\Omega_{43}^{(3)} = 6\pi(1-B^2l^2)B^2l^2[(3+7B^2l^2-2B^4l^4)E(k)-(9-B^2l^2)B^2l^2K(k)].$$

A.I.2.

The integrals I_t defined by (4.25a, b), occurring in the formulae for the wave drag coefficients $C_{dm}^{(n)}$ (4.26) and (4.35) for thin delta wings with subsonic edges:

$$I_0 = \frac{\pi}{2}, \quad I_1 = l$$

$$I_2 = \frac{\pi l^2}{4}, \quad I_3 = \frac{2l^3}{3},$$

$$I_4 = \frac{3\pi l^4}{16}, \quad I_5 = \frac{8l^5}{15},$$

$$I_6 = \frac{5\pi l^6}{32}, \quad I_7 = \frac{16l^7}{35}.$$

A.I.3.

The integrals J_t defined by (4.50) and the integrals $H_{k+q,\,k}$ and $H_{k+q,\,k+1}$ defined by (4.49), occurring in the formulae of the wave drag coefficients $C_{dm}^{(n)}$ (4.51) and (4.63) for delta wings with supersonic edges:

The integrals J_t

$$J_0 = \frac{\pi}{2B}, \quad J_1 = \frac{1}{B^2}, \quad J_8 = \frac{35\pi}{256B^9}, \quad J_9 = \frac{128}{315B^{10}},$$

$$J_2 = \frac{\pi}{4B^3}, \quad J_3 = \frac{2}{3B^4}, \quad J_{10} = \frac{63\pi}{512B^{11}}, \quad J_{11} = \frac{256}{693B^{12}},$$

$$J_4 = \frac{3\pi}{16B^5}, \quad J_5 = \frac{8}{15B^6}, \quad J_{12} = \frac{693\pi}{6144B^{13}},$$

$$J_6 = \frac{5\pi}{32B^7}, \quad J_7 = \frac{16}{35B^8}.$$

The integrals $H_{k+q,k}$ when k is an even number ($k = 0, 2, 4, \ldots$) and the integrals $H_{k+q,k+1}$ when $(k+1)$ is an even number ($k+1 = 2, 4, 6, \ldots$)

$$H_{0,k} = \frac{\pi}{2}\frac{\sqrt{B^2l^2-1}}{B},$$

$$H_{1,k} = H_{1,k+1} = \frac{\pi}{2}\frac{\sqrt{B^2l^2-1}}{2B^2}Bl,$$

$$H_{2,k} = H_{2,k+1} = \frac{\pi}{2}\frac{\sqrt{B^2l^2-1}}{3B^3}\left(B^2l^2+\frac{1}{2}\right),$$

$$H_{3,k} = H_{3,k+1} = \frac{\pi}{2}\frac{\sqrt{B^2l^2-1}}{4B^4}Bl\left(B^2l^2+\frac{1}{2}\right),$$

$$H_{4,k} = H_{4,k+1} = \frac{\pi}{2}\frac{\sqrt{B^2l^2-1}}{5B^5}\left(B^4l^4+\frac{1}{2}B^2l^2+\frac{3}{8}\right),$$

$$H_{5,k} = H_{5,k+1} = \frac{\pi}{2}\frac{\sqrt{B^2l^2-1}}{6B^6}Bl\left(B^4l^4+\frac{1}{2}B^2l^2+\frac{3}{8}\right),$$

$$H_{6,k} = H_{6,k+1} = \frac{\pi}{2}\frac{\sqrt{B^2l^2-1}}{7B^7}\left(B^6l^6+\frac{1}{2}B^4l^4+\frac{3}{8}B^2l^2+\frac{5}{16}\right),$$

$$H_{7,k} = H_{7,k+1} = \frac{\pi}{2}\frac{\sqrt{B^2l^2-1}}{8B^8}Bl\left(B^6l^6+\frac{1}{2}B^4l^4+\frac{3}{8}B^2l^2+\frac{5}{16}\right),$$

$$H_{8,k} = H_{8,k+1} = \frac{\pi}{2}\frac{\sqrt{B^2l^2-1}}{9B^9}\left(B^8l^8+\frac{1}{2}B^6l^6+\frac{3}{8}B^4l^4+\frac{5}{16}B^2l^2+\frac{35}{128}\right),$$

$$H_{9,k} = H_{9,k+1} = \frac{\pi}{2}\frac{\sqrt{B^2l^2-1}}{10B^{10}}Bl\left(B^8l^8+\frac{1}{2}B^6l^6+\frac{3}{8}B^4l^4+\frac{5}{16}B^2l^2+\frac{35}{128}\right),$$

$$H_{10,k} = H_{10,k+1} = \frac{\pi}{2}\frac{\sqrt{B^2l^2-1}}{11B^{11}}\left(B^{10}l^{10}+\frac{1}{2}B^8l^8+\frac{3}{8}B^6l^6+\frac{5}{16}B^4l^4\right.$$
$$\left.+\frac{35}{128}B^2l^2+\frac{63}{256}\right),$$

$$H_{11,k} = H_{11,k+1} = \frac{\pi}{2}\frac{\sqrt{B^2l^2-1}}{12B^{12}}Bl\left(B^{10}l^{10}+\frac{1}{2}B^9l^9+\frac{3}{8}B^6l^6+\frac{5}{16}B^4l^4\right.$$
$$\left.+\frac{35}{128}B^2l^2+\frac{63}{256}\right),$$

$$H_{12,k} = H_{12,k+1} = \frac{\pi}{2}\frac{\sqrt{B^2l^2-1}}{13B^{13}}\left(B^{12}l^{12}+\frac{1}{2}B^{10}l^{10}+\frac{3}{8}B^8l^8+\frac{5}{16}B^6l^6\right.$$
$$\left.+\frac{35}{128}B^4l^4+\frac{63}{256}B^2l^2+\frac{693}{3072}\right).$$

The integrals $H_{k+q,\,k}$ when k is an odd number ($k = 1, 3, 5, \ldots$) and the integrals $H_{k+q,\,k+1}$ when $(k+1)$ is an odd number ($k+1 = 1, 3, 5, \ldots$)

$$H_{0,k+1} = l\cos^{-1}\frac{1}{Bl},$$

$$H_{1,k} = H_{1,k+1} = \frac{l^2}{2}\cos^{-1}\frac{1}{Bl}+\frac{\sqrt{B^2l^2-1}}{2B^2},$$

$$H_{2,k} = H_{2,k+1} = \frac{l^3}{3}\cos^{-1}\frac{1}{Bl}+\frac{\sqrt{B^2l^2-1}}{3B^3}Bl,$$

$$H_{3,k} = H_{3,k+1} = \frac{l^4}{4}\cos^{-1}\frac{1}{Bl}+\frac{\sqrt{B^2l^2-1}}{4B^4}\left(B^2l^2+\frac{2}{3}\right),$$

$$H_{4,k} = H_{4,k+1} = \frac{l^5}{5}\cos^{-1}\frac{1}{Bl}+\frac{\sqrt{B^2l^2-1}}{5B^5}Bl\left(B^2l^2+\frac{2}{3}\right),$$

$$H_{5,k} = H_{5,k+1} = \frac{l^6}{6}\cos^{-1}\frac{1}{Bl}+\frac{\sqrt{B^2l^2-1}}{6B^6}\left(B^4l^4+\frac{2}{3}B^2l^2+\frac{8}{15}\right),$$

$$H_{6,k} = H_{6,k+1} = \frac{l^7}{7}\cos^{-1}\frac{1}{Bl}+\frac{\sqrt{B^2l^2-1}}{7B^7}Bl\left(B^4l^4+\frac{2}{3}B^2l^2+\frac{8}{15}\right).$$

A.I.4.

The integrals J' and $H'_{k+q,\,k}$, defined by (4.55) and (4.56 *a*, *b*), occurring in the formulae of the wave drag coefficients $C_{dm}^{(n)}$ (4.57) for the delta wing of symmetrical thickness:

The integrals J'_t

$$J'_0 = \frac{1}{B}\sin^{-1}Bl,$$

$$J'_1 = \frac{1}{B^2}\left[1-\sqrt{1-B^2l^2}\right],$$

$$J'_2 = \frac{1}{2B^3}\left[\sin^{-1}Bl-Bl\sqrt{1-B^2l^2}\right],$$

$$J'_3 = \frac{2}{3B^4}\left[1-\left(1+\frac{1}{2}B^2l^2\right)\sqrt{1-B^2l^2}\right],$$

$$J'_4 = \frac{3}{8B^5}\left[\sin^{-1}Bl-\left(1+\frac{2}{3}B^2l^2\right)Bl\sqrt{1-B^2l^2}\right],$$

$$J'_5 = \frac{8}{15B^6}\left[1-\left(1+\frac{1}{2}B^2l^2+\frac{3}{8}B^4l^4\right)\sqrt{1-B^2l^2}\right],$$

$$J'_6 = \frac{5}{16B^7}\left[\sin^{-1}Bl-\left(1+\frac{2}{3}B^2l^2+\frac{8}{15}B^4l^4\right)Bl\sqrt{1-B^2l^2}\right].$$

The integrals $H'_{k+q,\,k}$ when k is an even number ($k = 0, 2, 4, \ldots$)

$$H'_{0,k} = \frac{1}{B}\left[2Bl\cosh^{-1}\frac{1+Bl}{2\sqrt{Bl}}+\sqrt{1-B^2l^2}\sin^{-1}Bl\right],$$

$$H'_{1,k} = \frac{l}{2B}\sqrt{1-B^2l^2}\sin^{-1}Bl,$$

$$H'_{2,k} = \frac{1}{3B^3}\left[2B^3l^3\cos h^{-1}\frac{1+Bl}{2\sqrt{Bl}}+\left(B^2l^2+\frac{1}{2}\right)\sqrt{1-B^2l^2}\sin^{-1}Bl - \frac{1}{2}Bl(1-B^2l^2)\right],$$

$$H'_{3,k} = \frac{l}{4B^3}\left[\left(B^2l^2+\frac{1}{2}\right)\sqrt{1-B^2l^2}\sin^{-1}Bl-\frac{1}{2}Bl\sqrt{1-B^2l^2}\right],$$

$$H'_{4,k} = \frac{1}{5B^5}\left[2B^5l^5\cos h^{-1}\frac{1+Bl}{2\sqrt{Bl}}+\left(B^4l^4+\frac{1}{2}B^2l^2+\frac{3}{8}\right)\sqrt{1-B^2l^2}\sin^{-1}Bl - \frac{3}{4}Bl\left(B^2l^2+\frac{1}{2}\right)(1-B^2l^2)\right],$$

$$H'_{5,k} = \frac{l}{6B^5}\left[\left(B^4l^4+\frac{1}{2}B^2l^2+\frac{3}{8}\right)\sqrt{1-B^2l^2}\sin^{-1}Bl - \frac{3}{4}Bl\left(B^2l^2+\frac{1}{2}\right)(1-B^2l^2)\right].$$

The integrals $H'_{k+q,\,k}$ when k is an odd number ($k = 1, 3, 5, \ldots$)

$$H'_{1,k} = \frac{1}{2B^2}\left[B^2l^2\cosh^{-1}\frac{1}{Bl}+\sqrt{1-B^2l^2}-(1-B^2l^2)\right],$$

$$H'_{2,k} = \frac{l}{3B^2}\left[B^2l^2\left(\cosh^{-1}\frac{1}{Bl}-2\cosh^{-1}\frac{1+Bl}{2\sqrt{Bl}}\right)+\sqrt{1-B^2l^2}-(1-B^2l^2)\right],$$

$$H'_{3,k} = \frac{1}{4B^4}\left[B^4l^4\cosh^{-1}\frac{1}{Bl}+\left(B^2l^2+\frac{2}{3}\right)\sqrt{1-B^2l^2} - \frac{2}{3}(1+2B^2l^2)(1-B^2l^2)\right],$$

$$H'_{4,k} = \frac{l}{5B^4}\left[B^4l^4\left(\cosh^{-1}\frac{1}{Bl}-2\cosh^{-1}\frac{1+Bl}{2\sqrt{Bl}}\right)+\left(B^2l^2+\frac{2}{3}\right)\sqrt{1-B^2l^2} - \frac{2}{3}(1+2B^2l^2)(1-B^2l^2)\right],$$

$$H'_{5,k} = \frac{1}{6B^6}\left[B^6l^6 \cosh^{-1}\frac{1}{Bl} + \left(B^4l^4 + \frac{2}{3}B^2l^2 + \frac{8}{15}\right)\sqrt{1-B^2l^2} - \frac{1}{15}(8+14B^2l^2+23B^4l^4)(1-B^2l^2)\right],$$

$$H'_{6,k} = \frac{l}{7B^6}\left[B^6l^6\left(\cosh^{-1}\frac{1}{Bl} - 2\cosh^{-1}\frac{1+Bl}{2\sqrt{Bl}}\right) + \left(B^4l^4+\frac{2}{3}B^2l^2+\frac{8}{15}\right)\sqrt{1-B^2l^2} - \frac{1}{15}(8+14B^2l^2+23B^4l^4)(1-B^2l^2)\right].$$

APPENDIX II

THIS appendix refers to problems in Chapter V.

A.II.1.

The integrals $\mathcal{J}_t$, defined by (5.45), occurring in the formulae (5.36), (5.38), (5.40), (5.42,), (5.44), for the aerodynamic characteristics for yawed thin delta wings with subsonic edges:

$$\mathcal{J}_1 = \frac{\pi}{\sqrt{(1-l_1 \tan \delta)(1+l_2 \tan \delta)}},$$

$$\mathcal{J}_2 = \frac{\pi}{2} \frac{(1-1_1 \tan \delta)+(1+l_2 \tan \delta)}{[(1-l_1 \tan \delta)(1+l_2 \tan \delta)]^{3/2}},$$

$$\mathcal{J}_3 = \frac{\pi}{8} \frac{3(1-l_1 \tan \delta)^2+2(1-l_1 \tan \delta)(1+l_2 \tan \delta)+3(1+l_2 \tan \delta)^2}{[(1-l_1 \tan \delta)(1+l_2 \tan \delta)]^{5/2}},$$

$$\mathcal{J}_4 = \frac{\pi}{16}\left[\frac{5(1-l_1 \tan \delta)^3+3(1-l_2 \tan \delta)^2(1+l_2 \tan \delta)}{[(1-l_1 \tan \delta)(1+l_2 \tan \delta)]^{7/2}}\right.$$
$$\left.+\frac{3(1-l_1 \tan \delta)(1+l_2 \tan \delta)^2+5(1+l_2 \tan \delta)^3}{[(1-l_1 \tan \delta)(1+l_2 \tan \delta)]^{7/2}}\right],$$

$$\mathcal{J}_5 = \frac{\pi}{128}$$
$$\times\left[\frac{35(1-l_1 \tan \delta)^4+20(1-l_1 \tan \delta)^3(1+l_2 \tan \delta)+18(1-l_1 \tan \delta)^2(1+l_2 \tan \delta)^2}{[(1-l_1 \tan \delta)(1+l_2 \tan \delta)]^{9/2}}\right.$$
$$\left.+\frac{20(1-l_1 \tan \delta)(1+l_1 \tan \delta)^3+35(1+1_2 \tan \delta)^4}{[(1-l_1 \tan \delta)(1+l_2 \tan \delta)]^{9/2}}\right],$$

A.II.2.

The integrals $\mathcal{J}_t'$ and Λ_ν, defined by (5.57a, b), occurring in the formulae (5.54) and (5.56) for the aerodynamic characteristics for yawed thin delta wings having one subsonic and one supersonic edge:

$$\mathcal{J}_1' = \frac{\pi}{\sqrt{(1-l_1 \tan \delta)\left(1+\frac{\tan \delta}{B}\right)}},$$

$$\mathcal{J}_2' = \frac{\pi}{2}\frac{(1-l_1 \tan\delta)+\left(1+\dfrac{\tan\delta}{B}\right)}{\left[(1-l_1\tan\delta)\left(1+\dfrac{\tan\delta}{B}\right)\right]^{3/2}},$$

$$\mathcal{J}_3' = \frac{\pi}{8}\frac{3(1-l_1\tan\delta)^2+2(1-l_1\tan\delta)\left(1+\dfrac{\tan\delta}{B}\right)+3\left(1+\dfrac{\tan\delta}{B}\right)^2}{\left[(1-l_1\tan\delta)\left(1+\dfrac{\tan\delta}{B}\right)\right]^{5/2}},$$

$$\mathcal{J}_4' = \frac{\pi}{16}\left[\frac{5(1-l_1\tan\delta)^3+3(1-l_1\tan\delta)^2\left(1+\dfrac{\tan\delta}{B}\right)}{\left[(1-l_1\tan\delta)\left(1+\dfrac{\tan\delta}{B}\right)\right]^{7/2}}\right.$$
$$\left.+\frac{3(1-l_1\tan\delta)\left(1+\dfrac{\tan\delta}{B}\right)^2+5\left(1+\dfrac{\tan\delta}{B}\right)^3}{\left[(1-l_1\tan\delta)\left(1+\dfrac{\tan\delta}{B}\right)\right]^{7/2}}\right],$$

$$\mathcal{J}_5' = \frac{\pi}{128}\left[\frac{35(1-l_1\tan\delta)^4+20(1-l_1\tan\delta)^3\left(1+\dfrac{\tan\delta}{B}\right)}{\left[(1-l_1\tan\delta)\left(1+\dfrac{\tan\delta}{B}\right)\right]^{9/2}}\right.$$
$$\left.+\frac{18(1-l_1\tan\delta)^2\left(1+\dfrac{\tan\delta}{B}\right)^2+20(1-l_1\tan\delta)\left(1+\dfrac{\tan\delta}{B}\right)^3+35\left(1+\dfrac{\tan\delta}{B}\right)^4}{\left[(1-l_1\tan\delta)\left(1+\dfrac{\tan\delta}{B}\right)\right]^{9/2}}\right].$$

$$\Lambda_2 = \tfrac{1}{2}\frac{\sqrt{(l_1+l_2)\left(l_2-\dfrac{1}{B}\right)}}{1+l_2\tan\delta}\mathcal{J}_1',$$

$$\Lambda_3 = \tfrac{1}{4}\frac{\sqrt{(l_1+l_2)\left(l_2-\dfrac{1}{B}\right)}}{(1+l_2\tan\delta)^2}[\mathcal{J}_1'+(1+l_2\tan\delta)\mathcal{J}_2'],$$

$$\Lambda_4 = \tfrac{1}{6}\frac{\sqrt{(l_1+l_2)\left(l_2-\dfrac{1}{B}\right)}}{(1+l_2\tan\delta)^3}[\mathcal{J}_1'+(1+l_2\tan\delta)\mathcal{J}_2'+(1+l_2\tan\delta)^2\mathcal{J}_3'],$$

$$\Lambda_5 = \tfrac{1}{8}\frac{\sqrt{(l_1+l_2)\left(l_2-\dfrac{1}{B}\right)}}{(1+l_2\tan\delta)^4}[\mathcal{J}_1'+(1+l_2\tan\delta)\mathcal{J}_2'$$
$$+(1+l_2\tan\delta)^2\mathcal{J}_3'+(1+l_2\tan\delta)^3\mathcal{J}_4'].$$

A.II.3.

The integrals $\mathcal{J}_t''$, Λ_ν', Λ_ν'', defined by (5.67a, b, c), occurring in the formulae (5.64) and (5.66) for the aerodynamic characteristics for the yawed thin delta wings with supersonic edges:

$$\mathcal{J}_1'' = \frac{\pi B}{\sqrt{B^2-\tan^2\delta}},$$

$$\mathcal{J}_2'' = \frac{\pi B^2}{2}\,\frac{(B-\tan\delta)+(B+\tan\delta)}{(B^2-\tan^2\delta)^{3/2}},$$

$$\mathcal{J}_3'' = \frac{\pi B^3}{8}\,\frac{3(B-\tan\delta)^2+2(B^2-\tan^2\delta)+3(B+\tan\delta)^2}{(B^2-\tan^2\delta)^{5/2}},$$

$$\mathcal{J}_4'' = \frac{\pi B^4}{16}\,\frac{5(B-\tan\delta)^3+3(B-\tan\delta)^2(B+\tan\delta)}{(B-\tan^2\delta)^{7/2}}$$
$$+\frac{3(B-\tan\delta)(B+\tan\delta)^2+5(B+\tan\delta)^3}{(B^2-\tan^2\delta)^{7/2}}.$$

$$\Lambda_2' = \frac{1}{2B}\,\frac{\sqrt{B^2l_1^2-1}}{1-l_1\tan\delta}\,\mathcal{J}_1'',$$

$$\Lambda_3' = \frac{1}{4B}\,\frac{\sqrt{B^2l_1^2-1}}{(1-l_1\tan\delta)^2}\,[\mathcal{J}_1''+(1-l_1\tan\delta)\mathcal{J}_2''],$$

$$\Lambda_4' = \frac{1}{6B}\,\frac{\sqrt{B^2l_1^2-1}}{(1-l_1\tan\delta)^3}\,[\mathcal{J}_1''+(1-l_1\tan\delta)\mathcal{J}_2''+(1-l_1\tan\delta)^2\mathcal{J}_3''],$$

$$\Lambda_5' = \frac{1}{8B}\,\frac{\sqrt{B^2l_1^2-1}}{(1-l_1\tan\delta)^4}\,[\mathcal{J}_1''+(1-l_1\tan\delta)\mathcal{J}_2''$$
$$+(1-l_1\tan\delta)^2\mathcal{J}_3''+(1-l_1\tan\delta)^3\mathcal{J}_4''].$$

$$\Lambda_2'' = \frac{1}{2B}\,\frac{\sqrt{B^2l_2^2-1}}{1+l_2\tan\delta}\,\mathcal{J}_1'',$$

$$\Lambda_3'' = \frac{1}{4B}\,\frac{\sqrt{B^2l_2^2-1}}{(1+l_2\tan\delta)^2}\,[\mathcal{J}_1''+(1+l_2\tan\delta)\mathcal{J}_2''],$$

$$\Lambda_4'' = \frac{1}{6B}\,\frac{\sqrt{B^2l_2^2-1}}{(1+l_2\tan\delta)^3}\,[\mathcal{J}_1''+(1+l_2\tan\delta)\mathcal{J}_2''+(1+l_2\tan\delta)^2\mathcal{J}_3''],$$

$$\Lambda_5'' = \frac{1}{8B}\,\frac{\sqrt{B^2l_2^2-1}}{(1+l_2\tan\delta)^4}\,[\mathcal{J}_1''+(1+l_2\tan\delta)^2\mathcal{J}_2''$$
$$+(1+l_2\tan\delta)^2\mathcal{J}_3''+(1+l_2\tan\delta)^3\mathcal{J}_4''].$$

A.II.4.

The integrals, $\psi_t^{(\lambda)}$, $\Gamma_t^{(\lambda)}$, $\Phi_t^{(\lambda)}$, $\mathcal{J}_t^{(\lambda)}$ ($\lambda = 1, 2$), occurring in the formula (5.74) for the yawing moment coefficient for yawed delta wings of symmetrical thickness:

$$C_{n,q}\psi_t^{(\lambda)} = \text{Re}\, C_{n,q}\left\{(-1)^{\lambda+1}\frac{\cot\delta}{t-1}\left[\frac{1}{(1-y\tan\delta)^{t-1}}-\frac{1}{(1-l_1\tan\delta)^{t-1}}\right]\right.$$

$$\times\cosh^{-1}\sqrt{\frac{(1+Bl_1)(1-By)}{2B(l_1-y)}}\,\Bigg|_{-s}^{(-1)^{\lambda+1}l_\lambda}$$

$$\left.+\frac{\sqrt{1-B^2l_1^2}}{2(t-1)}\sum_{p=1}^{t-1}\frac{1}{(1-l_1\tan\delta)^{t-p}}\mathcal{J}_p^{(\lambda)}\right\},$$

$$G_{n,q}\Gamma_t^{(\lambda)} = \text{Re}\, G_{n,q}\left\{(-1)^{\lambda+1}\frac{\cot\delta}{t-1}\left[\frac{1}{(1-y\tan\delta)^{t-1}}-\frac{1}{(1+l_2\tan\delta)^{t-1}}\right]\right.$$

$$\times\cosh^{-1}\sqrt{\frac{(1+Bl_2)(1+By)}{2B(l_2+y)}}\,\Bigg|_{-s}^{(-1)^{\lambda+1}l_\lambda}$$

$$\left.+\frac{\sqrt{1-B^2l_2^2}}{2(t-1)}\sum_{p=1}^{t-1}\frac{1}{(1+l_2\tan\delta)^{t-p}}\mathcal{J}_p^{(\lambda)}\right\},$$

$$F_{n,q}\Phi_t^{(\lambda)} = \text{Re}\, F_{n,q}\left\{(-1)^{\lambda+1}\frac{\cot\delta}{t-1}\left[\frac{1}{(1-y\tan\delta)^{t-1}}-\frac{1}{(1+s\tan\delta)^{t-1}}\right]\right.$$

$$\times\cosh^{-1}\sqrt{\frac{(1+Bs)(1+By}{2B(s+y)}}\,\Bigg|_{-s}^{(-1)^{\lambda-1}l_\lambda}$$

$$\left.+\frac{\sqrt{1-B^2s^2}}{2(t-1)}\sum_{p=1}^{t-1}\frac{1}{(1+s\tan\delta)^{t-p}}\mathcal{J}_p^{(\lambda)}\right\},$$

$$\mathcal{J}_t^{(\lambda)} = \text{Re}\frac{\cot\delta}{(t-1)(1-B^2\cot^2\delta)}\left[(-1)^{\lambda+1}\frac{\sqrt{1-B^2y^2}}{(1-y\tan\delta)^{t-1}}\Bigg|_{-s}^{(-1)^{\lambda+1}l_\lambda}\right.$$

$$\left.-(2t-3)B^2(\cot\delta)^{t-1}\mathcal{J}_{t-1}^{(\lambda)}+(t-2)B^2(\cot\delta)^{t-2}\mathcal{J}_{t-2}^{(\lambda)}\right],$$

$$\mathcal{J}_1^{(\lambda)} = \text{Re}(-1)^{\lambda+1}\frac{2\cot\delta}{\sqrt{1-B^2\cot^2\delta}}$$

$$\times\cosh^{-1}\sqrt{\frac{(\tan\delta+B)(1-By)}{2B(1-y\tan\delta)}}\,\Bigg|_{-s}^{(-1)^{\lambda+1}l_\lambda},$$

$$\mathcal{J}_2^{(\lambda)} = \frac{\cot\delta}{1-B^2\cot^2\delta}\left[\operatorname{Re}(-1)^{\lambda+1}\frac{\sqrt{1-B^2y^2}}{1-y\tan\delta}\Bigg|_{-s}^{(-1)^{\lambda+1}l_\lambda} - B^2\cot\delta\,\mathcal{J}_1^{(\lambda)}\right],$$

$$\mathcal{J}_3^{(\lambda)} = \frac{\cot\delta}{2(1-B^2\cot^2\delta)^2}\left[\operatorname{Re}(-1)^{\lambda+1}(1-4B^2\cot^2\delta+3B^2y\cot\delta)\right.$$

$$\left.\times\frac{\sqrt{1-B^2y^2}}{(1-y\tan\delta)^2}\Bigg|_{-s}^{(-1)^{\lambda+1}l_\lambda} + (1+2B^2\cot^2\delta)B^2\cot\delta\,\mathcal{J}_1^{(\lambda)}\right],$$

$$\mathcal{J}_4^{(\lambda)} = \frac{\cot\delta}{6(1-B^2\cot^2\delta)^3}\left\{\operatorname{Re}(-1)^{\lambda+1}\left[2-5B^2\cot^2\delta+18B^4\cot^4\delta-3(1\right.\right.$$

$$\left.+9B^2\cot^2\delta)B^2y\cot\delta+(4+11B^2\cot^2\delta)B^2y^2\right]\frac{\sqrt{1-B^2y^2}}{(1-y\tan\delta)^3}\Bigg|_{-s}^{(-1)^{\lambda+1}l_\lambda}$$

$$\left.-3(3+2B^2\cot^2\delta)B^4\cot^3\delta\,\mathcal{J}_1^{(\lambda)}\right\}.$$

APPENDIX III

THIS appendix refers to problems in Chapter VII.

A.III.1.

The integrals $E^o_{t,q}$, P^o_t, $H^o_{t,q}$, R^o_t, defined by (7.41) and occurring in the formulae (7.40) for the wave drag coefficients C^{on}_{dm} for a double trapezoidal swept-back wing:

The integrals $E^o_{t,q}$ and P^o_t

$$E^o_{t,q} = \mathrm{Re}\frac{L}{(t-1)\sqrt{1-B^2l^2}}\left\{\left[(1+f)^{t-1}-\left(\frac{h}{h-f}\right)^{t-1}\right]\right.$$
$$\times\cosh^{-1}\sqrt{\frac{(1+Bl)(1+f-hBl)}{2B(1+f-h)l}}+(-1)^q\left[(1+f)^{t-1}-\left(\frac{h}{h+f}\right)^{t-1}\right]$$
$$\times\cosh^{-1}\sqrt{\frac{(1+Bl)(1+f+hBl)}{2(1+f+h)Bl}}-\tfrac{1}{2}\left[1+(-1)^q-\left(\frac{h}{h-f}\right)^{t-1}\right.$$
$$\left.\left.-(-1)^q\left(\frac{h}{h+f}\right)^{t-1}\right]\cosh^{-1}\frac{1}{Bl}\right\}+\frac{L}{2(t-1)}\sum_{p=1}^{t-1}\left[\left(\frac{h}{h-f}\right)^{t-p}\right.$$
$$\left.+(-1)^q\left(\frac{h}{h+f}\right)^{t-p}\right]\mathcal{J}^o_p,$$

$$P^o_t = \frac{L}{(t-1)B}\left[\mathrm{Re}(1+f)^{t-2}\sqrt{(1+f)^2-f^2B^2L^2}-1+B^2L^tJ^o_{t-1}\right.$$
$$\left.-B^2L^{t-1}\mathcal{J}^o_{t-2}\right],$$

where

$$\mathcal{J}^o_1 = \mathrm{Re}\frac{1}{\sqrt{1-B^2L^2}}\left[2\cosh^{-1}\sqrt{\frac{(1+BL)(1+f-fBL)}{2BL}}-\cosh^{-1}\frac{1}{BL}\right],$$

$$\mathcal{J}^o_2 = \frac{1}{L(1-B^2L^2)}\left[\mathrm{Re}\sqrt{(1+f)^2-f^2B^2L^2}-1-B^2L^2\mathcal{J}^o_1\right],$$

$$\mathcal{J}_3^o = \frac{1}{2L^2(1-B^2L^2)^2}\Big[\big(\mathrm{Re}(1+f-(4-f)B^2L^2)\sqrt{(1+f)^2-f^2B^2L^2}$$

$$-(1-4B^2L^2)+(1+2B^2L^2)B^2L^2\mathcal{J}_1^o\Big],$$

$$\mathcal{J}_4^o = \frac{1}{6L^3(1-B^2L^2)^3}\{\mathrm{Re}\,[2(1+f)^2-(5+13f+4f^2)B^2L^2+(18+9f$$
$$+2f^2)B^4L^4]\sqrt{(1+f)^2-f^2B^2L^2}-(2-5B^2L^2+18B^4L^4)$$
$$-3(3+2B^2L^2)B^4L^4\mathcal{J}_1^o\}.$$

The integrals $H_{t,q}^o$ and R_t^o

$$H_{t,q}^o = \mathrm{Re}\,\frac{1}{(t-1)l^{t-1}\sqrt{1-B^2l^2}}\left\{\left[\left(\frac{1+f}{h}\right)^{t-1}-1\right]\right.$$
$$\times\cosh^{-1}\sqrt{\frac{(1+Bl)(1+f-hBl)}{2(1+f-h)Bl}}+(-1)^q\left[\left(\frac{1+f}{h}\right)^{t-1}-(-1)^{t-1}\right]$$
$$\left.\times\cosh^{-1}\sqrt{\frac{(1+Bl)(1+f+hBl)}{2(1+f+h)Bl)}}-(-1)^q[1+(-1)^t]\cosh^{-1}\frac{1+Bl}{2\sqrt{Bl}}\right\}$$
$$+\frac{1}{2(t-1)l^t}\sum_{p=1}^{t-1}[1+(-1)^{t+q-p}]l^p\mathcal{J}_p^o,$$

$$R_t^o = -\frac{1}{(t-1)Bl^{t-1}}\mathrm{Re}\left[\sqrt{1-B^2L^2}-\frac{(1+f)^{t-2}}{h^{t-1}}\sqrt{(1+f)^2-h^2B^2l^2}\right.$$
$$\left.+B^2l^{t-1}\mathcal{G}_{t-2}^o\right],$$

where

$$\mathcal{G}_{2r+1}^o = -\frac{(2r)!}{2^{2r}(r!)^2l^{2r}}\mathrm{Re}\left\{(Bl)^{2r}\left[\cosh^{-1}\frac{1}{Bl}-\cosh^{-1}\frac{1+f}{hBl}\right]\right.$$
$$+\sum_{k=0}^{r-1}\frac{2^{2k}(k!)^2}{(2k+1)!}(Bl)^{2(r-k-1)}\left[\sqrt{1-B^2l^2}\right.$$
$$\left.\left.-\frac{(1+f)^{2k+1}}{h^{2k+2}}\sqrt{(1+f)^2-h^2B^2l^2}\right]\right\},$$

$$\mathcal{G}_{2r+2}^o = -\frac{2^{2r}(r!)^2}{(2r+1)!\,l^{2r+1}}\sum_{k=0}^{r}\frac{(2k)!}{2^{2k}(k!)^2}(Bl)^{2(r-k)}\left[\sqrt{1-B^2l^2}\right.$$
$$\left.-\frac{(1+f)^{2k}}{h^{2k+1}}\sqrt{(1+f)^2-h^2B^2l^2}\right].$$

A.III.2.

The integrals $H_{t,q}^F$, R_t^F, $E_{t,q}^F$, P_t^F defined by (7.43), occurring in the formulae (7.42) of the wave drag coefficients C_{dm}^{Fn} for a double trapezoidal swept-back wing:

The integrals $H_{t,q}^F$ and R_t^F

$$H_{t,q}^F = \mathrm{Re}\frac{1}{(t-1)L^{t-1}\sqrt{1-B^2L^2}}\left\{\left[1-\left(\frac{L}{a}\right)^{t-1}\right]\sinh^{-1}\sqrt{\frac{(1+BL)(1-Ba)}{2B(a-L)}}\right.$$

$$+(-1)^q\left[(-1)^{t-1}-\left(\frac{L}{a}\right)^{t-1}\right]\cosh^{-1}\sqrt{\frac{(1+BL)(1+Ba)}{2B(L+a)}}$$

$$\left.+(-1)^q[1-(-1)^{t-1}]\cosh^{-1}\frac{1+BL}{2\sqrt{BL}}\right\}+\frac{1}{2(t-1)l^t}$$

$$\times\sum_{p=1}^{t-1}[1+(-1)^{t+q-p}]l^p\mathcal{J}_p^F,$$

$$R_t^F = \mathrm{Re}\frac{1}{(t-1)BL^{t-1}}\left[\sqrt{1-B^2L^2}-\left(\frac{L}{a}\right)^{t-1}\sqrt{1-B^2a^2}-B^2L^{t-1}\mathcal{J}_{t-2}^F\right],$$

where

$$\mathcal{J}_{2r+1}^F = \mathrm{Re}\frac{(2r)!}{2^{2r}(r!)^2L^{2r}}\left\{(BL)^{2r}\left[\cosh^{-1}\frac{1}{BL}-\cosh^{-1}\frac{1}{Ba}\right]\right.$$

$$\left.+\sum_{k=0}^{r-1}\frac{2^{2k}(k!)^2}{(2k+1)!}(BL)^{2(r-k-1)}\left[\sqrt{1-B^2L^2}-\left(\frac{L}{a}\right)^{2k+2}\sqrt{1-B^2a^2}\right]\right\},$$

$$\mathcal{J}_{2r+2}^F = \mathrm{Re}\frac{2^{2r}(r!)^2}{(2r+1)!L^{2r+1}}\sum_{k=0}^{r}\frac{(2k)!}{2^{2k}(k!)^2}(Bl)^{2(r-k)}\left[\sqrt{1-B^2L^2}\right.$$

$$\left.-\left(\frac{L}{a}\right)^{2k+1}\sqrt{1-B^2a^2}\right].$$

The integrals $E_{t,q}^F$ and P_t^F

$$E_{t,q}^F = \mathrm{Re}\frac{(-1)^{t+1}l}{(t-1)\sqrt{1-B^2L^2}}\left\{\left[\left(\frac{l}{l-a}\right)^{t-1}-\left(\frac{f}{f-h}\right)^{t-1}\right]\right.$$

$$\times\sinh^{-1}\sqrt{\frac{(1+BL)(1-Ba)}{2B(a-L)}}+(-1)^q\left[(-1)^{t-1}-\left(\frac{L}{a}\right)^{t-1}\right]$$

$$\left.\times\cosh^{-1}\sqrt{\frac{(1+BL)(1+Ba)}{2B(L+a)}}\right\}+\frac{(-1)^tl}{2(t-1)}\sum_{p=1}^{t-1}\left[\left(\frac{f}{f-h}\right)^{t-1}\right.$$

$$\left.+(-1)^q\left(\frac{f}{f+h}\right)^{t-1}\right]\mathcal{J}_p^F,$$

$$P_t^F = \text{Re}\,\frac{(-1)^{t+1}l}{(t-1)B}\left[\left(\frac{l}{l-a}\right)^{t-1}\sqrt{1-B^2a^2} - B^2l^t\mathcal{J}_{t-1}^F + B^2l^{t-1}\mathcal{J}_{t-2}^F\right],$$

where

$$\mathcal{J}_1^F = \text{Re}\,\frac{-2}{\sqrt{1-B^2l^2}}\sinh^{-1}\sqrt{\frac{(1+Bl)(1-Ba)}{2B(a-l)}},$$

$$\mathcal{J}_2^F = \text{Re}\,\frac{-1}{l(1-B^2l^2)}\left[\frac{l}{l-a}\sqrt{1-B^2a^2} + B^2l^2\mathcal{J}_1^F\right],$$

$$\mathcal{J}_3^F = \text{Re}\,\frac{-1}{2l^2(1-B^2l^2)^2}\left[(1-4B^2l^2+3B^2la)\left(\frac{l}{l-a}\right)^2\sqrt{1-B^2a^2} - (1+2B^2l^2)B^2l^2\mathcal{J}_1^F\right],$$

$$\mathcal{J}_4^F = \text{Re}\,\frac{-1}{6l^3(1-B^2l^2)^3}\left\{[2-5B^2l^2+18B^4l^4-3(1+9B^2l^2)B^2la + (4+11B^2l^2)B^2a^2]\left(\frac{l}{l-a}\right)^3\sqrt{1-B^2a^2} + 3(3+2B^2l^2)B^4l^4\mathcal{J}_1^F\right\}.$$

A.III.3.

The integrals E_t^I, M_t^I, P_t^I, defined by (7.48), comprised in the formulae (7.47) of the wave drag coefficients C_{dm}^{In} for the double trapezoidal swept-back wing:

$$E_t^I = \frac{1}{2(t-1)}\left\{\text{Re}\left[1-\left(\frac{h}{h-f}\right)^{t-1}\right]\frac{L}{\sqrt{1-B^2l^2}}\cosh^{-1}\frac{1}{Bl} + \sum_{p=1}^{t-1}\left(\frac{h}{h-f}\right)^{t-p}L^p\mathcal{J}_p^I\right\},$$

$$M_t^I = \frac{1}{4(t-1)}\sum_{p=1}^{t-1}L^p\mathcal{J}_p^I,$$

$$P_t^I = \frac{L}{(t-1)B}[1+B^2L^t\mathcal{J}_{t-1}^I - B^2L^{t-1}\mathcal{J}_{t-2}^I],$$

where

$$\mathcal{J}_1^I = \text{Re}\frac{1}{\sqrt{1-B^2L^2}}\cosh^{-1}\frac{1}{BL},$$

$$\mathcal{J}_2^I = \frac{1}{L(1-B^2l^2)}[1-B^2L^2\mathcal{J}_1^I],$$

$$\mathcal{J}_3^I = \frac{1}{2L^2(1-B^2L^2)^2}[1-4B^2L^2+(1+2B^2L^2)B^2L^2\mathcal{J}_1^I],$$

$$\mathcal{J}_4^I = \frac{1}{6L^3(1-B^2L^2)^3}\,[2-5B^2L^2+18B^4L^4-3(3+2B^2L^2)B^4L^4\mathcal{J}_1^I].$$

A.III.4.

The integrals $E_{t,q}$ and P_t, comprised in the formulae (7.34) of the wave drag coefficients for the arrow-head wing:

$$E_{t,q} = \operatorname{Re}\frac{L}{2(t-1)\sqrt{1-B^2L^2}}\left\{[(1+f)^{t-1}-1]\,[1+(-1)^q]\cosh^{-1}\frac{1}{Bl}\right.$$

$$\left.+(-1)^q(1+f)^{t-1}\left[1-\frac{1}{(1+2f)^{t-1}}\right]\cosh^{-1}\sqrt{\frac{(1+Bl)(1+f+hBl)}{2(1+f+h)Bl}}\right\}$$

$$+\frac{L}{2(t-1)}\sum_{p=1}^{t-1}(1+f)^{t-p}\left[1+\frac{(-1)^q}{(1+2f)^{t-p}}\right]\mathcal{J}_p,$$

$$P_t = \operatorname{Re}\frac{L}{(t-1)B}\,[(1+f)^{t-2}]\,\sqrt{(1+f)^2-f^2B^2L^2}$$

$$-1+B^2L^t\mathcal{J}_{t-1}-B^2L^{t-1}\mathcal{J}_{t-2}],$$

where

$$\mathcal{J}_1 = \frac{1}{\sqrt{B^2L^2-1}}\left[2\cos^{-1}\sqrt{\frac{(1+BL)(1+f-fBL)}{2Bl}}-\cos^{-1}\frac{1}{BL}\right],$$

$$\mathcal{J}_2 = \frac{1}{L(1-B^2L^2)}\,[\operatorname{Re}\sqrt{(1+f)^2-f^2B^2L^2}-1-B^2L^2\mathcal{J}_1],$$

$$\mathcal{J}_3 = \frac{1}{2L^2(1-B^2L^2)^2}\,\{\operatorname{Re}\,[1+f-(4-f)B^2L^2]\sqrt{(1+f)^2-f^2B^2L^2}$$

$$-1+4B^2L^2+(1+2B^2L^2)B^2L^2\mathcal{J}_1\},$$

$$\mathcal{J}_4 = \frac{1}{6L^3(1-B^2L^2)^3}\,\{\operatorname{Re}\,[2(1+f)^2-(5+13f-4f^2)B^2L^2$$

$$+(18+9f+2f^2)B^4L^4]\sqrt{(1+f)^2-f^2B^2L^2}-(2-5B^2L^2+18B^4L^4)$$

$$-3(3+2B^2L^2)B^4L^4\mathcal{J}_1\}.$$

REFERENCES

A

1. Ackeret, J. Air forces on airfoils moving faster than sound. *Z. Flugtech. u. Motorluftschiffahrt* **16,** 72–4 (1925). Also NACA Technical Memorandum 317 (1925).
2. Ackeret, J. *Gasdynamik. Handbuch der Physik*, Band 7. Berlin, Springer, 1927.
3. Acum, W. Aerodynamic Forces on a Rectangular Wing Oscillating in a Supersonic Flow. Rep. Memor. Aero. Res. Coun., Lond., no. 2763 (1950).
4. Acum, W. Note on the Effect of Thickness and Aspect Ratio on the Damping of Pitching Oscillations of Rectangular Wings Moving at Supersonic Speeds. Aero. Res. Coun., Lond. Curr. Paper, no. 151 (1954).
5. Adams, M. C. and Sears, W. R. On an extension of slender wing theory. *J. Aero. Sci.* **19,** 424–5 (1952).
6. Adams, MacC. and Sears, W. R. Slender-body theory—review and extensions. *J. Aero. Sci.* **20,** 85–98 (1953).
7. Adams, M. C. Leading-edge separation from delta wings at supersonic speeds. *J. Aero. Sci.*, June 1953.
8. Andrews, L., Bergesen, J. and Porter, J. An Investigation of the Flow around Delta Wings with Leading-edge Separation. Princeton University Rep. no. 510, May 1960.
9. Arjanikov, N. S. and Maltev, V. N. *Aerodinamika (Aerodynamics)*. Moscow, 1952.

B

1. Baldwin, B. S. Triangular Wings Cambered and Twisted to support Specified Distributions of Lift at Supersonic Speeds. NACA Technical Note 1816, 1949.
2. Bardsley, O. The conditions at a sharp leading edge in supersonic flow. *Phil. Mag.* **42,** 255 (1951).
3. Bartels, R. C. F. and Laporte, O. An Investigation of the Exact Solutions of the Linearized Equations for the Flow past Conical Bodies. Engineering Research Institute. Univ. of Michigan, Ann Arbor, Bumblebee Report, 75, 1948.
4. Beastall, D. and Pallant, R. J. Wind Tunnel Tests on Two-dimensional Supersonic Airfoils at M = 1·86 and M = 2·48. Reports and Memoranda No. 2800. Ministry of Supply, Aeronautical Research Council, London: Her Majesty's Stationery Office, July 1950.
5. Behrbohm, H. The Lifting Trapezoidal Wing with Small Aspect Ratio at Supersonic Speed. SAAB Aircraft Co. Tech. Note 10, Linköping, Sweden, 1952.
6. Berler, I. and Nichols, S. *Interference between Wing and Body at Supersonic Speeds*. Part VI: Data report on pressure distribution tests of wing–body interference models at Mach number 2·0, phase two tests of June 1949. Cornell Aeronaut. Lab. Bumblebee Rept. CAL/CF-1569, May 1951.
7. Berler, I. and Nichols, S. *Interference between Wing and Body at Supersonic Speeds*. Part VII: Data report on pressure distribution tests of wing–body inter-

ference models at Mach number 2·0, phase three tests of August 1949. Cornell Aeronaut. Lab. Bumblebee Rept. CAL/CF-1570, June 1951.

8. Berman, J. Lift and Moment Coefficients for an Oscillating Rectangular Wing–aileron Configuration in Supersonic Flow. NACA Tech. Note 3644, 1956.
9. Berndt, S. B. On the Theory of Slowly Oscillating Delta Wings at Supersonic Speeds. Rep. Flygtekniska Försöksanstalten Stockholm, no. 43, 1952.
10. Beshkine, L. Forces aérodynamiques agissant sur les surfaces portantes aux vitesses supersoniques. *Les Cahiers d'Aérodynamique*, **6**, 1947.
11. Bolton-Shaw, B. W. Control on Delta Wings at Supersonic Speeds. College of Aeronautics, Report No. 36, May 1950.
12. Brandstatter, J, and Mortzschky, H. A Method for Calculating the Potential of an Oscillating Wing in Supersonic Flow. Rep. Lockheed Aircraft Co., no. MAM-245, 1954.
13. Britton, J. Pressure measurements on a cambered ogee wing ($p = 0{\cdot}45$) at Mach numbers up to 2·6. Unpublished M.o.A. Report.
14. Brown, C. E. Theoretical Lift and Drag of Thin Triangular Wings at Supersonic Speeds. NACA Report 839, 1946.
15. Brown, C. E. and Adams, MacC. Damping in Pitch and Roll of Triangular Wings at Supersonic Speeds. NACA Report 892, 1947.
16. Brown, C. E. and Michael, W. H. Effect of leading-edge separation on the lift of a delta wing. *J. Aero. Sci.*, October 1954.
17. Brown, Clinton E. and Michael, William H., Jr. On Slender Delta Wings with Leading-edge Separation. NACA Tech. Note 3430, 1955.
18. Browne, S. H., Friedman, L. and Hodes, I. A wing–body problem in a supersonic conical flow. *J. Aero. Sci.* **15**, 443–52 (1948).
19. Bulygina, V. E. K zadache o kryle zadannogo obiema s minimalnym volnovym soprotivleniem (On the problem of a minimum wave drag wing of given volume). *Prikl. Mat. Mekh.* **22**, 6 (1958).
20. Busemann, A. *Gasdynamik. Handbuch der Experimental Physik*, Band 4, Teil I. Leipzig, Akademischer Verlage, 1931.
    Busemann, A. Drücke auf kegelförmige Spitzen bei Bewegung und Überschallgeschwindigkeit. *ZAMM* **9**, 496–8 (1929).
21. Busemann, A. *Flüssigkeits und Gasbewegung. Handwörterbuch der Naturwissenschaften*, 2e. Auflage. Gustav Fischer, Jena, 1933.
22. Busemann, A. Aerodynamischer Auftrieb bei Überschallgeschwindigkeit. *Atti del Convegno della Fondazione Alessandro Volta, 1935*, 328–60. 1936. Also *Luftfahrtforschung* **12**, 210–20 (1935).
23. Busemann, A. Conical supersonic flow with axial symmetry. *Luftfahrtforschung* **19**, 134–44 (1942).
24. Busemann, A. Infinitesimale kegelige Überschallströmung. *Schr. Dtschen. Akad. Luftfahrt.* **7** B, 105–22 (1943); transl. in NACA Techn. Memo. 1100, 1947.

C

1. Carafoli, E. *Aerodinamica* (Aerodynamics). Editura Tehnică, Bucureşti. 1951.
   Also in German:
   Tragflügeltheorie. Verlag-Technik, Berlin, 1954.
   And in Russian:
   Aerodynamika kryla samoleta (Aerodynamics of aircraft wing). Izd-vo Akad. Nauk SSSR, Moscow, 1956.

2. Carafoli, E. *High-speed Aerodynamics*. Editura Technică, Bucureşti, 1956.
Also in Roumanian:
Aerodinamica Vitezelor Mari (High-speed aerodynamics). Edit. Acad. R.P.R., Bucureşti, 1957.
And in Russian:
Aerodinamika bolshih skorostey (High-speed aerodynamics). Izd-vo Akad. Nauk SSSR, Moscow, 1960.

3. Carafoli, E. *Conical Motions in Supersonic Flow* (in Chinese). Academia Sinica, Peking, 1955.

4. Carafoli, E. Asupra caracterului hidrodinamic al soluţiilor din mişăcrile conice aplicate la teoria aripilor poligonale (The hydrodynamic character of the solutions of conical flow applied to the theory of polygonal wings). *Comunicările Acad. R.P.R.* t. II, no. 9–10 (1952).

5. Carafoli, E. Extinderea analogiei hidrodinamice din mişcările conice la aripile groase şi cu incidenţă variabilă (Extension of the hydrodynamic analogy in the conical motions to thick wings and wings of variable incidence). *Communicările Acad. R.P.R.*, t. III, no. 3–4 (1953).

6. Carafoli, E. Asupra portanţei şi rezistenţei aripii triunghiulare la viteze supersonice (On the lift and drag of the triangular wing at supersonic speeds). *Comunicările Acad. R.P.R.*, t. IV, no. 5–6 (1954).

7. Carafoli, E. Teoria treugolnyh prostyh ili krestoobraznyh kryliev v sverhzvukovom potoke (The theory of simple or cruciform wings in supersonic flow). *Injenernyi Sbornik* **22** and **28,** 1960.

8. Carafoli, E. On a unitary formula for compression–expansion in supersonic–hypersonic flow. *Revue de Méc. Appl.* **7,** 5 (1962). Publishing House of the Academy of the R.P.R., Bucharest.

9. Carafoli, E. O raspredelenie davlenii na aerodinamicheskih profiliah v sverhzvukovom i ghiperzvukovom potokah. (On the pressure distribution on aerodynamic profiles in supersonic and hypersonic flows). *Izvestia Akad. Nauk SSSR, Otdelenie Techn. Nauk, Mechanika i Mashinostroenie*, no. 1, Moscow, 1962.

10. Carafoli, E. Extension of conical motions to quasi-conical motions. *Revue de Méc. Appl.* **10,** 2, 1965. Publishing House of the Academy of the Socialist Republic of Rumania, Bucharest.

11. Carafoli, E. and Atanasiu, Fr. Asupra unei clase de aripi în săgeată şi în coadă de rîndunică (On a class of arrow-head and swept-back wings). *Studii Cerc. Mec. Apl.* **18,** 3 (1965), Bucharest.

12. Carafoli, E. and Berbente, C. Aerodynamic characteristics of profiles in supersonic-hypersonic flow in the case of neglecting the pressure losses due to shock waves. *Revue de Méc. Appl.* **8,** 5 (1963). Publishing House of the Academy of the R.P.R., Bucharest.

13. Carafoli, E. and Berbente, C. Determination of the characteristics of aerodynamic profiles in supersonic–hypersonic flow, considering the pressure losses through shock waves. *Revue de Méc. Appl.* **9,** 5 (1964). Publishing House of the Academy of the R.P.R., Bucharest.

14. Carafoli, E. and Berbente, C. Calculation of pressures and aerodynamic characteristics of delta wings in supersonic–moderate hypersonic flow. *Revue de Méc. Appl.* **9,** 3 (1966). Publishing House of the Academy of the R.P.R., Bucharest.

15. Carafoli, E. and Dumitrescu, L. Asupra aripii delta subţiri dubluconice cu incidenţă variabilă, în curent supersonic (On the thin doubly-conical delta wing with variable angle of attack in supersonic stream). *Studii Cerc. Mec. Apl.* **6,** no. 5–4 (1955), Bucharest.

16. CARAFOLI, E. and DUMITRESCU, L. Rezistenţa de undă a unei clase de aripi delta subţiri dublu-conice cu indicenţă variabilă în curent supersonic (The wave drag of a thin doubly-conical delta wing with variable incidence in supersonic flow). *Studii Cerc. Mec. Apl.* **7**, 3 (1956), Bucharest.
17. CARAFOLI, E. and HOROVITZ, BEATRICE. Mişcarea supersonică în jurul aripii triunghiulare cu discuri marginale (Supersonic flow around a delta wing with end plates). *Comunicările Acad. R.P.R.*, t. III, no. 11–12 (1953), Bucharest.
18. CARAFOLI, E. and HOROVITZ, B. Influenţa discurilor axiale asupra mişcării supersonice în jurul aripilor unghiulare conice groase şi cu incidenţă variabilă (Influence of axial plates on supersonic flow around thick conical delta wings with varying incidence). *Comunicările Acad. R.P.R.*, t. IV, no. 5–6 (1954), Bucharest.
19. CARAFOLI, E. and HOROVITZ, B. L'écoulement supersonique autour d'une aile angulaire à disque axial. *Revue de Méc. Appl.* **1**, 1 (1956). Publishing House of the Academy of the P.R.R., Bucharest.
20. CARAFOLI, E. and HOROVITZ, B. Homogeneous flows of higher order round a cruciform wing, with application to harmonic oscillations. *The XIth International Congress of Applied Mechanics, Brussels*, Sept. 1956.
21. CARAFOLI, E. and HOROVITZ, B. Écoulements coniques d'ordre supérieur autour d'une aile triangulaire à plaque normale. *Revue de Méc. Appl.* **2**, 2 (1957). Publishing House of the Academy of the R.P.R., Bucharest.
22. CARAFOLI, E. and HOROVITZ, B. Problèmes mixtes des ailes triangulaires pourvues d'une plaque normale, en courant supersonique (Ailes cruciformes). *Revue de Méc. Appl.* **3**, 3 (1958). Publ. House of the Academy of the R.P.R., Bucharest.
23. CARAFOLI, E. and HOROVITZ, B. *Problèmes indirectes des ailes triangulaires cruciformes en courant supersonique.* Golden Jubilee, Calcutta Mathematical Society, vol. II, 1961.
24. CARAFOLI, E. and IONESCU, MIRCEA. Trainée d'ailes delta doublement coniques à pente variable, en régime supersonique. *Revue de Méc. Appl.* **1**, 2 (1956), and also in Rumanian in *Studii Cerc. Mec. Apl.* **5**, 3–4 (1954). Publishing House of the Academy of the R.P.R., Bucharest.
25. CARAFOLI, E. and IONESCU, M. Écoulements coniques d'ordre supérieur autour des ailes triangulaires où á épaisseur symétrique. *Revue de Méc. Appl.* **2**, 1 (1957). Publishing House of the Academy of the R.P.R., Bucharest.
26. CARAFOLI, E. and IONESCU, M. Théorie générale de l'aile triangulaire à distribution de pressions donnée (Le problème indirect). *Revue de Méc. Appl.* **2**, 2 (1958). Publishing House of the Academy of the R.P.R., Bucharest.
27. CARAFOLI, E. and MATEESCU, DAN. L'écoulement autour du système portant aile–fuselage conique. *Revue de Méc. Appl.* **4**, 3 (1959), and also in Rumanian in *Studii Cerc. Mec. Apl.* **10**, 2 (1959). Publishing House of the Academy of the R.P.R., Bucharest.
28. CARAFOLI, E. and MATEESCU, D. General method for determining the wing–conical body interference in supersonic flow. *Revue de Méc. Appl.* **5**, 2 (1960), and also in Rumanian in *Studii Cerc. Mec. Apl.* **11**, 1 (1960). Publishing House of the Academy of the R.P.R., Bucharest.
29. CARAFOLI, E. and MATEESCU, D. Supersonic flow around the cruciform wing–conical body system. *Revue de Méc. Appl.* **5**, 3 (1960), and also in Rumanian in *Studii Cerc. Mec. Apl.* **11**, 2 (1960). Publishing House of the Academy of the R.P.R., Bucharest.
30. CARAFOLI, E. and MATEESCU, D. High order conical flow around a delta wing fitted with conical body. *Proceedings of the Seminar on Aeronautical Sciences* (Nov. 27–Dec. 2, 1961, Bangalore, India), pp. 249–69.

31. Carafoli, E. and Mateescu, D. The wing–conical body interference in high order conical flows. *Revue de Méc. Appl.* **7**, 3 (1962), and also in Rumanian in *Studii Cerc. Mec. Apl.* **13**, 2 (1962). Publishing House of the Academy of the R.P.R., Bucharest.
32. Carafoli, E. and Mateescu, D. Sur une classe d'ailes delta à incidence et pente variables d'après des fonctions homogènes, en régime supersonique. *Revue de Méc. Appl.* **10**, 2 (1965), and also in Rumanian in *Studii Cerc. Mec. Apl.* **18**, 2, (1965). Publishing House of the Academy of the R.P.R., Bucharest.
33. Carafoli, E. and Mateescu, D. General method for determining the cruciform wing–conical body interference or tail–conical body interference in supersonic flow (unpublished).
34. Carafoli, E. and Mateescu, D. High order conical flow around the cruciform wing–conical body system (unpublished).
35. Carafoli, E. and Mateescu, D. Mouvement oscillatoire harmonique du système aile–fuselage conique en régime supersonique. *Revue de Méc. Appl.* **11**, 5 (1966), and also in Rumanian in *Studii Cerc. Mec. Apl.* **23**, 2 (1966). Publishing House of the Academy of the R.P.R., Bucharest.
36. Carafoli, E. and Mateescu, D. Application of quasi-conical motions to the study of wing–body lifting systems in supersonic flow. *Revue de Méc. Appl.* **12**, 1 (1967), and also in Rumanian in *Studii Cerc. Mec. Apl.* **24**, 3 (1967). Publishing House of the Academy of the R.P.R., Bucharest.
37. Carafoli, E. and Năstase, Adriana. Étude des ailes triangulaires minces à symétrie forcée. *Revue de Méc. Appl.* **3**, 4 (1958), and also in Rumanian in *Studii Cerc. Mec. Apl.* **9**, 4 (1958). Publishing House of the Academy of the R.P.R., Bucharest.
38. Carafoli, E. and Năstase, A. Détermination de la surface d'une aile à arête de separation de la distribution des pressions en courant supersonique. *Revue de Méc. Appl.* **5**, 1 (1960) and also in Rumanian in *Studii Cerc. Mec. Apl.* **10**, 4 (1959). Publishing House of the Academy of the R.P.R., Bucharest.
39. Carafoli, E. and Năstase, A. Trapetsievidnye krylia s minimalnym lobovym soprotivleniam v sverhzvukovom potoke (Trapezoidal wing of minimum drag in supersonic flow). *Mekhanika* **1**, 1961 (USSR, Moscow).
40. Carafoli, E. and Năstase, A. Minimum drag thin triangular wing in supersonic flow. *Revue de Méc. Appl.* **5**, 5 (1960) and also in Rumanian in *Studii Cerc. Mec. Apl.* **11**, 4 (1960). Publishing House of the Academy of the R.P.R., Bucharest.
41. Carafoli, E. and Năstase-Ghia, Adriana. On the minimum drag problem of the wing of symmetrical thickness in supersonic flow. *Revue de Méc. Appl.* **7**, 2 (1962) and also in Rumanian in *Studii Cerc. Mec. Apl.* **13**, 1 (1962). Publishing House of the Academy of the R.P.R., Bucharest.
42. Carafoli, E. and Năstase-Ghia, A. L'utilisation de la méthode des résidus dans le problème de résistance minimum de l'aile trapézoïdale en courant supersonique. *Arhiwum Mechaniki Stosowanej* **3**, 16 (1964) (Poland).
43. Carafoli, E. and Rădulescu, Silvia. Scurgerea supersonică in jurul aripii delta subţiri în derivă (Supersonic flow around the thin yawed delta wing). *Studii Cerc. Mec. Apl.* **19**, 4 (1965), Bucharest.
44. Carafoli, E. and Năstase, A. Optimum shape of the cruciform wing of symmetrical thickness in supersonic flow. *Revue de Méc. Appl.* **12**, 2 (1967).
45. Carafoli, E. and Săndulescu, Serban. Influenţa bracajului aripioarelor asupra caracteristicilor aerodinamice ale aripilor triunghiulare în regim supersonic (The influence of the deflection of ailerons on the aerodynamic characteristics of triangular wings in supersonic flow). *Studii Cerc. Mec. Apl.* **7**, 4 (1956), Bucharest.

46. CARAFOLI, E. and SĂNDULESCU, S. Oscillations harmoniques des ailerons de forme générale en régime supersonique. *Arhiwum Mechaniki Stosowanej* **10,** 6 (1958) (Poland).

47. CARAFOLI, E. and SĂNDULESCU, S. Caractéristiques aérodynamiques des ailerons ayant un mouvement oscillatoire harmonique en régime supersonique. *Revue de Méc. Appl.* **4,** 2 (1950) and also in Rumanian in *Studii Cerc. Mec. Apl.* **10,** 1 (1959). Publishing House of the Academy of the R.P.R., Bucharest.

48. CARAFOLI, E. and STAICU, ST. Sur un modèle théorique concernant l'aile delta mince avec décollement aux bords d'attaque. *Journ. de Méc.,* Vol. 6, No. 2, Juin 1967, Paris, and also in Rumanian in *Studii Cerc. Mec. Apl.* **22,** 4 (1966). Publishing House of the Academy of the R.P.R., Bucharest.

49. CARBSON, H. W. Pressure distribution at Mach number 2·05 on a series of highly swept arrow wings employing various degrees of twist and camber. NASA Tech. Note D-1264, May 1962.

50. CARRIER, G. F. The oscillating wedge in a supersonic stream. *J. Aero. Sci.* **16,** 150–2 (1949).

51. CHANG, C. C. and HOUSER, J. E. Correlation of Test Data and Comparison with Theoretical Results for Thin Wings in Supersonic Flow. Glenn L. Martin Co. Eng. Rep. 2729. 1947.

52. CHANG, C. C. The aerodynamic behavior of a harmonically oscillating finite sweptback wing in supersonic flow. NACA Tech. Note 2467. 1951: see *Math. Rev.* **13,** 730 (1952).

53. CHERNYI, G. G. *Techenia Gaza s Bolshoi Sverhzvukovoi Skorostiu* (Gas flow at high supersonic speed). Gos. Izd-vo Fiz.-Mat. Lit., Moscow, 1959. (*Introduction to Hypersonic Flow,* translated by Academic Press, New York and London, 1961.)

54. COHEN, D. Formulas and charts for the supersonic lift and drag of flat sweptback wings with interacting leading and trailing edges. NACA Tech. Note 2093, 1950.

55. COHEN, D. Formulas for the supersonic loading, lift, and drag of flat sweptback wings with leading edge behind the Mach lines. NACA Report 1050, 1951.

56. COHEN, D. and FRIEDMAN, M. D. Theoretical investigation of the supersonic lift and drag of thin, sweptback wings with increased sweep near the root. NACA Tech. Note 2959, 1953.

57. COLE, J. D. Note on non-stationary slender-body theory. *J. Aero. Sci.* **20,** 798 (1953).

58. COLEMAN, T. F. Supersonic lift solutions obtained by extending the simple linearized conical flow theory. North Amer. Aviation CM 440. 1948.

59. COURANT, R. and FRIEDRICHS, K. O. *Supersonic Flow and Shock Waves.* Interscience, 1948.

60. COURTNEY, A. L. A collection of data on the lift-dependent drag of uncambered slender wings at supersonic speeds. Unpublished M.o.A. Report.

61. COURTNEY, A. L. and ORMEROD, A. O. Pressure plotting and force tests at Mach numbers up to 2·8 of an uncambered slender wing of $p = \frac{1}{2}$, $s_T/c_0 = \frac{1}{4}$ (Handley Page Ogee). *A.R.C.* **23,** 109 (May 1961).

62. CRAMER, R. H. *Interference between Wing and Body at Supersonic Speeds.* Part V: Phase one wind tunnel tests correlated with the linear theory. Cornell Aeronaut. Lab. Bumblebee Report CAL/CM-597, Dec. 1950.

63. CRAMER, R. H. Interference between wing and body at supersonic speeds. Theoretical and experimental determination of pressures on the body. *J. Aero. Sci.* **18,** 629–32 (1951).

64. CRAMER, R. H. Wing–body interference at supersonic speeds. Recent experimental

results and their bearing on advances in theory. *Proc. U.S. Navy Symposium on Aeroballistics*, NAVORD Rept. 5146, May 1952. Also available as Cornell Aeronaut. Lab. special publication.

65. CUNNINGHAM, H. J. Total Lift and Pitching Moment on Thin Arrowhead Wings Oscillating in Supersonic Potential Flow. NACA Tech. Note 3433, 1955.
66. CARAFOLI, E. Theory of the wing and wing–body systems over a wide range of the supersonic–hypersonic regime. *Proceedings of the Vth I.C.A.S. Congress.*
67. CARAFOLI, E. and STAICU, ST. Antisymmetrical thin delta wing with flow separation at the leading edges. *Revue de Méc. Appl.* **11,** 6 (1966) and also in Rumanian in *Studii Cerc. Mec. Apl.* **23,** 6 (1966). Publishing House of the Academy of the R.S.R., Bucharest.

D

1. DARIE, GH. N. Aplicarea metodei Carafoli la calculul rezistenţei de undă a aripii delta dublu conice cu pantă dublă (Application of Carafoli's method to the calculus of the wave drag of the doubly-conical delta wing with double slope). *Studii Cerc. Mec. Apl.* **7,** 3 (1957), Bucharest.
2. DONOV, A. A plane wing with sharp edges in a supersonic stream. *Bull. Acad. Sci. URSS. Sér. Math. 1939*, 603 (1939).
3. DONOVAN, A. F. and LAWRENCE, H. R. (Editors). *Aerodynamic Components of Aircraft at High Speeds* (*High Speed Aerodynamics and Jet Propulsion*, vol. VII). London, Oxford University Press, 1957.
4. DÖRR, J. Les forces aérodynamiques sur une aile vibrant harmoniquement dans un écoulement supersonique. ONERA Rap. no. 37. 1949.
5. DÖRFNER, K. R. *Dreidimensionale Überschallprobleme des Gasdynamik*. Springer Verlag, Berlin, 1957.
6. DORRANCE, W. H. Two-dimensional airfoils at moderate hypersonic velocities. *J. Aero. Sci.* **19,** 9 (1952).
7. DUBS, F. *Hochgeschwindigkeits—Aerodynamik*. Birkhäuser, Verlag Basel/Stuttgart, 1961.
8. VAN DYKE, M. D. First Order and Second Order Theory of Supersonic Flow past Bodies of Revolution. Rand Corp. Rept. P-125, 1949.
9. VAN DYKE, M. D. The combined supersonic–hypersonic similarity rule. *J. Aero. Sci.* **18,** 499–500 (1951).
10. VAN DYKE, M. D. First- and second-order theory of supersonic flow past bodies of revolution. *J. Aero. Sci.* **18,** 161 (1951).
11. VAN DYKE, M. D. Supersonic Edges in Thin-wing and Slender-body Theory. NACA Tech. Note 3343, 1954.
12. VAN DYKE, M. D. A Study of Hypersonic Small Disturbance Theory. NACA Tech. Note 3173, May 1954.
13. DYE, F. E., Jr. *Interference between Wing and Body at Supersonic Speeds*. Part IX: Pressure distribution tests of wing–body interference models at Mach No. of 2·0. Phase V tests of Aug. 1951. Cornell Aeronaut. Lab. Rept. CAL/CF 1684, Dec. 1951.

E

1. EARNSHAW, A. B. An experimental investigation of the structure of leading-edge vortex. RAE Tech. no. 2740, Mar. 1961.
2. EBELING, P. and SCHWARM, H. Das Strömungsfeld ebener und rotationssymmetrischer Körper und Einläufe bei über- und hyperschallgeschwindigkeiten nach dem nicht linearen Charakteristikenverfahren. *Jahrbuch W.G.L.R.*, 1964.
3. EDWARDS, R. H. Leading-edge separation from delta wings. *J. Aero. Sci.*, Feb. 1954.

4. EDWARDS, J. B. W. Free-flight measurements of the zero-lift drag of a slender ogee wing at transonic and supersonic speeds. *A.R.C.* **24,** 448 (Oct. 1962).
5. ELLIS, M. C., JR. and HASEL, L. E. Preliminary Tests at Supersonic Speeds of Triangular and Swept-back Wings. NACA Res. Mem. 26 L 17, 1947.
6. EMINTON, E. Pressure Distributions at Zero Lift for Delta Wings with Rhombic Cross Sections. Aero. Res. Counc. Lond. Curr. Pap. 525, 1960.
7. EMINTON, E. Theoretical Comparison of the Flow over a Flat Delta Wing and a Rectangular Pyramid. Aeronaut. Res. Council, London, Curr. Pap. 637, 1963.
8. EVVARD, J. C. The Effects of Yawing Thin Pointed Wings at Supersonic Speeds. NACA Tech. Note 1429, Sept. 1947.
9. EVVARD, J. C. A Linearized Solution for Time-dependent Velocity potentials near Three-dimensional Wings at Supersonic Speeds. NACA Tech. Note 1699, 1948.
10. EVVARD, J. C. Use of Source Distributions for Evaluating Theoretical Aerodynamics of Thin Finite Wings at Supersonic Speeds. NACA Rept. no. 951, 1950.

F

1. FALKOVICI, S. V. K teorii kryla konechnogo razmaha v sverhzvukovom potoke (On the theory of wings of finite span in a supersonic stream). *PMM*, t. XI, fasc. 4, 1947.
2. FÉNAIN, M. Traînée d'ailes delta symétriques à incidence nulle en régime supersonique. *La Recherche Aéronautique*, no. 16, juillet–août 1950.
3. FÉNAIN, M. and VALLÉE DENISE. Application de la Théorie des Écoulements Homogènes à la Recherche de l'Adaptation de Certaines Ailes en Régime Supersonique. ONERA Mémo. Technique 1er et 2e partie, 14, 1959.
4. FÉNAIN, M. La théorie des écoulements à potentiel homogène et ses applications au calcul des ailes en règime supersonique. *Progress in Aeronautical Sciences*, vol. I (Edited by Ferri, A., Küchemann, D. and Sterne, L.H.F.). Pergamon Press, London, 1961.
5. FÉNAIN, M. Calculation of fixed wings of arbitrary form in supersonic regime (in French). *C.R. Acad. Sci. Paris* **253,** 23, p. 2634 (Dec. 1961).
6. FERRARI, C. Campo aerodinamico a velocità iperacustica attorno a un solido di rivoluzione a prora accuminata. *L'Aerotecnica* **16,** 121–30 (1936). Also Brown Univ. Transl. A 9-T-18.
7. FERRARI, C. Determination of the pressure exerted on solid bodies of revolution with pointed noses placed obliquely in a stream of compressible fluid at supersonic velocity. MOS R.T.P. Translation 1105. From *Atti R. Accad. Sci. Torino* **72,** 140–63 (1936).
8. FERRARI, C. Campi di corrente ipersonora attorna a solidi di rivoluzione. *L'Aerotecnica* **17,** 507–18 (1937). Also Brown Univ. Transl. A 9-T-29.
9. FERRARI, C. *Aerodinamica.* Arti Grafiche Viretto, Torino, 1939.
10. FERRARI, C. Interference between wing and body at supersonic speeds. Theory and numerical applications. *J. Aero. Sci.* **15,** 317–36 (1948).
11. FERRARI, C. Interference between wing and body at supersonic speeds. Note on wind-tunnel results and addendum to calculations. *J. Aero. Sci.* **16,** 542–6 (1949).
12. FERRARI, C. *Interference between Wing and Body at Supersonic Speeds.* Part III: Numerical application to the linear theory. Cornell Aeronaut. Lab. Bumblebee Rept. CAL/CM-463, Apr. 1951.
13. FERRARI, C. Sulla determinazione di alcuni tipi di campi di corrente ipersonora (On the determination of certain basic types of supersonic flow fields). *Rend. Reale*

Accad. Nazl. Linçei, Rome **1** (6), Ser, VIII (1949). Also available as NACA Tech Mem. 1381, Nov. 1954.

14. FERRARI, V. Calcolo della velocitá indotta da un'ala in corrente supersonica linearizzata (Induced velocities generated by a wing operating in supersonic flow, on the basis of the linear theory.) Encomium volume of *Aerotecnica,* in honour of Prof. G. A. Crocco, **33,** 65–69 (1953). English transl. Cornell Aeronaut. Lab. Rept. CAL-59, July 1953.
15. FERRI, A. Alcuni risultati sperimentali ottenuti alla Galleria ultrasonora di Guidonia. *Atti di Guidonia* **17,** 1939.
16. FERRI, A. Investigations and Experiments in the Guidonia Supersonic Wind Tunnel. NACA Tech. Mem. 901, 1939; Experimental Results with Airfoils tested in the High Speed Tunnel at Guidonia. NACA Tech. Mem. 946, 1940.
17. FERRI, A. *Elements of Aerodynamics of Supersonic Flows.* The Macmillan Company, New York, 1949.
18. FERRI, A. The Method of Characteristics for the Determination of Supersonic Flow over Bodies of Revolution at Small Angles of Attack. NACA Tech. Note 1809, 1949.
19. FERRI, A. Supersonic Flow Around Circular Cones at Angles of Attack. NACA Tech. Note 2236, 1950.
20. FERRI, A., NESS, N. and KAPLITA, T. Supersonic flow over conical bodies without axial symmetry. *J. Aero. Sci.* **20,** 563–71 (1953).
21. FIRMIN, M. C. P. The Pressure Distribution at Zero-lift on some Slender Delta Wings at Supersonic Speeds. Aeronaut. Res. Council, London, Curr. Pap. no. 774, 1964.
22. FLAX, A. (1949). Relations between the characteristics of a wing and its reverse in supersonic flow. *J. Aero. Sci.* **16,** 496–504 (1949).
23. FLAX, A. The reverse flow theorem for nonstationary flows. *J. Aero. Sci.* **19,** 352–3 (1952).
24. FOWELL, L. R. Exact and approximate solutions for the supersonic delta wing. *J. Aero. Sci.* **23,** no. 8, 709–20 (Aug. 1956).
25. FOWELL, L. R. An Exact Theory of Supersonic Flow around a Delta Wing. UTIA Rep. no. 30, Mar. 1955.
26. FRANKL, F. I. Osesimetrichnye sverhzvukovye techenya (Supersonic axial-symmetrical flow). *Izv. Artiler. Akad. RKKA,* no. 5 (1934).
27. FRANKL, F. I. and KARPOVITCH, E. A. *Gas Dynamics of Slender Bodies* (in Russian) Gos. Izd-vo Techn.-Theoret. Lit-ry, Moscow, 1948.
28. FROELICH, J. E. Nonstationary motion of purely supersonic wings. *J. Aero. Sci.* **18,** 298–310 (1951).
29. FÉNAIN, M. and VALLÉE, D. Effet d'incidence sur des plaques planes à pointe en delta. *La Recherche Aéronautique* 105 (1965).
30. FÉNAIN, M. and GUIRAUD-VALLÉE, D. Calcul numérique des ailes en régime supersonique stationaire ou instationaire, 1re partie: écoulement stationaire. *La Recherche Aéronautique* 115 (1966).

G

1. GALIN, L. A wing rectangular in plan in a supersonic flow. *Akad. Nauk SSSR, Prikl. Mat. Meh.* **2,** 465–74 (1947).
2. GARDNER, C. S. Time-dependent linearized supersonic flow past planar wings. *Comm. Pure Appl. Math.* **3,** 33–8 (1950).

3. GARDNER, C. S. A relation between time-dependent and steady linearized supersonic flows past conical bodies. *Comm. Pure Appl. Math.* **3,** 39–43.
4. GARRICK, I. E. and RUBINOW, S. I. Flutter and Oscillating Air Force Calculations for an Airfoil in a Two-dimensional Supersonic Flow. NACA Report 846, 1946.
5. GARRICK, I. E. and RUBINOW, S. I. Flutter and Oscillating Airforce Calculations for an Airfoil in Two-dimensional Supersonic Flow. NACA Report, 846, 1946.
6. GARRICK, I. E. and RUBINOW, S. I. Theoretical Study of Air Forces on an Oscillating or Steady Thin Wing in a Supersonic Main Stream. NACA Tech. Rept. 872, 1947.
7. GARRICK, I. E. and WATKINS, C. E. Some recent developments in the aerodynamic theory of oscillating wings. *NACA Univ. Conference on Aerodynamics, Construction and Propulsion*, Oct. 1954.
8. GARRICK, I. E. Aerodynamic Theory and its Application to Flutter. AGARD, Rept. 34, 1956.
9. GERMAIN, O. The General Theory of Conical Motions and its Application to Supersonic Aerodynamics. ONERA Publication 34, 1949.
10. GERMAIN, P. La théorie des mouvements homogènes et son application au calcul de certaines ailes delta en règime supersonique. *La Recherche Aéronautique* **7,** 3–16(1949)
11. GERMAIN, P. La Théorie des Mouvements Coniques et ses Applications à l'Aérodynamique Supersonique. Thèses, Paris, 1948, O.N.E.R.A. Publ. 34 (1949).
12. GERMAIN, P. and BADER, R. Quelques remarques sur les mouvements vibratoires d'une aile en régime supersonique. *La Recherche Aéronautique* **2,** 3–13 (1949).
13. GERMAIN, P. and VALLÉE, D. Effet de dièdre sur une aile delta en régime supersonique. *La Recherche Aéronautique* **52,** 13–20 (1956).
14. GERMAIN, P. Aile symétrique à portance nulle et de volume donné réalisant le minimum de trainée en écoulement supersonique. *Comptes Rendus* 244 (1957).
15. GERMAIN, P. Sur le minimum de traînée d'une aile de forme en plan donnée. *Comptes Rendus*, *Acad. Sci.*, févr. 1957.
16. GERMAIN, P. and FÉNAIN, M. Determination of Optimum Deltawings with Sonic Leading-edges (in French). O.N.E.R.A. Publ. 86, 3–20 (Jan./Feb. 1962).
17. GILBERT, J. Approximate Method for Computing the Pressure Distribution on Wings with Subsonic Leading Edges in a Steady Stream. English Electric Aero. Tech. Memo. AM18, 1956.
18. GOLDSTEIN, S. and WARD, G. N. The linearized theory of conical fields in supersonic flow, with applications to plane aerofoils. *Aeron. Quart.* **2,** 39–84 (1950).
19. GOODMAN, T. R. The Lift Distribution on Non-conical Flow Regions of Thin Finite Wings in a Supersonic Stream. Cornell Aeronaut. Lab. Rept., Oct. 1948.
20. GOODMAN, T. R. The lift distribution on conical and non-conical flow regions of thin finite wings in a supersonic stream. *J. Aero. Sci.* **16,** 365–74 (1949).
21. GOODMAN, T. R. Aerodynamics of a supersonic rectangular wing striking a sharp-edged gust. *J. Aero. Sci.* **18,** 519–26 (1951).
22. GOODMAN, T. R. The quarter-infinite wing oscillating at supersonic speeds. Cornell Aero. Lab. Report 36, 1951. Also *Quart. Appl. Math.* **10,** 189–92 (1952).
23. GOODMAN, T. R. Some Simplifications of Ferrari's Theory of Wing–Body Interference at Supersonic Speeds. Cornell Aeronaut. Lab. Bumblebee Rept. CAL/CM-687, Dec. 1954.
24. GRAHAM, W. E. The Calculation of Minimum Supersonic Drag by Solution of an Equivalent Two-dimensional Potential Problem. Douglas Aircraft Report, SM-22666, Dec. 1956.
25. GRAHAM, W. E. Note on the Use of Artificial Distribution of Singularities in Supersonic Minimum Drag Problems. Douglas Aircraft Report, SM-23.002.

26. GRAHAM, W. E. A geometric problem related to the optimum distribution of lift on planar wing in supersonic flow. *J. Aero. Space Sciences*, Dec. 1958.
27. GRAHAM, W. E. and LAGERSTROM, P. A. Reduction of Minimum Drag Problem in Supersonic Flow to Two Dimensions at Potential Problems by means of the Combined Flow fields. Douglas Aircraft Report SM-23.901, Mar. 1960.
28. GRIMMINGER, G., WILLIAMS, E. P. and YOUNG, G. B. W. Lift on inclined bodies of revolution in hypersonic flow. *J. Aero Sci.* **17,** no. 11 (Nov. 1950).
29. GRÖBNER, W. and HOFREITER, N. *Intergraltafel*. Teil I–II. Springer Verlag, Wien, 1949, 1950.
30. GUREVICH, M. I. The lift of a sweptback wing in supersonic flow (in Russian). *Prikl. Mat. Mekh.* **10,** no. 4 (1946).
31. GUREVICH, M. I. On the flow around the triangular wing in supersonic stream (in Russian). *Prikl. Mat. Mekh.* **11,** no. 2 (1947).
32. GUREVICH, M. I. On the problem of the thin triangular wing moving with supersonic speed (in Russian). *Prikl. Mat. Mekh.* **11,** no. 3 (1947).

H

1. HADAMARD, J. *Lectures on Cauchy's Problem in Linear Partial Differential Equations* Yale University Press, 1928.
2. HADAMARD, J. *Le Problème de Cauchy et les Équations aux Derivées Partielles Linéaires Hyperboliques*. Herman, Paris, 1932.
3. HAEFELI, R., MIRELS, H. and CUMMINGS, J. Charts for Estimating Downwash behind Rectangular, Trapezoidal, and Triangular Wings at Supersonic Speeds. NACA Tech. Note 2141, 1950.
4. HALL, G. and ROGERS, E. The Flow Pattern on a Tapered Sweptback Wing at Mach Numbers between 0·6 and 1·8. R. and M. 3271, 1957.
5. HALL, M. G. On the Vortex associated with Flow Separation from the Leading Edges of a Slender Wing. RAE Tech. No. 2629, June 1959.
6. HALL, M. G. A Theory of the Core of a Leading-edge Vortex. RAE Rep. 2644, Dec. 1960.
7. HALL, M. G. A theory for the core of a leading-edge vortex. *J. Fluid. Mec.* **11,** 209 (1961).
8. HAMAKER, F. M., NEICE, S. E. and EGGERS, A. J. JR. The Similarity Law for Hypersonic Flow about Three-dimensional Shapes. NACA Tech. Note 2443, 1951.
9. HANCOCK, G. J. On the transient motion of a slender delta wing with leading-edge separation. ARC no. 563, 1961.
10. HANTSCHE, W. and WENDT, H. *Jahrbuch 1942 der Deutschen Luftfahrtforschung*, Heft 7, 1942.
11. HANTSCHE, W. and WENDT, H. Cones in Supersonic Flow. NACA Tech. Mem. 1157, 1947.
12. HARMON, S. M. and JEFFREYS, I. Theoretical Lift and Damping in Roll of Thin Wings with Arbitrary Sweep and Taper at Supersonic Speeds. Supersonic Leading and Trailing Edges. NACA Tech. Note 2114, 1950.
13. HASKIND, M. D. and FALKOVICH, S. V. Vibration of a wing of finite span in a supersonic flow. *Akad. Nauk SSSR. Prikl. Mat. Meh.* **2,** 371–6 (1947). (Transl. in NACA Tech. Mem. 1257.)
14. HAYES, W. D., BROWNE, S. H. and LEW, R. J. Linearized Theory of Conical Supersonic Flow with Application to Triangular Wings. North Amer. Aviation Rept. NA-46-818, 1946.

15. HAYES, W. D. Linearized supersonic flows with axial symmetry. *Quart. Appl. Math.*, **4**, 255–61 (1946).
16. HAYES, W. D. On hypersonic similitude. *Quart. Appl. Math.* **5**, 105–6 (1947).
17. HAYES, W. D. Linearized Supersonic Flow. North American Aviation Report AL-222, 1947.
18. HAYES, W. D. Reversed flow theorems in supersonic aerodynamics. *Proc. VIIth Int. Congr. Appl. Mech. (London)*, **2**, 1948; Rep. North Amer. Aviation Inc., no. AL-755, 1948.
19. HAYES, W. D., ROBERTS, R. C. and HAASER, N. Generalized Linearized Conical Flow. NACA Tech. Note 2667.
20. HAYES, W. D. and PROBSTEIN, R. S. *Hypersonic Flow Theory*. Academic Press, New York and London, 1959.
21. HEASLET, M. A., LOMAX, H. and JONES, A. L. Volterra's Solution of the Wave Equation as applied to Three-dimensional Supersonic Airfoil Problems. NACA Report 889, 1947.
22. HEASLET, M. A. and LOMAX, H. The Use of Source-sink and Doublet Distributions extended to the Solution of Boundary-value Problems in Supersonic Flow. NACA Report 900, 1948.
23. HEASLET, M. A. and LOMAX, H. Two-dimensional Unsteady Lift Problems in Supersonic Flight. NACA Report 945, 1949.
24. HEASLET, M. A. and LOMAX, H. The Calculation of Pressure on Slender Airplanes in Subsonic and Supersonic Flow. NACA Tech. Note 2900, 1953.
25. HEASLET, M. A. and LOMAX, H. *Suporsonic and Transonic Small Perturbation Theory*. W. R. Sears (editor), Princeton, 1954.
26. HILTON, W. F. and PROUDEN, F. W. Subsonic and Supersonic High Speed Tunnel Tests of a Faced Double Wedge Airfoil. R. and M. 2057, 1943.
27. HILTON, W. F. *High-speed Aerodynamics*. Longmans, Green & Co., London, 1952.
28. HIPSCH, H. M. Harmonic Oscillations of a Narrow Delta Wing in Supersonic Flow. Ph.D. Thesis, Pasadena: California Inst. of Technology, 1951.
29. HOWARTH, L. (editor). *Modern Developments in Fluid Dynamics: High Speed Flow*. Oxford: Clarendon Press, 1953.
30. HUMMEL, D. Investigations on Vortex Breakdown on a Sharpedged Slender Delta Wing. Institut für Strömungsmechanik, Braunschweig, Rep. no. 64/24. Also *Jahrb. d. Wiss. Ges. Luft & Raumf.*, 1964.
31. HUNN, B. A. On the determination of the flutter forces on wings with supersonic leading edges. *Quart. J. Mech.* **8**, 293–309; see also *Math. Rev.* **17**, 314 (1955).

I

1. IACOB, C. *Introduction Mathématique à la Mécanique des Fluides*. Bucarest, Editions de l'Académie de la R.P.R., Paris, Gauthier-Villars, 1959.
2. IACOB, C. Teoria aripii unghiulare la viteze supersonice (Theory of angular wing at supersonic speeds). *Anal. Acad. R.P.R. (Seria mat., fiz., chim.)*, **3**, 15, 249 (1950).
3. IACOB, C. Cercetări asupra teoriei mişcărilor conice (Studies on the theory of conical motions). *Bul. ştiinţ. Acad. R.P.R. (Secţiunea de ştiinţe matematice şi fizice)*, **6**, 3, 603 (1954).
4. IONESCU, M. and DARIE, GH. Caracteristicile aerodinamice ale unei clase de aripi subţiri în scurgerea omogenă de ordin superior (Aerodynamic characteristics of a class of thin wings in high order homogeneous stream). *Studii Cerc. Mec. Apl.* **10**, 2 (1959), Bucharest.
5. IONESCU, M. and DARIE, HG. Asupra rezistenţei de undă şi a momentului de ruliu al

aripilor cu borduri de atac supersonice (On the wave drag and rolling moment of wings with supersonic leading edges). *Studii Cer. Mec. Apl.* **10,** 3 (1959), Bucharest.

6. IONESCU, M. and DARIE, GH. Asupra suprafeței aripii corespunzătoare unei distribuții de presiune dată (On the surface of the wing corresponding to a given pressure distribution). *Studii Cerc. Mec. Apl.* **10,** 4 (1959), Bucharest.
7. IVEY, R. H. Notes on the Theoretical Characteristics of Two-dimensional Supersonic Airfoils. NACA Tech. Mem. 1179, Jan. 1947.
8. IVEY, R. H., KLUNKER, E. B. and BOWEN, E. N. A Method for Determining the Aerodynamic Characteristics of two- and three-dimensional Shapes at Hypersonic Speeds. NACA Tech. Note 1613, 1948.

J

1. JIGULEV, N. V. and JILIN, IU. L. O neustanovivshemsya dvizhenii kryla ugolnoi dormy v plane (On the unsteady motion of an angular wing). *Priklad. Mat. Mekh.* **22,** 6 (1959).
2. JONES, J. P. The Breakdown of Vortices in Separated Flow. Univ. of Southampton Rep. 140. July 160.
3. JONES, R. T. Properties of Low Aspect-ratio Pointed Wings at Speeds Below and Above the Speed of Sound. NACA Rept. 855. Washington, 1946.
4. JONES, R. T. Thin Oblique Aifroils at Supersonic Speeds. NACA Tech. Note 1107, 1946.
5. JONES, R. T. Leading-edge singularities in thin airfoil theory. *J. Aero. Sci.* **17,** 307–10 (1950).
6. JONES, R. T. and COHEN, D. *High-speed Wing Theory*. Princeton University Press, Princeton, 1960.
7. JONES, R. T. and MARGOLIS, K. Flow over a Slender Body of Revolution at Supersonic Velocities. NACA Tech. Note 1081, 1956.
8. JONES, W. P. Supersonic Theory for Oscillating Wings of Any Planform. Rep. Memor. Aero. Res. Coun., Lond., no. 2655, 1948.

K

1. KAINER, J. H. and GRIJALVA, H. M. Unified Analysis of Aerodynamic Loads at Supersonic Speeds on Triangular Wings having Arbitrary Camber and Twist. CONVAIR (San Diego) Rep. ZA 251, Sept. 1956.
2. VON KÁRMÁN, TH. and MOORE, N. B. Resistance of slender bodies moving with supersonic velocities, with special reference to projectiles. *Trans. ASMW (Appl. Mech)*. **54,** 303–10 (1932).
3. VON KÁRMÁN, TH. and BURGERS, I. M. *Aerodynamic Theory*. W. F. Durand, Julius Springer, Berlin, vol. II, 1935.
4. VON KÁRMÁN, TH. The problem of resistance in compressible fluids. *Atti del Convergo della Fondazione Alessandro Volta 1935*, pp. 223–326 (1936).
5. VON KÁRMÁN, TH. Supersonic aerodynamics—principles and applications. *J. Aero. Sci.* **14,** 373–409 (1947).
6. KARPOVITCH, E. A. and FRANKL, F. I. The drag of delta wing in supersonic flow. *Prikl. Math. Mekh.* **11** (1947).
7. KELL, C. Free-flight measurements of the zero-lift drag of a slender wing at Mach numbers between 1·4 and 2·7. *A.R.C.* **23,** 511 (Aug. 1961).
8. KIRKBY, S. and ROBINSON, A. Wing–Body Interference at Supersonic Speeds. College of Aeronautics, Cranfield, Rept. 7, Apr. 1947.

9. Kochin, N. E., Kibel, I. A. and Roze, N. B. *Teoreticeskaia Ghidromekhanika* (Theoretical hydrodynamics). GITTL, Moscow, t. II, 1948.
10. Kogan, M. N. O telah minimalnogo soprotivlenia v sverhzvukovom potoke gaza. (On minimum drag bodies in supersonic flow). *Prikl. Math. Mekh.* **21,** 2 (1957).
11. Kopal, Z. Supersonic Flow of Air Around Cones. Mass. Inst. of Technol. Technical Report 1, 1947.
12. Kopal, Z. *Tables of Supersonic Flow around Cones.* Massachusetts, vol. I, 1947.
13. Krasilscikova, E. A. Disturbed motion of air caused by vibrations of a wing moving at supersonic speed. *Akad. Nauk SSSR, Prikl. Mat. Mekh.* **2,** 147–64 (1947); also *Dokl. Acad. Nauk SSSR* **56,** 571–4 (1947).
14. Krasilscikova, E. A. On the theory of the unsteady motion of a compressible fluid. *Dokl. Acad. Nauk SSSR* **72,** 23–6 (1950).
15. Krasilscikova, E. A. Supersonic flows over slender bodies. NACA Tech. Mem. 1383. 1956. Transl. from *Scientific Record of Moscow State University* **5,** 154, *Mechanics*, 181–239 (1951).
16. Krasilschikova, E. A. *Krylo Konechnova Nazhamua v Szhimaemom Potoke* (The wing of finite span in compressible stream). Gostehizdat, Moscow, 1952.
17. Krasilscikova, E. A. Unsteady motions of a wing of infinite span. *Dokl. Acad. Nauk SSSR* **94,** 397 (1954); *Izvetiya Akad. Nauk SSSR, Otd. Tech. Nauk* 25–41 (1954).
18. Krasilscikova, E. A. Unsteady motion of a wing of infinite aspect ratio. *Moscow Gos. Univ. Uc. Zap. 172, Meh.* **5,** 61–78 (1954).
19. Krasilschikova, E. A. Unsteady motion of finite wing span in a compressible motion. *Arch. Mech. Stos.* **16,** 285–90 (1964).
20. Küchemann, D. A Non Linear Lifting-surface Theory for Wings of Small Aspect-ratio with Edge Separation. R.A.E. Rep. 2540, April 1955.
21. Küchemann, D. Aircraft shapes and their aerodynamics for flight at supersonic speeds. *Advances in Aeronautical Sciences*, vol. 3, p. 221. Pergamon Press, 1961. *Proc. 2nd Int. Congr. Aero. Sci.*, Zürich, 12 to 16 Sept 1960.

L

1. Lagerstrom, P. A. Linearized Supersonic Theory of Conical Wings. NACA Tech. Note 1658, 1948.
2. Lagerstrom, P. A. and Graham, M. E. Downwash and Sidewash induced by Three-dimensional Lifting Wings in Supersonic Flow. Douglas Aircraft Co. Rept. SM 13007, Apr. 1947.
3. Lagerstrom, P. A. and Graham, M. E. Low Aspect Ratio Rectangular Wings in Supersonic Flow. Douglas Aircraft Co. Rept. SM 13110, 1947.
4. Lagerstrom, P. A. and Graham, M. E. The linearized theory of control surfaces. *J. Aero. Sci.*, Jan. 1949.
5. Lambourne, N. G. and Bryer, D. W. Some Measurements in the Vortex Flow Generated by Sharp Leading-edges. ARC 21·073, July 1959.
6. Lambourne, N. G. and Bryer, D. W. The Bursting of Leading-edge Vortices. R. and M. 3282, ARC 22775, Apr. 1961.
7. Laporte, O., Bartels, R. C. F. and Rourke, R. C. O. *An Investigation of Linearized Equations for the Flow Past Conical Bodies.* Part III. Supersonic flow past an elliptic cone at an angle of attack. Engineering Research Institute, University of Michigan, Ann Arbor, CM 575, 1949.
8. Lance, G. N. Some Problems of Supersonic Flight. Ph.D. Thesis, University of London, 1952.

9. LANCE, G. N. The delta wing in a non-uniform supersonic stream. *Aeronaut, Quart.* **5,** 55–72 (1954).
10. LANCE, G. N. The lift of twisted and cambered wings in supersonic flow. *Aeronaut. Quart.* **6,** 146–63 (1955).
11. LANDHAL, M., DROUGGE, G. and BEANE, B. Theoretical and experimental investigation of second-order supersonic wing–body interference. *J. Aero. Sci.* **27,** no. 9, Sept. 1960.
12. LAWRENCE, H. R. and FLAX, A. H. Wing–body interference at subsonic and supersonic speeds. Survey and new developments. *J. Aero. Sci.* **21,** 289–324, 328 (1954).
13. LEE, G. H. Note on the Flow around Delta Wings with Sharp Leading Edges. R. and M. 3070, 1958.
14. LEGENDRE, R. Écoulement au voisinage de la pointe avant d'une aile à forte flèche aux incidences moyennes. *8-me Congrès Int. de Méc. App. 1952* et R.A. no. 30, 1952–R.A. no. 12, 1953.
15. LEGENDRE, R. Écoulement au Voisinage de la Pointe avant d'une Aile à Forte Flèche aux Incidences Moyennes. R.A. no. 35, 1953.
16. LEGENDRE, R. Écoulement supersonique autour d'une aile à bords d'attaque subsoniques. *C.R. Acad. Sci., Paris* **245,** 889 (1957).
17. LEGENDRE, R. Nappes en Cornets aux Bords d'Attaque d'une Aile en Delta. R.A. no. 70, mai 1959.
18. LEGENDRE, R. Flow around a delta wing with subsonic leading edges (in French). *C.R. Acad. Sci., Paris* **256,** 11, 2287–90 (Mar. 1963).
19. LEGENDRE, R. Tourbillons en cornets des ailes delta. *C.R. Acad. Sci.*, dec. 1963.
20. LEGENDRE, R. Nappes de tourbillons en cornets. *C.R. Acad. Sci.*, janv. 1964.
21. LEGENDRE, R. Vortex sheets from leading edges of delta wings. *Progress in Aeronautical Sciences*, vol. 7, Pergamon Press.
22. LEGENDRE, R. Vortex sheets rolling up along leading edges of delta wings. *O.N.E.R.A., T.P.* no. 133, 1694.
23. MCLELLAN, CH. Exploratory wind-tunnel investigation of wings and bodies at M = 6·9 *J. Aero. Sci.* **18** (Oct. 1951).
24. LI, T. Y. Periodic rolling oscillations of a rectangular wing in supersonic flight. *J. Aero. Sci.* **18,** 191–8 (1951).
25. LI, T. Y. and STEWART, H. J. On an integral equation in the supersonic oscillating wing theory. *J. Aero. Sci.* **20,** 724–6 (1953) (see also references cited therein).
26. LI, TA. Aerodynamic influence coefficients for an oscillating finite thin wing in supersonic flow. *Inst. Aero. Sci.*, Preprint 500 (1955).
27. LI, TA. Aerodynamic influence coefficients for an oscillating thin wing in supersonic flow. *J. Aero. Sci.* **23,** 613–22 (1956).
28. LIEPMANN, H. W. and PUCKETT, A. E. *Introduction to the Aerodynamics of Compressible Fluids*. New York, John Wiley & Sons, 1947.
29. LIGHTHILL, M. J. The Conditions behind the Trailing Edge of the Supersonic Aerofoil. A.R.C.R. and M. 1930, 1944.
30. LIGHTHILL, M. J. The Supersonic Theory of Wings of Finite Span. R. and M., no. 2001.
31. LIGHTHILL, M. J. Supersonic Flow Past Bodies of Revolution. R. and M., no. 2003 (1945).
32. LIGHTHILL, M. J. The shock strength in supersonic "conical fields". *Phil. Mag.* **40,** 1202 (1949).
33. LIGHTHILL, M. J. Supersonic flow past slender bodies of revolution the slope of whose meridian section is discontinuous. *Quart. J. Mech. and Appl. Math.* **1,** 90 (1948).

34. Lighthill, M. J. Oscillating airfoils at high Mach number. *J. Aero. Sci.*, no. 6 (June 1953).
35. Lighthill, M. J. The wave drag at zero lift of slender delta wings and similar configurations. *J. Fluid. Mech.* **1,** Part 3, p. 337 (Sept. 1956).
36. Lighthill, J. *Mathematics and Aeronautics.* W. Wright Lect., 1960.
37. Lighthill, J. *Higher Appoximation s in Aerodynamic Theory.* Princeton Univ. Press, 1960.
38. Lilley, G. M., Alexander, A. J. and Craven, A. H. The Effect of Blowing from the Leading Edge of Slender Wings. Report Series of College of Aeronautics, Cranfield, 1964.
39. Linell, R. D. Two-dimensional airfoils in hypersonic flows. *J. Aero. Sci.* **16** (Jan. 1949).
40. Lomax, H., Sluder, L. and Heaslet, M. A. The Calculation of Downwash behind Supersonic Wings with an Application to Triangular Plan Forms. NACA Rept. 957, 1950.
41. Lomax, H. and Heaslet, M. A. Generalized Conical Flow Fields in Supersonic Wing Theory. NACA Tech. Note 2497, Sept. 1951.
42. Lomax, H. Lift Developed on Unrestrained Rectangular Wings entering Gusts at Subsonic and Supersonic Speeds. NACA Tech. Note 2925, 1953 .
43. Lomax, H. The wave drag of arbitrary configurations in linearized flow, as determined by areas and forces in oblique planes. NACA Res. Mem. A 55 A 18, Mar. 1955.
44. Love, E. S. Investigations at Supersonic Speeds of 22 Triangular Wings representing Two Airfoil Sections for each of 11 Apex Angles. NACA Res. Mem. L 9 DO 7, 1949.
45. Luke, Y. L. On the Evaluation of Generalized Aerodynamic Forces for Triangular Wings with Supersonic Leading Edges. Wright Air Development Center Rep. no. 56–525, 1956.
46. Luke, Y. L., Coombs, G. V. and Wagner, F. Solution of the Potential Equation for the Unsteady Flow about a Triangular Wing, etc. Wright Air Development Center Rep. no. 57–258, 1957.

M

1. Mabey, D. G. and Ilott, G. P. The characteristics of three slender "mild ogee" wings at Mach Numbers from 0·4 to 2·0. Unpublished M.o.A. Report.
2. Mangler, K. W. and Smith, J. H. A Theory of Slender Delta Wing with Leading Edge Separation. R.A.E. Rept. 2442, Apr. 1956.
3. Mangler, K. W. and Smith, J. H. Calculation of the Flow Past Slender Delta Wings with Leading-edge Separation. R.A.E. Rept. 2593, May 1957.
4. Mangler, K. and Smith, J. H. B. A theory of the flow past a slender delta wing with leading edge separations. *Proc. Roy. Soc.* **251,** 200–17 (1959).
5. Mangler, K. W. and Weber, J. An asymptotic solution for the inner part of the leading-edge vortex sheets. Unpublished R.A.E. Rept, 1964.
6. Mardsen, D. J., Simpson, R. W. and Rainbird, W. J. The flow over delta wings with leading-edge separation. *Col. of Aeron.*, no. 114 (Sept. 1958).
7. Martin, J. The Calculation of Downwash behind Wings of Arbitrary Plan Form at Supersonic Speeds. NACA Tech. Note 2135, 1950.
8. Maskell, E. C. Flow separation in three dimensions. Unpublished Ministry of Supply Report, 1955.

9. Mateescu, Dan. Sverhzvukovoe obtekanie trapetsoidalnovo kryla s dvumia conicheskimi obtekateliami (Supersonic flow around the trapezoidal wing fitted with two conical bodies). *Revue de Méc. Appl.* **6,** 3 (1961). Also in Rumanian in *Studii Cerc. Mec. Apl.* **12,** 1 (1961). Publishing House of the Academy of the R.P.R., Bucharest.
10. Mateescu, Dan. Tragflügel mit besonderer Abschlussfläche in Überschallströmung. *Revue de Méc. Appl.* **8,** 1 (1963). Also in Rumanian in *Studii Cerc. Mec. Apl.* **13,** 5 (1962). Publishing House of the Academy of the R.P.R., Bucharest.
11. Mateescu, Dan. Indirect problem of the wing–conical body system in supersonic flow (unpublished).
12. Mateescu, Dan. Supersonic flow around polygonal wings, both thin and of symmetrical thickness. *Revue de Méc. Appl.* **11,** 3 (1966). Publishing House of the Academy of the R.P.R., Bucharest.
13. Mateescu, Dan. Aerodynamic characteristics of the polygonal wings of symmetrical thickness in supersonic flow. *Revue de Méc. Appl.* **11,** 4 (1966) and also in Rumanian in *Studii Cerc. Mec. Appl.* **23,** 4 (1966). Publishing House of the Academy of the R.S.R., Bucharest.
14. Mateescu, Dan. Polygonal wings with diamond shaped profile in supersonic flow. *Revue de Méc. Appl.* **12,** 2 (1967) and also in Rumanian in *Studii Cerc. Mec. Apl.* **24,** 2 (1967). Publishing House of the Academy of the R.S.R., Bucharest.
15. Mateescu, Dan and Codoban, Al. Aerodynamic characteristics of the triangular wing fitted with triangular ailerons deflected antisymmetrically in supersonic flow (unpublished).
16. Michael, W. H. Flow Studies on Flat-plate Delta Wings at Supersonic Speeds. NACA Tech. Note 3472, July 1955.
17. Middleton, W. D. and Carlson, H. W. A Numerical Method for Calculating the Flat Plate Pressure Distributions on Supersonic Wings of Arbitrary Planform. NASA Tech. Note D-2570, Jan. 1963.
18. Miles, E. R. C. *Supersonic Aerodynamics*. McGraw-Hill Book Company, Inc., London, 1950.
19. Miles, J. W. Harmonic and transient motion of a swept wing in supersonic flow. *J. Aero. Sci.* **15,** 343–7 (1948).
20. Miles, J. W. The aerodynamic forces on an oscillating flap at supersonic speeds. *J. Aero. Sci.* **15,** 565–8; **16,** 442; **17,** 124 (1948).
21. Miles, J. W. On harmonic motion at supersonic speeds. *J. Aero. Sci.* **16,** 378 (1949).
22. Miles, J. W. On the oscillating rectangular airfoil at supersonic speeds. *J. Aero. Sci.* **16,** 381 (1949).
23. Miles, J. W. On the oscillating aileron at supersonic speeds. *J. Aero. Sci.* **16,** 511 (1949).
24. Miles, J. W. On nonsteady motion of delta wings. *J. Aero. Sci.* **16,** 568 (1949).
25. Miles, J. W. The Oscillating Rectangular Airfoil at Supersonic Speeds. U.S. Naval Ordnance Report, no. 1170, Inyokern, California, 1949.
26. Miles, J. W. On the reduction of unsteady supersonic flow problems to steady flow problems. *J. Aero. Sci.* **17,** 64 (1950).
27. Miles, J. W. On simple planforms in supersonic flow. *J. Aero. Sci.* **17,** 127 (1950).
28. Miles, J. W. On non steady motion of slender bodies. *Aeronaut. Quart.* **2,** 183–94 (1950).
29. Miles, J. W. Unsteady supersonic flow past slender pointed bodies. *Proceedings of*

Bureau of Ordnance Symposium, U.S. Naval Ordnance Report, no. 1651, pp. 429–84, 1950; no. 2031, 1953.
30. MILES, J. W. On slender wing theory. *J. Aero. Sci.* **18,** 770 (1951).
31. MILES, J. W. Slender body theory for supersonic unsteady flow. *J. Aero. Sci.* **19,** 280 (1952).
32. MILES, J. W. On the general solution for unsteady motion of a rectangular wing in supersonic flow. *J. Aero. Sci.* **19,** 421 (1952).
33. MILES, J. W. A note on the damping in roll of a cruciform winged body. *Quart. Appl. Math.* **10,** 276–7 (1952).
34. MILES, J. W. On the low aspect ratio oscillating rectangular wing in supersonic flow. *Aeronaut. Quart.* **4,** 231–44 (1953).
35. MILES, J. W. *Unsteady Supersonic Flow.* Baltimore: U.S. Air Force Research and Development Command, 1955.
36. MILES, J. W. *The Potential Theory of Unsteady Supersonic Flow.* Cambridge University Press, 1959.
37. MILNE-THOMSON, L. M. The circle theorem. *Proc. Camb. Philos. Soc.* 36 (1940).
38. MILNE-THOMSON, L. M. *Theoretical Hydrodynamics.* Macmillan, 1949.
39. MILNE-THOMSON, L. M. *Theoretical Aerodynamics.* Macmillan Co. Ltd., London, 1952.
40. MIRELS, H. Aerodyn amicsof Slender Wings and Wing–Body Combinations having Swept Trailing Edges. NACA Tech. Note 3105, 1954.
41. MIRELS, H. and HAEFELI, R. Line-vortex Theory for Calculation of Supersonic Downwash. NACA Rept. 983, 1950.
42. v. MISES, RICHARD. *Mathematical Theory of Compressible Fluid Flow.* Academic Press Inc., Publishers, New York, 1958.
43. MORIKAWA, G. K. The Wing–Body Problem for Linearized Supersonic Flow, Calif. Inst. Technol. Jet Propul. Lab. Progress Rep. 4–116, ORDCIT Project, Dec. 1949.
44. MULTHOPP, H. A Unified Theory of Supersonic Wing Flow, employing Conical Fields. Roy. Aircraft Establishment Rept. Aero. 2415, 1951.
45. MUNK, M. M. The reversal theorem of linearized supersonic airfoil theory. *J. Appl. Phys.* **21,** 159–61, 1950.
46. MURGULESCU, ELENA. Asupra unor probleme la limită din teoria mişcărilor conice (On the problems of the boundary conditions in the theory of conical motions). *Comunicările Acad. R.P.R.* **1,** 6, 447–53 (1951).
47. MURGULESCU, ELENA. Problema la limită din teoria mişcărilor conice în jurul aripii subţiri de secţiune dată (On the boundary conditions in the theory of conical motions around the thin wing of given section). *Comunicările Acad. R.P.R.* **2,** 9–10, 489–94 (1952).
48. MURGULESCU, ELENA. Mişcarea supersonică în jurul aripii delta cu fuzelaj conic (The supersonic motion around the delta wing with conical body). *Bul. ştiinţ. Acad. R.P.R. (Secţiunea de şt. mat. şi fiz.)* **6,** 4, 741–53 (1954).
49. MURGULESCU, ELENA. Asupra mişcării conice în jurul aripii subţiri, exterioară conului caracteristic al vîrfului (On the conical motion around the thin wing with supersonic leading edges). *Comunicările Acad. R.P.R.* **6,** 10, 1179–85 (1956).
50. MURGULESCU, ELENA. Asupra teoriei mişcărilor conice in cazul aripilor subţiri de secţiune dată (On the theory of conical motions în the case of thin wings of given section). (Thesis) *Studii Cerc. Mat.* **11,** 1 (1960). Publishing House of the Academy of the R.P.R., Bucharest.

N

1. NĂSTASE, ADRIANA. Asupra unor formule de recurenţă referitoare la calculul vitezei axiale de perturbaţie pentru mişcările omogene de ordin superior (On certain recurrence formulas for the calculus of the axial disturbance velocity for high order homogeneous motions). *Studii Cerc. Mec. Apl.* **10**, 1 (1959). Publishing House of the Academy of the R.P.R., Bucharest.
2. NĂSTASE, ADRIANA. The mixed problem of the wing of symmetrical thickness in conical motion (unpublished).
3. NĂSTASE, ADRIANA. Aerodynamic characteristics of triangular wings, both thin and of symmetrical thickness, in the indirect problem (unpublished).
4. NĂSTASE, ADRIANA. Determination of the axial velocity coefficient, of the characteristics and of the trapezoidal wing optimum symmetrical thickness in supersonic flow (unpublished).
5. NEICE, S. E. and EHRET, D. M. Simlarity laws for slender bodies of revolution in hypersonic flows. *J. Aero. Sci.* **18**, 725–31 (1951).
6. NELSON, H., RAINEY, R. and WITKINS, C. Lift and Moment Coefficients Expanded to the Seventh Power of the Frequency for Oscillating Rectangular Wings in Supersonic Flow. NACA Tech. Note 3076, 1945.
7. NELSON, H. C. Lift and Moment on Oscillating Triangular and Related Wings with Supersonic Edges. NACA Tech. Note 2494, 1951.
8. NICHOLS, S. *Interference between Wing and Body at Supersonic Speeds.* Part VIII: Data report on pressure distribution tests of wing–body interference models at Mach number 2.0, phase four tests of March 1950. Cornell Aeronaut. Lab. Bumblebee Rept. CAL/CF–1571, July 1951.
9. NIELSEN, J. N. and PITTS, W. C. Wing–Body Interference at Supersonic Speeds with an Application to Combination with Rectangular Wings. NACA Tech. Note 2677, 1952.

P

1. PAI, SHIH-I. *Introduction to the Theory of Compressible Flow.* D. van Nostrand Co., New York, 1959.
2. PATRAULEA, N. Mişcarea superonică în jurul unei aripi care traversează un con aproape circular (Supersonic flow around a wing which crosses an almost circular cone). *Bul. ştiinţ. Acad. R.P.R., secţiunea de ştiinţe tehnice şi chimice*, t. V, no. 1–4, 1953.
3. PATRAULEA, N. and PANTAZOPOL, D. Mişcarea conică în jurul unei aripi avînd la unul din capete un con aproape circular (Conical flow around a wing with an almost circular cone at one tip). *Bul. ştiinţ. Acad. R.P.R., secţiunea de ştiinţe tehnice şi chimice*, t. VI, no. 4, 1954.
4. PERKINS, E. W. and CANNING, T. N. Investigation of Downwash and Wake Characteristics at a Mach Number of 1·53. I: Rectangular Wing. NACA Res. Mem. A 8 L 16, 1949.
5. PERKINS, E. W. and CANNING, T. N. Investigation of Downwash and Wake Characteristics at a Mach Number of 1·53. II: Triangular Wing. NACA Res. Mem. A 9 D 20, 1949.
6. PICARD, CLAUDE. Caractéristiques aérodinamiques théoriques des ailes delta et des ailes en queue d'hirondelle pour $M < 3{,}5$. *La Recherche Aéronautique*, no. 39 (1954).
7. PINDZOLA, M. Symmetrical Double Wedge Airfoil Tests in the UAC Two-dimen-

sional Supersonic Wind Tunnel. Meteor Report, UAC-16, Research Department, United Aircraft Corporation, Jan. 1948.

8. PINDZOLA, M. Supersonic tests of conventional control surfaces on a double-wedge airfoil. *J. Aero. Sci.* **17** (Apr. 1950).
9. PINES, S. and DUGUNDJI, J. Application of Aerodynamic Flutter Derivatives to Flexible Wings with Supersonic and Subsonic Edges. Republic Aviation Corp. Rept. E-SAF-2, 1954.
10. PITTS, W. C., NIELSEN, J. N. and GIONFRIDDO, M. P. Comparison between Theory and Experiment for Interference Pressure Field between Wing and Body at Supersonic Speeds. NACA Tech. Note 2138, 1954.
11. PRANDTL, L. and BUSEMANN, A. Näherungsverfahren zur zeichnerischen Bestimmung von ebenen Stromungen und Überschallgeschwindigkeiten. *Stodola-Festschrift*, Zürich, 1929.
12. PUCKETT, A. E. Supersonic wave drag of thin airfoils. *J. Aero. Sci.* **13,** 9, 475–84 (1946).
13. PUCKETT, A. E. and STEWART, H. J. Aerodynamic performance of delta wings at supersonic speeds. *J. Aero. Sci.*, Oct. 1947.

R

1. RANDALL, D. A Theoretical Determination of the Flow and of the Air Forces on an Oscillating Delta Wing with Leading Edge Separation. RAE Rept. 284, Mar. 1963.
2. RANDALL, D. Oscillating Slender Wings in the Presence of Leading Edge Separation. RAE Rept. 286, Apr. 1963.
3. REESE, I. H. Notes on the Theoretical Characteristics of Two-dimensional Supersonic Airfoils. NACA Tech. Mem. 1179, Jan. 1947.
4. ROBINSON, A. Aerofoil Theory of a Flat Delta Wing at Supersonic Speeds. Brit. Aeronaut. Research Council Repts. and Mem. 2548, 1946.
5. ROBINSON, A. On Source and Vortex Distributions in the Linearised Theory of Steady Supersonic Flow. The College of Aeronautics, Cranfield, Rep. 9, 1947.
6. ROBINSON, A. Lift and Drag of a Flat Delta Wing at Supersonic Speeds. Roy. Aircraft Establishment Rept. 1791, 1946.
7. ROBINSON, A. Aerofoil Theory of a Flat Plate Delta Wing at Supersonic Speeds. Roy. Aircraft Establishment Rept. Aero. 2151, Farnborough, England, 1946.
8. ROBINSON, A. and HUNTER-TOD, J. Bound and Trailing Vortices in the Linearized Theory of Supersonic Flow, and the Downwash in the Wake of a Delta Wing. College of Aeronautics, Cranfield, Rept. 10, 1947.
9. ROPER, G. M. Formulae for Calculating the Camber Surfaces of Thin Swept-back Wings of Arbitrary Planform with Subsonic Leading Edges, and Specified Load Distribution. Aero. Res. Council, London, Rep. Mem. 3217, 1961.
10. ROTT, N. On the unsteady motion of a thin rectangular wing in supersonic flow. *J. Aero. Sci.* **18,** 775–6 (1951).
11. ROTT, N. On the viscous core of a line vortex. *ZAMP* **10,** 1 (1959).
12. ROTTY, R. M. *Introduction to Gas Dynamics*. John Wiley & Sons, 1962.
13. ROY, M. Caractères de l'écoulement autour d'une aile en flèche accentuée. *C.R. Acad. Sci., Paris* **234,** 2501 (1952).
14. ROY, M. Sur la théorie de l'aile delta. Tourbillons d'apex et nappes en cornets. *C.R. Acad. Sci.*, fév. 1957.
15. ROY, M. Remarques sur l'écoulement autour des ailes en flèche. *Zeits. f. ang. Math. w. Phys.* 554 (1958).
16 ROY, M. De la formation des zones tourbillonnaires dans les écoulements à faible viscosité. *Conf. L. Prandtl*, mai 1959 (and *Zeits. f. Flugwis.*, Aug. 1959).

17. Roy, M. On the rolling-up of the conical vortex sheets from a delta wing. *Progress in Aeronautical Sciences*, vol. 7, Pergamon Press.
18. Rush, C. K. The Icing of Delta Wings of Unity Aspect Ratio having Leading Edge Separation. N.R.C. (National Research Council) Rep. L.R. 268, Dec. 1959.

S

1. Săndulescu, S. Caractéristiques aérodynamiques des ailes triangulaires ayant un mouvement oscillatoire en régime supersonique. *Revue de Méc. Appl.* **4,** 4 (1959). Publishing House of the Academy of the R.P.R., Bucharest.
2. Săndulescu, S. Aerodynamicheskie charakteristiki mnogougolnyh eleronov so zvukovym sharnirom, imeischim garmonicheskoe kolebatelnoe dvizhenie (Aerodynamic characteristics of polygonal ailerons with sonic hinge, having a harmonic oscillating motion). *Revue de Méc. Appl.* **6,** 1 (1961). Publishing House of the Academy of the R.P.R., Bucharest.
3. Săndulescu, Serban and Năstase-Ghia, Adriana. Le problème mixte de l'aile triangulaire en mouvement conique d'ordre supérieur. *Revue de Méc. Appl.* **7,** 1 (1963). Publishing House of the Academy of the R.P.R., Bucharest.
4. Sankaranarayanan, R. Aerodynamic influence coefficients for finite wings oscillating in supersonic flow. *J. Aeronaut. Soc. India*, **16,** 1, 1–19 (Feb. 1964).
5. Sauer, R. *Theoretische Einführung in die Gasdynamik*. Berlin, Springer, 1943.
6. Sauer, R. Elementare Theorie des langsam schwingenden Überschallflügels. *Z. angew. Math. u. Phys.* **1,** 248–53 (1950).
7. Schlichting, H. Airfoil theory at supersonic speed. *Luftfahrtforschung* **13** (1936); also *Jahrbuch Deutschen Luftfahrtforschung*, 1937.
8. de Schwarz, M. J. Oscillazioni harmoniche di ali triangolari con bordo d'attaco supersonico. *Aerotecnica* **21,** 288–98, 306 (1951).
9. Sears, W. R. Operational methods in the theory of airfoils in non-uniform motion. *J. Franklin Inst.* **130,** 95–111 (1940).
10. Sears, W. R. On projectiles of minimum drag. *Quart. Appl. Math.* **14,** 361–6 (1947).
11. Sears, W. R. (editor). *General Theory of High Speed Aerodynamics* (Vol. VI of *High Speed Aerodynamics and Jet Propulsion*). Princeton University Press, Princeton, 1955.
12. Sears, W. R. *Small Perturbation Theory*. Princeton University Press, Princeton, 1960.
13. Sedov, L. I. *Ploskie Zadachi Ghidro i Aerodinamiki* (Two-dimensional problems of hydrodynamics and aerodynamics). Gostehizdat, Moscow, 1950.
14. Shapiro, A. H. *Compressible Flow* (2 vols.). The Ronald Press Company, New York, 1954.
15. Slezin, N. A. K zadachi ploskogo dvizhenia gaza (Contribution to the problem of two-dimensional flow in gases). *Tr. NGM*, 1935; *DAN*, 40, **3,** no. 9 (1936).
16. Smirnov. *Curs de Matematici Superioare* (Lectures on higher mathematics). Ed. Technică, București, vol. III, part II, 1955.
17. Smith, J. H. A Theory of the Separated Flow from the Curved Leading Edge of a Slender Wing. R. and M. 3116, 1959.
18. Snow, R. M. Aerodynamics of thin quadrilateral wings at supersonic speeds. *Quart. Appl. Math.* **5,** 417–28 (1948).
19. Spreiter, J. R. The Aerodynamic Forces on Slender Plane- and Cruciform-wing and Body Combinations. NACA Rept. 962, 1950.
20. Spreiter, J. R. On slender wing–body theory. *J. Aero. Sci.* **19,** 571–2 (1952).
21. Squire, H. B. An Example in Wing Theory at Supersonic Speed. A.R.C. Rep. and M. 2549 1947 (corrected, 1949).

22. Squire, L. C. An Experimental Investigation at Supersonic Speeds of the Characteristics of Two Gothic Wings, One Plane and One Cambered. A.R.C. R. and M. 3211, May 1959.
23. Squire, L. C. Some Applications of "not-so-slender-wing theory" to Wings with Curved Leading Edges. A.R.C. R. and M. 3278, July 1960.
24. Squire, L. C. The estimation of the non-linear lift of delta wings at supersonic speeds. *Journ. R.A.S.*, Aug. 1963.
25. Squire, L. C. The Characteristics of Some Slender Cambered Gothic Wings at Mach Numbers from 0·4 to 2·00. A.R.C. R. and M. 3370, 1964.
26. Squire, L. C. Pressure Distribution and Flow Patterns at M = 0·4 on some Delta Wings. A.R.C. R. and M. 3373, 1964.
27. Squire, L. C., Jones, J. G. and Stanbrook, A. An Experimental Investigation on the Characteristics of some Plane and Cambered 65° Delta Wings at Mach Numbers from 0·7 to 2·0 A.R.C. R. and M. 3305, July 1961.
28. Staicu, St. La détermination des caractéristiques aérodinamiques de l'aile cruciforme en régime supersonique. *Bul. Inst. Politehnic "Gh. Gheorghiu-Dej"*, no. 2, Bucharest, 1966.
29. Stanbrook, A. Lift-curve slope at subsonic and supersonic speeds. *Aircraft Engineering*, no. 306, vol. XXVI, Aug. 1954.
30. Stainback, P. C. Pressure Measurements on Two 60° Swept Delta Wings with Blunt Leading Edges and Dihedral Angles of 0° and 45° at a Mach Number of 4·95. NASA Tech. Note D-124b, Apr. 1962.
31. Stanbrook, A. and Squire, L. Possible types of flows at swept leading edges. *Aeronaut. Quart.* Feb. 1964.
32. Stanisic, M. M. On the aerodynamic forces acting on a delta wing in supersonic flow with supersonic leading edges (in English). *ZAMM* **41**, 9, 342–54 (Sept. 1961).
33. Stewart, H. J. The lift of a delta wing at supersonic speeds. *Quart. Appl. Math.* **4**, 246–54 (1946).
34. Stewart, H. J. and Puckett, A. E. Aerodynamic performance of delta wings at supersonic speeds. *J. Aero. Sci.* **14**, 567–8 (1947).
35. Stewart, H. J. and Li, T. Y. Periodic motions of a rectangular wing moving at supersonic speed. *J. Aero. Sci.* **17**, 529–39 (1950).
36. Stewart, H. J. and Li, T. Y. Source-superposition method of solution of a periodically oscillating wing at supersonic speeds. *Quart. Appl. Math.* **9**, 31–45 (1951). See *Math. Rev.* **13**, 86.
37. Stewart, H. J. and Li, T. Y. On an integral equation in the supersonic oscillating wing theory. *J. Aero. Sci.* **20**, 724–6 (1953).
38. Stewartson, K. On the linearized potential theory of unsteady supersonic motion. *Quart. J. Mech. Appl. Math.* **3**, 182–99 (1950).
39. Stewartson, K. Supersonic flow over an inclined wing of zero aspect ratio. *Proc. Camb. Phil. Soc.* **46**, 307–15 (1950).
40. Stewartson, K. On the linearized potential theory of unsteady supersonic motion. II. *Quart. J. Mech. Appl. Math.* **5**, 137–54 (1952).
41. Stocker, P. M. Supersonic flow past bodies of revolution with thin wings of small aspect ratio. *Aeronaut. Quart.* **3**, 61–79 (1951).
42. Stone, A. H. On supersonic flow past a slightly yawing cone. *MIT J. of Math. and Phys.* **27**, no. 1 (Apr. 1948).
43. Strand, F. Minimum wing wave drag with volume constraint. *J. Aero Space Sciences*, Aug. 1960.
44. Strang, W. Transient source, doublet and vortex solutions of the linearized equations of supersonic flow. *Proc. Roy. Soc.* A, **202**, 40–53 (1950).

45. Strang, W. Transient lift of three-dimensional purely supersonic wings. *Proc. Roy. Soc.* A, **202,** 54–80 (1950).

T

1. Taylor, G. I. Applications to aeronautics of Ackeret's theory of airfoils moving at speeds greater than that of sound. A.R.C. R. and M. 1467, 1932.
2. Taylor, G. I. and Maccoli, J. W. The air pressure on a cone moving at high speed. *Proc. Roy. Soc.* A, **139** (1933).
3. Taylor, G. I. and Maccoli, J. W. *The Mechanics of Compressible Fluids, Aerodynamic Theory.* Editor Durand, W. F. Berlin, Springer, vol. III, Div. H, 1935.
4. Temple, G. *Modern Developments in Fluid Dynamics. High Speed Flow.* Edited by L. Howarth. Oxford Univ. Press, 1953.
5. Temple, G. and Jahn, H. *Flutter at Supersonic Speeds.* Part I. Mid-chord derivative coefficients for a thin airfoil at zero incidence. Rep. Memor. Aero. Res. Coun., Lond., no. 2140, 1945.
6. Thomann, H. Measurements of Heat Transfer Coefficient, Temperature and Pressure Distribution on a Delta Wing at M = 3·0. R.A.E Lib. Trans. 964, Sept. 1961.
7. Truitt, R. W. *Hypersonic Aerodynamics.* Ronald Press Co., 1959.
8. Tsien, H. S. Supersonic flow over an inclined body of revolution. *J. Aero. Sci.* **5,** 480–3 (1938).
9. Tsien, H. S. Similarity laws of hypersonic flows. *J. Math. and Phys.* **25,** no. 3 (Oct. 1946).
10. Tsien, H. S. and Lees, L. *J. of Aero. Sci.,* **12,** 175 (1945).

V

1. Vincenti, W. G., Nielson, J. N. and Matteson, F. H. Investigation of Wing Characteristics at a Mach Number of 1·53. I: Triangular Wings of Aspect Ratio 2. NACA Res. Mem. A7110, 1947.
2. Vincenti, G. Walter. Comparison between theory and experiment for wings at supersonic speeds. *Joint. Meeting of J. Ae. S. and R. Ae. Soc.,* May 1949.

W

1. Walker, H. J. and Ballantyne, M. B. Pressure Distribution and Damping in Steady Pitch at Supersonic Mach Numbers of Flat Sweptback Wings having all Edges Subsonic. NACA Tech. Note 2197. 1950.
2. Walsh, J., Zartarian, G. and Voss, H. Generalized aerodynamic forces on the delta wing with supersonic leading edges. *J. Aero. Sci.* **21,** 739–48 (1954).
3. Ward, G. N. Supersonic flow past slender pointed bodies. *Quart. J. Mech. and Appl. Math.* **2,** 75–97 (1949).
4. Ward, G. N. *Linearized Theory of Steady High-speed Flow.* Cambridge University Press, 1955.
5. Watkins, C. E. Effect of Aspect Ratio on the Air Forces and Moments of Harmonically Oscillating Thin Rectangular Wings in Supersonic Potential Flow. NACA Tech. Note 2064, 1950.
6. Watkins, C. E. Air Forces and Moments on Triangular and Related Wings with Subsonic Leading Edges Oscillating in Supersonic Potential Flow. NACA Tech. Note 2457, 1951.

7. WATKINS, C. E. and BERMAN, J. Air Forces and Moments on Triangular and Related Wings with Subsonic Leading Edges Oscillating in Supersonic Potential Flow. NACA Rept. no. 1099, 1952.
8. WATKINS, C. E. and BERMAN, J. H. Velocity Potential and Air Forces associated with a Triangular Wing in Supersonic Flow, with Subsonic Leading Edges, and deforming Harmonically according to a General Quadratic Equation. NACA Tech. Note 3009. 1953.
9. WEBER, J. Some effects of flow separation on slender delta wings. *A.R.C.* **18,** 073, Nov. 1955.
10. WEBER, R. and RUPPEL, W. Étude du flutter en régime supersonique. *O.N.E.R.A. Rap.*, no. 35 (1953).
11. WERLÉ, H. Essai de vérification de la conicité de l'écoulement autour d'une aile delta mince avec incidence. R.A. no. 63, 1958.
12. WHITHAM, G. B. The behaviour of supersonic flow past a body of revolution far from the axis. *Proc. Roy. Soc.* **201,** 89–109 (1950).
13. WILBY, G. P. An Experimental and Theoretical Investigation of Second-order Wing–Body Interference at High Mach Numbers. F.F.A. Report 91, 1962.

Z

1. ZEOLDVICH. Theory of shock waves and introduction to gasdynamics. *Izdat. Akad. Nauk SSSR*, Moscow and Leningrad, 1946.

INDEX

OTHER TITLES IN THE SERIES IN AERONAUTICS AND ASTRONAUTICS

von Kármán – From Low-speed Aerodynamics to Astronautics

DIVISION I. SOLID AND STRUCTURAL MECHANICS

Vol. 1. Savin – Stress Concentration around Holes
Vol. 2. Gol'denveizer – Theory of Elastic Thin Shells
Vol. 3. Nowacki – Thermoelasticity
Vol. 4. Cox – The Buckling of Plates and Shells
Vol. 5. Morley – Skew Plates and Structures
Vol. 6. Mansfield – The Bending and Stretching of Plates
Vol. 7. Schuh – Heat Transfer in Structures
Vol. 8. Cox – The Design of Structures of Least Weight
Vol. 9. Washizu – Variational Methods in Elasticity and Plasticity

DIVISION II. AERODYNAMICS

Vol. 1. Scorer – Natural Aerodynamics
Vol. 2. Landahl – Unsteady Transonic Flow
Vol. 3. Guderley – The Theory of Transonic Flow
Vol. 4. Martynov – Practical Aerodynamics
Vol. 5. Irving – An Introduction to the Longitudinal Static Stability of Low-speed Aircraft
Vol. 6. Loitsyanskii – Mechanics of Liquids and Gases

DIVISION III. PROPULSION SYSTEMS INCLUDING FUELS

Vol. 1. Penner – Chemistry Problems in Jet Propulsion
Vol. 2. Ducarme, Gerstein and Lefebvre (Eds.) – Progress in Combustion Science and Technology, Vol. 1
Vol. 3. Ragozin – Jet Propulsion Fuels
Vol. 4. Kholshchevnikov – Some Problems in the Theory and Assessment of Turbo-jet Engines
Vol. 5. Zuyev and Skubachevskii – Combustion Chambers for Jet Propulsion Engines

DIVISION VI. FLIGHT TESTING

Vol. 1. Babister – Aircraft Stability and Control (2 vols.)

DIVISION VII. ASTRONAUTICS

Vol. 1. Alperin, Stern and Wooster (Eds.) – Vistas in Astronautics I
Vol. 2. Alperin and Gregory (Eds.) – Vistas in Astronautics II

DIVISION VIII. MATERIALS SCIENCE AND ENGINEERING

DIVISION IX. SYMPOSIA